World Energy Outlook 2019

www.iea.org/weo

INTERNATIONAL ENERGY AGENCY

The IEA examines the full spectrum of energy issues including oil, gas and coal supply and demand, renewable energy technologies, electricity markets, energy efficiency, access to energy, demand side management and much more. Through its work, the IEA advocates policies that will enhance the reliability, affordability and sustainability of energy in its 30 member countries, 8 association countries and beyond.

Please note that this publication is subject to specific restrictions that limit its use and distribution. The terms and conditions are available online at www.iea.org/t&c/

Source: IEA. All rights reserved.
International Energy Agency
Website: www.iea.org

IEA member countries:

Australia
Austria
Belgium
Canada
Czech Republic
Denmark
Estonia
Finland
France
Germany
Greece
Hungary
Ireland
Italy
Japan
Korea
Luxembourg
Mexico
Netherlands
New Zealand
Norway
Poland
Portugal
Slovak Republic
Spain
Sweden
Switzerland
Turkey
United Kingdom
United States

The European Commission also participates in the work of the IEA

IEA association countries:

Brazil
China
India
Indonesia
Morocco
Singapore
South Africa
Thailand

Foreword

More than 40 years after the International Energy Agency (IEA) published the first edition of the *World Energy Outlook*, the report's overarching aim remains the same – to deepen our understanding of the future of energy. It does so by examining the opportunities and risks that lie ahead, and the consequences of different courses of action or inaction. The *WEO* analyses the choices that will shape our energy use, our environment and our wellbeing. It is not, and has never been, a forecast of where the energy world will end up.

This year brings many changes. I would like to highlight two in particular. First, we have renamed the New Policies Scenario as the **Stated Policies Scenario**, making more explicit the intention to "hold up a mirror" to the plans and ambitions announced by policy makers without trying to anticipate how these plans might change in future.

Second, the **Sustainable Development Scenario** – which provides a strategic pathway to meet global climate, air quality and energy access goals in full – has been extended to 2050 and set out in greater detail. This delivers sharper insights into what is required for the world to move in this direction.

What comes through with crystal clarity in this year's *Outlook* is that there are no simple solutions to transform the world of energy. Multiple technologies and fuels have a part to play across all sectors of the economy. For this to happen, we need strong leadership from policy makers, as governments hold the clearest responsibility to act and have the greatest scope to shape the future.

The IEA is also acting on the insights contained in the *Outlook*. For instance, our analysis shows that the pace of energy efficiency improvements is slowing, but the potential for efficiency improvements to help the world meet its sustainable energy goals is massive. This has led us to set up a high-level Global Commission for Urgent Action on Energy Efficiency to recommend how progress can be rapidly accelerated through new and stronger policy action.

This year's *Outlook* underscores the crucial importance of Africa to the future of energy, and of energy to the future of Africa. In June 2019, the IEA and the African Union Commission co-hosted a first joint Ministerial Summit in Addis Ababa, bringing together high-level government representatives and other stakeholders to advance a positive, collaborative agenda for IEA engagement.

We are also acutely aware that while the ongoing transformation of the electricity sector is full of promise, it also has implications for the stability and reliability of power grids around the world. In response, we have introduced new initiatives, including co-organising with the German Federal Ministry for Economic Affairs and Energy the first Global Ministerial Conference on System Integration of Renewables in Berlin in October 2019 and undertaking a major new report on electricity security.

Another important issue is that global emissions of methane, a potent greenhouse gas, are rising alongside CO_2. This is why we recently launched a new online methane tracker to monitor the problem and identify ways to tackle it.

These are just four examples of how the *World Energy Outlook* provides strategic guidance to the energy community and results in real-world initiatives and solutions. The goal of this year's *Outlook*, once again, is to provide energy decision makers with the data and objective analysis that they need to pursue a more secure and sustainable future. Today, this mission is more urgent than ever.

I would like to thank every one of the many hard-working people who help make the *Outlook* better every year, in particular the dedicated *WEO* team, which is led by my colleagues Laura Cozzi and Tim Gould.

Dr. Fatih Birol
Executive Director
International Energy Agency

Acknowledgements

This study was prepared by the World Energy Outlook (WEO) team in the Directorate of Sustainability, Technology and Outlooks (STO) in co-operation with other directorates and offices of the International Energy Agency. The study was designed and directed by **Laura Cozzi**, Chief Energy Modeller and Head of Division for Energy Demand Outlook, and **Tim Gould**, Head of Division for Energy Supply and Investment Outlooks.

The special focus on Africa was co-ordinated by **Stéphanie Bouckaert**, also the overall lead on end-use modelling and analysis, **Tae-Yoon Kim**, also lead on petrochemicals, oil refining and trade, and **Kieran McNamara**, also lead on the energy efficiency and renewables chapter. **Brent Wanner** led the offshore wind analysis and also the power sector modelling and analysis. **Christophe McGlade** led the work on gas infrastructure and also on oil analysis. **Paweł Olejarnik** co-ordinated the oil, natural gas and coal supply modelling. Key contributions from across the *WEO* team were from: **Zakia Adam** (lead on data management, contributed to fossil fuel subsidies), **Lucila Arboleya Sarazola** (investment, Africa, Southeast Asia), **Yasmine Arsalane** (power, Africa, economic outlook), **Bipasha Baruah** (gender), **Simon Bennett** (hydrogen, innovation), **Michela Cappannelli** (oil, gas, low-carbon fuels), **Olivia Chen** (energy access and buildings), **Arthur Contejean** (energy access, Africa), **Haoua Cisse Coulibaly** (Africa), **Daniel Crow** (climate and environment), **Davide D'Ambrosio** (power and data management), **Amrita Dasgupta** (Africa, transport and agriculture), **John Connor Donovan** (power, offshore wind), **Marina Dos Santos** (Africa), **Livia Gallarati** (Southeast Asia[1]), **Timothy Goodson** (lead on buildings and demand-side response), **Lilly Yejin Lee** (transport), **Jianguo Liu** (China), **Wataru Matsumura** (lead on Southeast Asia, fossil fuel subsidies), **Yoko Nobuoka** (investment, coal, Southeast Asia), **Sebastian Papapanagiotou** (offshore wind), **Claudia Pavarini** (lead on energy storage, power), **Daniele Perugia** (power), **Apostolos Petropoulos** (lead on transport), **Arnaud Rouget** (Africa, energy access), **Marcela Ruiz de Chavez Velez** (industry, services, Africa), **Andreas Schröder** (lead on industry), **Glenn Sondak** (oil, gas), **Leonie Staas** (climate and environment), **Alberto Toril** (investment, offshore wind), **Michael Waldron** (lead on investment, contributed to Southeast Asia), **Molly A. Walton** (Africa, energy-water nexus), **Wilfred Yu** (Africa, offshore wind), and **Peter Zeniewski** (lead on natural gas). **Teresa Coon** and **Eleni Tsoukala** provided essential support.

Edmund Hosker carried editorial responsibility.

Debra Justus was the copy-editor.

Other key contributors from across the agency were: Ali Al-Saffar, Thibault Abergel, Chiara Delmastro, John Dulac, Carlos Fernández Alvarez, George Kamiya, Jinsun Lim, Raimund Malischek, Luis Munuera, Ryszard Pośpiech, Frances Reuland, Jacopo Tattini and Tiffany Vass.

---

[1] The Southeast Asia Energy Outlook 2019 is available at: www.iea.org/southeastasia2019.

The IEA is especially grateful for the guidance and valuable advice of H.E. Kandeh Yumkella on the Africa analysis and of Michał Kurtyka (Vice-Minister of Environment, Poland) and Peter Betts on climate issues.

Mechthild Wörsdörfer, Director of STO, provided encouragement and support throughout the project. Valuable comments and feedback were provided by other senior management and numerous other colleagues within the IEA. In particular, Paul Simons, Keisuke Sadamori, Dave Turk, Amos Bromhead, Timur Gül, Nick Johnstone, Laszlo Varro, Neil Atkinson, Joel Couse, Peter Fraser, Paolo Frankl, Brian Motherway, Aad van Bohemen, Rebecca Gaghen, Masatoshi Sugiura, Aya Yoshida, Christian Zinglersen, Heymi Bahar, Christophe Barret, Adam Baylin-Stern, Alessandro Blasi, Toril Bosoni, Jean-Baptiste Dubreuil, Jason Elliott, Kathleen Gaffney, Pharoah Le Feuvre, Peter Levi, Selena Lee, Peg Mackey, Samantha McCulloch, Sara Moarif, Kristine Petrosyan, Cedric Philibert and Andrew Prag.

Thanks go to the IEA's Communication and Digitalisation Office for their help in producing the report and website materials, particularly to Jad Mouawad, Jethro Mullen, Astrid Dumond, Jon Custer, Christopher Gully, Katie Lazaro, Magdalena Sanocka, Rob Stone and Sabrina Tan. Diana Browne and Ivo Letra provided essential support to the production process. IEA's Office of the Legal Counsel, Office of Management and Administration and Energy Data Centre provided assistance throughout the preparation of the report. Uğur Öcal also provided support.

Valuable input to the analysis was provided by: David Wilkinson (independent consultant); Markus Amann, Peter Rafaj, Gregor Kiesewetter, Wolfgang Schöpp, Chris Heyes, Zbigniew Klimont, Jens Borken-Kleefeld and Pallav Purohit (International Institute for Applied Systems Analysis); Iain Staffell (Imperial College London); Christopher Andrey and Maxime Chammas (Artelys) and Per Magnus Nysveen (Rystad Energy).

Dr. Andriannah Mbandi (Stockholm Environment Institute) and Jacqueline Senyagwa (University of Cape Town) provided valuable contributions to the special focus on Africa. Valuable modelling work for the special focus on Africa was contributed by Andreas Sahlberg, Babak Khavari, Alexandros Korkovelos and Mark Howells (KTH Swedish Royal Institute of Technology); Jose Ignacio Perez-Arriaga, Fernando de Cuadra-García, Andrés González-García, and Pedro Ciller-Cutillas (MIT-Comillas Universal Energy Access Lab).

Valuable input to the offshore wind analysis was provided by Kirsten Adlunger (German Environment Agency), Philipp Beiter (US National Renewable Energy Laboratory), Jesper Breinbjerg (Ministry of Energy, Utilities and Climate, Denmark), Karsten Capion (Danish Council on Climate Change), Sune Strøm (Ørsted), Lukas Wienholt (Federal Maritime and Hydrographic Agency, Germany) and Christoph Wolter (Danish Energy Agency).

Valuable input to the biogas and biomethane analysis was provided by Paul Hughes (independent consultant), and by the IEA Bioenergy TCP Task 37 (Jerry D. Murphy, School of Engineering University College Cork).

The work could not have been achieved without the support and co-operation provided by many government bodies, organisations and companies worldwide, notably: BHP Billiton; Danish Energy; Danish Energy Agency; Department of Energy, United States; Enel; Energy Market Authority, Singapore; Eni; Environmental Protection Agency; United States; Equinor; European Commission; Federal Ministry for Economic Affairs and Energy, Germany; Iberdrola; Ministry of Climate, Energy and Utilities, Denmark; Ministry of Economic Affairs and Climate Policy, Netherlands; Ministry of Economy, Trade and Industry, Japan; Ministry of Petroleum and Energy, Norway; Natural Resources Canada; Nexans; The Research Institute of Innovative Technology for the Earth, Japan; Schneider Electric; Snam; Temasek, Singapore; and Toshiba. Activities within the IEA Clean Energy Technologies Programme provided valuable support to this report.

Thanks also go to the IEA Energy Business Council, IEA Coal Industry Advisory Board, IEA Energy Efficiency Industry Advisory Board and the IEA Renewable Industry Advisory Board.

A number of events were organised to provide input to this report. The participants offered valuable new insights, feedback and data for this analysis.

- High-level workshop on Biogas and Biomethane, Paris, 19 February 2019
- High-level workshop on the Africa Energy Outlook, Paris, 17 April 2019
- High-level workshop on Offshore Wind Outlook, Paris, 13 May 2019

Further details on these events are available at www.iea.org/weo/events.

Peer reviewers

Many senior government officials and international experts provided input and reviewed preliminary drafts of the report. Their comments and suggestions were of great value. They include:

Thomas A. Frankiewicz	US Environmental Protection Agency (EPA)
Amani Abou-Zeid	African Union Commission, Ethiopia
Olalekan David Adeniyi	Chemical Engineering Department, Federal University of Technology, Nigeria
Keigo Akimoto	The Research Institute of Innovative Technology for the Earth, Japan
Safiatou Alzouma Nouhou	Africa Renewable Energy Initiative (AREI)
An Qi	Energy Research Institute, China
Venkatachalam Anbumozhi	Economic Research Institute for ASEAN and East Asia (ERIA)
Pedro Antmann	World Bank
Marco Arcelli	EPH Group
Edi Assoumou	Mines ParisTech, France
Peter Bach	Danish Energy Agency
Douglas K Baguma	Innovex, Uganda
Vicki Bakhshi	BMO Global Asset Management, United Kingdom
Rangan Banerjee	Indian Institute of Technology, Bombay
Marco Baroni	Independent consultant

Paul Baruya	Clean Coal Centre
Diana Bauer	US Department of Energy
Harmeet Bawa	ASEA Brown Boveri (ABB) Power Grids
Christopher Beaton	International Institute for Sustainable Development (IISD)
David Bénazéraf	Sahel and West Africa Club (SWAC), OECD
Christian Besson	Independent consultant
Murray Birt	DWS
Paul Bjacek	Accenture
Rina Bohla Zeller	Vestas, Denmark
Teun Bokhoven	Consolair, Netherlands
Jason Bordoff	Columbia University, United States
Nils Borg	European Council for an Energy Efficient Economy (ECEEE)
Edward Borgstein	Rocky Mountain Institute, Sustainable Energy for Economic Development (SEED) program (AFRICA)
Stephen Bowers	Evonik Industries AG
William Brent	Power for All
Tyler Bryant	Fortis British Columbia, Canada
Mick Buffier	Glencore
Nick Butler	Independent consultant
Irene Calvé Saborit	Sunkofa Energy
Guy Caruso	Center for Strategic and International Studies, United States
Pierpaolo Cazzola	International Transport Forum
Cho Ilhyun	Korea Energy Economics Institute (KEEI)
Drew Clarke	Australian Energy Market Operator
Ute Collier	Practical Action
Rebecca Collyer	European Climate Foundation
Emanuela Colombo	Politecnico di Milano, Italy
Francis Condon	UBS Asset Management
Anne-Sophie Corbeau	BP
Jon Lezamiz Cortazar	Siemens Gamesa
Fergus Costello	Siemens Gamesa
Ian Cronshaw	Independent consultant
Helen Currie	ConocoPhillips
Jostein Dahl Karlsen	Gas & Oil Technologies Collaboration Program (IEA GOT)
David Daniels	US Energy Information Administration
Francois Dassa	EDF
Ruud de Bruijne	Netherlands Enterprise Agency (RVO)
Christian de Gromard	Agence Française de Développement (AFD), France
Ralf Dickel	Oxford Institute for Energy Studies, United Kingdom
Giles Dickson	WindEurope
Dan Dorner	UK Department for Business Energy and Industrial Strategy
Loic Douillet	GE Power
Gina Downes	Eskom, South Africa
Jon Dugstad	Norwegian Energy Partners (NORWEP)

Joseph Essandoh-Yeddu	Energy Commission, Ghana
Simon Evans	Carbon Brief
Francesco Ferioli	DG Energy – European Commission
Capella Festa	Schlumberger
Nikki Fisher	Anglo American
Justin Flood	Sunset Power International – Vales Point Power Station
Fridtjof Fossum Unander	Research Council of Norway
Silvia Francioso	GOGLA
Nathan Frisbee	Schlumberger
David Fritsch	US Energy Information Administration
Mike Fulwood	Nexant
David G. Hawkins	Natural Resources Defense Council, United States
Jean-Francois Gagne	Department of Natural Resources Canada
Ashwin Gambhir	Prayas, Energy Group, India
Andrew Garnett	University of Queensland, Australia
Francesco Gattei	Eni
Peter George	Clean Cooking Alliance
Dolf Gielen	International Renewable Energy Agency (IRENA)
Olivia Gippner	DG Climate Action, European Commission
Craig Glazer	PJM Interconnection
Kazushige Gobe	Japan Bank For International Cooperation (JBIC)
Desmond Godson	Asia Biogas, Thailand
David L. Goldwyn	Atlantic Council, United States
Martin Graversgaard Nielsen	European Network of Transmission System Operators for Gas (ENTSOG)
Oliver Grayer	The Institutional Investors Group on Climate Change (IIGCC)
Andrii Gritsevskyi	International Atomic Energy Agency (IAEA)
Monica Gullberg	Swedish International Development Cooperation Agency (SIDA)
Han Wenke	Energy Research Institute, National Development and Reform Commission, China
Peter Handley	DG GROW, European Commission
Marc Hedin	ENERGIR
Jan Hein Jesse	JOSCO Energy Finance and Strategy Consultancy
Colin Henderson	Clean Coal Centre
James Henderson	Oxford Institute for Energy Studies, United Kingdom
Doug Hengel	German Marshall Fund of the United States
Andrew Herscowitz	Power Africa, US Agency for International Development
Gunnar Herzig	World Forum Offshore Wind
Martin Hiller	Renewable Energy and Energy Efficiency Partnership (REEEP)
Masazumi Hirono	Tokyo Gas
Kamiishi Hiroto	Japan International Cooperation Agency (JICA)
Neil Hirst	Imperial College London
Stéphane His	Agence Française de Développement

Nastassja Hoffet	Ministère de l'Europe et des Affaires Etrangères, France
Takashi Hongo	Mitsui Global Strategic Studies Institute, Japan
Christina Hood	Compass Climate
Didier Houssin	IFP Energies Nouvelles, France
Hu Jiang	Beijing Tianrun New Energy Investment
Thad Huetteman	US Energy Information Administration
Ole Hveplund	Nature Energy
Jan Hylleberg	Danish Wind Industry Association
Samuel Igbatayo	Afe Babalola University, Nigeria
Hans Ejsing Jorgensen	DTU Wind Energy, Germany
Emmanuel K. Ackom	UN Environment Programme (UNEP), Technical University of Denmark Partnership (DTU)
Sandholt Kaare	National Renewable Energy Centre, China
Sohbet Karbuz	Mediterranean Observatory for Energy (OME)
Yoichi Kaya	The Research Institute of Innovative Technology for the Earth, Japan
Talla Kebe	United Nations, Office of The Special Advisor On Africa
Daniel Ketoto	Government of Kenya/Office of the President
Kidong Kim	Korea Gas Corporation (KOGAS)
Robert Kleinberg	Columbia University, United States
Markus Klingbeil	Shell
David Knapp	Energy Intelligence Group
Oliver Knight	World Bank, Energy Sector Management Assistance Program
Hans Jorgen Koch	Nordic Energy Research, Norway
Lukasz Kolinski	DG Energy – European Commission
Pawel Konzal	Chevron
Christoph Kost	Institute for Transportation & Development Policy (ITDP)
Ken Koyama	Institute of Energy Economics, Japan
Masaomi Koyama	Ministry of Economy, Trade and Industry, Japan
Jim Krane	Baker Institute
Anil Kumar Jain	NITI Aayog, India
Atsuhito Kurozumi	Kyoto University of Foreign Studies
Michał Kurtyka	Ministry of Energy and Environment, Poland
Francesco La Camera	International Renewable Energy Agency
Sarah Ladislaw	Center for Strategic and International Studies (CSIS), United States
Susana Lagarto	European Investment Bank (EIB)
Glada Lahn	Chatham House
Per Landberg	Norwegian Agency for Development Cooperation (NORAD)
Richard Lavergne	Ministry for Economy and Finance, France
Francisco Laveron	Iberdrola
David Lecoque	Alliance for Rural Electrification
Andy Lewis	Cadent Gas Limited

Li Jiangtao	State Grid Energy Research Institute, China
Liu Wenke	China Coal Information Institute
Liu Xiaoli	Energy Research Institute, National Development and Reform Commission, China
Liu Yun Hui	China Energy Investment Group
Nikolaj Lomholt Svensson	Embassy of Denmark in Ethiopia
Giacomo Luciani	Sciences Po, France
Joan MacNaughton	The Climate Group
Domenico Maggi	SNAM
Emadeldin Ahmed Mahgoub	Agricultural Research Corporation (ARC), Sudan
Trieu Mai	National Renewable Energy Laboratory (NREL), United States
Mouhamadou Makhtar	Ministère du pétrole et des énergies, Senegal
Senatla Mamahloko	Council for Scientific and Industrial Research (CSIR), South Africa
Haigh Martin	Shell
Anne Marx Lorenzen	Danish Ministry of Climate, Energy and Utilities
Atef Marzouk	African Union Commission
Eric Masanet	Northwestern University, United States
Takeshi Matsushita	Mitsubishi Corporation
Felix Chr. Matthes	Öko-Institut – Institute for Applied Ecology, Germany
Dimitris Mentis	World Resources Institute (WRI), United States
Antonio Merino Garcia	Repsol
Bert Metz	European Climate Foundation
Michelle Michot Foss	University of Texas
Asami Miketa	International Renewable Energy Agency
Vincent Minier	Schneider Electric
Arthur Minsat	Organisation for Economic Co-operation and Development
Tatiana Mitrova	Energy Research Institute of the Russian Academy of Sciences
Simone Mori	ENEL
Peter Morris	Minerals Council of Australia
Charlotte Morton	Anaerobic Digestion and Bioresources Association (ADBA)
Isabel Murray	Department of Natural Resources, Canada
Rose Mutiso	Energy for Growth Hub
Steve Nadel	American Council for an Energy-Efficient Economy, United States
Sumie Nakayama	J-POWER
Joachim Nick-Leptin	Federal Ministry for Economic Affairs and Energy, Germany
Esben Baltzer Nielsen	Vattenfall
Susanne Nies	European Network of Transmission System Operators for Electricity (ENTSO-E)
Koshi Noguchi	Toshiba of Europe Ltd.
Ted Nordhaus	Breakthrough Institute, United States

Acknowledgements

Glory Oguegbu	Renewable Energy Technology Training Institute (RETTI)
Karin Ohlenforst	Global Wind Energy Council (GWEC)
Sheila Oparaocha	ENERGIA/HIVOS
Isaiah Owiunji	World Wide Fund for Nature (WWF), Uganda
Cathy Oxby	Africa Green
Pak Yangduk	Korea Energy Economics Institute (KEEI)
Adam Parums	CRU
Stefan Pauliuk	Faculty of Environment and Natural Resources University of Freiburg, Germany
Jose Ignacio Perez Arriaga	Comillas Pontifical University's Institute for Research in Technology, Spain
Glen Peters	CICERO
Gregor Pett	UNIPER
Marco Pezzaglia	Consorzio Italiano Biogas (CIB)
Stephanie Pfeifer	The Institutional Investors Group on Climate Change (IIGCC)
Jem Porcaro	Sustainable Energy for All
Elisa Portale	World Bank
Mark Radka	Economy Division, UN Environment Programme (UNEP)
Andrew Renton	Transpower
Christoph Richter	Solarway
Seth Roberts	Saudi Arabian Oil Company
Karen Roiy	Danfoss
Manuel Rudolph	German Environment Agency (UBA)
Amir Sadeghi Emamgholi	IHS Markit and George Washington University
Vineet Saini	Ministry of Science and Technology, India
Romain Saint Leger	Energy Pool
Papa Samba Ba	Ministère du Pétrole et des Énergies, Senegal
Hans-Wilhelm Schiffer	World Energy Council
Filip Schittecatte	ExxonMobil
Sandro Schmidt	Federal Institut for Geosciences and Natural Resources, Germany
Karl Schoensteiner	Siemens
Thomas Scurfield	National Resource Governance Institute, Tanzania
Shan Baoguo	State Grid Energy Research Institute, China
Adnan Shihab Eldin	Foundation for the Advancement of Sciences, Kuwait
Maria Sicilia Salvadores	Enagas
Katia Simeonova	United Nations Framework Convention on Climate Change
Stephan Singer	Climate Action Network International
Jim Skea	Imperial College London
Aaron Smith	Principle Power
Stuart Smith	National Offshore Petroleum Safety and Environmental Management Authority, Australia
Christopher Snary	UK Department for Business, Energy and Industrial Strategy
Takeshi Soda	Ministry of Economy, Trade and Industry, Japan

John Staub	US Energy Information Administration
James Steel	UK Department for Business, Energy and Industrial Strategy
Volker Stehmann	Innogy SE
Jonathan Stern	Oxford Institute for Energy Studies, United Kingdom
Robert Stoner	MIT Energy Initiative, United States
Bert Stuij	Netherlands Enterprise Agency
Dinesh Surroop	University of Mauritius
Minoru Takada	United Nations Department of Economic and Social Affairs
Yasuo Tanabe	Hitachi
Alban Thomas	grtGAZ
Wim Thomas	Shell
Johannes Trüby	Deloitte
Nikos Tsafos	Center for Strategic and International Studies (CSIS)
Sergey Tverdokhleb	Siberian Coal Energy Company (SUEK)
Charlotte Unger Larson	Swedish Wind Energy Association
Rob van der Hage	TenneT
Bob van der Zwaan	Energy Research Centre of the Netherlands (ECN part of TNO)
Noe Van Hulst	Ministry of Economic Affairs & Climate Policy, Netherlands
Tom Van Ierland	DG Climate Action, European Commission
Wim Van Nes	SNV Netherlands Development Organisation
Frank Verrastro	Center for Strategic and International Studies, United States
David Victor	UC San Diego School of Global Policy and Strategy, United States
Andreas Wagner	Stiftung Offshore-Windenergie
Andrew Walker	Cheniere Energy
Paul Welford	Hess Corporation
Paul Wendring	Prognos
Peter Westerheide	BASF
Akira Yabumoto	J-power
Masato Yamada	MHI Vestas Offshore Wind
Mel Ydreos	International Gas Union
Abdulmutalib Yussuff	Project Drawdown
Faruk Yusuf Yabo	Federal Ministry of Power, Works & Housing, Nigeria
William Zimmern	BP
Christian Zinglersen	Clean Energy Ministerial

The individuals and organisations that contributed to this study are not responsible for any opinions or judgments it contains. All errors and omissions are solely the responsibility of the IEA.

This document and any map included herein are without prejudice to the status of or sovereignty over any territory, to the delimitation of international frontiers and boundaries and to the name of any territory, city or area.

TABLE OF CONTENTS

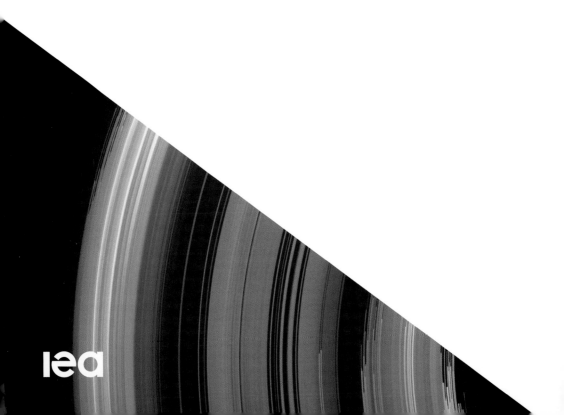

PART A: GLOBAL ENERGY TRENDS

Overview and key findings	1
Energy and the Sustainable Development Goals	2
Outlook for oil	3
Outlook for natural gas	4
Outlook for coal	5
Outlook for electricity	6
Outlook for energy efficiency and renewables	7

PART B: SPECIAL FOCUS ON AFRICA

Africa today	8
Urbanisation, industrialisation and clean cooking	9
Access to electricity and reliable power	10
Natural gas and resource management	11
Implications for Africa and the world	12

PART C: WEO INSIGHTS

Prospects for gas infrastructure	13
Outlook for offshore wind	14

ANNEXES

Foreword		3
Acknowledgements		5
Executive summary		23
Introduction		29

Part A: Global Energy Trends — 33

1 Overview and key findings — 35

Introduction		37
Scenarios		38
1.1	Overview	38
1.2	Primary energy demand by region	40
1.3	Final energy consumption and efficiency	42
1.4	Power generation and energy supply	44
1.5	Emissions	46
1.6	Trade	48
1.7	Investment	50
1.8	Differences between the *WEO-2019* and *WEO-2018*	52
Exploring the pace of change		54
1.9	Sustainability	54
1.10	Security	63
1.11	Affordability	70

2 Energy and the Sustainable Development Goals — 79

Introduction		81
Sustainable Development Scenario		82
2.1	Scenario overview	82
2.2	Scenario outcomes: Universal energy access	86
2.3	Scenario outcomes: Air pollution	87
2.4	Scenario outcomes: CO_2 emissions	88
2.5	Energy sector transformation in the Sustainable Development Scenario	89
2.6	Investment in the Sustainable Development Scenario	93
Key themes		96
2.7	How are we doing?	96
2.8	Where do we need to get to?	102
2.9	How much further can we go?	121

3 Outlook for oil — 129

Introduction		131

		Scenarios	132
	3.1	Overview	132
	3.2	Oil demand by region	134
	3.3	Oil demand by sector	136
	3.4	Oil supply by type	138
	3.5	Oil supply by region	140
	3.6	Oil product demand and refining	143
	3.7	Trade	145
	3.8	Investment	146
		Key themes	147
	3.9	Passenger cars: are we approaching the peak of the "ICE age"?	147
	3.10	Pushing the boundaries of US tight oil	155
	3.11	Can the world afford to relax about security of oil supply?	165

4 Outlook for natural gas — 175

		Introduction	177
		Scenarios	178
	4.1	Overview	178
	4.2	Natural gas demand by region and sector	180
	4.3	Natural gas production	183
	4.4	Trade and investment	186
		Key themes	188
	4.5	Associated gas: the upstream link between oil and gas markets	188
	4.6	How does innovation affect the outlook for LNG?	197
	4.7	Understanding the global potential for coal-to-gas switching	209

5 Outlook for coal — 219

		Introduction	221
		Scenarios	222
	5.1	Overview	222
	5.2	Coal demand by region and sector	224
	5.3	Coal production by region	226
	5.4	Trade	228
	5.5	Investment	229
		Key themes	230
	5.6	A view beyond power: industrial coal use	230
	5.7	Who will invest in coal supply?	236
	5.8	Coal mine methane	245

6 Outlook for electricity — 253

Introduction — 255
Scenarios — 256
- 6.1 Overview — 256
- 6.2 Electricity demand by region — 258
- 6.3 Electricity demand by sector — 261
- 6.4 Electricity supply by source — 264
- 6.5 Installed capacity by source — 266
- 6.6 Electricity supply by region — 268
- 6.7 Power sector investment — 271
- 6.8 Competitiveness of power generation technologies — 273
- 6.9 Power sector emissions — 275

Key themes — 278
- 6.10 Affordability of electricity — 278
- 6.11 Tackling emissions from coal-fired power plants — 283
- 6.12 Exploring the new frontiers of flexibility — 291

7 Outlook for energy efficiency and renewables — 299

Introduction — 301
Scenarios — 302
- 7.1 Energy efficiency overview — 302
- 7.2 Renewables overview — 304
- 7.3 Efficiency by sector and investments — 306
- 7.4 Renewables policies and investments — 308

Key themes — 310
- 7.5 Material efficiency in heavy industries — 310
- 7.6 Smart electricity use: the power of the hour in reducing emissions — 320
- 7.7 Biogas: turning organic matter into renewable energy — 328

Part B: Special Focus on Africa — 337

Introduction — 339

8 Africa today — 347

- 8.1 Context — 349
  - 8.1.1 Economic growth and industrialisation — 349
  - 8.1.2 Demographics and urbanisation — 351
  - 8.1.3 Infrastructure and investment — 352

8.2	Access to modern energy	355
	8.2.1 Clean cooking	357
	8.2.2 Electricity	361
	8.2.3 Affordability: energy prices and fossil fuel subsidies	367
8.3	Energy trends in Africa today	369
	8.3.1 Energy demand	369
	8.3.2 Power sector	375
	8.3.3 Fossil fuel resources and supply	386
	8.3.4 Renewable resources and supply	392
	8.3.5 Environment	395

9 Urbanisation, industrialisation and clean cooking — 399

9.1	Introduction	401
9.2	Urbanisation and industrialisation, drivers of growth	403
	9.2.1 Residential sector	408
	9.2.2 Transport sector	412
	9.2.3 Productive uses	416
9.3	Clean cooking: the role of cities and higher incomes	422
	9.3.1 Increasing access to clean cooking options	424
	9.3.2 Rapid urbanisation requires better use of charcoal	427
	9.3.3 Rural areas – how to unleash the potential of biogas?	430

10 Access to electricity and reliable power — 433

10.1	Introduction	435
10.2	Outlook for electricity access	436
10.3	Outlook for electricity demand	438
	10.3.1 Electricity demand growth by sector	440
	10.3.2 Electricity demand growth by region	442
10.4	Outlook for electricity supply	446
	10.4.1 On-grid supply	448
	10.4.2 Role of decentralised systems to reach universal access to electricity	454
10.5	Reliability	455
10.6	Affordability	459
10.7	Investment needs for reliable, sustainable and affordable power	461
10.8	Sources of finance for power investment in sub-Saharan Africa	463
	10.8.1 Investment framework and market structure	463
	10.8.2 Private financing is concentrated in IPPs, mostly in South Africa	465

10.9	Closing the investment and financing gap	467
	10.9.1 Improve the financial and operational performance of utilities	467
	10.9.2 Enhance policy and regulatory frameworks to improve bankability	469
	10.9.3 Create supportive enabling environments for rural electricity access	470
	10.9.4 Strengthen provision of long-term finance	472

11 Natural gas and resource management 475

11.1	Introduction	477
11.2	The role of natural gas in Africa's energy mix	479
	11.2.1 Prospects for gas in key regions	481
	11.2.2 Outlook for natural gas demand, production and infrastructure developments in Africa	488
	11.2.3 Conclusions	490
11.3	Maximising the value of Africa's resources	491
	11.3.1 Outlook for fossil fuel production	493
	11.3.2 Strategic responses for resource-holders in Africa	496
	11.3.3 Conclusions	504

12 Implications for Africa and the world 507

Introduction		507
Implications for the world		507
Regional and country profiles for Africa		524
12.1	Sub-Saharan Africa	526
12.2	Angola	530
12.3	Côte d'Ivoire	534
12.4	Democratic Republic of the Congo	538
12.5	Ethiopia	542
12.6	Ghana	546
12.7	Kenya	550
12.8	Mozambique	554
12.9	Nigeria	558
12.10	Senegal	562
12.11	South Africa	566
12.12	Tanzania	570

Part C: WEO Insights — 575

13. Prospects for gas infrastructure — 577

- 13.1 Introduction — 579
  - 13.1.1 Role of gas infrastructure today — 580
  - 13.1.2 Role of natural gas in energy transitions — 581
  - 13.1.3 Need for gas supply to evolve — 585
- 13.2 Low-carbon hydrogen — 587
  - 13.2.1 Hydrogen use today — 587
  - 13.2.2 Costs and potential to blend hydrogen into gas networks — 589
- 13.3 Biomethane — 594
  - 13.3.1 Biomethane use today — 595
  - 13.3.2 Blending biomethane into gas networks: costs and potential — 596
- 13.4 Outlook for low-carbon hydrogen and biomethane — 600
  - 13.4.1 Stated Policies Scenario — 600
  - 13.4.2 Sustainable Development Scenario — 601
- 13.5 Implications for emissions and energy security — 603
  - 13.5.1 Reducing CO_2 emissions — 603
  - 13.5.2 Avoiding methane emissions — 605
  - 13.5.3 Energy security — 607
- 13.6 Implications for policy makers and industry — 609

14. Outlook for offshore wind — 613

- 14.1 Introduction — 615
- 14.2 Offshore wind power today — 615
  - 14.2.1 Current status — 615
  - 14.2.2 Market size and key players — 617
  - 14.2.3 Offshore wind technology and performance — 619
  - 14.2.4 Offshore wind costs for projects commissioned in 2018 — 624
- 14.3 Offshore wind outlook to 2040 — 627
  - 14.3.1 Global outlook — 627
  - 14.3.2 Regional outlook — 628
  - 14.3.3 Offshore wind costs, value and competitiveness — 635
- 14.4 Opportunities for faster growth of offshore wind — 648
  - 14.4.1 Global technical potential for offshore wind — 648
  - 14.4.2 Improved economics for offshore wind — 654
  - 14.4.3 Increased demand for renewable hydrogen — 655
  - 14.4.4 Public acceptance — 656

14.5	Uncertainties that could slow offshore wind growth		657
	14.5.1	Developing efficient supply chains in new markets	657
	14.5.2	Environmental concerns	657
	14.5.3	Onshore grid development	658
14.6	Implications		661
	14.6.1	Achieving environmental goals	661
	14.6.2	Synergies with oil and gas activities	662
	14.6.3	Enhanced energy security and affordability	664

Annexes 667

Annex A. Tables for scenario projections	669
Annex B. Design of the scenarios	751
Annex C. Definitions	771
Annex D. References	785

Executive Summary

The energy world is marked by a series of deep disparities. The gap between the promise of energy for all and the fact that almost one billion people still do not have access to electricity. The gap between the latest scientific evidence highlighting the need for ever-more-rapid cuts in global greenhouse gas emissions and the data showing that energy-related emissions hit another historic high in 2018. The gap between expectations of fast, renewables-driven energy transitions and the reality of today's energy systems in which reliance on fossil fuels remains stubbornly high. And the gap between the calm in well-supplied oil markets and the lingering unease over geopolitical tensions and uncertainties.

More than ever, energy decision makers need to take a hard, evidence-based look at where they stand and the implications of the choices they make. The *World Energy Outlook* does not provide a forecast of what will happen. Instead, it provides a set of scenarios that explore different possible futures, the actions – or inactions – that bring them about and the interconnections between different parts of the system.

Understanding our scenarios

The Current Policies Scenario shows what happens if the world continues along its present path, without any additional changes in policy. In this scenario, energy demand rises by 1.3% each year to 2040, with increasing demand for energy services unrestrained by further efforts to improve efficiency. While this is well below the remarkable 2.3% growth seen in 2018, it would result in a relentless upward march in energy-related emissions, as well as growing strains on almost all aspects of energy security.

The Stated Policies Scenario, by contrast, incorporates today's policy intentions and targets. Previously known as the New Policies Scenario, it has been renamed to underline that it considers only specific policy initiatives that have already been announced. The aim is to *hold up a mirror to the plans of today's policy makers* and illustrate their consequences, not to guess how these policy preferences may change in the future.

In the Stated Policies Scenario, energy demand rises by 1% per year to 2040. Low-carbon sources, led by solar photovoltaics (PV), supply more than half of this growth, and natural gas, boosted by rising trade in liquefied natural gas (LNG), accounts for another third. Oil demand flattens out in the 2030s, and coal use edges lower. Some parts of the energy sector, led by electricity, undergo rapid transformations. Some countries, notably those with "net zero" aspirations, go far in reshaping all aspects of their supply and consumption. However, the momentum behind clean energy technologies is not enough to offset the effects of an expanding global economy and growing population. The rise in emissions slows but, with no peak before 2040, the world falls far short of shared sustainability goals.

The Sustainable Development Scenario maps out a way to meet sustainable energy goals in full, requiring rapid and widespread changes across all parts of the energy system. This scenario charts a path fully aligned with the Paris Agreement by holding the rise in global temperatures to "well below 2°C … and pursuing efforts to limit [it] to 1.5°C", and meets objectives related to universal energy access and cleaner air. The breadth of the world's energy needs means that there are no simple or single solutions. Sharp emission cuts are achieved across the board thanks to multiple fuels and technologies providing efficient and cost-effective energy services for all.

Energy security remains paramount, and oil stays in the spotlight

A fast-moving energy sector highlights the importance of a broad and dynamic approach to energy security. The attacks in Saudi Arabia in September 2019 underlined that traditional energy security risks have not gone away. Meanwhile, new hazards – from cybersecurity to extreme weather – require constant vigilance from governments. We estimate that almost one-fifth of the growth in global energy use in 2018 was due to hotter summers pushing up demand for cooling and cold snaps leading to higher heating needs.

Shale output from the United States stays higher for longer, reshaping global markets, trade flows and security. Annual US production growth slows from the breakneck pace seen in recent years, but updated official estimates of underlying resources nonetheless mean that the United States accounts for 85% of the increase in global oil production to 2030 in the Stated Policies Scenario, and for 30% of the increase in gas. This bolsters the position of the United States as an exporter of both fuels. By 2025, total US shale output (oil and gas) overtakes total oil and gas production from Russia.

Higher US output pushes down the share of OPEC countries and Russia in total oil production. This share drops to 47% in 2030, from 55% in the mid-2000s, implying that efforts to manage conditions in the oil market could face strong headwinds. Pressures on the hydrocarbon revenues of some of the world's major producers also underline the importance of their efforts to diversify their economies.

Whichever pathway the energy system follows, the world still relies heavily on oil supply from the Middle East. The region remains by far the largest net provider of oil to world markets, as well as an important exporter of LNG. This means that one of the world's busiest trade routes, the Strait of Hormuz, retains its position as a crucial artery for global energy trade, especially for Asian countries such as China, India, Japan and Korea that rely heavily on imported fuel. In the Stated Policies Scenario, 80% of international oil trade ends up in Asia in 2040, propelled in large part by a doubling of India's import needs.

Electricity moves to the heart of modern energy security

Cost reductions in renewables and advances in digital technologies are opening huge opportunities for energy transitions, while creating some new energy security dilemmas. Wind and solar PV provide more than half of the additional electricity generation to 2040 in the Stated Policies Scenario and almost all the growth in the Sustainable Development Scenario. Policy makers and regulators will have to move fast to keep up with the pace of technological change and the rising need for flexible operation of power systems. Issues such as the market design for storage, the interface between electric vehicles and the grid, and data privacy all have the potential to expose consumers to new risks.

The rise of the African energy consumer

Africa – the special focus of WEO-2019 – is increasingly influential for global energy trends. In the Stated Policies Scenario, the rise in Africa's oil consumption to 2040 is larger than that of China, while the continent also sees a major expansion in natural gas use, prompted in part by a series of large discoveries made in recent years. The big open question for Africa remains the speed at which solar PV will grow. To date, a continent with the richest solar resources in the world has installed only around 5 gigawatts (GW) of solar PV, less than 1% of the global total. Solar PV would provide the cheapest source of electricity for many of the 600 million people across Africa without electricity access today.

More than half a billion people are added to Africa's urban population by 2040. This is much higher than the growth seen in China's urban population between 1990 and 2010, a period in which China's production of materials such as steel and cement sky-rocketed. Africa's infrastructure development is not set to follow the same path, but the energy implications of African urbanisation trends are still profound. The expected growth in population in Africa's hottest regions also means that up to half a billion additional people would need air conditioners or other cooling services by 2040. Our Africa analysis underlines that the planning, design and governance of the world's growing cities, the industrial materials that are used in their construction, and the transport options that are available to their inhabitants are critical issues for the global outlook.

An urgent need to take full advantage of the world's "first fuel"

The faltering momentum behind global energy efficiency improvements is cause for deep concern. It comes against a backdrop of rising needs for heating, cooling, lighting, mobility and other energy services. Improvements in the energy intensity of the global economy (the amount of energy used per unit of economic activity) are slowing: the 1.2% improvement in 2018 was around half the average rate seen since 2010. This reflects a relative lack of new energy efficiency policies and of efforts to tighten existing measures.

A sharp pick-up in efficiency improvements is the single most important element that brings the world towards the Sustainable Development Scenario. The pursuit of all economically viable opportunities for efficiency improvement can reduce global energy intensity by more than 3% each year. This includes efforts to promote the efficient design, use and recycling of materials such as steel, aluminium, cement and plastics. This increased "material efficiency" could be enough in itself to halt the growth in emissions from these sectors. Innovative approaches also include the use of digital tools to shift electricity demand to cheaper and less emissions-intensive hours of the day, reducing electricity bills for consumers and helping with system balancing, while also helping to reduce emissions.

Critical fuel choices hang in the balance

A three-way race is underway among coal, natural gas and renewables to provide power and heat to Asia's fast-growing economies. Coal is the incumbent in most developing Asian countries: new investment decisions in coal-using infrastructure have slowed sharply, but the large stock of existing coal-using power plants and factories (and the 170 GW of capacity under construction worldwide), provides coal with considerable staying power in the Stated Policies Scenario. Renewables are the main challenger to coal in Asia's power sector, led by China and India. Developing countries in Asia account for over half of the global growth in generation from renewables. Demand for natural gas has been growing fast as a fuel for industry and (in China) for residential consumers, spurring a worldwide wave of investment in new LNG supply and pipeline connections. In our projections, 70% of the increase in Asia's gas use comes from imports – largely from LNG – but the competitiveness of this gas in price-sensitive markets remains a key uncertainty.

In the Stated Policies Scenario, global growth in oil demand slows markedly post-2025 before flattening out in the 2030s. Oil demand for long-distance freight, shipping and aviation, and petrochemicals continues to grow. But its use in passenger cars peaks in the late 2020s due to fuel efficiency improvements and fuel switching, mainly to electricity. Lower battery costs are an important part of the story: electric cars in some major markets soon become cost-competitive, on a total-cost-of-ownership basis, with conventional cars.

Consumer preferences for SUVs could offset the benefits from electric cars. The growing appetite among consumers for bigger and heavier cars (SUVs) is already adding extra barrels to global oil consumption. SUVs are more difficult to electrify fully, and conventional SUVs consume 25% more fuel per kilometre than medium-sized cars. If the popularity of SUVs continues to rise in line with recent trends, this could add another 2 million barrels per day to our projection for 2040 oil demand.

However fast overall energy demand grows, electricity grows faster

Electricity use grows at more than double the pace of overall energy demand in the Stated Policies Scenario, confirming its place at the heart of modern economies. Growth in electricity use in the Stated Policies Scenario is led by industrial motors (notably in China), followed by household appliances, cooling and electric vehicles. In the Sustainable Development Scenario, electricity is one of the few energy sources that sees growing consumption in 2040 – mainly due to electric vehicles – alongside the direct use of renewables, and hydrogen. The share of electricity in final consumption, less than half that of oil today, overtakes oil by 2040.

Solar PV becomes the largest component of global installed capacity in the Stated Policies Scenario. The expansion of generation from wind and solar PV helps renewables overtake coal in the power generation mix in the mid-2020s. By 2040, low-carbon sources provide more than half of total electricity generation. Wind and solar PV are the star performers, but hydropower (15% of total generation in 2040) and nuclear (8%) retain major shares.

Battery costs matter

The speed at which battery costs decline is a critical variable for power markets as well as for electric cars. India is the largest overall source of energy demand growth in this year's *Outlook*, and we examine how a cost-effective combination of cheaper battery storage and solar PV could reshape the evolution of India's power mix in the coming decades. Battery storage is well suited to provide the short-term flexibility that India needs, allowing a lunchtime peak in solar PV supply to meet an early evening peak in demand. In the Stated Policies Scenario, a major reduction in battery costs means that some 120 GW of storage are installed by 2040. We also examine the possibility that battery costs could decline even faster – an extra 40% by 2040 – because of greater industrial economies of scale or a breakthrough in battery chemistry, for example. In this case, combined solar and battery storage plants would be a very compelling economic and environmental proposition, reducing sharply India's projected investment in new coal-fired power plants.

Offshore wind is gathering speed

Cost reductions and experience gained in Europe's North Sea are opening up a huge renewable resource. Offshore wind has the technical potential to meet today's electricity demand many times over. It is a variable source of generation, but offshore wind offers considerably higher capacity factors than solar PV and onshore wind thanks to ever-larger turbines that tap higher and more reliable wind speeds farther away from shore. There are further innovations on the horizon, including floating turbines that can open up new resources and markets.

Increasingly cost-competitive offshore wind projects are on course to attract a trillion dollars of investment to 2040. Europe's success with the technology has sparked interest in China, the United States and elsewhere. In the Sustainable Development Scenario, offshore wind rivals its onshore counterpart as the leading source of electricity generation in the European Union, paving the way to full decarbonisation of Europe's power sector. Even higher deployment is possible if offshore wind becomes the foundation for the production of low-carbon hydrogen.

Tackling the legacy issues head on

If the world is to turn today's emissions trend around, it will need to focus not only on new infrastructure but also on the emissions that are "locked in" to existing systems. That means addressing emissions from existing power plants, factories, cargo ships and other capital-intensive infrastructure already in use. Despite rapid changes in the power sector, there is no decline in annual power-related CO_2 emissions in the Stated Policies Scenario. A key reason is the longevity of the existing stock of coal-fired power plants that account for 30% of all energy-related emissions today.

Over the past 20 years, Asia has accounted for 90% of all coal-fired capacity built worldwide, and these plants have potentially long operational lifetimes ahead of them. In developing economies in Asia, existing coal-fired plants are just 12 years old on average. We consider three options to bring down emissions from the existing stock of plants: to retrofit them with carbon capture, utilisation and storage (CCUS) or biomass co-firing equipment; to repurpose them to focus on providing system adequacy and flexibility while reducing operations; or to retire them early. In the Sustainable Development Scenario, most of the 2 080 GW of existing coal-fired capacity would be affected by one of these three options.

What's in the pipeline for gas?

Gas grids provide a crucial mechanism to bring energy to consumers, typically delivering more energy than electricity networks and providing a valuable source of flexibility. From an energy security perspective, parallel gas and electricity grids can be complementary assets. From an energy transitions perspective, natural gas can provide near-term benefits when replacing more polluting fuels. A key longer-term question is whether gas grids can deliver truly low- or zero-carbon energy sources, such as low-carbon hydrogen and biomethane. Low-carbon hydrogen is enjoying a wave of interest, although for the moment it is relatively expensive to produce. Blending it into gas networks would offer a way to scale up supply technologies and reduce costs. Our new assessment of the sustainable potential for biomethane supply (produced from organic wastes and residues) suggests that it could cover some 20% of today's gas demand. Recognition of the value of avoided CO_2 and methane emissions would go a long way towards improving the cost competitiveness of both options.

Shale and solar PV show that rapid change is possible, but the direction and speed is set by governments

Ten years ago, the idea that the United States could become a net exporter of both oil and gas was almost unthinkable. Yet the shale revolution – and over $1 trillion in upstream and midstream investment – is making this a reality. The foundations date back to a publicly funded research and development effort that began in the 1970s. This was followed by tax credits, market reforms and partnerships that provided a platform for private initiative, innovation, investment and rapid reductions in cost.

Today, solar PV and some other renewable technologies – mostly in the power sector – are similarly turning initial policy and financial support into large-scale deployment. Transforming the entire energy system will require progress across a much wider range of energy technologies, including efficiency, CCUS, hydrogen, nuclear and others. It will also require action across all sectors, not just electricity.

Meeting rising demand for energy services, including universal access, while cutting emissions is a formidable task: all can help, but governments must take the lead. Initiatives from individuals, civil society, companies and investors can make a major difference, but the greatest capacity to shape our energy destiny lies with governments. It is governments that set the conditions that determine energy innovation and investment. It is governments to whom the world looks for clear signals and unambiguous direction about the road ahead.

Introduction

The *World Energy Outlook (WEO)* does not aim to provide a view on where the energy world will be in 2030 or 2040. This will depend on hugely important choices that lie ahead. What the *WEO-2019* does aim to do is to inform decision makers as they design new policies or consider new investments or shape our energy future in other ways. It does so by exploring various possible futures, the ways that they come about, the consequences of different choices and some of the key uncertainties.

The scenarios in this *Outlook* are differentiated primarily by the assumptions they make about government policies, and they approach this task in two distinct ways. One type of scenario defines a set of starting conditions and then models where they lead the energy system, without aiming to achieve any particular outcome. Another approach does the opposite, defining a set of future outcomes and then working through how they can be achieved. The scenarios chiefly considered are the **Stated Policies Scenario** and the **Current Policies Scenario**, which are of the first type, and the **Sustainable Development Scenario,** which is of the second type.

Each of these approaches offers valuable insights, and the *WEO-2019* describes in detail their projections for all regions[1] as well as all fuels and technologies. In combination, they provide a powerful framework to understand the challenges facing today's decision makers. They highlight the gap between where the energy world appears to be heading, given certain starting assumptions on policy, and where it would need to go in order to meet key energy and environmental objectives in full.

All of the scenarios are generated by the IEA's World Energy Model, a large-scale simulation model developed over many years, using the latest data for energy demand and supply, costs and prices. A full description of the World Energy Model and the operation of its different modules is available online at: www.iea.org/weo/weomodel.

The aim of the **Stated Policies Scenario** (STEPS[2]), which occupies a central position in the *WEO* analysis, is to hold up a mirror to the actions and intentions of today's policy makers, and to provide a candid assessment of their implications for energy markets, energy security and emissions. The scenario reflects:

- The impact of energy-related policies that governments have already implemented.
- An assessment of the likely effects of announced policies, as expressed in official targets and plans.
- A dynamic evolution of the cost of energy technologies, reflecting gains from deployment and learning-by-doing.

The Stated Policies Scenario, previously called the New Policies Scenario, is not an IEA forecast. It takes into account policies that have already been announced ("stated"), but

---

[1] At the time of writing, the United Kingdom was a member of the European Union (EU) and for the purpose of this analysis, the United Kingdom is included in EU aggregates.

[2] STEPS: the **ST**ated (**E**nergy) **P**olicies **S**cenario.

does not speculate on how these might evolve in the future. The new name of this scenario in *WEO-2019*[3] has been chosen with the aim of avoiding misunderstanding on this point.

Policies announced by governments include some far-reaching commitments, including aspirations to achieve full energy access in a few years, to reform pricing regimes and, more recently, to reach net zero emissions in some countries and sectors. These ambitions are not automatically incorporated into the scenario: full implementation cannot be taken for granted, so the prospects and timing for their realisation are based upon our assessment of the relevant regulatory, market, infrastructure and financial constraints. Nonetheless, these targets and plans move the projections away from a business-as-usual trajectory, as a comparison with the **Current Policies Scenario**, in which such announcements are not considered, makes clear.[4]

The time horizon of the Stated Policies Scenario is to 2040. The design of this scenario, which relies on detailed bottom-up consideration of the impact of today's policies and plans, does not lend itself to very long-term time horizons.

The **Sustainable Development Scenario** (SDS) is an essential counterpart to the Stated Policies Scenario. It sets out the major changes that would be required to reach the key energy-related goals of the United Nations Sustainable Development Agenda. These are:

- An early peak and rapid subsequent reductions in emissions, in line with the Paris Agreement (Sustainable Development Goal [SDG] 13).
- Universal access to modern energy by 2030, including electricity and clean cooking (SDG 7).
- A dramatic reduction in energy-related air pollution and the associated impacts on public health (SDG 3.9).

The trajectory for emissions in the Sustainable Development Scenario is consistent with reaching global "net zero" carbon dioxide (CO_2) emissions in 2070. If net emissions stay at zero after this point, this would mean a 66% chance of limiting the global average temperature rise to 1.8 degrees Celsius (°C) above pre-industrial levels (or a 50% chance of a 1.65 °C stabilisation). In the light of the *Intergovernmental Panel on Climate Change Special Report on 1.5 °C*, we also explore what even more ambitious pathways might look like for the energy sector, either via "net negative" emissions post-2070 or by reaching the "net zero" point even earlier.

Analysis related to the Sustainable Development Scenario covers a longer time horizon than is the case for the Stated Policies Scenario: it stretches to 2050 and beyond.

---

[3] It takes over from the New Policies Scenario, but is identical in design.

[4] Increasing awareness by governments and other stakeholders of the implications of policy inaction mean that the Current Policies Scenario has become less prominent in recent editions of the *World Energy Outlook*. Nonetheless, it serves as a reminder of the effort required to reach the outcomes in the Stated Policies Scenario.

Inputs to the scenarios

Information on the inputs used to generate the scenarios, including the underlying assumptions for economic growth, population, policies and the trajectories for energy prices and (where applicable) carbon prices, is detailed in Annex B. Assumed rates of growth for global gross domestic product (an average of 3.4% per year to 2040) and population (an increase to just over 9 billion people in 2040) are constant across the scenarios, whereas assumed policies, costs and equilibrium prices differ substantially.

Our modelling, in all scenarios, incorporates a process of learning-by-doing that affects the costs of various fuels and technologies, including the cost of investing in energy efficiency. The analysis includes a number of sensitivity cases to discuss the implications of more rapid technology change and cost reductions, e.g. for battery storage. However, while technology learning is integral to the approach, this *Outlook* does not try to anticipate technology breakthroughs.

Structure of the WEO-2019

Part A begins with an overview of the projections and their implications, and explores the pace at which different parts of the energy system are changing. The subsequent chapters present updates to the scenario projections, starting with a dedicated chapter on the prospects for reaching energy-related sustainable development goals, and then examining the main elements of the outlook by fuel, including renewables and energy efficiency.

Part B looks at Africa, the special focus region in this *World Energy Outlook*. It provides data and robust, evidence-based analysis to guide energy decision makers across the continent. It relates the analysis to the sustainable development goals and different scenarios, including an Africa Case that reflects the vision of the Agenda 2063 that was adopted by African Union governments in 2015. This Africa focus provides individual detail for ten key countries in sub-Saharan Africa. It examines the opportunities to provide universal access to clean, affordable energy and how the energy sector can foster economic growth.

Part C includes two insight chapters. One addresses the future of natural gas infrastructure, exploring how the role of gas networks might evolve in a low-emissions energy system in which electricity takes a larger share of final consumption. The other examines the huge potential for offshore wind to contribute to a low-carbon transformation of the power sector.

Comments and questions are welcome and should be addressed to:

Laura Cozzi and Tim Gould

Directorate of Sustainability, Technology and Outlooks
International Energy Agency
9, rue de la Fédération
75739 Paris Cedex 15
France

E-mail : weo@iea.org

More information about the *World Energy Outlook* is available at www.iea.org/weo

PART A
GLOBAL ENERGY TRENDS

Part A of the World Energy Outlook takes an in-depth look at what impact policy, technology and investment choices may have on the future of energy. It does this for each fuel and technology for all sectors and regions, using the most up-to date market data.

The following chapters utilise scenario analysis to explore the possible pathways the energy sector could take and the attendant implications for climate change, energy security and the economy.

The analysis assesses how today's announced policies and energy technologies may shape the trajectory of the energy sector in the years ahead. It also considers how an integrated policy approach – centred on energy access, improving local air pollution and dramatic reductions in greenhouse-gas emissions – provides a way to make progress on multiple Sustainable Development Goals at once.

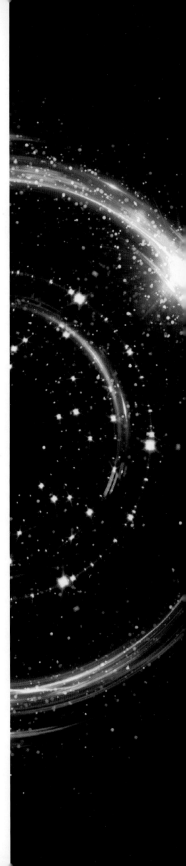

OUTLINE

Part A provides energy projections to 2040 for all energy sources, sectors and regions by scenario.

Chapter 1 provides an overview of the projections from our scenarios and the key insights from this year's *Outlook*, followed by a deep dive on the pace of change for different aspects of the energy sector in three key dimensions: sustainability, energy security and affordability.

Chapter 2 assesses if we are on track to meet the United Nations Sustainable Development Goals and lays out a pathway to reach three energy-related SDGs: universal energy access, reducing the impacts of local air pollution and tackling climate change in line with the Paris Agreement, including the pursuit of efforts to limit the rise in average global temperatures to 1.5 °C.

Chapter 3 focuses on the outlook for oil, going into detail on three topical areas that are critical for oil markets in the coming decades: the prospects for US tight oil, changing consumer preferences for vehicle ownership and the continued importance of oil security.

Chapter 4 tackles the outlook for natural gas and looks at the varying market and policy factors that will shape the competitive position of gas in different parts of the world. In this regard, it zooms in on three issues: the continued importance of associated gas; the scope for innovation to affect the costs and uses of liquefied natural gas; and the possibilities for gas to aid energy transitions by substituting for more polluting fuels.

Chapter 5 explores the outlook for coal, examining the dynamics of industrial coal use and the impact that constrained access to financing might have on coal investment and supply. It also provides new estimates on methane emissions from mining operations and considers the lifecycle emissions of coal versus gas.

Chapter 6 examines the outlook for power. Alongside the updated projections by scenario, it includes an in-depth look at electricity pricing and affordability, the options to bring down emissions from the existing coal-fired power fleet and the new frontiers of flexibility in power system operation, focusing on battery storage.

Chapter 7 concentrates on energy efficiency and renewables, detailing the potential for improved materials efficiency and recycling, and how policies can take advantage of variations in the carbon intensity of electricity generation. This chapter also presents a detailed assessment of the cost and supply potentials for biogas and the tools, technologies and policies that can exploit this potential.

Chapter 1

Overview and key findings
Understanding the pace of change

SUMMARY

- In the Stated Policies Scenario, which explores the implications of announced targets as well as existing energy policies, primary energy demand grows by a quarter to 2040 (Figure 1.1). The projected annual average demand growth of 1% per year in this scenario is well below the 2.3% seen in 2018. A continuation of 2018 growth trends would raise overall demand even higher than in our Current Policies Scenario, which does not consider announced targets, and would put considerable strain on all aspects of the global energy system.

- With the exception of coal, all fuels and technologies contribute to meeting demand growth in the Stated Policies Scenario, with the lead taken by renewables (50%) and gas (35%). After rising in the medium term, oil demand flattens by the 2030s under pressure from rising fuel efficiency and the electrification of mobility.

Figure 1.1 ▷ World primary energy demand by fuel and related CO_2 emissions by scenario

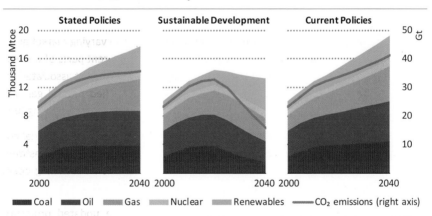

Existing policies and announced targets slow growth in global emissions to 2040, but they are not strong enough to force a peak in an expanding energy system

- The Sustainable Development Scenario identifies a pathway consistent with key energy-related sustainable development goals for emissions, access and air quality. Much more stringent energy efficiency policies mean that energy demand is lower in 2040 than today, even though the world economy grows by an average of 3.4% per year (the same economic outlook as in the other scenarios). Gas demand rises to 2030 before falling back; a rising share of low-carbon energy is accompanied by a dramatic reduction in coal use; and an oil market of 65 million barrels per day (mb/d) in 2040 returns to a level last seen in the early 1990s.

- China remains the world's largest energy consumer in all scenarios, but India is the largest source of demand growth. Africa – the focus region in this *Outlook* – is increasingly influential in determining global trends. How African decision makers manage the addition of almost 600 million people to the continent's urban population will have major demand implications for materials and mobility.

- There is no peak in global energy-related CO_2 emissions to 2040 in the Stated Policies Scenario, as the effects of an expanding economy and population on energy demand outweigh the strength of today's push for a more efficient and lower emissions energy system. Bridging the gap to the rapid emissions reductions seen in the Sustainable Development Scenario would require significantly more ambitious policy action in favour of efficiency and clean energy technologies, including decarbonised fuels and a major rebalancing of investment flows.

- Our scenarios point to the need for continued vigilance on a range of energy security issues. Oil security concerns do not disappear, and the risk of volatility and disruptions may intensify in scenarios where demand and prices are lower. Increased flexibility in global gas trade, thanks largely to liquefied natural gas (LNG), helps to underpin confidence in global gas supply, but the prospects for imported gas in price-sensitive import markets in Asia remain uncertain.

- A 60% increase in electricity demand to 2040 in the Stated Policies Scenario is supplied by an increasingly low-carbon generation mix. Solar photovoltaics (PV) and wind provide just over half of the growth in generation in this scenario. They help renewables to overtake coal in the power mix in the mid-2020s but require increasingly flexible operation of power systems. Cheaper battery storage plays an increasingly visible enabling role in this respect, notably in India. Offshore wind resources are huge and, although a variable source of generation, capacity factors are well above those for onshore wind and solar PV.

- With a higher assumed resource base than in last year's *Outlook*, the United States accounts for 85% of the increase in global oil production and around 30% of the increase in natural gas over the period to 2030 in the Stated Policies Scenario, expanding its position as an exporter of both fuels. The impacts on market shares and prices add to the pressure on some other major exporters. US shale output (oil and gas) overtakes total oil and gas production from Russia before 2025.

- The global value of fossil fuel consumption subsidies in 2018 was almost double the combined value of subsidies to renewable energy and electric vehicles and the revenue from carbon pricing schemes around the world. This imbalance greatly complicates the task of achieving an early peak in emissions.

- Some 60% of the existing coal fleet is 20-years old or less, and its emissions in the Stated Policies Scenario would take up most of the carbon budget to meet stringent climate goals. This year's *World Energy Outlook* outlines a cost-effective strategy to "retrofit, repurpose, retire" existing capacity to address coal-related emissions.

Introduction

2018 was an extraordinary year for energy. Global energy demand increased at almost twice the average rate for this decade, with consumption of all fuels and technologies – from coal to solar – up on the previous year. Renewables continued an impressive pace of growth, and the increase in their output covered roughly half of the increase in global electricity demand. Natural gas grew faster than any other fuel, with the United States and China leading the rise in consumption. More and more people got access to modern energy. India announced that all villages in the country had electricity connections, a huge step towards universal household access.

The environmental implications of these patterns of energy use are stark, and a groundswell of initiatives and pressure to mitigate these impacts has yet to make a decisive mark on the data. Energy-related carbon dioxide (CO_2) emissions hit another historic high in 2018, with a 1.9% rise marking the highest annual increase since 2013. The dissonance between the rising trend for CO_2 and the commitment of countries to reach an early peak in emissions was especially striking in the light of the latest scientific findings from the Intergovernmental Panel on Climate Change (IPCC, 2018). Poor air quality, another largely energy-related phenomenon, continued to have damaging effects on human health. Meanwhile, the IEA estimated that almost one-fifth of the growth in global demand in 2018 was due to unusual weather patterns: hotter summers in some countries pushing up demand for cooling, but also cold snaps leading to higher heating needs.

Each year about $1.8 trillion is invested in the energy sector, providing an important leading indicator of the way that the system is evolving, both on the supply and demand sides. Global spending on energy stabilised in 2018 after three years of decline, but a range of market, policy and geopolitical uncertainties could lead investment to fall short of what is needed to meet demand, to decarbonise and to provide energy access for all. This would be particularly damaging in parts of the world that face the highest economic and financial constraints, notably in Africa – the focus region for this *Outlook*.

This chapter starts with an overview of the projections from our scenarios, aiming to describe briefly the key trends and findings on energy demand, supply, end-use sectors, efficiency, emissions, trade and investment, and how these key indicators evolve under different assumptions about the world's energy future.

The remainder of the chapter digs down beneath the aggregate numbers to explore the pace of change in energy, looking at which elements of the energy system are really changing fast, which are moving much more slowly, and which might be the ones to watch. Analysis across three dimensions of energy – sustainability, energy security and affordability – illustrates some key opportunities ahead and also some of the huge challenges and risks that decision makers around the world are facing.

> Figures and tables from this chapter may be downloaded from www.iea.org/weo2019/secure/.

Scenarios

1.1 Overview

Table 1.1 ▷   World primary energy demand by fuel and scenario (Mtoe)

	2000	2018	Stated Policies 2030	Stated Policies 2040	Sustainable Development 2030	Sustainable Development 2040	Current Policies 2030	Current Policies 2040
Coal	2 317	3 821	3 848	3 779	2 430	1 470	4 154	4 479
Oil	3 665	4 501	4 872	4 921	3 995	3 041	5 174	5 626
Natural gas	2 083	3 273	3 889	4 445	3 513	3 162	4 070	4 847
Nuclear	675	709	801	906	895	1 149	811	937
Renewables	659	1 391	2 287	3 127	2 776	4 381	2 138	2 741
Hydro	225	361	452	524	489	596	445	509
Modern bioenergy	374	737	1 058	1 282	1 179	1 554	1 013	1 190
Other	60	293	777	1 320	1 109	2 231	681	1 042
Solid biomass	638	620	613	546	140	75	613	546
Total	10 037	14 314	16 311	17 723	13 750	13 279	16 960	19 177
Fossil fuel share	*80%*	*81%*	*77%*	*74%*	*72%*	*58%*	*79%*	*78%*
CO_2 emissions (Gt)	23.1	33.2	34.9	35.6	25.2	15.8	37.4	41.3

Notes: Mtoe = million tonnes of oil equivalent; Gt = gigatonnes. Other includes wind, solar PV, geothermal, concentrating solar power and marine. Solid biomass includes its traditional use in three-stone fires and in improved cookstoves.

Global primary energy demand grew by 2.3% in 2018, its largest annual increase since 2010. China, the United States and India accounted for 70% of the total energy demand growth. Despite the fact that growth in renewables has outpaced growth in all other forms of energy since 2010, the share of fossil fuels in global primary energy demand remains above 80% (Table 1.1).

The energy debate is often focused on the pace of change, but the forces of continuity in the energy sector should not be discounted. The **Current Policies Scenario** provides just such a "business-as-usual" picture, although its 1.3% average annual growth in energy demand to 2040 is well below the rate seen in 2018. Growth in line with this scenario would mean greater consumption of all fuels and technologies, leading to a continuous rise in energy-related emissions and increasing strains on almost all aspects of energy security.

In the **Stated Policies Scenario**, primary energy demand grows by one-quarter to 2040; the 1% annual average growth represents a slowdown compared with the 2% average seen since 2000. The global economy and the demand for energy move on diverging pathways due to structural shifts towards less energy-intensive output, energy efficiency gains and saturation effects, particularly in terms of vehicle use.

Low-carbon sources meet well over half of the increase in demand to 2040 in the Stated Policies Scenario, compared with 30% in 2017-2018. This is led by the power sector, where renewables dominate investment and capacity additions (Figure 1.2). However, demand for all sources of energy, except coal, continues to increase.

After rising strongly in the medium term, growth in oil demand slows markedly post-2025 in the Stated Policies Scenario before flattening in the 2030s. Oil use in passenger cars peaks in the late-2020s, despite the number of cars on the road increasing by 70% between 2018 and 2040. Coal demand in 2040 is slightly below today's level, and its share in the primary mix is overtaken by gas around 2030. Gas demand rises by 35%, with industrial use of gas increasing at more than twice the pace of gas in power generation.

In the **Sustainable Development Scenario**, a relentless focus on improving efficiency and a shift away from combustion for power generation (reducing losses from waste heat) means that the projected increase in the size of the global economy and population (the same in all scenarios) is accommodated without any rise in primary demand. With no overall increase in demand, the rise of low-carbon sources comes at the expense of coal and oil.

Global oil demand peaks within the next few years in the Sustainable Development Scenario. Much greater fuel efficiency and fuel switching, with almost half the global car fleet powered by near-zero carbon electricity, means that in 2040 oil use in transport is 40% lower than today; only the (non-combustion) use of oil, mostly as a feedstock for chemicals production, shows any increase. Natural gas use grows to 2030 and then falls back. Coal demand is hit hard in this scenario, declining at more than 4% per year.

Figure 1.2 ▷ Share of renewables in total capacity additions by region and scenario, 2019-2040

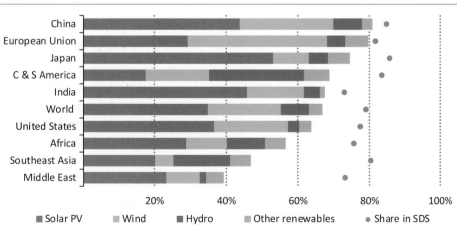

Renewable sources dominate the increase in global power generation capacity in the Stated Policies Scenario, with solar PV taking the lead

Notes: C&S America = Central and South America; SDS = Sustainable Development Scenario. Other renewables includes geothermal, concentrating solar power, bioenergy and marine.

1.2 Primary energy demand by region

Table 1.2 ▷ Total primary energy demand by region and scenario (Mtoe)

	2000	2018	Stated Policies 2030	Stated Policies 2040	Sustainable Development 2030	Sustainable Development 2040	Change 2018-2040 STEPS	Change 2018-2040 SDS
North America	2 678	2 714	2 717	2 686	2 377	2 087	-28	-627
United States	2 271	2 230	2 214	2 142	1 942	1 687	-89	-544
Central & South America	449	660	780	913	669	702	253	42
Brazil	184	285	342	397	299	312	112	27
Europe	2 027	2 000	1 848	1 723	1 689	1 470	-277	-530
European Union	1 692	1 613	1 414	1 254	1 311	1 101	-359	-512
Africa	489	838	1 100	1 318	698	828	480	-10
South Africa	108	134	133	139	112	107	5	-27
Middle East	365	763	956	1 206	802	880	443	117
Eurasia	742	934	980	1 031	858	807	97	-127
Russia	621	751	767	786	680	635	35	-116
Asia Pacific	3 012	5 989	7 402	8 208	6 232	6 085	2 218	96
China	1 143	3 187	3 805	3 972	3 226	2 915	785	-271
India	441	916	1 427	1 841	1 143	1 294	925	378
Japan	518	434	387	353	349	300	-80	-134
Southeast Asia	384	701	941	1 114	797	858	413	157
International bunkers	274	416	528	639	425	420	223	4
Total	10 037	14 314	16 311	17 723	13 750	13 279	3 409	-1 035

Notes: Mtoe = million tonnes of oil equivalent; STEPS = Stated Policies Scenario; SDS = Sustainable Development Scenario. International bunkers include both marine and aviation fuels.

Developing countries, led by Asia, are the main engines of global growth in our projections. India, where energy demand doubles, is the single largest source of demand growth to 2040 in the Stated Policies Scenario (Table 1.2). China remains the world's largest energy consumer in this scenario, but its demand growth slows considerably and is increasingly met by low-carbon sources (Figure 1.3). China's oil demand flattens out from 2030 at a per capita level far below those reached in advanced economies. Southeast Asia is another region where energy demand grows strongly.

Energy use in advanced economies, which already have high per capita consumption, generally declines. But energy demand rises rapidly in the Middle East, where per capita consumption is already high today, spurred in part by increased demand for cooling and desalination. Rising incomes and population growth underpin strong growth in Africa, from a very low base; reduced reliance on solid biomass as a cooking fuel – especially in the Sustainable Development Scenario – also implies significant gains in efficiency.

Figure 1.3 ⊳ Change in energy demand and average annual GDP growth rate by region in the Stated Policies Scenario, 2018-2040

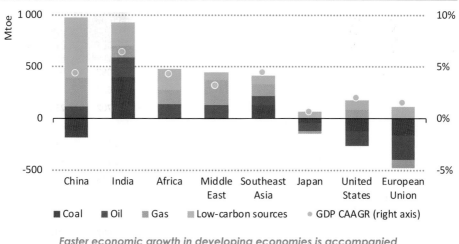

Faster economic growth in developing economies is accompanied almost everywhere by rising demand for all fuels and technologies

Notes: GDP CAAGR = gross domestic product, compound average annual growth rate. Low-carbon sources includes nuclear and renewables.

In most developing economies, all fuels and technologies are mobilised to satisfy rapidly growing economies and energy demand; the focus is on new infrastructure, in particular for the 70 million people each year that are added to the urban population. The design of the world's growing cities and towns, the industrial outputs that are used in their construction, and the mobility services available to their inhabitants, are critical issues for the energy outlook.

China is a somewhat separate case, given the remarkable build-out of infrastructure there over the last two decades: in this case, the challenge is to manage the implications of a development phase that relied heavily on heavy industry and coal. In many advanced economies, declining or stagnant energy demand alters the investment dynamic: the pace of change is dictated by the need to replace or upgrade existing energy-using infrastructure.

In the Sustainable Development Scenario, choices on all new energy-using infrastructure incorporate a systematic preference for increased efficiency and reduced emissions. But there is also much greater attention across the board to reducing the environmental footprint of the existing stock of assets, whether by efficiency retrofits of existing buildings, the introduction of more efficient and low-emissions industrial processes, or the use of low-carbon electricity and, increasingly, low-carbon fuels as well such as biofuels, biomethane and hydrogen.

1.3 Final energy consumption and efficiency

Table 1.3 ▷ Final energy consumption by sector, fuel and scenario (Mtoe)

	2000	2018	Stated Policies 2030	Stated Policies 2040	Sustainable Development 2030	Sustainable Development 2040	Change 2018-2040 STEPS	Change 2018-2040 SDS
Industry	1 881	2 898	3 460	3 839	2 949	2 904	940	5
Transport	1 958	2 863	3 327	3 606	2 956	2 615	742	-249
Buildings	2 446	3 101	3 455	3 758	2 735	2 709	657	-391
Other	758	1 092	1 365	1 470	1 264	1 272	378	180
of which: feedstock	433	549	743	825	681	707	277	158
Electricity	1 092	1915	2 503	3 061	2 349	2 902	1 146	988
District heat	248	296	313	312	264	224	16	-72
Hydrogen*	0	0	1	2	6	65	2	65
Direct use of renewables	272	482	696	876	887	1 142	395	660
of which: modern bioenergy	263	430	592	718	729	873	288	443
Natural gas	1 127	1 615	2 032	2 360	1 816	1 719	746	105
Oil	3 124	4 043	4 469	4 561	3 695	2 838	518	-1 205
Coal	542	984	979	954	746	533	-29	-451
Solid biomass	638	620	613	546	140	75	-74	-545
Total	7 043	9 954	11 607	12 672	9 904	9 500	2 718	-455

* Includes hydrogen only used as a fuel and produced from dedicated hydrogen production facilities.

Notes: STEPS = Stated Policies Scenario; SDS = Sustainable Development Scenario. Solid biomass includes its traditional use in three-stone fires and in improved cookstoves.

The increasing role of electricity among energy end-uses is a common feature of our projections for final energy consumption (Table 1.3). The global share of electricity rises from 19% today to 23% in the Current Policies Scenario, to 24% in the Stated Policies Scenario in 2040, and to more than 30% in the Sustainable Development Scenario, confirming the importance of this vector in decarbonisation strategies. The total electricity used in the latter two scenarios is quite similar, but the consumption in the Sustainable Development Scenario takes place in the context of a much more efficient system.

Industry accounts for the largest share of growth (35%) in final consumption in the Stated Policies Scenario, nearly all of which is in the form of natural gas and electricity. Less than half of the growth in total transport demand is met by oil, a dramatic change from previous trends. Considering only road transport demand, electricity and biofuels together account for half the growth, considerably higher than the 30% from oil products. Rising demand for space cooling, and higher ownership levels for a range of electric appliances and digitised devices, mean higher energy use in buildings: 50% of the growth comes from developing Asian economies.

Figure 1.4 ▷ Change in final energy consumption by sector 2000-2018 and by scenario to 2040

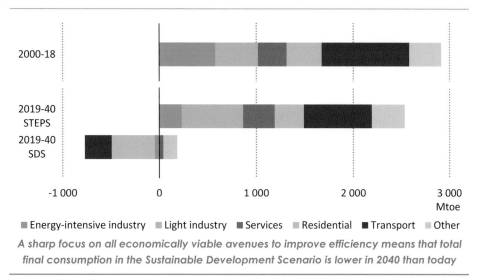

A sharp focus on all economically viable avenues to improve efficiency means that total final consumption in the Sustainable Development Scenario is lower in 2040 than today

Notes: STEPS = Stated Policies Scenario; SDS = Sustainable Development Scenario. Other includes agriculture and non-energy use.

In the Sustainable Development Scenario, increased material efficiency (see Chapter 7) is a key reason for flat industrial energy demand to 2040 (Figure 1.4), despite a slight rise in energy used by chemical industries. Energy consumption in transport declines due to much higher fuel efficiency and half the global car fleet running on electricity. Stronger building codes and standards means energy consumption in buildings falls by more than 10% by 2040.

Electricity, the direct use of renewables (for example biofuels, biogas and solar thermal heating) and hydrogen are the elements that see rising shares of final energy use in 2040, and electricity overtakes oil around 2040 to become the largest component of final consumption (it is less than half the level of oil today). Oil consumption falls by around 1.5% per year on average, and coal use by 3%. Gas demand in final consumption rises slightly through to the early 2030s, largely because of increased industrial demand, but then declines under pressure from efficiency improvements (including deep buildings retrofits), low-carbon gases such as hydrogen and biomethane, and electrification.

The energy intensity of the global economy decreases by 2.3% each year on average to 2040 in the Stated Policies Scenario as the links between gross domestic product (GDP) growth and energy demand growth continue to weaken. This is well above the 1.6% rate of change seen since 1990. Significant improvements occur in China, India and Africa. All economically viable opportunities for efficiency are pursued in the Sustainable Development Scenario, and energy intensity improves at a much faster rate, decreasing by around 3.6% each year on average.

1.4 Power generation and energy supply

Table 1.4 ▷ World electricity generation by fuel, technology and scenario (TWh)

	2000	2018	Stated Policies 2030	Stated Policies 2040	Sustainable Development 2030	Sustainable Development 2040	Change 2018-2040 STEPS	Change 2018-2040 SDS
Coal	5 995	10 123	10 408	10 431	5 504	2 428	307	-7 695
Oil	1 207	808	622	490	355	197	-319	-611
Natural gas	2 760	6 118	7 529	8 899	7 043	5 584	2 781	-534
Nuclear	2 591	2 718	3 073	3 475	3 435	4 409	757	1 691
Hydro	2 613	4 203	5 255	6 098	5 685	6 934	1 895	2 731
Wind and solar PV	32	1 857	5 879	9 931	7 965	15 503	8 073	13 645
Other renewables	217	739	1 344	2 020	1 785	3 628	1 281	2 889
Total generation	15 436	26 603	34 140	41 373	31 800	38 713	14 770	12 110
Electricity demand	*13 152*	*23 031*	*29 939*	*36 453*	*28 090*	*34 562*	*13 422*	*11 531*

Notes: TWh = terawatt-hours. STEPS = Stated Policies Scenario; SDS = Sustainable Development Scenario. Total generation includes other sources. Electricity demand equals total generation minus own use (for generation) and transmission and distribution losses.

Global electricity generation grows by around 55% (14 800 terawatt-hours [TWh]) between 2018 and 2040 in the Stated Policies Scenario (Table 1.4). Growth is led by Asia: China alone constitutes one-third of growth followed by India at 20% and other developing countries in Asia at 13%. Despite the fact that full electricity access is achieved in the Sustainable Development Scenario, a more efficient energy system means that electricity generation grows at a slightly slower pace, increasing by a (still impressive) 12 100 TWh over the same period.

Wind and solar photovoltaics (PV) provide over half of additional power generation to 2040 in the Stated Policies Scenario; coal provides just 2% although this varies widely by region. As a result, soon after 2025, renewables account for a greater share of generation than coal. Hydropower remains the largest low-carbon source of electricity, but by 2040 wind gets close to equalling its share on the back of a rapidly expanding offshore wind market. By the early 2030s, the installed capacity of solar PV overtakes coal and by 2040 its share of generation rises above 10%. The availability of cheap natural gas in some regions, allied with a rising need for system flexibility, means that gas-fired generation grows in all regions except Europe and Japan.

The Sustainable Development Scenario sees an even more dramatic transformation. Low-carbon technologies provide almost 85% of generation in 2040; 40% of generation comes from wind and solar PV. Coal-fired generation plummets as almost all coal-fired power plants without carbon capture are repurposed to provide flexibility, or retired. Although gas-fired power provides important balancing services, especially to help meet seasonal variability in demand, its use peaks in the mid-2020s as battery storage and demand-side management take on a much larger role in meeting short-term flexibility needs.

Figure 1.5 ▷ Change in fossil fuel production and demand in selected regions in the Stated Policies Scenario, 2018-2040

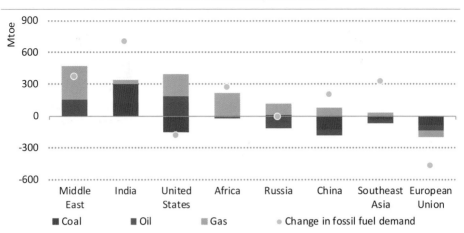

Gas output rises to 2040 in all major regions except the EU, oil growth is concentrated in the Middle East and US, and India is one of the few countries where coal production grows

Global oil output rises by around 10% and natural gas production by almost 40% to 2040 in the Stated Policies Scenario. Over the period to 2030, the United States accounts for 85% of the increase in oil and 30% of the increase in gas. The rise in US output, together with increases from Norway and Brazil, mean that the Organization of Petroleum Exporting Countries' (OPEC) share of global oil production declines to 36% by 2025, a level not seen since 1990.

US tight oil production levels off and starts falling back from the mid-2030s in the Stated Policies Scenario, resulting in a steady decline in total US oil output through to 2040. As a result, major resource-holders in the Middle East start to regain market share, and eventually see that share return to today's levels.

Around 60% of the growth in global gas production to 2030 comes from unconventional sources, predominantly shale gas in the United States. As unconventional production in the United States begins to plateau, Canada, China and Argentina pick up the pace. Russia sees the largest increase in conventional output. Mozambique overtakes Nigeria to become the largest gas producer in sub-Saharan Africa by 2040. Coal production declines globally, but higher output (mainly from India) almost offsets declines elsewhere, including a peak in production in China (Figure 1.5).

Aggregate fossil fuel production falls precipitously by 2040 in the Sustainable Development Scenario. Natural gas output rises by around 8% to 2030 before falling below today's levels in 2040. Oil production peaks in the next few years and drops to 65 million barrels per day (mb/d) (a level not seen since 1990) as the shift to alternative modes in the transport sector takes away its main demand base. Coal production in 2040 is more than 60% lower than today.

1.5 Emissions

Table 1.5 ▷ World energy-related CO_2 emissions by fuel and scenario (Mt)

			Stated Policies		Sustainable Development		Change 2018-2040	
	2000	2018	2030	2040	2030	2040	STEPS	SDS
Coal	8 946	14 664	14 343	13 891	8 281	3 424	-773	-11 240
Oil	9 640	11 446	12 031	12 001	9 436	6 433	555	-5 012
Natural gas	4 551	7 134	8 486	9 697	7 464	6 032	2 563	-1 102
Total CO_2	23 137	33 243	34 860	35 589	25 181	15 796	2 345	-17 448

Notes: Mt = million tonnes; STEPS = Stated Policies Scenario; SDS = Sustainable Development Scenario. Total CO_2 accounts for captured emissions from bioenergy with carbon capture, utilisation and storage (CCUS).

Global energy-related carbon dioxide (CO_2) emissions increased again in 2018, reaching a record high and marking out the 2014-2016 flattening in emissions as a pause in a continued upward climb rather than a turning point. This recent uptick is due to a range of factors including robust economic growth, a slowdown in efficiency gains and the fact that deployment of cleaner energy sources has not kept pace with rising energy demand.

A growing number of countries, together with the European Union, are actively considering or have announced that they will set long-term "net-zero" CO_2 or greenhouse-gas (GHG) emissions goals. Only a few of these have yet been formally adopted but, where they have, we assess the specific policies identified to achieve these long-term goals and include them in the Stated Policies Scenario. The Sustainable Development Scenario is consistent with the net-zero goals being reached in full; in this scenario, they constitute the leading edge of a much wider effort in line with reaching global net-zero CO_2 emissions by 2070.

The **Stated Policies Scenario** does not see a peak in energy-related CO_2 emissions, which rise by just over 100 million tonnes (Mt) CO_2 per year on average between 2018 and 2040 (Table 1.5 and Figure 1.6). This is slower than the average rate of increase seen since 2010 (emissions have increased on average by 350 Mt CO_2 per year), but still very far from the emissions reductions necessary to achieve the goals of the Paris Agreement.

Electricity generation increases by more than half in the Stated Policies Scenario, but annual power sector CO_2 emissions in 2040 are broadly the same as today. This occurs because of the rising share of renewables and related improvements to networks, as well as because of improvements in the efficiency of fossil fuel power plants and some additional fuel switching from coal to gas.

The **Sustainable Development Scenario** is consistent with limiting the rise in average global temperatures to below 1.8 degrees Celsius (°C) at a 66% probability, or a 50% probability of a 1.65 °C stabilisation, without any recourse to net-negative emissions. Energy sector CO_2 emissions in this scenario peak immediately at around 33 gigatonnes (Gt) and then fall to less than 10 Gt by 2050.

Figure 1.6 ▷ Cumulative energy-related CO₂ emissions (since 1890) and annual emissions by fuel and scenario

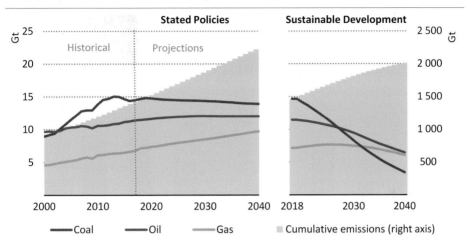

Existing and announced policies do not produce a peak in global CO₂ emissions by 2040. The Sustainable Development Scenario is on course for net-zero emissions in 2070.

Note: Gt = gigatonnes.

The emphasis on strong early action in the Sustainable Development Scenario and the subsequent rapid reduction in emissions means that this scenario is fully in line with the Paris Agreement objective to hold the temperature rise to "well below 2 °C" while "pursuing efforts to limit the temperature increase to 1.5 °C". If technologies were to be deployed at scale in the second-half of the century to remove carbon from the atmosphere then the Sustainable Development Scenario would provide a 50% chance of a 1.5 °C outcome. The cumulative level of net-negative emissions required to do this would be less than the median level in 1.5 °C scenarios assessed by the IPCC (IPCC, 2018).

In the Sustainable Development Scenario, emissions in advanced economies fall at an average of 5.6% per year to 2050, compared with 3.2% in developing economies. Energy efficiency becomes the primary option for policy makers to reduce emissions in most sectors. Policies that accelerate the deployment of renewables extend beyond electricity generation and incorporate the direct use of renewables in buildings and industry for heat, and of biofuels and biomethane in transport.

Another key technology is carbon capture, utilisation and storage (CCUS). By 2030, there are around 0.7 Gt of CO_2 emissions captured each year; this rises to almost 2.8 Gt in 2050. CCUS deployment is equally split between the power and industry sectors. Around 170 GW of coal plants are retrofitted with CCUS by 2050, mostly in China: without this, the majority of these plants would need to be repurposed to provide back-up capacity or closed before the end of their technical lifetimes to be consistent with the necessary emissions reductions.

1.6 Trade

Figure 1.7 ▷ Net oil and gas imports to Asia by scenario

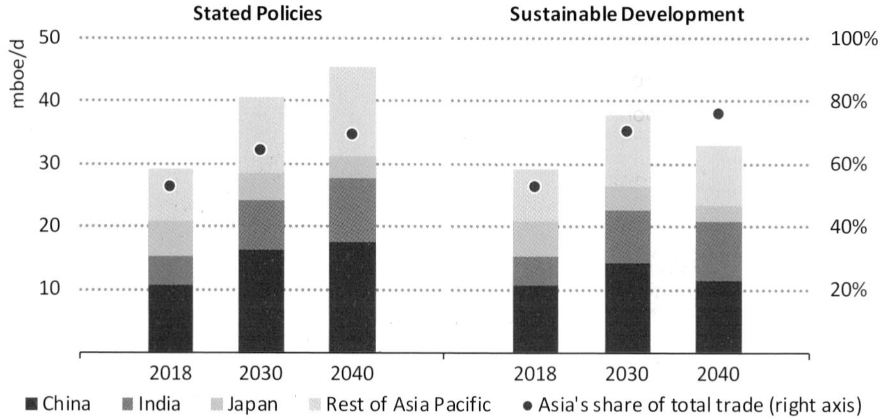

Some 70% of global net oil and gas imports flow to Asia by 2040 in both scenarios

Note: mboe/d = million barrels of oil equivalent per day.

In the Stated Policies Scenario, China soon overtakes the European Union and cements its position as the world's largest net oil importer through to 2040, even though its demand flattens out after 2030. India's net oil import requirements more than double between 2018 and 2040 and its level of import dependency reaches roughly 90%, one of the world's highest (Table 1.6). India's reliance on imported fuels becomes a major factor in global trade and energy security.

The United States becomes a net oil exporter soon after 2020 and North America becomes the world's second-largest oil exporter by 2030. However the United States also remains a major importer of oil given the configuration of its refineries – the US tends to produce and export light crude oil and to import heavier crude – and more oil flows into and out of US ports than before the shale boom.

With North America becoming a major exporting region, European net imports declining and Asian demand for imports rising, around 80% of the world's oil trade in 2040 ends up in Asia (Figure 1.7).

In the Stated Policies Scenario, LNG overtakes pipeline gas as the dominant method of inter-regional trade by the late 2020s. Australia, Qatar and the United States jostle for position as the world's largest LNG exporter, but a number of other players – notably Canada, Russia and Mozambique – strengthen positions as major LNG exporters over the period to 2040.

China is the destination for most of the major new inter-regional pipeline projects that are anticipated to 2040, notably the Power of Siberia pipeline from Russia and a fourth link (along a different route) from Turkmenistan. However, a majority of China's gas imports by 2040 are in the form of LNG. All in all, over 70% of gas demand growth to 2040 in emerging Asian markets is met by imports.

China overtook Japan to become largest gas-importing country in 2018. In the Stated Policies Scenario, its imports reach the level of those of the European Union by 2040.

Table 1.6 ▷ **Net import (shaded) and export shares by fuel, region and scenario (%)**

	Oil			Natural Gas			Coal			Total		
	2018	2040		2018	2040		2018	2040		2018	2040	
		STEPS	SDS		STEPS	SDS		STEPS	SDS		STEPS	SDS
North America	1	30	43	3	10	12	15	13	31	4	15	11
United States	20	17	39	2	13	14	14	10	17	3	10	7
C & S America	11	30	11	5	10	11	44	20	51	21	24	18
Brazil	16	42	27	29	23	17	90	89	99	18	30	26
Europe	76	76	70	55	67	61	49	61	72	42	40	34
European Union	87	91	89	76	90	85	49	66	72	50	47	39
Africa	49	8	15	33	37	47	29	27	25	45	33	50
Middle East	77	70	73	20	21	23	74	88	76	61	51	47
Eurasia	71	64	55	34	40	39	45	53	42	49	47	34
Asia Pacific	78	86	87	26	43	42	1	5	2	17	27	30
China	70	83	85	43	57	57	6	3	0	18	27	34
India	83	92	92	49	59	74	29	27	38	18	35	38
Japan	100	99	99	74	98	97	99	100	100	87	67	55
World	45	44	45	20	24	26	21	20	20	25	24	21

Notes: STEPS = Stated Policies Scenario; SDS = Sustainable Development Scenario; C & S America = Central and South America. Shaded orange cells indicate net imports; white cells indicate net exports. Import shares for each fuel are calculated as net imports divided by primary demand. Export shares are calculated as net exports divided by production. Total also includes bioenergy, hydropower, nuclear and renewables.

China's net imports of coal peak in the next few years and decline substantially towards 2040. China is overtaken by India as the world's largest coal importer in the mid-2020s. However, there is substantial uncertainty over future import volumes: small changes in the supply-demand balance in either China or India could quickly have substantial implications for traded coal.

In the Sustainable Development Scenario, the volumes of oil traded between regions in 2040 are almost 40% lower than in the Stated Policies Scenario, and fall below 30 mb/d. By contrast, trade in natural gas falls by roughly 20% due to continued demand for imports in a number of carbon-intensive Asian economies.

1.7 Investment

Following three consecutive years of decline, energy investment stabilised in 2018 at over $1.8 trillion. A rise in upstream oil and gas spending was partially offset by lower capital expenditure in power. Nevertheless, for the fourth year in a row, the power sector exceeded oil and gas supply as the largest investment sector. Investment in energy efficiency remained flat due to sluggish spending on energy efficient buildings. Two-of-every-ten dollars invested in energy worldwide currently goes to electricity provision in the Asia Pacific region with roughly another two dollars divided between oil and gas and power in North America.

Table 1.7 ▷ Global average annual energy investment by type and scenario ($2018 billion)

		Stated Policies		Sustainable Development		Change 2031-40 vs. 2014-18	
	2014-18	2019-30	2031-40	2019-30	2031-40	STEPS	SDS
Fossil fuels without CCUS	1 063	1 017	1 063	749	555	0	-508
Renewables	308	356	398	548	703	89	395
Electricity networks	291	354	455	345	631	164	340
Nuclear and other	44	64	74	99	141	29	97
Fuels and power	**1 706**	**1 792**	**1 989**	**1 741**	**2 030**	**282**	**323**
Fuels	*55%*	*52%*	*50%*	*42%*	*28%*	*-5%*	*-27%*
Power	*45%*	*48%*	*50%*	*58%*	*72%*	*5%*	*27%*
Energy efficiency	238	445	635	625	916	397	678
Renewables and other	127	220	308	332	950	181	824
End-use	**365**	**665**	**943**	**957**	**1 866**	**578**	**1 501**
Total	**2 071**	**2 457**	**2 931**	**2 697**	**3 896**	**860**	**1 825**
		2019-40		2019-40			
Cumulative total		58 795		71 329			
Average annual total		2 673		3 242			

Notes: STEPS = Stated Policies Scenario; SDS = Sustainable Development Scenario. Other fuels and power includes battery storage and power plants equipped with CCUS. Other end-use includes CCUS in industry sector, and electric vehicles and EV slow chargers.

Compared with the past decade, average annual investment in the **Stated Policies Scenario** rises by more than one-third to almost $2.7 trillion over the period to 2040 (Table 1.7). The overall split in energy supply investment between fuel supply and power is around 50:50. On the fuel side, the balance of spending in oil and gas investment shifts more towards gas, while coal supply investment falls back. Within the power sector, there is a major shift in generation spending away from coal and towards renewables, and also towards upgrading networks and storage. Investment in energy efficiency triples by 2040. The European Union, China and the United States together account for more than half of all energy efficiency investments in 2040.

In the **Sustainable Development Scenario**, around 20% more investment is needed than in the Stated Policies Scenario and investment rises to an annual average of $3.2 trillion to 2040. There is a significant reallocation of investment away from fossil fuels and towards renewables, energy efficiency and low-carbon technologies. The investment split on the supply side moves decisively in favour of power, which absorbs two-thirds of overall spending (Figure 1.8). Within the power sector, spending on renewables almost doubles while spending on nuclear rises by nearly 80%; investment in electricity grids and batteries also rises strongly, in part to accommodate rising shares of wind and solar PV. Spending on energy efficiency nearly quadruples by 2040, 45% of which is in buildings.

In both scenarios, developing countries account for nearly 60% of investment over the next two decades. China and the United States remain the two largest destinations for investment, but there is a shift in spending towards India and other developing economies, especially in the Sustainable Development Scenario. Regions where investment has been lagging – for example Southeast Asia and sub-Saharan Africa – see a boost, raising the question of how investment frameworks will evolve to mobilise these capital flows in practice.

Comparisons between the investment needs in the two scenarios are instructive but do not tell the whole story. The Sustainable Development Scenario has a larger share of spending on technologies that have higher initial capital costs, but significantly lower operating costs.

Figure 1.8 ▷ Global average annual energy supply investment by type and scenario

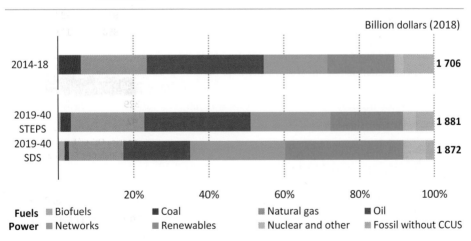

Investment in fuels and power needs rise in both scenarios; a major capital reallocation from fuels to power would be needed to meet sustainability goals

Notes: STEPS = Stated Policies Scenario; SDS = Sustainable Development Scenario. Nuclear and other includes nuclear, battery storage and power plants equipped with CCUS.

1.8 Differences between the WEO-2019 and the WEO-2018

Figure 1.9 ▷ US tight oil production and assumed resources in the announced policies scenarios of the *WEO-2018* and *WEO-2019*

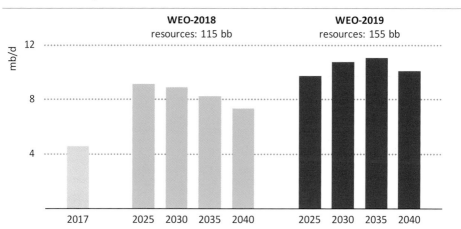

Upward revisions to US tight oil numbers means tight oil production peaks later in WEO-2019 than WEO-2018 and does not decline as steeply in the later years of the Outlook

Notes: bb = billion barrels. Compares the Stated Policies Scenario in the *WEO-2019* and the New Policies Scenario in the *WEO-2018*.

Oil demand in 2040 in the Stated Policies Scenario is similar to that projected in the New Policies Scenario of the *World Energy Outlook-2018 (WEO-2018)* (a difference of just over 0.1 mb/d), but there is a significant change in supply dynamics. Tight oil resources in the United States have been revised upwards this year by 35%, based on the latest official assessments of each play, and this means that tight oil production peaks at a later stage and does not decline by as much in later years (Figure 1.9). Conventional production and heavy oil output from Venezuela are lower as a result. The increase in tight oil resources also means that the oil price in 2040 is 10% lower than in the *WEO-2018*.

For power, the largest differences between the *WEO-2019* projections and those in the 2018 edition are in wind and solar PV generation (Figure 1.10). For wind, this is a mainly the result of an increase this year in projected offshore wind capacity. Six US states announced new targets for deploying offshore wind capacity in 2018 and early 2019; auction results in Europe have reconfirmed the maturity of the market there, improving financing arrangements. In China, there have been positive policy signals on wind and the supply chain there is strengthening. The reversal of the 2018 decision to slash financial support for new solar PV in China also boosted prospects for deployment. In India, the outlook for solar PV has benefited from the prospect of pairing new projects with lower cost battery storage.

Figure 1.10 ▷ Differences in oil supply and power generation in the announced policies scenarios of the *WEO-2019* and *WEO-2018*

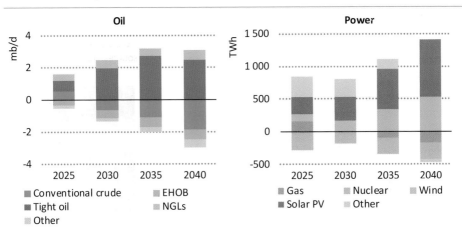

Tight oil production and NGLs are higher in WEO-2019, at the expense of conventional oil: projected generation from wind and solar PV is also higher than WEO-2018

Notes: EHOB = extra heavy oil and bitumen; NGLs = natural gas liquids. Other in oil includes kerogen oil, coal-to-liquids, gas-to-liquids and additives. Other in power includes oil, coal, hydro, bioenergy and other renewables. Compares the Stated Policies Scenario in the *WEO-2019* and the New Policies Scenario in the *WEO-2018*.

Global coal demand in 2040 has been revised downwards by 30 million tonnes of oil equivalent (Mtoe) compared with last year as air quality and climate policies, coal phase-out plans, declining costs of renewables and (in some regions) abundant supplies of natural gas put additional pressure on coal use. The upward revision in the Asia Pacific region stems mainly from a revision of demand in the base year. Coal prices have also been revised downwards slightly, partly because of the reduction in demand, but also because a lower oil price results in a reduction in coal production costs in some regions.

Global gas demand in 2040 is slightly higher than in *WEO-2018* (an increase of nearly 10 Mtoe). An upward revision to the use of gas in industry is largely offset by a downward adjustment to power generation. On a regional basis, consumption in the United States has been revised upwards, but this is counterbalanced by lower demand in the European Union and slightly slower demand growth in China.

Gas prices have been revised down slightly. The Henry Hub price has been reduced because of a larger shale gas resource base in the United States and because increases in tight oil production mean larger volumes of associated shale gas. This price is an increasingly important benchmark for gas prices in other regions, and its reduction has led to gas prices in all other regions also being revised downwards. Anticipated reductions in the average cost of LNG liquefaction and a slight acceleration of the move away from oil indexation have also contributed to this downwards revision.

Exploring the pace of change

One of the topics on which it is difficult to find consensus is the pace of change in today's energy sector. Many see incipient signs of an energy revolution, with ever-cheaper clean energy technologies accelerating transitions at the expense of incumbent fuels. Others emphasise the political and social challenges of forcing rapid change upon a system characterised by inertia and increasing demand for energy services, and underscore how poorly managed transitions could affect the affordability and reliability of supply. Another view might identify the expansion of access to modern energy as the critical vector for change, or recognise the way that technology can disrupt the existing order but point to the shale revolution as the most pertinent example. As well, there is a position that focuses not on the pace of change that we can see today, but on the urgency to move much faster towards carbon neutrality given the increasingly visible threat of climate change.

Our aim in the discussion that follows is to dig deeper into these questions, using the new *World Energy Outlook* scenarios to examine the forces of technology, market and policy change in more detail, in order to identify the elements that are moving quickly, as well as those that are not, and to consider their implications for the future. The sections that follow also highlight a few elements that may not yet be very visible today in our projections, but which may emerge to shape energy developments in the coming years. We divide this analysis in three ways across the different dimensions of energy: sustainability, security and affordability.

1.9 Sustainability

Moving fast: solar PV and electric mobility

Solar PV becomes the single largest source of installed generation capacity worldwide by 2040 in the Stated Policies Scenario, overtaking coal capacity in the early-2030s. This reflects continued cost reductions as well as policy support in some key markets. Year-on-year capacity additions flattened out in 2018 at just below 100 GW (Box 1.1), although the indications are that 2019 is seeing a resumption of strong growth. A renewed acceleration in annual solar PV deployment, alongside enhanced efforts to ensure smooth integration of the resulting solar generation into power systems, is essential to reach climate targets and other sustainable development goals.

Measured in terms of generation, solar PV and wind (onshore and, increasingly, offshore) lead the way in terms of growth in the Stated Policies Scenario, followed by natural gas, hydropower and biomass. As a result, the overall share of renewables in electricity rises from around one-quarter today to almost 45% in 2040 (of which solar contributes 11%, up from the current 2%). Together with output from nuclear plants and plants equipped with CCUS, this pushes the overall share of low-carbon electricity slightly above 50%.

Box 1.1 ▷   How much is 100 gigawatts of solar each year?

After several years of explosive growth, new global solar PV deployment plateaued in 2018 at the 2017 level, close to 100 GW. Preliminary data indicate that 2019 will see a welcome resumption of growth with more than 110 GW of additions. Renewed momentum is essential if the world is to meet its climate targets, but reaching the 100 GW mark nonetheless provides an opportunity to reflect on this level of annual deployment.

Adding 100 GW of solar PV each year means large-scale industrial activity. Even with the recent impressive cost declines, solar absorbed $135 billion in investment in 2018, almost equal to the total capital investment in the entire Swedish economy. Around 800 000 cargo containers were packed with solar panels in the factories of China and Southeast Asia for delivery around the world. Manufacturing is on a massive scale: the largest factories in China employ 15 000 workers each and churn out a 2 m^2 solar module every two seconds. For utility-scale projects, which represent the majority of global solar deployment, today's annual deployment could cover the equivalent of 200 000 football fields.

The output from a given PV panel depends upon where it is installed, and the average is rising as deployment shift towards countries like India and Mexico. Even so, it would take 200 years at an annual deployment rate of 100 GW to reach the scale of 2018 global electricity demand. And electricity demand is not standing still, but increasing rapidly: 100 GW of new solar each year would meet around 20% of the annual growth in power demand projected in the Stated Policies Scenario, so the effect would be merely to dampen growth in other generation sources. Moreover, with time, an increasing share of new solar would be required to replace retiring capacity.

Electricity accounts for around 20% of final energy consumption today. The success of solar is a major reason for optimism about electrification – that is, the potential for increasingly cost-efficient low-carbon electricity to replace fossil fuels for mobility and heat. The impact of 100 GW of solar in these cases varies depending on the end-use. It would be enough electricity for more than 50 million electric cars to be added to the global fleet every year. But if converted to gas (power-to-gas), the conversion losses would make this around 1% of the gas used today for heating in buildings. If used to electrify industrial processes, 100 GW of solar could produce sufficient hydrogen to produce 3% of global primary steel production via direct iron ore reduction.

These numbers underline that, despite the rapid growth to date of the solar industry, deployment at today's scale is not enough to usher in a "solar century", during which solar becomes the central pillar of a rapid decarbonisation. For that to happen, solar deployment would need to maintain the dynamic growth pattern observed in 2011-17, but at a much larger scale. This is undoubtedly possible, as our Sustainable Development Scenario shows, but would require a broad range of policy, infrastructure, financing and market design obstacles to be overcome.

This rise of renewables in the Stated Policies Scenario comes largely at the expense of coal-fired power, whose share drops from 38% today to 25% in 2040. The outlook for gas-fired power varies widely by region, but overall it maintains its share in global generation at around 23%. With ample gas supply currently exerting strong downward pressure on prices, gas generation could continue to play a near-term role in displacing coal-fired power. We estimate in Chapter 4 that, if policies and relative prices were sufficiently favourable, a further 1.2 Gt of emissions reduction could be achieved immediately by running existing gas-fired plants harder and reducing coal use commensurately. This would be enough to bring emissions back down to where they were in 2013. By 2040, the overall carbon intensity of global power generation drops by one-third from the current average of 480 grammes of CO_2 per kilowatt-hour (g CO_2/kWh), which brings it below the typical carbon intensity of a new combined-cycle gas turbine (at 350-400 g CO_2/kWh). There are significant regional differences: in the European Union, carbon intensity drops to less than 70 g CO_2/kWh in 2040, below the projected global average for that year in the Sustainable Development Scenario, whereas in Southeast Asia the continued prevalence of coal-fired power means that the average kWh of electricity generated emits 550 g CO_2 in 2040.

Figure 1.11 ▷ Global installed power generation capacity by scenario

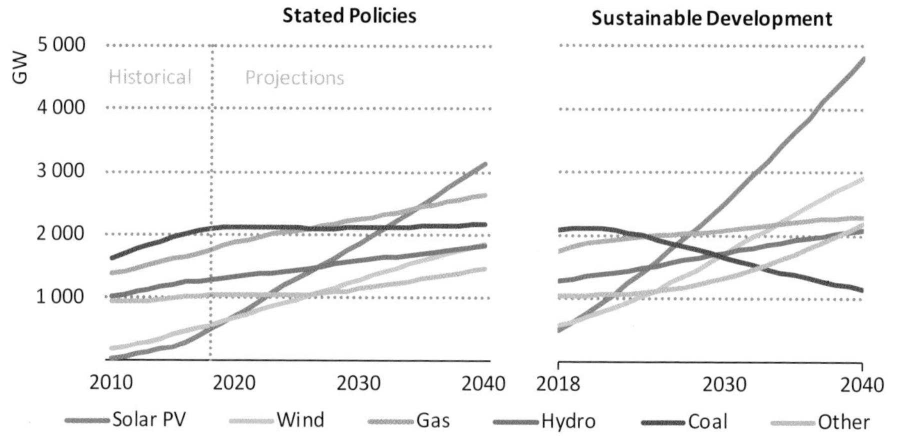

Solar PV accounts for more global installed capacity than any other energy source by the late 2020s to mid-2030s, depending on the scenario

In the Sustainable Development Scenario, global deployment of solar PV accelerates to reach 300 GW per year by 2040, alongside roughly 170 GW of annual additions of wind power (Figure 1.11). To illustrate what this would mean, the additional energy generated from the solar PV and wind added between 2030 and 2040, expressed in terms of primary energy, would be almost exactly equivalent to the additional output that has come from US tight oil and shale gas over the ten years since 2008. Translated into useful energy, i.e. taking into account the efficiency with which these outputs are eventually used by consumers, the projected growth from solar PV and wind over this period in the Sustainable Development Scenario would be far greater than the recent impact of shale.

Huge potential for solar, as well as the need for supportive policies to help its deployment, is a major theme of our special focus on Africa in Part B. Solar PV could provide the continent with far more than its projected electricity needs, but so far deployment has been slow: one of the most solar resource-rich parts of the world has installed a total of 4 GW, less than 1% of the global total. For the moment there is limited capacity within governments to help get projects off the ground, and a lack of scale and competition: transaction costs are also high. Yet solar is central to the achievement of the aims of our Africa Case, developed as part of the focus in this *WEO*, which meets not only the UN Sustainable Development Goals in relation to energy but also the economic growth goals of African economies. The Africa Case, which reflects the vision of the Africa 2063 Agenda adopted by the African Union in 2015, has around 315 GW of solar capacity by 2040 – making solar the largest source of power generation capacity in Africa.

Electric mobility is also on a fast upward trajectory. The global electric car fleet now stands at well over 5 million vehicles, helped by 2 million new sales in 2018. Growth in electric two-wheelers and buses has also been impressive, albeit so far heavily concentrated in China. In the Stated Policies Scenario, annual electric car sales rise from 2 million today to reach 10 million by 2025 and more than 30 million in 2040. Key factors putting electric mobility in the fast lane are subsidies (at least in the near term), increasingly strict fuel economy targets, restrictions or penalties on the sale or use of conventional cars or their circulation, fleet procurement decisions by public authorities and large private logistics companies, and investment in new recharging infrastructure. Alongside these policies are the plans (examined in Chapter 3) of today's manufacturers to step up production of battery electric or plug-in hybrid vehicles and/or to re-tool existing production lines. Lower battery costs are an important part of the story: battery costs fall to less than $100 per kilowatt-hour (kWh) by the mid-2020s, down from $650/kWh five years ago, meaning that electric cars in several key markets become cost competitive on a total cost of ownership basis with conventional cars.

What does this mean for sales of conventional internal combustion engine cars? Rising electric vehicles sales, together with a potential shift in consumer preferences away from personal ownership of vehicles, raise profound questions about the future of conventional cars. Set against this, however, is the rising demand for mobility in developing countries, concerns about the adequacy of recharging sites for electric cars, and also – a trend that we examine in detail – the growing appetite among consumers for bigger and heavier cars. The share of sport-utility vehicles (SUVs) in global car sales has more than doubled over the last ten years to reach 40%, and this trend is visible not only in the United States (where almost 50% of new car sales are SUVs) but also in Europe, China, India and many other countries. The rise of the SUV is significant because larger, heavier cars are more difficult to electrify fully, and because a conventional SUV consumes about 25% more fuel to travel a given distance than a medium-size car.

Passenger cars account for around a quarter of today's oil use and so do not tell the whole story for oil demand. In the Stated Policies Scenario, oil consumption continues to grow in other sectors such as trucks, aviation, shipping and petrochemicals. However, the increase in electric car sales and more stringent fuel economy standards together do lead to a peak in oil demand from passenger cars in the late-2020s. By 2040, the 330 million electric cars on the road in this scenario displace around 4 mb/d of oil use.

This projection is very sensitive to what happens to SUVs. On the one hand, if their popularity continues to rise in line with recent trends, then this could add another 2 mb/d to 2040 oil demand. If, on the other, the share of SUVs in new sales levels off at around 50% (as it does in the Stated Policies Scenario) and if in addition a higher share of the SUV fleet is electrified by 2040, the effect would be in the opposite direction.

Electrification proceeds much more quickly in the Sustainable Development Scenario, where the electric car fleet reaches almost 900 million by 2040. However, electrification alone is not sufficient to secure the deep decline in transport oil demand seen in this scenario: it is accompanied by additional improvements in fuel efficiency, as well as by the use of other alternative transport fuels, such as advanced biofuels and hydrogen.

Moving slow: emissions, coal demand, improvements in efficiency

While solar PV and electric mobility are moving fast, their impact at the pace projected in the Stated Policies Scenario is not sufficient to force a peak in global emissions. Energy-related CO_2 emissions in the Stated Policies Scenario remain on a slight upward trajectory, in essence because energy and climate policies in this scenario are not strong enough to overcome the effects of population and economic growth on energy consumption. There is no single solution that can be deployed to turn the emissions trend around. Multiple technologies and policy approaches are required in different parts of the energy system. What is certain, though, is that any scenario that involves deep cuts in emissions has to address a core question: what to do about Asia's large and youthful fleet of coal-fired power plants.

Coal-fired power generation is the single largest source of CO_2 emissions, accounting for 10 Gt out of the 33 Gt emitted today by the energy sector. Over the past 20 years, Asia accounted for 90% of all coal-fired capacity built worldwide, and these plants have potentially long operational lifetimes ahead of them, locking in emissions for many years to come. Moreover, even though final investment decisions for new coal-fired plants have slowed dramatically from almost 90 GW in 2015 to 20 GW in 2018, the size of the global coal-fired fleet continues to grow. Additions to the fleet in 2018 (primarily in Asia) came in just ahead of retirements (concentrated in advanced economies). Coal consumption is projected to see a long plateau in the Stated Policies Scenario, though the use of coal in power generation falls slightly to 2040 (the specificities of coal use in industry are discussed in Chapter 5). Under these circumstances, emissions just from the existing coal fleet would make sustainable energy targets very hard to reach.

In Chapter 6, we consider three options to bring down emissions from the existing stock of coal-fired plants: to retrofit them with CCUS or biomass co-firing equipment; to repurpose them to focus on providing system adequacy and flexibility while reducing operations; or to retire them early. In the Sustainable Development Scenario, most of the 2 080 GW of existing coal-fired capacity would be affected by one of these three options. Around 600 GW reaches 50 years of age and retires (as in the Stated Policies Scenario). In addition, around 240 GW of existing coal-fired capacity is retrofitted with CCUS or biomass co-firing equipment; around a third of the existing fleet reduces operations, limiting energy output but still providing system adequacy and flexibility; and an additional 550 GW retires by 2040 before reaching 50 years of age (Figure 1.12). The investment cost of the new power capacity that would be needed to replace these low marginal cost but high-carbon assets falls with time. This steadily closes the gap with the operating costs of existing coal plants, but operating costs for existing coal can be very low – $20 per megawatt-hour or less – making it challenging to replace existing facilities without raising the costs of electricity supply.

Figure 1.12 ▷ Reducing CO₂ emissions from existing coal-fired power capacity by measure

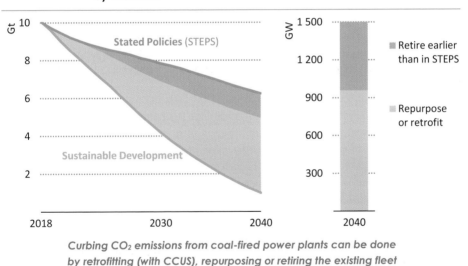

Curbing CO₂ emissions from coal-fired power plants can be done by retrofitting (with CCUS), repurposing or retiring the existing fleet

There is also much more that could be done on the demand side to accelerate energy transitions. **Energy efficiency** contributes the largest share (37%) of the additional emissions savings that the Sustainable Development Scenario provides over those in the Stated Policies Scenario. This is the main reason why, even with 3.4% annual GDP growth and an additional 1.6 billion people on the planet by 2040, the world's consumers use less energy in 2040 in this scenario than we do today.

For the moment, however, efficiency policies are fighting a losing battle against rising demand for energy services. Improvements in the energy intensity of the global economy (the amount of energy used per unit of economic activity) slowed to 1.2% in 2018, roughly half the annual average improvement seen since 2010. A key factor is the lack of dynamism both in the implementation of new energy efficiency policies and in efforts to increase the stringency of existing measures. There are many areas of global energy use that could be transformed by greater attention to efficiency across the end-use sectors of industry, transport and buildings.

One example comes from the recent *Southeast Asia Energy Outlook: World Energy Outlook Special Report* (IEA, 2019a), which addresses the future of cooling in the region and highlights challenges found in many other parts of the world. Space cooling is the fastest growing use of electricity in buildings in Southeast Asia, propelled by rising incomes and high cooling needs in a very hot part of the world. For the moment, less than 20% of households across the region have air conditioning: in Indonesia, the most populous country, around 10% do. Cooling demand is poised for very strong growth, yet our detailed market analysis shows that the average efficiency of air conditioning units sold is well below the global average, even though much more efficient units are available at comparable cost. A key problem is the ineffectiveness of labelling programmes that put the majority of products in the top category, preventing consumers from differentiating between them. In the Stated Policies Scenario, appliance ownership and cooling demand skyrocket, not only raising overall electricity demand but accentuating strains on power systems (as the share of cooling in peak power demand rises towards 30%). By contrast, enhanced efforts to improve building and equipment efficiency in the Sustainable Development Scenario reduce the growth in cooling demand in 2040 by around half.

Energy efficiency policies can also be targeted to take into account variations in the emissions intensity of power generation. As examined in Chapter 7, each electric appliance or piece of equipment has a different contribution to power sector emissions, depending on the time of use. In many countries, as the share of solar PV rises in the generation mix, the cleanest period of generation will be around the middle of the day, and the most carbon-intensive during the evening peaks once solar PV output falls: this would mean, for example, that programming a washing machine to run during the day would lead to lower emissions than running it in the evening or night (Figure 1.13). In India, ever-higher shares of solar mean a projected reduction in the emissions intensity of generation by up to 95% in the middle of the day, compared with the typical levels seen today. Policies can help by focusing efficiency efforts on the electricity uses that occur once people return home in the evening: air conditioners, lighting, televisions, computers and other household appliances. Digital tools can help shift some electricity demand to less emissions-intensive hours of the day and reduce electricity bills for consumers.

Another area where joined-up thinking is essential is material efficiency, especially for key energy-intensive materials such as steel, aluminium, cement and plastics; this is a key component of the drive for a more "circular" economy. Demand for these materials grows in the Stated Policies Scenario, with significant implications for energy use and emissions. This demand growth is undercut in the Sustainable Development Scenario by changes in the way that these materials are produced via more efficient, lower carbon process routes and by changes in the way that these materials are used. In the transport sector, for example, the Sustainable Development Scenario incorporates lightweighting in vehicle design in order to improve efficiency, which increases the demand for aluminium, plastics and composites at the expense of steel: demand for steel for road vehicles falls by 30% between the two scenarios, although total car sales in 2040 are lower by only 12% (due to increased utilisation of the fleet and an assumed shift in favour of mass transit: the latter is the key element of behavioural change that is incorporated in this scenario). The shift towards electric mobility also requires more aluminium for electric vehicles, as well as other elements such as lithium and cobalt for batteries – the latter also bringing in security of supply considerations.

Figure 1.13 ▷ Average CO₂ emissions intensity of hourly electricity supply in India and the European Union, 2018, and by scenario, 2040

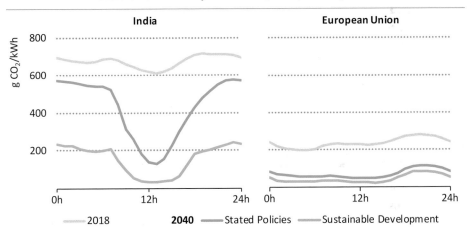

Variability of hourly emissions intensity of electricity generation increases as the share of solar and wind expands. Certain hours are almost completely decarbonised by 2040

Note: g CO_2/kWh = grammes of CO_2 per kilowatt-hour.

To watch: a new round of pledges

Interconnected commitments to a more sustainable energy system are coming from four main sources. There are the energy and climate policies adopted by different levels of government, primarily at national level but also by states in federal jurisdictions, and by cities and municipalities. There are private sector initiatives to reduce the environmental impact of their operations, ranging from individual corporate commitments to renewable electricity to sector-wide emissions targets. There are commitments by investors and the financial community that favour funding for clean energy projects, or that place restrictions on funding to more polluting sectors (the effect on such actions on coal supply is discussed in Chapter 5). And there is the bottom-up pressure exerted by society, typically felt in all of the above areas but also manifested in community-level commitments, including in sustainable communities that rely on locally produced renewable electricity and, in some cases, biogas.

These **clean energy initiatives** are a key area to watch for the future. They are also a crucial element of the architecture of the Stated Policies Scenario. The focus of the scenario analysis is mainly on the impacts of national policies, but an assessment of the impact of all of these types of commitment is incorporated, as far as it is possible to do so: so, for example, the scenario takes account of the efforts made by industry to reduce emissions in individual sectors such as aviation, shipping and steel.

One particularly dynamic area is targets for economy-wide "net-zero" emissions. An increasing number of countries, together with the European Union, have announced or are actively considering setting long-term net-zero emissions targets. Only a few of these have yet been formally adopted, but, where they have, we assess the specific policies that have been identified to achieve these long-term goals: we look at them across different elements of energy supply, transformation and end-use, and include these policies in the Stated Policies Scenario.

Figure 1.14 ▷ **Effects of including announced net-zero carbon pledges on CO_2 emissions in the Stated Policies Scenario** (STEPS)

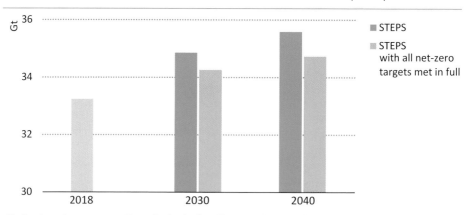

Today's net-zero commitments, including those under consideration, do not yet transform the global emissions outlook, but they are setting the terms for debate in many countries

In Chapter 2, we consider the incremental impact of all of the economies in question, including all member states of the European Union, meeting a net-zero target by 2050. The weight of these countries in global economic and energy affairs is considerable: today they account for 21% of global GDP and nearly 13% of energy-related CO_2 emissions. Full implementation of these net-zero commitments, on their own, would make a measurable, but not decisive difference to the Stated Policies Scenario (Figure 1.14). The role of these countries in terms of leadership and of technology innovation in hard-to-abate sectors would be at least as valuable. This underlines that achieving the goals of the Sustainable Development Scenario and the Paris Agreement will require not only an unprecedented effort, but also a broad one.[1]

---

[1] The Sustainable Development Scenario is consistent with the net-zero goals being reached in full.

1.10 Security

Moving fast: shale, LNG, offshore wind

The experience of the United States with the shale revolution shows how a concentrated shift in investment flows can change energy dynamics very rapidly. Ten years ago, the idea that the United States could become a net exporter of oil was almost unthinkable, yet it is on the verge of becoming reality (a consideration which should give pause to those discounting the potential for rapid changes elsewhere in the energy sector). The shale revolution has stimulated more than $1 trillion in new investment over the last ten years – nearly $900 billion in the upstream, and the rest on new pipelines and other infrastructure, including LNG export terminals. This has not, for the moment, been a profitable business for many of the companies involved: as of 2018, the upstream shale industry as a whole has yet to achieve positive free cash flow. But the industry has developed with remarkable speed, with global implications for prices, trade flows, emissions and energy security. And our projections suggest that the shale race is not yet run; many of the most profound impacts of the shale revolution still lie ahead (Figure 1.15). Higher tight oil output in the United States is the main reason why our equilibrium oil price in the Stated Policies Scenario is 10% lower than in last year's edition, adding to the pressures on many of the traditional oil producing and exporting countries (see below).

Figure 1.15 ▷ **Tight oil and shale gas output in the United States, 2010-2018, and in the Stated Policies Scenario**

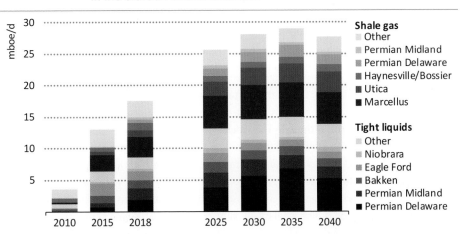

United States shale production alone (oil and gas) is projected to exceed the entire oil and gas output of Russia before 2025

Note: mboe/d = million barrels of oil equivalent per day.

Tight oil production in the United States was 0.6 mb/d in 2010. Today it exceeds 6 mb/d – a rise that already matches the fastest ever seen previously in oil markets (Saudi Arabia in the late 1960s). The speed at which tight oil has risen has created problems, especially given the recent concentration of activity in the Permian Basin. Infrastructure has struggled to keep pace with oil output: there has been insufficient pipeline capacity to transport associated natural gas to buyers (gas flaring in the Permian increased to an estimated 17 million cubic metres (mcm) per day in the first-half of 2019) and it has also proved difficult to scale adequate housing and services for the influx of workers.

The assumed size of the resource base for US tight oil, based on the latest play-by-play assessments and estimates, is 35% higher than in the *WEO-2018*. As a result, output stays higher for longer; the largest difference comes not in the high point reached, but the extent to which it is sustained over the latter part of our projection period. In Chapter 3, we examine the sensitivity of this projection to changes in our assumptions about the pace of technology learning, the size of remaining recoverable resources, the oil price and the cost of capital. All of the factors (the last to a lesser extent than the others) have significant impact on the projected level of US production. The sensitivity to price provides something of a buffer against large price swings in the international market and therefore represents an energy security asset, although our analysis suggests that this response is asymmetrical: the loss from a fall in production due to a lower price is larger than the gain from a similar size upswing. The other important factor in play is the possibility that social or policy headwinds may constrain upstream activity or the ability to get production to market.

Thanks to tight oil, the US becomes a major actor in international oil markets. Gross exports of lighter crude and refined products reach 12 mb/d by 2030: imports of oil, mainly heavier crudes fall to 7.5 mb/d. The increased availability of US exports brings energy security gains as well, although their extent needs to be kept in context. Most US export barrels are light and sweet (only just over 20% is light and medium-sour) and are not a one-to-one substitute for the sour grades, primarily from the Middle East, that are the main feedstock for Asian refineries.

Tight oil also moves relatively fast in the Sustainable Development Scenario, reaching a high point of above 10 mb/d in the 2030s before tailing off (because of the lower demand and prices in this scenario). Although the social and environmental performance of the upstream would come under even greater scrutiny, the characteristics of tight oil are reasonably well suited to such a scenario. Decline rates are high and so, faced with huge uncertainties about the future, there is less need for a long-term outlook on demand and prices: operators need just enough market visibility to know when to increase or throttle back on drilling. Tight oil is also a relatively light crude oil that is well suited to provide the kinds of products most in demand in this scenario.

Production of **shale gas** is bolstered in the near term by the increased volumes of associated gas produced along with tight oil. By the mid-2020s, almost one-quarter of the gas produced worldwide is coming from the United States. The domestic US market and the pipeline connections to neighbouring markets in Canada and Mexico can only absorb a part of this increase. The main outlet for shale comes in the form of **liquefied natural gas** (LNG) which confirms its status as the preferred way of moving gas over long distances: LNG trade grows at 3.5% per year in the Stated Policies Scenario on the back of rising exports from

North America, East Africa (mainly Mozambique) and the Middle East (Qatar), whereas long-distance pipeline trade expands at a much more modest 1% per year as additional import capacity to China comes online from Russia and Central Asia. Rising exports of US LNG, in particular, act as a catalyst for a much more interconnected, diverse and liquid global gas market. While this should give comfort to importers worried about the security of gas supply, questions remain about the extent to which LNG can cultivate new buyers and markets in emerging economies such as China (Box 1.2).

Box 1.2 ▷ LNG is big business in China, but can it also be a small business?

Trade in LNG has traditionally been conducted by buyers and sellers that are large enough to manage the costs and risks of a very capital-intensive supply chain. Size still matters, especially among sellers: recent final investment decisions offer some signs of consolidation in LNG supply, as they have been dominated by international majors and national oil companies. But the buyer side is starting to look a lot more fragmented as a range of new, smaller companies jostle alongside the established importers for access to cargoes and terminals. They are often less creditworthy than the established buyers and have less visibility on medium-term needs, but nonetheless are being catered to by LNG suppliers that are keen to develop new outlets for gas.

China is already the world's second-largest LNG importer and total gas imports are on track, in the Stated Policies Scenario, to reach those of the European Union by 2040. In addition to the large national oil companies, the gradual opening of China's gas markets is allowing a growing cast of mid-size utilities and private enterprises to contract for new LNG supplies, develop transport infrastructure and facilitate trade, turning China into a focal point for innovation in the use of LNG. Prices are unregulated in some of these market segments, and the most attractive commercial opportunities are often to displace oil product use, rather than coal.

Over the past few years, China has built the world's largest small-scale LNG system, in response to growing demand and pipeline infrastructure bottlenecks. Truck-loading capacity at LNG terminals stands at over 20 Mt, serving nearly 350 000 LNG-fuelled trucks which service a network of 3 000 refuelling stations. New bunkering facilities are also being built to support a government target to reach at least a 10% market share for LNG in inland waterways by 2025. Several companies have been experimenting with delivering LNG using container ships. This aspect of China's gas story may turn out to be significant for the prospects of LNG in other emerging economies, where there are still plenty of commercially accessible, relatively small-scale opportunities to compete with oil, but where the scope to displace coal is contingent to a large extent on policy developments.

Oil and gas may appear to be heading along divergent tracks in our scenarios, but the links between our US tight oil and shale gas outlooks are a reminder that, even as downstream ties are loosened, there are still strong upstream connections between the two fuels. On average, every barrel of oil produced today comes with around 60 cubic metres of gas.

The situation varies widely across different parts of the world – as examined in case studies of the Gulf countries, United States and Brazil in Chapter 4 – but in total more than 850 billion cubic metres (bcm) of gas was produced in 2018 along with oil. Only 60% of this associated gas ended up being sold to consumers, while 15% was utilised by the energy industry, primarily for reinjection to maintain reservoir pressure. The remaining 200 bcm – roughly equivalent to current LNG imports to Japan and China combined – was either flared (140 bcm) or vented (an estimated 60 bcm) – a significant and particularly wasteful source of emissions.

The *WEO-2019* also includes a chapter dedicated to the outlook for **offshore wind**[2], which has become a rising force in renewable power generation. Spurred by initial deployment in Europe's North Sea, technology gains are now driving down costs towards those for fossil fuel-based power and other renewables, demonstrated by several projects in the pipeline in Europe. There are further innovations on the horizon, including floating turbines that have the potential to open new markets and resource potentials. Offshore wind already contributes 60 TWh, or around 2% of the European Union's electricity generation, and this rises to over 560 TWh by 2040 in the Stated Policies Scenario; in the Sustainable Development Scenario, it rivals onshore wind as the European Union's single largest source of electricity. Europe's success is sparking increasing interest from elsewhere: China now stands among the market leaders and the United States is poised for growth with the first contracts awarded for large-scale projects in 2019.

The technical potential for offshore wind to grow is enormous: our detailed country-by-country estimate suggests that tapping just the most attractive potential, in shallow waters near to shore (while avoiding shipping lanes and environmentally sensitive areas), could produce close to 36 000 TWh globally per year, an amount nearly equal to global electricity demand in 2040 in the Stated Policies Scenario. At these levels, offshore wind could also be a crucial input to the production of low-carbon hydrogen (IEA, 2019b). The degree of deployment worldwide will depend on how risks to development are managed and mitigated, including immature or inefficient supply chains (outside the established areas in Europe), difficulties with permitting or transmission connections, and challenges relating to public acceptance, relationships with other maritime industries and the marine environment.

The key asset that offshore wind brings to the transformation of the power sector is the scalability and dependability of output. It remains a variable source of generation like onshore wind and solar PV, but offshore wind offers significantly higher capacity factors than either of these (Figure 1.16). This is thanks to ever-larger turbines that can access higher and more reliable wind speeds further from shore: individual turbines can reach annual capacity factors of well over 50% or more. This gives offshore wind characteristics that might be termed "variable baseload", a halfway house between solar PV and onshore wind, on the one hand, and dispatchable sources of generation on the other.

---

[2] *Offshore Wind Outlook 2019: World Energy Outlook Special Report* (IEA, 2019b) is available at: http://www.iea.org/offshorewind2019.

Figure 1.16 ▷ Average annual capacity factors for various power generation technologies by region/country

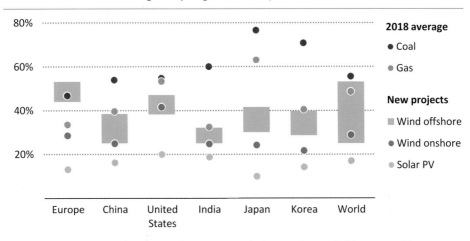

Offshore wind offers capacity factors well above other variable renewables, and levels comparable to efficient gas-fired power plants in several regions

Moving slow: integrated thinking; reform in producer economies

The concept of energy security is much broader and more dynamic today than it has been in the past, which is why the IEA has made this a key element of its own modernisation strategy. The attacks on Saudi Arabia in September 2019 made it clear that traditional energy security risks have not dissipated (see next section regarding oil), but there is a host of new considerations that arise as energy transitions gather pace, particularly in the electricity sector, as well as new physical risks to energy infrastructure from a changing climate and extreme weather. From a policy perspective, there are twin hazards: that decision makers fail to keep up with the pace of change (Box 1.3); and that a "silo mentality" means they do not account for interactions and trade-offs across different parts of the energy sector. The need for integrated thinking about energy security is a core theme of this *Outlook*.

There are difficult questions that are not yet being adequately addressed about the respective contributions of electricity and **natural gas networks** to least-cost energy transitions. The policy focus in many countries is on the potential for low-carbon electricity to play a much greater direct role, but there are limits to how quickly and extensively electrification can happen in practice. Where they exist already, gas networks are typically the largest and most flexible energy delivery mechanism to consumers, and there are plans to expand gas grids in many of the world's fastest growing economies, notably China and India. However, emissions from the combustion of gas, together with methane leaks along the supply chain, mean that the long-term role of this infrastructure in a low-emissions future is not clear, even if gas brings near-term benefits when replacing more polluting fuels. From an energy security perspective, maintaining overlapping infrastructure can be an important asset, compared with an approach relying exclusively on a single electricity network. But this means thinking through some complex interactions between electricity and gas security; it also requires the gas grid eventually to be repurposed or retooled to deliver low or zero-carbon energy. In Chapter 13, we consider two key options: low-carbon hydrogen and biomethane.

Box 1.3 ▷ Staying ahead of the digital revolution

The pace of technological change presents a challenge. For decades energy has been a relatively slow moving, technologically predictable industry and this was often reflected in the legal and regulatory environment. Today, a new phase of technology-driven transformation is emerging, propelled in large part by advances in digital technologies, and it can be difficult for policy makers to stay ahead of the curve. Notable regulatory gaps in the electricity sector include the market design treatment of storage, the contractual and data management aspect of demand response, the interface between electric vehicles and the grid, and the location and production incentives for solar "prosumers". Moreover, the confluence of digitalisation and electrification, if not managed well, could make systems more vulnerable to cyber-attacks and also lead to significant privacy concerns.

Some jurisdictions have made progress in addressing these issues, but there is unfinished business everywhere. The discrepancy between the speed of change in technology on the one hand and regulation on the other could lead to major electricity security risks.

Changing energy dynamics are also exerting considerable strain on many of the world's **traditional oil and gas producers**. This was highlighted in a *WEO-2018* special report (IEA, 2018), but the pressures to reform and diversify these producer economies are even more visible in our new projections. The higher US tight oil outlook and a lower equilibrium oil price in the Stated Policies Scenario represent a double blow to the net income of incumbent producers. The long-term response to climate change also underlines the need for producer economies to prepare for a world in which hydrocarbons may no longer be the main source of revenue. This pressure is felt in many large resource-holders in the Middle East and beyond. Given this year's focus on Africa, we highlight in Chapter 11 the pressure on Nigeria, where a combination of a less favourable international market setting (especially for producers of light, sweet crude oil in direct competition with tight oil) and continued uncertainty over regulatory frameworks has meant a downward revision in projected output.

To watch: oil security; Africa

The rise of US tight oil and the increasing presence of the United States as an exporter offer some reasons for comfort on the outlook for **oil security**. As the attacks on tanker traffic and on Saudi Arabia in 2019 illustrated, however, there are also ample reasons for caution.

Whichever pathway the energy system follows, the world will still be relying heavily for some time yet on the adequacy of investment and supply from the Middle East. This means that one of the world's busiest trade routes, the Strait of Hormuz, is set to remain firmly in the spotlight. The Strait currently carries some 16 mb/d of crude and another 4 mb/d of oil products primarily from Saudi Arabia, Iraq and Iran to major importers in Asia, but also to Europe and Africa. All of Qatar's LNG exports go via the same route.

In the projection period to 2040, either a shortfall in conventional spending or an impediment to trade through the Strait of Hormuz would materially tighten markets and leave the world facing significantly higher prices for purchases of oil. All would be affected, but direct vulnerabilities would be concentrated among the key emerging economies of Asia, with potentially serious knock-on effects on economies (as well as intensified moves to reduce reliance on oil). In the Stated Policies Scenario, oil demand growth in China grinds to a halt in the 2030s, largely due to a very rapid rise in electric vehicles and improvements in fuel efficiency, but oil use in Asia as a whole nonetheless rises by 8 mb/d between 2018 and 2040. With declining domestic production, Asia's import requirements grow by 10 mb/d. Asia also sees rapid gas demand growth in this scenario, with the majority of demand in developing economies met by rising seaborne imports. Rising Asian import needs increase flows through the Strait of Malacca, a key stretch of water between Malaysia and Indonesia that connects exporters in the Middle East and Africa with Asian importers.

The development pathway that **Africa** follows will be an increasingly influential factor in the future of energy. The global population without access to modern energy, the most extreme form of energy insecurity, increasingly becomes concentrated in rural areas in sub-Saharan Africa (see next section). At the same time, almost 600 million people are likely to be added to Africa's urban population over the period to 2040. This is significantly higher than the growth that China saw during the twenty years from 1990, a period when China's production of iron and steel and cement skyrocketed (Figure 1.17). We do not anticipate that Africa's future infrastructure development will follow the same path as China, but the implications for energy of Africa's urbanisation trends are still profound: as discussed in Chapter 9, these are felt not only in the demand for materials, but also in industry and agriculture, as well as in increased mobility of people and goods. The expected growth in population in Africa's hottest regions also means that up to half a billion additional people will be living in regions that need cooling, a challenge for electricity provision and for public policy in setting efficiency standards and designing Africa's future built environment.

The development of Africa's resources and minerals is also set to be a major factor in energy markets over the coming decades. In the past, the focus has been on Africa's oil production and exports, but in our projections the focus switches towards natural gas and to the mineral and metal deposits that can play an important role in energy transitions. Outside North Africa, infrastructure constraints and low purchasing power have so far limited gas use: at 5%, the share of gas in the energy mix in sub-Saharan Africa is one of the lowest in the world. However, Africa has seen major gas discoveries in recent years in

Mozambique, Tanzania, Egypt, Mauritania, Senegal and South Africa, and the continent's evolving energy needs are starting to create room for gas to grow. This is particularly visible in our projections for the Africa Case where a renewables-plus-gas approach provides the bedrock for growth. Gas investment also supports an increasing presence in international LNG, and Mozambique becomes the largest gas producer in sub-Saharan Africa.

Figure 1.17 ▷ Urban population and cement demand growth in China (historical) India (Stated Policies Scenario) and Africa (Africa Case)

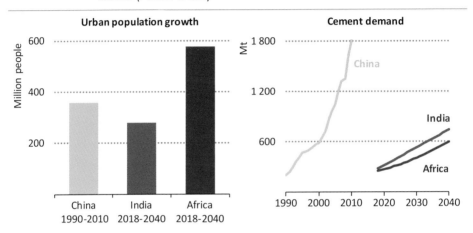

Africa is set to experience unprecedented growth in its urban population, a critical factor in demand for materials, industrialisation and economic growth

1.11 Affordability

Moving fast: battery storage; pathways to electricity access

Keeping energy affordable for consumers and industry while achieving challenging policy goals remains at the forefront of policy makers' minds in all countries. As energy bills reflect the dynamic interplay of prices, energy efficiency and energy service demand, often averaged annually across all different times of day, a systemic approach to considering affordability is essential. Batteries and digital applications are examples of enabling technologies that are improving rapidly and thereby raising confidence that high peak electricity prices can be avoided, energy efficiency services can be commoditised and renewable mini-grids can be cost-effective options in certain regions.

Battery cost reductions are changing how the electricity system accommodates the rise of wind and solar in the power mix. In Chapter 6 we examine how this becomes a key variable in the outlook for India's power supply, especially in the mid-2020s once continued strong growth in electricity demand requires a wave of investment decisions in new generation capacity. In the Stated Policies Scenario, most of this investment goes to solar, battery storage plants and coal.

Battery storage is one of the technologies best suited to provide the short-term flexibility that India needs to balance electricity supply and demand, and 120 GW is installed by 2040 in the Stated Policies Scenario. The fortunes of the battery industry in India are linked to global market developments and also to the policy push for accelerated deployment of more affordable electric vehicles. The scale of the industrial opportunity, the economies of scale, the pace of learning by doing and battery chemistry improvements are all uncertain, and there is the possibility that costs could decline even more quickly. In this *Outlook*, we consider a "Cheap Battery Case" in which the costs of a four-hour battery system reach $120/kWh by 2040, 40% lower than in the Stated Policies Scenario, resulting in around 200 GW of battery deployment (Figure 1.18). This makes combined solar and storage plants a compelling proposition for India, meaning that the number of new coal-fired plants would fall sharply once the existing project pipeline runs dry (avoiding almost 80 GW of coal-fired capacity by 2040 relative to the Stated Policies Scenario).

Figure 1.18 ▷ **Installed power generation capacity by source and CO_2 emissions from electricity generation in India in the Stated Policies Scenario and Cheap Battery Case**

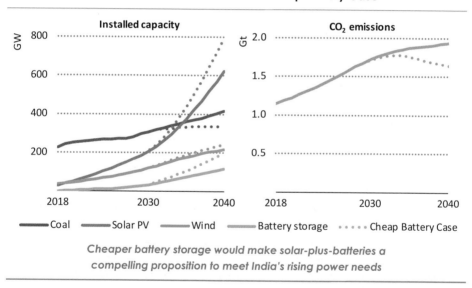

Cheaper battery storage would make solar-plus-batteries a compelling proposition to meet India's rising power needs

Technology cost reductions and innovative business models are also opening new and more affordable **pathways to electricity access**, especially for communities far from the existing grid. Three-quarters of the people without access to electricity in the world live in sub-Saharan Africa, but there are signs of progress. The number of people gaining access to electricity has accelerated over the past five years, more than doubling from 9 million per year between 2000 and 2013 to 20 million per year between 2014 and 2018, and outpacing population growth. Much of this recent dynamism has been in East Africa, in part due to continued grid extensions but also to a rapid rise in access via solar home systems: Kenya,

Tanzania and Ethiopia accounted for around half of the 5 million new solar home systems that we estimate were deployed in sub-Saharan Africa in 2018 (up from only 2 million installations in 2016) (Figure 1.19).

Figure 1.19 ▷ Electricity access by sub-Saharan region in the Stated Policies Scenario

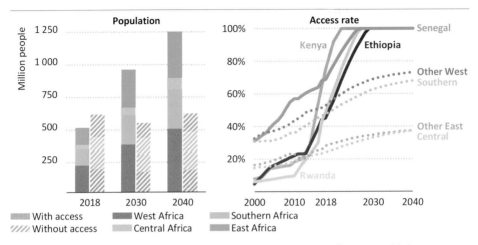

A combination of affordable technologies and strong policy support is key to the rapid increase in access rates in East African countries

In the Stated Policies Scenario, more than 500 million people in sub-Saharan Africa remain without access to electricity in 2030; achieving full electricity access by this date (as achieved in the Sustainable Development Scenario and the Africa Case) would mean tripling the rate of connections seen in the Stated Policies Scenario to reach more than 60 million people each year. A variety of routes play a role in scaling up access, but our analysis suggests that mini-grids will be an important part of the solution, especially for settlements far from existing grids that have relatively high population densities. A key strength of mini-grids is that they can integrate household needs with the provision of other services to support, for example, schools, health services and light industries. Digital technologies, allied with better data collection and management, help to underpin more effective business models for mini-grids by improving billing and lowering financing costs.

Moving slow: pricing reform; access to affordable capital

Prices are a critical motor for change in any economic system, and energy is no exception. Given the expressed priority to act on emissions, one might imagine that the bias in energy **pricing policies** around the world would be tilted in favour of investment in cleaner and more efficient energy technologies, but this is not the case (Figure 1.20). The data for 2018 show a one-third increase in the estimated value of global fossil fuel consumption subsidies, to more than $400 billion, as a higher international oil price increased the gap between market-based pricing and the artificially low prices paid by some consumers. This was almost double the combined value of subsidies to renewable energy and electric vehicles and the revenue from carbon pricing schemes around the world. Approaches to energy pricing and subsidies vary widely by country, but this overall imbalance greatly complicates the task of achieving an early peak in global emissions.

Figure 1.20 ▷ **Estimated value of subsidies to fossil fuel consumption, renewables and electric vehicles, and carbon pricing, 2010-2018**

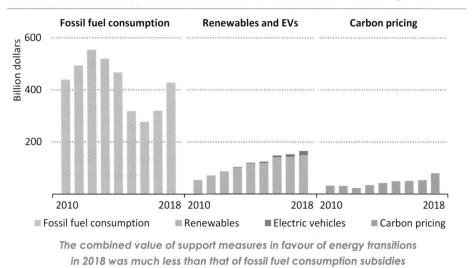

The combined value of support measures in favour of energy transitions in 2018 was much less than that of fossil fuel consumption subsidies

There can be good reasons for governments to make energy more affordable, particularly for the poorest and most vulnerable groups. However, many fossil fuel consumption subsidies are poorly targeted, disproportionally benefiting wealthier segments of the population that make much more use of the subsidised fuel. Such untargeted subsidy policies encourage wasteful consumption, pushing up emissions while also straining government budgets and leading to market distortions. Their removal would promote sound energy policy and energy transitions.

Efforts at pricing reform have intensified in recent years following the fall in the oil price in 2014. The fall in price offered an opportunity for reform to oil importers such as India, Indonesia and Malaysia, and made reform very desirable in fiscal terms for many exporters such as those in the Middle East that have traditionally provided very cheap energy to their populations. However, in 2018 the oil price trended higher for much of the year before falling back in the last quarter. This became a source of strain in importing countries where consumers were exposed to rising retail prices, while relieving some of the pressure on oil exporters. Reforms in many cases were partially reversed (this was the case in both Indonesia and Malaysia) or postponed.

On the other side of the equation, estimated subsidies to renewables continue to grow in absolute terms: the position varies from technology to technology, but overall their relative importance is set to decline as renewable costs fall and governments move towards auctions or competitive tenders as the most cost-effective way to promote deployment. One of the most important subsidy initiatives for electric vehicles, China's New Energy Vehicle programme, is being scaled back to encourage innovation rather than reliance on state support. The scope of carbon pricing in China, though, is set to increase substantially with the envisaged launch of a national emissions trading system in 2020. Most of the existing application of carbon prices is in Europe, which in 2019 showed that meaningful carbon pricing can be very effective: the price of European emissions allowances has risen to the $25-30/tonne range, after languishing in single figures for many years. Carbon pricing schemes becomes increasingly widespread and stringent in the Sustainable Development Scenario, eventually covering all advanced economies together with Brazil, China, Russia and South Africa by 2040.

Getting pricing right is fundamental to the prospects for adequate **investment** in low-carbon technologies. This is particularly difficult for countries where existing subsidies to fossil fuel consumption and poorly developed financial systems constrain access to affordable capital. Annual energy investment spending by developing economies would need to increase by more than two-thirds over the next two decades, to $1.8 trillion, to provide what is required by the Sustainable Development Scenario (Figure 1.21). The conditions that would facilitate these flows of investment and finance are becoming a major focus area for *WEO* analysis.

Figure 1.21 ▷ Energy investment indicators by economy and scenario

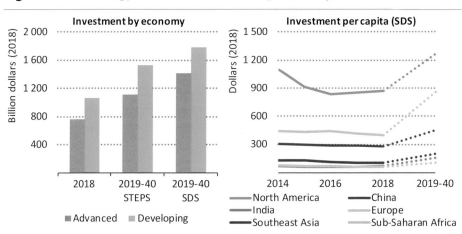

Investment needs in both advanced and developing economies pick up considerably. Spending on a per capita basis rises in all regions in the Sustainable Development Scenario

Note: Investment shown for the Stated Policies Scenario (STEPS) and the Sustainable Development Scenario (SDS) is the average annual investment from 2019 to 2040.

The mix between public and private sources of finance is a key consideration. In China, which accounts for 20% of investment in the Sustainable Development Scenario, public finance institutions have made ample low-cost financing widely available, but few other emerging economies have the balance sheets to replicate this approach. The investment challenge in many emerging economies lies in matching the potential catalytic effect of (limited) public funds with market reforms that simultaneously crowd-in larger amounts of affordable capital from private sources (see Chapter 2).

This year's *WEO* highlights opportunities to close investment gaps in several key regions, with particular attention to Southeast Asia and sub-Saharan Africa. In Southeast Asia, the cost and availability of capital vary widely, as do the investment frameworks for fast-growing power systems (IEA, 2019a). Policy priorities range from improving the financial position of the state utility in Indonesia through to enhancing the bankability of contracts in Viet Nam, and supporting markets for efficiency upgrades in Thailand. In sub-Saharan Africa, access to long-term finance is low outside South Africa, and investment decisions on around two-thirds of power generation capacity in the past five years relied on public funding (see Chapter 10). Investments are hampered by power purchase risks from cash-strapped utilities, weak or unclear power procurement policies and volatile currencies. Some countries have implemented successful procurement programmes (Senegal, South Africa, Zambia), but reform momentum needs to pick up fast to accelerate investment.

To watch: decarbonised gases

Natural gas grids in many countries are the primary network to bring energy to consumers. In Europe and the United States, for example, they deliver far more energy to end-users than electricity networks. They also provide a valuable source of flexibility, scaling up deliveries as necessary to meet winter demand peaks. In the context of energy transitions, replicating the energy services that gas grids provide with electricity networks looks likely to be challenging and expensive. If there is, instead, an option to fill existing pipelines with decarbonised gases, then a valuable existing asset could be used through energy transitions and beyond.

There are two main candidates for this role, which are examined in detail in Chapter 13: **low-carbon hydrogen and biomethane** (biogas supply is also covered in Chapter 7). Low-carbon hydrogen can be produced either from fossil fuels (with CCUS) or by electrolysis from water using low-carbon electricity. Our analysis shows that there is considerable supply potential for both hydrogen and biomethane. The issue is affordability (Figure 1.22). We estimate that today there is around 15 bcm of biomethane, mostly landfill gas, that is competitive with the wholesale price of natural gas in its respective regions. However, the supply costs of most biomethane and all hydrogen production are not competitive with natural gas supply today.

This changes somewhat to 2040 in the Sustainable Development Scenario, in part because natural gas prices are higher but also because of cost reductions for decarbonised gases. In the case of hydrogen, there is strong interest from policy makers and industry in scaling up new low-carbon routes for its production, especially for the (currently more expensive) electrolysis route. The first waves of deployment at scale will – as with almost all technologies – bring technology learning and cost reductions, which are incorporated into our modelling framework. The situation with biomethane is different: production technologies are mature and, absent a technology breakthrough, the cost benefits of increased deployment are likely to be more modest. The cost-competitiveness of biomethane relies instead on policy. If CO_2 prices are applied to the combustion of natural gas, then biomethane becomes a more attractive proposition. If policy recognises the value of avoided methane emissions that would otherwise take place from the decomposition of feedstocks, then an even larger quantity would be cost competitive. Methane is such a potent greenhouse gas that attaching a value to these avoided emissions makes a dramatic difference to its overall supply cost profile.

Figure 1.22 ▷ Supply costs of natural gas, biomethane and hydrogen in the Sustainable Development Scenario, 2018 and 2040

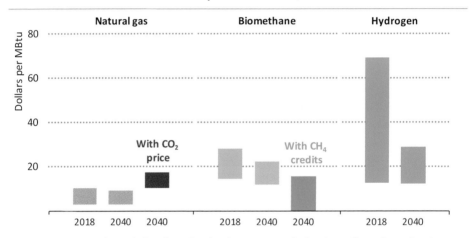

By 2040, the supply costs for decarbonised gases are closing in on those for natural gas, especially if externalities are taken into account

Notes: MBtu = million British thermal units; CH_4 = methane.

Conclusions

The global energy system is large and typically slow moving, but it is not immune to change. Our analysis in this *WEO-2019* highlights some profound shifts that are already visible and some that may become more so. It also highlights both areas of great dynamism and areas where it is in short supply. Each of these areas, and the asymmetry between them, change the nature and distribution of risks to energy security, requiring constant vigilance from energy policy makers.

Perspectives on the future of energy can also differ greatly depending on which part of the energy system is being examined. For example, it makes a huge difference whether the focus is on new flows of energy or on the existing stock of energy-using infrastructure. Flow indicators could include the rising number of electric cars sold each year, or the growth in

US shale output, the efficiency of new appliances sold, or the investment into solar PV. Flows are often what matter to investors, or to incumbents whose revenues are squeezed. But they do not tell the whole story. It can take a long time to renew the energy capital stock, and for new flows to overcome the accumulated weight of decades of previous investments. And ultimately it is the stock of existing assets and infrastructure that matters most for emissions. Dealing with legacy stock issues, such as the existing buildings in advanced economies or the existing coal-fired power generation fleet in Asia, is a tough proposition for policy makers interested in a rapid transformation.

It likewise makes a difference if the focus is on electricity, which accounts for around 20% of final energy consumption around the world, or on other aspects of energy that may be more resistant to electrification and to change. The discussion about an advanced economy, where energy demand tends to be flat or in decline and the means to change course are greater, also differs significantly from one about parts of the developing world where – as our Africa analysis underlines – energy needs are growing fast or are unmet today, and affordability is an even more critical consideration.

The changes projected in the Stated Policies Scenario cannot be taken for granted: the rate and composition of growth in energy demand in 2018, if maintained, would put energy and emissions on a higher growth trajectory than in the business-as-usual Current Policies Scenario. And there is still a large gap between where the world appears to be heading in the Stated Policies Scenario, based on today's announced policies and targets, and what would need to happen in order to reach sustainable development goals and avoid severe impacts from climate change.

Closing this gap means addressing both flows and stocks, transforming not just the electricity sector but the whole energy system, and requires actions in advanced and developing economies alike. And it is not enough, in that context, that a handful of technologies are moving very quickly. As the IEA has emphasised repeatedly, there is no single or simple solution to turn emissions around. Multiple approaches and technologies – including much greater efficiency – are required across all parts of the energy system, alongside a clear-eyed appreciation of where emissions occur and what the abatement options are in each area.

Today's speed of change is already creating uncertainties for some players in the energy sector, especially where incumbents are faced with technologies with potential for rapid progress. This is the case for conventional resource-holders in a world where shale output is growing fast, or for thermal generators facing the rise of wind and solar PV, or nuclear operators in a world of cheaper natural gas and ever-cheaper renewables (the latter example underlining that there is disruption, too, for some low-carbon technologies). It also applies to oil companies faced with the rise of electric vehicles – although here perceptions are important too: thus far, the impact of electric cars on investor sentiment towards oil has been considerably larger than their impact on oil demand.

Reactions to this market and technological uncertainty have included a systematic preference for shorter cycle investments: in many parts of the world, committing to any project with a long lead time or extended payback period is seen as risky. But these types of projects have also been a mainstay, in the past, of adequate supply. There are also large, capital-intensive undertakings that are essential to the shift to a low-emissions future: CCUS, nuclear (in some countries), hydrogen, integrated smart-city infrastructure and many more. As the IEA's *World Energy Investment* has highlighted, there is a risk of getting caught between two worlds, the old and the new, and failing to succeed in either of them (IEA, 2019c). As things stand, there is not enough investment in traditional forms of supply to maintain today's consumption patterns, nor enough in clean energy technologies and efficiency to change track.

The faster the transformation required – and the scientific evidence shows that this push needs to be very rapid indeed – the greater the risk of poor co-ordination or unintended consequences for the reliability or affordability of supply. This puts extraordinary responsibility on governments. Private initiative and capital is vital to the task of creating a more secure, sustainable and affordable energy future, but the vast majority of energy investment around the world is still made in response to conditions set by governments. The need for clear and effective signals and unambiguous direction is greater than ever.

Chapter 2

Energy and the Sustainable Development Goals
Are we on the path to achieve the SDGs?

SUMMARY

- The Sustainable Development Scenario lays out a pathway to reach the United Nations Sustainable Development Goals (SDGs) most closely related to energy: achieving universal energy access (SDG 7), reducing the impacts of air pollution (SDG 3.9) and tackling climate change (SDG 13). It is designed to assess what is needed to meet these goals, including the Paris Agreement, in a realistic and cost-effective way.

- Recent progress has been mixed, and so are prospects for improvement given proposed policies. The number of people without access to electricity fell from 980 million in 2017 to 860 million in 2018, but the lack of electricity access in sub-Saharan Africa remains acute. Current and planned policies deliver universal electricity access in many parts of the world, but are insufficient to fully electrify Africa by 2030. Annual premature deaths linked to household and outdoor air pollution stand at around 5.5 million globally. This figure is set to rise to around 7 million by 2050.

Figure 2.1 ▷ Energy-related CO_2 emissions and reductions by source in the Sustainable Development Scenario

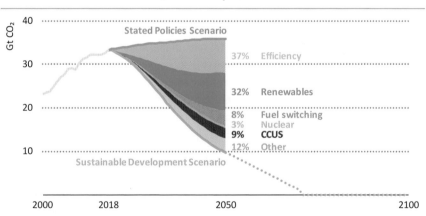

Efficiency and renewables provide most emissions reductions, but more technologies are needed as emissions become increasingly concentrated in hard-to-abate sectors

Note: CCUS = carbon capture, utilisation and storage.

- Energy-related carbon dioxide (CO_2) emissions grew by 1.9% in 2018. A peak in these emissions is not in sight: they are set to reach nearly 36 gigatonnes (Gt) annually by 2050 in the Stated Policies Scenario. Jurisdictions accounting for nearly 13% of total current emissions have put forward net-zero emissions targets, an important step in increasing global ambition, but insufficient to significantly alter the trajectory of global energy-related CO_2 emissions.

- In our Sustainable Development Scenario, universal access to electricity and clean cooking facilities is achieved by 2030 with net reductions in greenhouse gas emissions, as increased CO_2 emissions from the power sector and use of liquefied petroleum gas (LPG) for cooking are more than offset by the reduction in methane emissions from reduced biomass burning for cooking. Premature deaths from air pollution are 3 million lower in 2050 than in the Stated Policies Scenario.

- Thanks to efficiency improvements, electrification and fuel switching, energy demand remains broadly stable, despite a growing global economy. However, the fuel mix changes dramatically. Electricity demand grows steadily in the Sustainable Development Scenario, as mobility and heating are electrified. By 2040, wind and solar become the top two sources of power generation and by 2050 the power sector is mostly decarbonised. Electric cars make up three-quarters of all cars sold. Hydrogen and biomethane are used in gas grids. Energy efficiency, material efficiency and carbon capture, utilisation and storage (CCUS) together decarbonise heavy industries. A host of new low-carbon technologies move from low market shares to wide deployment at a speed as fast as anything in the history of the energy sector.

- By 2050, oil use falls to 50 million barrels per day (mb/d), 40% of which is used as a feedstock to produce plastics and asphalt (i.e. non-emitting uses). Demand for natural gas hovers just above 4 000 billion cubic metres (bcm) until the late 2030s, before entering a pronounced decline. Coal demand declines in both absolute and relative terms, and by 2050 accounts for 8% of total energy use. By 2050, low-carbon technologies, most of which are renewables, but also nuclear and CCUS, support well over half of global energy demand, from less than 20% today, a reversal of the stable share of fossil fuels at over 80% for the past three decades.

- Global energy-related CO_2 emissions decline rapidly, in line with the objectives of the Paris Agreement. They reach 25 Gt in 2030, reduce to under 10 Gt in 2050, and are on track for net zero in 2070 (Figure 2.1). This trajectory is consistent with a 66% chance of limiting the temperature rise to 1.8 °C or a 50% chance of a 1.65 °C stabilisation, and does not rely on large-scale negative emissions. If negative emissions technologies were deployed in the second-half of the century at levels similar to those reported in the IPCC's *Special Report on Global Warming of 1.5 °C*, then the Sustainable Development Scenario would provide a 50% chance of limiting the temperature increase to 1.5 °C. For this temperature target to be achieved without negative emissions, global energy-related emissions would need to fall to zero by 2050. The changes required to set such a course are rapid, deep and unprecedented and their implications would be far-reaching and extend well beyond the energy sector.

Introduction

This chapter illustrates our Sustainable Development Scenario, depicting an energy future that simultaneously delivers on the United Nations Sustainable Development Goals (SDGs) most closely related to energy: universal energy access (SDG 7), reducing impacts of air pollution (part of SDG 3) and tackling climate change (SDG 13). The time horizon[1] for this analysis has been extended this year for the first time to 2050 to assess the potential of new technologies (such as hydrogen and renewable gases) in supporting the global energy transition and to reflect the announcements that several countries have made about plans to reach carbon neutrality by 2050.

The first part of this chapter provides an overview which shows that there has been mixed progress on meeting the SDGs most closely related to energy. It then describes the scenario outcomes in terms of achieving the SDGs, before looking at the energy sector transformation and the investment needs implied by the scenario.

The chapter continues with three focus areas that take a long-term perspective towards 2050 (and beyond). They cover:

- **How are we doing?** This section assesses where stated policies will take us in meeting the SDGs. It also looks at current plans to increase ambitions and the impact of those on carbon dioxide (CO_2) emissions.

- **Where do we need to get to?** This section explores in detail the energy sector transformation in the Sustainable Development Scenario, shedding light on critical gaps, including finance and technology innovation. It also explains how the Sustainable Development Scenario has evolved, capturing recent technology progress as well as science and investment trends.

- **How much further can we go?** This section, more exploratory in nature, looks at the scope to do more in the light of proposals for even more ambitious temperature goals. It depicts a future pathway for the energy system compatible with limiting the global temperature increase to 1.5 degrees Celsius (°C) without the use of net-negative emissions, and discusses the standing of the Sustainable Development Scenario in the pursuit of this more ambitious target.

> Figures and tables from this chapter may be downloaded from www.iea.org/weo2019/secure/.

---

[1] The *World Energy Outlook* scenarios extend to 2050, with inputs (see Annex B) quoted to 2040 for the analyses, while this chapter looks to 2050 and beyond.

Sustainable Development Scenario

2.1 Scenario overview

There are three pillars to the Sustainable Development Scenario. These are to ensure universal energy access for all by 2030; to bring about sharp reductions in emissions of air pollutants; and to meet global climate goals in line with the Paris Agreement (see section 2.4). The scenario meets them simultaneously while reflecting national priorities and the latest technology and market developments. Taking those changes into account, the scenario outcome has evolved over successive editions of the *World Energy Outlook (WEO)* (see Box 2.1).

In many cases the policies necessary to achieve the multiple SDGs covered in the Sustainable Development Scenario are complementary. For example, energy efficiency and renewable energy significantly reduce local air pollution, particularly in cities, while access to clean cooking facilitated by liquefied petroleum gas (LPG) also reduces overall greenhouse gas (GHG) emissions by reducing methane emissions from incomplete combustion of biomass as well as by reducing deforestation (Masera, Bailis and Drigo, 2015; IEA, 2017a).

Yet there can be trade-offs. For example, while electric vehicles (EVs) reduce local air pollution from traffic, the overall CO_2 footprint of EVs can actually exceed that of combustion engine vehicles if there is not a parallel effort to decarbonise the power sector (Box 2.4). To take another example, retrofitting coal-fired power plants with pollution controls may be the cheapest option in the short term to deal with local pollution, but may lead to long-term emissions not aligned with climate goals (see Chapter 6).

Dichotomies between short- and long-term policy actions can also add to the challenge of accomplishing the Sustainable Development Scenario. Ultimately, the balance of potential synergies or trade-offs depends on the route chosen to achieve the energy transition. The Sustainable Development Scenario portrays an energy future which emphasises the co-benefits of the measures needed to simultaneously deliver energy access, clean air and climate goals.

This chapter shows that, despite positive movement in many countries and sectors, we are not on track to meet the Sustainable Development Goals, and sets out some possible ways to address the shortfall. On the basis of current stated policies, energy-related CO_2 emissions are set to continue to rise, premature deaths linked to air pollution are set to increase, and in 2030 there would still be around 620 million people without access to electricity and around 2.3 billion people cooking with primitive stoves or without access to cleaner fuels. Consequently, the transformation required in the energy sector to deliver the SDG goals is profound (Table 2.1).

Table 2.1 ▷ Key energy indicators in the Sustainable Development and Stated Policies scenarios

	2018	Sustainable Development 2030	Sustainable Development 2050	Stated Policies 2030	Stated Policies 2050
SDG 7: Access (million people)					
Population without access to electricity	862	0	0	623	736
Population without access to clean cooking	2 651	0	0	2 302	1 538
Related premature deaths	2.5	0.6	0.8	2.4	1.8
SDG 13: Energy-related GHG emissions					
CO_2 emitted (Gt)	33.2	25.2	9.75	34.9	35.9
CO_2 captured with CCUS (Mt)	32	763	2 776	71	154
Methane (CH_4) (Mt)	127	51	30	116	108
of which from oil and gas operations	77	20	14	66	63
SDG 3: Air pollution (million people)					
Premature deaths from energy-related outdoor air pollution	3.0	2.7	3.0	3.6	5.1
Primary energy supply					
Total primary energy supply (Mtoe)	14 314	13 750	13 110	16 311	18 832
Share of low-carbon supply	19%	30%	61%	23%	29%
Energy intensity of GDP (toe/$1 000)	106	67	37	79	53
average annual reduction from 2018		3.8%	3.2%	2.4%	2.1%
Power generation					
CO_2 intensity of power (g CO_2/kWh)	476	237	23	370	262
Share of low-carbon generation	36%	61%	94%	46%	57%
of which renewables	71%	80%	84%	80%	86%
Final consumption					
Total final consumption (Mtoe)	9 954	9 904	9 225	11 607	13 555
of which renewables	10%	21%	44%	14%	21%
Industry					
Share of electricity	28%	31%	40%	29%	31%
CO_2 intensity (t CO_2/$1 000 VA)	0.26	0.15	0.06	0.20	0.15
Transport					
Electric cars % of new car sales	2%	47%	72%	15%	27%
Carbon intensity of new PLDVs (g CO_2/v-km)	175	62	21	121	90
Carbon intensity of truck fleet (g CO_2/t-km)	82	55	21	65	50
Shipping emissions (Mt CO_2)	878	832	435	1 064	1 276
Aviation emissions (Mt CO_2)	982	925	625	1 217	1 661
Buildings					
Energy intensity: residential (toe/dwelling)	1.04	0.72	0.59	0.94	0.88
Energy intensity: services (toe/$1 000 VA)	0.016	0.012	0.007	0.014	0.010

Notes: Mt = million tonnes; CCUS = carbon capture, utilisation and storage; GDP = gross domestic product; $PM_{2.5}$ = particulate matter with a diameter of less than 2.5 micrometres; Mtoe = million tonnes of oil equivalent; toe = tonnes of oil equivalent; t = tonnes; $ = USD (2018); g = grammes; kWh = kilowatt-hour; VA = value added; PLDVs = passenger light-duty vehicles; v-km = vehicle-kilometre; t-km = tonne-kilometre. Buildings: services include public and commercial buildings. See Annex B for details of the scenarios.

Chapter 2 | Energy and the Sustainable Development Goals

Box 2.1 ▷ The future ain't what it used to be – a decade of WEO energy transition scenarios

The *World Energy Outlook* introduced a detailed energy transition scenario in 2009. Back then, climate policy discussions focused on the targeted stabilisation level of CO_2 concentration. The scenario got its 450 name from 450 parts per million (ppm), the CO_2 concentration that was seen at that time to be consistent with a 50% likelihood of keeping average global temperature rise below 2 °C. Since then the global goalposts have shifted, technological progress has been uneven, and emissions have continued to grow. How has the *WEO* energy transition scenario changed?

A tougher starting point. Energy-related CO_2 emissions in 2018 reached a record high 33 gigatonnes (Gt), only marginally lower than the figure projected in the Reference Scenario of 2009, but a huge 2.5 Gt above what was set out in the 450 Scenario (which, as with all *WEO* energy transition scenarios, embodies the principle of early action). Higher emissions are associated with a larger carbon-intensive capital stock: the first version of the 450 Scenario in the *WEO-2009* saw no more than 1 700 GW of unabated coal capacity in 2020, but in the years since that publication, the world has continued to build unabated coal-fired power plants and is on track to have over 2 100 GW in operation in 2020. This investment wave completely dwarfed the impact of positive developments from solar panels to EVs. In fact, without the slower than expected recovery after the financial crisis, global CO_2 emissions today would probably be higher than projected in the 2009 Reference Scenario.

Figure 2.2 ▷ Average annual post-peak CO_2 emissions reductions and power sector mix in various WEO scenarios

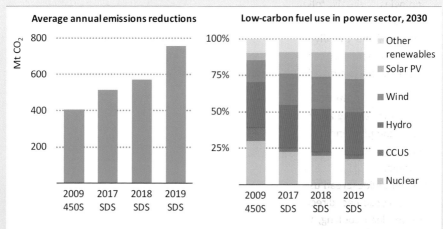

Annual WEO updates of energy transition scenarios have changed to reflect continued emissions growth, uneven technology progress and more ambitious long-term targets

Note: 450S = 450 Scenario in *WEO-2009*; SDS = Sustainable Development Scenario in *WEO* editions 2017-2019; CCUS = carbon capture, utilisation and storage.

Higher ambition. The 450 Scenario was compatible with reaching net-zero CO_2 emissions towards the end of the century. The emission trajectory of the 2019 Sustainable Development Scenario (Figure 2.5) combined with the higher starting point result in 730 million tonnes (Mt) average emissions decline per year (with peaks of yearly reduction of 1 Gt) compared with just 400 Mt reductions per year in the *WEO-2009* (Figure 2.2).

Uneven technological progress. The Sustainable Development Scenario relies much more on solar and wind in the power sector, and less on carbon capture, utilisation and storage (CCUS) and nuclear in the power sector than the 450 Scenario. The 2040 low-carbon system described by the 2014 edition of the 450 Scenario would have been dominated by baseload low-carbon technologies with nuclear at roughly the same level as wind and solar combined. Five years later, the Sustainable Development Scenario continues to rely on nuclear power and CCUS in 2040, but wind and solar contribute 2.5-times their combined generation.

This is a reflection of technology progress, or lack of it, principally determined by policy preferences. In the European Union, for example, the revision of transport-related policies strongly benefited electric cars over biofuels. Korea, which in 2014 had a robustly pro-nuclear policy stance, has since introduced a moratorium on new nuclear construction and an ambitious renewable energy investment drive and – apart from the recent 45Q tax credit reforms in the United States designed to incentivise carbon capture projects – there has been relatively little policy progress on CCUS.

Hydrogen provides another example of how shifting technology developments and preferences are reflected in the *Outlook*. The inclusion of new hydrogen-based technologies is timely, following the IEA's first comprehensive hydrogen study and it reflects unprecedented interest in hydrogen (IEA, 2019a). *WEO-2019* includes the modelling of the full diversity of possible contributions of hydrogen to sustainable development across all sectors, as well as an assessment of the technical potential and costs of biogas and biomethane supply, together with projections for the future.

Today, around 330 million tonnes of oil equivalent (Mtoe) of hydrogen is used globally, mainly in the refining and chemicals sectors, with future interest focused on hydrogen's ability to complement variable renewable electricity and enable deep decarbonisation. In the Sustainable Development Scenario, smart policies combine to support the development of the hydrogen industry and bring down costs as a springboard for subsequent widespread use in a wider variety of sectors. Globally, around 150 Mtoe of biogas is consumed directly in 2050, and this contributes significantly to improving access to clean cooking, with around 240 million people using biogas to move away from the traditional use of biomass. In addition, the uptake of biomethane (or renewable low-carbon gas) rises strongly to 280 Mtoe in 2050, equal to around 10% of natural gas demand in that year.

2.2 Scenario outcomes: Universal energy access

In the Sustainable Development Scenario, universal access to both electricity and clean cooking facilities is achieved by 2030, in line with target 7.1 of SDG 7 (Figure 2.3). Given expected strong population growth over that period, particularly in countries where many people still lack access, this means a cumulative total of around 1 billion additional people with access to electricity by 2030, and more than 2.5 billion people moving away from the traditional use of biomass by the same date. This reduces the health impact of air pollution, promotes gender equality and is achieved without increasing GHG emissions (IEA, 2017a).

A cost-effective way to achieve electricity access in many areas is with renewable energy sources, thanks to the declining costs of small-scale solar photovoltaics (PV) for off-grid and mini-grid electricity as well as batteries, and to the increasing use of renewables for grid-connected electricity. This is especially the case in rural areas in sub-Saharan Africa, home to around 60% of the global population still deprived of electricity access (see Chapter 10).

The means of achieving clean cooking depends on cultural and economic factors, as well as on resources available locally and infrastructure. LPG delivers access to clean cooking in about half of all cases – particularly in urban areas - with improved and more energy-efficient biomass cookstoves playing a more significant role in rural communities. Overall, the additional CO_2 emissions resulting from increased electricity consumption and use of LPG are more than offset by the reduction in methane and nitrous oxide emissions from incomplete combustion of biomass as well as by reduced deforestation. Taking into account the higher equivalent warming effect of methane and nitrous oxide relative to CO_2, even a conservative calculation shows a net climate benefit from simultaneously achieving universal access to both electricity and clean cooking solutions.

Figure 2.3 ▷ Pathways to universal access in the Sustainable Development Scenario

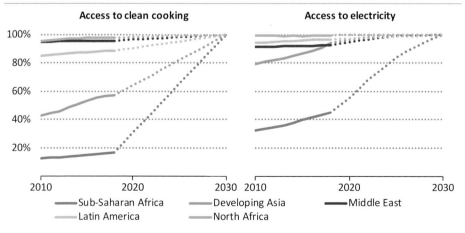

Acceleration in access is particularly needed in sub-Saharan Africa and developing Asia

2.3   Scenario outcomes: Air pollution

Air pollution has major implications for health: nine-out-of-ten people breathe polluted air every day, causing more than 5 million premature deaths each year. Around 3 million people die prematurely each year from diseases linked to breathing polluted air outdoors and around 2.5 million from smoky household air pollution: the biggest burden is on developing economies, where more than 2.6 billion people still do not have access to clean cooking.

While premature deaths due to outdoor air pollution are projected to increase and reach 5 million a year by 2050 in the Stated Policies Scenario, the higher level of ambition on air quality in the Sustainable Development Scenario translates into more than 2 million premature deaths avoided each year by 2050, compared to the Stated Policies Scenario (Figure 2.4). This is driven by reductions in emissions from the three major air pollutants – sulphur dioxide (SO_2), nitrogen oxides (NO_x) and fine particulate matter ($PM_{2.5}$). Improvements in power plant and industrial facilities – especially reduced coal use – cause total energy-related SO_2 emissions to be two-thirds lower in 2050 compared to the Stated Policies Scenario in that year; these measures, together with stricter emissions standards in the transport sector, cause NO_x emissions to drop by more than 70%. Reducing reliance on polluting fuels for cooking by 2030 reduces $PM_{2.5}$ emissions by more than 80% compared to the Stated Policies Scenario: premature deaths due to household air pollution fall to 0.8 million a year[2] in the Sustainable Development Scenario by 2050, compared to 1.8 million a year in the Stated Policies Scenario.

Figure 2.4 ▷   Global premature deaths attributable to air pollution

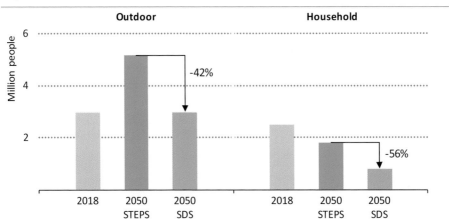

Lower pollutant emissions from power plants, cars and industries could avoid 2.2 million premature deaths a year, and broader access to clean cooking another 1 million a year

Note: STEPS = Stated Policies Scenario; SDS = Sustainable Development Scenario.
Source: International Institute for Applied Systems Analysis (IIASA).

---

[2] While improvements in biomass cookstoves reduce indoor pollution by around 75%, they do not completely eliminate health risks.

2.4 Scenario outcomes: CO_2 emissions

The Paris Agreement has an objective of "holding the increase in the global average temperature to well below 2 °C above pre-industrial levels and pursuing efforts to limit the temperature increase to 1.5 °C above pre-industrial levels". Article 4 of the Agreement sets out related aims on how to achieve this temperature goal including the need to: reach a global peak in GHG emissions as soon as possible (recognising that peaking will take longer for developing economies); undertake rapid reductions thereafter; and achieve a balance between anthropogenic emissions by sources and removals by sinks of greenhouse gases in the second-half of this century.

The Sustainable Development Scenario is constructed on the basis of limiting the temperature rise to below 1.8 °C with a 66% probability without the implied reliance on global net-negative CO_2 emissions, or 1.65 °C with a 50% probability. Because emissions do not turn net negative, this means that there is no "overshooting" of the 1.8 °C temperature rise (see section 2.9). However the emissions trajectory of the Sustainable Development Scenario to 2050 leaves open the possibility that – if emissions were to turn net negative during the second-half of the century – the temperature rise could be limited to 1.5 °C with a 50% probability (Figure 2.5). The Sustainable Development Scenario adopts a principle of early action and sees energy sector CO_2 emissions peak immediately at around 33 Gt, and then fall at an average of 3.8% per year to less than 10 Gt by 2050, on course to net zero by 2070.

In the Sustainable Development Scenario, the cumulative level of CO_2 that is emitted between 2018 and 2070 is 880 Gt. After taking into account emissions from land-use change and industrial processes, this gives a remaining energy-related CO_2 budget of around 800 Gt.

Figure 2.5 ▷ Energy-related CO_2 emissions in the Sustainable Development Scenario to 2050 and extended pathway to 2100

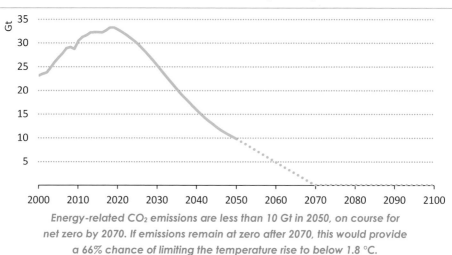

Energy-related CO_2 emissions are less than 10 Gt in 2050, on course for net zero by 2070. If emissions remain at zero after 2070, this would provide a 66% chance of limiting the temperature rise to below 1.8 °C.

2.5 Energy sector transformation in the Sustainable Development Scenario

Total final consumption

Households and industries are at the heart of the energy transitions in the Sustainable Development Scenario. Households opt for very efficient appliances, buildings renovations and electric options. Industrial processes become very efficient (including through material efficiency, see Chapter 7), fuel switching, electrification and CCUS. Innovations are made in sectors such as cement, and ammonia (with 50 Mtoe produced via electrolytic hydrogen in 2050) is utilised widely (Figure 2.6).

Efficiency, including through electrification, has a major role to play and is a common denominator across sectors: efficiency gains mean that final energy use stays flat, despite a near tripling of the economy and an additional 2 billion people. Conventional cars sold in 2050, for example, consume around 50% of the fuel consumed by the average car sold today. This is accompanied by strong electrification: in 2050, three-quarters of cars sold globally are electric, as are all scooters, all urban buses, and 50% of trucks.

In the Sustainable Development Scenario, both oil and coal use peak imminently. Oil use declines steeply for passenger cars and for other uses. The use of oil for non-energy and non-emitting uses (for example in plastic feedstock and asphalt) rises to 40% of the final consumption of oil. Coal use declines steeply, but it continues to be used in industry, mainly for the production of cement, iron and steel, with a third of these emissions captured via CCUS. Natural gas use increases through to the late 2020s, as it replaces more polluting fuels, but consumption declines after this as deep retrofits, biomethane, hydrogen and electrification provide less polluting alternatives. Hydrogen starts to be used in gas grids and shipping. Inefficient use of biomass is phased out in buildings.

Figure 2.6 ▷ Total final consumption by sector and fuel in the Sustainable Development Scenario

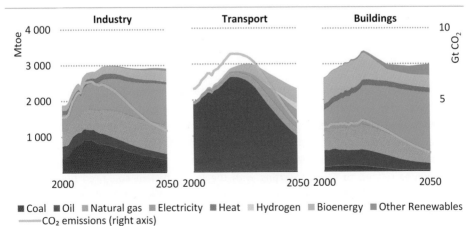

A shift to low emitting energy sources cuts end-use emissions by 57%, but the drop in total final consumption in buildings flattens out due to expanding electricity access

Power generation

In the Sustainable Development Scenario, the entire global population gains access to electricity by 2030 and a number of sectors – including transport – see a shift from other fuels to electricity. Electricity increases its share in the fuel mix across all sectors, doubling its current share of 19% by 2050. By 2050 the electricity system grows to more than 45 000 terawatt-hours (TWh), some 70% larger than today (Figure 2.7). Global electricity supply in the Sustainable Development Scenario moves rapidly from unabated fossil fuels towards low-carbon sources. By 2030, 60% of global electricity production comes from low-carbon sources, increasing to 94% in 2050 (up from 36% today). The global carbon intensity of electricity supply falls to just 23 grammes of CO_2 per kilowatt-hour kWh (g CO_2/kWh) on average in 2050, from 475 g CO_2/kWh in 2018.

Figure 2.7 ▷ **Electricity generation by source and carbon intensity of electricity in the Sustainable Development Scenario**

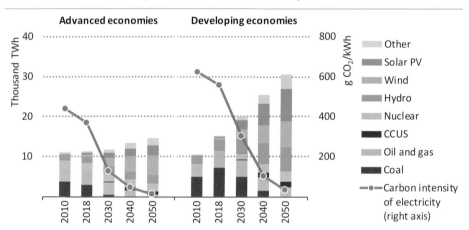

Both advanced and developing economies move towards full decarbonisation of electricity supply by 2050

Note: CCUS = carbon capture, utilisation and storage.

Wind and solar become the two main sources of generation by 2040, and supply half of global electricity generation by 2050. Hydro keeps its current share of around 17% of generation, and bioenergy grows to supply 7% of generation. Nuclear supplies about 10% of total generation, and coal and natural gas with CCUS an additional 5%. Coal-fired generation declines sharply, although CCUS mitigates the drop (see IEA, 2017b for uncertainties related to CCUS). Starting with the least efficient coal-fired power plants, virtually all generation from coal-fired power plants without carbon capture is phased out by 2030 in advanced economies and by 2045 in developing economies. Natural gas-fired power generation increases through to the late 2020s, providing important flexibility (in some regions), but then declines as renewables become cheaper and as flexibility needs are increasingly met by battery storage.

Box 2.2 ▷ The risk of nuclear power fading away in advanced economies

Nuclear power can play an important role in clean energy transitions. Today, it provides 18% of electricity supply in advanced economies, where it is the largest low-carbon source of electricity. Alongside renewable energy and CCUS technologies, nuclear power will be needed for clean energy transitions around the world. Nuclear power also contributes to electricity security as a dispatchable source.

Policy and regulatory decisions remain critical to the fate of reactors, particularly in advanced economies, where the average age of reactors is 35 years. Lifetime extensions offer a low-cost source of clean baseload energy, at $40-60 per megawatt-hour (MWh), competitive in most cases with the falling costs of renewables. At the same time, hurdles to investment in new nuclear projects are daunting, as cost overruns and delays raise doubts of future development, although advanced nuclear technologies, such as small modular reactors, could offer new opportunities.

Without investment in lifetime extensions or new projects, operational nuclear capacity in advanced economies would decline by two-thirds from 2018 to 2040 (Figure 2.8), with important implications for sustainability and affordability. Achieving the clean energy transition with less nuclear power is possible but would require more to be done to reduce emissions in other ways, adding to the difficulty of delivering ambitious emissions goals. It would also be very likely to cost more: offsetting less nuclear power with more renewables would raise overall power investment needs by some $1.6 trillion over the period to 2040, resulting in 5% higher electricity bills for consumers in advanced economies (IEA, 2019b).

Figure 2.8 ▷ Operational nuclear power capacity in advanced economies absent further investment

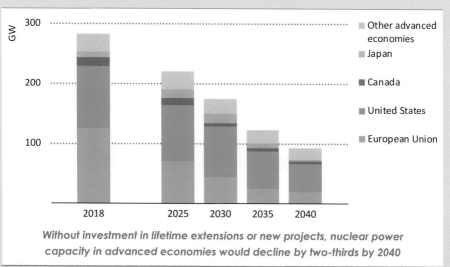

Without investment in lifetime extensions or new projects, nuclear power capacity in advanced economies would decline by two-thirds by 2040

Primary energy demand

Thanks to efficiency improvements, electrification and fuel switching, energy demand remains broadly stable, despite a growing economy. However the fuel mix changes dramatically. Low-carbon technologies (including CCUS) grow to supply 60% of primary energy by 2050 from less than 20% today, a reversal of a stable share of fossil fuels at over 80% for the past three decades (Figure 2.9).

Coal demand declines in both absolute and relative terms, and by 2050 accounts for just 8% of total energy uses, with industrial processes (iron, steel and cement production) and transformations accounting for the largest component. Around half of the remaining energy-related coal emissions (for industry and power) are captured via CCUS.

By 2050 natural gas and oil account for 18% and 20% of primary energy demand respectively. The share of natural gas rises to 26% in 2030 and then slowly falls back. Across the energy system, half of the decline in the use of natural gas is compensated for by strong growth in the use of biomethane and hydrogen (see Chapter 13).

Although oil demand for non-combustion uses, such as petrochemical feedstock, increases until 2050, oil is still used in 2050 to fuel 17% of the car fleet and 40% of the truck fleet. It is also used in aviation and shipping, where substitutes are more difficult to find.

Figure 2.9 ▷ Primary energy mix and fuel use by sector in the Sustainable Development Scenario, 2018 and 2050

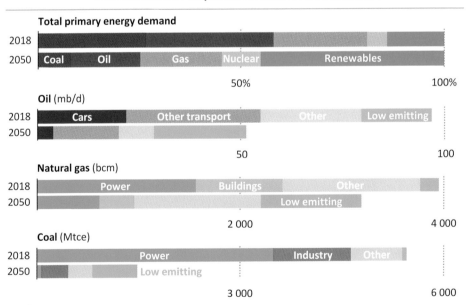

While growth in renewables and nuclear power contribute most to decarbonising the energy mix, the carbon footprint of fossil fuels also improves

Notes: mb/d = million barrels per day; bcm = billion cubic metres; Mtce = million tonnes of coal equivalent. Low emitting includes feedstocks and fuel use equipped with CCUS.

2.6 Investment in the Sustainable Development Scenario

The Sustainable Development Scenario sees an increase in overall investment compared to the Stated Policies Scenario of around 25% over the period 2019-50. This additional investment cost is partially counterbalanced by reduced fuel costs. The Sustainable Development Scenario brings considerable benefits in terms of energy access, health/air quality and mitigating the impacts of climate change. It also requires a different approach to financing. Average annual supply-side investment in power increases by three-quarters through to 2050, whereas investment in fuels decreases by about 40% from today's level. This marks a significant shift away from fossil fuels to renewables and other low-carbon sources as well as to electricity (Figure 2.10).

Figure 2.10 ▷ Average annual energy investment in the Sustainable Development Scenario, 2014-2018 and 2019-2050

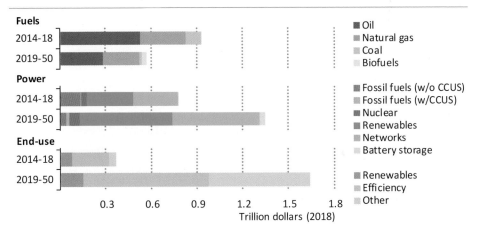

Investment in fuels and power is marked by a major reallocation of capital towards renewables and electricity networks; demand-side investment increases substantially

Notes: Other end-use includes CCUS in industry, spending to meet the incremental cost of EVs and investment in EV charging infrastructure. w/o CCUS = without CCUS, w/CCUS = with CCUS.

Despite this shift, some investment in oil supply – in both currently producing fields and new fields – is still required to meet demand (Box 2.3). While investment in natural gas supply is lower in 2050 than it is today, it rises over the next decade to meet increasing demand, as natural gas increasingly substitutes for coal, and to maintain and develop the gas infrastructure that helps support the use of low-carbon gases.

The largest increase in supply investment comes from renewables-based power, which doubles from today's levels to nearly $610 billion a year on average. This ramp-up is supported by additional spending on electricity grids and battery storage. While there continues to be investment in fossil fuel power generation, about half of this spending is associated with plants equipped or retrofitted with CCUS technology after the late 2020s.

Spending on more efficient buildings, industrial processes and transport accounts for half of demand-side investment needs. In buildings, this includes spending on more efficient appliances as well as on efficiency measures such as thermal insulation and efficient lighting. In transport, it includes spending that supports the shift towards EVs and associated charging infrastructure, as well as the costs of more efficient internal combustion engines.

The investment needed to achieve universal energy access amounts to some $45 billion per year between 2019 and 2030, the lion's share of it for electricity access. While this is more than double the amount in the Stated Policies Scenario, it is less than 2% of the total annual energy sector investment in the Sustainable Development Scenario.

Box 2.3 ▷ Oil and gas investment in the Sustainable Development Scenario

Fossil fuel producers are used to making long-term investment decisions despite the range of uncertainties faced in markets. But the lower demand trajectory of the Sustainable Development Scenario, coupled with a prolonged period of low prices would represent a new set of pervasive risks for the oil and gas industry.

For oil, demand in the Sustainable Development Scenario peaks within the next few years and then falls to just over 50 million barrels per day (mb/d) in 2050 from close to 97 mb/d today. This is an average drop of around 1.4 mb/d every year, but this decline is not spread equally across all sectors. Consumption in sectors where the oil is not combusted and does not produce CO_2 emissions see continued growth in oil use. For example:

- There is a significant increase in plastic recycling rates in the Sustainable Development Scenario (from a global average of 15% today to over 40% in 2050), but oil use as a petrochemical feedstock still grows to almost 15 mb/d in 2050.
- A further 5 mb/d is used in 2050 for non-energy products such as lubricants, bitumen, asphalt and paraffin waxes.

Because of the rise in the relative share of non-emitting[3] uses, the average emissions from using a barrel of oil fall globally by nearly 30% between 2018 and 2050 (these figures exclude emissions from the production, processing and transport of the oil).

Despite these increases, overall oil demand drops on average by 2% every year between 2018 and 2050. This is well short, however, of the decline in production that would occur if all capital investment in currently producing fields were to cease immediately, which would lead to a loss of over 8% of supply each year. If investment were to continue in currently producing fields but no new fields were developed, then

---

[3] The IPCC 2006 Guidelines for National Greenhouse Gas Inventories as well as the IEA statistics on *CO_2 Emissions from Fuel Combustion 2018* exclude all non-energy use of fuel from energy sector emissions calculations and thus apply a zero emissions factor to oil use as a feedstock (IPCC, 2006; IEA, 2018d).

the average annual loss of supply would be around 4.5%. Continued investment in both new and existing oil fields, even as overall production declines in line with climate goals, is therefore a necessary part of the energy transition envisaged in the Sustainable Development Scenario.

Demand for natural gas grows by around 10% between 2018 and the late 2020s. The rise stems mainly from the need to offset the major drop in coal consumption. Huge increases in generation from variable renewable electricity technologies plug most of the gap left by the decline in coal generation, but increased natural gas use also plays a role. During the 2030s, natural gas consumption falls slowly as it becomes too emissions intensive to be consistent with the emissions reductions required. Decline rates from existing gas fields are similar to those for oil, and investment in new gas assets continues to be necessary even as the use of gas declines (Figure 2.11). Investment in maintaining gas infrastructure is also important as the gas grid helps to support the uptake of low-carbon gases such as biomethane and hydrogen (see Chapter 13).

The need for continuing investment in oil and gas fields in the Sustainable Development Scenario is an important point. However, it is just as important that global decarbonisation plans are fully and clearly integrated into resource development strategies, so that future investment takes account of them and that resources are not developed in the expectation of much higher trajectories for oil and natural gas demand and prices. It is also important that the oil and natural gas industries should minimise the emissions impacts of these fuels to the fullest extent possible. In particular, reducing methane emissions from oil and natural gas operations is an essential component of action to address climate change.

Figure 2.11 ▷ Average annual upstream oil and gas investment in the Stated Policies and Sustainable Development scenarios

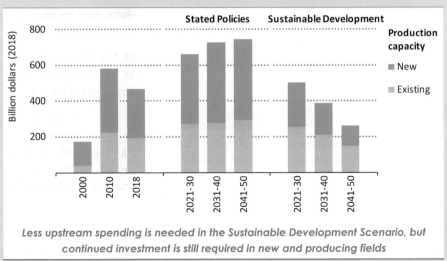

Less upstream spending is needed in the Sustainable Development Scenario, but continued investment is still required in new and producing fields

Key themes

2.7 How are we doing?

Status of emissions trajectories

Current country commitments, the Nationally Determined Contributions (NDCs), made under the Paris Agreement and domestic energy policy plans fail to bring about the rapid, far-reaching changes required to avert dangerous and irreversible changes in the global climate system. These are assessed in our Stated Policies Scenario and lead to total global energy-related CO_2 emissions growing steadily from today's levels before plateauing around 36 Gt after the mid-2040s. This trajectory is consistent with limiting the temperature increase to below 2.7 °C above pre-industrial averages with a 50% probability (or below 3.2 °C with 66% probability).

Figure 2.12 ▷ Energy-related CO_2 emissions by region in the Stated Policies Scenario

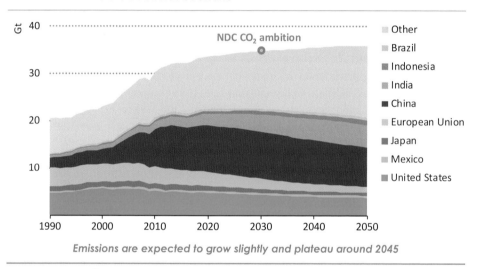

Emissions are expected to grow slightly and plateau around 2045

Note: NDC = Nationally Determined Contribution.

In the Stated Policies Scenario, trends in emissions see significant regional variations (Figure 2.12). In advanced economies in aggregate they are set to decline by 3.6 Gt to 2050, albeit with rates of reduction varying significantly depending on domestic circumstances. In the European Union they are in steady decline and are on course to be cut by more than half by 2050 under stated policies, reaching 1.3 Gt in 2050 (2.2 Gt in 2030) from 3.1 Gt in 2018. Some countries, such as the United Kingdom, Sweden and France have set very ambitious targets of net carbon neutrality by 2050 or earlier, and a qualified, sector-by-sector interpretation of these targets, which takes into consideration likely barriers to their full realisation, has been incorporated in the Stated Policies Scenario. The European Union's power sector leads the emissions reductions, with generation 90% decarbonised by

2050. Coal phase-out plans, carbon pricing and increasingly competitive renewables all play a role. As the power sector decarbonises it is overtaken by transport as the largest emitter in the next five years. By 2050 it is one of the least emitting sectors. Drops in emissions are also seen in road passenger segments. Emissions also decline in Japan, where they are on course almost to halve compared with today. In the United States, emissions are on a gently declining trend through to 2050 when they reach 3.8 Gt, compared with 4.9 Gt today. Cheaper natural gas and renewables push out coal from the power sector, with its share of generation set to decline from 28% today to 12% in 2050.

Given its stated policies, China's emissions are expected to increase slightly to the late 2020s, reaching a peak of 10 Gt, and then decline to around 8 Gt in 2050. Strong policy support for low-carbon technologies in the power sector, a switch from coal towards electricity and gas for heating in buildings, and a strong push for electrification of vehicles explain this trend. Demographics also play a large role: population is expected to peak and decline on a similar timescale.

In contrast to China, other Asian developing countries in aggregate are expected to increase emissions by more than 4 Gt. Under its stated policies, India would see national CO_2 emissions more than double from 2.3 Gt today to 4.8 Gt in 2050, 3.3 Gt of which would come from coal. Increases are also seen in the Middle East and Africa.

These overall CO_2 trends mask significant differences between particular sectors and technologies. The cost decline in solar and wind and the expected cost decline in offshore wind are transforming the power sector at an unprecedented rate. Power sector emissions remain broadly stable to 2050, while electricity generation almost doubles, as 85% of new additions are low carbon or natural gas. Emissions are also set to decline in the buildings sector by 0.3 Gt as rising electrification offsets oil and coal use, which today account for 15% of energy use in the sector. The road passenger segments see a similar trend with EVs accounting for more than 40% of all sales by 2040 and continued efficiency improvements bringing about a deep decoupling between activity and emissions. However emissions rise in industry (+1 Gt in 2040 compared to today), trucks (+0.6 Gt), aviation and shipping (+0.8 Gt), all contributing to significant emissions growth.

Status of energy access and air pollution

Our latest country-by-country assessment shows that in 2018, the number of people without access to electricity had dropped to 860 million, a record in recent years (Figure 2.13). India continues to make remarkable progress towards its target of universal electrification, with almost 100 million people gaining access in 2018 alone. In March 2019, the government announced that it had provided access to all willing households after connecting 26 million households between October 2017 and March 2019 through the Saubhagya scheme, with 99% of them through the grid. Access to electricity in sub-Saharan Africa remains low, and around 600 million people – more than half of the population – are still without access to electricity. These numbers improve only slightly under current and announced policies: the total number of people without access to electricity reaches

620 million by 2030 and then increases to 740 million by 2050 as progress in expanding electricity access fails to keep pace with population growth.

Progress on access to clean cooking facilities has been gradual and limited compared to progress on electricity access. More than 2.6 billion people continue to rely on the traditional use of biomass, coal or kerosene as their primary cooking fuel. This has damaging consequences for health and productivity, especially for women. The challenge remains particularly acute in sub-Saharan Africa, where less than one-person-in-five has access to clean cooking fuels and technologies. Even though a number of countries, mostly located in developing Asia, have shown signs of improvement in recent years through dedicated policies supporting LPG, there are still 1.5 billion people without access to clean cooking in the Stated Policies Scenario in 2050.

Figure 2.13 ▷ Population without modern energy access and premature deaths due to air pollution in the Stated Policies Scenario, 2018 and 2050

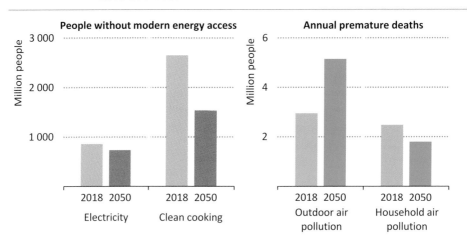

More effort is needed to achieve universal access to modern energy and to bring about major reductions in air pollution

Sources: IIASA and IEA analyses.

The lack of access to clean cooking facilities is directly linked to household air pollution and high numbers of premature deaths. In the Stated Policies Scenario, 1.8 million premature deaths are still linked to pollutant emissions from cooking in 2050. The Stated Policies Scenario leads to some reductions in outdoor air pollution, but these remain insufficient to prevent major threats to human health. Due to the complex relationship between emission levels, atmospheric conditions, exposure levels and timing, the number of premature deaths from outdoor air pollution is actually set to rise to 5 million in 2050.

Recent policy developments

Some governments have recently announced new, more ambitious targets. For example, by the end of September 2019, at least 65 countries, together with the European Union, had set or were actively considering long-term net-zero carbon targets.[4] The economies of these countries together accounted for 21% of global GDP and nearly 13% of emissions in 2018.

Figure 2.14 ▷ **Net-zero carbon or GHG emissions reduction announcements**

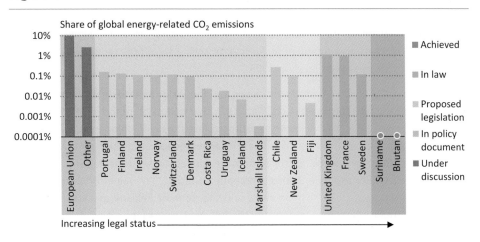

More than 65 jurisdictions accounting for 13% of global CO_2 emissions have announced net-zero CO_2 or GHG emissions commitments for 2050

Notes: The other category includes countries not shown that have recently signalled their intent to put forward net-zero targets (UNFCCC, 2019a). The 13% total share of emissions covers all jurisdictions in the figure, without double counting. Under discussion category means that consultations to develop a net-zero target are ongoing. The policy document category means that a net-zero target has been put forward, however, without legally binding status. The proposed legislation category means that the target has been proposed to parliament to be voted into law. The In law category means that a net-zero target has been approved by parliament and is legally binding. The achieved category means that the jurisdiction absorbs more CO_2 than it emits, e.g. through afforestation. Not included in the figure are the efforts of some state and local entities, such as California, though they would increase the share of emissions covered to over 16%.

Net-zero targets serve an important role in shifting the centre ground of global ambition. By stimulating and testing innovations, regulations and markets that can be replicated elsewhere, they play a role in accelerating progress towards climate targets around the world. This is likely to be particularly the case in hard-to-abate sectors, to which net-zero targets give a clear signal about the need to plan for technological change. As the necessary technologies and business models will take time to develop, first-mover regions - especially those as large as the European Union – will generate vital knowledge. Overall, this

---

[4] The United Kingdom, Ireland, Fiji, Bhutan and the Marshall Islands have put forward net-zero GHG targets for 2050. It is likely that net-zero CO_2 emissions will be achieved several years before then.

contribution may be as large as the absolute CO_2 reductions that would result from the announcements in Figure 2.14 if all proposals are fully implemented (Figure 2.15).

Non-state actors are also increasingly bringing forward pledges of action. City networks, such as C40[5], account for 2.4 Gt of emissions and are required to have plans to deliver their contribution towards the Paris Agreement. If target setting and actions were to become widespread, these could potentially reduce emissions further given that cities account for 70% of global emissions. Several companies have also announced pledges to make their business compatible with the Paris Agreement. Such pledges can be important because companies can drive innovation and learning. Global shipping giant Maersk has, for example, committed to go carbon neutral by 2050, while the International Maritime Organization reached an agreement in April 2018 calling for a 50% reduction in shipping emissions by 2050, relative to levels in 2008.

Recent years have seen an increased focus on the risks to investment from energy transitions and climate change impacts. The 2019, G20 endorsed, report of the Task Force on Climate Related Financial Disclosure encouraged investors and major asset owners to be more transparent about the risks of: climate change impacts; transition to a low-carbon world; and of litigation. Some jurisdictions have moved to set expectations of greater disclosure of risks. For example, France's recently updated 2017 Value for Climate Action Law required disclosure of transition risks, and the Bank of England in its guidance has encouraged company boards to understand the full range of risks they face.

Figure 2.15 ▷ Effects of including announced net-zero carbon pledges on CO_2 emissions in the Stated Policies Scenario

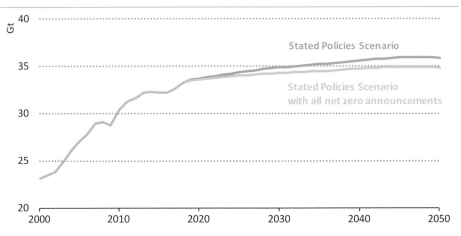

If all the announced net-zero pledges were implemented by 2050, the direct impact on CO_2 trends would be limited, but the indirect impacts could be significant

---

[5] The C40 Cities Climate Leadership group focuses on reducing GHG emissions from cities as well as reducing risks to the urban environment from climate change.

SPOTLIGHT

Chile's Decarbonisation Plan

Chile will host the United Nations Framework Convention on Climate Change (UNFCCC) climate talks (COP 25) in December 2019, where countries will meet to discuss, among other things, increasing their climate ambition. In advance of COP 25, the president of Chile announced that the country aims to achieve carbon neutrality by 2050. The draft bill has been presented to congress and is expected to receive approval during the climate summit in December. This is a significant increase in ambition compared to the NDC in which Chile committed to reducing the intensity of its emissions relative to its GDP by 30% by 2030 from 2007 levels.

Chile is responsible for less than 1% of global GHG emissions. However, the country is highly vulnerable to changes in climate conditions and ranks among the 16 countries most affected by climate variability (GCRI, 2019). It is home to 82% of Andean glaciers, most of which are in retreat. Highly unusual tornadoes, heatwaves and forest fires are among the other impacts already being felt by Chile (UNFCCC, 2019b).

Among the key actions announced, Chile is planning to phase out coal by 2040 and to generate 70% of electricity from renewables by 2030. Chile's decarbonisation plan focuses on a phase-out of coal in two stages: by 2024, it will close eight of the oldest coal units, which account for 20% of its current coal electricity capacity, and the second stage will phase out the remaining 20 coal units by 2040.

Chile's power generation mix is led by coal, which accounted for 35% of generation in 2018. Hydropower was the second-largest source of electricity, natural gas accounted for 16%, bioenergy for 7%, solar PV for 6% and wind for 5%. Chile has excellent solar and wind resources, and recent competitive tenders brought record low offers for solar PV (at $29 per megawatt-hour). In the Sustainable Development Scenario, the share of renewables generation in Chile expands to almost 80% in 2030 and 94% in 2050, with coal-fired generation phased out by 2040. Chile's strategy is not confined to the power sector; it aims to electrify 40% of the private vehicle fleet and 100% of public urban transport by 2050.

While decarbonising power generation in Chile is within reach, given its large renewable resources and the phase-out plan for coal, additional policies and technologies also will be needed to set end-use sectors, such as industry, on course to meet its stated carbon neutrality goal.

In the Sustainable Development Scenario, policies targeted at the electrification of end-use sectors, the expansion of bioenergy and hydrogen as well as enhanced energy efficiency standards put total emissions from transport and buildings on a declining trend.

2.8 Where do we need to get to?

An energy sector transformation of the scale and pace required to achieve the Sustainable Development Scenario depends upon fundamental changes to the way energy is produced and consumed. This would compel all sectors of the economy to significantly accelerate the uptake of low-carbon technologies, including energy efficiency and renewables as well as nuclear and CCUS (Figure 2.16).

Figure 2.16 ▷ CO₂ emissions reductions by measure in the Sustainable Development Scenario relative to the Stated Policies Scenario

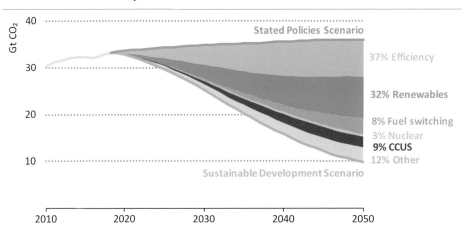

All clean energy technologies are needed in the Sustainable Development Scenario; energy efficiency is the main contributor to emissions savings to 2050

Notes: CCUS = carbon capture, utilisation and storage. Reduced thermal losses in power generation account for 15% of efficiency improvements.

In the Sustainable Development Scenario, the relative contributions of clean energy technologies differ for a variety of reasons, but generally consider least-cost opportunities as well as national circumstances such as the age of the existing capital stock. Energy efficiency is the primary "fuel" of choice in most regions, because of its cost-effectiveness. Energy efficiency measures generally offer an attractive payback, and the barriers to their deployment such as access to finance or lack of information are successfully addressed through policy measures in the Sustainable Development Scenario. No decarbonisation pathway is achievable without rapid and significant deployment of energy efficiency measures. But energy efficiency alone cannot deliver the emissions reductions required for achieving the SDGs. Absent further breakthroughs in energy-efficient technologies beyond what is considered in the Sustainable Development Scenario, energy efficiency alone approximately stabilises global energy-related CO_2 emissions at slightly below 30 Gt by around 2040 and beyond.

The second key option for reducing CO_2 emissions is the deployment of renewables. The cost of solar PV and wind in particular have fallen significantly in recent years and are set to decline further (see Chapter 1). Their deployment in the Sustainable Development Scenario is supported by a host of measures that further strengthens their competitiveness *vis-à-vis* fossil fuel power plants (such as carbon prices) and allows for their successful integration into the power generation mix. The uptake of renewable energy technologies in sectors such as industry and buildings (for heating purposes) and transport (advanced biofuels) has been limited to date, given high costs and lack of sufficiently widespread policy support. Targeted policy measures such as fuel blending mandates and renewable energy quotas are assumed in the Sustainable Development Scenario to overcome these hurdles. In tandem with renewables, the Sustainable Development Scenario sees nuclear energy play an important role in decarbonising the power sector in countries that seek to support its future deployment, especially given the cost-effectiveness of lifetime extensions (Box 2.2).

CCUS needs to be more widely deployed in order to capture an annual average of 1.5 Gt CO_2 between 2019-50 to put the world on track to meet the objectives of the Paris Agreement. As the decarbonisation of hard–to-abate sectors becomes more pressing over time, the volume of carbon captured increases to 2.8 Gt in 2050, or 28% of total CO_2 emissions in that year. Governments would have to take steps to enable a framework to foster the uptake of CCUS in order to achieve such levels of captured and stored emissions (IEA, 2017b).

In the Sustainable Development Scenario, CCUS is almost equally split in 2050 between the power and industry sectors (including cement, iron and steel, upstream oil and gas, and refineries). In the power sector, CCUS is concentrated in a handful of countries, most notably China and the United States. Around 215 GW of coal plants are equipped with CCUS by 2050, predominantly in China where the fleet is very young and the potential for CCUS deployment is high. A similar amount of natural gas power plants are equipped with CCUS, led by deployment in the United States where natural gas prices remain low and a young fleet of natural gas plants fitted with CCUS provide cheap and flexible power generation. The use of CCUS in industrial applications is widespread, as emissions from energy-intensive sectors are typically hard-to-abate, and CCUS constitutes one of the few currently available technology options to achieve deep levels of decarbonisation. For example, today the iron and steel sector emits around 2 Gt of emissions each year. Currently, 92% of primary steel is produced in blast furnaces (primarily fuelled by coal) while 7% is produced via the direct reduced iron route (mainly fuelled by natural gas), or in some cases coal (e.g. India). For existing blast furnaces, CCUS is the main decarbonisation option. Similarly, CCUS is the main option under consideration in the cement sector, where process emissions account for two-thirds of the 2.5 Gt CO_2 the sub-sector emits today (IEA, 2018a).

The net result of these changes is a shift in the way energy is produced and consumed (Figure 2.17). Emissions of CO_2 from coal drop by 90% in 2050 in the Sustainable Development Scenario compared to today, and those that remain mostly stem from the iron and steel and cement sub-sectors. Oil is the main source of remaining CO_2 emissions in the transport sector, despite strong inroads made by electrified cars (including electric and hydrogen fuel cell cars) which make up three-quarters of the global car fleet by 2050, about

1.3 billion cars. Among the main contributors to transport-related emissions in 2050 in the Sustainable Development Scenario are trucks, airplanes and ships: modes for which key alternatives to oil such as clean hydrogen or advanced biodiesel and bio-kerosene are not yet available at commercial scale. Demand for natural gas hovers just over 4 000 bcm throughout much of the 2030s as significant near-term reductions of methane emissions from natural gas production further enhance its ability to contribute to bringing down CO_2 emissions by switching away from coal (see Chapter 4). However, from the late 2030s, natural gas demand falls away, and its use is increasingly concentrated in industrial sub-sectors (e.g. petrochemicals) as well as non-energy intensive sectors.

Figure 2.17 ▷ Global fossil fuel demand by CO_2 content in the Sustainable Development Scenario, 2018 and 2050

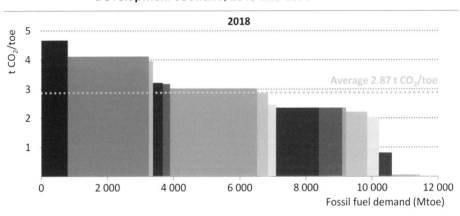

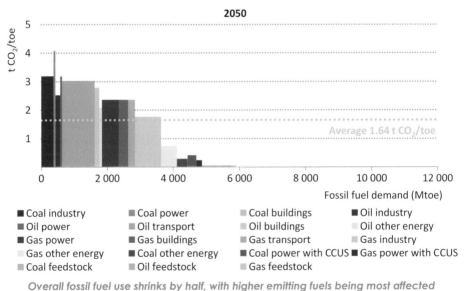

Overall fossil fuel use shrinks by half, with higher emitting fuels being most affected

Notes: t CO_2/toe = tonnes of carbon dioxide per tonne of oil equivalent; CCUS = carbon capture, utilisation and storage. Fossil fuel demand does not include demand from agriculture.

The Sustainable Development Scenario assumes that all countries immediately significantly scale up their clean energy ambitions. That does not mean that emissions in all countries peak at the same time. Aggregate CO_2 emissions from advanced economies have already peaked; emissions in advanced economies have fallen by 0.8% per year on average since 2010 (Figure 2.18). In the Sustainable Development Scenario, the pace of decline accelerates to 5.6% per year through to 2050. In developing economies, emissions have increased on average by 2.3% per year since 2010. In the Sustainable Development Scenario, they peak in aggregate by around 2020 even if some regions (such as India and Southeast Asia) would not be expected to reach peak emissions until later; emissions in developing economies fall by 3.2% on average per year through to 2050.

Figure 2.18 ▷ CO_2 emissions in advanced and developing economies in the Sustainable Development Scenario

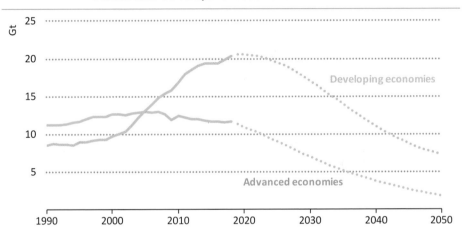

All countries accelerate their clean energy transitions, with the timing of the peak in their emissions depending on each individual country's particular circumstances

A closer look at the transformation by sector

The pace at which energy sectors decarbonise in the Sustainable Development Scenario differs widely, reflecting the availability, scale and cost-competitiveness of low-carbon technology options by sector. The **power sector** far and away shows the steepest decline in emissions in the Sustainable Development Scenario; CO_2 emissions fall by around 90% until 2050, relative to today, reflecting the scale up in the use of renewables and nuclear, which are commercially available today. The clean energy transition in the end-use sectors is more complex given its diversity and the number of actors involved. In this section, we describe the main transformational changes involved in the pathway described by the Sustainable Development Scenario.

The **transport sector** sees significant growth in activity to 2050, particularly for trucks (120% increased activity) and aviation (200%). Even with the rise in activity, CO_2 emissions in the transport sector decline by around 60% to 3.5 Gt by 2050 in the Sustainable

Development Scenario, relative to today (Figure 2.19). The transport sector is the second-largest contributor to overall emissions reductions, but given the fast pace of decline in the power sector, transport becomes the largest source of CO_2 emissions by the early 2030s. By 2050, transport constitutes 35% of global energy-related CO_2 emissions, compared with 25% today. Around 60% of the emissions reductions are from passenger cars alone due to the combined effects of incentivising the use of public transport, significantly enhancing the efficiency of conventional road vehicles and increasing electrification. Electrifying the global car fleet leads to its decarbonisation given that 94% of electricity is generated from low-carbon sources in the Sustainable Development Scenario by the middle of the century (Box 2.4).

Figure 2.19 ▷ **Oil demand in transport by mode (left) and change in transport energy use by scenario in 2050 relative to today**

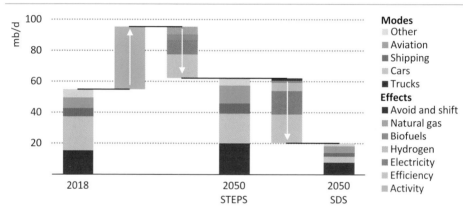

Oil demand in transport falls to 20 mb/d by 2050 in the Sustainable Development Scenario and increasingly concentrates in transport modes where low-carbon options are limited

By 2050, there are around 10% fewer cars on the road than in the Stated Policies Scenario, due to a modal shift from private vehicles to public transport. Three-out-of-four cars still on the road are electrified (including electric cars, plug-in hybrids and fuel cell cars). The remaining conventional vehicles are around 50% more efficient than today, broadly representing what is technically achievable including through hybridisation. Oil demand from passenger cars is already set to decline under stated policies (see Chapter 3). Yet the decline required to meet the objectives of the Sustainable Development Scenario is much steeper: oil demand from passenger cars is around 15 mb/d lower by 2050 than in the Stated Policies Scenario.

Emissions reductions across other transport modes are more modest, owing partly to the more limited suite of commercially available low-carbon options today (especially for trucks, ships and airplanes), but also to the limited amount of CO_2 emissions related to the use of other transport modes. For example, two/three-wheelers are nearly entirely electrified by 2050 in the Sustainable Development Scenario, and so are around 40% of buses. Today their aggregate contribution to transport emissions is around 10% and so the impact of these transitions, although important, is much smaller than for other modes of transport.

The slower pace of emissions decline in trucks, ships and airplanes does not mean that a clean energy transition is absent in these areas in the Sustainable Development Scenario. By 2050, trucks are close to 50% more efficient on average; ships use about 60% less fuel to transport a tonne of goods a kilometre, and airplanes use about 60% less fuel per revenue passenger-kilometre. Alternative fuels make inroads as well, in particular advanced biofuels. In aviation, around 60% of global fuel use in 2050 is bio-kerosene; in shipping, more than one-third of fuel use is advanced biodiesel by 2050, and around 20% is natural gas, hydrogen and hydrogen-based fuels. In aggregate, emissions in these three transport modes fall by 43% to 2050, relative to today, with the largest contribution from trucks (55%), with shipping (25%) and aviation (20%) making significant contributions as well. The emissions reductions in shipping are in line with the targets of the International Maritime Organization.

Box 2.4 ▷ How clean is your car?

The fuel economy of new cars improved in all regions during the last decade, and while their stringency varies, fuel economy standards today cover around 85% of global car sales. While cars within each segment are becoming more efficient, a slowdown in the rate of global average improvement has been recorded over the last three years, caused in part by the boom in sports utility vehicle (SUV) ownership (see Chapter 3, section 3.9).

In the Sustainable Development Scenario, a conventional car sold in 2050 consumes around half the energy of one sold today, as the technical potential for fuel efficiency improvements in internal combustion engines (ICEs) is maximised. Gains in fuel efficiency make an even larger contribution to curbing oil demand than the expansion of EVs.

Efficient hybrid petrol cars available on the market today emit nearly 110 g CO_2/km, while EVs running on electricity with carbon intensity[6] close to 600 g CO_2/kWh emit around 130 g CO_2/km (Figure 2.20). The difference in terms of emissions is marginal in regions where coal- or oil-fired plants are the dominant sources of electricity generation. In the Stated Policies Scenario, nearly half of car sales take place in markets in which the gains are less than 10 g CO_2/km by 2030.

Strategies that tap both the remaining potential for improvements in ICEs and the deployment of EVs while decarbonising power will be key to meet the Paris Agreement goals. EV deployment requires strong inter-sectoral co-ordination to maximise emissions abatement such as smart-charging (see Chapter 7, section 7.6).

---

[6] For comparison, today's global average intensity of the power sector is 475 g CO_2/kWh.

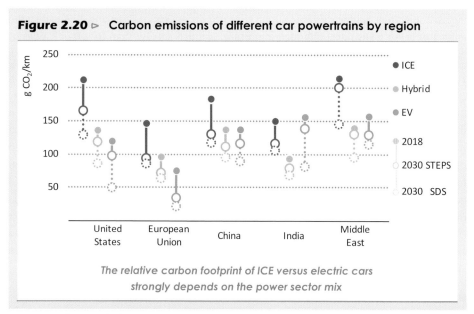

Figure 2.20 ▷ Carbon emissions of different car powertrains by region

The relative carbon footprint of ICE versus electric cars strongly depends on the power sector mix

Note: EV = electric vehicle; ICE = internal combustion engine; STEPS = Stated Policies Scenario; SDS = Sustainable Development Scenario.

In the **industry sector**, emissions from the combustion of fossil fuels decline by around 50% to 3.0 Gt in 2050, relative to today (Figure 2.21). Emissions related to industrial processes (mostly cement production) decline by nearly 40% to around 1.4 Gt in 2050. The majority of the overall emissions savings (i.e. process and energy-related) are from the production of cement, iron and steel and petrochemicals; energy-intensive industries that together account for around two-thirds of total industry sector CO_2 emissions today. Because the transport sector electrifies more quickly than the industry sector, the relative contribution of the industry sector to global energy- and process-related CO_2 emissions rises from 25% today to nearly 40% by 2050 due to the rapid decarbonisation of power.

A variety of technology opportunities are being deployed to achieve the projected emissions reductions. The largest near-term options are in energy efficiency, material efficiency and fuel switching. Energy efficiency and fuel switching account for 37% and 28% of emissions reductions in industry. They reduce oil and coal consumption by almost a third in 2050, with electricity, natural gas and bioenergy stepping in as substitutes and some use of hydrogen in the iron and steel industries, where pilot projects start around the mid-2020s. In light industry sub-sectors and chemicals, process heat requirements in the low-temperature segment allow for high shares of electrification and fuel switching at reasonable cost, for instance through the use of heat pumps. In China, for example, a marked shift to electricity in light industries helps cut coal use in industry by almost 80% in 2050, relative to today. In 2050, electricity accounts for half of industrial end-use energy demand in China, almost double the share of today.

Figure 2.21 ⊳ Savings in energy-related CO_2 emissions in industry by measure and scenario

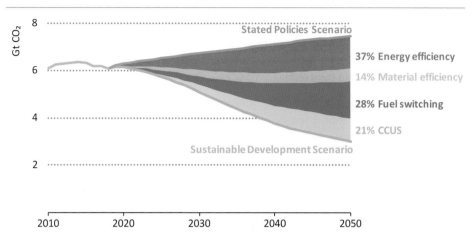

The industry sector is slow to decarbonise given the long lifetime of its capital stock; energy efficiency, material efficiency and fuel switching make the most difference in the near term

Notes: CCUS = carbon capture, utilisation and storage. Excludes all fuel transformation and industrial process emissions.

Material efficiency in the industry sector contributes around 14% of cumulative emissions reductions in the Sustainable Development Scenario, relative to the Stated Policies Scenario (see Chapter 7). Material efficiency stems from a variety of sources; it comprises direct strategies such as reducing yield losses and other process improvements in the aluminium and iron and steel sub-sectors. Yet, the majority of savings come from systemic strategies across the energy sector (IEA, 2019c). For example, in the Sustainable Development Scenario, iron and steel demand in 2050 is 15% less than in the Stated Policies Scenario as a result of lightweighting strategies for reducing the fuel consumption of cars and trucks and of lifetime extension for capital stock in the buildings and power sectors; in the chemicals sector, recycling reduces the need to produce plastics.

As discussed, a clean energy transition in the industry sector at the pace and scale depicted in the Sustainable Development Scenario is very difficult to envisage without the use of CCUS given the long lifespans of much of the capital stock and related lock-in effects, as well as the general absence of commercially available alternatives. In the Sustainable Development Scenario, about 1 Gt CO_2 from the combustion of fossil fuels is captured in the industry sector in 2050, and a further 0.7 Gt from process-related emissions, on the assumption that price uncertainties are effectively addressed by governments (IEA, 2017b). The majority of total CO_2 capture in the industry sector is in cement production with much of the remainder in iron and steel production. Additional CO_2 capture occurs in the refining sub-sector, and in oil and gas extraction.

Direct CO_2 emissions from the **buildings sector** fall from 3.1 Gt today to 1.1 Gt in 2050 in the Sustainable Development Scenario, two-thirds of which are in the residential sector. Emissions from the production of electricity and heat used in buildings, the largest source of buildings-related CO_2 emissions today, fall by 5.7 Gt (around 85%) due to the fast pace of power sector decarbonisation. In aggregate, this means that by 2050 direct and indirect emissions from the buildings sector account for just one-fifth of global energy-related CO_2 emissions, compared to a third today.

The rapid decline in CO_2 emissions from the buildings sector in the Sustainable Development Scenario should not hide the significance of the challenge. Unlike in some other sectors, there is no lack of viable technological and economic options that are generally available to the market today. The complexity of the transition to low-emissions in the buildings sector relates to the large variety of actors involved, the split incentives they face and the complexity of considerations involved. In the Sustainable Development Scenario, a host of policies are introduced to facilitate the transition to clean energy use in buildings, with energy efficiency being the most prominent. The rate of renovations of buildings rises to about 4% per year through to 2050, compared with less than 1% today (and around 2% in the Stated Policies Scenario), helping to curb demand for space heating and cooling.

As a result of these renovations and stringent minimum energy performance standards (MEPS), electricity demand for cooling – a major driver of electricity demand growth in the buildings sector today – is over 30% lower by 2050 than in the Stated Policies Scenario. This decline is achieved despite the increased need for cooling associated with rising global temperatures. Overall energy demand for space and water heating is lowered by a similar rate. More efficient refrigerators, cleaning appliances, TVs and computers, light bulbs and appliances further help to curb energy demand in buildings. Overall, energy demand from the buildings sector falls steadily through efficiency improvements, and is around 10% lower by 2050 than today, despite supporting a projected total floor space that is a two-thirds larger than today and a level of GDP per capita that is twice as large.

The decrease in energy demand in buildings in the Sustainable Development Scenario comes with a change in the way energy demand is satisfied (Figure 2.22). Worldwide, the use of coal in buildings, most of which is linked to space heating in China today, all but disappears by 2050, further accelerating a trend that is already reflected in the Stated Policies Scenario. Demand for oil in the buildings sector also declines rapidly in the Sustainable Development Scenario as its use for space and water heating (the source of 70% of oil demand in buildings today) declines by three-quarters to 2050. Oil demand for cooking (much of the remaining use of oil in buildings today) roughly stabilises at today's level. Oil use for cooking is nearly phased out by 2050 in advanced economies in the Sustainable Development Scenario, but remains an important means of providing access to clean cooking in developing countries. Around 2.3 mb/d of oil are still used for cooking purposes in 2050 in the Sustainable Development Scenario, mostly in the form of LPG, and mostly in Africa, India, Southeast Asia and Latin America.

Figure 2.22 ▷ **Change in energy demand by end-use in the buildings sector in the Sustainable Development Scenario relative to the Stated Policies Scenario, 2050**

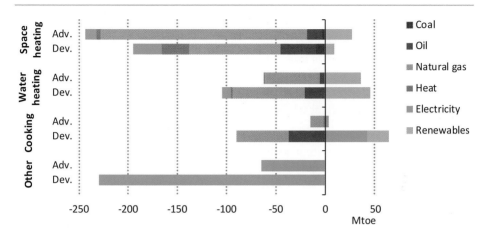

Sustainable Development Scenario sees significant improvements in buildings efficiency as well as fuel switching; gas is the hardest hit, with demand cut by two-thirds

Note: Adv = advanced economies; Dev = developing economies.

The use of natural gas in buildings in 2050 is cut by 60% in the Sustainable Development Scenario, relative to today. Demand for space and water heating – at more than 80% the primary source of gas demand in buildings today – contracts by 75% to 2050 as a result of both reduced overall demand (through improved buildings insulation) and a shift to electric heat pumps. The use of gas for cooking purposes also falls by around 25%. As with oil for cooking, the use of gas for cooking is increasingly concentrated in developing countries. The use of gas for desalination provides access to clean water in the Sustainable Development Scenario and is the only growth area for natural gas in the services portion of the buildings sector; three-times more gas is used for desalination in 2050 than today.

Electricity is the primary means for decarbonising the buildings sector. At 19 000 TWh, electricity demand from buildings in 2050 in the Sustainable Development Scenario is around 3 000 TWh lower than in the Stated Policies Scenario thanks to stringent MEPS for appliances and cooling equipment. But its use is much more widespread: in 2050, electricity demand accounts for nearly 60% of all energy use in buildings, more than ten percentage points above the level of the Stated Policies Scenario. Electricity becomes the primary source for space and water heating as well as cooking, breaking into the long-held domain of oil and natural gas. In developing countries, the achievement of full electricity access by 2030 further boosts electricity demand.

How to finance the Sustainable Development Scenario?

Investment in clean energy assets would need to rise substantially to meet the goals of the Sustainable Development Scenario, which require nearly $115 trillion of investments over the next three decades. This is around 25% more than in the Stated Policies Scenario. While energy investment as a share of GDP rises initially in the Sustainable Development Scenario, it falls to close to 1.3% by 2050, comparable to today's level, and there are additional operational savings in terms of reduced fuel spending. With greater emphasis on capital-intensive low-carbon assets, the Sustainable Development Scenario is more sensitive to the cost of capital than the Stated Policies Scenario, so factors affecting financing play a more important role in the pace and affordability of the transition.

In the Sustainable Development Scenario, governments play an increasingly important role in influencing the overall institutional, regulatory and market environment, and this influences the willingness and ability of the financial community and industry to mobilise clean energy investment at scale. Some jurisdictions, notably in Europe, are starting to address these challenges in a more holistic way, with policy linkages between the energy and finance sectors. There are particular challenges in developing economies, where investment can be affected by the structure and maturity of local financial sectors and by perceived risks.

Figure 2.23 ▷ Financial flows in energy investment

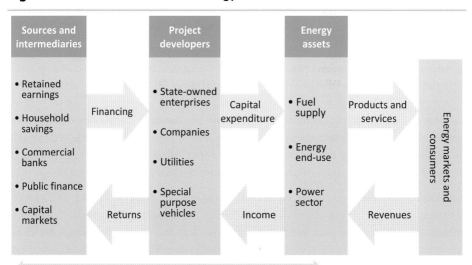

Financial pathways to a more sustainable energy system are multifaceted, involving a range of actors and vehicles at different stages of the investment value chain

How a particular investment is financed depends on the business model of the developer, the source of funds, the available financial instruments and the risk-return profile of the asset, among other factors, all of which impact the cost of capital. The framework in Figure 2.23 shows project developers as the primary actors investing in energy assets, but their success depends on a having robust inter-connected system of secondary financial sources and intermediaries, diverse investment vehicles to facilitate flows and clear signals for action, based on profit expectations and risk profile.

There is substantial investor appetite for more sustainable investments. As the financial community increasingly seeks strategies for allocating capital in way that is consistent with the SDGs, green bond issuance has surged to $650 billion cumulatively since 2007. Actors are paying more attention to climate-related risks and institutions responsible for trillions of dollars in investment funds have announced divestments from fossil fuel holdings.

These trends and ambitions raise fundamental questions about potential trade-offs between increased financing for sustainable energy and long-term returns, and about potential new risks and financing models investors will need to navigate along the way. Another key uncertainty is the extent to which current investors have the necessary skills, incentives and products to fund the clean energy transition adequately. While low-carbon projects vary considerably in terms of risk profile, lead times, useful life and level of complexity, they tend to raise several common key issues (Table 2.2).

Table 2.2 ▷ Key financing issues in the Sustainable Development Scenario

	Potential challenges	Potential options
Managing risks for low-carbon power and infrastructure	Risk-return profiles for renewables and flexible assets with changing government support.	Financial strategies (e.g. insurance, contracts, hedging instruments), beyond subsidies, to manage potential exposure to short-term market pricing (in competitive markets).
Attracting capital to developing economies	Underdeveloped financial systems, investment risks and high cost of capital.	Provision of low-cost debt and guarantees from public sources coupled with reforms that reduce risks and crowd-in private capital (in regulated markets).
Financing efficiency and distributed resources	Small transaction sizes, limited consumer balance sheets and complex cash flow evaluation based on energy savings.	Repayment through energy bills or property taxes, pay-for-performance markets and third-party finance from energy service companies, plus better measurement and verification.
Broader participation of capital markets	Limited routes and higher transaction costs for direct investment in projects by institutional investors.	Aggregation of projects from the balance sheets of developers into portfolios that can be securitised and issued as debt or equity.

First, the risk and return profiles of low-carbon investment mostly differ considerably from fossil fuels, raising questions over the willingness of traditional developers to shift capital allocations based on pure profit motives. Over time, the top oil and gas companies have earned, on average, higher returns on invested capital than the power companies who have led investment in solar PV and wind (IEA, 2019d). Lower risk – due to policy frameworks that have supported revenue certainty for renewables and grids – has partly compensated for lower returns in the power sector. There are however questions over how these policies will evolve and what this might mean for risk allocation between public and private actors. A balance of continued government support and increased use of more market-based solutions to manage risk will probably be required for there to be enough capital for the renewable energy and flexibility investments foreseen in the Sustainable Development Scenario.

Second, power investment depends on risk perceptions and the capacity and willingness of banks to make available long-term, low-cost debt. In Europe, a combination of supportive long-term policies, improved technology, participation of public finance institutions to reassure private investors and low interest rates has helped to halve debt costs for offshore wind (see Chapter 14). In India, similar dynamics have improved the confidence of banks to lend at lower rates for renewables (Box 2.5). However, new challenges are emerging to securing bank debt, which is less able to absorb uncertainty and market volatility than equity finance. Large project sizes and persistent risks make financing nuclear increasingly difficult without state-backed capital or guarantees (IEA, 2019b). Battery storage costs have fallen considerably, but few banks are willing to lend without a long-term capacity contract due to the complexity of the revenue model. As newer technologies (e.g. CCUS, hydrogen) enter the mix, financiers will need to continually navigate new risks and revenue models.

Third, in emerging economies, domestic banking constraints make the facilitation of domestic capital markets and the attraction of international capital particularly important. Domestic public finance institutions in China have made ample low-cost financing available for a range of technologies, but few countries have such balance sheets. Moreover, the provision of attractively priced public finance is often most effective when coupled with reforms that simultaneously crowd-in private sources. In South Africa, the implementation of a transparent framework for competitive procurement for renewables has been successful in attracting considerable finance from private international lenders, but there has been less activity in other sub-Saharan Africa markets without the involvement of debt or guarantees from development banks or governments (see Chapter 10).

Box 2.5 ▷   Improved risk perceptions and financing for renewables in India[7]

India's renewable power investment has doubled over the past five years, reaching nearly $20 billion in 2018, and now exceeds that for coal power. Ambitious targets, supportive policies with competitive bidding and falling costs have lowered risks for investors and led to reductions in power purchasing tariffs for utility-scale solar PV and wind. Better financing terms also played a key role. Domestic banking rules force renewables, coal and gas power to compete for the same pool of debt finance (all categorised as power sector lending, with exposures capped by the central bank), which makes risk perceptions central to determining the share of capital flowing to renewables.

---

[7] This analysis has been developed in collaboration with the Council on Energy, Environment and Water (CEEW), Centre for Energy Finance in New Delhi. For further information, see CEEW and IEA (2019f).

An analysis of final investment decisions on solar PV and wind projects between 2014 and 2018 shows improvements in key risk parameters. Banks have been willing to fund projects with more debt-heavy capital structures, and have offered longer duration loans. The debt risk premium to benchmark bank lending rates has come down by around 75 basis points (Figure 2.24). Combined with lower economy-wide interest rates, improved debt financing makes over a 15% difference to the levelised cost of electricity (LCOE) for solar PV. Future financing cost pathways continue to have an impact on electricity prices and the affordability of the transition in the power sector (see Chapter 6).

Figure 2.24 ▷ Evolution of debt risk premiums for solar PV and wind in India and sensitivity of solar PV LCOE 2018 to debt financing

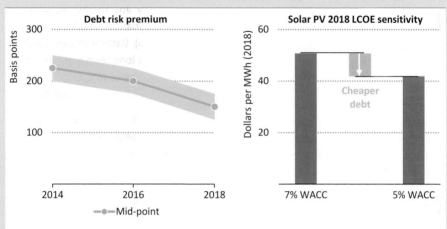

Risk perceptions and debt financing terms have improved for solar PV and wind in India, enabling investment in energy transition at lower costs

Notes: LCOE = levelised cost of electricity; WACC = weighted average cost of capital. The debt risk premium is the excess interest charged by banks over benchmark rates, reflecting the risks to a specific project or company. The LCOE is expressed in real terms and the sensitivity analysis assumes constant capital cost levels at $796 per kilowatt (average of commissioned plants in 2018) and capacity factor of 19%, with only the cost of debt and capital structure varying, which results in a 4.7% WACC compared with 7.0% under base assumptions.

Sources: Debt risk premiums are based on CEEW and IEA (2019f); LCOEs are based on IEA analysis.

Financing the clean energy transition also requires more focus on investments made at the consumer level, which tend to be much smaller and depend on the creditworthiness of households and small and medium-size enterprises. Traditional financing approaches can make an impact here – for example, the size of outstanding auto loans in the United States would support the purchase of 20 million EVs. However, additional measures may be needed to encourage the turnover of the capital stock in line with sustainability goals, while also balancing potential economic risks from consumers potentially taking on too much debt.

Obtaining finance for efficiency measures in the buildings and industry sectors is a bigger difficulty, given the challenge for banks to evaluate financial models based on energy savings. Measures that improve capital recovery, help improve the risk profile of investments. In 20 US states it is now possible to make repayments on upgrades through electricity bills or property taxes. Commercial arrangements with developers, such as energy service companies, provide an additional channel, which is strongest when backed by energy savings and performance guarantees. Pay-for-performance markets are emerging in California and other US states, where programme administrators (e.g. utilities) offer incentives (that help reduce cash flow risks) to customers or project developers in exchange for energy savings measured during pre-agreed periods, though the attractiveness of this practice depends on the ability of the project implementer to manage and insure against potential performance risks.

In terms of financing energy access, pay-as-you-go models for solar home systems – supported by digital finance – have facilitated access for electricity in East Africa by addressing creditworthiness concerns for a class of consumers with severe balance sheet constraints (see Chapter 10).

Newer financial solutions will need to be scaled up to crowd-in larger pools of capital. The aggregation of small-scale projects from the balance sheets of developers into portfolios that can be securitised, i.e. issued as debt or equity in the capital markets, can spread the risks among more investors and reduce financing costs However, structuring and evaluating the credit profiles of diverse small-scale assets remains challenging, and requires further co-operation between developers, banks and ratings agencies.

Investors such as pension funds and insurance companies can provide long-term capital at lower cost than developers and banks. Their demand for vehicles such as green bonds continues to rise as investors and companies pay more attention to climate-related risks in their companies. Some of these investors have also been particularly active in announcing financial restrictions on investment in fossil fuel assets, particularly coal (see Chapter 5).

Governments will play a fundamental role in setting the policies and regulations that influence who invests and the risks and returns that shape the allocation of energy capital: around 40% of energy capital is in any case owned by state-owned enterprises (IEA, 2018b). Policy makers also set the broader financial rules, and some countries are trying to facilitate financial decision making that supports the sustainable energy transition. For example, the United Kingdom issued a Green Finance Strategy in 2019 that seeks to align investment decision making with sustainability through a combination of long-term policy frameworks, new financial standards and products, and better industry analytics. The European Union has meanwhile proposed legislation seeking to channel private sector investment into sustainable development through better classification of aligned economic activities, enhanced corporate disclosure and development of benchmarks for investors. However the appropriate balance between private investment and state-directed capital mobilisation remains an area of debate, with calls in some quarters for states to play a more direct role in financial pump-priming and directing energy capital in order to achieve deeper levels of decarbonisation.

Which technology portfolio for the Sustainable Development Scenario?

The energy sector is technology-intensive, and it has passed through several innovation-led transformations in the last two centuries. We are now entering a new phase of technology-driven transformation propelled by advances in digital technologies, mass manufacturing and environmental awareness. Meeting the goals of the Sustainable Development Scenario will require the deployment of a range of technologies, some of which may evolve rapidly in this new phase. The markets for these technologies will be largely shaped by policy goals, which means that governments have a central role to play.

History tells us that the adoption of new technologies takes time. Experience shows that it can take two to three decades to move from first commercialisation of energy technologies to 2.5-3% market share, and more decades to reach maximum deployment (Bento, Wilson and Anadon, 2018; Gross et al., 2015). It took almost 25 years from the introduction of the first market-based feed-in tariff for renewable electricity to reach the point where solar PV made up 1% of global electricity output. Fortunately, most of the clean energy technologies deployed in the Sustainable Development Scenario are already well established. Many of the technologies that play a major role over the outlook period can be directly traced back to the 1950s. In that decade, gas turbines, hydraulic fracturing, liquefied natural gas (LNG), nuclear, solar PV and three-bladed alternating current wind turbines all made their debuts on the energy scene. The primary focus therefore is on how to accelerate their market uptake and improve performance and costs in the most efficient manner.

Table 2.3 ▷ **Examples of technologies that scale from low levels today to over 3% market share in the Sustainable Development Scenario**

	By 2030	After 2030
Liquids supply	Bio-liquids (jet fuel, diesel and gasoline)	CCUS on refinery hydrogen supply.
Natural gas supply	Remote, continuous methane leak detection	Biomass gasification. Clean hydrogen in gas supply to buildings.
Power generation and supply	Solar PV Smart meters	Small modular nuclear reactors. Battery storage. Biomass-fired power generation.
Buildings	Heat pumps Near-zero emissions buildings Home biogas digesters for clean cooking	Hydrogen fuel cells and boilers.
Transport	Electric cars and trucks	Alternative drivetrains for ships. Hydrogen-powered heavy trucks.
Industry	CCUS for iron and steel and cement	Electrolytic hydrogen feedstock for industrial processes. Heat pumps for industrial heat.

Notes: Market share defined as share of global sales or additions of equipment for the provision of equivalent energy services. For CCUS and hydrogen supply, shares refer to the emissions captured and gas delivered.

Of the critical clean energy technologies required to meet the Sustainable Development Scenario, only 7 of 45 are "on track" in terms of deployment, indicating the need for performance and cost improvements (IEA, 2019e). The next 30 years will witness dramatic improvements in a wide variety of energy technologies, many of which have components still in the laboratory today. Most will result in incremental changes to existing ways of doing things, but some could lead to radical and barely foreseeable new approaches. In the Sustainable Development Scenario, a number of key technologies move from very low levels of market penetration today to widespread commercialisation in the next decade, and in subsequent years (Table 2.3). As they do so, their costs decline in line with learning rates. A few of the technologies deployed in the Sustainable Development Scenario have not yet been commercialised, for example steel smelting with integrated CCUS and low-carbon cement production. However, the Sustainable Development Scenario does not assume any breakthroughs that would lead to deployment of technologies that have not yet been demonstrated.

The technologies that need to be improved and costs reduced throughout the Sustainable Development Scenario period have very diverse characteristics. They range from the physical sizes of individual units and the types of owners or operators to the types of materials and engineering involved. New nuclear designs, CCUS and low-carbon industrial processes are similar in many ways to the types of technologies that have dominated energy supply over the past century: each unit is designed for 50 megawatts (MW) to 2 GW of energy throughput; deployment is up to around 50 units per year; and much of the innovation is in materials and chemical engineering. In contrast, fuel cell, battery and solar PV units are designed for energy throughput of up to 0.1 kW to 100 kW, are deployed at rates of 100 000 to more than 400 million units per year worldwide in the Sustainable Development Scenario, and benefit from innovations in mass production, electronics and standardised installation. Other technologies, such as wind turbines, electrolysers and new ship drivetrains can be mass produced, but their unit size is larger and fewer units are expected to be deployed (Figure 2.25).

The policy and market dynamics that drive performance improvements and cost reductions in these types of technologies are very different. Mass produced consumer energy products are much more responsive to changes in prices, policies and social preferences than traditional energy technologies. More factories are needed to meet global demand for these items and industrial competition leads to faster turnover of products. New products with improved features hit the market every few years, and sometimes mismatched investment and consumption cycles can lead to oversupply and intensive competition for market share. These dynamics were at play in the case of solar PV over the last decade. This was the first time that mass produced, small unit size energy products and technologies became a significant feature of the energy supply landscape. Furthermore, because capital costs dominate for solar PV, costs reductions from manufacturing scale up had a disproportionate impact on electricity supply costs compared to cost reductions in traditional energy technologies requiring fuel inputs.

Figure 2.25 ▷ Low-carbon technologies by unit size and average annual installations in the Sustainable Development Scenario

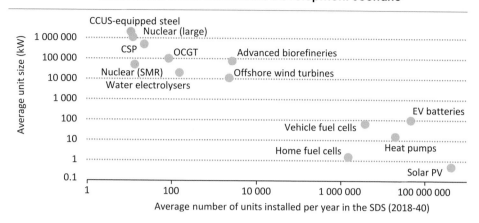

Products with large unit sizes are installed in double-digits per year, while small products are installed at annual rates over 400 million enabling faster innovation

Notes: CCUS = carbon capture, utilisation and storage; CSP = concentrating solar power; SMR = small modular reactor; EV = electric vehicles; OCGT = open-cycle gas turbine; PV = photovoltaics; SDS = Sustainable Development Scenario. Capacities refer to rated maximum energy output. For technologies that do not have output rated in energy terms, energy throughput for the relevant technology component is used.

The link between policy, learning and cost reductions is particularly instructive in the case of solar PV. As described in prior editions of the *World Energy Outlook*, the capital costs of this technology have been decreasing with a learning rate[8] of around 20% for over a decade. In practice, this means that each time the cumulative amount of added PV capacity doubles, the costs fall by 20% as a result of economies of scale and innovations, especially by manufacturers. As in previous *WEOs*, the Stated Policies and Sustainable Development scenarios both follow this trend over the outlook period, but with different outcomes (Figure 2.26). In the Sustainable Development Scenario, policies and markets drive a higher level of installation overall, resulting in more progress along the learning curve: total capital costs fall towards $600/kW by 2030, which is around 10% lower than in the Stated Policies Scenario. However, the absolute level of cost decline per year in the outlook is much less pronounced than in the period from 2010 to 2018 for two reasons: it takes longer to double the cumulative production of PV panels from 1 terawatt (TW) to 2 TW than from 0.25 TW to 0.5 TW; and a 20% reduction translates to fewer dollars per kilowatt when applied to a cost that is getting smaller over time.

There is however no guarantee that this learning rate will be maintained in the future. If total capital costs are falling towards $600/kW by 2030, they will be similar to the current costs of the basic hardware alone in the cheapest locations in the world. As a minimum, incremental innovations in technology, manufacturing and installation will be required, yet

---

[8] The capital cost reduction for a doubling of cumulative installed capacity.

there are indications that patenting and venture capital activity in these areas are slowing (Cárdenas Rodríguez, Haščič and Johnstone, 2019; IEA, 2019d). This is in part related to the success of mass manufacturing in establishing a dominant type of PV with low prices, which makes it harder for innovative approaches with significant future potential to enter the market. There are two main ways to counter this with the aim of facilitating faster cost declines. The first is to put more effort into researching new approaches that could potentially yield a step-change in cost reduction, especially if they enable much lower local installation costs. Thin, flexible, printed PV cells would fall into this category. The second is to ensure that advances in techniques are shared as quickly and widely as possible. Some elements of the solar PV learning rate are global, and some are local. International collaboration and trade can raise the global component of learning, and therefore bend the curve towards lower costs.

Figure 2.26 ▷ Evolution of capital costs of solar PV in the Stated Policies and Sustainable Development scenarios

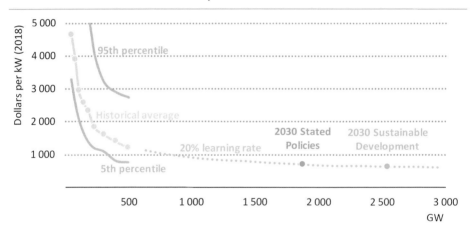

Policies drive more unit additions in the Sustainable Development Scenario, leading to lower costs that approach the expected limits of current technology

Sources: IEA analysis; IRENA (2019).

In the Sustainable Development Scenario, some of the biggest changes in the energy system relate to the diffusion of technologies that have small unit size and are for energy end-uses. These include electric heat pumps, EVs, electrolysers and fuel cells that are modular even in large installations. Governments have an important role to play in establishing a market framework that helps to make these technologies attractive to end-users, thus promoting uptake and innovation and enabling manufacturers and installers to raise finance at low cost. As the size and turnover of the market rises, the private incentives for improving the technologies increase. This is especially the case where developments can piggyback on progress in other fast-evolving sectors, such as communications and smart technology, as is the case for batteries and sensors for energy efficiency and demand response. Private investment is also likely to flow more readily where consumer energy products can be differentiated for various users, as with cars, or where end-user data can be commercialised.

The Sustainable Development Scenario, however, cannot be achieved without a range of complex, large unit size technologies that require associated infrastructure and generally involve a high degree of investment risk. These technologies require more capital to be put at risk in an early stage of the innovation chain and often face regulatory uncertainties. CCUS, nuclear, hydrogen and integrated smart city solutions fall into this category because of their costs and complexity. In the case of CCUS, for example, the first commercial projects can cost around $1 billion, take five or more years to move from engineering designs to results, and generally produce products – such as lower emissions cement, gasoline or power – with market values below their levelised and marginal costs today. If these technologies are to thrive, governments around the world will have to take on a significant proportion of the costs and risks of early commercial projects, sometimes for well over a decade, and provide strong signals that they will be supported in the future. Given the reduced appetite for governments to become involved in long-term, large-scale projects in some countries, reducing unit size and diversifying potential applications could be helpful: small modular nuclear reactors, CCUS for hydrogen production and even direct air capture all offer potential in this perspective.

The technologies that move from low market shares to wide deployment in the Sustainable Development Scenario mostly do so at a speed that is as fast as anything in the history of the energy sector.[9] Regardless of the technology type, the Sustainable Development Scenario requires an integrated approach to technology support which seeks to identify gaps in innovation, promote research that advances key technologies, establish market frameworks that support new clean energy technologies, penalise polluting technologies and encourage entrepreneurship. In this context, governments may wish to consider the scope for redirecting revenues from pollution charges to clean energy research grants, loans and other instruments, which would tackle two market failures at the same time.

2.9 How much further can we go?

The Paris Agreement goal to "pursue efforts towards 1.5 °C", together with the Intergovernmental Panel on Climate Change (IPCC) *Special Report on Global Warming of 1.5 °C* (*IPCC SR1.5*) released in 2018, has prompted a robust debate in some quarters about what a trajectory compatible with 1.5 °C might look like (IPCC, 2018). A salient point in this debate is that many of the physical impacts of climate change scale in a non-linear fashion with temperature rise: the impacts of 2.0 °C of warming are far worse than those of 1.5 °C of warming. This section looks at the Sustainable Development Scenario in the context of this debate and explores the implications for the energy system of limiting the temperature rise to 1.5 °C.

---

[9] Among transport options, the government-directed penetration of compressed natural gas into the vehicle fleets of Pakistan and Iran, and the adoption of LPG in Turkey, are the fastest to date, gaining around 5% fleet share per year.

Emissions trajectories consistent with 1.5 °C

A useful starting point to frame the energy-related consequences of limiting the temperature rise to 1.5 °C is the remaining CO_2 budget commensurate with this target. For a 50% chance of limiting the temperature rise to 1.5 °C, the *IPCC SR1.5* report provides remaining CO_2 budgets of 580 Gt and 770 Gt.[10] These different budgets reflect different views about the extent to which warming has already taken place, but both assume accompanying wholesale reductions in non-CO_2 emissions from methane and other gases, which have a strong, short-term effect on the temperature rise. Without such reductions the remaining CO_2 budgets would be lower.

Figure 2.27 ▷ Emissions trajectories for total CO_2 emissions in the Sustainable Development Scenario and to limit warming to 1.5 °C

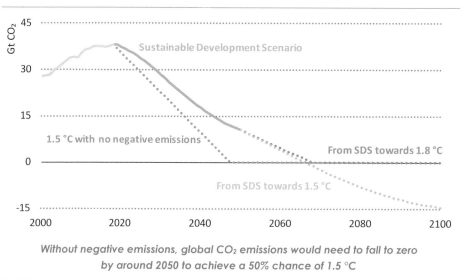

Without negative emissions, global CO_2 emissions would need to fall to zero by around 2050 to achieve a 50% chance of 1.5 °C

Note: SDS = Sustainable Development Scenario.

In the Sustainable Development Scenario, energy-related emissions fall to 10 Gt in 2050 and are on course to reach net zero in 2070. If emissions were to remain at exactly zero after 2070, then cumulative emissions between 2018 and 2100 would be 880 Gt and this would provide a 66% chance of keeping the temperature rise below 1.8 °C, and a 50%

---

[10] These budgets are from the start of 2018. The different budgets included in the *IPCC SR1.5* are associated with differences over the temperature increase today relative to pre-industrial times. The 770 Gt budget is associated with a lower current temperature rise today (0.87 °C above pre-industrial times) and the 580 Gt budget with a higher current temperature rise (0.97 °C above pre-industrial times). There are a number of additional uncertainties, such as the assumed future level of non-CO_2 GHG emissions and their impact on the temperature rise, that could also have a substantial impact on the remaining budget (although these are not included in the choice of 580 Gt or 770 Gt as a remaining budget).

chance of staying below 1.65 °C.[11] If emissions were to turn net negative, and around 300 Gt of CO_2 were to be absorbed from the air cumulatively by 2100, this would lead to a 50% chance of limiting the temperature rise to below 1.5 °C (using the smaller of the two IPCC budgets).

There are uncertainties about the scale of negative emissions that may be possible, and about their impacts and costs. However, 88 of the 90 scenarios in the *IPCC SR1.5* report which have at least a 50% chance staying below 1.5 °C warming in 2100 rely on net negative emissions, and the median level of cumulative net-negative emissions in these scenarios is around 420 Gt. In other words, an assumption about net-negative emissions which is well below the median in terms of the scenarios in the *IPCC SR1.5* report would make the Sustainable Development Scenario compatible with at least a 50% chance of keeping the temperature rise below 1.5 °C in 2100 (Figure 2.27).

SPOTLIGHT

What is the role of negative emissions in 1.5 °C scenarios?

One of the key characteristics of nearly all scenarios that aim to limit the temperature rise to 1.5 °C is their reliance on net negative CO_2 emissions in the second-half of the century.

There are four main options that could provide large-scale negative CO_2 emissions: afforestation and reforestation, sequestration of biochar[12], bioenergy used in conjunction with carbon capture and storage (often called "BECCS"), and direct air capture.

While it is technically conceivable that the world will reach a point where large quantities of CO_2 are absorbed from the atmosphere, there are uncertainties about what may be possible and about the likely impacts. Many of the technologies or methods involved are unproven at scale, and could have negative consequences outside the energy system related to land use, biodiversity and food security (IPCC, 2019; Anderson and Peters, 2016).

Negative emissions could help in particular to offset emissions from hard-to-abate sectors, such as aviation or the manufacturing of iron, steel and cement. Without future technological solutions to decarbonise these sectors or else negative emissions at sufficient scale, it would be necessary to curtail activity in order to reduce emissions from these sectors.

---

[11] A 66% chance of a less than 1.8 °C temperature rise is broadly equivalent to a 50% chance of a less than 1.65 °C temperature rise (IPCC, 2018).

[12] Biochar is a carbon-rich soil amendment made from biomass by pyrolysis (burning with limited oxygen), which could be used to sequester carbon in soil.

The range of sustainable bioenergy potential globally is estimated to be 130-240 exajoules (EJ) (3 100-5 700 Mtoe) per year (IEA, 2017c), similar to the level of current oil demand globally (4 450 Mtoe). The Sustainable Development Scenario uses around 80 EJ (1 900 Mtoe) of bioenergy in total in 2050, and around 0.25 Gt CO_2 is absorbed from the atmosphere in that year through the use of BECCS, compared to a median of 4.7 Gt from scenarios in the *IPCC SR1.5* database of scenarios (IPCC, 2018).

Figure 2.28 ▷ Cumulative net-negative CO_2 emissions between 2018 and 2100 in 1.5 °C scenarios assessed by the IPCC

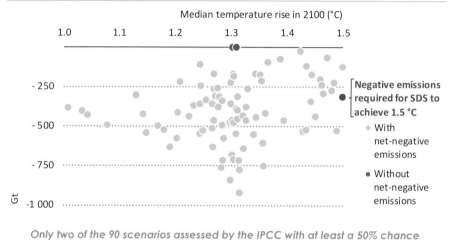

Only two of the 90 scenarios assessed by the IPCC with at least a 50% chance of 1.5 °C warming in 2100 achieve this without recourse to negative emissions

Notes: SDS = Sustainable Development Scenario. Cumulative from the point at which energy sector emissions reach net zero until 2100.

Implications for the energy sector of achieving zero CO_2 emissions in 2050

A 1.5 °C scenario which does not rely on negative emissions implies achieving zero CO_2 emissions around 2050 (Figure 2.27). This in turn implies a reduction in emissions of around 1.3 Gt CO_2 every year from 2018 onwards. The 1.3 Gt CO_2 is roughly equivalent to the emissions that would be avoided by shutting down around 290 GW of coal-fired power capacity (14% of the global installed coal capacity) or by replacing 490 million ICE passenger cars (around 40% of the global passenger car fleet) with electric cars running on zero carbon electricity.

We have not modelled this scenario in detail, but on the assumption of a regional pathway similar to that used in the Sustainable Development Scenario, a 1.5 °C scenario would imply that advanced economies reach net-zero energy sector CO_2 emissions around 2045, and developing economies around 2050 (Figure 2.29). The year in which net-zero emissions are achieved may vary depending on domestic circumstances, stage of development and energy sector characteristics.

Figure 2.29 ▷ Illustrative trajectory of energy-related CO_2 emissions to achieve a 50% chance of 1.5 °C in advanced and developing economies

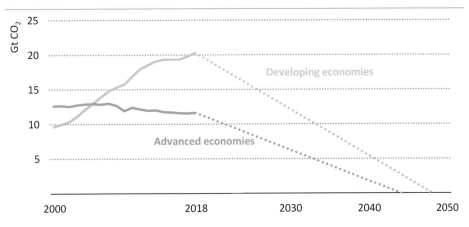

Achieving a 1.5 °C stabilisation would imply that advanced economies reach net-zero energy-related CO_2 emissions around 2045 and developing economies around 2050

Power sector

There are technologies available today to move towards full decarbonisation of electricity generation, but the challenges and costs of achieving this by 2050 would be considerable.

- A zero carbon power system would need to pre-date an economy-wide decarbonisation goal by at least a few years. This implies moving to a zero emissions electricity system in the 2030s for advanced economies and around 2040 for developing economies.

- Adding only zero carbon generation would not be sufficient: existing assets would have to be repurposed, retrofitted with CCUS or retired. In Chapter 6, we examine the practicalities and costs of this for coal-fired power plants. The costs of the possible strategy for coal outlined in Chapter 6 are significant; bringing this forward would inevitably increase these costs considerably.

- Building sufficient flexibility into a zero emissions power system would be challenging given that the practical constraints – such as planning and permitting a major expansion of the transmission grid – would need to be overcome on a shorter timeline than in the Sustainable Development Scenario.

End-use sectors

In end-use sectors, the starting point today is that 5% of kilometres driven are low carbon (that is, they involve the use of biofuels or EVs that use low-carbon electricity), 11% of energy used for heating buildings comes from low-carbon sources, and 17% of industry energy demand is low carbon. In some sectors technical solutions for decarbonisation are yet to be developed, for instance there are no commercially viable low-carbon planes.

There are some end-use sectors where technologies are now available that would allow decarbonisation. One example is electric cars, if coupled with zero carbon electricity. As the lifetime of cars is also relatively short[13], at least compared with other assets and infrastructure in the energy sector, to achieve zero CO_2 emissions in 2050 the sale of internal combustion engine cars would probably need to fall to zero in advanced economies in 2030 or soon after and in developing economies a few years later. This implies that sales of EVs would need to reach around 100 million in 2030 (from around 2 million a year today). This would increase electricity demand and the need for the deployment of low-carbon technologies in the power sector.

The focus for any economy-wide 100% decarbonisation target eventually narrows to the hard-to-abate sectors other than power generation and personal mobility. This includes decarbonisation of buildings, energy-intensive industries, aviation and freight transport. The Sustainable Development Scenario sees about 1.2 Gt CO_2 emissions from these sectors in advanced economies in 2045 and in developing economies around 3.4 Gt in 2050. Nearly all of the technical potential for energy efficiency measures is exploited in the Sustainable Development Scenario. Some of the key difficulties relate to:

- **Infrastructure constraints**: Most of the existing buildings stock would need to be retrofitted to become net-zero energy and be heated using zero carbon means. For example, in the United Kingdom, the Climate Change Committee has calculated that meeting a net-zero GHG target in 2050 implies that new homes would not be connected to the natural gas grid from 2025, that almost all heating systems for existing homes would be low carbon or ready for hydrogen by 2035, and that the share of low-carbon heating would need to increase from 4.5% today to 90% in 2050 (CCC, 2019). These types of changes would need to apply to many of the 2 billion houses existing today, as well as the additional one billion houses built to 2050 in order to reach net-zero emissions globally.

- **Social acceptance and behavioural changes**: Some of the changes that would be necessary would have wide-ranging effects, implying the need for a broad measure of public acceptance. For example, it would be hard to renovate and retrofit all buildings by 2050 without a high degree of support from the public. Another example is air travel, where it is uncertain that low-carbon aviation will be widely available before 2050. Currently, aviation emits around 1 Gt CO_2 per year and passenger activity is expected to at least double over the next 20 years. In the Sustainable Development Scenario, we assume significant efficiency improvements in aircraft and widespread use of advanced biofuels, but aviation still emits over 600 Mt CO_2 in 2050. Eliminating these emissions entirely could entail measures to limit air travel either via behavioural changes, administrative constraints or higher prices. However, there are areas where behavioural changes not associated with any absolute reductions in demand for goods and services could also help achieve deep decarbonisation. For example, dietary shifts towards less meat are often stipulated in 1.5 °C compatible scenarios. These go

---

[13] Assumes an average of 16 years in advanced economies and 18 years in developing economies.

beyond the energy sector but become equally essential as the share of emissions from agricultural and land use increases in importance. Digitalisation could help to enable behavioural changes, and make it easier to accept changes to the way that energy services are accessed, for example with the use of EV smart-charging as part of demand-side response (see Chapter 7, section 7.6).

- **Capital stock replacement**: Getting to zero emissions by 2045 for the hard-to-abate sectors in advanced economies would require the development and deployment of new technologies for all production – and in many cases the replacement of all capital stock – within the next 25 years. This process would need to take place in parallel in all developing economies within an additional five years, in a setting where the extent of the capital stock involved is much larger and, in most cases, the capital available to effect such a transformation is much more limited.

In terms of the fuel mix, all zero carbon energy sources would need to ramp up even faster than envisaged in the Sustainable Development Scenario. Three-quarters of energy demand for steel production today is met using coal. This would mean that in 20 years the majority of primary steel plants (around 80% of global production) in the world would need to be refurbished or retired and replaced.

Natural gas use may need to increase in the short term in countries that already have a well-established natural gas grid to help replace unabated coal and to allow for the extra flexibility needed by the very steep ramp-up of variable renewables. For unabated natural gas, stringent year-on-year emissions reductions mean that this increase would be temporary, but low carbon gases could make use of existing infrastructure and moderate the decline.

Assuming that oil could continue to be used for long-haul air travel (around 3 mb/d in 2050, with marked efficiency gains and use of alternative fuels), and for non-energy and non-emitting purposes, such as the manufacture of asphalt or chemicals (20 mb/d in 2050), then oil demand would fall sharply through to 2050, following a trajectory closer to the decline in supply from fields already producing today, assuming continued investment in these fields. However, given both the geographical locations and differing characteristics of fields around the world, not all the fields that are producing today would be well placed to reliably satisfy the ongoing demand for oil, meaning that some new field developments would still find a place even in a world of rapidly falling demand.

Summary and implications

More than ten years ago, the 2008 edition of the *World Energy Outlook* introduced an ambitious new climate scenario, warning that the "the consequences for the global climate of policy inaction are shocking" and underlining that "strong co-ordinated action is needed urgently to curb the growth in greenhouse gas emissions and the resulting rise in global temperatures".[14] Today, emissions continue to grow, and the emissions cuts required to avoid the worst effects of climate change get steeper every year.

---

[14] Executive Summary of *World Energy Outlook-2008* (IEA, 2008).

The Sustainable Development Scenario shows a pathway that would keep the temperature rise below 1.8 °C with a 66% probability without any implied reliance on global net-negative emissions. It is fully consistent with the Paris Agreement, and it sees energy sector CO_2 emissions fall to around 10 Gt by 2050, on course for net-zero emissions by 2070. If an assumption were to be made that emissions would turn net negative in the second-half of the century, it would take 300 Gt of negative emissions for the scenario to be consistent with at least a 50% chance of limiting warming to 1.5 °C – less than the median amount in the 90 scenarios included in the *IPCC SR1.5 report*.

There is a significant gap between the outcomes of the Stated Policies Scenario, which reflects the policies that governments have put in place or announced, and the Sustainable Development Scenario, which takes its starting point as the achievement of the energy-related SDGs and then shows what would be necessary to deliver this. This gap is set to grow in the future – an indication of the scale and pace of the transformation set out in the Sustainable Development Scenario.

There are uncertainties associated with net-negative emissions, and it would be possible in the light of concern about these to construct a scenario that goes further than the Sustainable Development Scenario and delivers a 50% chance of limiting warming to 1.5 °C without any reliance on net-negative emissions on the basis of a zero carbon world by 2050. This analysis does not reflect detailed modelling, but it nevertheless shows that eliminating the 10 Gt CO_2 emissions remaining under the Sustainable Development Scenario in 2050 would not amount to a simple extension of the changes to the energy system described in the Sustainable Development Scenario. The additional changes involved would pose challenges that would be very difficult and very expensive to surmount.

This is not something that is within the power of the energy sector alone to deliver. It would be a task for society as a whole, and likely involve widespread behavioural changes. It bears repeating that there is no single or simple solution to turn emissions around. The "moon shot" analogy – a concentrated dedication of resources, leadership and effort in favour of a single, visible outcome – is flawed. Change on a massive scale would be necessary across a very broad front, and would impinge directly on the lives of almost everyone.

Chapter 3

Outlook for oil
Summit fever?

SUMMARY

- Global oil demand continues to grow in the Stated Policies Scenario, but it loses momentum over the next two decades. From 97 million barrels per day (mb/d) in 2018, demand rises by around 1 mb/d on average every year to 2025. There is a material slowdown after 2025, but this does not lead to a definitive peak in oil use. Demand increases by 0.1 mb/d each year on average during the 2030s and ends up at 106 mb/d in 2040. By contrast, oil demand peaks very soon in the Sustainable Development Scenario and falls back to 67 mb/d by 2040, a level last seen in 1990.

- Thanks to determined policy action, China's period of rapid oil demand growth is set to end soon in the Stated Policies Scenario. Demand in China reaches a peak of 15.7 mb/d in the early 2030s. China nonetheless becomes the world's largest oil consumer just before 2040 as consumption falls steadily in the United States. Demand in India nearly doubles from today's level to 9 mb/d in 2040.

- US tight crude oil production grows in the Stated Policies Scenario from 6 mb/d in 2018 to a maximum of 11 mb/d in 2035. The majority of this growth comes from the Permian Basin, which by itself produces more crude oil than the continent of Africa soon after 2030.

- The outlook for US tight oil is not set in stone. Alongside uncertainties on the demand side, changing assumptions on future technology development and resource availability could each increase or decrease production in 2030 by more than 2 mb/d. Concerns about social and environmental issues, such as the level of flaring in the Permian Basin, and the ability of operators to raise affordable finance are other important variables that could affect production prospects.

- The increase in US output accounts for 85% of the global increase in production to 2030 in the Stated Policies Scenario. There are also notable increases in deepwater output in Brazil (the third-largest source of growth globally after the United States and Iraq) and in Guyana. These changes have profound implications for the share of OPEC countries plus Russia in total production, which drops to 47% in 2030 (from 55% in the mid-2000s). Continued investment in new sources of supply in OPEC and Russia nonetheless remains important to long-term oil market stability in this scenario, and their projected share of the oil market rises in the 2030s.

- In the Stated Policies Scenario, more than 30 million electric cars are sold each year by 2040, and the 330 million electric cars on the road in 2040 avoid 4 mb/d of oil demand. By the late-2020s, sales of cars with traditional internal combustion engines are in decline globally and oil use in passenger cars has peaked. This scenario assumes that the rising popularity of sports utility vehicles (SUVs) levels off: if the consumer preference for SUVs were to continue to increase, it would boost projected 2040 oil demand by nearly 2 mb/d, since these vehicles are larger, generally less fuel-efficient and more difficult to electrify than others.

Figure 3.1 ▷ Change in oil demand, supply and net trade position in the Stated Policies Scenario, 2018-2040

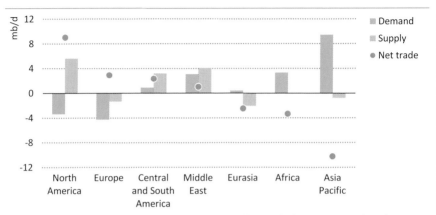

There is a marked shift in the geography of global oil trade towards Asia; North America becomes the world's second-largest oil exporter by 2030

Notes: Positive net trade values are increases in net exports, negative values are increases in net imports. Demand figures here include international aviation and marine bunkers.

- The changing geography of oil supply and demand in the Stated Policies Scenario transforms global oil trade. Asia takes an increasing share of global imports, and gross oil exports from the United States overtake those from Saudi Arabia by the mid-2020s. Refineries steadily adapt to the additional supply of light crudes. However, the United States also remains a major importer of heavier crudes: gross crude oil imports in 2030 are only one-third lower than in 2010.

- Growing trade volumes and rising geopolitical risks surrounding key chokepoints in oil markets highlight the need for policy makers to keep a close watch on oil security. By 2040, nearly 26 mb/d of oil passes through the Strait of Malacca in the Stated Policies Scenario and around 20 mb/d through the Strait of Hormuz. Any impediment to shipments could materially tighten markets. Emergency oil stocks continue to play an important role in helping to weather potential disruptions.

- Some traditional producers and exporters are seeing increasing pressure on their development model as a result of changing oil market dynamics. They face the prospect of a world where markets for their ample oil resources are not guaranteed, and where reduced income from hydrocarbons hampers their ability to maintain upstream spending and constrains the investments necessary to diversify their economies. A shift towards the lower demand and lower price environment of the Sustainable Development Scenario would further underscore the urgency of economic reform and diversification.

Introduction

What does a regular year for oil markets look like? The year 2018 saw a steady rise in demand of 1.2 million barrels per day (mb/d), but in other ways it confounded expectations. The leading source of consumption growth was not China or India, but rather the United States. Sales of electric vehicles set new records in 2018, though are yet to make a very visible dent in oil consumption trends. The main increases in oil product demand came from gasoline and diesel but there were also sizeable contributions from ethane, liquefied petroleum gas (LPG) and naphtha as the use of oil as a petrochemical feedstock continues to grow in importance.

On the supply side, US tight oil defied talk of infrastructure constraints and its growth to date matches the fastest pace ever seen in the history of oil markets. Geopolitics also came to the fore, as economic, political and security issues affected supply from Venezuela, Libya and Iran. Attacks on oil tankers and processing infrastructure in the Middle East have heightened awareness about risks to oil supply from the region, including via some key choke points in international trade. The oil price has generally remained subdued despite these events. The level of new resource developments appears to be rebounding from its post-2014 slump, with notable prospects for new offshore projects in Latin America.

So where do we go from here? As ever, the *World Energy Outlook* sets out a number of pathways. In the Stated Policies Scenario, demand growth is robust to 2025, but growth slows to a crawl thereafter and demand reaches 106 mb/d in 2040, while the Current Policies Scenario sees continued annual growth in line with historical averages. In the Sustainable Development Scenario, the unprecedented scale, scope and speed of changes in the energy landscape paints a very different picture for oil markets: demand soon peaks and drops to under 67 mb/d in 2040. Against this backdrop, we examine three key issues that shape the future of oil markets:

- **Impact of changes in vehicle ownership:** While passenger cars remain the largest single element of oil demand today, there are some signs that the world is moving beyond traditional modes of car ownership and oil-powered cars. However, recent years have also seen consumers opting for heavier and less efficient cars. We explore the impact of these trends on oil demand.

- **Outlook for US tight oil:** The growth in tight oil in the United States has arguably been the biggest story for oil markets over the past decade. We examine how tight oil operators fared in the wake of the 2014 oil price crash, discuss the outlook for US tight oil and explore some key uncertainties and sensitivities that could affect this outlook.

- **Outlook for oil security:** The consequences of the rise in US production have been felt well beyond North America and beyond the energy sector. At the same time, recent events have highlighted the risks of physical disruption to oil supply. We explore how these various dynamics influence patterns of oil trade and oil supply security: we also look at how they impact, and are impacted by, crude oil quality.

> Figures and tables from this chapter may be downloaded from www.iea.org/weo2019/secure/.

Scenarios

3.1 Overview

Table 3.1 ▷ Global oil demand and production by scenario (mb/d)

	2000	2018	Stated Policies 2030	Stated Policies 2040	Sustainable Development 2030	Sustainable Development 2040	Current Policies 2030	Current Policies 2040
Road transport	30.1	42.2	45.5	44.5	36.7	22.8	48.9	53.4
Aviation and shipping	8.3	12.0	14.5	16.8	11.2	9.4	15.7	19.2
Industry and petrochemicals	14.4	18.3	21.5	22.9	18.9	18.5	21.5	23.0
Buildings and power	14.3	12.3	10.5	9.2	8.4	5.8	11.5	11.1
Other sectors	10.2	12.0	13.3	13.1	12.0	10.4	13.8	14.2
World oil demand	77.4	96.9	105.4	106.4	87.1	66.9	111.5	121.0
Asia Pacific share	25%	33%	36%	37%	37%	38%	36%	37%
World biofuels	0.2	1.9	3.5	4.7	6.3	7.7	2.8	3.6
World liquids demand	77.6	98.8	108.9	111.1	93.4	74.6	114.3	124.6
Conventional crude oil	64.5	67.1	65.1	61.9	52.7	36.9	68.5	70.6
Existing fields	64.5	67.1	39.6	25.9	39.6	25.9	39.6	25.9
New fields	-	-	25.5	36.0	13.1	11.0	28.9	44.7
Tight oil	-	6.3	12.0	13.4	10.1	9.2	13.1	15.5
Natural gas liquids	9.0	17.3	20.4	21.7	17.7	14.8	21.2	23.1
Extra-heavy oil and bitumen	1.0	3.8	4.0	4.9	3.3	2.9	4.3	6.3
Other production	0.6	0.8	1.3	1.6	1.2	1.2	1.5	2.2
World oil production	75.1	95.4	102.8	103.5	85.0	65.1	108.7	117.7
OPEC share	41%	39%	37%	39%	37%	37%	37%	39%
World processing gains	1.8	2.3	2.6	2.9	2.2	1.8	2.8	3.3
World oil supply	76.9	97.7	105.4	106.4	87.1	66.9	111.5	121.0
IEA crude oil price ($2018/barrel)	40	68	88	103	62	59	111	134

Notes: Other production includes coal-to-liquids, gas-to-liquids, additives and kerogen oil. Historical supply and demand volumes differ due to changes in stocks. See Annex C for definitions.

In the **Stated Policies Scenario**, global oil demand rises by around 1 mb/d on average every year until 2025 (Table 3.1). Oil use in passenger cars peaks in the late-2020s and during the 2030s demand increases by only 0.1 mb/d on average each year (Figure 3.2). There is no definitive peak in oil use, given increases in petrochemicals, trucks and the shipping and aviation sectors. The largest increases in production between 2018 and 2040 come from the United States, Iraq and Brazil. The share in oil production from countries in the Organization of the Petroleum Exporting Countries (OPEC) plus Russia falls to 47% for much of the 2020s, a level not seen since the 1980s. The oil price required to balance supply and demand in this scenario edges higher to nearly $90/barrel in 2030 and $103/barrel in 2040.

The **Current Policies Scenario** provides a reminder that, if the effects of new policies and alternative technologies are discounted, there is scope for more rapid growth in oil demand. In this scenario, global oil demand rises by 1.1 mb/d on average every year to 2040, similar to the average increase seen since 2000. Without strengthened policies on fuel efficiency or the use of alternative fuels, there is little restraint on the pace of oil demand growth. Growth is led by road transport, accounting for nearly half of the increase to 2040, and there are major increases in petrochemicals and aviation. The oil price rises steadily to just under $135/barrel in 2040.

Figure 3.2 ▷ Global oil demand and crude oil price by scenario

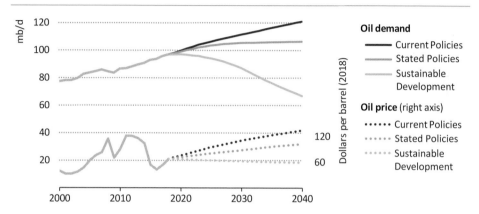

In the Stated Policies Scenario, demand growth slows substantially after 2025, while the Sustainable Development Scenario paints a very different picture for demand and prices

In the **Sustainable Development Scenario**, determined policy interventions lead to a peak in global oil demand within the next few years. Demand falls by around 2 mb/d each year on average during the 2030s and is 30 mb/d below today's level by 2040. Demand falls by more than 50% in advanced economies between 2018 and 2040, and by 10% in developing economies. Reductions in oil use in road transport are particularly significant. By 2040, 50% of cars are electric (with 900 million electric cars on the road), as are most of the world's urban buses; almost 2 million barrels of oil equivalent (mboe) per day of biofuels are consumed in the aviation and shipping sectors; and almost 20% of the fuel used by trucks worldwide is low-carbon. The only sector to see demand growth is petrochemicals: while the rate of plastics recycling more than doubles (from around 15% today to 35% in 2040), demand increases by almost 3 mb/d to 2040.

Tight oil production in the United States continues to grow for a number of years: it has a rapid decline rate and so is relatively well suited to a world where the outlook for demand and prices is uncertain. In very challenging market conditions, OPEC is assumed to continue to pursue efforts at market management, meaning that its production falls by 5 mb/d over the period to 2025. The oil price remains below $70 per barrel and falls to just under $60/barrel in 2040.

3.2 Oil demand by region

Table 3.2 ▷ Oil demand by region and scenario (mb/d)

	2000	2018	\multicolumn{4}{c}{Stated Policies}	\multicolumn{2}{c}{Sustainable Development}
			2025	2030	2035	2040	2030	2040
North America	23.5	22.8	22.5	21.5	20.3	19.1	17.7	11.7
United States	19.6	18.5	18.4	17.4	16.3	15.1	14.2	9.1
Central and South America	4.5	5.8	6.1	6.2	6.4	6.5	4.9	3.8
Brazil	1.9	2.4	2.6	2.7	2.8	2.8	2.1	1.6
Europe	14.9	13.2	12.4	11.1	9.7	8.7	9.2	5.0
European Union	13.1	11.1	10.1	8.8	7.4	6.3	7.3	3.5
Africa	2.2	3.9	4.9	5.5	6.2	7.0	4.9	5.2
South Africa	0.4	0.5	0.6	0.6	0.7	0.7	0.6	0.5
Middle East	4.3	7.5	8.4	8.8	9.6	10.2	6.7	6.3
Eurasia	3.1	3.9	4.3	4.3	4.2	4.2	3.8	3.1
Russia	2.6	3.2	3.4	3.4	3.3	3.2	3.0	2.4
Asia Pacific	19.4	31.6	35.8	38.0	38.9	39.2	32.4	25.4
China	4.7	12.5	14.5	15.6	15.6	15.5	12.7	9.3
India	2.3	4.7	6.4	7.5	8.4	9.0	6.5	6.1
Japan	5.1	3.6	3.0	2.7	2.3	2.0	2.3	1.2
Southeast Asia	3.1	5.3	6.3	6.6	6.9	6.9	5.8	4.7
International bunkers	5.4	8.2	9.3	10.0	10.7	11.4	7.6	6.4
World oil	77.4	96.9	103.5	105.4	106.0	106.4	87.1	66.9
World biofuels	0.2	1.9	2.8	3.5	4.1	4.7	6.3	7.7
World liquids	77.6	98.8	106.4	108.9	110.1	111.1	93.4	74.6

Note: See Annex C for definitions.

In the Stated Policies Scenario, there is a 0.9 mb/d average annual increase in demand in developing economies between 2018 and 2030 and a 0.4 mb/d average annual decrease in demand in advanced economies (Figure 3.3). This swing has major consequences for oil trade flows and oil security dynamics. There is a marked slowdown in global oil demand growth during the 2030s: demand fluctuates on an annual basis but there is no definitive peak in overall use. The average annual increase in the 2030s is 0.1 mb/d.

Road transport accounts for half of total oil demand in advanced economies today, and changes in this segment are the key determinant of overall demand trends. In the United States, reductions in the use of oil in passenger cars account for over 90% of the overall decline in demand between 2018 and 2040. The implementation of efficiency standards is central to this decline. Electric cars also play an increasingly important role: on average there are nearly 2.5 million electric cars sold each year during the 2030s and in 2040 there are nearly 30 million electric cars on the road (just over 10% of the total US car stock).

Figure 3.3 ▷ Change in oil demand by region in the Stated Policies Scenario

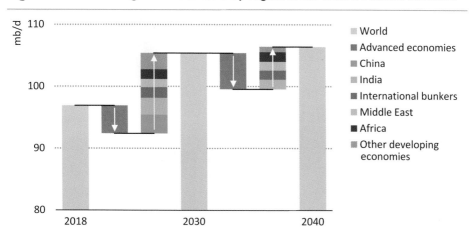

Demand growth in many developing countries stalls after 2030 and global demand growth tails off, however the first peak in oil demand will not necessarily be the last

In Europe, oil demand falls by around 0.2 mb/d each year on average to 2040. Most of this decline comes from passenger cars, but there is also a decrease in oil demand for buildings given stated policies to ban the installation of oil boilers in new houses and phase out existing equipment.

In China, oil demand growth remains robust until the early 2020s but grinds to a halt in the 2030s. Oil use in petrochemicals rises steadily between 2018 and 2040, as does demand for aviation, stemming from increasing domestic tourism. However, with widespread electrification (150 million cars in China are electric in 2040) and an ever more efficient car fleet, oil use in cars reaches a peak in 2030 and then slowly declines. The use of oil in buildings also falls and total oil demand peaks in the early 2030s at 15.7 mb/d.

In India, demand nearly doubles between 2018 and 2040, the largest absolute growth of any country. One-third of the growth comes from trucks; another quarter comes from passenger cars, with the Indian car fleet growing by a factor of seven between 2018 and 2040; a further 15% of the growth comes from the use of oil as a petrochemical feedstock.

In the Middle East, the largest contributor to demand growth is the petrochemical sector, which many countries see as having the potential to help them gain more value from their oil reserves. There is also an increase in oil use in power as large volumes of heavy fuel oil are displaced from the shipping sector around 2020, though the use of oil then declines as the heavy fuel oil market rebalances and prices start to rise.

In Africa, demand growth is largest in transport, but demand also increases for diesel (to use in back-up generators) and for LPG, which, by 2040, helps to provide clean cooking facilities to over 260 million people previously relying on traditional uses of biomass (see Chapter 9).

3.3 Oil demand by sector

Figure 3.4 ▷ Annual average change in global oil demand by sector in the Stated Policies Scenario

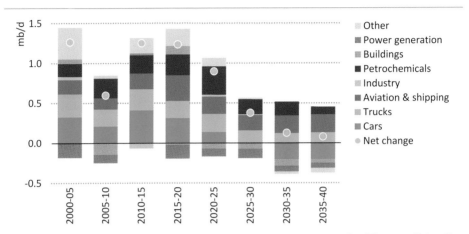

Petrochemicals, aviation & shipping and trucks account for the bulk of the growth in oil demand, offsetting declines in other sectors. Oil use in cars peaks in the late-2020s.

In the Stated Policies Scenario, oil use as a **petrochemical** feedstock is particularly strong in the period to 2025 given the announced plans of many companies to add new capacity, most notably in the United States, China and the Middle East (Figure 3.4). Today around 15% of plastics are recycled globally and an increasing number of countries aim to curb single-use plastics. However, most future growth in plastics comes from developing countries with relatively low levels of recycling and so the global average remains less than 20% in 2040. If the average global recycling rates were instead to reach 35%, this would avoid around 1.7 mb/d of oil demand in 2040; even in this case, however, the use of oil as a feedstock would still increase by 3 mb/d between 2018 and 2040.

Around 15 mb/d of oil is consumed by **trucks** today and the increase in demand to 2040 is the second-largest of any sector (after petrochemicals). The 25% increase in oil demand, however, is much less than the 90% increase in global freight activity over this period. A number of alternatives to the use of oil gain ground: there is an increase in the use of natural gas and biofuels, primarily in China and the United States, and there is some electrification of light commercial vehicles. In total these fuels avoid around 1.5 mb/d of oil demand growth between 2018 and 2040. However, efficiency measures play a bigger role, avoiding a further 4.5 mb/d increase in demand to 2040.

Oil demand for **passenger cars** peaks in the late-2020s. This occurs despite the number of cars on the road globally increasing by 70% between 2018 and 2040. In 2040, there are 330 million electric cars on the road – 15% of the global car fleet – which avoid 4 mb/d of oil demand. The average gasoline or diesel-based car on the road in 2040 is also 25% more efficient than today: this avoids nearly 9 mb/d of oil demand in 2040. This increase in efficiency occurs despite the recent trend, seen across all major car markets, for purchasing larger and heavier cars such as sports utility vehicles (SUVs) (see section 3.9).

There is a 50% increase in oil use in **aviation** between 2018 and 2040 and it accounts for 10% of total oil demand in 2040 (from 7% in 2018). Biofuels are one of the few alternatives to oil use in aviation and they comprise around 5% of liquids demand in the sector in 2040.

The International Maritime Organization (IMO) regulation to limit the sulfur content of oil use in **shipping** to no more than 0.5% by 2020 leads to an 2 mb/d drop in high sulfur fuel oil consumption after the regulation comes into force, together with offsetting increases of 1 mb/d in marine gasoil (a type of diesel), and 1 mb/d in "very low sulfur fuel oil". Over time, there is an increase in the use of liquefied natural gas (LNG) as a bunker fuel (which avoids 0.9 mb/d of oil in 2040) and a modest increase in biofuels. However, total oil use increases by nearly 1.5 mb/d between 2018 and 2040.

Oil combustion in **industry** is mainly used to generate steam and process heat in the chemicals and cement sub-sectors and to power equipment in manufacturing. There are many alternatives to oil for these processes: however, oil is often preferred if it is subsidised or if there is an established distribution network, since site-specific or process-specific constraints can make fuel switching an expensive proposition. Oil consumption in industry remains broadly constant at just over 6 mb/d between 2018 and 2040.

Today, around 65% of oil use in **buildings** occurs in developing economies, with LPG and kerosene used for cooking and lighting, and heavy fuel oil used in some countries for heating. The use of oil in buildings increases modestly in Africa, India and Southeast Asia, but this is offset by losses elsewhere: the United States and Europe see falls of around 70% and 80% respectively.

Oil is generally used in **power generation** only in countries with major subsidy regimes or where there is a legacy fleet of stations. The new IMO regulation on shipping fuels leads to a surplus of heavy fuel oil, which in turn leads to a near-term boost in oil use in power. In the longer term, subsidies are assumed to be phased out, and, as renewable-based electricity becomes an increasingly attractive option, the consumption of oil in power generation falls by 40% globally between 2018 and 2040.

Half of the energy used by **trains** today is oil (0.6 mb/d), which is mostly used for freight rather than passenger trains. Oil use in trains declines as electricity becomes the preferred option and demand falls by 0.1 mb/d to 2040. Electric **bus** sales have soared recently in China, which accounted for 99% of global electric bus sales in 2018. There is a 35% increase in the number of buses on the road globally between 2018 and 2040, and so while 25% of urban buses in 2040 are electric, oil use in buses increases by 0.1 mb/d to 2040.

There are limited viable alternatives to oil in sectors such as agriculture, petroleum refineries, oil extraction and non-energy uses like asphalt, bitumen and lubricants. Oil consumption across these sectors increases by just over 1 mb/d over the *Outlook* period.

3.4 Oil supply by type

Figure 3.5 ▷ Oil production by type in the Stated Policies Scenario

- Onshore conventional crude oil
- Offshore shallow water
- Offshore deepwater
- NGLs
- Tight oil
- EHOB & other

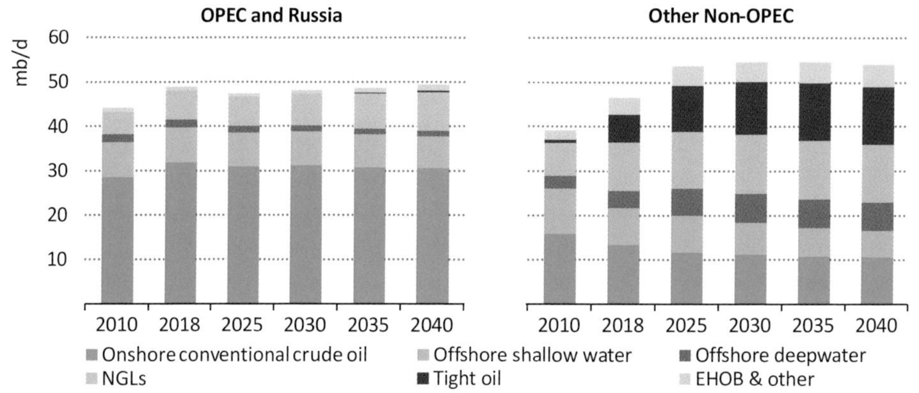

OPEC and Russia's share of the oil market remains below 50% to 2040; NGLs and unconventional oil comprise more than two-thirds of other non-OPEC production by 2030

Notes: NGLs = natural gas liquids; EHOB & other = extra-heavy oil and bitumen, kerogen oil, coal-to-liquids, gas-to-liquids and additives.

Conventional crude oil is by far the largest source of oil production today (67 mb/d) (Figure 3.5). However many currently producing fields are mature (almost 40% of production today comes from fields older than 40 years) and replacing declining production from these fields is difficult. Globally, conventional production drops in the Stated Policies Scenario by around 5 mb/d by 2040 and its share in global supply falls from 70% today to 60% in 2040.

Onshore conventional crude oil comes mostly from the Middle East and Russia, which today account for 65% of the 45 mb/d total produced globally. Most of the onshore oil in Russia comes from the Western Siberian Basin. This is a mature province and production declines steadily to 2040. There are other onshore resources elsewhere in Russia, but the harsh environment and lack of infrastructure make these difficult to develop. Total onshore crude oil production falls slightly but stabilises just above 41 mb/d after 2030, with new production coming online from Iraq and North Africa in particular.

Offshore conventional crude oil production remains broadly at today's level (22 mb/d) to 2040. Shallow water oil production from the Middle East grows as Saudi Arabia ramps up production, but continued declines elsewhere, especially in Europe and Southeast Asia, mean that global shallow water production falls by 3 mb/d in the period to 2040. This decline is offset by a near 2 mb/d increase in deepwater production in the same period. There has been a recent uptick in deepwater investment, and most of the increase in production is set to take place before 2030, with notable increases in the United States, Brazil, Mexico and the newly discovered giant fields in Guyana. Deepwater production in Africa falls to around 1.5 mb/d to 2040.

Figure 3.6 ▷ Composition of global production in the Stated Policies Scenario

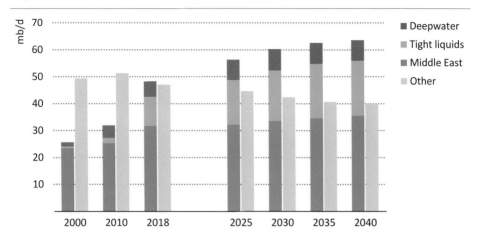

Investment and growth is increasingly concentrated in a handful of areas: US tight liquids dominate the near term, but the Middle East and deepwater production also play vital roles

Note: Tight liquids include tight crude oil, condensate and NGLs.

Natural gas liquids (NGLs) play an important role in the economics of gas developments and production increases by 4.5 mb/d to reach 22 mb/d in 2040. Growth in NGLs production generally follows increases in gas production, with the largest increases in the United States, Saudi Arabia and Russia.

Tight crude oil production globally expands from 6.3 mb/d in 2018 to around 13.5 mb/d by 2040 (Figure 3.6). Production in the United States reaches a maximum of 11 mb/d in 2035 but declines slightly in later years (see section 3.10). Production also increases in Argentina, China, Canada, Mexico and Australia. The Vaca Muerta Basin in Argentina has received more attention in recent years, and its production grows to more than 1 mb/d by 2040. Countries outside of the United States produce almost 25% of global tight oil by 2040. The growth in tight oil leads to a large increase in the volumes of light crude oil available. This is not projected to cause major issues for markets: refiners are assumed gradually to adjust the configuration of refineries to accept these volumes and to meet the steady shift in global oil demand towards lighter products (see section 3.11).

Extra-heavy oil and bitumen production increased by 9% each year on average between 2000 and 2015, but growth has stagnated recently. In Canada, Alberta imposed a production curtailment order from January 2019 to bring supplies more closely in line with available pipeline capacity. In Venezuela, production has been severely impacted by its economic and political difficulties. Concerns about the wider social and environmental impacts of production weigh on prospects for major new greenfield developments: extra-heavy oil and bitumen production nevertheless increases in Canada by 0.6 mb/d to 3.7 mb/d in 2040, and in Venezuela by 0.4 mb/d to 1.2 mb/d in 2040.

3.5 Oil supply by region

Table 3.3 ▷ Non-OPEC oil production in the Stated Policies Scenario (mb/d)

	2000	2018	2025	2030	2035	2040	2018-2040 Change	CAAGR
North America	14.2	23.0	28.4	29.6	29.7	28.6	5.6	1.0%
Canada	2.7	5.4	5.7	5.8	6.0	6.0	0.6	0.5%
Mexico	3.5	2.1	1.9	2.1	2.4	2.8	0.8	1.4%
United States	8.0	15.5	20.9	21.7	21.3	19.8	4.3	1.1%
Central and South America	3.2	4.5	6.3	6.8	7.1	7.8	3.3	2.5%
Argentina	0.9	0.6	0.7	0.9	1.1	1.6	1.0	4.7%
Brazil	1.3	2.7	3.9	4.1	4.4	4.7	2.0	2.6%
Europe	7.1	3.7	4.0	3.3	2.8	2.6	-1.0	-1.5%
European Union	3.6	1.7	1.5	1.1	0.9	0.7	-0.9	-3.6%
Norway	3.3	1.9	2.4	2.0	1.8	1.7	-0.1	-0.3%
Africa	1.3	1.5	1.8	1.9	1.7	1.6	0.1	0.4%
Middle East	3.0	3.2	3.3	3.5	3.8	4.0	0.8	1.0%
Qatar	0.9	2.0	2.1	2.3	2.6	2.8	0.8	1.6%
Eurasia	7.9	14.5	14.1	13.6	13.0	12.4	-2.1	-0.7%
Kazakhstan	0.7	1.9	2.0	2.2	2.2	2.3	0.4	0.8%
Russia	6.5	11.5	11.1	10.6	9.9	9.4	-2.1	-0.9%
Asia Pacific	7.8	7.6	6.9	6.5	6.4	6.4	-1.2	-0.8%
China	3.2	3.8	3.4	3.2	3.1	3.0	-0.9	-1.2%
India	0.8	0.8	0.9	0.8	0.8	0.8	-0.1	-0.5%
Southeast Asia	2.8	2.3	2.0	1.8	1.6	1.5	-0.8	-1.9%
Conventional crude oil	37.0	35.9	36.0	34.2	32.0	30.6	-5.3	-0.7%
Tight oil	-	6.3	10.5	11.9	13.1	13.2	6.9	3.4%
United States	-	5.9	9.8	10.8	11.1	10.1	4.2	2.5%
Natural gas liquids	6.4	11.9	13.9	14.5	14.7	14.5	2.6	0.9%
Canada oil sands	0.6	3.1	3.4	3.4	3.5	3.7	0.6	0.8%
Other production	0.5	0.7	1.0	1.1	1.3	1.4	0.7	3.1%
Total non-OPEC	44.5	58.0	64.8	65.2	64.6	63.4	5.5	0.4%
Non-OPEC share	59%	61%	64%	63%	63%	61%	0%	n.a.
Sustainable Development			59.2	53.8	46.7	40.8	-17.2	-1.6%

Notes: CAAGR = compound average annual growth rate; n.a. = not applicable. See Annex C for definitions.

Non-OPEC production grows by nearly 7 mb/d to 2025, led by increases in the United States. While OPEC provides nearly 40% of global production today, with a continuing strategy of moderating production to support prices, its share of the market falls to a low of 35% in the early 2020s. Russia has joined OPEC's market management efforts, and Russia and OPEC members together see a fall in market share to a low of 47% in 2025 (from a recent high of 55% in the mid-2000s). While OPEC and Russia's share of the market rises in later years, as US tight oil production reaches a peak, it stays below 50% through to 2040.

Table 3.4 ▷ OPEC oil production in the Stated Policies Scenario (mb/d)

	2000	2018	2025	2030	2035	2040	2018-2040 Change	CAAGR
Middle East	20.5	28.5	28.9	30.1	30.8	31.6	3.1	0.5%
Iran	3.8	4.6	3.8	4.1	4.3	4.5	-0.1	-0.1%
Iraq	2.6	4.7	5.3	5.8	6.2	6.5	1.8	1.5%
Kuwait	2.2	3.1	3.1	3.2	3.2	3.2	0.1	0.2%
Saudi Arabia	9.3	12.4	12.6	12.9	13.0	13.1	0.8	0.3%
United Arab Emirates	2.6	3.8	4.0	4.1	4.2	4.4	0.6	0.6%
Non-Middle East	10.1	8.9	7.3	7.5	7.9	8.5	-0.4	-0.2%
Algeria	1.4	1.5	1.3	1.3	1.3	1.3	-0.2	-0.8%
Angola	0.7	1.5	1.3	1.3	1.2	1.2	-0.3	-1.0%
Congo	0.3	0.3	0.1	0.1	0.1	0.1	-0.3	-6.7%
Ecuador	0.4	0.5	0.4	0.4	0.3	0.3	-0.2	-2.4%
Equatorial Guinea	0.1	0.2	0.1	0.1	0.1	0.1	-0.1	-5.2%
Gabon	0.3	0.2	0.1	0.1	0.1	0.1	-0.1	-2.1%
Libya	1.5	1.0	1.0	1.1	1.3	1.5	0.5	1.9%
Nigeria	2.2	2.1	2.1	2.1	2.2	2.3	0.2	0.5%
Venezuela	3.2	1.5	0.8	0.9	1.2	1.6	0.1	0.2%
Conventional crude oil	27.5	31.1	30.1	30.9	31.0	31.3	0.1	0.0%
Natural gas liquids	2.7	5.4	5.5	5.9	6.6	7.2	1.8	1.3%
Venezuela extra-heavy oil	0.4	0.8	0.5	0.6	0.8	1.2	0.4	2.1%
Other production	0.1	0.1	0.1	0.2	0.3	0.4	0.3	6.7%
Total OPEC	30.6	37.4	36.2	37.6	38.7	40.1	2.7	0.3%
OPEC share	41%	39%	36%	37%	37%	39%	0%	n.a.
Sustainable Development			32.7	31.2	27.6	24.3	-13.1	-1.9%

Notes: CAAGR = compound average annual growth rate. See Annex C for definitions.

In the **United States**, total oil production grows to a peak of 22 mb/d in 2030 and it accounts for 85% of the total global increase in this period. In 2030, production in the United States is almost 70% greater than the next largest country (Saudi Arabia). Tight oil production peaks in 2035 and, alongside a drop in deepwater and NGLs production, this leads to a 1.5 mb/d decline in US production between 2035 and 2040. The tax credits offered for enhanced oil recovery (EOR) using carbon dioxide (CO_2) lead to a doubling in today's CO_2-EOR production to 0.8 mb/d in 2040.

The focus in **Mexico** has returned to developing shallow water and onshore projects, with less immediate emphasis on deepwater projects. Total production declines to 2025 until deepwater production starts to pick up. There is also some tight oil development.

Brazil registers the second-largest increase of any country over the period to 2040. The majority of this growth occurs before 2025 as production ramps up from new developments and extensions at the deepwater Buzios, Mero, Iara and Lula fields.

Guyana has seen a number of major deepwater discoveries in recent years. These lead to large increases in production over the *Outlook* period. Production grows from negligible levels today to 0.8 mb/d in 2025 and, with further discoveries, to 1 mb/d in 2040.

The tight oil resources of **Argentina** have received a great deal of attention recently. Tight oil production today is less than 60 kb/d, but major investment helps this to increase to 1 mb/d in 2040, and this accounts for 70% of total oil production in Argentina at that time.

Production in **Norway** increases to 2025 as major projects such as Johan Sverdrup come online: this project alone accounts for more than one-quarter of total production in 2025. Despite some further new oil field developments and expansions, total production then declines by around 2% each year on average between 2025 and 2040.

Production in **China** peaked in 2015 at 4.3 mb/d and has since declined by 3.5% on average each year. This slows to an annual average 1% decline through to 2040.

Iraq nearly doubled its oil production over the past decade to 4.7 mb/d despite formidable challenges. Future growth is contingent on a stable political and security environment, attracting sufficient foreign capital and the successful development of new projects to provide the vast quantities of water needed for reinjection.

Production in **Saudi Arabia** remains largely flat until 2025 but then grows by around 0.5 mb/d to reach 13 mb/d in 2040. Around a third of Saudi Arabia's current oil production comes from offshore fields and there are notable increases at the Marjan and Berri fields.

In **Iran**, production fell in 2018 following the reimposition of US sanctions. While Iran has vast oil resources, for the moment we project only a gradual recovery in production, meaning that Iran does not exceed its previous output peak (6 mb/d in 1974) any time before 2040.

In **Venezuela**, the production outlook is strongly affected by political uncertainty and profound economic difficulties. In the absence of conditions that allow for a pickup in investment, we assume that output declines to a low of 0.8 mb/d in 2025 before the situation stabilises, inward investment picks up and production slowly recovers.

Russia's agreement to join OPEC's market management efforts, along with the maturing of many of its largest fields, means that oil production there does not grow materially above today's level. Nonetheless, production remains above 11 mb/d until the mid-2020s before declining by around 1.5 mb/d by 2040. Near-term prospects for developing new tight oil resources and projects in the Arctic are constrained by sanctions and high costs.

Production in **Africa** falls in the coming years because of declines in Angola and Nigeria. But after 2025, production grows in Nigeria as new deepwater fields are developed while Angola stems declines. Production in non-OPEC African countries creeps upwards to 2030: there are increases from smaller producers such as Uganda, Kenya and Senegal. Additions from these countries are relatively small at a global level, but the revenue generated can make a significant difference domestically (see Chapter 11).

3.6 Oil product demand and refining

Table 3.5 ▷ World liquids demand by scenario (mb/d)

	2018	2025	2030	2035	2040	Sustainable Development 2030	Sustainable Development 2040
		Stated Policies					
Total liquids	98.8	106.4	108.9	110.1	111.1	93.4	74.6
Biofuels	1.9	2.8	3.5	4.1	4.7	6.3	7.7
Total oil	96.9	103.5	105.4	106.0	106.4	87.1	66.9
CTL, GTL and additives	0.8	1.0	1.2	1.4	1.5	1.1	1.2
Direct use of crude oil	1.0	0.6	0.4	0.2	0.1	0.2	0.1
Oil products	95.1	101.9	103.8	104.4	104.8	85.8	65.6
LPG and ethane	11.9	14.2	14.8	15.3	15.2	13.5	11.6
Naphtha	6.7	7.2	8.0	8.5	9.1	7.6	8.3
Gasoline	24.6	25.2	25.4	24.4	23.4	20.2	11.0
Kerosene	7.6	8.4	9.0	9.8	10.7	6.8	5.8
Diesel	27.8	30.5	30.6	30.5	30.7	24.3	17.8
Fuel oil	6.6	5.7	5.7	5.8	5.8	4.1	3.0
Other products	10.0	10.6	10.4	10.2	9.9	9.3	8.0
Fractionated products from NGLs	10.6	12.1	12.6	12.8	12.8	11.7	9.5
Refinery products	84.5	89.8	91.2	91.6	92.0	74.1	56.1
Refinery market share	*86%*	*84%*	*84%*	*83%*	*83%*	*79%*	*75%*

Notes: CTL = coal-to-liquids; GTL = gas-to-liquids; NGLs = natural gas liquids; LPG = liquefied petroleum gas. See Annex C for definitions.

Petrochemical feedstocks (ethane, LPG and naphtha) and kerosene account for over 90% of the net increase in total oil product demand to 2040. This contrasts with the trend since 2000, when gasoline and diesel provided two-thirds of the growth in total oil products. Gasoline and diesel remain very important in developing economies, however, with their combined demand growing by nearly 40% between 2018 and 2040.

Gasoline demand peaks globally in the late-2020s and is 1.2 mb/d lower than today by 2040. Diesel demand is boosted around 2020 due to the IMO sulfur regulation, which prompts shippers to look for alternatives to heavy fuel oil. But the pace of growth slows later in the *Outlook* period as a result of the rise of alternative fuels and efficiency improvements in trucks. Demand for heavy fuel oil gradually recovers after 2020 as refineries adapt to produce compliant fuels and scrubbers are installed on large vessels.

The amount of new refining capacity coming online in 2019 is set to be the largest since 2010. This suggests greater competition in the refining sector in the future, especially as demand growth slows after 2025, and as biofuels and NGLs make a growing contribution to liquids demand. These non-refinery sources meet some 40% of the incremental liquids demand between 2018 and 2040. As a result, refineries see their market share of liquids demand decline from 86% today to 83% in 2040.

Table 3.6 ▷ Refining capacity and runs by region in the Stated Policies Scenario (mb/d)

	Refining capacity			Refinery runs			Capacity at risk
	2018	2030	2040	2018	2030	2040	2040
North America	22.7	22.4	22.1	19.2	19.6	18.3	1.7
Europe	16.2	15.0	14.5	13.5	10.7	9.6	5.3
Asia Pacific	35.2	41.3	43.0	29.3	33.8	35.9	3.9
Japan and Korea	7.0	6.6	6.0	6.1	5.2	4.3	1.9
China	15.7	19.2	19.2	12.0	14.4	14.8	2.1
India	5.2	6.3	7.7	5.1	6.1	7.5	-
Southeast Asia	5.1	6.8	7.6	4.2	6.1	7.0	-
Middle East	9.3	12.2	12.7	7.9	10.8	11.4	-
Russia	6.6	6.4	6.4	5.7	4.9	4.6	1.5
Africa	3.5	4.4	4.8	2.0	3.5	4.0	0.2
Brazil	2.2	2.5	2.5	1.7	2.2	2.2	-
Other	4.8	4.8	4.8	2.6	3.0	3.2	1.1
World	100.4	109.0	110.7	82.1	88.6	89.1	13.7
Atlantic Basin	55.4	55.0	54.6	44.5	43.5	41.4	9.8
East of Suez	45.0	54.0	56.2	37.7	45.1	47.8	3.9
Sustainable Development	100.4	96.1	77.9	82.1	71.9	54.3	44.2

Notes: Capacity at risk is defined as the difference between refinery capacity and refinery runs, with the latter including a 14% allowance for downtime. Projected shutdowns beyond those publicly announced are also counted as capacity at risk.

The immediate challenge facing the refining industry is to adapt to the major shift in product demand triggered by the IMO regulation which will require a move away from high sulfur fuel oil for shipping. Refiners are responding in diverse ways, which include investing in residue cracking or desulfurisation units, adjusting configurations to increase the yield of diesel at the expense of gasoline, and introducing a new low sulfur fuel by blending gasoil and high sulfur fuel oil (IEA, 2019a).

The long-term challenge is to secure assets that can maintain competitiveness. Many developing Asian countries are actively expanding refining capacity in the light of rising domestic demand. Saudi Arabia is trying to extract more value from its oil by pursuing refining and petrochemical investment opportunities in Asia. Other countries are integrating petrochemical facilities with refining capacities to adapt to changing demand patterns. In the Stated Policies Scenario, some 15 mb/d of new capacity comes online between 2018 and 2040, primarily in developing economies in Asia and the Middle East. This upends the traditional order of the global refining industry. China and the Middle East both overtake Europe in terms of refining activity and the combined share of developing Asia and the Middle East in global refinery runs increases from 37% today to 48% in 2040.

In the Sustainable Development Scenario, refinery runs in 2040 are around 40% lower than in the Stated Policies Scenario. Higher shares of biofuels and lighter products demand mean that the market share of refiners in liquids demand falls from 86% today to 75% in 2040.

3.7 Trade[1]

Table 3.7 ▷ Oil trade by region in the Stated Policies Scenario

Net importer in 2040	Net imports (mb/d)				As a share of demand			
	2000	2018	2030	2040	2000	2018	2030	2040
China	1.7	9.4	12.9	13.3	34%	71%	78%	79%
Other Asia Pacific	2.2	7.1	9.8	10.7	36%	70%	79%	79%
India	1.5	3.7	6.7	8.4	64%	77%	88%	90%
European Union	10.8	10.9	9.6	7.5	74%	83%	88%	89%
Japan and Korea	7.3	6.1	5.2	4.2	98%	96%	97%	97%
Rest of world	-1.6	0.3	0.4	0.7	n.a.	11%	14%	28%

Net exporter in 2040	Net exports (mb/d)				As a share of production			
	2000	2018	2030	2040	2000	2018	2030	2040
Middle East	18.9	23.5	24.2	24.6	80%	74%	71%	68%
North America	-9.6	-0.1	7.6	8.9	n.a.	n.a.	25%	30%
Russia	3.9	8.1	6.9	5.8	59%	69%	64%	61%
Central and South America	2.2	0.5	1.5	2.8	31%	7%	18%	28%
Caspian	0.8	2.3	2.1	2.0	59%	75%	69%	66%
Africa	5.4	4.2	2.2	0.8	68%	50%	28%	10%

Global oil trade becomes increasingly centred on Asia in the Stated Policies Scenario. A combination of rising demand and declining domestic production leads to growing import dependency, rising from 76% today to 83% in 2040. China soon overtakes the European Union as the world's largest oil importer and holds that position to 2040, despite the flattening of its oil demand in the 2030s. India's oil import requirements more than double between 2018 and 2040 and its level of import dependency reaches 90%, one of the world's highest. The growing concentration of trade flows to Asia increases the amount of oil passing through major global chokepoints, with implications for global oil security (see section 3.11).

The United States becomes a net oil exporter soon after 2020 and North America becomes the world's second-largest oil exporter by 2030 (overtaking Russia). Gross oil exports (crude oil and refined products) from the United States are five-times higher in 2030 than in 2010, but because its refineries are geared towards heavy crude oil while its production consists mostly of light crude oil, US gross oil imports in 2030 are only one-third lower. More oil will be flowing into and out of US ports than before the shale boom and the United States therefore becomes more connected with global markets.

Net oil exports from Russia decline through to 2040 due to Russia's subdued production outlook. While Africa and Central and South America remain net crude oil exporters, they increasingly rely on imports for refined products.

---

[1] Unless otherwise stated, trade figures in this chapter reflect volumes traded between regions modelled in the *WEO* and therefore do not include intra-regional trade.

3.8 Investment

Table 3.8 ▷ Cumulative oil and natural gas supply investment by region in the Stated Policies Scenario, 2019-2040 ($2018 billion)

	Total oil and gas	Upstream oil and gas	Transportation		Refining oil	Annual average upstream oil and gas
			Oil	Gas		
North America	5 492	4 547	141	665	139	207
Central and South America	1 817	1 558	115	103	40	71
Europe	1 618	1 201	18	320	80	55
Africa	1 911	1 623	68	167	54	74
Middle East	2 711	2 098	183	291	140	95
Eurasia	2 496	2 089	36	329	42	95
Asia Pacific	3 284	2 052	80	747	405	93
Shipping	402	n.a.	286	116	n.a.	n.a.
World	19 730	15 167	927	2 737	899	689
Sustainable Development	13 227	10 085	332	2 272	538	458

Total oil and gas upstream investment in 2018 was $475 billion. This is 42% below the peak levels seen in 2014. The difference is less stark when adjusted for the decline in upstream costs over this period (based on the IEA Upstream Investment Cost Indices [IEA, 2019a]): on this basis, cost-adjusted investment in 2018 was 16% lower than the peak level in 2014.

In the *World Energy Outlook-2018*, we warned of a possible shortfall in supply in the mid-2020s to meet rising demand if US tight oil did not maintain its record-breaking pace of growth or if there was not a pickup in conventional crude oil approvals. Tight oil production has been revised upwards in this *Outlook* and there are some signs of an increase in project approvals in 2019. We continue to monitor carefully the long-term adequacy of oil supply.

In the Stated Policies Scenario, annual upstream oil and gas spending averages $650 billion between 2019 and 2030 and $730 billion thereafter. Between 2019 and 2030, around $380 billion (60% of total upstream investment) is spent on average each year developing new capacity; the remainder is spent at already producing fields.[2] On average $90 billion is spent globally on tight oil between 2019 and 2040.

In the Sustainable Development Scenario, fewer new developments are required, but continued investment in both new and existing oil fields is an essential element of energy transitions, even as overall production declines in line with climate goals (see Chapter 2). Around $510 billion is spent on average each year on upstream oil and gas between 2019 and 2030, while $390 billion is spent between 2030 and 2040.

---

[2] This *World Energy Outlook* includes a new method for estimating investment levels based on capital spending over time rather than "overnight" capital expenditure. This provides further insight on spending differences over time and between types of field. The cumulative outcome remains the same as in previous editions.

Key themes

3.9 Passenger cars: are we approaching the peak of the "ICE age"?

There are over 1.1 billion passenger cars on the road today, nearly 50% more than the number just a decade ago, and cars currently account for just under one-quarter of global oil demand.[3] The level of private car ownership in developing economies is still far below that of most advanced economies, and so continued growth in the global car fleet cannot be ruled out.

The latest data show that the number of electric cars has grown rapidly in recent years. There were over 5 million electric cars on the road in 2018, an increase of nearly 65% from 2017. Sales continued to rise during the first-half of 2019, although the rate of increase has declined from 65% to around 45%, mainly because of an overall shrinking car market and because of reforms that have reduced subsidies for electric cars in China.

Many countries aim to phase out conventional vehicles in the coming decades, and various car manufacturers have announced plans to roll out new electric car models. Electric car sales therefore look set to continue to grow. This growth is likely to spread to other vehicle types (trucks, buses and two/three-wheelers), and thus to encompass road transport as a whole, which accounts for nearly half of today's global oil demand.

While the number of electric car sales is still relatively small, there are signs of a slowdown in sales of conventional cars, and this has given rise to the idea that a peak in sales might come sooner than expected. The year 2018 was particularly challenging for car manufacturers: in several key markets, including China, United States and Europe, the total number of cars sold either stagnated or fell for the first time since the 2008 recession. This trend has continued so far into 2019: passenger car sales have dropped by 14% in China, 10% in India, 3% in Europe and 2% in the United States.

What does all of this mean for oil demand from passenger cars? The answer is not straightforward. Many factors influence the decisions people make about whether to buy a car and what type of car to purchase. Consumers do not make their choices based only on economic factors, and trends visible today may or may not persist in the coming decades. For example, the increasing prospects for ride-sharing, vehicle digitalisation and automation as well as improving public transport systems could spell a wholesale shift in how many, and how much, cars will be used in the future. Range anxiety and slow progress in building up recharging infrastructure could slow the pace at which electric mobility grows. There could be a rebound in conventional car sales as the world's population grows and becomes richer. Finally, there could be a continuation of the recent steady rise in consumers choosing to buy larger and heavier cars. There has been a marked rise in sales of sports utility vehicles, which on average consume around 25% more fuel than medium-size

---

[3] Passenger cars here include passenger vans. Electric cars include both plug-in hybrid electric and battery electric cars. Conventional cars are any car that primarily relies on an internal combustion engine (ICE) including hybrid vehicles.

cars and can be challenging to electrify fully because of power and battery size requirements (IEA, 2019b). This trend could bolster future oil demand from passenger cars.

The following section looks at these countervailing forces and assesses their effect on the outlook for passenger cars and the evolution of future oil demand. We start by examining recent developments in electric car and SUV sales, and what they might mean for the future. We examine how these trends, along with announced transport sector policies and fuel economy targets, could impact global oil demand in the Stated Policies Scenario, focusing in particular on whether the rise of SUVs could offset gains from the electrification of the car fleet.

Rise of electric vehicles

There are many reasons to believe that electric vehicle (EVs) sales will continue to increase. The largest vehicle manufacturers have announced ambitious deployment targets (Box 3.1). Many governments have also announced policies that support the deployment of electric cars, including by: increasing the stringency of fuel economy or emissions targets, introducing restrictions or penalties on the use of conventional cars (such as low emission zones in urban environments), committing to phase out conventional cars sales (as discussed in the *World Energy Outlook-2018*), and investing in new EV recharging infrastructure. Plus battery costs continue to fall. Today, the battery in electric cars costs less than $180 per kilowatt-hour (kWh), down from around $650/kWh five years ago. In the Stated Policies Scenario, this falls to less than $100/kWh in the mid-2020s, by which time electric cars in several key markets are cost competitive with conventional cars on a total cost of ownership basis.[4]

Box 3.1 ▷ Announced plans of vehicle manufacturers for electric cars

The largest manufacturer of electric cars globally today is Tesla, a dedicated electric car company. There are also a number of other EV-only manufacturers in China and elsewhere looking to expand operations. In China, many car manufacturers were quick to embrace electric models and some already focus exclusively on producing electric cars. In 2018, China accounted for 55% of global electric car sales. Car manufacturers in China aim to produce around 12 million electric cars within the next decade.

However, for the largest auto companies worldwide, EVs make up only a small fraction of their overall sales. How quickly this share grows is a crucial variable for the future. Over the past two years, these manufacturers have been turning their attention to electric cars. The number of models and the share of electric car sales are expected to rise significantly (Table 3.9).

The plans of the 20 largest car manufacturers, which account for 75% of current global car sales, suggest a tenfold increase in electric car sales from 2 million today to more

---

[4] Regional variations occur depending on fuel taxation levels, consumer preferences on car size and driving habits i.e. average daily trip distances (IEA, 2018a).

than 20 million in 2030. This would increase the sales of electric cars as a share of total car sales from around 2% to 15% over the same period. The automotive industry also plans to spend $300 billion on electro-mobility research, development and demonstration over the next decade, and by 2025 there will be over 350 electric car models available globally. However there is a large degree of uncertainty about what will actually happen. While the pace of change could accelerate, few of the car manufacturers have provided concrete plans on how their targets will be achieved in practice.

Table 3.9 ▷ Electric car targets of the world's 20 largest car manufacturers

Manufacturing alliance or brand	Total car sales in 2018 (million)	Electric car share of sales in 2018 (%)	Automaker plans for electric cars for given year		
			Sales (million)	Share of company sales	Number of models
VW group	9.4	0.7%	0.4 (in 2020) 22* (by 2030)	25% (in 2025)	80 (by 2025)
Volkswagen	6.5	0.8%			
Audi	1.8	0.9%	0.8 (in 2025)		
Skoda	1.2	0.0%		25% (by 2025)	10 (by 2022)
Toyota	8.0	0.6%	1 (by 2030)		more than 10 (by early 2020s)
Renault-Nissan	6.8	2.2%			12 (by 2022)
Nissan	4.7	2.1%	1 (by 2022)		
Renault	2.2	2.4%		20% (in 2022)	
Hyundai-Kia	6.8	1.2%			12 (by 2020)
Hyundai	4.2	1.1%			
Kia	2.6	1.5%			
Ford	5.2	0.2%			40 (by 2022)
Honda	5.1	0.4%		15% (in 2030)	
Chevrolet (GM)	3.7	1.3%			20 (by 2023)
Suzuki	3.1	0.1%	1.5 (in 2030)		1 (in 2020)
FCA group	2.8	0.1%			28 (by 2022)
Jeep	1.4	0.0%			14 (by 2022)
Fiat	1.4	0.2%			2 (by 2022)
Mercedes (Daimler)	2.5	1.5%		15-25% (by 2025)	50 (by 2022)
SAIC	2.4	4.1%	0.6 (by 2020)		
BMW	2.0	6.4%		15-25% (in 2025)	25 (by 2025)
Geely	1.5	4.7%	1 (in 2020)	32% (in 2020)	30 (by 2020)
Peugeot (PSA group)	1.5	0.3%	0.9 (in 2022)		40 (by 2025)
Mazda	1.4	0.0%		5% (by 2030)	1 (in 2020)

* Target refers to cumulative sales.

Notes: GM = General Motors; FCA = Fiat Chrysler Automobiles; PSA = Peugeot Société Anonyme; SAIC = Shanghai Automotive Industry Corporation. Figures show only the top-20 car brands globally; targets are given for brands within manufacturing alliances where specified.

Sources: IEA analysis based on IHS Markit (2018); EV Volumes (2019); China Association of Automobile Manufacturers (2019); IEA (2019b); car manufacturer announcements.

Policy support and declining costs for electric cars does not necessarily mean the end of conventional car sales. A consumer may still prefer to purchase a conventional car despite higher lifetime expenses, perhaps because of the lower upfront cost or anxiety concerning the driving range of an electric car. Further, car sales have historically been driven by population and economic growth, and there are major differences in the number of cars per household between various markets, depending on the availability of public transport, geography and cultural preferences. A peak in conventional car sales in one market does not necessarily mean a peak in all markets, especially given the shift in global car markets towards Asia.

Shared mobility services could also impact car ownership and mobility patterns, particularly in cities. Services such as ride-sourcing (e.g. Uber) and micro-mobility (e.g. electric scooters) have emerged and spread rapidly in just a few years. There are already more than 1 billion active users of ride-sourcing services worldwide, taking over 50 million rides per day. Survey data from major cities in the United States indicate that around 40% of ride-sourcing trips are displacing private cars and taxis, and around 50% of trips are displacing public transit, walking or cycling (Schaller, 2018). Preliminary findings for micro-mobility suggest that around 30% of e-scooter rides today replace trips in cars and 20% of trips connect to public transit (Bird, 2019; Lime, 2019). However, usage patterns may change over time and it is still too early to discern the long-term impact of these trends on car ownership.

Rise of SUVs

While an increasing number of consumers are choosing to buy electric cars, there is also a growing appetite for bigger and heavier cars. This is not simply a phenomenon confined to North America or advanced economies: the share of SUVs has also increased very significantly in both India and China in recent years (Box 3.2). In 2018, the number of SUV sales reached a record high, accounting for around 40% of global car sales, more than double the share they had ten years ago (Figure 3.7). The boom in SUVs came mainly at the expense of small cars, whose share of the market fell by ten percentage points over the same period (IEA, 2019c).

Box 3.2 ▷   Why buy an SUV?

Today nearly 50% of the cars sold in the United States and one-third of the cars sold in Europe are SUVs (Figure 3.8). There has been a surge in the number of SUV models available to attract new customers, particularly in the past three years. This preference for SUVs has increased both in countries with low gasoline prices and in regions where more and more consumers can afford the higher upfront cost. Cost-effectiveness does not seem to be what dictates a consumer's choice when purchasing such vehicles: SUVs are more expensive and tend to have higher running costs than smaller cars. The available evidence suggests that geographical, social and behavioural factors are all relevant. For example, SUVs are more likely to be purchased by male and younger consumers, and by households that lease rather than purchase a new vehicle, and these consumers often attach a high value to attributes such as safety and engine power (Kitamura et al., 1999). Another factor is that SUVs tend to provide higher profit margins for manufacturers, and nearly all car manufacturers have increased advertising for SUVs.

Figure 3.7 ▷   Share of SUV sales in key car markets

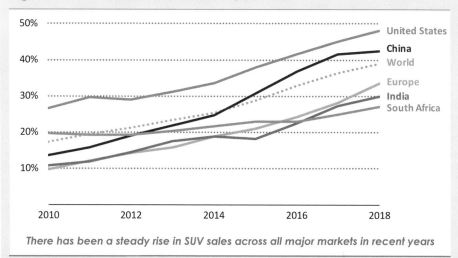

There has been a steady rise in SUV sales across all major markets in recent years

Source: IEA analysis based on IHS Markit (2018).

In recent years, there has been a noticeable upturn in SUV sales in developing economies. In China, SUVs are considered symbols of wealth and status. In India, sales are currently lower, but consumer preferences are changing as more and more people can afford SUVs. Similarly in Africa, the rapid pace of urbanisation and economic development means that demand for premium and luxury vehicles is relatively strong; the poor condition of roads in many African countries also increases the attractiveness of SUVs (Automotive World, 2018).

A SUV today consumes around 25% more oil than a medium-size car per kilometre travelled. The rise of SUV sales therefore has important implications for oil use in the future. Efficiency standards to moderate future oil demand cover around 85% of global car sales, but they often differentiate the required level of improvement by car size. Even if the required rate of improvement were the same for different car segments, a shift to SUVs would increase oil use. In some regions such as the European Union, policy makers have committed to monitor trends closely in an effort to achieve targets for improving the average fuel economy of new car sales, but major uncertainties remain about the overall development of the market and the implications for oil use.

Figure 3.8 ▷ Historical global trends in car sales by size

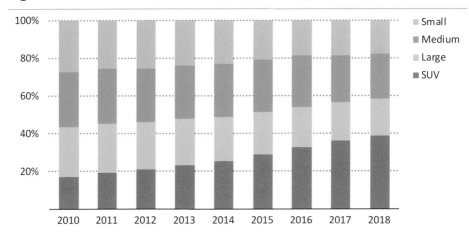

The share of SUVs in total sales globally has increased significantly in recent years

Note: Cars sizes based on Euro NCAP segment classification: small = A & B; medium = C; large = D-F & S; SUV = J & M.
Source: IEA analysis based on IHS Markit (2018).

Outlook for car sales and oil demand

Oil consumption in passenger cars grew by 3.3 mb/d between 2010 and 2018. Most of this growth was caused by an increase in the size of the car fleet, but the rise in the share of SUVs on the road caused around 15% (or 0.5 mb/d) of the overall increase. This effect is five-times larger than the oil displaced by electric cars over the same period (less than 0.1 mb/d). Besides electrification, fuel efficiency standards avoided a further 2 mb/d of additional oil demand increase over this period.

Total car sales grow steadily to 2040 in the Stated Policies Scenario, but with differences by region (Figure 3.9). Car sales remain on a plateau in advanced economies at around current levels (just under 50 million a year), while sales in developing economies double from 40 million today to 80 million in 2040. Even with this increase, car ownership in 2040 in developing countries (160 cars per 1 000 people) remains well below that of advanced economies (540 cars per 1 000 people).

In the Stated Policies Scenario, conventional global car sales rebound in the early 2020s from the declines occurring in 2018 and 2019, and peak by the late-2020s. Annual electric car sales grow from 2 million today to around 20 million in 2030, and over 30 million in 2040. This is broadly in line with the existing plans of car manufacturers. Electric cars account for more than three-quarters of the increase in sales between 2018 and 2040. In 2040, there are 1 700 million conventional cars on the road compared with 330 million electric cars. The increase in electric car sales in combination with fuel economy improvements leads to a peak in oil demand from passenger cars in the late-2020s.

By 2040, the 330 million electric cars on the road displace around 4 mb/d oil use, while more stringent fuel economy standards displace nearly 9 mb/d. In total, oil use in cars falls slightly from 22 mb/d in 2018 to 21 mb/d in 2040.

Figure 3.9 ▷ Passenger car sales in the Stated Policies Scenario

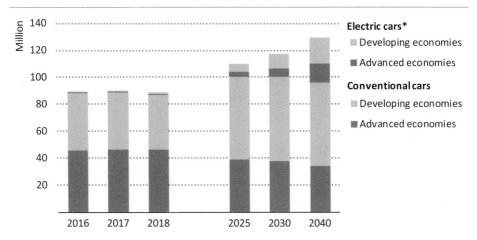

Conventional car sales peak by the late-2020s and electric cars account for over three-quarters of the growth in sales between 2018 and 2040

* Includes battery electric, plug-in hybrids and fuel cell cars.

There is a steady increase in the market share of SUVs in the Stated Policies Scenario from around 40% today to just over 50% in 2040. This increase is slower than the rate of increase recorded in recent years as the appetite of consumers for heavier cars is assumed to saturate in key markets over the course of the next decade and fuel economy standards put pressure on the heavier car segments.

The rise of SUVs in the Stated Policies Scenario is accompanied by a partial electrification of this segment, mitigating some of the upward pressure on oil demand from SUVs. Plug-in hybrids are the main electric option for SUVs in most markets today, reflecting the fact that they can be difficult to electrify fully. Powering a heavier car for long ranges needs a larger battery, and the additional technical and financial costs involved mean that non-SUVs currently have around double the electrification share of SUVs (IEA, 2019c). Nonetheless, car manufacturers in many countries are now looking into fully electric SUV models (IEA, 2019b). In 2018, China reported sales of 1.1 million electric cars, of which 280 000 were SUVs (China Passenger Car Association, 2019). With the roll out of more electrified SUV models, we project that around 20% of SUVs will be electrified globally in 2040, narrowing the electrification gap between the two segments and contributing nearly 40% of the growth in total electric car sales.

There are two key uncertainties in this *Outlook* with implications for the future trajectory of passenger car oil demand. The first is that a stronger-than-expected continuation of consumer preferences for SUVs could increase oil demand. The second is that a faster electrification of SUVs could have the opposite effect.

On the one hand, the share of SUVs in total car sales could increase at a similar pace to that seen in recent years (the "Rise of SUVs" case in Figure 3.10). In this case, SUVs would reach close to 80% of total car sales by 2040, and the average new conventional car sold in 2040 would consume around 10% more fuel than the level projected in the Stated Policies Scenario. This would increase total oil consumption by nearly 2 mb/d in 2040, offsetting the oil displacement impact of close to 150 million electric cars. The increase in global oil demand between 2018 and 2040 would therefore be 20% higher than the 9.5 mb/d increase in the Stated Policies Scenario.

Figure 3.10 ▷ Share of SUVs in total sales and oil demand in the Stated Policies Scenario and two cases examining differences in the SUV market

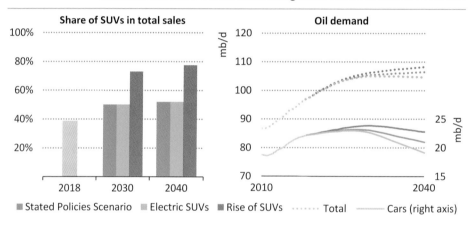

A continuation of historic trends in SUV sales would add nearly 2 mb/d to oil demand in 2040, but further electrification of SUVs would curb the oil demand increase beyond 2030

Note: The Electric SUVs case assumes 20% of the SUV fleet is electrified in 2040 compared to 10% in the Stated Policies Scenario; the Rise of SUVs case assumes SUVs sales reach close to 80% of total sales.

On the other hand, the SUV fleet could be electrified more quickly than projected in the Stated Policies Scenario (the "Electric SUVs" case in Figure 3.10). If SUVs were to follow a similar electrification trajectory to that of small and medium cars, this would add over 100 million electric cars by 2040, helping the total number of electric cars to reach around 450 million. The result would be a reduction in oil demand by nearly 2 mb/d over the same period of time, relative to the Stated Policies Scenario.

More transformational changes to passenger car transport are possible. In the Sustainable Development Scenario, a much stronger push for electrification leads to a peak in conventional car sales in the mid-2020s, followed by a much steeper decline than in the

Stated Policies Scenario. As a result, over 80 million electric cars are sold each year by 2040, accounting for three-quarters of total car sales, and conventional car sales are only 30% of today's level. By 2040 there are nearly 900 million electric cars on the road. In this scenario, all small, medium and large cars sold are electric by 2040, as are 40% of all SUVs. The average conventional car sold in 2040 consumes nearly 45% less fuel than an average car sold today (compared with a 25% reduction in the Stated Policies Scenario). The combination of the rapid rise in electric cars, downsizing of the car fleet, and efficiency improvements in conventional cars mean that oil demand in passenger cars falls from 22 mb/d today to 8 mb/d in 2040.

One conclusion to draw from this is that, while discussions surrounding the perspectives for oil demand for cars often focus on fleet electrification and fuel economy standards, the segmentation of the car fleet is an equally important consideration. Unless there is a major change in consumer preferences, the recent boom in SUV sales could be a major obstacle towards developing cleaner car fleets and reducing overall oil demand.

3.10 Pushing the boundaries of US tight oil

The rise of tight oil in the United States has been remarkable. From less than 0.6 mb/d production in 2010, tight crude oil production rose to nearly 6 mb/d in 2018 – an increase that already matches the fastest rise seen previously in oil markets (Saudi Arabia in the late 1960s). This was initially spurred by publicly funded research and development efforts, and then by the ingenuity of hundreds of small, independent companies, fuelled by ample external sources of capital. However the tight oil race is only half run. If tight oil alone is to satisfy the 8.5 mb/d increase in global oil demand to 2030 in the Stated Policies Scenario, the pace of growth to date will need to be maintained for a prolonged period. This possibility should not be ruled out: resource estimates in many of the major areas of activity continue to be revised upwards, while the larger international oil companies, initially hesitant to devote large levels of capital to shale production, have significantly increased their exposure to tight oil.

But there are also some signs of fatigue: the steep decline rates of individual tight oil wells mean that, despite continued drilling, some of the largest shale plays may have already reached their production peak. There are a number of distribution infrastructure bottlenecks in key regions. Concerns over water use, local disruption and flaring of associated gas could also impede production activities. In addition, with less than 35% of US shale companies currently achieving positive cash flows, investors are demanding more capital discipline, and there has been a high degree of consolidation within the sector, with many companies acquired by larger operators (Lukash, 2019).

This section examines the dynamics of tight oil production since the oil price crash in 2014. It describes the outlook for tight oil in the Stated Policies and Sustainable Development scenarios, assesses the impact of key sensitivities on these outlooks, and discusses the implications for global oil markets.

Box 3.3 ▷   Some key features of tight oil

Tight oil is generally considered an "unconventional" source of production. Tight oil rocks have low permeability (fluids cannot easily flow) and low porosity (there is limited void space within the rock). They need to be hydraulically fractured using a mixture of water, proppant (generally sand) and chemicals to extract the oil. The vast majority of tight oil wells are horizontal. Lateral lengths of a single well can be as long as 4.5 kilometres (km), but they average around 3 km. The total lateral length of all tight oil wells drilled in 2017 and 2018, if set end-to-end, would circle the Earth's circumference.

Production from a tight oil well tends to decline by between 60% and 70% within 12 months of first production (compared with a decline of around 6% for a conventional well). This means that high levels of drilling are required to maintain production levels: nearly 12 000 tight oil wells were drilled in 2018, 40% of which were needed just to keep production constant. Over time the decline rate of tight oil wells slows, and they often have a long tail of low-level production. The large number of wells drilled to date – over 80 000 horizontal tight oil wells have been drilled in the United States since 2000 – means that these "tails" can provide an important long-term baseload of production. Tight oil wells mainly produce a light, sweet crude oil, together with some condensate and associated gas. In addition to the 6 mb/d tight crude oil production in 2018 in the United States, there was 0.5 mb/d tight condensate and 3.1 mb/d NGLs produced from shale gas wells.

A key characteristic used to assess the productivity of a tight oil well is its estimated ultimate recovery (EUR). The EUR is an estimate of the total volume of oil and gas that will be extracted over the lifetime of the well (generally taken to be 30 years). It is a function of the specific geology of the "play",[5] the length of the laterals that have been drilled and the completion technique.

There are at least 30 different tight oil plays in the United States, but the top-five plays account for nearly 90% of tight crude oil and condensate production today and are the focus of this analysis. Recently, the largest share of tight oil drilling has taken place in the Permian Basin in Texas and New Mexico. There are many geological layers containing tight oil prospects within the Permian Basin; it is common to differentiate resources by geography into the Permian-Midland and Permian-Delaware. The other most important plays are the Bakken (in North Dakota), the Eagle Ford (in Texas) and the Niobrara (mainly in Colorado and Wyoming).

Estimates of the remaining recoverable tight oil resources in the United States have been revised upwards on a number of occasions. For example, the United States

---

[5] A play is strictly one or more geologically related prospects that must be evaluated by drilling to determine whether it contains commercial quantities of oil or gas (AAPG, 2000). In this *World Energy Outlook*, we model 23 individual tight oil plays each separated into four areas according to productivity.

Geological Survey (USGS) recently estimated that the Permian-Delaware contains around 50 billion barrels of technically recoverable tight oil (USGS, 2018), an area whose technical potential had not previously been assessed. For this edition of the *WEO*, the estimate for remaining US tight crude oil and condensate is 155 billion barrels, a major increase from the 115 billion barrels estimate in the *WEO-2018* (IEA, 2018a).[6]

Evolution in tight oil production costs

Prior to 2015, the cost of drilling, hydraulically fracturing and connecting a tight oil well to distribution infrastructure, and paying taxes and royalties meant that operators required, on average, an oil price over $80/barrel to generate a reasonable rate of return. The crash in the oil price in 2014 from over $100/barrel to less than $50/barrel in the space of nine months presented a major challenge for tight oil operators. The response took a number of forms (Figure 3.11):

- Operators focussed activity on the most productive areas they owned to generate larger and faster returns on investment.[7] Across the five main plays, 40% of wells drilled in 2012 were in these core areas; by 2018 this had increased to 67%.
- Operators increased the volumes of oil and gas produced from a given well. The horizontal length of the average tight oil well drilled today is 50% greater than the average well drilled in 2012, and it uses three-times more water and proppant. There have also been continuous improvements in completion strategies such as the placement and mechanism for producing fractures.
- Operators significantly reduced the unit costs of drilling and completing tight oil wells, for example by drilling multiple wells from a single location, and by drilling multiple wells so as to produce oil simultaneously from different shale layers. They were helped by reductions in the cost of oil field services. The cost of hiring a drilling rig, for example, fell by almost 50% between 2012 and 2016.

The aggregate effect of these changes is that the wellhead breakeven price for tight oil production is now around 60% lower than it was in 2012.

Can these impressive rates of improvement be maintained in the future? There are certainly further improvements that could be made (Box 3.4). However, there are also signs that rates of improvement are slowing and that costs cannot be expected to continue to fall in the future in the same way as the past. The costs of drilling and completion services have risen in recent years, although they remain below 2012 levels, and we expect future

---

[6] In previous editions of the *World Energy Outlook*, we directly used the resource estimates from the US Energy Administration Information (EIA), but this year we shift to a composite approach, taking the most recent estimate coming from either the USGS or EIA for each play.

[7] These are often referred to as the play "sweet spots". These are defined here to be an area where well breakeven prices are less than the play median. We assume that this area remains fixed geographically.

Figure 3.11 ▷ Changes in the average efficiency and economics of US tight oil production, 2012-2018

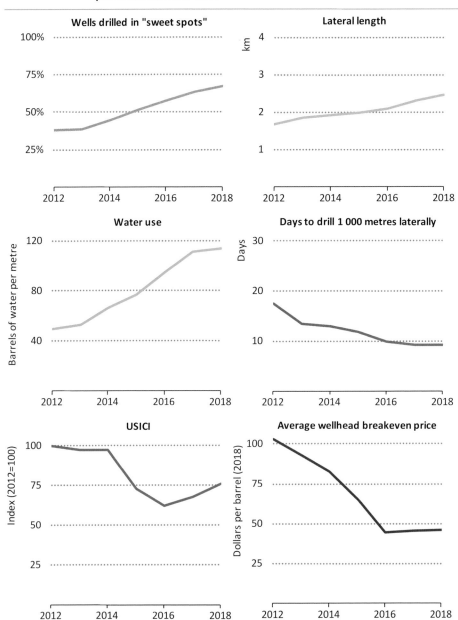

There were major technological improvements and changes in operating procedures following the oil price crash, but the rate of improvement appears to have slowed recently

Note: USICI = IEA Upstream Shale Investment Cost Index.
Source: IEA analysis based on Rystad Energy (2019).

costs broadly to follow changes in the oil price, as is the case for conventional resources. It is becoming increasingly difficult to continue to concentrate drilling only in sweet spots: although the spacing between adjacent wells could be reduced, this runs the risk of having a negative impact on production. There is also evidence of a slowdown in the rate of reduction in the number of days to drill a well, and in the rate of increase in volumes of water and proppant that are used to complete a tight oil well.

Box 3.4 ▷ Possible future technology breakthroughs for tight oil operations

Currently only around 5-15% of the oil in the ground can be extracted during tight oil operations, whereas conventional resources tend to produce between 15% and 40% after primary and secondary recovery. For conventional resources this percentage can be increased by applying tertiary recovery methods such as enhanced oil recovery (IEA, 2018b). While the use of these methods for tight oil wells is still in its infancy, more than 150 tight oil wells used enhanced oil recovery in 2018 (either injecting CO_2 or natural gas); this was found to approximately double the recovery factor (Jacobs, 2019).

Another option to increase productivity and extend the life of existing wells is through "re-fracking". This applies newer and more effective fracking technologies to older wells that are typically one-third to one-half of the way through their expected decline. While there is significant potential for re-fracking, the major hurdle is cost. Re-fracking an old well costs around 25-35% of the cost of drilling and completing a new well. Although re-fracking increases the amount of oil recovered by around 60% (Oruganti et al., 2015), a large share of this additional oil is often not recovered until many years later, meaning that payback periods can be long. Wells therefore need to be assessed carefully prior to re-fracking. To date more than 800 wells have been re-fracked in the United States (Rystad Energy, 2019).

In addition, wider adoption of digitalisation could lead to cost reductions. To take one example, the use of AI-enabled horizontal drilling equipment that can interpret data autonomously in real time could boost productivity and reduce drill wear-and-tear. It could also allow operators to spend less time on data interpretation and to oversee a higher number of wells than would otherwise be possible.

Outlook for US tight oil and implications for other producers

In the Stated Policies Scenario, US tight crude oil production rises from close to 6 mb/d today to a maximum of 11 mb/d in 2035 before declining to 10 mb/d in 2040 (Figure 3.12). The majority of the growth in this period comes from the Permian-Delaware and Permian-Midland: soon after 2030, crude oil production from the Permian Basin surpasses crude oil production from the continent of Africa. Total tight liquids production, which also includes tight condensate and tight NGLs, grows to a maximum of 16.5 mb/d in 2035, a level that is 30% higher than that of the next largest oil producer globally.

Figure 3.12 ▷ US tight crude oil production in the Stated Policies Scenario

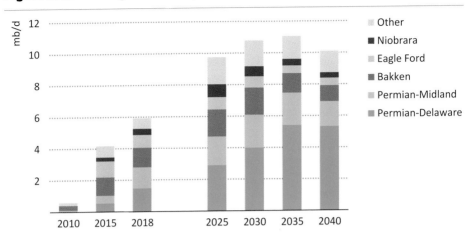

US tight crude oil rises to a maximum of 11 mb/d in 2035. Production from the Permian Basin alone surpasses crude oil production from the continent of Africa in the early 2030s.

An increasing number of wells need to be drilled to realise this rate and level of growth, but we assume that operators exercise a degree of capital discipline that ensures financially sustainable returns. By 2030, over 16 000 tight oil wells are drilled annually, around 40% more than were drilled in 2018. As tight oil production grows, an increasing share of activity is required simply to maintain production (Figure 3.13). By the mid-2020s, if drilling were to stop, tight crude oil production would fall by over 2.5 mb/d within 12 months. This rate of decline drops slightly after the 2030s as the long tail of production from older wells provides an increasing bedrock of production and as the overall level of production begins to decline.

Our projection for US tight crude oil in this *Outlook* in 2025 is 0.6 mb/d higher than in the *WEO-2018*, in which production reached around 9.2 mb/d in 2025. Subsequent years show a larger difference. The 35% increase in the resource base from the *WEO-2018* means that, even though the oil price in the Stated Policies Scenario is lower to 2040, production reaches a plateau later and only declines slowly thereafter. In 2040, tight crude oil production is 3 mb/d higher than in the 2018 projection.

This trajectory poses a stark challenge for many of the world's producer economies, especially those that rely heavily on oil and gas revenues (IEA, 2018b). In the Stated Policies Scenario, members of OPEC are assumed to adjust production in an attempt to manage markets. What this means in practice is that the aggregate production of OPEC countries only exceed its 2018 level in 2030, by which time demand growth has slowed markedly compared with today. An alternative approach for some resource-rich countries could be to prioritise their share of the oil market and therefore boost production over this period. This would lead to a major drop in the oil price and would probably curtail the levels of other sources of production, including tight oil. Members of OPEC chose to do this in 2014: while this temporarily stemmed the rise in tight oil, the drop in the oil price also led to 40-70% fall in net income in these countries. Whether OPEC countries would be able to maintain this strategy for a prolonged period hinges, in many cases, on the extent of their structural economic reliance on hydrocarbon revenues and on the steps taken to address this.

Figure 3.13 ▷ Year-on-year growth from new wells and underlying declines from existing wells in the Stated Policies Scenario

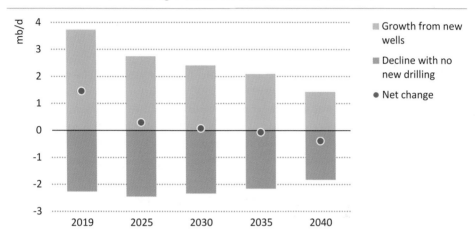

By 2030, over 16 000 tight oil wells are drilled each year, the vast majority of these are needed just to offset underlying declines in production

Note: Declines are the 12-month drop in production within that year if drilling stops on 1 January; growth is the 12-month increase from wells completed within that year.

It is possible that new policy measures and accelerated technology developments in the future will reduce oil use beyond what is projected in the Stated Policies Scenario; the Sustainable Development Scenario provides an illustration of such a case. US tight crude oil production grows at a slower pace in this scenario but it still exceeds 9 mb/d in 2030. Faced with uncertainty about the future, tight oil could be a logical choice for many looking to produce oil, as its high decline rates reduce the need for a long-term outlook on demand and prices.

Whether or not there are such new policy measures, the extended growth in US tight oil to the 2030s is set to squeeze the oil income for some of today's major conventional producers and exporters. Alongside the pressing need in many cases to create employment opportunities for a large and youthful population, this reinforces the case for strengthening and accelerating their programmes of diversification and economic reform.

Technical and economic uncertainties in US tight oil production

There are a number of technical and economic uncertainties that could affect the long-term outlook for US tight oil. Here we quantify the impact on tight crude oil production in 2030 of four key issues by running individual sensitivity cases that examine a "low" and "high" variant of each:

- **Technology improvement:** There is uncertainty over the permanency of the cost reductions seen to date and the extent to which future technological innovation can continue to reduce costs, while some more advanced drilling and completion strategies have been demonstrated but not yet adopted by all operators. In the Stated Policies Scenario, we include a variety of technology improvements that reduce costs or increase recovery.[8] For our sensitivity cases we assume either no future technology learning or a doubling in these rates of improvement.

- **Resources:** Remaining technically recoverable tight oil resources have increased substantially in recent years. The *WEO-2011* (IEA, 2011) assumed 24 billion barrels of recoverable tight crude oil and tight condensate resources; we now assume 155 billion barrels. For our sensitivity cases, we assume either the lowest current publicly available estimate for each shale play (totalling 100 billion barrels) or the highest estimate, with an allowance for increased deployment of new and emergent technologies (totalling 200 billion barrels).

- **Oil price:** The oil price affects the number of rigs operating, production costs (via inflation in the oil field services), and the number of wells that can be drilled economically. For the sensitivity cases, we examine the impact of trajectories for the oil price reaching $65/barrel and $110/barrel in 2030 (compared with $88/barrel in the Stated Policies Scenario).

- **Cost of capital:** Low interest rates and the ability of tight oil operators to access low-cost capital has undoubtedly helped underpin tight oil production growth to date. There is no guarantee that this will continue indefinitely. A related factor is that the major international oil companies also plan to increase investment levels in tight oil areas. These companies have very different costs of capital than the smaller, independent companies that have been responsible for the majority of spending to date. In the Stated Policies Scenario, we assume that tight oil operations must achieve a 12% rate of return to be economic; for our sensitivity cases we examine the impact of changing this to 8% and 16%.

The rate of future technology availability, resource availability and the oil price all have a major impact on the level of US tight crude oil production in 2030 (Figure 3.14). If the rate of technology learning is double the level assumed in the Stated Policies Scenario, then tight crude oil production in 2030 would be more than 2 mb/d higher (at around 13 mb/d).

---

[8] These factors include: a 20% reduction in the cost of drilling and completing a well for every doubling in cumulative production from 2018 across each play. A 1% increase per year in the EUR of a given well and a 1% increase per year in the number of wells that can be drilled by a rig. These factors broadly match the rates of learning observed historically and are independent of any changes in drilling location or the oil price.

However, the sensitivities examined here are slightly weighted to the downside. If remaining technically recoverable resources are 100 billion barrels, then production would peak much sooner and by 2030 would be nearly 3 mb/d lower than in the Stated Policies Scenario. Changes in the cost of capital have a smaller, but nevertheless important, impact on production.

Figure 3.14 ▷ **Sensitivity of US tight crude oil production in 2030 to technical and economic uncertainties**

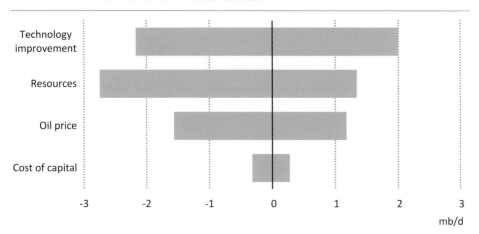

There are major uncertainties in the outlook for US tight oil. Future technology development and resource availability could each affect production by more than 4 mb/d in 2030.

Note: Changes in production are relative to the 2030 level in the Stated Policies Scenario (10.8 mb/d).

Policy uncertainties, social and environmental issues

There are also potential policy uncertainties affecting tight oil, especially in relation to social and environmental issues. This is a fast-moving area and there have been significant changes in both regulation and industry performance in recent years. Future trends in policy and regulation naturally are difficult to assess but, if social and environmental concerns are not adequately addressed, they could have a significant impact on long-term production prospects.

- **Regulatory oversight**: In 2018, over 80% of US tight oil production came from privately owned lands and just over 10% from federal lands. Hydraulic fracturing is currently exempt from many federal environmental laws, and applicable rules can differ from state to state, which can have a major impact on production prospects in different regions. For example, the State of Colorado recently passed a bill to expand the authority of local communities over drilling operations, including the issuing of permits and the setting of minimum distances between drilling operations and homes or schools.

- **Water use:** Tight oil operations require a lot of water. Today around 0.3 million barrels of water are needed on average to drill and complete a new well, and water use for hydraulic fracturing has grown by a factor of five since 2010. Although still a small percentage of overall water use, around 50% of wells drilled in the Permian are located in areas at high risk of water stress (Freyman, 2016).

- **Handling produced water:** Up to ten barrels of water are extracted from the ground with every barrel of tight oil (Wood Mackenzie, 2018). This water requires careful handling because it is usually saline and because it contains some of the chemicals used in the fracking process as well as metals, minerals and hydrocarbons from the reservoir rock. Since 2016, the discharge of wastewater into publicly owned water treatment plants has been prohibited, and today most wastewater is injected back underground using disposal wells (US EPA, 2019). In some locations this has led to a major increase in seismic activity (Magnani et al., 2017). An alternative option is to recycle it by using it to fracture new wells. This has the added benefit of reducing water extraction in water-stressed areas. There has been some progress in this area, but on average only 10% of the water produced today from tight oil wells is recycled (GWPC, 2019), implying that additional regulations or incentives are likely to be needed to boost recycling.

- **Local consent for infrastructure projects:** The steep rise in production from the Permian Basin caused distribution bottlenecks and led to major discounts in local crude oil prices. This was relieved only as new pipelines were rapidly commissioned and constructed and new infrastructure will be crucial to the future of tight oil growth. The prospects for such infrastructure can be significantly influenced by public and political support or opposition.

- **Air quality:** Local air pollution can be caused by flaring, truck traffic, and noise and fumes from the diesel generators required to power drilling and completion equipment. Between 2 000 and 6 000 truck journeys are required to transport drilling and completion materials equipment for a single well (Quiroga et al., 2016). In addition to negative effects on air quality, this can lead to congestion, wear-and-tear on local infrastructure and, if not managed carefully, can increase road accidents.

- **Flaring:** Large volumes of associated natural gas produced alongside tight oil are currently being flared. In the Permian region, for example, pipeline infrastructure for associated gas has not kept pace with the expansion in oil production. Although flaring is only allowed if a permit has been issued, applications for permits are rarely (if ever) turned down. The volumes of gas flared have soared, particularly in the second-half of 2018. Flaring across the whole of the United States jumped by 50% in 2018 compared with 2017 to over 14 billion cubic metres (bcm) (World Bank, 2019). Another concern is the level of methane emissions. Methane is a much more powerful greenhouse gas than CO_2 and the incomplete combustion of flares can result in a large level of methane emissions to the atmosphere (see Chapter 5). The combination of flaring and methane leaks means that tight oil can have relatively high lifecycle emissions intensity (IEA, 2018b).

In 2012, the IEA provided a set of principles – the *Golden Rules for a Golden Age of Gas* – to help policy makers, regulators, operators and members of the public understand and address the environmental and social impacts of shale operations (IEA, 2012). Our judgement at the time was that there is a critical link between the way that governments and industry respond to these impacts and the prospects for production. Seven years on, this link is even more visible: a continuous drive to improve performance remains essential if public confidence is to be earned or maintained.

3.11 Can the world afford to relax about security of oil supply?

Changing oil market dynamics offer some grounds for an upbeat assessment about the prospects for oil security. In the Stated Policies Scenario, US tight oil production continues its upward climb and the growth in global oil demand slows markedly. These developments could alleviate some concerns over security of supply.

But there are also reasons for caution. Despite the relentless growth of US tight oil, shrinking contributions from other non-OPEC countries after 2025 in the Stated Policies Scenario mean that the world continues to rely heavily on investments made by traditional resource holders, notably those in the Middle East. On the demand side, a flattening of demand in China contributes to the slowing global pace of growth, but there is still a marked shift in demand towards Asia. Oil consumption in Asia rises by 8 mb/d between 2018 and 2040 (close to 80% of the growth in global oil demand). With declining domestic production, Asia's import requirements grow by 10 mb/d, raising import dependency across the board (Figure 3.15).

Figure 3.15 ▷ Oil net imports and import dependency in selected developing Asian economies in the Stated Policies Scenario

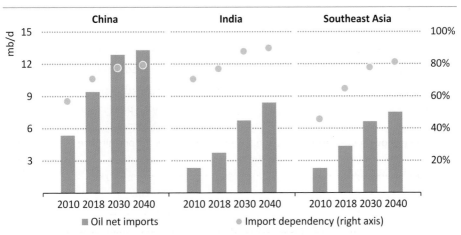

By 2040, over 80% of oil demand in developing economies in Asia is met by imports, stimulating a strong acceleration in oil trade flows to Asia

Note: Import dependency = net oil imports divided by demand including international bunkers.

Asian importers tap into a wider variety of supply sources, and there is a major increase in flows from North and South America to Asia. However major suppliers in the Middle East maintain their position of primacy (Figure 3.16). In the Stated Policies Scenario, despite the major changes in oil markets over the period to 2040, the share of seaborne crude oil trade from the Middle East to Asia remains remarkably consistent.

Figure 3.16 ▷ Seaborne crude oil trade by route in the Stated Policies Scenario

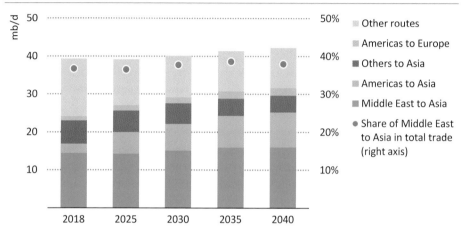

Despite the rapid growth of tight oil output, trade flows from the Middle East to Asia are set to remain the mainstay of global crude oil trade

Growing pressure on trade chokepoints

Escalating geopolitical tensions in the Middle East have brought the risks of physical disruption in the world's busiest trade passage, the Strait of Hormuz, back into the spotlight. The Strait of Hormuz is a narrow bend of water between Oman and Iran that connects oil producers around the Gulf with global markets. It carries some 16 mb/d of crude oil and 4 mb/d of oil products (around one-third of global seaborne oil trade): primarily from Saudi Arabia, Iraq and Iran to major importers in Asia, Europe and Africa (Figure 3.17). In 2018, around 80% of crude oil imports to Japan came through the Strait, as did 40% of China's oil imports. In addition, over a quarter of global LNG trade flows through the Strait: a large share of Asia's gas needs is met by LNG imports, primarily from Qatar and the United Arab Emirates (UAE).

In the Stated Policies Scenario, the volume of crude oil and oil products passing through the Strait of Hormuz remains high through to 2040. LNG trade via the Strait also grows, reflecting the strong position of Qatar as a low-cost exporter, although its share in global LNG trade gradually declines as Asian countries diversify their sources of imports, notably from the United States and Australia.

Any impediment to shipments through the Strait of Hormuz could materially tighten markets and leave importers rushing to find alternatives sources of supply. It could also render unavailable the vast majority of OPEC's spare production capacity. Only Saudi Arabia and the UAE have pipelines that can circumvent the Strait. These two countries have a combined pipeline capacity that bypasses the Strait of around 6.8 mb/d, of which around 3.8 mb/d is currently unused.[9] The complete closure of the Strait, while unlikely, would therefore result in a blockage of around 16 mb/d of crude oil and oil products, or around 17% of global oil supply. The recent attack on Saudi Arabia's oil facilities in Adqaiq and Khurais that led to a temporary loss of 5.7 mb/d of processing capacity (making it the largest supply disruption in history) led to a short-lived 20% spike in the oil price. Impacts on gas markets would also be significant, especially if any closure were to occur during the winter season: gas consumption in Asia is typically highest in winter months when underground gas storage in other regions (e.g. Europe) is also being drawn down.

Figure 3.17 ▷ **Major exporters and importers of crude oil traded via the Strait of Hormuz, 2018**

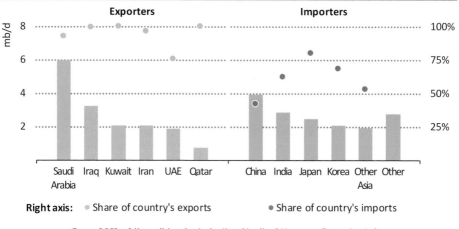

Over 80% of the oil traded via the Strait of Hormuz flows to Asia; most of the world's exporters and importers rely heavily on trade flows via the Strait

Notes: UAE = United Arab Emirates. Other importers include: Europe (with 8% of total imports flowing through the Strait of Hormuz), Africa (6%), Americas (2%) and other Middle Eastern countries.

The Strait of Hormuz is not the only chokepoint for oil trade. The Strait of Malacca is a key stretch of water between Malaysia and Indonesia that connects exporters in the Middle East and Africa with Asian importers. Around 19 mb/d of crude oil and oil products pass through the Strait of Malacca today. It is also a crucial location for fuel storage, blending and ship refuelling. In the Stated Policies Scenario, the volume of oil flowing through the

---

[9] Saudi Arabia plans to expand the capacity of its East-West pipeline from 5 mb/d to 7 mb/d although the timing of completion is uncertain. It is also assessing a reopening of the Ipsa pipeline and another 1 mb/d link from Iraq's Basrah oil hub to the Red Sea. If completed, these would increase the capacity that bypasses the Strait of Hormuz, although it would raise the reliance on another chokepoint, the Bab al-Mandab. Iran also plans to build a pipeline from Goreh to a port terminal at Jask, located just outside the Strait of Hormuz.

Strait of Malacca increases strongly, underpinned by robust demand growth in developing economies in Asia. LNG flows also increase as a result of supplies going to both traditional and emerging LNG importers in Asia. As a result, the Strait of Malacca becomes the largest chokepoint for global oil and gas trade by 2030 (Figure 3.18).

Figure 3.18 ▷ **Oil and gas trade volumes via major chokepoints in the Stated Policies Scenario**

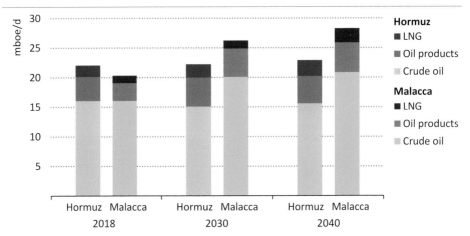

The Strait of Malacca takes over from the Strait of Hormuz as the largest potential chokepoint for global oil and gas trade

Note: mboe/d = million barrels of oil equivalent per day.

Growing traffic and the narrow nature of the Strait of Malacca pose potential risks of physical congestion, collision or attacks, which could have major implications for global oil and LNG markets. Finding alternative routes is not a straightforward task. The Sunda Strait, which separates the Indonesian islands of Java and Sumatra, is the closest alternative, but is too shallow and narrow to accommodate large oil vessels. The Lombok Strait between the Bali and Lombok islands represents the most viable alternative due to its width and depth, but is constrained by the lack of adequate infrastructure, notably port facilities and refuelling stations. Such a diversion would also increase voyage times between the Middle East and Asia, causing delays and incurring additional costs.

Quality matters: US tight oil is helpful, but not a panacea

Crude oil exported from the Middle East consists mainly of light and medium sour crude oil.[10] Since Asian refiners have been importing Middle Eastern oil for many years, many of their refineries are configured precisely to process these crude oil grades. For example, over 70% of the crude oil processed in refineries in Japan and Korea is light and medium

---

[10] Light and medium sour crude oil has a density greater than 24 °API and sulfur content more than 1% (see Spotlight).

sour crude oil. There is also a large appetite for these grades from refiners in China and India, although they process a slightly more diverse range of different crude oil grades (Figure 3.19). A potential output disruption either in the Middle East or in the chokepoints discussed would therefore have a particularly large impact on the global supply of the oil most wanted by Asian refiners.

Figure 3.19 ▷ Crude oil quality of selected producers and refiners, 2018

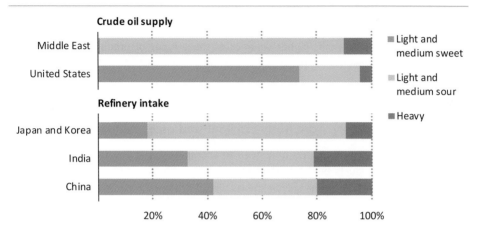

Asian refineries are configured to produce crude oil grades produced in the Middle East; quickly shifting to US crude oil could pose major short-term issues

If a sudden disruption in the Middle East were to occur, these supplies in theory could be replaced by increased output from other regions. A key candidate would be the United States, where tight oil production could likely ramp up relatively quickly in the event of a disruption. However, because of differences in crude quality, this is more difficult than it initially looks. Crude oil produced in the United States is mostly light and sweet: only 20% of the crude oil it produces is the light and medium sour crude oil preferred by Asian refiners. This would limit the ability of additional US crude oil to fill a sudden drop in the supply of medium sour grades. Asian refiners could switch to refine a lighter, sweeter crude oil feedstock, but this would take time. It would also require a careful assessment of a number of technical and economic factors. A change in feedstock would result in a major shift in product yield, which would affect the economics of refinery operations and, for complex refineries, would come at the cost of lower utilisation of upgrading units. The forced intake of light oil in the event of a supply disruption would therefore be likely to incur additional costs, weighing on already thin refining margins.

There are also major differences between US tight oil and OPEC spare capacity. Unlike OPEC's spare capacity, US tight oil production is determined by the responses of numerous market participants and is therefore not amenable to central direction. The scalability of US tight oil and its ability to respond to market signals should not be taken for granted given the numerous uncertainties that can impact production levels (see section 3.10).

The dynamics would be different if any disruption in oil supply from the Middle East were to extend for a longer period. If US tight oil were to respond to market signals, Asian refiners would adapt over time to make use of the higher volumes of US crude oil available on the market. Refineries in Asia have been processing growing volumes of US crude oil in recent years and they are already the largest export market for US crude oil (IEA, 2019d). This trend continues in the Stated Policies Scenario, with US crude oil comprising a growing share of Asian refiners' crude intake, particularly in China. In addition, global oil demand increasingly shifts towards lighter and sweeter products as a result of rising demand for petrochemicals and the need to meet tightening sulfur specifications (such as the IMO regulation). US tight oil yields higher volumes of these types of products and so is well positioned to help meet this changing demand pattern.

SPOTLIGHT

Crude quality: from heavy to light and onward to where?

To stay competitive, refiners not only need to anticipate and adapt to changes in the demand for oil products but also adapt to changes in the types of crude oil that are available. The question of changes in the quality of crude oil has rapidly risen up the agenda as a result of the rise in tight crude oil and NGLs production in the United States, escalating geopolitical and economic tensions in several producer economies, as well as the tightening regulations on sulfur content in oil products (such as the Euro emissions standards on road vehicles and the IMO regulation on marine fuels).

There are two factors that are widely used to define the quality of crude oil: density and sulfur content. Density is generally measured using American Petroleum Institute (API) gravity (an inverse proportional measure of petroleum density relative to that of water). Light crude oil is lighter than 32 °API; medium crude oil lies between 32 °API and 24 °API; and heavy crude oil is less than 24 °API. Crude oil is "sour" if it has sulfur content higher than 1% and "sweet" if it contains less.

Heavy crude oil tends to produce more low-value residue streams than light crude oil unless it is upgraded in a more complex refinery. Sour crude requires energy-intensive processing to remove the sulfur. Light and sweet crude oil therefore has historically been valued more highly than heavy and sour crude oil.

In the past, it was anticipated that crude oil available on the market was going to become increasingly heavy because major production increases were expected from Canada, Mexico and Venezuela. Refiners invested heavily in upgrading capacity that could produce a larger proportion of valuable products from lower priced heavy crude oil. Complexity became a mantra of the refining industry.

The rapid growth of US tight oil output turned this expectation on its head. Over the past decade, light and medium sweet crude oil supply accounted for almost 60% of the increase in crude oil supply; heavy crude oil represented just 15% (Figure 3.20). This was compounded by a major increase in very light NGLs. This trend looks set to continue, even though there is still uncertainty over how long and to what level US production can grow.

Figure 3.20 ▷ Change in global oil production by type, 2008-2018

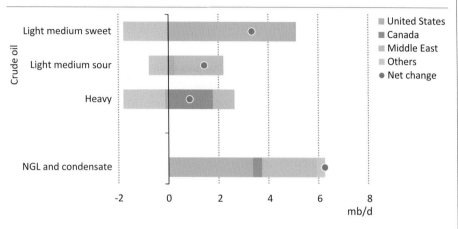

Light and medium sweet crude oil, and NGLs accounted for the majority of oil supply growth over the past decade

Note: Crude oil includes conventional crude oil, tight crude oil and extra-heavy oil and bitumen.

In the Stated Policies Scenario, US tight oil production nearly doubles between 2018 and 2030. In response, refiners are likely to strengthen efforts to adapt to changes in the quality of crude oil coming onto the market. This could undermine the attractiveness of heavy crude oil and weaken the case for related new investments.

The sulfur content inherent in crude oil is another important element. Refineries today need to remove some 70-75% of the sulfur from crude oil. They are likely to need to remove even higher proportions in the future. Many countries, including China, have already reduced the permissible sulfur content of road transport fuels to under 15 parts per million (ppm). Constraints are also extending beyond road transport, with the IMO stipulating that, by 2020, 40% less sulfur will be allowed in oil products than in 2005.

Traditionally, crude quality has been a constant in the decision making processes of refiners, and refiners have focussed on optimising operations for the specific type of crude oil processed. In the future, crude quality is likely to be a variable as the grades of crude oil that are available change; the range of products demanded by consumers is also likely to change. The ability to process a wider range of crude oil and optimise operations depending on market conditions looks set to be an increasingly important element in refinery competitiveness. For refiners, the era of complexity could be shifting towards an era of flexibility.

Producer economies matter for consumers

The enduring importance of traditional producers as a key source of global oil supply means that any social, economic or geopolitical turmoil in these regions will have material impacts on oil consumers. Many of these producers rely on hydrocarbon revenue to finance a significant proportion of their national budgets. In the *WEO-2018 Special Report: Outlook for Producer Economies*, we highlighted the vulnerabilities that these economies might face in a changing energy system and the case for fundamental changes to their development models (IEA, 2018b).

One year on, the factors that put pressure on producer economies to reform their resource development models have all heightened. In this *Outlook*, US tight oil production remains higher for longer and there are longer term uncertainties over the strength of oil demand. Despite this, some of the momentum behind reform seems to have waned with higher oil prices in 2018.

When oil prices were low, many governments in producer economies were keen to implement energy sector reforms and promote economic diversification. There was some positive progress in this area. Many countries made notable reductions in fossil fuel subsidies by raising regulated prices, which helped ease the strain on public finances. Their deployment of renewable technologies also gathered pace and ambitious targets have been introduced in a number of countries. In Saudi Arabia, broader economic reforms (such as the introduction of a value-added tax) have helped spur an increase in non-oil revenues.

However, history suggests that short-term pressure for reform tends to lessen when the oil price increases. Net income from oil and gas production in major producer economies almost doubled between 2016 and 2018. Some countries delayed proposed cuts to subsidies, leading to an increase in subsidy volume and energy intensity. In Nigeria, longstanding uncertainties around the Petroleum Industry Bill, which have been constraining upstream investment, remain unresolved. The pace of growth in non-oil gross domestic product has also moderated in many countries.

The changing dynamics of global energy markets pose major long-term questions about the durability of this relief. In this *Outlook*, cumulative net income from oil and gas production in major producer economies is around 10% lower over the period to 2040 than in the *WEO-2018* (Figure 3.21). This would reduce to an even greater degree if countries were to accelerate efforts to tackle climate change. Producer economies are increasingly heading towards a future world where markets for their ample resources are not guaranteed.

A long-term reduction in income would not only hamper the ability of governments to make the investments necessary to diversify their economies but also limit their capacity to respond to any potential periods of economic disruption. This could risk short-term supply interruptions ballooning into more sustained outages, with obvious implications for oil importers. While many producer economies achieved notable progress on reforms in recent years, the easing pressure for further reform matters for consumers as well as for the producers themselves.

Figure 3.21 ▷ Net income from oil and gas production in selected producer economies in the Stated Policies Scenario

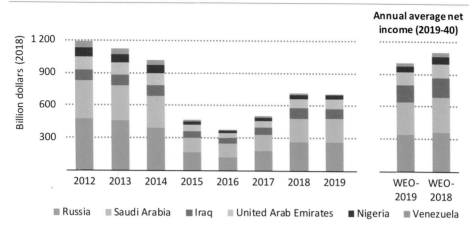

Recent increases in net income have diminished the pressure for producer economies, but the rapidly changing dynamics of global energy raises increasing questions going forward

Conclusion

Despite some positive developments in recent years, there are plenty of reasons for policy makers to continue to pay close attention to oil market security. While there is a marked slowdown in the pace of overall global oil demand growth in the Stated Policies Scenario, demand continues to grow briskly in Asia and much of the supply to meet it flows through major chokepoints. The rise of US tight oil output offers Asian importers opportunities for supplier diversification, but increases the pressure on producer economies, some of which are facing escalating geopolitical tensions.

Against this backdrop, the role of emergency oil stocks to help weather sudden supply disruptions remains vital. The effectiveness of such stocks will be greater with broader participation and with increased attention to changes in crude quality and product demand. It will also be important for refiners to improve the flexibility of their operations; for importing countries to promote energy efficiency and alternative technologies to moderate the pace of growing import dependency; and for producer economies to expedite their efforts to reform and diversify their economies. The changing market environment requires a broad and sustained approach to oil security.

Chapter 4

Outlook for natural gas
Does cheap mean cheerful?

SUMMARY

- Over the next two decades, global demand for natural gas grows more than four-times faster than demand for oil in the Stated Policies Scenario. Natural gas sees broad-based growth across the energy economy, in contrast to oil where growth is concentrated in parts of the transport sector (trucks, shipping and aviation) and petrochemicals.

- Developing Asian economies account for half of the global growth in natural gas demand and almost all of the increase in traded volumes. By 2040, they consume one-quarter of the world's gas production, much of it sourced from other regions. By 2040, the average gas molecule travels over 5 000 kilometres to reach consumers in developing Asian markets, nearly twice as far as today.

Figure 4.1 ▷ Change in gas supply balance by region in the Stated Policies Scenario, 2018-2040

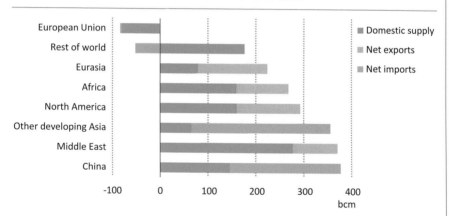

Most of the gas produced is consumed within the region where it is extracted. Developing countries in Asia underpin most of the growth in gas imports.

- China's natural gas consumption increased by 33% in just two years (2017-18). This massive expansion is having wide-ranging effects on the global liquefied natural gas (LNG) balance. It is underpinning major investments in new liquefaction capacity despite low gas prices in both Europe and Asia. By 2040, China imports almost twice as much LNG as the next largest importing country, India, and the share of gas in China's energy mix rises from 7% today to 13% by 2040.

- The United States adds nearly 200 billion cubic metres (bcm) to global natural gas production by 2025, over half of which is destined for export. In the Stated Policies

Scenario, the United States produces more natural gas than the whole of the Middle East over the period to 2040. Iraq with mostly associated gas and Mozambique with vast new offshore discoveries emerge as large gas producers from a low base.

- LNG dominates growth in global gas trade. Technological and financial innovations are making LNG more accessible to a new generation of importers. The combination of a growing spot market and more destination flexibility are accelerating a move toward market-based pricing of LNG and away from oil-indexed pricing. In the Stated Policies Scenario, the share of "pure" oil-indexed import contracts in Asia declines from around 80% today to less than 20% in 2040.

- There is significant uncertainty as to the scale and durability of demand for imported LNG in developing markets around the world. LNG is a relatively high-cost fuel; investment in liquefaction, transportation and regasification adds a considerable premium to each delivered gas molecule. Competition from other fuels and technologies, whether in the form of coal or renewables, loom large in the backdrop of buyer sentiment and appetite to take volume or price risk.

- Associated gas – a by-product of oil production – reached 850 bcm in 2018, around 20% of the world's gas output. Globally, only 75% of this gas is used by the industry or brought to market. We estimate that 140 bcm was flared and 60 bcm released into the atmosphere in 2018, more than the annual LNG imports of Japan and China combined. This significant source of emissions represents 40% of the total indirect emissions from global oil supply.

- In 2018, on a lifecycle basis, natural gas resulted in 33% fewer carbon dioxide (CO_2) emissions on average than coal per unit of heat used in the industry and buildings sectors, and 50% fewer emissions than coal per unit of electricity generated. Coal-to-gas switching can therefore provide "quick wins" for global emissions reductions. Theoretically, up to 1.2 gigatonnes (Gt) of CO_2 could be avoided using existing infrastructure in the power sector. Doing so would bring down global power sector emissions by nearly 10%. With coal prices in 2019 in a $60-80 per tonne range, most of this potential would require gas prices below $4 per million British thermal units (MBtu). Such prices are below the long-run marginal cost of delivering gas for many of the world's suppliers, implying the need for additional policy support (in the form of carbon prices or regulatory intervention) to realise these emissions savings.

- Near-term efforts by member countries of the International Maritime Organization (IMO) to reduce sulfur emissions from shipping may provide a new market for LNG. The window of opportunity is relatively narrow and clouded by uncertainty about the lifecycle emissions intensity of LNG. The IMO long-term strategy envisages an overall emissions reduction of at least half by 2050, compared with 2008; in this case, the maximum amount of emissions that could come from ships using LNG in 2050 is estimated to be 290 million tonnes of CO_2.

Introduction

Natural gas had a remarkable year in 2018, with a 4.6% increase in consumption accounting for nearly half of the increase in global energy demand. In 2011, a *World Energy Outlook* (*WEO*) special report asked whether the world might be poised to enter a "Golden Age of Gas", based on supportive assumptions about gas availability and price, as well as policies on the demand side that could promote its use in certain countries, notably China (IEA, 2011). A few years on, global gas consumption is now very close to this 2011 projection. Since 2010, 80% of the growth has been concentrated in three key regions: United States, where the shale gas revolution is in full swing; China, where economic expansion and air quality concerns have underpinned rapid growth; and the Middle East, where gas is a gateway to economic diversification from oil. Liquefied natural gas (LNG) is the key to more broad-based growth; 2019 is already a record year for investment in new LNG supply, even as regional spot gas prices have fallen to record lows.

Natural gas continues to do far better than either coal or oil in both the Stated Policies Scenario (where gas demand grows by over a third) and the Sustainable Development Scenario (where gas demand grows modestly to 2030 before reverting to present levels by 2040). The stage appears to be set for natural gas to thrive, at least in relative terms, over the coming decades. However, the headline findings should not obscure some important commercial and environmental challenges facing the gas industry, as well as some major variations in the storyline in different parts of the world. We examine some of these uncertainties in three in-depth sections of this chapter:

- Market linkages between natural gas and oil are gradually loosening, at least when it comes to pricing arrangements. However, there are upstream ties between the fuels that will be much more difficult to undo. In this section, we examine the continued importance of gas produced as a by-product of oil – associated gas – in shaping market developments, drawing on case studies of the United States, Middle East and Brazil.

- The outlook for natural gas relies heavily on LNG as a way to connect regional markets and bring gas to new consumers, especially in fast-growing parts of Asia. In a world where innovation is bringing down costs in many areas, we ask how innovation might affect the outlook for LNG. We explore four issues: the costs of LNG supply; the way that LNG is contracted; its environmental credentials; and the potential for demand-side innovation to unlock new markets.

- We consider how natural gas contributes to energy transitions when it competes with, and substitutes for, more polluting fuels, in particular coal. Our analysis examines to what extent, and in which sectors and timeframes, this substitution reduces emissions compared with other policy approaches, and how the calculation changes when methane leaks are considered. It also highlights regional variations in the case for fuel switching and reflects on the role of natural gas in the Sustainable Development Scenario.

Figures and tables from this chapter may be downloaded from www.iea.org/weo2019/secure/.

Scenarios

4.1 Overview

Table 4.1 ▷ Global gas demand, production and trade by scenario (bcm)

	2000	2018	Stated Policies		Sustainable Development		Current Policies	
			2030	2040	2030	2040	2030	2040
Power	908	1 571	1 708	1 936	1 580	1 248	1 823	2 197
Industrial use	644	909	1 229	1 474	1 108	1 114	1 243	1 527
Buildings	651	846	945	998	740	557	1 011	1 131
Transport	70	137	200	295	268	330	181	249
Other sectors	257	490	639	701	568	605	681	788
World natural gas demand	2 530	3 952	4 720	5 404	4 264	3 854	4 940	5 891
Asia Pacific share	*12%*	*21%*	*26%*	*28%*	*29%*	*34%*	*26%*	*28%*
Low-carbon gases	-	4	53	90	138	269	29	52
World total gases	2 530	3 956	4 773	5 494	4 402	4 123	4 968	5 943
Conventional gas	2 318	3 004	3 293	3 694	3 004	2 689	3 433	3 926
Existing fields	2 318	3 004	2 200	1 659	2 200	1 659	2 200	1 659
New fields	-	-	1 094	2 035	804	1 030	1 234	2 266
Tight gas	148	274	267	238	262	141	253	232
Shale gas	3	568	1 020	1 290	863	871	1 113	1 532
Coalbed methane	38	88	103	129	101	103	102	143
Other production	-	3	36	54	34	50	38	58
World natural gas production	2 507	3 937	4 720	5 404	4 264	3 854	4 940	5 891
Shale gas share	*0%*	*14%*	*22%*	*24%*	*20%*	*23%*	*23%*	*26%*
LNG	136	352	598	729	608	636	633	768
Pipeline	378	436	528	549	463	358	589	704
World natural gas trade	514	788	1 126	1 278	1 071	993	1 222	1 472
Share of production that is traded	*20%*	*20%*	*24%*	*24%*	*25%*	*26%*	*25%*	*25%*
Henry Hub price ($2018/MBtu)	6.1	3.2	3.3	4.4	3.2	3.4	3.8	5.1

Notes: Low-carbon gases are biomethane and low-carbon hydrogen injected into the gas grid. Historical production and demand volumes differ due to stock changes. World trade reflects volumes traded between regions modelled in the *WEO* and therefore excludes intra-regional trade. See Annex C for definitions.

In the **Stated Policies Scenario**, overall global gas demand in 2040 is broadly similar to the level projected in the *World Energy Outlook-2018*, as a slight upward revision to the use of gas in industry compensates for a downward adjustment to gas consumption for electricity generation (IEA, 2018). Demand in the United States has edged higher, but this is offset by a sharper decline in the European Union, as well as by slightly slower projected growth in China. Production growth is dominated by shale gas, which grows at a rate of almost 4% each year, four-times faster than conventional gas. Natural gas prices have been revised down slightly: this is a consequence of lower oil prices, a larger shale gas resource base in the United States putting pressure on the Henry Hub price (which increasingly influences prices elsewhere), anticipated declines in the costs of LNG liquefaction and an acceleration in the move away from oil indexation.

In the **Current Policies Scenario,** higher overall demand for energy pushes up natural gas consumption: demand for gas in this scenario increases by 2 trillion cubic metres (tcm) by 2040, a level 50% higher than in 2018. In the absence of further support for renewables or efficiency policies, gas satisfies a third of total energy demand growth, and more than any other energy source. The Middle East, North America and Eurasia each provide a fifth of the total additional volumes required in this scenario, with around 45% of global incremental production consumed in developing Asian markets, primarily in China.

In the **Sustainable Development Scenario**, natural gas consumption increases over the next decade at an annual average rate of 0.9% before reaching a high point by the end of the 2020s. After this, accelerated deployment of renewables and energy efficiency measures, together with a pickup in production of biomethane and later of hydrogen, begins to reduce consumption (decarbonised gases are modelled in detail for the first time in this *World Energy Outlook*). By 2040, natural gas demand in advanced economies is lower than current levels in all sectors apart from transport, where demand remains broadly similar to the level reached in the Stated Policies Scenario. In developing economies, gas growth in the power sector rises to 2030 but falls back due to a growing share of renewables, while growth in industrial demand is around half the level of the Stated Policies Scenario (Figure 4.2). Although absolute consumption falls, natural gas gains market share at the expense of both coal and oil in sectors that are difficult to decarbonise, such as heavy-duty transport and the use of heat in industry. Even though natural gas-fired electricity generation declines, capacity grows compared with today as gas expands its role as a provider of power system flexibility.

Figure 4.2 ▷ Change in gas demand by region and scenario, 2018-2040

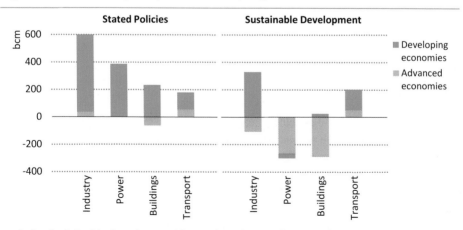

In the Sustainable Development Scenario, advanced economies consume much less gas than today; in developing economies growth is more subdued, especially in power

4.2 Natural gas demand by region and sector

Table 4.2 ▷ Gas demand by region and scenario (bcm)

			Stated Policies				Sustainable Development	
	2000	2018	2025	2030	2035	2040	2030	2040
North America	800	1 067	1 163	1 183	1 195	1 221	1 052	791
United States	669	860	936	947	949	957	870	646
Central and South America	97	172	178	198	224	257	168	169
Brazil	10	36	34	37	46	57	30	40
Europe	606	607	621	593	578	557	519	380
European Union	487	480	477	442	416	386	387	266
Africa	58	158	185	221	265	317	176	200
South Africa	2	5	5	7	8	9	6	8
Middle East	186	535	559	646	739	807	550	507
Eurasia	471	598	628	639	652	674	551	471
Russia	388	485	505	506	506	514	438	363
Asia Pacific	313	815	1 071	1 218	1 374	1 522	1 234	1 322
China	28	282	454	533	598	655	508	497
India	28	62	103	131	166	196	199	303
Japan	81	120	102	90	90	89	92	62
Southeast Asia	89	163	203	231	264	295	212	240
International bunkers	-	0	11	21	34	50	14	15
World natural gas	2 530	3 952	4 415	4 720	5 060	5 404	4 264	3 854
World low-carbon gases	-	4	27	53	72	90	138	269
World total gases	2 530	3 956	4 442	4 773	5 132	5 494	4 402	4 123

In 2018, over 45% of the growth in the world's energy demand was met by natural gas, with rising consumption in the United States and China together accounting for 70% of the increase. Gas use also increased in some established markets such as Korea.

In the Stated Policies Scenario, natural gas demand reaches 5 400 billion cubic metres (bcm) in 2040, a level nearly 40% higher than today. There is broad-based growth across all sectors, with the share of gas in global primary energy demand increasing from 23% today to 25% by 2040. Gas overtakes coal by 2030 but trails oil (with a 28% share in 2040) in the global energy mix. In the Sustainable Development Scenario, the rapid fall in oil use means that gas becomes the main fuel in the global mix by the mid-2030s.

Industry accounts for almost half of the projected growth in natural gas use in the Stated Policies Scenario. Gas is increasingly used in steel making and petrochemical production (primarily fertilisers), as well as in a broad range of medium- and small-scale manufacturing (e.g. textiles, food processing, glass and ceramics). Gas is well suited to provide adjustable levels of process heat for industrial boilers and furnaces. Most of this growth occurs in developing economies, particularly in China, India and the Middle East.

In **electricity generation**, demand for natural gas differs by region. Several advanced economies, such as Europe and Japan, see a peak in gas use for power before the mid-2020s, after which demand flattens and in some cases falls below 2018 levels. In Europe, gas use in power rises to fill part of the gap left by declining nuclear and coal capacity, before falling in the second-half of the projection period as renewables continue to grow strongly. Japan lowers its gas consumption in power as nuclear plants gradually return to service and renewable capacity is added to the mix.

Projected growth in natural gas for power is more durable in several developing economies, where it increases by more than 350 bcm (a 40% increase over current levels) to meet strong electricity demand growth in parts of Asia as well as the Middle East and sub-Saharan Africa. However, gas is not the main challenger to coal in China or India, where its share in generation remains well below 10% in 2040. Overall, the share of gas in fossil fuel-based electricity generation globally rises from 35% in 2018 to 45% in 2040, while efficiency gains of 0.5% per year reduce average fuel requirements for gas-fired power plants by 11% compared to 2018.

Compressed natural gas (CNG) and LNG both make inroads in the **transport** sector. CNG is primarily used for passenger vehicles while LNG enters markets for shipping and large road vehicles such as trucks and buses. Demand starts from a low base of around 140 bcm today, but grows to nearly 300 bcm by 2040, spurred by increased use in China and the United States, and to a lesser extent in India and the Middle East.

In the **buildings** sector, around 45% of the global net increase in natural gas demand is due to efforts to close the water supply gap in the Middle East through thermal desalination, which requires over 90 bcm of gas in 2040, primarily to provide drinking water. Much of the remaining growth comes from greater energy demand for cooking, where gas maintains an overall market share of around 40%. However, there is limited scope for gas to expand its role in providing residential heat, except in China. In advanced economies the use of gas for heating is curbed by improved efficiency, the direct use of renewables and increased electrification. This is particularly visible in the European Union, where overall gas demand in buildings declines by around two-thirds.

The **United States** remains by far the world's largest natural gas consumer. Underpinned by ample domestic supply, demand continues to grow strongly until the late-2020s before levelling off at around 950 bcm; the scope for large-scale switching in the power sector starts to dwindle, as does the scope for additional gas-intensive industrial development, while efficiency measures and electrification of heat in some parts of the country start to reduce demand in the buildings sector.

In **China**, natural gas demand more than doubles over the next two decades, rising by 370 bcm, more than the rest of developing Asia combined (Figure 4.3). There is an ongoing strong drive to use gas to reduce residential and industrial coal demand to improve air quality and reduce CO_2 emissions. Considerable potential remains: natural gas use currently accounts for 7% of total industrial energy demand compared with a global average of 22%, while only 12% of residential heating demand is met by gas in China. Although gas competes with electricity and the direct use of renewables in displacing coal in these sectors, its market share in industry and heat demand for buildings more than doubles over the period to 2040. By then, China consumes 650 bcm of gas, with the share of gas in the total energy mix reaching 13% (still well below the global average).

Figure 4.3 ▷ Change in gas supply and demand in developing Asian markets in the Stated Policies Scenario, 2018-2040

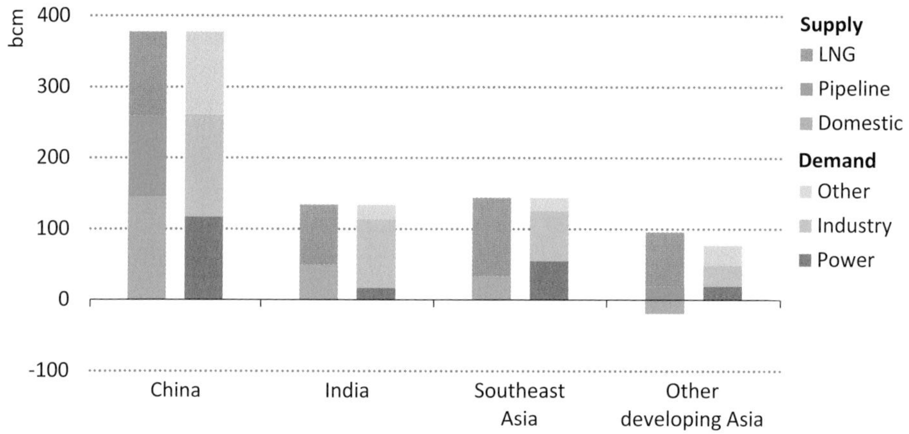

In China, gas demand increases more than the rest of developing Asia combined and is supplied by LNG, pipelines and domestic production; LNG underpins growth elsewhere

In **India**, the prospects for natural gas are limited by supply constraints and affordability issues, as well as by the lack of infrastructure. Most of the projected growth in electricity demand is met by a combination of renewables (especially solar) and coal, with gas largely confined to a balancing role. Demand grows more quickly in other sectors, supported by efforts to expand the distribution grid to make gas more accessible to industry (including for new fertiliser plants), to households for cooking and water heating, and to new CNG infrastructure in urban areas. Overall gas consumption triples to nearly 200 bcm by 2040.

In **Southeast Asia**, coal and renewables are in a strong position to serve growing energy demand, but natural gas use also rises, nearly doubling to nearly 300 bcm by 2040 even as the region switches from a net exporter to a net importer of natural gas.

In **sub-Saharan Africa**, natural gas demand rises threefold; spurred by economic growth, expanding population and development of the large-scale gas resources discovered over the last decade (see Part B). Though the majority of production is exported, investments in domestic infrastructure stimulate demand across a broad range of industries as well as for household cooking.

4.3 Natural gas production

Table 4.3 ▷ Natural gas production by region in the Stated Policies Scenario (bcm)

	2000	2018	2025	2030	2035	2040	2018-2040 Change	2018-2040 CAAGR
North America	763	1 083	1 254	1 336	1 358	1 376	293	1.1%
Canada	182	190	193	199	201	217	27	0.6%
Mexico	37	31	22	25	34	45	15	1.8%
United States	544	862	1 040	1 111	1 122	1 114	252	1.2%
Central and South America	102	177	188	209	244	285	108	2.2%
Argentina	41	45	58	78	101	126	81	4.8%
Brazil	7	26	28	37	55	75	49	5.0%
Europe	338	277	236	206	191	188	-89	-1.7%
European Union	265	120	66	47	44	40	-79	-4.8%
Norway	53	126	120	108	97	95	-31	-1.3%
Africa	124	240	287	372	435	508	268	3.5%
Algeria	82	96	96	104	112	125	29	1.2%
Egypt	18	59	81	92	95	98	39	2.3%
Mozambique	0	5	19	54	66	78	74	13.7%
Nigeria	12	44	41	45	56	65	22	1.8%
Middle East	198	645	721	787	912	1 016	371	2.1%
Iran	59	231	251	257	286	302	71	1.2%
Iraq	3	9	37	51	76	111	102	12.1%
Qatar	25	171	188	216	260	289	118	2.4%
Saudi Arabia	38	97	106	118	135	154	57	2.1%
Eurasia	691	918	1 021	1 054	1 105	1 143	224	1.0%
Azerbaijan	6	19	28	34	38	40	21	3.4%
Russia	573	715	797	798	834	853	138	0.8%
Turkmenistan	47	81	105	129	142	158	77	3.1%
Asia Pacific	290	598	708	757	816	889	291	1.8%
Australia	33	118	164	174	184	199	81	2.4%
China	27	160	224	250	274	306	146	3.0%
India	28	32	44	54	69	82	50	4.4%
Indonesia	70	72	75	79	85	95	22	1.2%
Rest of Southeast Asia	89	140	137	142	147	152	12	0.4%
World	2 507	3 937	4 415	4 720	5 060	5 404	1 467	1.4%
Sustainable Development			4 265	4 264	4 146	3 854	-83	-0.1%

Notes: CAAGR = compound average annual growth rate. See Annex C for definitions.

In the Stated Policies Scenario, natural gas production goes through two distinct phases. In the period to 2025, nearly 70% of the increase in global gas production comes from unconventional resources, largely driven by the continued ramping up of shale gas in the

United States, which is responsible for around 40% of total global gas production growth to 2025. Shale gas plays are found across the United States, but growth is primarily concentrated in the Marcellus and Utica shale plays, which together hold 17 tcm, or 40% of the country's total shale gas resources. Alone, these two plays are projected to add 200 bcm to total gas production in the United States over the next decade. The other major source of gas supply in the near term is associated gas from the Permian Basin.

The second period, from 2025-40, sees a shift in momentum back towards conventional natural gas, with accelerating production growth in the Middle East and several emerging exporters in sub-Saharan Africa. Meanwhile, shale gas production becomes more broad-based, as the peak in tight oil production in the United States contributes to a levelling off and subsequent decline in associated gas production. After 2025, 80% of growth in shale gas comes from outside the United States, primarily due to growth in Canada, Argentina and China, as well as smaller quantities in Australia, Algeria, Saudi Arabia and India.

Russia continues to be the second-largest natural gas producer after the United States, producing 850 bcm by 2040, a level nearly 20% higher than today. It also remains the largest gas exporter. The launch of the Bovanenkovo field in 2012 marked the start of a major new gas province in the Yamal Peninsula, which gradually becomes a mainstay for westward deliveries. The development of the Yamal and Arctic LNG projects underpins LNG export growth, while dramatic production growth in Eastern Siberia facilitates pipeline exports to China, initially via the Power of Siberia pipeline.

Qatar sees off competition from the United States and Australia to reclaim its position as the world's largest LNG exporter by the end of the projection period. An integrated LNG project to expand capacity by 45 bcm brings total liquefaction capacity above 150 bcm.

Mozambique sees the development of major offshore resources in the Rovuma Basin support a significant ramp up in LNG export capacity. The Golfinho and Atum fields within Area 1 block provide the feed gas for the recently approved 18 bcm LNG project (the largest in Africa), with further projects supported by the development of the adjacent Area 4. These projects enable Mozambique to overtake Nigeria in the late 2020s to become the largest gas producer in sub-Saharan Africa.

Iraq, barely in the top-fifty natural gas producers today, becomes the fastest-growing producer in the Middle East. Proven gas reserves stand at nearly 3.7 tcm, the majority of which is associated with oil. Two-thirds of the country's gas output, around 16 bcm, is currently flared, but capture rates increase and, as oil production expands to over 6 mb/d by 2040, additional associated gas is brought to market. As the political situation stabilises, particularly after 2030, the country sees a more than ten-fold increase in marketed production, from 9 bcm today to over 100 bcm by 2040.

Australia faces declining production from some mature basins, which brings some near-term challenges to maintain supplies for both domestic consumption growth and LNG exports. Over the long term, an additional 15 bcm of coalbed methane (CBM) is brought online by 2040, solidifying Australia's position as the world's largest CBM producer.

Although restrictions on gas development remain in place in some states, the lifting of a ban on hydraulic fracturing in the Northern Territories gives some impetus to shale gas exploration and eventual production, which increases to 30 bcm by 2040. In total, natural gas production reaches 200 bcm in 2040, a near doubling from today, but this relies on significant additional investment being unlocked by an alignment of state and federal hydrocarbon policies.

In **Europe,** natural gas production drops by 25% over the next decade, resulting in large part from the cessation of production at the earthquake-prone Groningen field in the Netherlands by 2022, and gradual resource depletion in the offshore North Sea. Norwegian production remains close to today's record highs before tailing off gradually. For Europe as a whole, marginal productivity gains from existing fields and some new offshore developments limit the decline in production to 90 bcm to 2040 (compared with a decline of 250 bcm – or 90% – if there were no further upstream investments).

Despite important shifts in the global gas trade balance, two-thirds of production growth over the projection period is consumed within each respective region. This results in a close relationship between the average annual change in production and the average annual change in demand (Figure 4.4).

Figure 4.4 ▷ Annual average change in gas demand and production in selected regions in the Stated Policies Scenario, 2018-2040

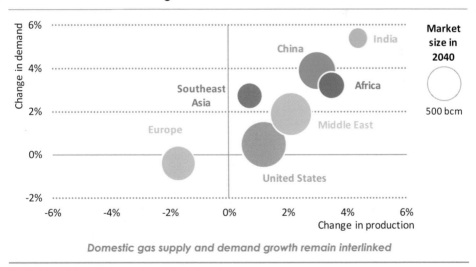

Domestic gas supply and demand growth remain interlinked

There are wide regional variations in the percentage of gas produced that is exported, with the range running from 20% in the United States to over 60% in sub-Saharan Africa. In some regions, countries increasingly require imports to satisfy demand growth, despite increases in production. China and India are the two prime examples: domestic output in both countries doubles, but consumption still runs ahead of production and therefore dependence on imports in both cases rises above 50% in the Stated Policies Scenario.

4.4 Trade[1] and investment

Table 4.4 ▷ Natural gas trade by region in the Stated Policies Scenario

Net importer in 2040	Net imports (bcm)				As a share of demand			
	2000	2018	2030	2040	2000	2018	2030	2040
European Union	221	360	400	356	46%	75%	89%	90%
China	1	122	286	353	5%	43%	53%	54%
Other Asia Pacific	-65	-27	88	181	n.a.	n.a.	24%	38%
Japan and Korea	97	170	145	153	97%	98%	99%	99%
India	-	30	78	115	-	48%	59%	58%
Other Europe	46	-29	-8	24	39%	n.a.	n.a.	14%

Net exporter in 2040	Net exports (bcm)				As a share of production			
	2000	2018	2030	2040	2000	2018	2030	2040
Russia	185	230	290	336	32%	32%	36%	39%
Middle East	12	109	138	203	6%	17%	18%	20%
North America	-37	16	150	149	n.a.	1%	11%	11%
Australia	10	78	126	148	31%	67%	73%	74%
Sub-Saharan Africa	5	36	97	133	33%	50%	62%	55%
Caspian	36	91	123	130	30%	45%	48%	45%
North Africa	61	45	53	57	57%	27%	25%	22%
Central and South America	5	5	10	26	5%	3%	5%	9%

World	Trade (bcm)				As a share of production			
	2000	2018	2030	2040	2000	2018	2030	2040
LNG	136	352	598	729	5%	9%	13%	13%
Pipeline	378	436	528	549	15%	11%	11%	10%
World	514	788	1 126	1 278	20%	20%	24%	24%
Sustainable Development			1 071	993			25%	26%

Notes: n.a. = not applicable. See Annex C for definitions.

In the Stated Policies Scenario, gas trade between regions expands by nearly 500 bcm, at an annual rate of around 2% through to 2040, running ahead of global gas demand growth. Imports continue to shift towards developing economies in Asia (Figure 4.5). China overtook Japan as the largest gas-importing country in 2018 and its imports are projected to reach the level of the entire European Union by 2040. China, India and other developing Asian markets account for most of the increase in global gas trade to 2040.

LNG emerges as the preferred way of moving gas over long distances, growing by more than 3% per year through to 2040. Pipeline gas increases at a more modest rate of 1% per year, with the increase primarily due to a tripling of China's pipeline imports; this offsets declines in net pipeline imports into the European Union in the latter half of the projection period, as the region takes higher volumes of LNG.

---

[1] Unless otherwise stated, trade figures in this chapter reflect volumes traded between regions modelled in the *WEO* and therefore exclude intra-regional trade.

The European Union's dependence on imported gas grows to nearly 90% by 2030 and the region retains a balancing role in global gas trade, thanks to a well-integrated gas market and a diversified mix of LNG, pipeline imports and storage infrastructure. Emerging importers on the other hand increasingly rely on the liquidity and responsiveness of the global LNG market to meet seasonal or short-term gas supply requirements (see section 4.6).

Russia and the Middle East retain their role as the largest gas exporting regions, but emerging exporters increase their market share. Australia briefly overtakes Qatar as the world's largest LNG exporter, while the United States becomes the second-largest gas exporter after Russia by 2025. Africa provides nearly 40% of export growth in the period 2025-40, primarily due to growing LNG exports from East Africa.

Figure 4.5 ▷ Natural gas net trade by region in the Stated Policies Scenario

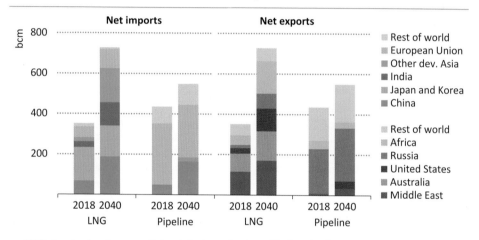

LNG dominates the growth in global gas trade, with developing Asia its main recipient

Note: Other dev. Asia = other developing Asia.

In the Stated Policies Scenario, on average around $370 billion of annual investment in natural gas is needed between 2019 and 2040: $240 billion for developing upstream resources and $130 billion for infrastructure including transmission and distribution pipelines, shipping, and LNG liquefaction and regasification facilities. The last three years have seen comparatively low levels of gas upstream investment (IEA 2019a), but mid-stream investments are at a high level. Several major pipeline projects are nearing completion and 2019 is already a record year for new LNG liquefaction project approvals.

In the Sustainable Development Scenario, LNG trade remains robust through to the mid-2030s, in part due to increased imports in coal-intensive economies in developing Asia. Trade levels then start to reduce as overall gas demand begins to decline.

Key themes

4.5 Associated gas: the upstream link between oil and gas markets

Are natural gas and oil going separate ways? Spending on large-scale, capital-intensive LNG projects has been increasing. This is in contrast to the oil industry's focus on smaller, less complex short-cycle projects (IEA 2019a). Rising gas-on-gas competition is also challenging the rationale for oil indexation. However, because oil rarely rises to the surface without gas, the two fuels are unlikely ever to be completely delinked. Associated gas from oil fields is also the main source of flaring as well as a major source of gas that is vented directly to the atmosphere – both major sources of greenhouse gas (GHG) emissions – and it is important to understand the environmental footprint of this source of gas supply. This section examines the extent of global associated gas production and use, and explores in three cases how upstream linkages between oil and gas continue to affect the outlook for gas in the Middle East, the United States and Brazil.

Associated gas today

Most wells that are drilled to target oil formations also yield a mixture of other hydrocarbons such as condensates, natural gas liquids and natural gas. The latter is known as "associated gas". In essence, associated gas comes as a by-product from an oil well or field and non-associated gas comes from a well or field that is primarily geared for gas production.[2]

Associated gas has often been seen as an inconvenient by-product of oil production: it is generally less valuable than oil per unit of output and is costlier to transport and store. It is often used on-site as a source of power or heat. It can also be reinjected into oil wells to create pressure for secondary liquids recovery (as is common, for example, in Norway, Iran and Venezuela). Under the right geological conditions, it can also be stored and sold to the market at a later stage. Associated gas is usually collected via a network of gathering pipelines for further processing or direct injection into gas grids. When the gas is rich in natural gas liquids (NGLs), extra processing is required to separate out the heavier hydrocarbons such as ethane, butane and propane.

When there is no on-site use for the gas and a lack of infrastructure prevents it from reaching nearby markets, it is vented or flared. Associated gas can also be unintentionally released to the atmosphere as fugitive methane emissions. Together, such non-productive uses of gas have significant environmental consequences, making up around 40% of the indirect emissions associated with oil production. They also represent a wasted economic

---

[2] There is no single definition of the boundary between associated and non-associated gas. Here, if the amount of gas is more than 60% of the hydrocarbons extracted in energy terms (or around 250 cubic metres per barrel of oil), it is classified as a gas well that produces non-associated gas; otherwise it is classified as an oil well that also produces associated gas. Changing this to 50% would reclassify around 30 bcm from associated to non-associated gas.

opportunity: the 200 bcm that was flared (140 bcm) or escaped into the atmosphere or vented (60 bcm) in 2018 was greater than the annual LNG imports of Japan and China combined. Globally, only 75% of associated gas is used or brought to market (Figure 4.6).

Figure 4.6 ▷ Use of associated gas by region, 2018

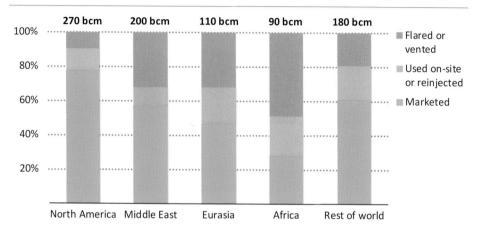

Only 75% of associated gas is put to productive use; flaring and venting is more common in remote areas of production which often lack nearby markets

While gas is often flared temporarily while operators test equipment and undergo pre-commissioning works or for safety reasons, "routine" flaring occurs when there is a failure to put associated gas to productive use. This may be because of the remoteness of fields or the topography of the surrounding area or because the local price of gas discourages operators from developing costly gas transportation infrastructure to reach existing or potential new markets. Associated gas also often comes with a combination of water vapour, hydrogen sulphide, nitrogen or carbon dioxide, and the cost of separating out these unwanted elements may be higher than the potential profits from the gas. Although several countries have imposed regulatory measures to restrict flaring, these are often inadequately enforced. Even in countries with well-developed gas markets, such as the United States, around 10% of associated gas extracted today is flared or vented.

The United States is the largest producer of associated gas, accounting for over a third of the global total, nearly equal to the next five countries combined. In most countries, associated gas constitutes less than a fifth of total marketed gas production, but in some large oil and gas-producing countries, such as Mexico, Saudi Arabia, Brazil and Nigeria, it makes up a much larger share. On average, gas makes up around 10% of the energy content of an oil field, but this is subject to wide variation depending on geological conditions, well design and production method (Figure 4.7).

Figure 4.7 ▷   Associated gas volumes in the total output from oil fields, 2018

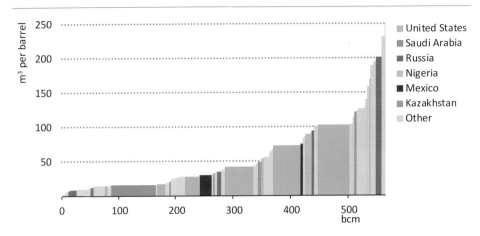

On average, every barrel of oil produced today comes with around 60 cubic metres of gas, but this is subject to wide variations across the world's oil- and gas-producing regions

Middle East

The Middle East holds nearly 40% of global proven gas reserves, but these are not spread evenly across the region. In the two largest gas-producing countries, Iran and Qatar, natural gas and condensate resources have been developed independently of oil. However, more than 80% of the gas in Saudi Arabia and Kuwait is associated gas, and there are also significant volumes in Iraq and Oman. Associated gas has underpinned the rise of gas demand in many of these countries since it has largely been available at close to zero cost as a by-product of oil production. It has also provided a basis for economic diversification away from oil.

In recent years, associated gas production in the Middle East has struggled to keep pace with soaring domestic demand, which has tripled since 2000. Gas has been used as a substitute for oil in the power sector in producer countries because it frees up additional volumes of crude for export. It has also become a crucial fuel for water desalination plants. However, these end-uses cause significant peaks in consumption in summer months and, in the absence of significant storage capacity, associated gas has struggled to accommodate this seasonal variability. Unable to keep pace with demand, Kuwait and the United Arab Emirates have resorted to seasonal LNG imports, while Oman has had to cut back LNG exports to redirect supply to the domestic market.

The shortfall in associated gas, combined with the soaring pace of gas demand, has driven several oil-rich countries to develop non-associated gas fields, particularly those containing NGLs. Since 2000, the Middle East has seen a fourfold growth in non-associated gas production. This has supported the development of new gas value chains and has underpinned Qatar's rise to become the world's largest LNG exporter (helped by the liquids-rich North Field). It has also led to integrated gas and NGL projects in the United Arab Emirates and Saudi Arabia that have spurred the development of heavy industries and petrochemical complexes.

Some countries are facing a need to raise gas prices to support further upstream development of more complex non-associated gas projects. They face a delicate balancing act on price given their overall macroeconomic reliance on low-cost gas supply. The Saudi government is planning a pricing regime that differentiates associated from non-associated gas to reflect the higher development cost of non-associated gas. End-user prices for non-associated gas are likely to be between $1.5-3.9/MBtu, compared with less than $0.50/MBtu for associated gas (Saudi Aramco, 2019). Even prices at the higher end of this range are unlikely to cover the long-run production costs of these resources, which are located in difficult-to-develop fields. The Saudi Arabian government has also granted tax and fiscal incentives to Saudi Aramco for gas production, a sign of the priority assigned to domestic gas development. Oman, Bahrain and the United Arab Emirates have put similar measures in place.

Figure 4.8 ▷ Associated and non-associated gas production in selected countries in the Middle East in the Stated Policies Scenario

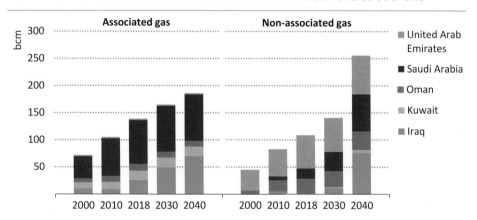

Several Middle Eastern countries are diversifying into non-associated gas fields to support further development of domestic gas markets

Note: Associated gas volumes include flaring.

In the Stated Policies Scenario, several countries in the Persian Gulf squeeze some marginal gains out of existing associated gas production, as oil output grows by 3.9 million barrels per day (mb/d) over the period to 2040. However, the majority of growth comes from non-associated gas resources, with production nearly doubling to reach 250 bcm by 2040 in the countries in Figure 4.8. Saudi Arabia, for example, derives most of its incremental production from non-associated gas, which allows the country more or less to keep pace with demand growth in the power, petrochemical and desalination sectors. The main exception is Iraq, which is a relative latecomer in the development of its non-associated gas resources. In the medium-term, most of its gas production growth comes in the form of associated gas from a ramp up in oil production and a reduction in the rate of flaring, allowing it to meet significant latent demand for electricity generation and removing the need to import gas from Iran.

Overall, natural gas's share of total marketed oil and gas production in the region rises from 26% today to nearly 33% by 2040. The brisk pace of non-associated gas production allows the Middle East to develop upstream oil and gas supply chains that are increasingly separate from each other. It also accelerates the displacement of oil in electricity generation. At the same time, oil and gas remain tied together by continued investment in large-scale petrochemical and refining complexes in several parts of the Middle East, since these require an integrated hydrocarbon processing chain consisting of gas, oil and NGL feedstocks. Looking ahead, the development of non-associated gas has important implications for the region's ability to adapt to the demands of the Sustainable Development Scenario, where oil production begins to decline very soon while gas demand continues to grow until 2030. The divergent paths in this scenario raise questions about how investment is divided between associated and non-associated gas, and about how countries in the Middle East reconcile declining crude oil exports with continued robust growth in domestic gas demand.

United States

Associated gas production in the United States has doubled over the last decade, reaching nearly 200 bcm in 2018. It accounts for around 20% of the gas produced in the United States today.

More than half of associated gas production comes from tight oil plays and significant further growth is expected (Figure 4.9). The Bakken formation in North Dakota and the Eagle Ford play in Texas, developed in the early 2010s, were the first basins to yield significant quantities of both gas and oil from horizontal drilling. Since 2017, the Permian Basin in southwest Texas has been the main source of growth reflecting a huge ramp up in drilling activity and major productivity gains that have increased the amount of oil and associated gas output per well (see Chapter 3). As drilling has increased, so have estimates of remaining technically recoverable resources: the Permian is now estimated to hold nearly 7 tcm of gas, alongside 80 billion barrels of oil.

Drilling in the Permian primarily targets more valuable liquids; gas is essentially a very low-cost by-product of this activity. This has created a number of dilemmas for US producers. While crude oil can be transported by road or rail, associated gas must be transported through dedicated pipelines; the strong production growth seen over the past three years has outpaced mid-stream processing and pipeline transport capacity, creating bottlenecks that have put strains on upstream operations. Companies have responded by leaving drilled wells uncompleted and selling off associated gas at very low, and at times even negative, prices. They have also increased the amount of flaring; in the Permian alone, levels of flaring have risen more than twenty-fold since 2011, when total flared volumes were estimated at 250 million cubic metres. Recent data suggest that as much as 7 bcm could be flared in the Permian in 2019 (Oil and Gas Journal, 2019). As earlier experience in the Bakken formation showed (where as much as 30% of extracted gas was flared in the early 2010s, and where levels have come down only slightly since), flaring on this scale invites a regulatory response and is extremely damaging to the reputation of the upstream industry at a time when there is ever-increasing scrutiny of its environmental performance.

Figure 4.9 ▷ Associated gas production in the United States in the Stated Policies Scenario, 2010-2025

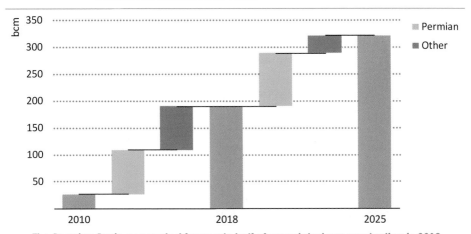

The Permian Basin accounted for nearly half of associated gas production in 2018

Constraints on pipeline capacity to transport associated gas away from where it is extracted are visible in widening price differentials between associated gas-producing basins and the Henry Hub benchmark. These differentials send a clear economic signal about the case for investment in new connecting pipelines, but the scale of projected associated gas production growth in the Permian in the Stated Policies Scenario would require a doubling of take-away capacity (currently around 100 bcm) by 2025.

It is not clear which demand centres have the greatest capacity to absorb this low-cost gas. Pipeline exports to Mexico have increased more than fourfold since 2010, but the prospects for exporting additional quantities are limited by the need for further development of pipeline infrastructure within Mexico, as well as by Mexico's own gas production prospects. In the Stated Policies Scenario, exports to Mexico rise modestly to just above 55 bcm by 2025. There is some potential for gas demand growth in the power sector in the southern parts of the United States, as coal-fired capacity is increasingly retired, but this is tempered by expanding deployment of wind and solar capacity. Switching all coal-fired electricity generation that remains in the region in the Stated Policies Scenario in 2025 to natural gas would require an additional 30 bcm. Taken together, additional exports to Mexico and further coal-to-gas switching would only absorb around 40% of the growth in associated gas from the Permian Basin.

Chapter 4 | Outlook for natural gas

LNG export terminals are expected to be the main outlet for associated gas. Around 80% of liquefaction terminal capacity in operation in the United States today is on the coast of the Gulf of Mexico. New pipelines that link low-cost Permian gas to these export facilities, such as the recently completed 20 bcm Gulf Coast Express, have a critical role to play in relieving the pressure on upstream producers by providing a route to market for their associated gas. Several proposed LNG export projects have made sourcing lower cost gas from the Permian a key part of their business model, which means that new pipelines will also help support the commercial case for the next wave of US LNG export expansion.

The implications of a possible rise in oil prices are likely to reinforce the desire of producers to send associated gas to LNG terminals for export. Any increase in the oil price would be likely to stimulate additional tight oil production, which would lead to more low-cost associated gas, thereby putting downward pressure on domestic gas prices. At the same time, higher oil prices would increase oil-indexed LNG prices, enhancing the competitiveness of Henry Hub linked LNG exported from the United States. A $20/barrel increase in the oil price compared with our projection in 2025 would push tight oil production in the United States up by 0.6 mb/d. This, in turn, would increase associated gas production by 10 bcm, pushing higher volumes of lower cost gas to LNG export terminals.

Brazil

Associated gas has a crucial place in the future energy mix of Brazil. The development of offshore pre-salt fields has led to increasing quantities of associated gas production. Today, every barrel of oil extracted from pre-salt fields is accompanied on average by 20 cubic metres of gas, which means that associated gas accounts for around 10% of total output in energy terms. In the Stated Policies Scenario, offshore gas production rises from 21 bcm today to over 60 bcm by 2040.

Despite growing production from this significant resource base, Brazil today relies in part on imports to meet its domestic gas demand. It could potentially make much more use of its own gas, but it lacks the necessary pipeline infrastructure. Pipeline capacity is sufficient to bring onshore around 8.5 bcm of associated gas, while gross production is running at around 32 bcm. Producers reinject over a third of the associated gas into the pre-salt fields and flare the remaining 3%.

There is also a mismatch between associated gas supply and demand. Gas demand for electricity generation varies considerably from one year to the next in Brazil because gas is used to balance the annual availability of large-scale hydropower. As a result, gas demand for power over the last decade has varied from as little as 3 bcm to nearly 20 bcm (Figure 4.10). This means that producers of associated gas do not at present have a relatively constant source of demand to justify building additional pipeline capacity to bring gas onshore. Significant storage capacity could potentially provide such a constant source of demand, but it would be difficult to obtain finance for this precisely because of the annual unpredictability of gas demand. New power projects linked to associated gas that would typically run as baseload supply could also potentially provide a stable source of demand. Currently, however, Brazil's power market is oversupplied, meaning that the delivered cost of associated gas to such power projects would need to be low enough to justify new investment in plants that need to run as baseload. Meanwhile expanding wind and solar capacity adds a further challenge to the investment case.

As oil and associated gas output grows, the question of how best to market the gas beyond the power sector becomes more pressing. Industry may provide the relatively constant year-round demand required by associated gas production, but demand growth in industry may not keep pace with the expected growth in gas production. Petrobras, Brazil's oil and gas incumbent company, is exploring other technologies that potentially avoid the need for costly offshore-to-onshore pipeline infrastructure, such as floating LNG, gas-to-liquids processes or CNG for the transport sector. It is not yet clear whether these technologies represent cost-effective alternatives to conventional infrastructure.

Figure 4.10 ▷ Natural gas demand and production in Brazil in the Stated Policies Scenario

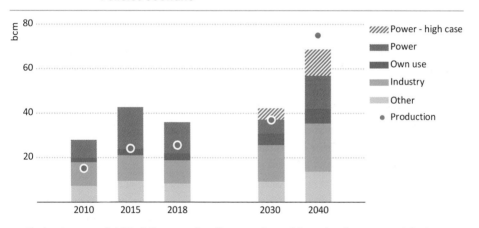

Hydropower variability influences Brazil's annual need for natural gas; associated gas production creates new challenges for balancing supply with this variable demand

Note: "Power - high case" illustrates the additional gas demand if gas-fired power plants ran at higher load factors in order to compensate for reduced hydro availability.

In the Stated Policies Scenario, Brazil continues to import gas on a flexible, short-term basis to complement the variability of its hydro generation. Some baseload gas demand for power is met by associated gas, but the main source of growth is demand for industry, at a rate of 3.4% per year to 2040. Supply still runs ahead of demand, and so we anticipate the development of LNG export capabilities by the end of the next decade.

Outlook for associated gas

These case studies suggest that the key hurdles to finding markets for associated gas are the need to build additional infrastructure and the need to match a stable supply of output with demand in situations where many sources of demand are variable. In most parts of the world, in the absence of determined action by regulators, operators have greater incentives to flare associated gas than to curb more valuable oil output.

In the Stated Policies Scenario, efforts to reduce this practice bear fruit, with flaring rates declining by half even as oil production increases. The United States dominates the rise in associated gas over the next decade, accounting for 75% of total growth. Production becomes more evenly distributed after 2030, reflecting a greater diversity in oil supply during this period. Associated gas increases from 565 bcm to 680 bcm by 2040, but its share of total marketed gas production drops to 13%, as global oil demand levels off while natural gas demand continues to rise.

In the Sustainable Development Scenario, flaring rates decline faster (as do methane emissions) and the share of total gas output accounted for by associated gas drops more quickly. Global demand for crude oil declines in the early 2020s, while demand for gas continues to rise throughout most of the rest of the decade; demand for natural gas then falls away more slowly than demand for oil (Figure 4.11). The divergent trajectories in this scenario create a variety of challenges for operators: in a world of reduced revenue from oil, they face an even stronger imperative to invest in gas capture, and minimise flaring and venting. Some emerging technologies, such as small-scale LNG, may offer a commercially viable alternative to the reduction of flaring and venting of associated gas.

Figure 4.11 ▷ Change in global gas and oil production in the Sustainable Development Scenario

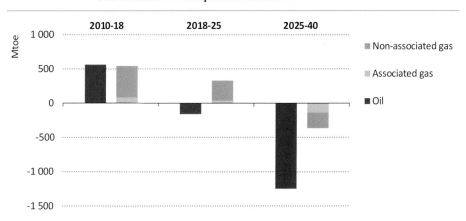

In the Sustainable Development Scenario, falling demand for oil reduces associated gas production, while increasing pressure on operators to ensure its productive use

4.6 How does innovation affect the outlook for LNG?

In the Stated Policies Scenario, LNG overtakes pipeline gas as the main way of trading gas over long distances by the late 2020s. Developing economies in Asia are the main engines of LNG growth, with the market share of LNG in total gas demand growing from 20% in 2018 to 40% by 2040.

There is significant uncertainty, however, as to the scale and the durability of demand for imported LNG in developing markets around the world. The price sensitivity of demand is one key uncertainty: another is the extent of competition from other fuels and technologies, whether in the form of coal or renewables. LNG is a relatively high-cost fuel, with investment in liquefaction, transportation and regasification adding significantly to the cost of the gas itself. As shown in Figure 4.12, emerging markets in Asia are facing significantly higher costs for imports than for domestically produced gas. Even though spot gas prices fell to record lows in 2019 on the back of ample LNG supplies, over the long-term end-user prices generally seem set to rise: unless they do, LNG suppliers will be unable to recover their long-term investment costs or governments will have to continue to subsidise the cost of LNG imports. The LNG industry faces a struggle to gain a strong foothold in developing markets where affordability is a key consideration.

Figure 4.12 ▷ Domestic natural gas production costs, LNG import prices and industry gas prices in developing Asian import markets, 2018

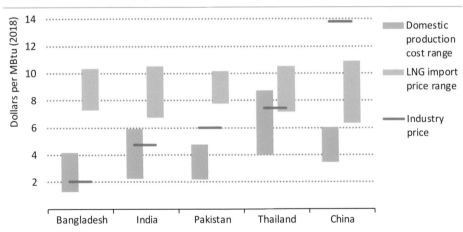

End-user prices are generally high enough to sustain domestic production in developing Asian gas markets, but below the costs of imported LNG in several markets

Note: LNG import price ranges are based on data from Argus Media (2019).

LNG has compensating advantages that have the potential to justify its premium price. It can be provided flexibly and relatively quickly, which are helpful qualities from the point of view of energy security. In some cases LNG also has a competitive advantage over oil. In addition, although LNG supply chains tend to produce more emissions per unit of gas than pipeline gas because of the additional energy requirements for liquefaction, LNG still has significant environmental benefits when it substitutes for more polluting fuels such as coal or oil (see section 4.7). The limited air pollutants emitted by natural gas make it particularly attractive for developing economies concerned about air quality. Policy support for building gas infrastructure and tapping into LNG supply, including through small-scale infrastructure, has been a critical part of its evolution.

The business model for buying and selling LNG is changing. As gas-on-gas competition increases, fewer LNG export projects are likely to be able to rely in the future on a return on investment by delivering volumes under fixed oil-indexed long-term contracts with re-export restrictions and take-or-pay conditions. Sellers will therefore have to innovate to attract new buyers. Technology is opening new opportunities to supply LNG directly to various end-use sectors, notably transport. We examine how innovation in the LNG supply chain affects the outlook for liquefaction costs, contracting patterns, environmental performance and demand-side technologies.

Where are liquefaction costs heading?

Liquefaction is the most capital-intensive part of the LNG supply chain, in most cases accounting for nearly 50% of the delivered cost of gas. While there are common elements, the costs of liquefaction terminals are dictated ultimately by project-specific factors, such as their location, size and complexity.

Figure 4.13 ▷ Capital costs of liquefaction projects

Liquefaction costs have come down from their pre-2014 highs

Note: Mtpa = million tonnes per annum; FID = final investment decision.

The evolution of LNG liquefaction costs since 2000 can be separated into three phases (Figure 4.13). At the start of the century investment costs of below $500/tonne were achieved at a number of projects through competitive bidding by engineering, procurement and construction companies and an increased focus on cost. Size also matters, as demonstrated in the case of Qatar, where efforts to develop integrated projects using the world's largest LNG trains yielded important economies of scale. With the rise in oil prices and the move toward more remote, technically complex sites over the 2008-12 period, liquefaction costs began to escalate beyond $2 000/tonne. Most of these high-cost projects were built in Australia, where investment decisions were compressed into a relatively short space of time (and in the case of projects in Queensland, a relatively small geographical area). This created labour and service bottlenecks as well as higher material costs, which were compounded by a strong local currency. Since 2014, around two-thirds of investment in new liquefaction has been made in the United States, where projects were primarily based on converting regasification terminals. With some of the necessary infrastructure already in place, a large and competitive LNG construction industry, and declining oil prices, costs have reverted to the previous range of $500-1 000/tonne.

Due to the scale, location and complexity of its projects, the LNG industry has found it difficult to deliver projects on time and on budget. Cost overruns and delays in the period between 2009-14 were particularly common, with over half of projects sanctioned experiencing delays of a year or more or exceeding the budget estimates made at the time when the final investment decision was taken. Delays appear to have continued since 2014: out of 22 projects that have come online, 11 have seen significant delays in commissioning dates.

Despite these downside risks, LNG remains one of the few parts of the oil and gas sector that has continued to see traditional large-scale, capital-intensive infrastructure projects. After a three-year lull in project approvals, the market for new liquefaction capacity appears to have turned a corner, with 2019 seeing a record number of project approvals. There is a long list of competing projects around the world that are seeking to advance towards a final investment decision. A number of brownfield projects are also being pursued. These are additions to existing facilities that can tie in to new nearby gas resources, while making use of existing infrastructure and drawing on established operational knowledge and relationships with contractors. Some existing LNG export facilities also seem set to undergo "debottlenecking" to optimise plant facilities and squeeze out additional capacity.

In addition to the "brownfield advantage", there are several emerging business model innovations that have important implications for costs. In the United States, a supportive regulatory environment and an unbundled, competitive gas industry has helped to reduce the weighted average cost of capital for new projects (Box 4.1). Some projects – particularly in North America – are also phasing in the development of large-scale liquefaction terminals, by using mid-scale train sizes of around 1-2 million tonnes per annum (Mtpa). This kind of project structure theoretically requires lower upfront capital and hence makes financing more accessible. It also allows for the monetisation of gas resources incrementally and can be coupled with a modular design, whereby standardised equipment and prefabricated liquefaction modules are built off-location and delivered when complete.

There are several design configurations available for modular, mid-scale projects and increasing competition between nearly a dozen major vendors. In some cases, projects have also benefited from concessionary tax breaks provided by governments keen to encourage the significant capital inflows brought by these large-scale projects.

Figure 4.14 ▷ Investment cost ranges for liquefaction capacity and long-run marginal costs in the Stated Policies Scenario, 2018-2040

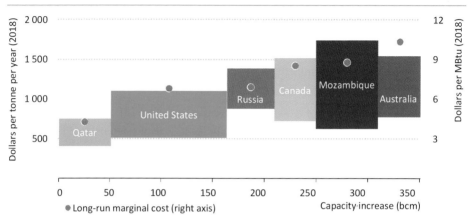

There is a wide range of investment costs for liquefying gas; along with feed gas and shipping costs, LNG project economics set the marginal price of gas in several markets

Notes: Long-run marginal costs equal the weighted average costs in each country of developing gas resources, building liquefaction terminals and shipping the total LNG volumes delivered over the projection period. Shipping costs reflect the volume-weighted average cost of delivery to importing regions in the Stated Policies Scenario. Assumed asset lifetime is 30-years with a cost of capital in the range of 5-10%.

In the Stated Policies Scenario, the costs of liquefaction are maintained in a relatively wide range of $400-1 200/tonne per year, with some projects exceeding the upper value towards the end of the projection period. Around 430 bcm of liquefaction capacity is developed at a total investment cost exceeding $300 billion, with 80% of total capacity being built in just six countries (Figure 4.14). Ultimately, gas prices rise to meet long-run production costs. Taking into account the cost of feed gas, liquefaction costs and shipping costs to regasification facilities, this scenario sees a weighted average long-run marginal cost of LNG of $7.50/MBtu over the projection period, with marginal projects exceeding $10/MBtu by 2040. This projection takes into account different types of projects and designs, but does not build in any cyclical element to the market for new LNG infrastructure. It assumes that the industry shows strong discipline on costs and project management. Given the competitive pressures facing LNG, there is very little margin to absorb cyclical upswings in costs or cost overruns.

Box 4.1 ▷   Innovation in LNG financing and marketing strategies

LNG projects are among the most capital-intensive in the energy world. Sponsors have typically relied on project finance to raise the necessary funds to build liquefaction capacity and procure dedicated upstream feed gas. This type of finance involves securing loans from banks and other large lenders such as export credit agencies, backed by long-term contractual commitments from large creditworthy buyers, which agree to a minimal offtake of LNG with re-export restrictions and take-or-pay conditions. Final investment decisions are typically made when at least 80% of the proposed output from an LNG terminal is contracted to long-term buyers. The lenders are paid back from the cash flows generated by the project.

This business model has been in place for decades, but it has come under threat from the emergence of new LNG suppliers and market participants, especially in the United States. Projects there were built under either tolling or merchant models, without dedicated upstream supplies or downstream transport assets. Project finance was secured by having buyers sign up to long-term capacity rights at the terminal (whether it is used or not). Much of the gas has been sold to portfolio players, who take ownership of LNG at the liquefaction facility and are free to deliver it to a diverse set of buyers through a combination of spot transactions and short- and long-term contracts. Such models have unlocked bigger volumes of destination-flexible LNG supply.

Growing confidence in an LNG spot market has encouraged large players to move away from project finance and toward balance sheet financing, which involves using retained earnings as well as raising debt or equity on the strength of a project sponsor's own creditworthiness. This route is essentially only open to international oil companies and national oil companies with significant financial resources. It implies higher financing costs and involves higher market risk, as it means fewer guarantees from long-term buyers and more exposure to global gas price volatility, but it offers the potential to capture larger returns. LNG Canada, Golden Pass in the United States and the Tortue floating LNG project off the cost of Mauritania and Senegal are examples of recently approved projects that combine balance sheet financing with a portfolio marketing model. The risks involved in balance sheet financing can be managed, for example by entering into joint ventures or shifting some of the construction risk onto engineering contractors through lump sum turnkey contracts. Some projects in the United States are testing new models of risk allocation by offering prospective partners equity stakes in the liquefaction terminal itself.

Balance sheet financing implies that greater amounts of the volumes from LNG terminals are bought by intermediaries rather than end-users, thereby increasing market liquidity. However, it is not yet clear how it might impact the delivered cost of LNG. We estimate that decreasing the cost of financing by half could decrease the delivered cost of LNG by up to 10%. However, this decrease may be offset by the need for higher risk-adjusted returns. In the Stated Policies Scenario, $14 billion on average is spent every year on new LNG capacity to 2040. Although this scenario sees changing financing patterns, long-term contracts that commit buyers to significant deliveries of gas remain an important element in the approval of new projects.

Growth in contractual innovation

Over the last decade, the number of companies purchasing LNG jumped from 40 to nearly 100, with total contracted volumes more than doubling to 360 Mtpa. Fifteen new LNG buyers have emerged in the last two years alone, a reflection of the growing supply and accessibility of LNG. In this more crowded marketplace, LNG contracts are becoming much more diverse and the terms of trade between buyers and sellers are evolving.

The changes underway are partly a consequence of utility buyers in both established and emerging markets no longer being certain of their long-term gas requirements. Their investment horizons have been clouded by the declining costs and rising deployment of renewable energy sources, and in some cases by increased competition from market opening. New LNG buyers in developing markets tend to have a different risk profile and to be more willing to contract gas on a more speculative basis. At the same time, a growing secondary market for LNG has given buyers more confidence in the short-term availability of supply. Together, these forces have translated into buyer demands for contracts with lower volumes and greater delivery flexibility.

These demands have become increasingly visible in recent contracting trends. The number of new LNG deals signed for volumes of 2 Mtpa or less has grown from 40% of the market in the 2010-14 period to more than two-thirds in the 2015-19 period. Around half of companies buying LNG have a portfolio size no bigger than 2 Mtpa; these buyers represent over 15% of LNG contracts signed in 2015-19, up from about 6% during the previous five years.

The rise in short-term contracts and spot trading has been partly facilitated by floating storage and regasification units (FSRUs), which tend to have relatively low upfront capital requirements, as they are usually leased on a short-term basis rather than purchased. In particular, FSRUs appeal to buyers uncertain of their long-term gas requirements. They also enable near-term needs to be met relatively quickly: Bangladesh and Pakistan are good examples of where FSRUs were used to provide LNG to plug an electricity generation deficit or as a stopgap in the face of a decline in indigenous production. The chartering of FSRUs provides fertile ground for contractual innovation, given the short-term nature of LNG procurement, the involvement of smaller and less creditworthy buyers than in the past, and the need for contract terms to reflect bundled services such as LNG-to-power. In the past three years, however, FSRU chartering activity has slowed, with several projects in emerging markets cancelled or postponed for diverse reasons: some countries are developing their own gas resources and therefore no longer require imports, while others have struggled to secure financing or favourable terms of trade.

More traditional long-term LNG contracts are also gradually being reshaped. Half of all existing LNG contracts, involving some 200 bcm of LNG, are due to expire over the next decade (Figure 4.15). Negotiations for contract extensions seem likely to involve demands for enhanced flexibility, with buyers pointing out that the initial costs of the projects have now been written off, against the background of a regulatory push to remove destination clause restrictions in major LNG importing regions (with continuing efforts in Europe since 2000 and more recent initiatives by Japan and Korea).

Changes in LNG contracts are not without challenges. Buyer demands for seasonal supply conflict with seller needs to market the entire output of a terminal continuously on a long-term basis (although large sellers may be able to balance different seasonal needs across a global portfolio). Meanwhile a growing number of LNG buyers have lower credit ratings than has traditionally been the case, increasing the risks to the seller: nearly 45% of new LNG buyers have a non-investment grade rating or no rating for their default risk.

Figure 4.15 ▷ **LNG trade volumes by contract type and assumed oil indexation levels in the Stated Policies Scenario**

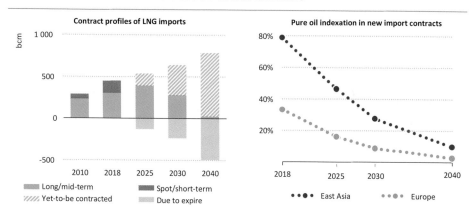

LNG traded on a spot or short-term basis is increasing as the market opens to new buyers including portfolio traders; pricing increasingly moves away from oil linkage

Notes: Long/mid-term contracts are those longer than three years. Contract volumes include those sold by LNG suppliers on the primary market, net of secondary sales e.g. by portfolio players. "Pure" indexation refers to a situation where over 80% of the price of gas sold under a sales contract is determined through a linkage to the price of crude oil or oil products.

LNG pricing trends

Historically, gas has been priced in relation to oil, providing a reliable reference price for large-scale investments in upstream projects, transport pipelines and LNG terminals. As discussed, there are forces at work to separate gas and oil markets, but around 45% of internationally traded gas today remains based on some form of oil indexation (IGU, 2019). The use of oil indexation varies considerably by region. In North America, gas prices are dictated by the fundamentals of supply and demand as a result of competition between multiple sellers and buyers interacting in a spot market and continuously trading on both physical and virtual hubs. This is commonly referred to as "gas-on-gas competition". European prices, by contrast, are determined by a mixture of gas-on-gas competition and oil indexation, while the dominant pricing mechanism in Asia remains oil indexation, with only a small portion of gas traded on the short-term spot market.

Oil indexation levels are expected to diminish over the outlook period, due to a number of mutually reinforcing developments. The growth of destination-flexible, hub-priced LNG exports from the United States is already providing the catalyst for a more liquid global gas market. The concurrent rise in traders and portfolio players, along with the relaxation of destination clause restrictions, facilitates short-term optimisation of LNG volumes based on arbitrage opportunities. This further supports the rise of a spot market, which becomes increasingly important as a backstop for contractual surpluses and deficits as the supply of international LNG grows.

Over time, the Stated Policies Scenario sees spot gas prices in Europe and Asia settle in a range between the short-run marginal cost of importing LNG from the United States ($4-7/MBtu over the projection period) and the global long-run marginal costs of developing new LNG export projects around the world (at a weighted average cost of $7.50/MBtu, rising above $10/MBtu by the end of the projection period). As the oil price rises, buyers locked into oil-indexed gas contracts pay $9-13/MBtu for their gas imports (depending on contract "slopes" that dictate the strength of the oil-gas price link), and this is likely to expedite contractual renegotiations that incorporate a more diverse set of pricing benchmarks. The rise in such hybrid pricing structures linked to multiple indices allows for a more even spread of market risk between buyers, sellers and traders. As interregional LNG trade grows and increasingly pivots toward Asia, there is further impetus to develop more liquid and transparent pricing references that can help financing for new LNG infrastructure. This, in turn, encourages an acceleration of domestic gas market reforms in several emerging Asian economies, allowing for third-party access to infrastructure to promote downstream competition.

These conditions give rise to a scenario in which the price of LNG, and gas more generally, is increasingly determined by its own supply-demand fundamentals, rather than linked to the price of oil or any competing fuel. Over time, the dominance of oil indexation in new import contracts recedes (Figure 4.15). In the Stated Policies Scenario, less than 20% of LNG trade worldwide remains based on oil indexation by 2040, mostly representing legacy contracts concluded before the mid-2020s.

The LNG business has evolved substantially over the past years, with an increasing roster of suppliers and buyers. However, there are limits to the extent to which LNG, and natural gas more generally, can become a global commodity similar to oil. LNG requires large-scale shipping and costly liquefaction and regasification infrastructure, and gas is more expensive to store than oil or coal. In the Stated Policies Scenario, around 10% of global natural gas supply is liquefied and transported by sea, and two-thirds of this gas remains contracted under long-term, point-to-point delivery arrangements. The scale of upstream gas resources and the financial resources required to sanction an LNG project biases the playing field in favour of larger, well-capitalised sellers and buyers.

However, a key element in LNG's growth story is the ability of sellers to offer more flexible terms to a new wave of buyers, and to allow for innovation in contracting and pricing arrangements to accommodate the rising liquidity and availability of LNG. Several traditional elements of the LNG business seem likely to endure, but are set to co-exist with these more novel elements as the sector continues to evolve.

Box 4.2 ▷ Technologies to minimise emissions from LNG supply

In the Stated Policies Scenario, 80% of the growth in global gas trade to 2040 comes in the form of LNG, with the majority making its way to Asia. Therefore, the indirect emissions arising from the production, transport and delivery of LNG are set to rise significantly.

Liquefying gas is an energy-intensive process and therefore often emissions intensive. Around 10% of gas is consumed as part of the liquefaction process, with most of this being used to power the equipment used to cool that gas to minus 162 degrees Celsius. Pipeline transport also results in emissions (gas is used in compressor stations along the pipeline and in some cases the losses arising from ageing transmission lines are significant), but globally LNG transport, on average, is more emissions intensive than pipeline transport.

A key way to reduce indirect emissions arising along the LNG supply chain (apart from efforts to minimise methane leaks, which are very important, but not specific to LNG) is to electrify the liquefaction process using low-carbon electricity. This would eliminate nearly all of the emissions associated with liquefaction, and lead to a 40% average reduction in greenhouse gas (GHG) emissions from coal-to-gas switching for the production of heat, compared with a 30% reduction if these mitigation strategies were not in place. There is one electric LNG plant currently in operation (the Snøhvit facility in the Norwegian Sea) and others are under construction or under consideration (Freeport LNG in Texas, as well as a number of projects in Canada, such as LNG Canada, Woodfibre and Kitimat). If all the world's existing liquefaction facilities were electrified using zero-carbon electricity, it would reduce annual emissions from the LNG supply chain by 80 million tonnes of CO_2.

Can new LNG technologies create "unconventional demand"?

In recent years, there has been growing interest in the potential for small-scale LNG to unlock new markets and applications for natural gas. Small-scale LNG refers to the use of LNG as a liquid fuel in a number of niche applications for which pipeline gas is unsuitable. Close to 30 Mtpa of small-scale LNG capacity is estimated to be in place, with the vast majority in China (White and Brooks, 2018).

Small-scale LNG has potential uses as a marine fuel for ships and bunkers (see Spotlight), as a liquid fuel for trucks and rail, and as a fuel for use in remote locations that are not served by gas infrastructure (a prime example being oil-based electricity generation in remote off-grid locations). Small-scale LNG also provides an opportunity to underpin a broader transition to natural gas by building new customer bases for the use of LNG that in time could justify the construction of large-scale pipeline infrastructure.

The use of small-scale LNG in various forms is growing in a number of places. In Europe, regasification facility operators have doubled the number of LNG truck-loading services and tripled the number of bunkering facilities over the last five years. More than three-quarters of European regasification terminals now possess truck-loading capabilities, and this has encouraged the expansion of associated infrastructure such as refuelling stations (Gas Infrastructure Europe, 2018). China has quickly developed LNG transportation infrastructure in the form of both trucks and inland bunkering facilities. There are now nearly 300 000 LNG-fuelled trucks in China, a ten-fold increase in four years. Nearly 25 million tonnes of small-scale LNG was delivered in China in 2018 (accounting for around 50% of total Chinese LNG demand), around half of which was sourced from regasification terminals and the rest from small-scale inland liquefaction facilities.

Elsewhere in the world, developing supply chains for small-scale LNG is more challenging, but the potential is high. Figure 4.16 compares the competitiveness of small-scale LNG versus oil products in medium-size industries and for baseload electricity generation. The costs of small-scale LNG lie in a relatively wide range of $2.5-8.5/MBtu, and largely depend on the presence of existing LNG infrastructure, in particular LNG liquefaction or regasification terminals with truck-loading capabilities. Receiving terminals would need to be either FSRUs or otherwise offer "break bulk" capabilities, which could parcel out smaller quantities for local industrial demand; storage tanks would also enable peak shaving in the power sector. Trucking LNG further inland would entail additional costs as well as logistical challenges: for example, a 100 megawatt (MW) baseload power plant would require, on average, around 20 daily deliveries from tanker trucks.

Using the average price of crude oil in 2018 of $70/barrel, up to 60 Mtpa worth of small-scale LNG could be delivered at a lower cost to end-users than oil products. While the substitution potential is only a fraction of oil demand in these sectors, it represents a sizeable 25% of the global LNG market. Around 40% of the global potential lies in developing Asian markets, where gas is around 20-40% cheaper than heavy fuel oil and diesel for industrial consumers. Substitution is costlier in the Middle East, where small-scale LNG fails to compete with highly subsidised oil product prices. Moreover, although the region has significant oil-fired electricity generation, less than a third of plants there are commercially suited to small-scale LNG (i.e. have a capacity less than 300 MW).

Although the potential for cost-effective substitution is high, there are logistical and other hurdles to using small-scale LNG. Lower fuel costs and improved efficiency might be offset by the additional capital expenditure required for converting equipment, which is site- and process-specific. Relatively high storage costs and boil-off rates are also important considerations, which makes the economics of delivering small-scale LNG-to-power plants challenging for plants running intermittently, e.g. as back-up or for peaking purposes. In developing markets, affordability and logistical concerns have discouraged some potential buyers from contracting what are relatively small quantities of flexible supply that need to be delivered on short-term contracts. In addition, while a growing spot market and an expanding list of intermediaries such as trading houses can support small-scale LNG projects, contracting and financing challenges have led to projects being cancelled or delayed.

Figure 4.16 ▷ Fuel cost competitiveness of small-scale LNG versus oil products for stationary uses, 2018

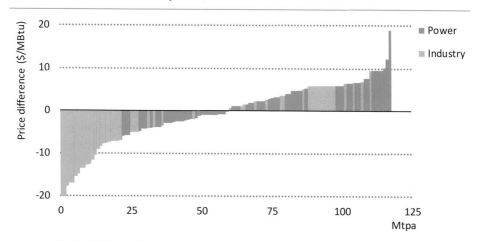

Up to 60 Mtpa of small-scale LNG could be delivered at a lower cost than oil products that are used for electricity generation and in medium-size industries

Note: Price differences based on an assessment of end-user prices for both gas and oil products that are paid by industrial customers and power plant operators in different countries around the world, taking into account the costs of building small-scale LNG infrastructure, which is a crucial prerequisite for substitution.

In the Stated Policies Scenario, natural gas captures almost 40% of the growth in industrial energy demand in emerging economies, reflecting the economic case for natural gas to satisfy incremental demand that otherwise would be met by costlier oil products, particularly in small- and medium-scale manufacturing subsectors. Traditional large-scale onshore regasification facilities and the build out of transmission and distribution networks underpin the majority of this growth: small-scale LNG is likely to remain a niche part of the global gas market, with its development potential driven by smaller players serving peripheral demand in end-use markets.

SPOTLIGHT

Can LNG sink emissions in maritime shipping?

The International Maritime Organization (IMO) sulfur cap, which enters into force in January 2020, has generated interest in using LNG as an alternative to high-sulfur fuel oil in maritime transport. LNG emits almost no sulfur dioxide or particulate matter, and contains up to 90% fewer nitrogen impurities than heavy fuel oil. For shipping companies, LNG is one choice among a range of options to comply with the IMO sulfur regulation, with scrubbers or low-sulfur fuels seen as the main alternatives. There are currently around 130 orders for new LNG-fuelled ships: two-thirds of these are due to be based in Europe, where bunkering infrastructure is currently the most developed. While this would double the global size of the existing fleet, it represents only around 4% of the total order book for new vessels.

In the Stated Policies Scenario, the use of LNG in international shipping reaches 50 bcm by 2040 from less than 1 bcm today, and accounts for 13% of shipping fuel mix. In the Sustainable Development Scenario, whether LNG in shipping has a role to play depends on its ability to reduce GHG emissions. The international shipping sector was responsible for around 700 million tonnes (Mt) of CO_2 emissions in 2018 (2% of global energy-related CO_2 emissions). The IMO has adopted an initial strategy to cut GHG emissions in 2050 by at least half compared to 2008 levels. While the Stated Policies Scenario includes future IMO mandates on sulfur, nitrogen oxide emissions and energy efficiency, it does not include its GHG target for 2050 because details of its implementation have yet to be defined. However, with vessels expected to be in service for up to 30 years, shippers are already thinking about ways to make substantial reductions to emissions intensity.

Burning LNG as a bunker fuel is estimated to achieve at best a 20% reduction in CO_2-equivalent direct emissions compared to fuel oils (Speirs et al., 2019). Assuming scrappage rates consistent with historical averages, "locked-in" carbon emissions from the existing fleet (including vessels on order) would be around 30 Mt CO_2 in 2050. This is some 290 Mt CO_2 less than the maximum level of emissions consistent with the IMO target. In the highly unlikely event of all new ships being powered either by zero-carbon fuels or by LNG, then all of the 290 Mt CO_2 could potentially be emitted by LNG-fuelled ships (Figure 4.17). This translates into a hypothetical upper-bound for annual LNG bunkering of around 100 bcm, or roughly 30% of global LNG trade in 2018.

The extent to which LNG-fuelled ships can maximise this potential, however, is clouded by uncertainty about the overall emission benefits. A comprehensive assessment must consider "well-to-wake" emissions, which combines GHG emissions that can occur upstream before the LNG reaches the ship (i.e. "well-to-tank"), and emissions from methane slip, where gas is not fully combusted in a ship engine and escapes into the atmosphere (i.e. "tank-to-wake"). We assess the average upstream emissions intensity of LNG to be around 160 grammes of CO_2 per kilowatt-hour (g CO_2/kWh), while the extent of methane slip can vary considerably depending on engine performance and dynamic conditions (such as weather). We estimate that, on average, the GHG benefits of utilising LNG as a maritime fuel compared to marine diesel are neutralised once total fugitive methane emissions exceed 4% of a ship's gas consumption.

Figure 4.17 ▷ Locked-in emissions from international shipping and the maximum potential remaining emissions from LNG

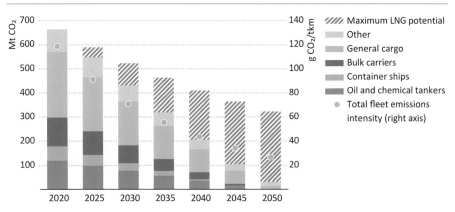

As the current shipping fleet is scrapped, demand growth for LNG as a bunker fuel would be limited to around 100 bcm if the 2050 IMO target is to be met

Notes: Mt CO_2 = million tonnes of carbon dioxide; g CO_2/tkm = grammes of carbon dioxide per tonne kilometre.

In the Stated Policies Scenario, the focus for ship operators is principally on the 2020 sulfur cap, for which LNG offers an established and proven option, as well as a reasonable hedge against a tightening of future environmental regulations. In the Sustainable Development Scenario, the focus is more on long-term emissions reduction targets, and here the contribution of LNG is less certain. In this scenario, the overall fleet must accommodate a projected doubling of shipping activity to 2050, while simultaneously achieving a near 80% reduction in the average amount of CO_2 emitted per tonne kilometre. Maximising the potential for LNG in this context depends on a significant uptake of zero-carbon options to fill the sizeable gap between maximum allowable emissions and total shipping activity (as shown by the total fleet emissions intensity in Figure 4.17). Overall, LNG is not a silver bullet, but it could play a role along with other options in helping to reduce emissions in maritime shipping.

4.7 Understanding the global potential for coal-to-gas switching

The substitution of one fuel by another is a fundamental part of energy system change. The shares of fuels in the global energy balance have undergone several important shifts throughout history, notably the transition from biomass to coal during the Industrial

Revolution, and from coal to oil and gas in the 20th century. A new transition is now underway to low- or zero-carbon fuels. It is in this context that we examine the global potential for coal-to-gas switching.

Natural gas is the cleanest burning fossil fuel. Combustion results in around 40% fewer CO_2 emissions relative to coal and 20% fewer than oil for each unit of energy output. Natural gas also emits fewer air pollutants, giving it the potential to improve air quality rapidly when substituting for other combustible fuels. Methane leaks to the atmosphere along the oil, gas and coal value chains are a very important part of the picture, but we find that in the majority of cases, gas is preferable to both oil and coal even when taking into account the full spectrum of emissions arising from extraction, transport and end-use (IEA, 2019b).

Some of the prospects for gas to challenge oil in both stationary applications and in maritime and road transport are explored in the previous section. The focus here is on the opportunities and limits for gas to gain ground at the expense of coal, and in which timeframes and sectors this might apply.

Near-term opportunities for natural gas

The clearest case for switching from coal to gas arises when there is the possibility of using existing infrastructure to provide the same energy services but with lower emissions. The power sector – where 40% of gas and 60% of coal is consumed – is the main arena for competition between the two fuels. Nearly 10 gigatonnes (Gt) of CO_2 emissions, around one-third of global energy sector emissions, come from coal-fired electricity generation, making this by far the largest single category of CO_2 emissions.

Since 2010, we estimate that over 500 Mt of CO_2 emissions have been avoided due to coal-to-gas switching (Figure 4.18). Two-thirds of these savings have occurred in the power sector, reflecting the lower emissions from gas-fired electricity (which is around half that of coal on a lifecycle basis). The largest savings occurred in the United States, where the rise of shale gas has pushed down natural gas prices and underpinned large-scale switching in the power sector, where emissions have dropped by a fifth since 2010. The shale revolution has also had implications for switching in other regions, with rising US exports of LNG helping to push spot gas prices in Europe from over $8/MBtu in 2018 to below $4/MBtu, at times in 2019, thereby improving the economics of gas-fired electricity generation.

Switching from coal to gas in the European Union (EU) has also been supported by policy interventions. In the United Kingdom, coal-to-gas switching has contributed to a 55% drop in the emissions intensity of electricity generation following the introduction of a carbon price floor in 2013: this imposed a minimum cost to generators of GBP 9/tonne CO_2, which was doubled in 2015. The EU's Emissions Trading System (ETS) has been reformed to eliminate surplus allowances from the market: this has pushed up the ETS price from an average of EUR 5/tonne CO_2 in 2016 to over EUR 20/tonne CO_2 in late 2018, helping to tip the balance of short-run generating costs in favour of gas over coal. Coal-fired power plants have also been affected by stricter EU directives governing pollution from large combustion plants, and national policies to phase out coal generation in several European countries.

Figure 4.18 ▷ CO_2 emissions reductions since 2010

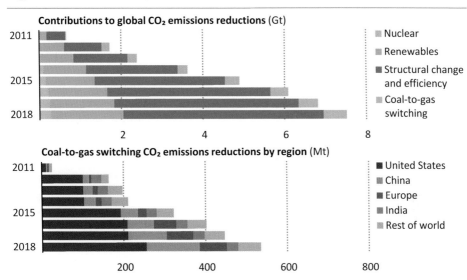

Coal-to-gas switching has prevented over 500 Mt CO_2 of emissions since 2010. More than two-thirds of the savings have taken place in the United States and China.

Note: Structural change and efficiency shows emissions savings compared to a baseline that assumes no further improvements in the energy and CO_2 intensities of gross domestic product since 2010.

In 2019, coal, gas and carbon prices in the European Union have at times been at levels that place more than three-quarters of the EU's gas-fired generation capacity within a competitive range for coal-to-gas switching. Ultimately, the extent to which gas displaces coal in practice depends on daily electricity demand profiles, fuel contracting patterns as well as on the extent of renewable-based generation: the accelerating pace of wind and solar output in several key markets is set to shrink the space in which coal and gas compete for market share.

Policies are also driving coal-to-gas switching in China, where the main motivation is to improve air quality and the main sectors involved are industry and buildings rather than power. Given the continued growth in energy demand, a high priority is being given to building new gas infrastructure, even though imported gas is relatively expensive. Nearly 9 million additional households switched from coal to natural gas or electricity for heating in 2017/18, with a further 3 million conversions planned to be completed by 2021. China's National Energy Administration is also seeking to expand biogas production to promote rural coal-to-gas switching. It aims for production to reach 30 bcm by 2030 (from around 10 bcm today).

The example of China highlights that, in many gas-importing countries in Asia, gas needs a supporting policy push in order to displace coal. It also highlights that, in a world of ever-cheaper renewables, gas does not necessarily represent the major challenger to baseload coal in Asia's electricity generation. In India, gas currently has a low share of the energy mix (as in China), and there has been very little evidence of switching to gas so far. Plans to expand city gas distribution grids in India, if realised, would make gas far more widely available, but the economics would not appear to favour replacing coal with gas in power and industry; instead, gas may displace liquefied petroleum gas (LPG) as a cooking fuel, and oil as a fuel in some parts of the transport and industry sectors.

In many countries in Southeast Asia, the immediate gains from switching from coal to gas are only feasible with very low gas prices. Given the relatively young and modern fleet of coal-fired plants and the low cost of indigenous coal, gas prices would need to drop to a range between \$2-4/MBtu to create a commercial case for switching – which in many cases is below the long-run marginal cost of domestic supply – and well below the cost of LNG imports (Figure 4.12).

Box 4.3 ▷ How does carbon pricing affect the switching calculation?

Commodity prices for coal and gas paid by utilities today mean that most of the switching potential identified in our analysis is out of reach. However, this could change if a higher price was put on CO_2 emissions. The size, location and relative efficiencies of existing coal and gas-fired power plants are key determinants of the technical and economic potential for a carbon price to lead to a switch from higher to lower emitting units. Power systems designed around a merit order whereby plants are dispatched according to their short-run marginal costs of generation are a crucial prerequisite for allowing gas plants to run ahead of coal if relative prices are favourable.

Globally, a price of \$60/tonne of CO_2 could trigger nearly 700 Mt CO_2 in avoided emissions from coal-to-gas switching in electricity generation using existing infrastructure and our commodity price assumptions in the Stated Policies Scenario in 2025 (Figure 4.19). In the United States, significant gains from switching could be achieved with an average carbon price of \$40/tonne CO_2. In Europe, savings from switching could be unlocked once carbon prices exceed \$22/tonne CO_2 (EUR 20/tonne CO_2); the ETS has been around this level since early 2019, supporting an increase in gas-fired generation.

China's emerging carbon market may offer a similar opening for natural gas to gain market share. However, with a cost gap between coal and gas-fired electricity generation of \$30-40 per megawatt-hour, a CO_2 price in the \$60-80/tonne CO_2 range would be needed to provide enough support for switching to gas. Similar conditions prevail in Southeast Asia, where a carbon price averaging \$70/tonne CO_2 could result in savings of some 120 Mt CO_2, or around 20% of Southeast Asia's total power sector CO_2 emissions. In markets such as India, where high gas prices have led to stranded gas-fired power plants, carbon prices would have to average \$90/tonne CO_2 to stimulate switching.

Carbon prices can also affect the investment case for new natural gas plants as compared with renewables or other technologies such as nuclear, battery storage or carbon capture, all of which present alternatives to the use of gas to reduce emissions in the power sector. The costs of wind and solar technologies in particular are falling considerably, bolstering the economic case for switching directly from coal to renewables. However, there are other factors that bear on the investment calculation. Power systems typically need a certain proportion of baseload, mid-merit and peaking plants to meet variable levels of demand: the extent to which gas, renewables or other power technologies can meet these demands, in addition to providing services such as flexibility and standby capacity, is an important consideration. Moreover, while the cost of carbon stimulates investment by influencing the wholesale power price, new gas-fired power plants and renewable energy projects often derive much of their revenue from outside the wholesale power market. A carbon price therefore is one of a range of considerations that will determine the role of gas in future power systems.

Figure 4.19 ▷ Average cost of potential emissions savings from coal-to-gas switching in the power sector, 2025

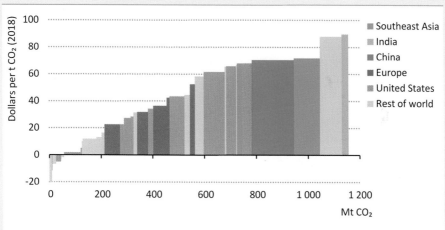

Carbon prices needed to trigger switching differ around the world; most low-cost potential is in Europe and the US which have older coal plants and spare gas capacity

Notes: Average delivered cost to power plants (gas / coal per MBtu) with no carbon price applied: United States: $9 / $3; European Union: $35 / $11; China: $44 / $17; India: $41 / $10. The United States is broken down by regional independent system operators.

Taking full advantage of near-term opportunities for switching in the power sector, based on existing infrastructure would yield global CO_2 savings of around 1 200 Mt CO_2 (Figure 4.20). Most of the realisable potential is in the European Union and the United States. With relatively slow electricity demand growth, ample gas and power infrastructure and significant spare gas capacity, these markets could displace around half of their respective coal-fired power output.

Elsewhere in the world most of the switching potential is not economic at current commodity prices. The overall economic potential for switching in much of Asia is limited by the efficiency of the relatively young coal-fired power fleet. In China, for example, the share of supercritical and ultra-supercritical coal plants within the fleet increased from 3% in 2005 to almost 40% in 2018, and plants have modernised quickly, adapting to stricter regulations governing air pollutant emissions. Since the gas fleet is less than one-tenth the size of the coal fleet in China, the current savings potential from switching (around 100 Mt CO_2) is in any case small relative to its overall power sector emissions (4 900 Mt CO_2). With the absence of a carbon price and a coal price in the range of $60-80/tonne, the delivered cost of gas to most utilities in Asia would need to be below $4/MBtu to stimulate switching, compared with an average price for imported LNG in 2018 of $9.50/MBtu. Falling spot prices in 2019 have placed some gas plants in a more competitive position, but a positive differential for gas would need be sustained over a longer period to effect a more durable shift away from coal.

Figure 4.20 ▷ **Potential CO₂ savings from coal-to-gas switching at various gas prices using existing power plants**

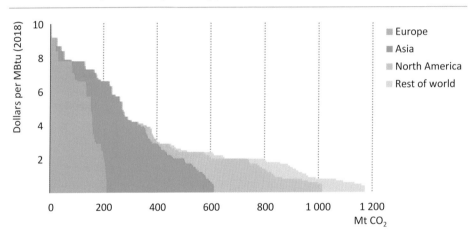

Rising carbon prices mean gas-fired power plants in Europe are competitive with coal, whereas most Asian markets require much lower gas prices – less than half of 2018 levels

Is coal-to-gas switching a viable long-term route to emissions reductions?

In the Stated Policies Scenario, gas use increases by an average of 1.4% each year through to 2040 and helps to meet existing energy policy commitments and ambitions. However, this scenario puts energy-related CO_2 emissions on an upward trend to 2040, far from the emissions trajectory required to tackle climate change. In the Sustainable Development Scenario, renewables and efficiency measures are the most important drivers of the energy sector transition. Natural gas plays a role in this scenario, although the extent varies by country, sector and timeframe.

In mature natural gas markets, like the United States and Europe, coal-to-gas switching is a compelling near-term option for reducing emissions, given existing infrastructure and spare capacity. Gas can also contribute to security of electricity supply by balancing variable renewables and meeting peaks in demand. However, given the need for decarbonisation efforts to intensify in the Sustainable Development Scenario, a role for unabated gas in the energy mix becomes increasingly challenging in this scenario: by 2040, gas demand is 40% lower than today's levels in Europe and 25% lower in the United States (Figure 4.21). The case for natural gas is also challenged by increasing investments in battery storage and grid management capabilities, which, if scaled up, could fulfil the same short-term flexibility functions as gas-fired power plants.

Figure 4.21 ▷ Change in gas and coal demand by scenario, 2018-2040

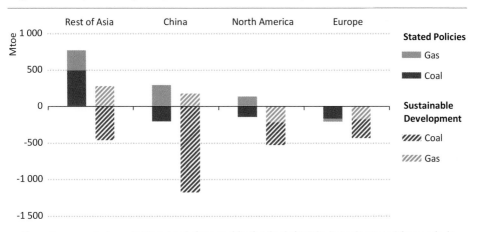

Gas plays a role in reducing coal demand in the Sustainable Development Scenario in Asia, while in North America and Europe both gas and coal are pushed out of the mix

Gas plays a more pronounced role in developing economies that are very carbon-intensive today, helping to push more polluting fuels out of the energy system. Gas demand in China is lower in the Sustainable Development Scenario, but still helps to displace coal demand in both power and industry, while in India gas demand is even higher than in the Stated Policies Scenario as gas replaces coal as a baseload source of electricity generation.

The long-term growth opportunities for gas in the electricity generation mix in Asia depend heavily on policy priorities in each country. Southeast Asia, where the demand for electricity is rising robustly, offers an illustration of the challenges facing gas as a lower emissions option than coal for providing power. Until recently, natural gas was the largest source of electricity generation in the region. However, several countries are now turning to coal, partly due to slowing or declining domestic gas production. Coal is plentiful and available at low cost and foreign direct investment has supported new coal capacity additions. In the Stated Policies Scenario, power sector emissions in Southeast Asia double, as natural gas cedes market share to coal while renewables, despite tripling, covers barely 40% of total electricity demand growth in the period to 2040.

While renewables offer a compelling pathway to long-term emissions reductions, investing in gas-fired power as a baseload source of generation in place of coal can also help reduce emissions. Figure 4.22 compares the emissions and generation performance for different power technologies under an illustrative $5 billion investment in Southeast Asia, taking into account capital, operating and fuel costs over a 30-year asset lifetime. If the priority is to maximise electricity output, then coal comes out on top (as it does for this region in the Stated Policies Scenario). Subcritical coal plants in particular remain among the cheapest generation options given lower fuel costs. More expensive ultra-supercritical coal plants yield lower emissions than subcritical plants, but their emissions per unit of power output are higher than the average emissions intensity of total generation in 2018 for the region as a whole (around 510 g CO_2/kWh, which is the baseline in Figure 4.22). If the priority is to lower this average, then wind or utility-scale solar offer larger long-term emissions savings than gas-fired combined-cycle gas turbine (CCGT) plants. Gas is therefore a second-best option, both in terms of environmental performance and power output.

Figure 4.22 ▷ Cumulative effects on electricity generation and emissions in Southeast Asia of spending $5 billion in the power sector

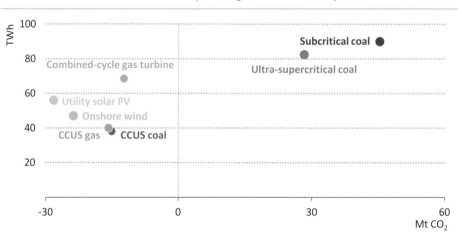

The highest efficiency coal plants still negatively impact average emissions in the region; gas brings improvements, but wind and solar provide more than twice the savings

Notes: TWh = terawatt-hours; CCGT = combined-cycle gas turbine; CCUS = carbon capture, utilisation and storage. Spending equals total capital and operating expenditure over a thirty-year investment period starting in 2020 and including construction lead times. Load factors for thermal power plants = 60%; wind = 24%; solar PV = 13%. CCUS capture rates assumed to be 90%. The change in emissions is compared to generating the same amount of electricity with the average emissions intensity of electricity generation in Southeast Asia in 2018.

The investment case for wind and solar capacity, however, is complicated by the variability of their generation profiles, which creates challenges for ensuring electricity system stability. Gas and coal plants have an advantage since they are dispatchable sources of electricity generation. In this regard, they can provide standby capacity to balance the variability of renewable output or, if equipped with carbon capture, utilisation and storage (CCUS), ensure stable baseload supply while also delivering significant emissions reductions.

A crucial variable for the future that could change the prospects for both gas and coal is the extent to which CCUS technologies are deployed. However, retrofitting existing plants would require large upfront capital costs and relatively long construction lead times. As for new plants, the finite amount of capital available for new generation capacity in the illustrative investment case would cover the installation of 250 MW of gas-based CCUS capacity compared with 500 MW of conventional gas turbine capacity. Taking into account the average emissions intensity for the power fleet as a whole, this means overall CO_2 emissions savings for new, conventional gas plants would be similar to CCUS units over the 30-year investment horizon.

While coal-to-gas switching is not the long-term answer to climate change, it is one of a portfolio of options to reduce emissions across a broad range of sectors, helping to bring down emissions in the short term, while providing valuable flexibility to power systems with growing shares of variable renewable energy sources. In the Sustainable Development Scenario, renewables and efficiency do the heavy lifting, while coal-to-gas switching contributes around 8% of the overall emissions savings required in the 2018-40 period.

Chapter 5

Outlook for coal
It is not just about power

SUMMARY

- Global coal demand increased for the second straight year in 2018, driven by strong electricity demand in developing Asian countries. This gradually rising trajectory is maintained in the Current Policies Scenario, but in the Stated Policies Scenario, demand is essentially flat, ending up in 2040 at around 5 400 million tonnes of coal equivalent (Mtce). Declines in China (-9%), United States (-40%) and European Union (-73%) are offset by rising demand in India (+97%) and Southeast Asia (+90%). Overall, coal use in power generation decreases and its industrial use expands.

- In the Sustainable Development Scenario, coal consumption falls steeply as a result of much more stringent environmental policies across the board. By 2040, unabated coal use comes under intense pressure, and global coal use is 60% lower than in the Stated Policies Scenario. As a result, coal's share in the primary energy mix falls towards 10% (Figure 5.1).

Figure 5.1 ▷ Global coal demand by scenario

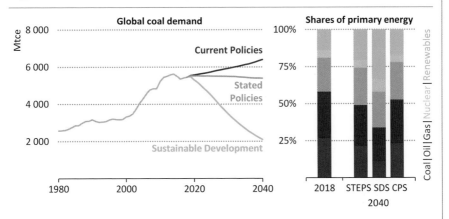

Coal's future is heavily dependent on the breadth and stringency of environmental policies across the world

Note: Mtce = million tonnes of coal equivalent; STEPS = Stated Policies Scenario; SDS = Sustainable Development Scenario; CPS = Current Policies Scenario.

- In many advanced economies, coal demand for power in the Stated Policies Scenario is in deep structural decline, hastened by specific phase-out commitments in some countries, the continued rise of renewables, competition from natural gas in the United States and higher carbon dioxide (CO_2) prices in the European Union. Coal demand drops too in China – by far the world's largest coal consumer – due in large

- part to a strong policy push to improve air quality. However, in other parts of developing Asia countries look to increase their use of coal to satisfy fast-rising demand for electricity and for industrial development.

- The main source of coal production growth in the Stated Policies Scenario is India, which doubles its production by 2040 on the strength of government production targets and efforts to reduce supply chain bottlenecks. The additional supply, however, is not enough to meet growing domestic demand, and India overtakes China by the mid-2020s to become the world's largest coal importer.

- Industrial coal use which today accounts for around one-third of consumption, increases by some 225 Mtce over the period in the Stated Policies Scenario: coal remains the backbone of the iron and steel, and cement sub-sectors and its use in the chemicals sub-sector keeps rising, particularly in China. In the Sustainable Development Scenario, overall use drops significantly, but coal remains important to several industrial processes, reflecting the difficulty and expense of finding substitutes for coal in these processes.

- A crucial variable in the Sustainable Development Scenario, and for the future of coal, is the extent to which carbon capture, utilisation and storage (CCUS) technologies are deployed in power generation and industry. In the industry sector, CCUS can provide a cost-competitive decarbonisation option for key industrial processes. The current pipeline of projects however is far short of what is required under this scenario to abate emissions from key industrial sectors of the economy.

- Global investment in coal supply today stands at about $80 billion per year, half the level reached at the 2012 peak. In the Stated Policies Scenario, nearly $1 135 billion is spent over the projection period, although investment sees a gradual, broad-based decline, resulting in an annual average of $50 billion to 2040.

- Financing restrictions from an increasing number of lenders add to the headwinds facing coal supply projects. This is not yet an issue affecting projects in China and India, which are the main countries investing in coal in the Stated Policies Scenario. But projects elsewhere that cannot be financed from the balance sheets of larger companies could struggle to move ahead. Among exporters this could provide an opening for Russian producers to increase market share.

- Discussions of energy-related methane emissions have focused on oil and natural gas operations, but coal mining also releases methane. We estimate that global emissions of coal mine methane (CMM), the methane naturally contained in coal seams, were around 40 million tonnes in 2018, equal to around 1 200 million tonnes of CO_2-equivalent (Mt CO_2-eq). The concentration of CMM is often low and can fluctuate in quality and quantity, making it technically and economically difficult to abate. More active regulatory regimes are needed to incentivise operators to install CMM abatement technologies where technically possible.

Introduction

Global coal use in 2018 rose for the second straight year, although it remained some 160 million tonnes of coal equivalent (Mtce) below the level of the peak in 2014. A shift in consumption towards Asia was again visible, as coal use rose in China, India, Indonesia and some other countries in South and Southeast Asia. Demand for electricity in Asia has continued to grow and coal remains by some distance the largest source of electricity generated in Asia, and among the cheapest.

Meanwhile coal is being steadily squeezed out of the energy mix in many advanced economies by a mixture of environmental policies and competitive pressures from increasingly cost-competitive renewables and, in some markets, also from natural gas. The United Kingdom, whose industrial revolution was built on coal, now goes for extended periods without any coal-fired power. Germany, a stronghold of coal demand in Europe, plans to phase out coal by 2038, and the latest data show coal-fired power generation falling sharply in 2019.

In our projections, there is a stark variation in the coal outlook between the Stated Policies Scenario, in which coal demand is essentially flat, and the Sustainable Development Scenario, in which it falls rapidly. A major consideration in this *Outlook* is that coal plays a central role in much of Asia, given a large coal supply industry and the young average age of the coal-fired fleet (see Chapter 6). Large-scale deployment of CCUS technologies could yet allow for a distinction to be made between coal use and the emissions from its combustion, and this is an important feature of the Sustainable Development Scenario, alongside a significant reduction in overall coal demand.

This chapter delves into three aspects of this complex picture in more detail:

- The future of coal is often seen as synonymous with the future of power generation, but around one-third of coal is used outside the power sector, mostly in industry. We explore the dynamics of industrial coal use, the scope for finding substitutes for coal across the main industrial sub-sectors, the outlook for industrial coal use in China, and the role of CCUS in reducing coal-related industrial emissions.

- Access to finance for new coal investment is becoming constrained, as an increasing number of banks and other financial institutions announce that they will no longer support any coal-related investment. We take stock of this move away from coal financing and ask what impact it might have on coal supply.

- The issue of methane emissions from oil and natural gas operations has received more attention so far than the extent and nature of methane emissions in coal supply. We provide new estimates of methane released during mining operations, describe the abatement possibilities and examine the implications for lifecycle emissions comparisons of coal versus natural gas.

Figures and tables from this chapter may be downloaded from www.iea.org/weo2019/secure/.

Scenarios

5.1 Overview

Coal demand is essentially flat in the **Stated Policies Scenario**, ending up in 2040 at around 5 400 Mtce, some 60 Mtce below where it is today (Table 5.1). This represents a slight downward revision compared with the *World Energy Outlook (WEO)-2018* (IEA, 2018). Flat demand in an expanding energy system means that the share of coal in the global energy mix declines from 27% in 2018 to 21% in 2040, falling behind natural gas in the process.

The strength of the economic and policy headwinds facing coal vary widely by scenario and, within each scenario, across different countries and sectors. The net effect in the Stated Policies Scenario is that global coal use in power generation decreases slightly, while its industrial use grows modestly. The **Current Policies Scenario**, in which energy demand is stronger and policy pressure on coal is weaker, sees coal use rise in both areas.

Table 5.1 ▷ Global coal demand, production and trade by scenario (Mtce)

	2000	2018	Stated Policies 2030	Stated Policies 2040	Sustainable Development 2030	Sustainable Development 2040	Current Policies 2030	Current Policies 2040
Power	2 233	3 500	3 470	3 395	1 872	858	3 789	4 156
Industrial use	869	1 680	1 852	1 903	1 461	1 206	1 926	2 075
Other sectors	207	279	175	100	137	36	220	168
World coal demand	3 309	5 458	5 498	5 398	3 471	2 101	5 934	6 399
Asia Pacific share	*47%*	*75%*	*81%*	*83%*	*86%*	*84%*	*79%*	*81%*
Steam coal	2 504	4 342	4 393	4 394	2 672	1 515	4 753	5 266
Coking coal	449	955	857	790	676	497	885	854
Lignite and peat	302	270	247	214	123	89	297	280
World coal production	3 255	5 566	5 498	5 398	3 471	2 101	5 934	6 399
Asia Pacific share	*48%*	*73%*	*78%*	*79%*	*80%*	*83%*	*77%*	*78%*
Steam coal	310	859	733	726	381	197	888	964
Coking coal	175	319	314	371	258	247	332	404
World coal trade	471	1 169	1 039	1 087	633	413	1 206	1 355
Trade as share of production	*14%*	*21%*	*19%*	*20%*	*18%*	*20%*	*20%*	*21%*
Coastal China steam coal price ($2018/tonne adjusted to 6 000 kcal/kg)	34	106	89	92	74	76	98	105

Notes: Mtce = million tonnes of coal equivalent; kcal/kg = kilocalories per kilogramme. Unless otherwise stated, industrial use in this chapter reflects volumes also consumed in own use and transformation in blast furnaces and coke ovens, petrochemical feedstocks, coal-to-liquids and coal-to-gas plants. Historical supply and demand volumes differ due to changes in stocks. World trade reflects volumes traded between regions modelled in the *WEO* and therefore does not include intra-regional trade. See Annex C for definitions.

Figure 5.2 ▷ Growth in global GDP, coal demand and related CO_2 emissions by scenario

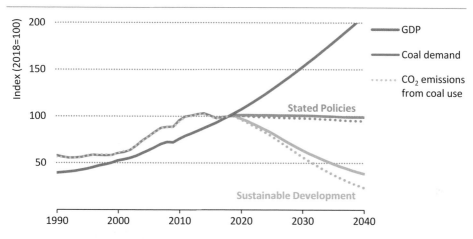

Coal demand has now decoupled from global GDP, largely due to changes in China; the relationship between future demand and emissions depends mainly on CCUS

Notes: GDP = gross domestic product. The divergence in decline rates for coal demand and CO_2 emissions from coal use over the period stems from efficiency gains and the uptake of CCUS.

The outlook for coal is very different in the **Sustainable Development Scenario**. With a much more stringent focus on reducing emissions, coal use decreases steeply at an annual rate of 4.2%. By 2040, world coal use is 60% lower than in the Stated Policies Scenario and coal's share in the primary energy mix falls towards 10%.

Until the early 2010s, coal demand was aligned with economic growth. That is not the case in the future in either the Stated Policies or Sustainable Development scenarios (Figure 5.2). In advanced economies, e.g. European Union, United States and Japan, the trend in coal demand becomes detached from the overall economic outlook. By contrast, strong growth in incomes and energy needs in parts of developing Asia continues to go hand-in-hand with higher coal demand. China's position moves progressively closer to that of the advanced economy group, exerting a strong influence on the global decoupling of coal demand from economic growth.

With coal demand growth levelling off, CO_2 emissions from coal combustion flatten in the Stated Policies Scenario, but they do not reduce significantly. In the Sustainable Development Scenario, the deployment of CCUS and improvements in plant efficiencies result in coal-related CO_2 emissions falling faster than coal demand. By 2040, almost 160 gigawatts (GW) of coal-fired plants are equipped with CCUS, accounting for 40% of the electricity generated from coal, although today's policies fall far short of those which could stimulate needed investment in CCUS.

5.2 Coal demand by region and sector

The flat outlook worldwide for coal demand in the Stated Policies Scenario masks significant contrasts between regions (Table 5.2).

In the United States, coal demand reached a 39-year low in 2018 and continues to fall over the outlook period, as coal-to-gas switching continues in the power sector.

In a growing number of other **advanced economies**, such as Canada, Germany and United Kingdom, demand has decreased and continues to do so over the period to 2040, reflecting the policy drive to reduce CO_2 emissions.

China sees a modest reduction in consumption of 0.4% per year on average from 2018 to 2040, while remaining, by some way, the largest consumer of coal worldwide. The outlook for coal use in China from year to year remains contingent on prevailing policy preferences and economic conditions. Final investment decisions for new coal-fired plants fell from 60 GW in 2015 to less than 6 GW in 2018 when, for the first time, more gas-fired power was approved than coal. However, China has a stock of more than 1 000 GW of coal-fired capacity, much of it recently commissioned and highly efficient (see Chapter 6).

Table 5.2 ▷ Coal demand by region and scenario (Mtce)

			Stated Policies				Sustainable Development	
	2000	2018	2025	2030	2035	2040	2030	2040
North America	818	492	369	328	304	285	81	50
United States	763	451	350	314	291	272	71	41
Central and South America	29	46	48	47	47	49	30	23
Brazil	19	24	23	23	23	23	14	12
Europe	578	447	314	263	219	203	129	84
European Union	459	319	204	157	113	87	84	59
Africa	129	159	165	160	160	161	113	92
South Africa	117	142	133	117	107	97	92	56
Middle East	2	6	9	10	12	14	7	6
Eurasia	202	229	225	212	203	199	136	74
Russia	171	166	160	144	136	129	88	50
Asia Pacific	1 551	4 079	4 385	4 476	4 502	4 487	2 976	1 771
China	955	2 834	2 934	2 845	2 710	2 568	2 065	1 154
India	208	586	771	938	1 063	1 157	546	395
Indonesia	17	72	97	114	130	148	67	39
Japan	139	165	141	124	115	107	70	43
Rest of Southeast Asia	28	132	172	194	219	241	91	38
World	3 309	5 458	5 515	5 498	5 446	5 398	3 471	2 101

Note: See Annex C for definitions.

Figure 5.3 ▷ Global coal demand by key sector and scenario

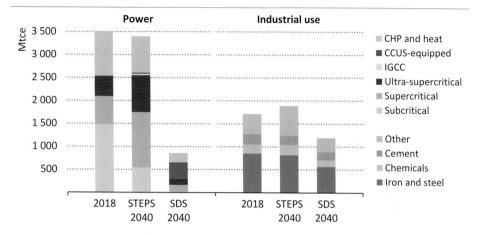

In the Sustainable Development Scenario, coal demand for unabated coal-fired power in 2040 is 90% lower than today. Industrial demand decreases across the sectors.

Notes: CCUS = carbon capture, utilisation and storage; IGCC = integrated gasification combined-cycle; CHP = combined heat and power; STEPS = Stated Policies Scenario; SDS = Sustainable Development Scenario.

Coal demand remains strong across much of **developing Asia**. India sees the largest increase, although the 3.5% annual rate of growth observed over the past five years is dampened by a large-scale expansion of renewables and the use of supercritical and ultra-supercritical technologies in new plants. Consumption also increases in Southeast Asia, notably in Indonesia: 40% of the projected rise in the region's electricity demand is met by coal.

Almost 65% of global coal demand today comes from the **power sector** (Figure 5.3). In the Stated Policies Scenario, slightly less coal is used in the power sector in 2040 than today (down 105 Mtce), but this generates slightly more electricity (up 305 terawatt-hours [TWh]) because more efficient technologies are deployed while old plants are retired: the average efficiency of coal-fired generation improves from 42% today to 43% in 2040.

Industrial coal use is a growth area for demand in some regions. The 225 Mtce growth in this sector in the Stated Policies Scenario is underpinned by the less energy-intensive industrial sub-sectors, and by the use of coal as a feedstock for coal-to-gas and coal-to-liquids projects, particularly in China (see section 5.6).

In the Sustainable Development Scenario, coal demand decreases across the majority of sectors. Coal-fired power drops to 6% of total generation, but industrial demand is more resilient because the substitution possibilities are more limited for some industrial processes than for power (see Chapter 6). The availability of CCUS at scale becomes a crucial variable in this scenario. CCUS and hydrogen are important tools to address emission abatement in industry (Box 5.1).

5.3 Coal production by region

Table 5.3 ▷ Coal production by region in the Stated Policies Scenario (Mtce)

	2000	2018	2025	2030	2035	2040	2018-2040 Change	CAAGR
North America	824	576	432	385	358	329	-247	-2.5%
United States	767	526	399	358	333	302	-223	-2.5%
Central and South America	48	82	75	75	62	62	-20	-1.3%
Colombia	36	77	70	70	58	58	-19	-1.3%
Europe	397	230	174	131	105	80	-150	-4.7%
European Union	307	163	117	79	57	30	-134	-7.5%
Africa	187	225	217	199	210	221	-4	-0.1%
South Africa	181	209	198	174	176	173	-37	-0.9%
Middle East	1	1	2	2	2	2	0	0.7%
Eurasia	234	414	408	409	415	424	10	0.1%
Russia	184	338	330	330	341	350	11	0.2%
Asia Pacific	1 564	4 039	4 208	4 297	4 295	4 282	243	0.3%
Australia	235	411	386	402	426	453	42	0.5%
China	1 019	2 668	2 751	2 710	2 589	2 481	-188	-0.3%
India	187	415	558	681	773	844	429	3.3%
Indonesia	65	416	380	367	368	359	-57	-0.7%
World	3 255	5 566	5 515	5 498	5 446	5 398	-168	-0.1%
Sustainable Development			4 473	3 471	2 682	2 101	-3 466	-4.3%

Notes: CAAGR = compound average annual growth rate. See Annex C for definitions.

China continues to be the largest coal producer over the projection period but, after a slight increase in the 2020s, the Stated Policies Scenario sees a gradual decline in output (Table 5.3). Authorities in China have been actively engaged in a programme of restructuring that has seen the closure of small, less efficient mines, but the expansion of capacity in the west of China has more than offset these closures.

India's production grows by 3.3% per year, reflecting the ambitious output targets set by the authorities (mainly for Coal India Limited, which accounts for around 80% of the country's output), together with efforts to reduce bottlenecks in the rail transportation network and overhaul the coal allocation system to increase the reliability of deliveries to consumers. Steam coal accounts for 90% of output and almost all the output growth. India's share in global production increases from 7% today to 16% in 2040 (Figure 5.4).

Russia has some of the lowest mining costs in the world, but output has to be transported over long distances from coal mining regions to ports. Transport costs are diluted by subsidies, but the share of transport in delivered costs is still higher than for most other producers. Russia has been investing in new mining capacity and infrastructure with a view to expanding exports to Asia. In our projections, coal production increases slightly from 338 Mtce in 2018 to 350 Mtce in 2040, while exports rise from 166 Mtce to around 220 Mtce.

After an uptick in 2017, coal production in the **United States** declined slightly in 2018 as higher exports were offset by falling domestic demand. In our projections, the decline in output continues through the period. Coal production of 305 Mtce in 2040 is less than 40% of the peak level reached in 2008. Weakening demand in the power sector and shrinking export markets, especially for steam coal in Europe, lead to additional falls in US output.

In the **European Union**, near-term hard coal production remains steady in Poland and lignite production continues in Germany, Czech Republic and other Eastern European countries. Over time, however, coal production is reduced in line with plans to reduce domestic consumption, and coal output is down 80% in 2040 compared with today.

Even as domestic demand falls, **Australia** is one of the few major producers projected to increase coal production to 2040 in the Stated Policies Scenario. Our projections are consistent with the development of some new mines, although there are large uncertainties over the extent of import demand in Asia and the financing environment for greenfield projects (see section 5.7).

Indonesia's coal production reduces over the period to 2040. The share of output serving domestic markets rises steadily, from less than 20% today to 40% in 2040, and coal exports fall significantly. Indonesia serves as something of a swing producer in Asia, as companies wishing to export can increase production quickly in response to price signals from international markets.

Figure 5.4 ▷ Share of coal production by key country in the Stated Policies Scenario

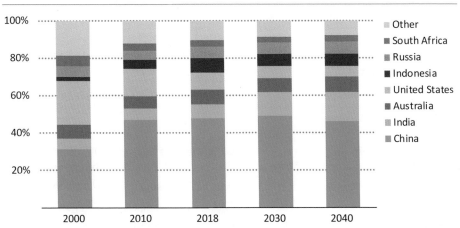

While overall world coal production remains flat to 2040, India increases its share by more than doubling its level of production, offsetting declines in China and the United States

5.4 Trade[1]

In the Stated Policies Scenario, traded coal volumes remain broadly flat through to 2040 at around 1 100 Mtce, equivalent to about 20% of global coal production (Table 5.4). Import demand for coal slows in China and advanced economies in Europe and Asia, but it is partially offset by increasing imports of coal in India and in some other developing Asian economies.

Table 5.4 ▷ Coal trade by region in the Stated Policies Scenario

Net importer in 2040	Net imports (Mtce)				As a share of demand			
	2000	2018	2030	2040	2000	2018	2030	2040
India	20	187	256	313	10%	32%	27%	27%
Other Asia Pacific	52	122	181	257	56%	53%	60%	66%
Japan and Korea	192	280	219	178	97%	99%	100%	100%
China	-58	228	135	87	n.a.	8%	5%	3%
European Union	139	158	78	58	30%	50%	50%	66%
Rest of world	24	81	92	111	12%	33%	39%	40%
Net exporter in 2040	**Net exports (Mtce)**				**As a share of production**			
	2000	2018	2030	2040	2000	2018	2030	2040
Australia	173	353	359	418	74%	86%	89%	92%
Russia	14	166	186	221	7%	49%	56%	63%
Indonesia	48	341	253	211	74%	82%	69%	59%
South Africa	66	66	57	76	36%	32%	33%	44%
Colombia	33	76	62	48	93%	99%	88%	82%
United States	40	95	44	30	5%	18%	12%	10%
World	**Trade (Mtce)**				**As a share of production**			
	2000	2018	2030	2040	2000	2018	2030	2040
Steam coal	310	859	733	726	12%	20%	17%	17%
Coking coal	175	319	314	371	39%	33%	37%	47%
Stated Policies	471	1 169	1 039	1 087	14%	21%	19%	20%
Sustainable Development			633	413			18%	20%

Notes: n.a. = not applicable. World trade is the sum of net exports for all WEO regions and may not match the sum of steam and coking coal as a region could be a net exporter of one coal type but a net importer of another. See Annex C for definitions.

China's net import of coal is projected to peak in the next few years and then to decline substantially. By the mid-2020s, China is overtaken by India as the world's largest coal importer. However, there are uncertainties over how the supply-demand balance in Asia plays out. Among exporters, Australia and Russia could take advantage of any new export opportunities, as Indonesian exports decline.

---

[1] Unless otherwise stated, trade figures in this chapter reflect volumes of coking and steam coal traded between regions modelled in the WEO and therefore do not include intra-regional trade.

5.5 Investment

Investment in coal supply (mining and transport infrastructure) today stands at about $80 billion per year, half the level of the historical investment peak in 2012. Some 40% of this is spent on maintaining output at existing mines. Falling investment in China accounted for some three-quarters of this decline, but investment in China still accounts for more than half the global total, largely as a result of a continuing process of industrial restructuring involving the closure of smaller, inefficient mines and their replacement with more modern facilities, often in the west of the country. A westward shift in the location of mining areas in China increases the distance to the main demand hubs, implying higher transport costs and investment plans to reduce them.

In the Stated Policies Scenario, total global investment falls to an annual average of $50 billion between today and 2040, as suppliers calibrate their investment in line with demand trends (Table 5.5). Once China's restructuring process is complete, investment there in new facilities falls substantially, as existing mines are sufficient to meet the bulk of the country's needs at least until they start to reach the end of their operational lifetimes from the late 2020s.

Over the period to 2040, India is a key country investing in both existing mines and in new capacity, reflecting its goal of meeting a higher share of its rising demand from domestic sources and limiting reliance on imported coal. In North America and Europe, demand decreases, and public opposition and an unfavourable financing environment weigh on investment in new coal supply (see section 5.7).

Table 5.5 ▷ Cumulative coal supply investment by region in the Stated Policies Scenario, 2019-2040 ($2018 billion)

	Total	Mining			Ports and rail	Total annual average
		Capacity additions	Maintenance	Total		
North America	44	17	24	41	3	2
Central and South America	16	8	6	14	1	1
Europe	16	4	6	10	6	1
Africa	43	17	18	35	8	2
Middle East	1	0	0	0	1	0
Eurasia	68	27	28	55	12	3
Asia Pacific	898	379	424	803	94	41
Shipping	47	n.a.	n.a.	n.a.	47	2
World	1 133	453	506	959	174	51
Sustainable Development	461	149	242	391	70	21

Note: See Annex C for definitions.

Key themes

5.6 A view beyond power: industrial coal use

The future of coal is often seen as synonymous with the future of power generation. However, around one-third of coal is used in industry, mostly in the iron and steel, cement and chemicals sub-sectors. The relative importance of industrial coal use increases in the Stated Policies Scenario as coal continues to power these sectors and as more of it is used as a feedstock for coal-to-gas and coal-to-liquids projects (Figure 5.5).

Overall coal use drops sharply in the Sustainable Development Scenario, but coal still remains an important part of the energy mix because of the difficulty in finding substitutes for it in some industrial processes. The share of industrial use in total coal demand almost doubles to 2040 as coal use in power generation declines, and innovative technologies like CCUS or the use of hydrogen become important levers for deep decarbonisation of industry over the outlook period and beyond.

Figure 5.5 ▷ Share of global coal demand by sector and scenario

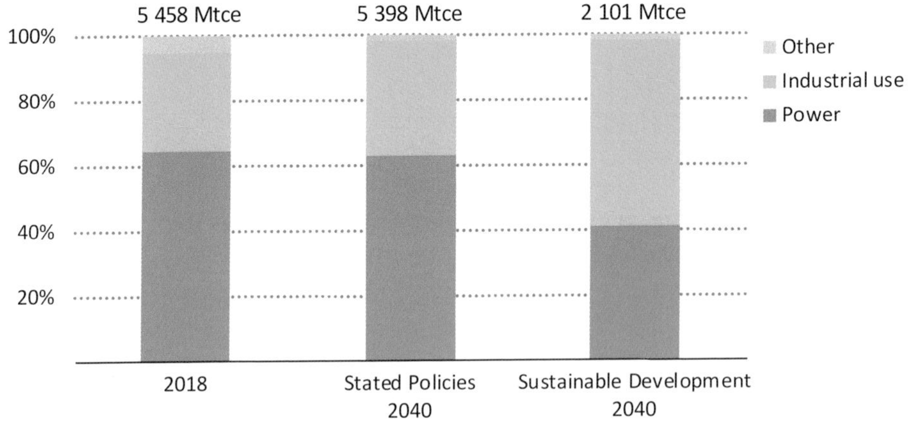

Industrial coal use accounts for an increasing share of coal demand in both scenarios

A series of technical and economic challenges differentiate the industry sector from other parts of the energy system, and they largely determine emission abatement opportunities and the scope for finding substitutes for coal. Current and planned industrial assets are important in this context because they have the potential to lock in production pathways for the future.

Current status

Coal is the largest fuel source in industry, ahead of electricity, natural gas and oil. It is the dominant fuel in the iron and steel (74%), and cement (61%) sub-sectors, and has a substantial share of 13% in the chemicals sub-sector (Figure 5.6). Coal's share in industrial energy use increased from 24% in 2000 to around 31% in 2018 due to the growth in coal consumption of nearly 93% over this period. This increase was in large part due to growth in industrial production in China. Increasing demand for cement, steel and chemicals has historically coincided with economic and population growth. However, as economies develop, urbanise, consume more goods and build up infrastructure, material demand per capita tends to increase significantly. Once industrialised, material demand per capita may level off and even begin to decline. At that stage, materials like steel and cement tend to be used primarily for replenishing and renovating rather than building the capital stock.

Figure 5.6 ▷ Global industrial energy use by fuel (left) and energy mix for selected sub-sectors (right)

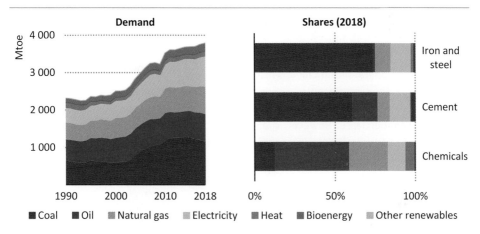

Coal is the largest source of industrial energy use, dominating demand in the cement and iron and steel sub-sectors

In the iron and steel sub-sector, 70% of global crude steel is produced through the blast furnace-basic oxygen furnace (BF-BOF) route which is heavily dependent on coking coal for the production of coke. Scrap-based electric arc furnaces (EAF) account for most of the remaining production. The scope to shift away from coal by making greater use of scrap-based or direct reduction of iron (DRI)-based EAFs is limited by the availability and cost of scrap, as well as the cost competitiveness of electricity.

Coke serves three functions in the BF-BOF route: first, as a fuel (providing heat); second, as a chemical-reducing agent (reducing iron oxides); and third, as a permeable support. While coke is essential as a permeable support for the blast furnace charge, its two other functions can be substituted by oil, natural gas, coal or alternative fuels. Due to the scarcity and relatively high price of good quality coking coal, the ratio of coke to hot metal has increasingly been reduced in recent decades by the pulverised coal injection (PCI) method which injects pulverised coal into the furnace together with hot air.

Coal is the most widely used fuel in cement production, although used tyres, mineral oil and industrial waste are also increasingly used as fuel. Kilns, predominantly fuelled by coal, heat raw materials to about 1 450 degrees Celsius (°C) to create clinker, the main ingredient in cement. While kilns could in principle also be fuelled by oil or natural gas, this is not economically viable in most regions. In any case, around two-thirds of CO_2 emissions from making cement come from the chemical reactions that take place when clinker is created through the calcination of limestone rather than from the coal that is used to provide heat.

Coal also plays an important role in the chemicals sub-sector, in particular in China where coal is the dominant feedstock in methanol and ammonia production. China has the largest coal conversion sector in the world, in which coal is converted into other fuels or chemical products. The majority of these are coal-to-liquids and coal-to-chemicals conversions. There are also some coal-to-gas projects, reflecting the push for natural gas in China's "Winning the battle for blue skies" policy, though poor economics and technical problems have plagued many of these projects.

Coal conversion operations in China, which can bring benefits in terms of air pollution, tend to be seen by their proponents as a way to increase energy security and to promote economic development in certain regions by monetising coal reserves that might otherwise be hard to commercialise. However, coal conversion projects require high upfront capital expenditure and substantial water consumption, and they produce significant CO_2 emissions.

While coal is the largest fuel in industry globally, there are marked regional differences (Figure 5.7). Coal is by far the main fuel used in industry in China and India. Outside Asia, natural gas tends to take larger shares in the industrial energy mix, in particular outside the iron and steel sector.

Figure 5.7 ▷ Energy mix in industrial use for selected regions, 2018

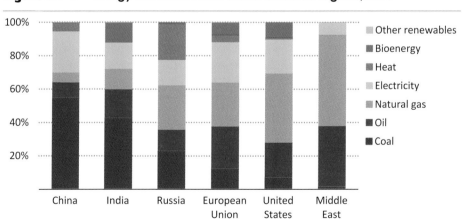

Coal is the dominant fuel in industrial energy use in China and India, while other fuels tend to account for larger shares elsewhere

Outlook and key uncertainties

In the Stated Policies Scenario, industrial coal use increases by some 225 Mtce over the period to 2040 (Figure 5.8). In the context of a rise of almost one-third in overall industrial energy consumption, this means that the share of coal in energy used for industrial processes declines from 31% today to 26% in 2040. The share of electricity in final energy use in this sector rises from 21% in 2018 to 23% in 2040. Pronounced regional differences in industrial production underlie these global trends. Most notably, while iron and steel and cement production in China declines over the outlook period, production in India and other developing Asian economies picks up strongly.

Figure 5.8 ▷ Global industrial use of energy by fuel and scenario

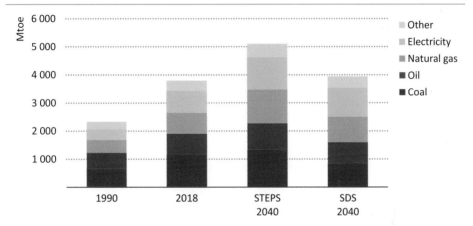

The challenge to find cost-effective substitutes for coal in industrial energy use means coal remains relatively strong even in the Sustainable Development Scenario

Coal use in the iron and steel industry declines in the Stated Policies Scenario by around 30 Mtce by 2040, reflecting efficiency gains and the gradual rise in the use of electricity-based routes for steel production. Coal use in cement production decreases by a similar amount, due to process improvements (a reduction in the clinker-to-cement ratio), energy efficiency improvements and a switch in some cases to alternative fuels such as waste or biomass. Together, these measures result in a reduction of around 13% in the CO_2 intensity of global cement production. There is, however, an increase of more than 180 Mtce in the use of coal as a feedstock for coal-to-gas and coal-to-liquids projects as well as in chemicals production, largely due to projected project start-ups in China.

In the Sustainable Development Scenario, coal demand decreases in all sectors, but coal remains an important fuel in industry, accounting for 21% of energy use in this sector in 2040. CCUS and hydrogen become important tools to address some of the most difficult to abate emissions in the energy sector. Together with cleaner production routes (via fuel switching), energy efficiency improvements and process improvements (via a reduction in the clinker-to-cement ratio) they reduce the CO_2 intensities of the iron and steel, and cement sub-sectors by an average 3% per year. The vigorous application of material efficiency strategies contributes by reducing the amount of iron and steel and cement produced from what it would otherwise have been (see Chapter 7). As a result, there are notable decreases in coal demand, especially in iron and steel (-285 Mtce) and in cement production (-50 Mtce). The scenario's trajectory is critically dependent on the implementation of policies to reduce the carbon footprint of industry, and to promote the deployment of technology innovations and the creation of markets for premium lower-carbon materials.

Box 5.1 ▷ **What role can CCUS play in mitigating emissions from industry?**

Industrial CO_2 emissions are among the hardest to abate, one-quarter of these are non-combustion process emissions that result from chemical or physical reactions, and therefore cannot be avoided by a switch to alternative fuels. This presents a particular challenge for the cement sub-sector, where 65% of emissions result from the calcination of limestone, a chemical process underlying cement production. Furthermore, one-third of industrial energy demand is used to provide high-temperature heat. Switching from fossil to low-carbon fuels or electricity to generate this heat would require facility modifications and substantially increase electricity requirements (see *World Energy Outlook 2018*, Box 5.1 [IEA, 2018]). Plus, industrial facilities are long-lived assets of up to 50 years, and so have the potential to "lock in" emissions for decades. Exposure to highly competitive, low-margin international commodity markets accentuates the challenges faced by firms and policy makers.

A portfolio of technologies and approaches will be needed to address the decarbonisation challenge while supporting industry sustainability and competitiveness. CCUS technologies can play a critical role in reducing industry sector CO_2 emissions. CCUS is one of the most cost-effective solutions available to reduce emissions for some industrial and fuel transformation processes, especially those that inherently produce a relatively pure stream of CO_2, such as natural gas and coal-to-liquids processing, hydrogen production from fossil fuels and ammonia production. CCUS can be applied to these facilities in some cases for as little as $15-25 per tonne of CO_2.

In the Sustainable Development Scenario, around 10 gigatonnes of carbon dioxide (Gt CO_2) is captured from industrial processes and fuel transformation in the period to 2040, the vast majority of it from the cement, iron and steel, and chemical sub-sectors. Around half of this comes from coal-based facilities. CCUS makes significant inroads in the late 2020s, capturing around 450 million tonnes (Mt) CO_2 in 2030, and expands rapidly thereafter.

Focus: industrial coal use in China

China is the world's largest producer of steel and cement, accounting for around 50% of iron and steel and 55% of cement production. In addition, a significant share of global chemicals production takes place in China. As a result of this dominance in global materials manufacturing, China also accounts for the world's largest shares of global industrial energy use (35%) and industrial CO_2 emissions (nearly 45%).

Industrial production in China has been based largely on domestic coal, but its reliance on coal gradually decreases over the outlook period in the Stated Policies Scenario. The share of coal in its industrial energy mix decreases from 55% today to 40% in 2040. Overall coal demand in industry decreases by some 0.8% per year, mainly due to lower coal demand in the iron and steel, and cement sub-sectors (Figure 5.9). Recent policies, often aimed at improving air quality, have already resulted in production cuts at inefficient steel plants and in plant closures. Electrification of steel-making, increased use of scrap steel, process improvements and general efficiency improvements are important drivers behind coal demand reductions.

Figure 5.9 ▷ Industrial use of energy by fuel and key sub-sector in China in the Stated Policies Scenario

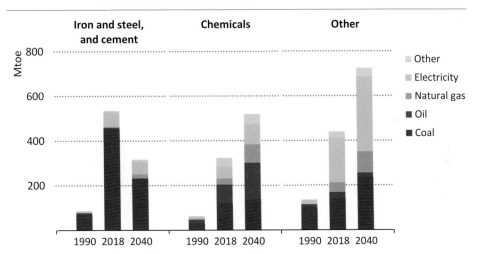

A deep rebalancing across industrial activities is underway in China, with large effects on the industrial energy mix

The anticipated rebalancing of the Chinese economy towards lighter and less energy-intensive manufacturing leads to a reduction in high-temperature heat demand over the projection period (Figure 5.10). This reduces the need for coal in industry and opens the possibility of other fuels to provide lower temperature heat. There has been some substitution in recent years of small industrial coal-based distributed boilers by gas or electric boilers, or combined heat and power (CHP) systems, especially for industries close to urban areas. This has resulted in improved air quality and a reduction in coal demand.

The main drivers of industrial coal demand in China in the Stated Policies Scenario are chemicals production and the use of coal as a feedstock for coal-to-gas and coal-to-liquids projects (+165 Mtce to 2040). Around 100 Mt of coal are used for dedicated hydrogen production using coal gasification in China today, with the resultant hydrogen used mainly to make ammonia – a key primary chemical and the chemical base of all nitrogen fertilisers – and methanol (see Chapter 11). The development of methanol-to-olefins and methanol-to-aromatics technology has opened an indirect route from methanol to high-value chemicals (HVCs), and thus to plastics. Methanol-to-olefins technology is currently deployed at commercial scale in China, accounting for 9 million tonnes per year or 18% of domestic HVC production in 2018. Methanol-to-aromatics, which is used to produce more complex HVC molecules, is currently in the demonstration phase.

Figure 5.10 ▷ Change in heat demand by temperature in China in the Stated Policies Scenario, 2018-2040

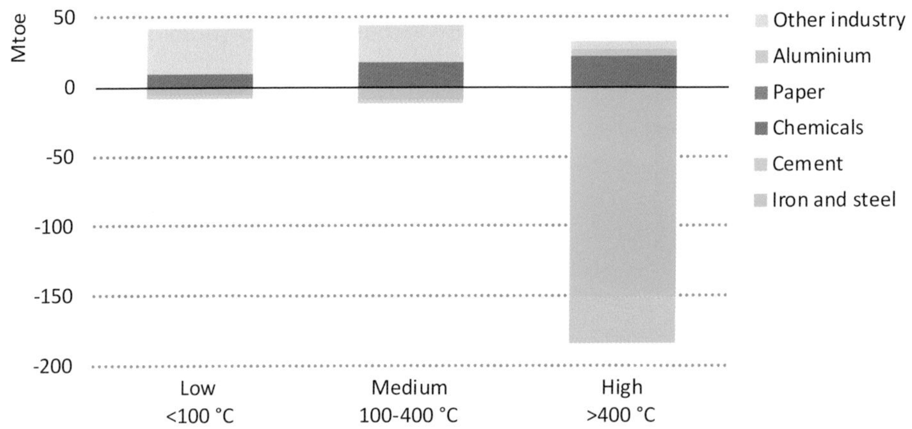

High-temperature heat demand in China is falling due to lower demand in the iron and steel, and cement sub-sectors

5.7 Who will invest in coal supply?

In 2018, for the first time since 2012, investment in coal supply increased slightly at 2% pushing the global estimate to $80 billion: growth was due to increased spending on existing mines, rather than capacity expansion, aided by the uptick in prices and revenues (Figure 5.11). The overall figure includes investment in mining and related infrastructure to bring coal to market, but excludes spending on coal-fired power plants, and represented 4% of overall capital spending on energy supply in 2018. The share of coal supply investment in total investment has been steadily decreasing in recent years; in 2012, coal supply investment of $163 billion was double the 2018 figure and made up more than 8% of the global total. In this section, we analyse the dynamics of coal supply investment, look at the increasingly restrictive conditions for access to finance for many coal supply projects and consider the implications for the scenarios.

The headwinds facing new coal supply projects start with uncertainties over future demand. Climate policies are hitting coal hard in many advanced economies. Air pollution policies, often overlooked, have been effective at reducing coal use, both in Europe and United States in the power sector, and in China in the industrial and residential sectors. In addition, an increasing number of countries are adopting phase-out policies and setting deadlines for the end of unabated coal power generation. This threefold policy action, alongside the fall in costs of renewables, and lower natural gas prices in some markets, is raising questions about the long-term profitability of coal assets. In addition, an increasing number of banks, insurance companies, institutional and private investors, utilities and mining companies are restricting, reducing or giving up investment in coal. The risk of reputational damage is becoming an increasingly important consideration, especially among companies for whom coal is not their core business. Last but not least, public opposition to coal, including through legal litigation, is an impediment for new coal developments across most of the world, from the developed world to emerging countries in Asia and Africa.

Figure 5.11 ▷ Coal supply investment

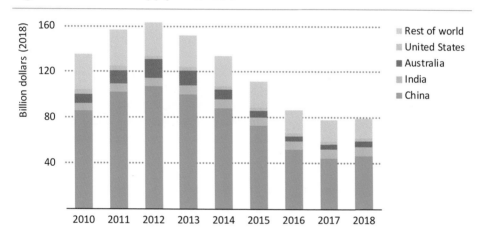

Investment in coal supply rose slightly in 2018, although the global total remains at around half the level seen at the start of the decade

The allocation of coal investment along the value chain is very different to that of oil or gas. The power sector, the largest user of coal, accounts for the bulk of the investment in the coal supply chain. Although the figure varies from country to country and from plant to plant, a 1 000 megawatt (MW) plant typically involves investment of somewhere between $700 million and $2.5 billion, whereas the mine to fuel the plant typically involves investment one order of magnitude lower. This is a very different allocation from the gas supply chain for electricity, in which the upstream (exploration and extraction) and midstream (transportation) typically account for the bulk of investment. In case of seaborne traded coal involving rail and ports, additional investment is needed for transport, but transportation costs for coal are still much cheaper than natural gas per unit of energy.[2]

Given that coal-fired power plants are the most capital-intensive part of the value chain, analysis of the implications of restricted access to finance for coal has typically focused on power. However, with the increasing constraints on coal finance and the long lifetime of coal consuming assets, the question also arises whether securing coal supply might become more difficult or expensive in the future.

The policy and financing setting is only part of the story. Coal is a commodity, and hence subject to standard commodity cycles, and affected by the broader economic and financial environment. The experience of the commodity price downturn in 2015-16 has made mining companies cautious about investing in capacity expansion. At present mining companies are generally avoiding big capital expenditures. Capital efficiency has overtaken growth as the key management principle.

The position of coal in different parts of today's energy sector is an important part of the context, including the significance of coal use outside the power sector, notably in industry (see section 5.6). Moving away from coal is not easy anywhere, but it is easier in some circumstances and countries than others, especially when existing assets are older and substitutes are readily available. In the power sector, around 70% of today's global coal power capacity and coal-based electricity generation is in Asia, where electricity demand is rising fast and coal plants are around 12 years old on average, more than two decades younger than those in North America or Europe. Reducing emissions from this stock of coal-fired assets poses difficult social and economic questions for policy makers (see Chapter 6); this helps to explain why global coal use in the Stated Policies Scenario in 2040 is only just below today's level.

Changing landscape for coal investment and financing

Opposition to coal is not new, but in recent years there has been an increasing number of announcements by various players, whether governments or business organisations, of restrictions or prohibitions on coal financing and/or investment. Even if some have come from countries or investors with limited interests in coal, these announcements create momentum, open dialogue and can point a direction to others. Within the coal industry, companies leaving coal mining or power generation typically sell their assets to other companies, so there may not be any reduction in overall capacity. However, the new owners may have different strategies or possibilities: for example, when a big mining company sells coal assets to a small player, the expansion potential decreases significantly, owing to weaker balance sheets, higher capital cost and less favourable access to finance,

---

[2] In significant cases like China, Russia and India, coal transport tariffs are not necessarily cost-reflective. This can cut both ways: in Russia, the railways subsidise coal exports, whereas in India, coal transport cross-subsidises passenger rail.

in particular when complex infrastructure is needed to develop a project. The cumulative impact of multiple decisions can also be very significant: one bank moving away from coal has little impact, but when one hundred banks decide not to finance coal, it is far from irrelevant.

One of the more consequential moves away from coal has come from multilateral development banks. In 2013, the World Bank Group (WBG) decided to stop financing coal power plants except in very exceptional circumstances. Other banks followed suit, including the European Investment Bank (EIB) and the European Bank for Reconstruction and Development. Institutions such as the Asian Development Bank or the Asian Infrastructure Investment Bank, although they have not excluded coal as explicitly as WBG or EIB, are not in practice financing coal projects. The African Development Bank has recently financed the Sendou coal power plant in Senegal, but it has not been involved in coal production for more than a decade. This strategy shift among the main multilateral lending institutions has constrained available financing for coal projects in developing countries and made private finance for projects more expensive, and many other banks and institutions are mirroring their stance. Export credit agencies (ECAs) represent a similar case. In November 2015, Organisation for Economic Co-operation and Development (OECD) countries agreed to strengthen the rules for ECAs and most have announced that they will not finance any new coal projects.

Among the country-led initiatives, the most prominent has been the Powering Past Coal Alliance (PPCA), launched by United Kingdom and Canada in November 2017 during the Conference of the Parties[3] (COP) 23 in Bonn, Germany. By joining the PPCA, governments commit to phase out unabated existing coal power generation and not to build any new capacity. Non-government members commit to powering their operations without coal. As of September 2019, the Alliance had 91 members: 32 country governments, 25 sub-national governments and 34 companies (PPCA, 2019). Among the country signatories, Germany is the largest coal power generator, followed by Canada, Mexico and Italy. South Chungcheong Province in Korea, home to half of the country's coal capacity and power generation, also joined the Alliance.

Overall, the coal used for power generation in the national and sub-national members of PPCA accounts for about 4% of global consumption, although in some of the countries concerned, such as Germany, Israel, Netherlands and Portugal, coal accounts for a significant share of their overall electricity generation. But the indirect effect on coal investment decisions nonetheless may be powerful, particularly since the members, both governments and businesses, commit to restricting financing for unabated coal power generation.

Another important trend is in the financial community. Many institutional investors and pension funds, banks, insurance companies and others have committed to reduce or end their involvement in coal in one way or another (Table 5.6). The early commitments came

---

[3] COP of the United Nations Framework Convention on Climate Change.

from Europe and North America, but recent announcements from China, Australia, Japan and Singapore underline the scale of this movement both in terms of the volumes involved and in geographical spread.

Table 5.6 ▷ Selected financial and investment institutions committed to reduce or end involvement in coal supply and coal-fired power

Country	Institutions
Australia	QBE
China	State Development & Investment Corporation
France	AXA, Société Générale, Crédit Agricole, BNP Paribas, CNP
Germany	Allianz, Deutsche Bank
Italy	Generali
Japan	Marubeni, Mitsui, Itochu, Sojitz, MUFG
Korea	Teachers Pension System, Government Employees Pension System
Norway	Wealth Pension Fund
Singapore	Oversea Chinese Banking Corp, United Overseas Bank
South Africa	Standard Bank, Nedbank
Spain	Mapfre, BBVA, Banco Santander
Switzerland	Zurich, Swiss Re
United Kingdom	Lloyds, Aviva, Barclays
United States	Chubb Ltd.

Others prefer engagement rather than divestment, as is the case of Japan's Government Pension Investment Fund, the world's largest pension fund by volume. The principle is that this gives investors more opportunities to have an impact. When Glencore, the world's largest coal exporter, pledged in February 2019 to cap coal production and look for growth in other commodities necessary for energy transitions, this came in part as a result of engagement with investor signatories of the Climate Action 100+ initiative. The commitment may not seem very stringent, but it establishes a principle and could be further tightened in the future.

Last but not least, any coal project has to deal with opposition from local communities and international civil society groups. This public opposition typically includes legal litigation, which can make projects more difficult and more costly, and can lead to cost overruns, delays and, in some cases, cancellation of projects.

Focus: the main global coal producers

A review of the main global producing companies shows how their composition has changed over time (Table 5.7).

The shift in recent years to Asia and, in particular, to China is striking. Coal India Limited, a state-owned company, remains as the top producer in the world by tonnes. China now accounts for six of the top-ten largest producers. In 2009, only two Chinese companies, Shenhua Group and China National Coal Group, both owned by the central government, were among the world's top-ten. China Energy Investment Corporation, its new name after

Shenhua merged with Guodian Group, remained in second place in 2017 (although if measured in terms of the value or energy content of production, rather than tonnes of output, it would be the largest producer), and China National Coal Group moved up to fourth. In addition, four companies owned by Chinese provinces joined the list of the top ten producers: Shandong Energy and Yankuang Group, both owned by Shandong Province; Shaanxi Coal and Chemical Industry Group, owned by Shaanxi Province; and Datong Coal Mining Group, owned by Shanxi Province. The share of China in global coal production was stable over this period: 46% in 2009 and 47% in 2017, and the changes to the top-ten list reflect the recent process of consolidation in the coal industry in China rather than a rise in its share of global coal production. The only new non-Chinese company in the list is SUEK, a Russian private company, which is also the world's fourth-largest coal exporter.

Table 5.7 ▷ **Top-ten global coal producing companies**

Main countries of operation	Company	Output (Mt)
2009		
India	Coal India	431
China	Shenhua Group	328
United States	Peabody Energy	221
United States / Australia	Rio Tinto	140
China	China National Coal Group	125
United States	Arch Coal	114
Colombia / South Africa / Australia	BHP Billiton	105
Germany	RWE	100
Colombia / South Africa / Australia	Anglo American	96
Colombia / South Africa / Australia	Xstrata	95
2017		
India	Coal India	561
China	China Energy Investment Corporation	513
United States / Australia	Peabody Energy	172
China	China National Coal Group	164
China	Shandong Energy	141
China	Shaanxi Coal and Chemical Industry Group	140
China	Yankuang Group	135
China	Datong Coal Mining Group	127
Colombia / South Africa / Australia	Glencore	121
Russia	SUEK	108

Among the big diversified global mining companies, Rio Tinto left the coal business definitively in 2018, after a divestment process which started with its US assets almost a decade ago and finished with the sale of its last coal assets in Australia. BHP, which spun off part of its coal assets to South32, and Anglo American, which also reduced its coal exposure, are still the world's second- and third-largest exporters. Glencore, which merged with Xstrata in 2013, remains in the top-ten, the only big diversified global mining company in this ranking. Meanwhile US companies, facing declining demand on the domestic market, are now more streamlined, with more focus on financial discipline and less on international expansion. Peabody, still the largest private producer and the only US company on the 2017 top-ten list, announced a joint venture with Arch Coal to share assets producing 185 Mt in 2018 in order to maximise value and compete in a difficult environment.

A related question is who owns the coal mining companies. In China, state-owned companies produce most of the coal. The same is true in India, where Coal India accounts for 80% of domestic production and other public companies produce a further 10%. Analysis of recent trends in the ownership of listed companies shows that large institutional investors have maintained their positions in the big diversified mining companies, and there are no new large investors, apart from Anil Agarwal's Conclave, which bought 20% of Anglo American. The main shift in terms of ownership has occurred in US coal companies, following the bankruptcy and recovery process that almost all of the big US companies have followed. The large institutional investors have reduced their positions, with some hedge funds and other asset managers stepping in. In the case of pure coal export-oriented companies, the story is quite different: these companies are much smaller than the big diversified miners and the participation of international institutional investors is rather limited. Individuals or families are the main owners of some of them, with increasing participation from Chinese state-owned enterprises. These two types of companies, big global diversified mining groups and pure coal export-oriented players, face very different financing constraints (Box 5.2).

Box 5.2 ▷ Increasing cost of capital for coal producers

Access to capital is becoming more difficult for coal producers. An increasing number of financial institutions and investors are turning away from coal, and this is contributing to higher financing costs.

Historically, coal mining has been characterised by high volatile returns and a higher cost of capital than other extractive energy industries, such as upstream oil and gas. Figure 5.12 shows the weighted average cost of capital (WACC) and return on invested capital (ROIC) for big global diversified mining groups and for a group of selected export-oriented coal companies. The ROIC trend closely follows the commodity price cycle for the diversified companies and the coal price for the coal-only companies.

In addition, coal-only companies have historically seen higher financing costs than diversified companies. This continues to be the case. From 2015, the cost of capital has increased for both diversified miners and coal-only companies, but the increase has been bigger for the coal-only companies, reflecting investor perceptions that the risks associated with coal supply businesses have increased for the reasons described. Coal-only companies have had considerable difficulty in creating value for shareholders over the past decade, and this may constitute another hurdle for investment in the sector in the current environment of rising pressure against coal, although coal proved to be a profitable commodity in 2017.

Figure 5.12 ▷ Return on invested capital and after-tax weighted average cost of capital for selected coal companies

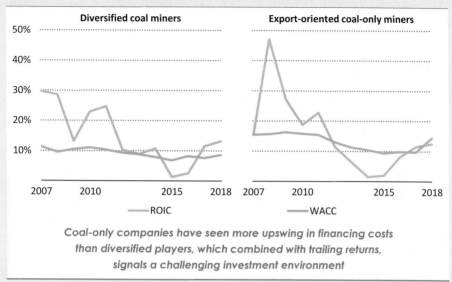

Coal-only companies have seen more upswing in financing costs than diversified players, which combined with trailing returns, signals a challenging investment environment

Notes: Diversified coal miners include Glencore, BHP Billiton, Rio Tinto and Anglo American. Export-oriented coal-only miners include Adaro Energy, Bumi Resources, Exxaro Resources and Whitehaven Coal. ROIC measures the ability of a company's core business investments to generate profits, expressed as operating income adjusted for taxes divided by invested capital. The WACC is expressed in nominal terms and measures the company's required return on equity and the after-tax cost of debt issuance, weighted according to its capital structure. The tax rate is assumed at 35% for all companies.

Sources: IEA analysis based on data from Thomson Reuters Eikon (2019) and Bloomberg (2019).

Outlook and key uncertainties

The decline in coal supply investment since the high point in 2012 – notwithstanding the slight uptick in 2018 – is set to continue in our projections in both scenarios. From $80 billion in 2018, average annual investment declines to around $60 billion in the Stated Policies Scenario by the 2020s (averaged for the decade as a whole) and to around a third of this in the Sustainable Development Scenario. These sums also represent a smaller share of the overall amount of capital expenditure in the energy sector, which edges higher towards $2.5 trillion per year from about $1.8 trillion today. Investment in coal supply increasingly bifurcates into two worlds – one in which financing constraints start to bite and the other in which financing does not yet appear to be such a hard constraint. Most of the projected coal supply investment is concentrated in the latter.

China and India account for a large portion of projected expenditure on coal supply. Cumulative investment in coal supply in China to 2040 is $585 billion in the Stated Policies Scenario (50% of the global total and $27 billion per year on average), and $220 billion in the Sustainable Development Scenario (just under half of the global total). The comparable figures for India are $170 billion in the Stated Policies Scenario and $50 billion in the

Sustainable Development Scenario. National policies are set to play a defining role in determining investment outlooks in both cases, and the predominance of state-owned companies in production means that external financial constraints may not loom large in these decisions.

Figure 5.13 ▷ Global coal production by type in the Stated Policies Scenario

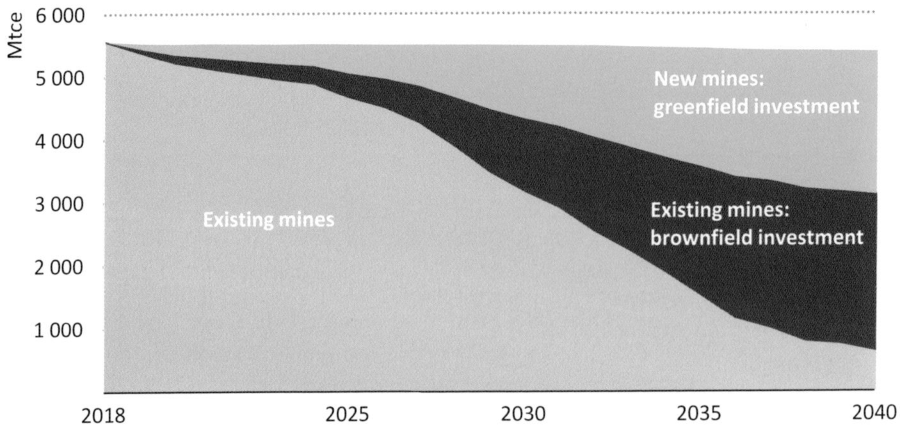

As production from existing mines declines, so demand in the Stated Policies Scenario requires investment either to expand existing mines or to open new ones

In China, a pivotal moment for coal supply investment arrives in the mid-2020s, when the wave of mines that were opened in the early 2000s start to reach the end of their operational lifetimes. In the Stated Policies Scenario, we assume a decision to reinvest in new coal production capacity, in line with future projected domestic needs. But this could also be a point at which there is a significant move away from coal, if alternative technologies and fuels are deemed to provide adequate substitutes and the implications for employment appear manageable. Much will depend on the timing of the retirement of existing coal-fired power plants and the extent to which any move away from coal is slowed by widespread deployment of CCUS technologies (see Chapter 6).

India too faces crucial decisions in the 2020s, once the current over-capacity in the power sector is absorbed by rising demand for electricity. In the Stated Policies Scenario, the next wave of investments in generation capacity involves a balance of renewables, battery storage and coal, the latter stimulating new investment in domestic coal supply in turn. However, India could decide at this point to opt for a more dramatic transformation of the power mix. Much is likely to depend on developments between now and then in renewables and storage technologies (see sensitivity analysis on battery storage in Chapter 6). Financing considerations could also play into these power sector choices.

Outside China and India, there is a distinction to be made between sustaining capital expenditure to maintain production from existing mines and funding new mines (Figure 5.13). In the Stated Policies Scenario, half of the $275 billion in coal mining investment outside China and India goes to sustaining capital expenditure; this share is higher in the Sustainable Development Scenario (60%). The distinction is important because the expansion of existing mines is likely to be financed by the incumbent, whereas greenfield projects are likely to require external financing, which could be challenging to obtain.

Russian producers are likely to be least affected by constraints on investment finance. Big producers have access to finance from Russian and Asian banks, and expansion plans are backed by the government strategy of developing infrastructure to increase market share. In part for this reason, our projections in the Stated Policies Scenario show Russia continuing to expand exports and to increase its share in the seaborne traded coal market.

A visible trend in recent years has been for companies to downsize big supply projects. Carmichael, in Australia, was first planned as a 60 million tonnes per annum (Mtpa) mine with external project finance and has been downscaled to 10 Mtpa to be self-financed by Adani, the owner. Boikarabelo mine in South Africa, initially planned for about 20 Mtpa, has been downscaled to 6 Mtpa and is still working on a lending syndicate. Market conditions may partially explain this downsizing trend, but the increasing restrictions on external finance for major projects have also played an important role.

The apparent market oversupply in 2019 following the high prices in 2017-18 suggests that restrictions affecting external financing have not so far had much impact on curtailing coal production, but increased restrictions imply a higher cost of capital in the future, pushing up the long-term supply cost curve and making the profitability of projects more uncertain. It is difficult to know how this will play out. Producers in countries such as the United States and Indonesia may be ready to step in if the market starts to signal a shortfall in supply. Conversely, the tougher environment for investment and difficult conditions for market entry may reduce the possibility of oversupply, one of the biggest risks in the commodity business.

5.8 Coal mine methane

The energy sector is responsible for around 40% of anthropogenic methane emissions. This is a major cause for concern because methane is a potent greenhouse gas: over a 100-year timeframe, one tonne of methane absorbs around 30-times more energy than a tonne of CO_2 (the 100-year global warming potential [GWP]); over a 20-year timeframe, one tonne absorbs as much as 85-times more energy.

Previous *WEO* analysis looked in detail at the levels of methane emissions from oil and gas operations, and the costs and opportunities of avoiding the emissions (IEA, 2017). This analysis found that around 45% of the current 80 Mt of methane that is emitted by the oil and gas sector could be avoided at no net cost, since the value of the captured methane is greater than the cost of installing the abatement measures.

Methane emissions play a critical role both in the lifecycle emissions intensity of different sources of oil and gas and in the emissions savings that would result from switching from one fuel to another (see Chapter 4). Methane emissions also occur during coal mining operations: coal seams naturally contain methane, referred to as coal mine methane (CMM) and this can be released during or after mining operations (Box 5.3). The extent of these emissions is relevant to coal-to-gas switching. Coal combustion results in around 75% more CO_2 emissions than gas per unit of energy, but a more complete comparison between coal and gas must also look at the methane emissions from getting the coal or gas from the ground to the end-user.

Box 5.3 ▷   What is the difference between CMM and CBM?

Coal mine methane is trapped in coal seams that are suitable for mining. CMM is a hazard to mining activities and must therefore be dealt with carefully. It can be released in various ways during coal mining operations, including:

- Seepage of methane from coal seams exposed in surface or open pit mines.
- Degasification and ventilation systems that are used to extract and dilute methane found in underground coal mines to create safe working conditions and avoid explosions.
- Post-mining activities such as processing, storage and transport when quantities of methane still trapped in the matrix of the coal seeps out.
- Abandoned mines, since there may still be large volumes of coal remaining in a mine even after operations have ended; methane contained in this coal may continue to escape to the atmosphere.[4]

Coalbed methane (CBM)[5] is natural gas, predominantly methane, which is deliberately extracted from underground coal seams that generally are not suitable for normal mining. However in some countries – most notably in China – CBM can also include CMM emissions that are captured and subsequently marketed. CBM is considered an "unconventional" source of natural gas with nearly 90 billion cubic metres produced globally in 2018.

If coal is used in conjunction with CCUS, this would reduce the associated combustion emissions and open longer term possibilities for coal. But it would still be important to minimise the methane emissions that occur before combustion.

As with oil and gas, there is a high degree of uncertainty in estimating the level of CMM emissions that occur today. Few direct measurements of methane emissions occur at coal mines: baseline or background levels of emissions were often not established prior to the start of operations and the emissions levels that do exist are based on sparse and sometimes conflicting data.

---

[4] This analysis focuses on CMM emissions from active coal mines since we aim to estimate the full emissions intensity of current sources of coal production. Abandoned mines are therefore not included.

[5] CBM is also sometimes called coal-seam gas.

This section describes our new approach to estimating methane emissions from coal operations, provides an overview of some of these estimates and examines how the full lifecycle emissions of coal compare with gas.

Method for estimating methane emissions from coal supply

Our new estimate of CMM is based on deriving mine-specific methane emissions intensities for all major coal producing countries.[6] The starting point was to generate emission intensities for coal mines in the United States using the US Environmental Protection Agency's Greenhouse Gas Reporting Program and 2018 US Greenhouse Gas Inventory (US EPA, 2018). This was supplemented by data sources that provided disaggregated CMM data for China (Wang et al, 2018; Zhu et al, 2017) and India (Singh and Sahu, 2018; India Ministry of Coal, 2018). The mine-level CMM estimates generated in this way were then aggregated and verified against the country-level estimates for China and India taken from satellite-based measurements (Miller et al., 2019). Based on these data, coal quality (e.g. the ash content or fixed carbon content of coal produced by individual mines), mine depth and regulatory oversight[7] were used as key factors to estimate CMM emission intensities for mines in other countries for which there are no reliable direct estimates.

An important factor in estimating CMM emissions is the depth in metres (m) and age of various mines in operation today (Figure 5.14). Deeper coal seams tend to contain more methane than shallower seams, while older seams have higher methane content than younger seams. In the absence of any mitigation measures, methane emissions to the atmosphere will therefore tend to be higher for underground mines than for surface mines.

Figure 5.14 ▷ Depth of coal production in selected countries, 2018

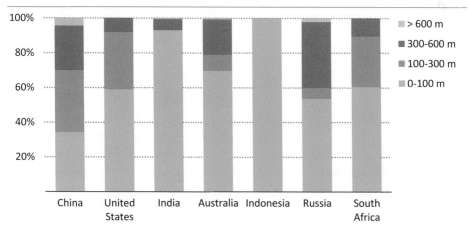

There is wide variation in the depths of coal mines across countries; deeper mines tend to have higher methane concentrations

---

[6] Mine-level data is provided in the coal mine database produced by the CRU Group (CRU, 2019).

[7] This incorporates government effectiveness, regulatory quality and the rule of law as given by the Worldwide Governance Indicators compiled by the World Bank (World Bank, 2019).

Methane emissions from coal supply

We estimate that global CMM emissions in 2018 were just under 40 Mt. Assuming that one tonne of methane is equal to 30 tonnes of CO_2, this is equivalent to around 1 200 Mt CO_2-eq, which is broadly similar to the current level of total annual emissions from international aviation and shipping combined. On average, the production of coal results in just over 0.3 tonnes of CO_2-eq indirect emissions for every tonne of coal equivalent (tce) produced. CMM emissions are responsible for two-thirds of these emissions. Coal combustion emissions average around 2.9 tonne CO_2-eq/tce globally, which means that indirect emissions account for around 10% of the lifecycle emissions of a tonne of coal.

The level of CMM emitted to the atmosphere is the main determinant of how the indirect emissions intensities of different mines compare (Figure 5.15). The lowest 10% production has a total average emissions intensity of 80 kilogrammes (kg) of CO_2-eq/tce, while the highest 10% has an emissions intensity of 1 000 kg CO_2-eq/tce. In other words, the most emissions intensive coal produces more than ten-times more indirect emissions than the least emissions intensive.

Figure 5.15 ▷ Indirect CO_2 and methane emissions intensity from global coal supply, 2018

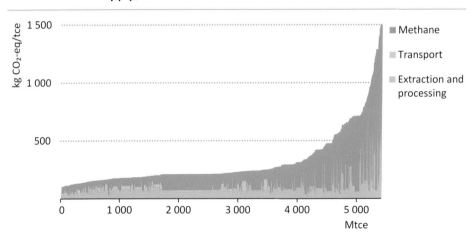

There is wide distribution in the emissions associated with different coal mines; CMM is the main determinant of where coal sits in the indirect emissions spectrum

It is no surprise that the geographic distribution of total indirect emissions is clearly weighted towards overall levels of production (Figure 5.16). Coal operations in China alone result in over 900 Mt CO_2-eq today. The indirect emissions intensity of coal production in China is marginally higher than the global average, largely because of the number of deep underground mines in operation in China, despite the use of a number of mitigation technologies to reduce or make productive use of these methane emissions (Box 5.4).

Figure 5.16 ▷ Indirect CO₂ and methane emissions, and emissions intensities for the ten-largest coal producing countries, 2018

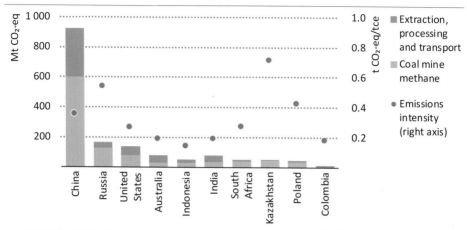

Countries with deeper coal mines and less regulatory oversight have higher indirect emissions intensities, but China dominates the absolute level of emissions

Box 5.4 ▷ Other indirect emissions from coal mining

The indirect emissions intensity of coal includes all sources of greenhouse gas (GHG) emissions from the point where coal is extracted to, but not including, where and how it is consumed. Indirect emissions come from CMM, but they also from the energy required to extract, process and transport the coal. The level of CO_2 emissions produced depends on the method of extraction, coal quality, and transport methods and distances. In surface mines, for example, diesel is usually used to power the earth-moving equipment such as draglines, shovels and trucks. Underground mining tends to rely on more mechanised processes powered by electricity, which is also required to power belt conveyors, crushers and other auxiliary equipment at or near production sites. Data for the energy use across mines is available from the CRU database, which we convert into emissions based on the emissions intensity of diesel and electricity within each country.

Depending on its quality, coal also often undergoes some processing prior to use. Coal "washing", for example, aims to reduce or remove ash, impurities and contaminants, and requires the use of electric-powered systems and large quantities of water. Plus, the coal needs to be transported from where it is produced to consumers; this is usually done by rail or ship. Our estimates of transportation energy use and emissions take into account the two modes of transport and corresponding trade routes.

In the Stated Policies Scenario, there are very few policies in place globally that aim to reduce the emissions intensity of coal production and so CMM emissions remain broadly constant at around 40 Mt to 2040. In the Sustainable Development Scenario, the 60%

decline in coal demand to 2040 is the primary factor in reducing CMM, but there are also some broader efforts to reduce emissions from mines, and total methane emissions fall to less than 15 Mt in 2040.

Reducing CMM

A key problem with mitigating CMM is that the methane concentration of emissions is often very low and can fluctuate in quality and quantity. The lower the concentration of methane, the more technically and economically difficult it is to abate. The same applies to methane emitted during the mining process. For example, air from the ventilation and degasification systems of underground mines (called ventilation air methane) contains less than 1% methane.

Processes such as blending or oxidation are required to make the recovered methane usable as an energy source (e.g. for heating mine facilities or drying coal). These can be expensive, are not always fully effective and can also pose considerable safety risks. However, there may be opportunities to produce higher concentration sources of methane if these are carefully planned prior to the start-up of mining operations either in a new mine or a new area. Degasification wells and drainage boreholes can produce methane and this can be used for small-scale power generation or, if concentrations exceed 95%, injected into a local gas grid.

A lack of gas infrastructure or nearby consumers can preclude the business case for onward sale of recovered methane. In the absence of a viable recovery project, methane can either be destroyed by thermal oxidisation or flared. However there are relatively few projects globally that have installed the equipment necessary to do this, with safety concerns and a lack of regulatory incentives cited as reasons for the lack of progress.

Much more active policies and regulatory regimes are clearly needed to incentivise or require mine operators to install CMM abatement technologies. This would need to involve requiring any new mines to develop plans to handle methane emissions before operations commence (or expand to a new area) and to make the necessary capital investments in drainage technologies, pipelines networks, and auxiliary and monitoring equipment.

SPOTLIGHT

How does the environmental performance of coal and gas compare?

The supply and use of coal and natural gas has a number of environmental impacts. Coal is responsible for over 40% of global emissions of energy-related sulfur dioxide (SO_2) emissions – a cause of respiratory illness and a precursor of acid rain – and nearly 15% of particulate emissions. In contrast, SO_2 and particulate emissions from the use of natural gas are close to zero. Combustion of natural gas does produce nitrogen oxides (NO_X), which can trigger respiratory problems and the formation of other hazardous particles and pollutants, but it accounts for less than 10% of global energy-related NO_X emissions compared with around 15% for coal.

With our new estimates for indirect emissions of coal, it is possible to provide a full comparison of the relative GHG emissions performance of coal and gas. Processing and transport are generally more energy intensive for gas than for coal, but gas extraction is generally less energy intensive than coal production. While there is a wide range in emissions from different sources of both coal and gas, on average the indirect emissions intensity of gas is around 50% higher than coal (0.67 tonne CO_2-eq per tonne of oil equivalent (toe) and 0.45 tonne CO_2-eq/toe respectively).

Indirect emissions intensity is only one element in this comparison. The emissions associated with the use of gas and coal also need to be factored in. There is some variation in the combustion emissions of coal depending on its quality (lignite has a higher emissions intensity than steam coal for example), but on average emissions are around 75% higher than gas. Gas-fired power plants are also generally more efficient in producing electricity than coal plants (the global average efficiency of gas-fired plants today is 52% while for coal-fired plants it is 42%).

Figure 5.17 ▷ Lifecycle emission intensities of coal and natural gas used for heat and electricity generation, 2018

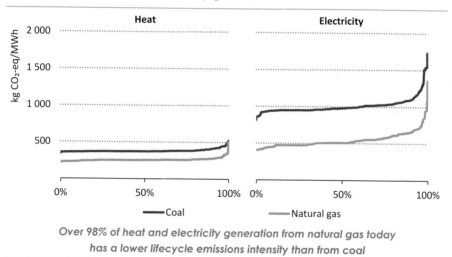

Over 98% of heat and electricity generation from natural gas today has a lower lifecycle emissions intensity than from coal

Notes: MWh = megawatt-hour. For heat, includes 1 300 Mtoe natural gas and 1 000 Mtoe coal used in the residential and industrial sectors in 2018. For electricity, includes 1 280 Mtoe natural gas and 2 400 Mtoe coal converted to electricity using region-specific electric plant efficiencies.

Overall we find that around 98% of heat generated in the residential and industrial sectors using gas has lower lifecycle emissions intensity than the heat provided from coal; similarly, 98% of electricity generated from gas has lower lifecycle emissions intensity than coal. Therefore, it is clear that the vast majority of gas use results in fewer emissions than coal (Figure 5.17). This finding highlights the importance both of minimising methane emissions from coal operations and reducing combustion emissions through the use of CCUS.

Chapter 6

Outlook for electricity
Lighting the way?

> **SUMMARY**
>
> - Electricity is at the heart of modern economies, powering communications, healthcare, industry, education, comfort and entertainment. In the Stated Policies Scenario, global electricity demand grows at 2.1% per year to 2040, twice the rate of primary energy demand. This raises electricity's share in total final energy consumption from 19% in 2018 to 24% in 2040. Electricity plays a larger role in the Sustainable Development Scenario, reaching 31% of final energy consumption.
>
> - Electricity demand follows two distinct regional paths. In advanced economies, growth linked to increasing digitalisation and electrification is largely offset by energy efficiency improvements. In developing economies, rising incomes, expanding industrial output and a growing services sector push demand firmly up. Developing economies contribute nearly 90% of global electricity demand growth to 2040, but their demand per person remains 60% lower than in advanced economies.
>
> - Industry and buildings account for over 90% of global electricity demand today, while transport makes up less than 2%. In the Stated Policies Scenario, the leading drivers of global electricity demand growth are industrial motors (over 30% of the total growth to 2040), space cooling (17%), and large appliances, small appliances and electric vehicles (10% each). Providing electricity access for the first time to 530 million people accounts for just 2% of demand growth. In the Sustainable Development Scenario, electric vehicles become the main source of demand growth.
>
> - Electricity supply continues a major shift towards low-carbon sources. In the Stated Policies Scenario, the share of renewables rises from 26% in 2018 to 44% in 2040, as they provide three-quarters of the growth in electricity supply to 2040. Solar photovoltaics (PV) and wind rise from 7% to 24% of supply, thanks to falling costs and supportive government policies. At the same time, coal-fired output plateaus and its share declines from 38% today to 25% in 2040, while gas holds at 20% and nuclear at 10%. The centre of gravity for nuclear power shifts towards Asia.
>
> - Solar becomes the largest source of installed capacity around 2035 in the Stated Policies Scenario, surpassing coal and gas (Figure 6.1). Global coal-fired capacity plateaus, with the project pipeline of 710 gigawatts (GW), mainly in Asia, just exceeding coal plant retirements, mainly in advanced economies. Renewables make up two-thirds of all capacity additions to 2040 globally. Wind power capacity triples, with offshore wind taking off in Europe, China and the United States. Gas-fired capacity grows in most markets for reliability purposes and battery storage skyrockets from 8 GW today to 330 GW in 2040. Annual power sector investment averages over $900 billion to 2040, including $500 billion for power plants, $400 billion for networks and more than $15 billion for battery storage.

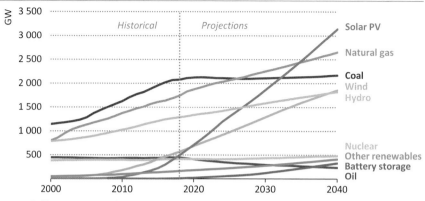

Figure 6.1 ▷ Global power capacity by source in the Stated Policies Scenario

Falling costs and increasing competitiveness of solar PV push its installed capacity beyond all other technologies in the Stated Policies Scenario

- Global power sector carbon dioxide (CO_2) emissions stabilise at today's level in the Stated Policies Scenario, while pollutant emissions decline thanks to wider use of pollution controls. In the Sustainable Development Scenario, the global average CO_2 intensity of electricity generation falls to 80 grammes per kilowatt-hour by 2040, about 80% below today's level. This is the result of 90% more growth for low carbon sources of electricity to 2040 compared with the Stated Policies Scenario, including renewables, nuclear power and carbon capture, utilisation and storage (CCUS).

- Coal-fired power has long provided reliable, flexible and affordable electricity, but it is the single largest source of greenhouse gas emissions today. Under current and proposed investment plans and policies, coal-fired generation would use up most of the remaining carbon budget consistent with meeting the stringent climate goals agreed at the United Nations. This reflects the fact that 60% of the 2 080 GW existing coal fleet is 20-years old or less. We outline a cost-effective strategy to "retrofit, repurpose, retire" existing capacity to address coal-related emissions. It includes retrofitting about 240 GW with CCUS or biomass co-firing equipment, with $225 billion of CCUS investment, and efforts to repurpose plants for more flexibility.

- Power system flexibility needs grow faster than electricity demand in the Stated Policies Scenario to 2040, due to rising shares of variable renewables and growing demand for cooling and electric vehicles. Flexibility needs grow even faster in the Sustainable Development Scenario. Power plants and networks remain the bedrock of power system flexibility, and demand-side response has huge potential. As the fastest growing flexibility option, battery storage capacity rises 40-fold by 2040 in the Stated Policies Scenario, due to its falling costs, short construction periods, widespread availability and scalability. Flexibility needs increase especially in India, where cheap batteries could offer a cost-effective flexibility option and eliminate the need for new coal-fired capacity after 2030.

Introduction

Electricity is at the heart of modern economies and it is providing a rising share of energy services. Demand for electricity is set to increase further as the world gets wealthier, as its share of energy services continues to rise, and as new sources of demand expand, such as digital connected devices, air conditioning and electric vehicles. Electricity demand growth is set to be particularly strong in developing economies.

The commercial availability of a diverse suite of low emissions technologies puts electricity at the vanguard of efforts to combat climate change and pollution. Decarbonised electricity, in addition, could provide a platform for reducing carbon dioxide (CO_2) emissions in other sectors through electricity-based fuels such as hydrogen or synthetic liquid fuels. Renewable energy also has a major role to play in providing access to electricity for all. Rising electricity demand caused global CO_2 emissions from the power sector to reach a record high in 2018, signalling the need for the transformation that is underway to speed up.

The first part of this chapter describes the projections for the Stated Policies Scenario and the Sustainable Development Scenario, analysing electricity demand and supply, and looking at drivers of demand growth, the changing electricity generation mix, investment needs, and CO_2 and pollutant emissions.

The second part of the chapter looks at three key topics that build on the first in-depth focus on electricity presented in last year's *World Energy Outlook (WEO)* (IEA, 2018a):

- The affordability of electricity and of energy more generally. This continues to be a key concern for consumers and policy makers. We examine the outlook for electricity prices by scenario, explore the impact of uncertainties on fossil fuel prices, wind and solar photovoltaics (PV) costs and the cost of capital, and analyse how growing shares of electricity in final energy demand affect the average household energy bill.

- Emissions from coal-fired power plants, which pose a serious threat to achieving environmental goals. The existing fleet of coal power plants also represents more than $1 trillion in capital investment that has yet to be recovered. We examine a possible way forward.

- Rising needs for flexibility. It is clear that flexibility will be a cornerstone of electricity security in a changing power mix. We look at how it might be provided, and explore how low-cost batteries could turn India into a large-scale launch pad for battery storage technology.

Figures and tables from this chapter may be downloaded from www.iea.org/weo2019/secure/.

Scenarios

6.1 Overview

Table 6.1 ▷ Global electricity demand and generation by scenario (TWh)

	2000	2018	Stated Policies		Sustainable Development		Current Policies	
			2030	2040	2030	2040	2030	2040
Electricity demand[1]	13 152	23 031	29 939	36 453	28 090	34 562	30 540	37 418
Industry	5 398	9 333	11 843	13 525	10 751	12 169	11 998	13 874
Transport	218	377	1 025	2 012	1 374	4 065	725	1 091
Buildings	6 738	11 755	15 198	18 893	14 264	16 606	15 835	20 176
Share of population with electricity access	*73%*	*89%*	*93%*	*93%*	*100%*	*100%*	*93%*	*93%*
Electricity generation[2]	15 427	26 607	34 140	41 373	31 800	38 713	34 988	42 824
Coal	5 994	10 123	10 408	10 431	5 504	2 428	11 464	12 923
Natural gas	2 750	6 122	7 529	8 899	7 043	5 584	8 086	10 186
Nuclear	2 591	2 718	3 073	3 475	3 435	4 409	3 112	3 597
Renewables	2 863	6 799	12 479	18 049	15 434	26 065	11 627	15 485

Note: TWh = terawatt-hours.

In the **Stated Policies Scenario**, electricity demand grows by 2.1% per year, resulting in over 13 000 terawatt-hours (TWh) more demand in 2040 than today (Table 6.1). Developing economies account for almost 90% of demand growth, of which two-thirds is in Asia where demand is rising in particular for electric motors, space cooling and household appliances. About 530 million people gain access to electricity around the world by 2040, mainly in Africa and developing Asia, accounting for 2% of global electricity demand growth.

Renewables, led by wind and solar PV, provide three-quarters of the increase in electricity supply, underpinned by policy support and declining technology costs. The share of generation from renewables increases from 26% today to 44% in 2040, with solar PV and wind together rising from 7% to 24%. The share from nuclear power decreases, but its output nevertheless rises in absolute terms, with growth in China and more than twenty other countries more than offsetting reductions in advanced economies.

The share of fossil fuels in electricity supply falls below 50% in 2040, down from two-thirds (where it has been for decades). Coal remains the largest source of electricity, though its share of overall generation declines from 38% to 25%, and its share of generation in advanced economies falls by more than half over the period to 2040. Gas-fired generation grows steadily, maintaining roughly its current share of generation, thanks to the availability of cheap gas in some regions and the role of gas in supporting flexibility.

---

[1] Electricity demand is defined as total gross electricity generated less own use generation, plus net trade (imports less exports), less transmissions and distribution losses.

[2] Other sources are included in total electricity generation.

There have been a number of significant developments since the *WEO-2018*. An additional 30 million electric cars, reflecting policy targets, are now projected by 2040, adding to electricity demand growth (see Chapter 3.9). For electricity supply, solar PV projections have been boosted by some 20% to 2040, mainly reflecting policy changes: in China, where there has been a partial reversal of a previous decision to reduce subsidies; in India, where an ambition to reach 450 gigawatts (GW) of renewables-based capacity by 2030 (excluding hydropower) has been announced; and in the United States, where state-level policies have been strengthened. Projections for battery storage capacity have been raised by close to 50%, in part due to the increases for solar PV (see section 6.12), and offshore wind projections have been revised upward by some 80% with new policies and technology gains (see Chapter 14).

In the **Current Policies Scenario**, electricity demand increases by 2.2% per year, and by 2040 demand is nearly 1 000 terawatt-hours (TWh) higher than in the Stated Policies Scenario. This figure hides important differences in the structure of electricity demand growth between the two scenarios; without the implementation of proposed increases in the coverage and stringency of energy efficiency policies, electricity demand in buildings is nearly 1 300 TWh higher, while the difference in industry is 350 TWh, largely due to higher demand related to motor systems. Uptake of electric vehicles continues to accelerate under current policies, although not as rapidly as in the Stated Policies Scenario, as a result, transport electricity demand is nearly 1 000 TWh lower. Without the implementation of proposed policies, the pace of change for the power mix is slower than in the Stated Policies Scenario. Renewables provide half of the increase in electricity supply over the next two decades, though their share of global electricity generation to 2040 remains below 40%. At the same time, under current policies, fossil fuels continue to play a large role to 2040: coal-fired electricity generation still accounts for 30% of electricity supply and gas-fired generation for about 25%. Overall, power sector emissions rise by some 20% by 2040.

In the **Sustainable Development Scenario** the share of electricity in final consumption grows faster than in the Stated Policies Scenario, rising from 19% today to 31% in 2040. Increased energy efficiency dampens demand growth, and total electricity demand is just below 35 000 TWh in 2040. Full electricity access is achieved by 2030 and contributes 5% of demand growth. The growth of renewables generation exceeds electricity demand growth by almost 8 000 TWh, raising their share of generation to two-thirds by 2040. Wind and solar PV together provide 40% of generation in 2040. Solar PV and other renewables also play a critical role in providing electricity access to all, particularly in sub-Saharan Africa (see Chapter 10). Nuclear and power plants equipped with carbon capture, utilisation and storage (CCUS) supplement renewables, raising the global low-carbon share of generation to about 85% in 2040. Generation from fossil-fuelled power plants without carbon capture declines sharply. Coal-to-gas switching provides a bridge to a low-carbon future in the near term, though, in the longer term, the role of gas moves increasingly towards the provision of flexibility.

6.2 Electricity demand by region

In all scenarios, electricity accounts for the largest growth in final energy consumption and grows at double the pace of final energy demand, with today's global electricity demand of 23 000 TWh rising by 13 400 TWh to 2040 in the Stated Policies Scenario (+2.1% per year), and 11 500 TWh in the Sustainable Development Scenario (+1.9% per year) (Table 6.2).

In advanced economies, electricity demand growth has already started to flatten or even decline (18 out of the 30 IEA member countries have seen their electricity demand fall since 2010). The future level of demand in these countries is dictated by the balance between the increasing electrification of mobility and heat, speed of digitalisation and extent of further energy efficiency gains. In the Stated Policies Scenario, electricity demand flattens or declines in some countries (e.g. Japan); in others it rebounds (e.g. the United States). (For further analysis on the impacts of energy efficiency, electrification and digitalisation on electricity demand see the *WEO-2018 Special Focus on Electricity* [IEA, 2018a]).

Table 6.2 ▷ Electricity demand by region and scenario (TWh)

			Stated Policies		Sustainable Development		Change 2018-2040	
	2000	2018	2030	2040	2030	2040	STEPS	SDS
North America	4 260	4 786	5 160	5 626	4 966	5 602	840	816
United States	3 589	4 011	4 226	4 517	4 099	4 573	506	563
Central & South America	660	1 081	1 445	1 837	1 331	1 660	757	579
Brazil	327	517	675	845	619	745	328	228
Europe	3 114	3 631	3 975	4 346	3 926	4 724	715	1 093
European Union	2 604	2 884	3 045	3 243	3 050	3 645	359	761
Africa	380	703	1 086	1 653	1 073	1 696	950	993
South Africa	190	211	252	319	210	249	108	38
Middle East	361	954	1 309	1 817	1 189	1 621	863	667
Eurasia	809	1 084	1 302	1 474	1 132	1 220	390	137
Russia	677	893	1 043	1 149	916	971	256	77
Asia Pacific	3 569	10 792	15 662	19 699	14 474	18 038	8 907	7 246
China	1 174	6 330	9 127	10 912	8 415	10 052	4 582	3 723
India	376	1 243	2 417	3 718	2 254	3 263	2 475	2 020
Japan	962	994	980	989	926	942	-4	-52
Southeast Asia	323	935	1 510	2 091	1 346	1 888	1 156	953
World	13 152	23 031	29 939	36 453	28 090	34 562	13 422	11 531

Note: TWh=terawatt-hour; STEPS = Stated Policies Scenario; SDS = Sustainable Development Scenario.

In developing economies, electricity demand has almost tripled since 2000. It continues to grow, but at a slower pace, with demand increasing by 3% annually. Demand growth continues to be driven by economic fundamentals: increases in industrial output, rising incomes and an expanding services sector.

These two different pathways result in a continuing shift of global electricity demand towards developing economies. **Advanced economies** account for 44% of global electricity demand today, down from a share of 66% in 2000. In the Stated Policies Scenario, this share drops to 32% in 2040, with electricity demand seeing annual growth of 0.7%, which is four-times lower than the rate of growth in developing economies. Developing economies account for nearly 90% of global demand growth to 2040, this share is 83% in the Sustainable Development Scenario, with higher demand in advanced economies.

The United States and the European Union account for nearly 70% of electricity demand in advanced economies today, a share broadly maintained through to 2040 in the Stated Policies Scenario with demand increases of 510 TWh and 360 TWh respectively. The switch from gas to more efficient electric heat pumps in buildings and industry reduces total energy consumption, but drives up electricity demand. As a result, the share of electricity in final energy consumption reaches 29% in 2040 in the European Union, up from 21% today. In the United States, electricity's share rises from 21% to 24% in 2040. Demand in these economies rises even higher in the Sustainable Development Scenario as the impact of pushing energy efficiency towards its maximum potential is more than offset by increasing electrification, notably in transport. By 2040, 270 million additional electric vehicles (EVs) (cars, buses, trucks and two/three-wheelers) relative to the Stated Policies Scenario push up demand by over 900 TWh in advanced economies,[3] 190 million of these EVs are on the road in the United States and European Union.

The share of electricity in final energy demand is currently 29% in Japan and 25% in Korea, and it rises in both countries in the Stated Policies Scenario. In Japan, increasing electrification offsets the impact of higher efficiency and declining population, keeping demand in 2040 around today's level; in Korea, it helps push demand almost 130 TWh higher by 2040. In the Sustainable Development Scenario, additional energy efficiency sees lower electricity demand in 2040 in both countries relative to the Stated Policies Scenario.

In **developing economies**, the 3% rate of average annual demand growth to 2040 is half the rate seen in the period between 2000 and 2018, but it still leads to electricity demand in the Stated Policies Scenario nearly doubling from today's levels by 2040. Demand in China, the largest user of electricity today, increases by more than 70% from today to 2040, accounting for 40% of global demand growth. Further economic development and rising standards of living also see demand almost triple in both India and Indonesia, with particularly strong demand growth in buildings. Despite this, per capita demand for electricity in India in 2040 is still less than a fifth of that in the United States today (Figure 6.2). Demand also increases across South America, with the largest growth in Brazil, where economic growth leads to a rise in electricity demand of almost 330 TWh.

In the Sustainable Development Scenario, electricity demand in developing economies in 2040 is 9% below that in the Stated Policies Scenario (2 200 TWh). Efficiency gains, notably in industry, are the primary driver of lower demand in major industrial centres, and their

---

[3] The additional electricity demand has been estimated based on regional annual average mileage of EVs.

impact is biggest in China (-860 TWh). These improvements more than offset the impact of over 500 million additional EVs in developing economies, of which 300 million are in China.

Figure 6.2 ▷ Per capita electricity demand and share of electricity in total final consumption in advanced and developing economies

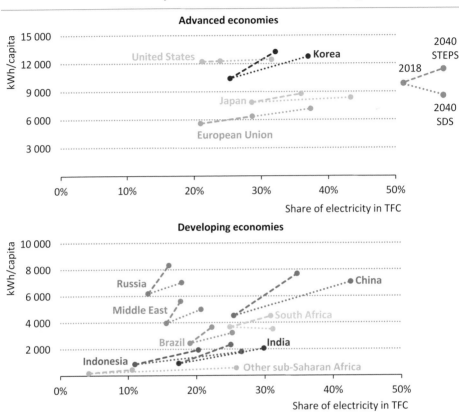

All regions see higher electrification in the Sustainable Development Scenario, while levels of demand relative to the Stated Policies Scenario vary according to regional contexts

Notes: TFC = total final consumption; kWh = kilowatt-hour; STEPS = Stated Policies Scenario; SDS = Sustainable Development Scenario. Other sub-Saharan Africa excludes South Africa.

Achieving universal access to electricity adds 470 TWh to demand in the Sustainable Development Scenario by 2040, with the biggest impact in sub-Saharan Africa, where over 900 million people gain access to electricity. The additional connections create opportunities for economic development and improve access to basic services, while increasing demand in 2040 by only 150 TWh more than the Stated Policies Scenario, which sees progress but does not deliver universal access (see Chapter 10). Despite universal access to electricity, per capita electricity demand in 2040 averages less than 700 kWh in sub-Saharan Africa, under a quarter of today's global average.

6.3 Electricity demand by sector

Buildings and industry are the two largest electricity consuming sectors today, accounting respectively for nearly 11 800 TWh and 9 300 TWh, while transport (mostly rail today) accounts for less than 2% of electricity consumption. The share of electricity in total final consumption (TFC) today is 19%; by 2040, further electrification of all sectors drives up the share of electricity in both the Stated Policies (24%) and Sustainable Development scenarios (31%). Given the higher efficiency of equipment using electricity, compared with equipment using fuels directly, the share of electricity in meeting useful energy demand[4] is even higher, surpassing 43% by 2040 in the Sustainable Development Scenario.

Table 6.3 ▷ Electricity demand by sector and scenario (TWh)

	2000	2018	Stated Policies 2030	Stated Policies 2040	Sustainable Development 2030	Sustainable Development 2040	Change 2018-2040 STEPS	Change 2018-2040 SDS
Total electricity demand	13 152	23 031	29 939	36 453	28 090	34 562	13 422	11 531
Share in TFC	16%	19%	22%	24%	24%	31%	4.9%	11.3%
Share in useful energy	21%	24%	30%	33%	34%	43%	9.0%	19.7%
Industry	5 406	9 346	11 857	13 540	10 768	12 188	4 194	2 842
Electric motors	3 414	6 930	9 427	11 418	8 160	9 290	4 488	2 359
Heat pumps	-	-	85	215	476	1 138	215	1 138
Share in industry final consumption	19%	21%	23%	24%	24%	28%	2.1%	6.1%
Share in useful energy	20%	22%	26%	27%	29%	36%	5.2%	14.0%
Buildings	6 738	11 755	15 198	18 892	14 264	16 606	7 137	4 851
Space and water heating	165	231	272	281	249	262	50	31
Cooling	948	1 850	2 904	4 072	2 530	3 040	2 222	1 190
Share in buildings final consumption	24%	33%	38%	43%	45%	53%	10.6%	20.1%
Share in useful energy	36%	45%	59%	60%	65%	68%	15.2%	23.7%
Transport	218	377	1 025	2 012	1 374	4 065	1 635	3 688
Electric vehicles	3	57	541	1 347	890	3 393	1 290	3 336
Share in transport final consumption	1.0%	1.1%	3%	5%	4%	13%	3.7%	12.2%
Share in useful energy	2%	3%	6%	10%	9%	26%	7.7%	23.1%

Note: STEPS = Stated Policies Scenario; SDS = Sustainable Development Scenario; TFC = total final consumption.

Most of the electricity used in the industry sector is for motor-driven systems (including pumps, fans and compressors). Motor systems currently account for 30% of global

---

[4] Useful energy refers to the energy that is available to end-users to satisfy their needs. This is also referred to as energy services demand. As a result of transformation losses at the point of use, the amount of useful energy is lower than the corresponding final energy demand for most technologies. Equipment using electricity often has a higher conversion efficiency than equipment using other fuels, meaning that for a unit of energy consumed, electricity can provide more energy services.

electricity demand, increasing to 31% in the Stated Policies Scenario by 2040 as demand grows by almost 4 500 TWh. Improving the efficiency of motors and wider motor systems (for example by installing variable speed drives, ensuring correct motor sizing and using energy management systems) plays an important role in slowing electricity demand growth. Today, low efficiency motors represent two-thirds of the current stock. In the Stated Policies Scenario, their use declines rapidly, and motors with an equivalent to or higher than IE3[5] standard represent close to 90% of the stock. In the Sustainable Development Scenario, increasing sales of motors compliant with the IE4+ standard avoids over 2 100 TWh of consumption compared with the Stated Policies Scenario (Table 6.3).

Buildings (including households and the services sector) continue to be both the biggest contributor to electricity demand and to its growth in the Stated Policies Scenario. Increasing ownership of cooling systems is one of the main drivers of growth. Today, only 8% of the 3 billion people that live in hot places[6] own a cooling system, compared with 90% of people in United States and Japan (IEA, 2018b). In China, air conditioner ownership has grown rapidly in recent years and is now around 60%. In Africa and India ownership levels for air-conditioners and household appliances remain very low. By 2040, increases in incomes, electricity access rates and ownership of appliances and air conditioners drives up electricity demand in the buildings sector by 4 200 TWh in developing economies, fourteen-fold the growth projected in advanced economies. Household electricity consumption in developing economies increases 60% in in the Stated Policies Scenario, although where implemented, minimum energy performance standards (MEPS) for air conditioners and large appliances help to slow electricity demand growth.

More stringent MEPS for air conditioners and large household appliances in the Sustainable Development Scenario avoid 1 350 TWh of electricity demand compared with the Stated Policies Scenario in 2040. Demand growth is slowed further in the Sustainable Development Scenario by more efficient heat pumps, better insulation and greater use of renewables for heating (biomass stoves, solar thermal).

Among the end-use sectors, **transport** uses the least amount of electricity today (about 2%), most of it in rail. Yet, road transport electricity use is growing rapidly as the global electric vehicle fleet expands, underpinned by policy measures in many countries. In 2018, worldwide sales of electric cars grew 70%, led by China, the United States, Canada and the European Union. Growth in EV sales contrasts with a decline in overall car sales in many major economies. The total number of electric cars rises to 330 million by 2040 in the Stated Policies Scenario (one-out-of-six cars), and the total number of EVs (cars, buses, trucks and two/three-wheelers) rises to over 1 billion, corresponding to 30% of the total fleet. By the mid-2020s, EVs overtake railways as the largest consumer of electricity in transport, although there is strong potential for growth from rail (IEA, 2019a).

---

[5] There are a number of standards and classifications for motors. Many can be benchmarked to the International Electro-technical Commission's "International Efficiency" standards, which range from low (IE0) to super premium (IE4), with minimum efficiency requirements based on size and number of poles.

[6] Hot places are defined as experiencing average perceived daily temperature over 25 °C for the whole year.

Although electrification efforts still need to be bolstered in many countries, the IEA's *Tracking Clean Energy Progress* recognises EVs as one of the very few energy technologies that are on track to meet the requirements of the Sustainable Development Scenario (IEA, 2019b). In the Sustainable Development Scenario, the total number of EVs reaches 1.9 billion by 2040, with almost 900 million electric cars, leading demand growth (Figure 6.3). Heavier freight vehicles are more difficult to electrify than their passenger counterparts, and only 8% of freight vehicles are electric in the Stated Policies Scenario by 2040, although in the Sustainable Development Scenario this figure rises to 26%.

Figure 6.3 ▷ Electricity demand growth by end-use and scenario in advanced and developing economies, 2018-2040

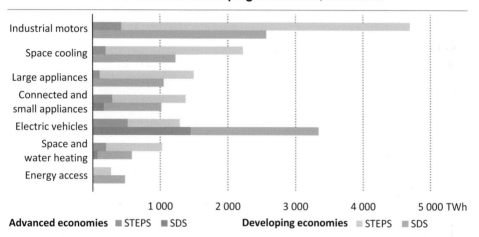

Electricity demand growth from industrial motors and cooling is nearly halved in the Sustainable Development Scenario, while most growth is from the almost 1 billion EVs

Note: STEPS = Stated Policies Scenario; SDS = Sustainable Development Scenario.

Box 6.1 ▷ What if the future is electric?

Electricity demand growth is uncertain and could be accelerated by policy actions beyond those in the Stated Policies Scenario. In recognition of this, the Future is Electric Scenario was introduced in the *WEO-2018* (IEA, 2018a). It achieves the economic potential for the electrification of heat and transport, and sees an increasingly digital economy and the provision of universal access to electricity. In the Future is Electric Scenario, the share of electricity in total final consumption reaches 31% by 2040, and its share of useful energy demand reaches 48%. The buildings sector sees a share in useful energy of almost three-quarters, whereas electrification of some modes of transport (for example trucks) makes limited progress for economic or technical reasons. The Sustainable Development Scenario takes account of the assessment in the Future is Electric Scenario on electrification potential.

6.4 Electricity supply by source

In the **Stated Policies Scenario,** electricity generation from renewables increases rapidly, surpassing coal by 2026. Renewables contribute three-quarters of electricity supply growth to 2040, underpinned by policy support in nearly 170 countries (REN21, 2019), as well as their improving competitiveness (see section 6.8). Their share of generation rises from 26% in 2018 to 44% in 2040 (Table 6.4). Hydro remains the largest source of renewable power, at about 15% of generation through to 2040. Wind and solar PV together provide over half of the growth in electricity supply, raising their share from 7% in 2018 to 24% in 2040. Bioenergy adds to the renewables total, offering dispatchable low-carbon electricity. Rising shares of variable renewables transform power systems in a number of ways, including by increasing the need for flexibility (see section 6.12). They also raise concerns over whether some current electricity market designs will deliver efficient and timely investment.

Table 6.4 ▷ Global electricity generation by source and scenario (TWh)

	2000	2018	Stated Policies				Sustainable Development	
			2025	2030	2035	2040	2030	2040
Coal	5 994	10 123	10 291	10 408	10 444	10 431	5 504	2 428
of which CCUS	-	0	1	16	43	69	246	994
Natural gas	2 750	6 118	6 984	7 529	8 165	8 899	7 043	5 584
of which CCUS	-	-	-	0	0	1	220	915
Oil	1 207	808	724	622	556	490	355	197
Nuclear	2 591	2 718	2 801	3 073	3 282	3 475	3 435	4 409
Renewables	2 863	6 799	9 972	12 479	15 204	18 049	15 434	26 065
Hydro	2 613	4 203	4 759	5 255	5 685	6 098	5 685	6 934
Bioenergy	164	636	916	1 085	1 266	1 459	1 335	2 196
Wind	31	1 265	2 411	3 317	4 305	5 226	4 453	8 295
Solar PV	1	592	1 730	2 562	3 551	4 705	3 513	7 208
Geothermal	52	90	125	182	248	316	282	552
CSP	1	12	28	67	124	196	153	805
Marine	1	1	2	10	25	49	14	75
Total	15 427	26 603	30 803	34 140	37 682	41 373	31 800	38 713

Note: TWh = terawatt-hours; CCUS = carbon capture, utilisation and storage; PV = photovoltaics; CSP = concentrating solar power.

Global coal-fired electricity generation plateaus after more than a century of growth, but remains the largest source of electricity through to 2040. From 1970 to 2013, it grew fivefold, an average of 3.8% per year, but slowed to 1% in the last five years. In 2018, final investment decisions for new coal plants were at their lowest level in a century (IEA, 2019b). To 2040, coal-fired generation rises by just 0.1% per year. As a result, coal's share in electricity generation declines from 38% in 2018 to 25% in 2040 (Figure 6.4). Without additional efforts to develop the technology, CCUS in coal-fired power remains limited.

Natural gas-fired generation tripled over the past 22 years, and is set to increase by nearly 50% to 2040, thanks in part to its role in providing flexibility to power systems. The global share of gas in total electricity supply remains stable at about one-fifth, though its share declines in Europe and Japan. The availability of large, low-cost gas resources in the United States is having an impact on gas markets around the world and has spurred coal-to-gas switching: in total such switching avoided around 500 million tonnes (Mt) of CO_2 emissions from 2010 to 2018, and it has the potential to avoid more in the future (IEA, 2019c).

Nuclear power generation increases by 30% worldwide to 2040, though its share of supply declines slightly. Two regional pathways emerge. In advanced economies, nuclear power is the largest low-carbon source of electricity today (18% of generation), but falls to the third-largest in 2040, behind wind and hydro, and just ahead of solar PV. This trajectory for nuclear power includes many planned lifetime extensions as well as new construction, but there is a risk that nuclear power will fade away faster in advanced economies (IEA, 2019d). In developing economies, the role of nuclear power is maintained at about 5% of electricity supply to 2040, though electricity production increases by 150% over the period to 2040.

Figure 6.4 ▷ Global electricity generation mix by scenario

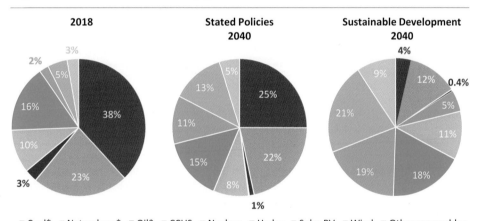

Electricity supply shifts towards renewable energy under current and proposed policies, but all low-carbon technologies are needed to support clean energy transitions

* Excludes capacity equipped with CCUS.

In the **Sustainable Development Scenario**, renewables provide two-thirds of electricity supply worldwide by 2040: solar PV and wind together provide 40%, with a further 25% from dispatchable renewables, including hydro and bioenergy. Nuclear power makes a moderate additional contribution, and close to 320 GW of coal- and gas-fired capacity is equipped with CCUS. Unabated coal-fired generation is cut in half by 2030, and its share falls from 38% in 2018 to less than 5% by 2040. The role of gas moves away from supplying bulk energy and towards providing system flexibility, with its share of generation cut by one-third by 2040 as a result.

6.5 Installed capacity by source

In the **Stated Policies Scenario**, global power generation capacity expands by 80% by 2040, to over 13 000 GW from 7 220 GW in 2018. Renewables' share in global power capacity rises from 35% today to 55% in 2040. Solar PV becomes the largest source of power capacity around 2035 overtaking wind in 2020, hydro in 2027, coal in 2033 and gas in 2035 (Figure 6.5). Wind power also rises steadily to over 1 850 GW of capacity by 2040, on par with hydropower, with an increasing emphasis on offshore wind (see Chapter 14).

Figure 6.5 ▷ Global power generation capacity by source and scenario

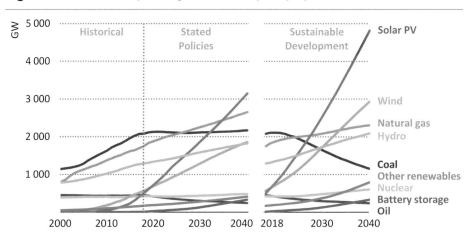

Policy support and improving competitiveness of solar PV lead to rapid growth, and by 2040 its total capacity is higher than any other technology

Global coal-fired power capacity flattens out in the near term, ending a century of growth. At the start of 2019, the global pipeline for new coal-fired capacity totalled 710 GW, with 170 GW under construction and 540 GW of planned projects. This compares with coal-fired capacity additions of 690 GW by 2040 in the Stated Policies Scenario, with more than 90% of this capacity being built in developing economies, principally in India, China and Southeast Asia. Cumulative retirements of coal-fired capacity by 2040 total nearly 600 GW, reflective of ageing fleets, challenging market conditions and phase-out commitments. The majority of retirements occur in advanced economies. As a result, the share of coal in global power capacity falls from 29% in 2018 to 17% in 2040.

Nuclear power capacity worldwide increases to 2040, declining by one-quarter in advanced economies but more than doubling in developing economies. At the start of 2019, 60 GW (gross) of nuclear power projects were under construction in nearly 20 countries. Increased capacity to 2040 is driven by programmes in China, India, Russia and the Middle East. China is set to become the leader in nuclear power capacity around 2030, overtaking the United States and Europe. Several new projects in China were announced in 2019, after two years without a new start. In most advanced economies, nuclear power capacity is set to decline as a result of existing plants reaching the end of their life and of concerns about the economic feasibility of new projects. However, small modular reactors are gaining interest because they would require less upfront capital investment and could offer a dispatchable source of low-carbon electricity, with a potential market of over 20 GW (OECD/NEA, 2016).

Natural gas-fired capacity and battery storage are very important sources of flexibility to support the integration of variable renewables and for electricity security. In the Stated Policies Scenario, combined-cycle gas turbines (CCGTs) account for the vast majority of the average of 60 GW of gas-fired capacity built each year, though lower efficiency gas-fired capacity takes up an increasing share to meet rising flexibility needs. About 330 GW of battery storage are deployed by 2040, reducing the need for new thermal capacity. The future costs are uncertain, but battery storage has significant potential and new developments could change the landscape of electricity supply (see section 6.12).

Figure 6.6 ▷ Renewables share in capacity additions by region in the Stated Policies and Sustainable Development scenarios, 2019-2040

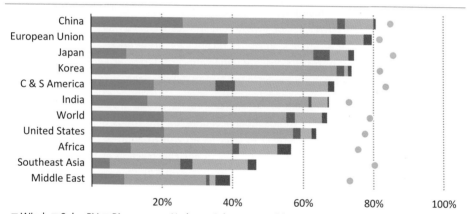

Renewables account for the majority of new capacity built to 2040 in most regions in the Stated Policies Scenario and rise to higher shares in the Sustainable Development Scenario

Note: C & S America = Central and South America.

In the Stated Policies Scenario, nearly 8 500 GW of new power capacity is added globally by 2040, of which two-thirds is renewables (Figure 6.6). Renewables account for the majority of capacity additions in most regions. This includes about 80% of additions in the European Union and China, but they provide less than half of additions in Southeast Asia and the Middle East. Solar PV provides the largest share of renewable capacity additions in most regions, including China and India, helping the global solar market return to growth after a pause in 2018. Wind power provides the second-largest share of renewables capacity additions in many regions and leads the way in the European Union.

In the **Sustainable Development Scenario**, the deployment of all low-carbon technologies accelerates. Renewables make up the vast majority of capacity additions in all regions, and about 80% of all capacity additions worldwide through to 2040. The pace of solar PV deployment increases from 97 GW in 2017 and 2018 to nearly 210 GW in 2030 and over 300 GW in 2040. Wind power additions average close to 130 GW per year through to 2040, two-and-a-half times the additions in 2018. Nuclear power capacity additions average about 15 GW per year, up from the 11 GW completed in 2018. Starting in the late 2020s, the deployment of power generation capacity equipped with CCUS averages over 20 GW per year to 2040, with two-thirds as retrofits that extend the operational life of relatively young and efficient coal- and gas-fired capacity. Addressing the CO_2 emissions from the existing fleet of coal-fired power plants calls for a multifaceted strategy (see section 6.11).

6.6 Electricity supply by region

The transformation of electricity supply is underway around the world, but the extent of change varies widely. Regional outlooks depend critically on four factors: policy environment; technology preferences; available resources; and the relative economics of different power generation technologies. The policy environment is of central importance, and more than 95% of power sector investment in recent years was directly related to policies or regulation (IEA, 2018c). Regional technology preferences, mainly expressed through financial support or restrictions, are important considerations and vary widely. This section focuses on the outlook envisaged in the Stated Policies Scenario.

China is the largest power market by far and poised for strong growth, making it a key determinant of global electricity trends. Coal accounts for two-thirds of China's electricity supply today, but efforts are underway to limit its growth to address pollution concerns and limit CO_2 emissions. Even with over 60 GW under construction at the start of 2019, coal-fired capacity and generation are set to plateau (Figure 6.7). China is already the global leader in a wide array of low-carbon technologies and maintains this position: it accounts for 30-40% of the global market to 2040 for solar PV, wind, hydro and nuclear power. The solar PV outlook improved in 2019 when new subsidy allocations for utility-scale solar PV were announced that partially reversed sudden cuts to support in the previous year.

In the **United States**, cheap natural gas has reshaped electricity supply in recent years, helping to cut the share of coal by 18 percentage points from 2010 to 2018. Wind and solar PV have been gaining momentum thanks to falling costs, state-level renewable energy targets and federal tax incentives. Growth is set to continue due to improved competitiveness and corporate demand, overcoming the impact of expiring federal tax credits. The growth in renewables and low gas prices are contributing to challenging market conditions for all generators, trimming the margins for existing capacity and making new investment difficult. Natural gas prices are set to remain low for some time, contributing to the retirement of about 90 GW of coal-fired capacity by 2030. Nuclear power is set to decline as the existing fleet is nearly 40 years old on average and no large-scale projects are in view after two new units are completed at Plant Vogtle. Nuclear power's share of generation is set to be cut from 19% in 2018 to 11% in 2040, even with recent decisions to support nuclear power in several states.

In the **European Union**, there is increasing emphasis on developing renewables. The "Clean Energy for All Europeans" package sets a renewable energy target of 32% of gross final consumption by 2030. In effect, this calls for the share of renewables in electricity to exceed 50%. Wind power is set to become the European Union's leading source of electricity around 2025, overtaking gas and nuclear, offshore wind is poised for rapid growth with country-level targets targeting at least a tripling of installed capacity by 2030 (see Chapter 14). Thirteen member states have plans to phase out all coal-fired power: together, these plans cover close to 100 GW of coal and aim to phase out more than 80% of it by 2030. In 2018, nuclear power provided 25% of electricity in the European Union, with several member states getting more than half their supply from nuclear. However, phase-out plans have been announced in Germany, Belgium and Spain (in 2019) and France aims to reduce its nuclear share of generation to 50%.

The future of electricity supply in **Japan** is guided by the 5th Strategic Energy Plan, published in 2018. It sets targets for the energy mix in 2030 and considers the path towards 2050. One key issue is the restarting of the nuclear fleet, with many plants having remained idle since the accident at the Fukushima Daiichi Nuclear Power Plant in 2011. As of May 2019, nine reactors had restarted operations and six more had received approval to do so from the nuclear regulator. A new order calling for anti-terrorism facilities at nuclear power plants is adding uncertainty about the pace of the nuclear restart. Japan is also targeting expansion of renewable energy to 22-24% in 2030, led by hydro and solar PV. Both nuclear and renewables would help Japan to reduce its reliance on imported coal and natural gas.

In **India**, three-quarters of electricity today comes from coal. Coal-fired capacity of 40 GW is currently under construction. Further capacity additions are planned to meet increasing demand, including peak loads and to ensure reliability with more dispatchable capacity. India is looking to diversify its power mix away from coal with a target of 175 GW of renewable energy capacity by 2022 (excluding large hydro), including 100 GW of solar PV. A longer term target of 500 GW of renewables capacity by 2030 (including large hydro) has also been announced. Battery storage, among other options, could support diversification away from coal, if costs can fall far enough (see section 6.12).

Expanding populations and rising incomes in developing economies are driving strong electricity supply growth in **Africa**, the focus region in this *World Energy Outlook*. Africa is rich in energy resources but poor in energy supply and aims to provide reliable and affordable electricity to a population that is set to grow to over 2 billion by 2040. Electricity demand is set to more than double to 2040 and many countries are actively developing renewable energy resources to meet rising needs (see Chapter 10).

Southeast Asia is set to achieve access to electricity for all in the near term. Countries in the region are seeking to develop a variety of sources of energy and expand regional trade to meet rising demand while maintaining affordability (IEA, 2019e).

Figure 6.7 ▷ Electricity generation by region in the Stated Policies (STEPS) and Sustainable Development (SDS) scenarios

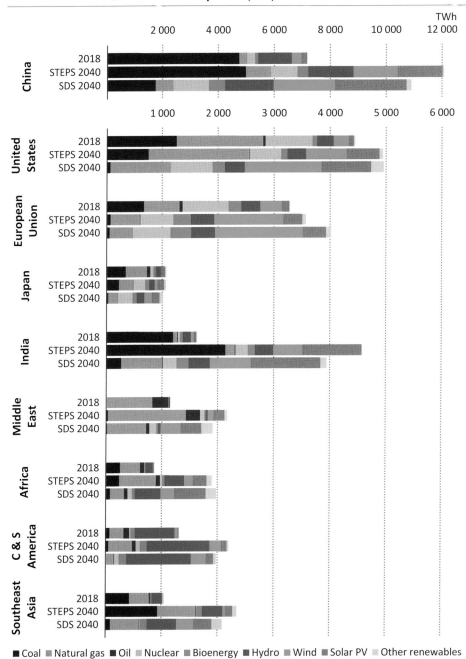

A suite of low-carbon sources play a larger role in the Stated Policies Scenario, but all regions need to accelerate efforts to drive clean energy transitions

6.7 Power sector investment

In the Stated Policies Scenario, global power sector investment totals $20 trillion over the period to 2040, on average this is 20% higher than annual spending from 2010 to 2018 (Figure 6.8). The power sector represents 50% of total energy supply investment worldwide and includes spending on new power plants, transmission and distribution lines, as well as refurbishments and upgrades. Renewables account for about $360 billion per year, most of it for solar PV and wind power, an increase of about one-fifth from recent levels, plus falling costs mean that this investment also buys more capacity per unit of spend than in the past. Investment in transmission and distribution networks represents nearly 45% of total power sector investment, with average spending of about $400 billion per year, about 3% of which is for fast chargers for EVs. Network spending also rises from recent levels to support replacements, upgrades and extensions to assure the security of electricity supply and to integrate rising shares of variable renewables. Spending is cut by half for coal-fired power plants and by 90% for oil-fired capacity. The vast majority of new investment in coal-fired power plants is in developing economies, mainly in India, China and Southeast Asia, but alternatives are rapidly emerging that are more compatible with environmental goals including reducing air pollution (see section 6.11).

Figure 6.8 ▷ Global annual average power sector investment, historical and by scenario, 2019-2040

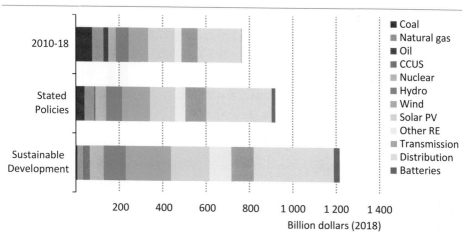

Solar PV and wind make up half of power plant investment and grid spending rises in both scenarios. Overall power sector investment rises by 60% in the SDS from recent levels

Note: CCUS in this figure represents coal- and gas-fired plants equipped with carbon capture, utilisation and storage equipment; RE = renewable energy technologies.

In the Sustainable Development Scenario, power sector investment reaches nearly $1.2 trillion per year on average, or $27 trillion in total to 2040. This is around 60% higher than recent spending levels. In this scenario, power sector investment accounts for two-thirds of total energy supply investment, compared with almost half today. Annual spending on renewables doubles as the primary means of decarbonising electricity supply in markets around the world. Annual spending on wind power increases to $210 billion, followed by solar PV at $180 billion, hydropower at $100 billion and other renewables at $100 billion (Figure 6.9). Spending on nuclear power doubles from recent levels, with more than 10% to extend the lifetime of existing reactors: such extensions are one of the most cost-effective ways of sourcing low-carbon electricity. After the fossil fuel power projects currently under construction (including around 170 GW of coal-fired and 120 GW of gas-fired generation) are completed, investment in fossil-fuelled power plants without CCUS slows dramatically to around $40 billion per year or 75% less than average annual spending from 2010 to 2018, while annual spending on fossil-fuelled plants equipped with CCUS rises to almost $30 billion per year, with most of this increase coming in the second-half of the outlook period. Close to 40% of annual spending, some $475 billion per year, is for electricity networks, with 80% for distribution networks.

Figure 6.9 ▷ Average annual power sector investment by region, 2019-2040

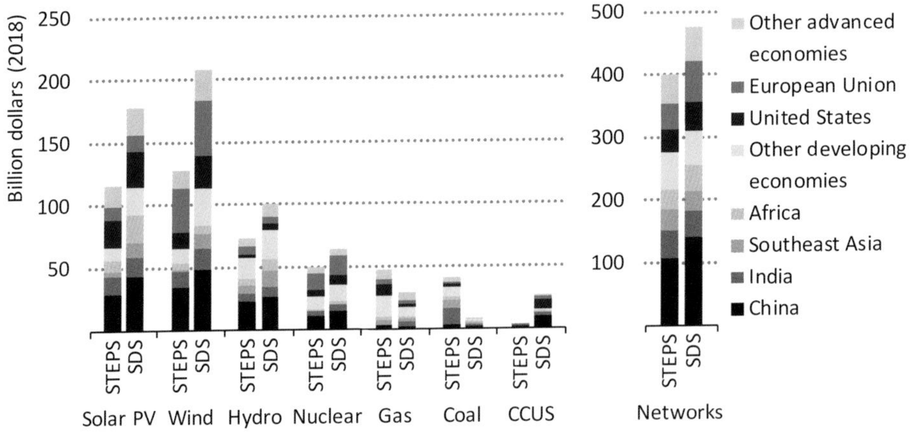

In both scenarios, developing economies make up over 60% of investment in power plants, accounting for 80% of hydropower and coal, and over 60% in solar PV and nuclear

Notes: STEPS = Stated Policies Scenario; SDS = Sustainable Development Scenario. CCUS in this figure represents coal- and gas-fired plants equipped with carbon capture, utilisation and storage equipment.

In the Sustainable Development Scenario, total power plant investment is over 40% higher than in the Stated Policies Scenario. Over 60% of global power sector investments are made in developing economies, with China leading the way in solar PV, wind and hydro, and coal plants equipped with CCUS. About two-thirds of investment in electricity networks is in developing economies in both scenarios. Among advanced economies, the United States leads investment in solar PV in both scenarios and remains the largest investor in gas-fired power plants. Europe leads investment in wind, with particularly strong development of offshore wind in the Sustainable Development Scenario (see Chapter 14).

6.8 Competitiveness of power generation technologies

Power systems are undergoing unprecedented changes, including rising shares of variable renewables and an increasing need to source power system flexibility. This calls for new metrics to assess the competitiveness of various power generation technologies. The levelised cost of electricity (LCOE) is the most commonly used metric for competitiveness and it remains useful in certain cases, but it needs to be complemented with additional system-specific information. The value-adjusted LCOE, introduced in the *WEO-2018*, provides a single metric for evaluating competitiveness under market conditions, considering both the costs and value for each technology in a system at a given time (IEA, 2018a). The value-adjusted LCOE does not include the costs of environmental externalities where they are not priced in the market, nor does it include network integration costs that are extremely site-specific. Other metrics, including the system LCOE (Ueckerdt et al., 2013) and value-cost ratio (US EIA, 2019) also offer approaches to capture both costs and value.

As a result of continued cost reductions, solar PV becomes the most competitive source of electricity in 2020 in China and India, and largely closes the gap with other sources by 2030 in the European Union and United States (Figure 6.10). In the Stated Policies Scenario, the global average LCOE of solar PV declines by about 50% from 2018 to 2030. After 2030, solar PV has the lowest LCOE of any power generation technology in many markets, including China and India (see Annex B).[7] However, the competitiveness of solar PV and resulting deployment should not be taken for granted, because the system value of solar PV tends to decline relative to the system average as its share of generation rises and this offsets further cost reductions in many cases. The output of all solar PV projects in a region is concentrated in the same hours of the day, so as the share of solar PV increases, then the value of additional output in those hours tends to decline (Hirth, 2013). This means that exponential growth for solar PV is not necessarily an obvious conclusion (Wanner, 2019). However, energy storage, such as batteries, can help by shifting solar power output to when it is valuable. As a result, solar PV plus storage becomes an increasingly competitive option in several leading markets, including India (see section 6.12).

Other renewables can also provide competitive new sources of electricity. Hydropower has long been a low-cost source of electricity, where high quality resources were able to offset relatively high upfront costs. Some new hydro projects are underway, mostly in developing economies. Onshore wind power has been one of the cheapest sources of electricity in mature markets for several years, and offshore wind is catching up as a result of recent technology gains (see Chapter 14). Geothermal can also be a competitive power generation technology where high quality resources are available at shallow depths, as for example in the East Africa Rift Valley. Bioenergy for electricity production and concentrating solar power are not at present generally competitive, though the technologies involved may well be developed further, and as dispatchable renewable energy technologies they could become more important.

---

[7] For additional power technology cost assumptions and projections, see iea.org/weo/weomodel/#power.

Figure 6.10 ▷ Value-adjusted levelised cost of electricity by technology in selected regions in the Stated Policies Scenario, 2020-2040

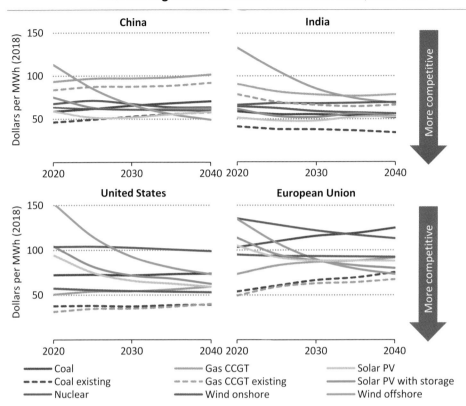

Considering both the changing technology costs and system value, solar PV is set to become one of the most competitive new sources of electricity by 2030

Note: CCGT = combined-cycle gas turbine.

The competitiveness of conventional power plants depends heavily on the availability of low-cost resources and market conditions. Coal-fired power plants remain one of the most competitive new sources of electricity wherever CO_2 or pollutant emissions have low or zero prices. Efficient gas-fired power plants are competitive especially where natural gas prices are very low, as they are in the United States. New nuclear power plants have proved to be high-cost endeavours in recent years, often involving delays and cost overruns, with notable exceptions in China, Korea and the United Arab Emirates. However, lifetime extensions for existing nuclear reactors are one of the most competitive sources of clean energy (IEA, 2019d). In general, existing conventional power plants are more competitive with renewables than new builds, because they usually need to cover only their short-run fuel and maintenance costs to continue operations, and capital investment for refurbishment is lower than for new construction. Just as with new builds, both the costs and value of existing projects need to be considered.

The competitiveness of power generation technologies is not fixed and is influenced by decisions taken by policy makers, regulators and the industry. For example, pricing CO_2 and other pollutant emissions would fundamentally change the comparisons, improving the competitiveness of all low-carbon sources compared with fossil fuels. A lower cost of capital would also improve the case for most low-carbon technologies. For comparison purposes, the results shown in Figure 6.10 assume the same weighted average cost of capital (WACC) for all technologies within a given region (7-8% in real terms depending on the stage of economic development). Where market risks are relatively low, cheaper financing might be available and this would improve their competitiveness. Supportive policy frameworks, for example reducing revenue risk by guaranteeing prices, can also enable access to cheap financing. A sensitivity analysis on the impact of lower costs of capital is included in the offshore wind outlook (see Chapter 14). Demand-side response can also improve the case for variable renewables: to the extent that it enables demand to be shifted to align with available supply, it reduces the effect of declining value of wind and solar PV. The development of transmission and distribution networks also strongly influences the value of additional power plants: congestion in the network for example could otherwise limit the ability to tap high quality renewable energy resources.

6.9 Power sector emissions

CO_2 emissions

Table 6.5 ▷ Annual CO_2 emissions from the power sector by scenario (Mt)

			Stated Policies				Sustainable Development	
	2000	2018	2025	2030	2035	2040	2030	2040
Electricity generation	8 247	12 655	12 608	12 643	12 703	12 759	7 544	3 129
Coal	5 920	9 357	9 262	9 263	9 175	9 048	4 665	1 323
Natural gas	1 341	2 656	2 770	2 897	3 099	3 330	2 594	1 732
Oil	986	641	575	483	429	381	285	168
Bioenergy with CCUS	-	-	-	-	-	-	-1	-94
Heat production	1 055	1 163	1 151	1 135	1 110	1 075	916	651
Coal	532	708	679	657	627	593	461	228
Natural gas	415	403	426	435	443	445	415	391
Oil	108	51	45	43	40	37	40	32
Total	9 302	13 818	13 759	13 777	13 813	13 834	8 460	3 780
CO_2 captured with CCUS	-	1	1	14	37	59	309	1 323

Note: Mt = million tonnes; CCUS = carbon capture, utilisation and storage equipment.

In the Stated Policies Scenario, global CO_2 emissions from the power sector remain stable to 2040 (Table 6.5), even though electricity generation rises by almost 60%. This is not only due to the rising share of renewables but also to continuing efficiency improvements in fossil fuel power plants, especially natural gas-fired power plants. Electricity generation makes up more than 90% of total power sector emissions, with the rest coming from the production of heat. Stable power sector CO_2 emissions to 2040 compares with a 50% increase over the period from 2000 to 2018.

Coal-fired power plants are the single largest source of energy-related CO_2 emissions today and represent about three-quarters from the power sector. To 2040, coal-fired power emissions decline by less than 5% compared with today. Tackling the emissions from the existing coal fleet is central to meet global climate objectives (see section 6.11). Although it is the least carbon-intensive fossil fuel, the 25% increase in the use of natural gas leads to an increase in emissions, and gas accounts for almost 30% of total emissions from the power sector in 2040. However, gas-fired power plants run more efficiently in 2040 producing 55% more electricity than now, while emitting 25% more emissions.

Power sector emission trends differ significantly by region. India overtakes United States before 2030 to become the second-biggest emitter behind China in terms of power-related CO_2 emissions, while Southeast Asia overtakes the European Union even earlier. In the United States, CO_2 emissions in the power sector decline by more than 25% to 2040. China sees its power sector emissions peak by 2030 and then falls by about 5% to 2040.

Figure 6.11 ▷ CO_2 intensity of electricity generation by region and scenario

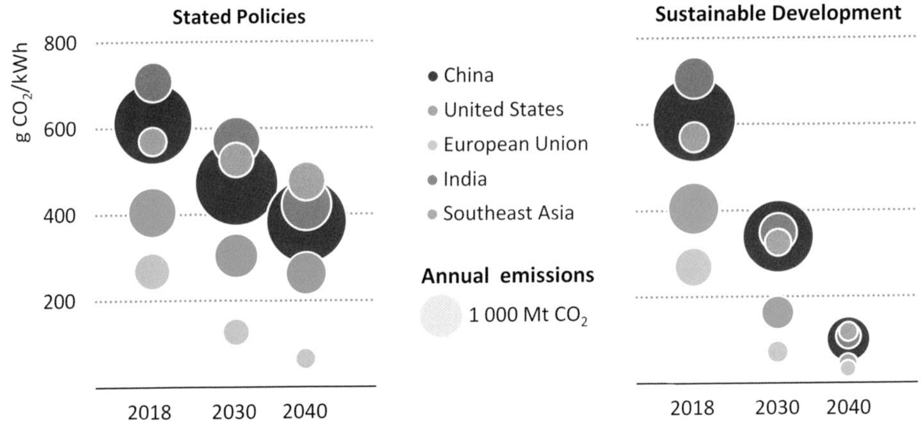

Carbon intensity of electricity generation declines in each region and scenario, though to a much greater extent in the Sustainable Development Scenario

The global carbon intensity of electricity generation declines by more than one-third on average from today to 2040, from 475 grammes of carbon dioxide per kilowatt-hour (g CO_2/kWh) to 310 g CO_2/kWh in the Stated Policies Scenario. This is the result of falling intensities in all regions, including a 75% reduction in the European Union and about 40% declines in China, India and the United States (Figure 6.11). In the Sustainable Development

Scenario, faster adoption of low-carbon technologies and addressing emissions from the existing coal fleet means that all regions are converging to CO_2 intensities below 100 g CO_2/kWh, with reductions of about 90% in the European Union, the United States, Japan and Korea, and at least 75% in China, India, Southeast Asia, Middle East and Africa.

Pollutant emissions

In the Stated Policies Scenario, global emissions of sulfur dioxide (SO_2) fall by 40%, and nitrogen oxides (NO_X) by 25% and particulate matter ($PM_{2.5}$) emissions decrease by one-third. These outcomes are driven by a shift to renewables and by a reduction in the use of coal in power (Figure 6.12). Most regions see reductions in these air pollutants, including the United States, the European Union, China and India. The main exceptions are Southeast Asia and other developing economies in Asia, where the use of coal in power increases significantly over the next two decades.

In the Sustainable Development Scenario, SO_2, NO_X and $PM_{2.5}$ emissions decline dramatically in all regions as a result of reduced use of fossil fuels and enhanced end-of-pipe measures. These reductions contribute to improve air quality around the world and reduce negative impacts on human health.

Figure 6.12 ▷ SO_2, NO_X and $PM_{2.5}$ emissions in the power sector by region and scenario, 2018-2040

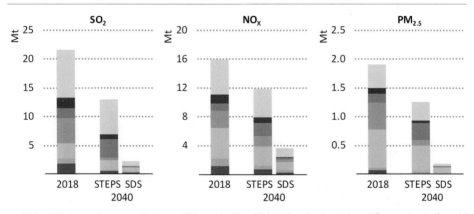

Pollutant emissions from the power sector, mainly driven by coal use in Asia, are projected to reduce by 2040, most significantly in the Sustainable Development Scenario, where end-of-pipe technologies and lower fossil fuel use drive the change

Note: Mt = million tonnes; STEPS = Stated Policies Scenario; SDS = Sustainable Development Scenario.
Source: International Institute for Applied Systems Analysis.

Key themes

6.10 Affordability of electricity

As electrification of the economy continues, the affordability of electricity is an increasingly central concern of consumers and policy makers. Residential electricity prices are the most visible metric for most consumers by which affordability is considered. There are significant differences in average annual electricity prices across regions (Figure 6.13). The production costs of electricity hinge on fuel costs and technology investments, and these can vary from one country to another, while the final price paid by the consumer is also affected by taxes, CO_2 prices, network costs and subsidies.

Figure 6.13 ▷ Residential electricity prices in selected regions by scenario

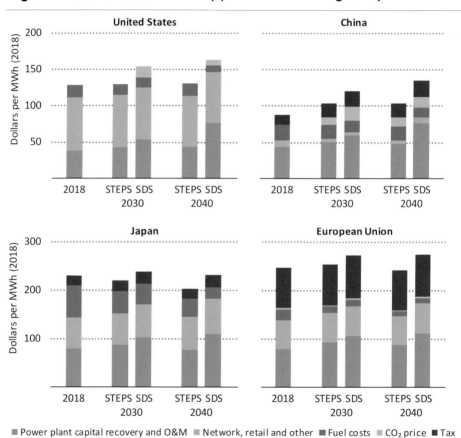

■ Power plant capital recovery and O&M   ■ Network, retail and other   ■ Fuel costs   ■ CO_2 price   ■ Tax

Electricity prices are set to rise in most regions, but depend on many regional factors

Note: MWh = megawatt-hours; O&M = operation and maintenance; STEPS = Stated Policies Scenario; SDS = Sustainable Development Scenario.

Residential electricity prices in the United States today are lower than most other advanced economies. This is thanks to relatively abundant and affordable indigenous resources like coal and gas, a well-established fleet of existing power plants, low investment needs in the face of subdued demand growth, and very low taxes on average. Electricity prices in Japan are notably higher today, having increased following the Fukushima Daiichi accident and the subsequent suspension of output from nuclear plants that made the country more reliant on more expensive liquefied natural gas imports. In the European Union, residential electricity prices are relatively high, in large part due to high taxes, which account for about one-third of the final price paid by consumers. In China, electricity prices are very low for households. This reflects both low electricity production costs that are a result of abundant and relatively cheap domestic coal resources and the fact that residential electricity prices benefit from cross-subsidies from the industry sector.

The cost structure of electricity becomes more capital-intensive across almost all regions in the Stated Policies Scenario because of increased reliance on technologies with low or zero fuel costs but higher upfront costs per unit of electricity produced. Where they are introduced, CO_2 prices push costs up in regions that rely heavily on coal-fired power generation, with a smaller impact in regions more reliant on gas. Despite cross-subsidies from industry, China sees growing electricity prices for households due to the introduction of a CO_2 price and to continuing investment in capital-intensive technologies. Electricity prices decline in Japan where the gradual restart of nuclear plants drives prices down by reducing costs from imported fuels. Residential electricity prices are relatively stable in the United States and the European Union in the Stated Policies Scenario.

In the Sustainable Development Scenario, residential electricity prices tend to be higher as a result of a more rapid move away from reliance on existing fossil-fuelled power plants and a faster roll-out of new capital-intensive investments in low-carbon technologies. The costs of renewable energy technologies decline faster than in the Stated Policies Scenario with accelerated deployment, but they are not enough to offset the expanded investment needs. In this scenario, residential electricity prices increase by about 10% in the European Union and 25% in the United States, while staying broadly stable in Japan. Accelerated efforts to reduce CO_2 emissions in China bring an increase of about one-half although prices in China still remain below today's US prices.

There are some areas in which action could help to improve the affordability of electricity. Any reduction in policy uncertainty would help to boost investor confidence, reduce risks and lower the cost of finance. Support for innovation through research, development, demonstration and deployment activities could facilitate new and cheaper technologies. Bills could also be lowered by subsidies for low-income households to buy more energy-efficient equipment.

Sensitivity analysis

Market dynamics, technology costs and policy decisions all have to be considered in any assessment of projected electricity prices. This section examines how sensitive residential electricity prices are to three of the largest uncertainties in our scenarios: fossil fuel prices, the cost of wind and solar PV, and the cost of capital (i.e. financing costs) for all power generation technologies (excluding transmission). No single factor yields changes higher than 12%, but all have a significant impact on electricity prices in the Stated Policies and Sustainable Development scenarios (Figure 6.14).

Figure 6.14 ▷ Sensitivity analysis of residential electricity prices by scenario, 2040

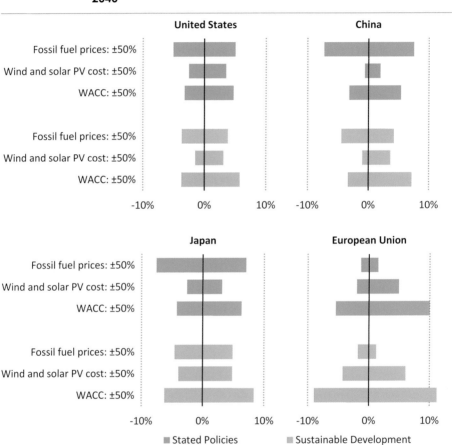

All factors influence electricity prices and their impact is determined by regional circumstances

Notes: WACC = weighted average cost of capital. All sensitivities are compared with the projected residential electricity price in 2040. Additional information on fossil fuel prices and power generation technology costs is available in Annex B.

The variability of fossil fuel prices has the biggest impact in countries like Japan that rely heavily on fossil fuel energy imports and such as China where fossil-fuelled power provides the biggest share of the electricity supply. However, even in these countries, 50% lower fossil fuel prices in 2040 in the Stated Policies Scenario would reduce residential electricity prices by only 7-8%. In the Sustainable Development Scenario, consumers are less exposed to changes in fossil fuel prices.

The sensitivity of electricity prices to wind and solar PV costs is relatively low, less than a 5% reduction for 50% lower costs, despite the fact that they dominate power plant investment over the next two decades in both the Stated Policies and Sustainable Development scenarios. In part, this reflects the fact that the existing power plant fleet represents a large part of electricity supply costs for years to come, as power plants have operational lifetimes measured in decades. The cost of wind and solar is particularly important in the European Union, where wind alone accounts for 40% of new capacity additions in both scenarios, more than one-third of which takes the form of offshore wind. The United States and Japan are less sensitive to the cost of wind and solar PV because these technologies account for a smaller share of their total capacity additions than is the case in the European Union. In China, the cost of wind and solar PV has an even smaller effect on the final electricity price than elsewhere because its technology costs are among the lowest in the world.

The WACC for power generation technologies affects the affordability of electricity across all regions and is particularly important in those characterised by a shift towards more capital-intensive technologies such as renewables and nuclear power. Reducing the WACC by 50% would reduce residential electricity prices by 3-6% in the Stated Policies Scenario in all four regions. In the Sustainable Development Scenario, the shift away from fossil fuels increases the effect on electricity bills due to variations in the WACC and of the cost of wind and solar PV, especially in the European Union where the impact would reach nearly 10%.

Role of electricity in household energy spending

The level of consumer energy spending today varies substantially across regions, with marked differences between advanced and developing economies attributable to a combination of different consumption patterns, tax policies and fossil fuel subsidies (Figure 6.15). While energy bills are highest in advanced economies, they tend to account for a similar share of disposable income in most countries, although taxes and subsidies mean that there are plenty of exceptions to this. Expenditure on oil products generally constitutes the largest component of household spending on energy and accounts for between 40% and 60% of the total.

In advanced economies, overall energy spending is set to decline over the projection period as a share of disposable income in the Stated Policies Scenario, mainly thanks to more energy-efficient cars and heating systems. Expenditure on electricity increases as a proportion of the total and expenditure on fossil fuels declines. In developing economies, total household energy bills are projected to increase in all regions, as households become increasingly affluent and demand more energy services. These increases more than offset energy efficiency improvements. Both electricity and fossil fuel related bills are set to rise, though electricity tends to increase as a share of overall household energy spending.

In the Sustainable Development Scenario, stronger energy efficiency measures reduce overall consumer energy spending even further in advanced economies and mitigate the effects of increased consumption in developing economies, thus reducing the impact of energy purchases on disposable income.

Electricity bills today account for between one-third and one-half of consumer energy payments in most regions. The electricity share of overall consumer energy expenditure is set to rise in both scenarios. Electricity spending increases on average to over 40% of total household energy spending in the Stated Policies Scenario and to two-thirds in the Sustainable Development Scenario, where there is more electrification of end-uses.

Figure 6.15 ▷ Household energy bill by fuel, 2018 and 2040 by scenario

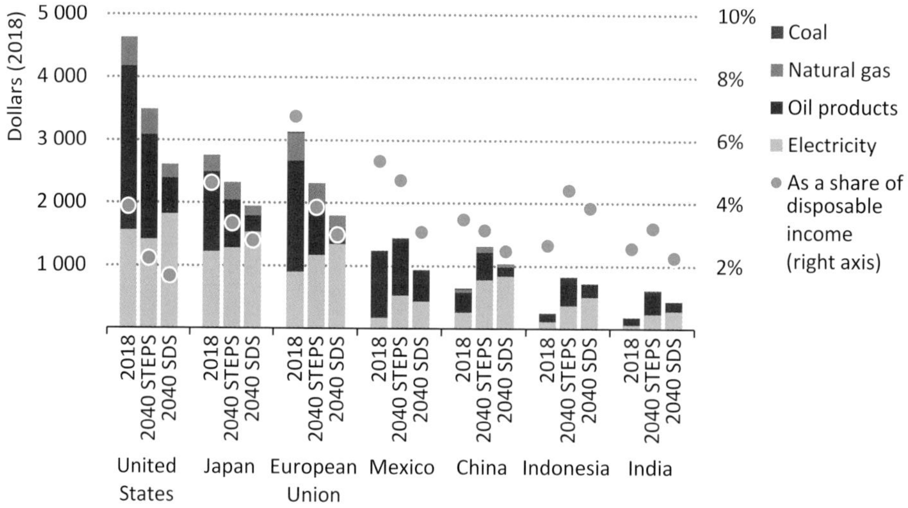

Electrification and energy efficiency reduce the impact of energy bills on household disposable income

Note: STEPS = Stated Policies Scenario; SDS = Sustainable Development Scenario.

This increase in spending on electricity as a result of increasing electrification is combined with a reduction in spending on other fuels, for example as consumers turn from conventional to electric cars. This shift has implications for costs. In the case of cars, for instance, it costs more at the outset to buy an electric car than it does to buy a conventional one, but there are significant subsequent savings in annual fuel bills. In the Stated Policies Scenario, payback periods are five to six years in China, European Union and United States in 2025: these reduce by half by 2040.

6.11 Tackling emissions from coal-fired power plants

Overview

Coal-fired power plants have been providing reliable, flexible and affordable electricity for well over a century. In recent decades they have helped to bring access to electricity to hundreds of millions of people across China, India and Southeast Asia, and have been a catalyst for economic development.

Coal-fired power generation, however, is the single largest source of greenhouse gas (GHG) emissions. It accounts for 30% of energy-related CO_2 emissions and around a quarter of energy-related GHG overall. Addressing these emissions is central to meet global climate objectives. The continued operation of the world's existing coal fleet would "lock in" a significant amount of CO_2 emissions and contribute to air pollution, potentially putting sustainable development targets out of reach.

The Paris Agreement and other commitments look to sharply reduce global CO_2 emissions. For example, the Paris Agreement aims to hold "the increase in the global average temperature to well below 2 degrees Celsius above pre-industrial levels and pursuing efforts to limit the temperature increase to 1.5 degrees Celsius above pre-industrial levels". Keeping the temperature rise below such limits implies a global budget for CO_2 emissions, beyond which climate goals become unattainable. If operations continued at current levels, emissions from coal-fired power plants in the period to 2050 would use up most of the remaining carbon budget consistent with meeting the stringent climate goals agreed at the United Nations.

Retrofitting coal-fired power plants with CCUS technologies, repurposing coal-fired plants to provide flexibility and, in some cases, gradually phasing out coal-fired plants where CCUS is not possible would create considerable room to manoeuvre to achieve climate goals. Such a strategy would also require rapidly scaling up other low-carbon sources, led by renewables (see Chapter 2).

New coal-fired power plants are still being built today. Over 170 GW of coal-fired power capacity was under construction at the start of 2019. Their completion would expand the global coal fleet by close to 10% and risk locking-in another 15 Gt of CO_2 emissions over the next two decades.

This section provides an analysis of the potential emissions from existing coal-fired power plants around the world. It then assesses region specific least-cost options for making the existing coal-fired power plant fleet compatible with global goals to tackle climate change, reduce air pollution and increase energy access.

Where are CO_2 emissions from coal-fired power generation locked in?

There are 2 080 GW of coal-fired power plants in operation worldwide, accounting for 38% of global electricity generation, more than any other source. Coal's share today is the same as it was throughout the 1970s. Almost 60% of the existing coal fleet is 20 years old or younger (Figure 6.16). In developing economies in Asia, existing coal plants are on average just 12 years old, meaning they are likely to operate for decades to come. Over the past 20 years, Asia accounted for 90% of all coal-fired capacity built worldwide, including 880 GW in China, by far the most of any country, followed by India (173 GW) and Southeast Asia (63 GW). Elsewhere there were smaller additions of coal-fired capacity in Europe (45 GW), Korea (28 GW), United States (25 GW), Japan (20.5 GW) and Africa (10 GW).

Figure 6.16 ▷ Global coal-fired power capacity by plant age, 2018

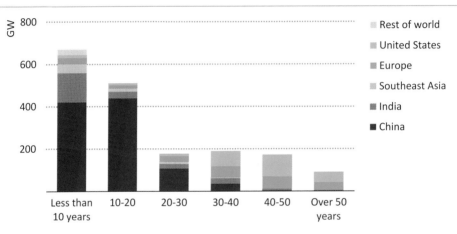

Asia is home to 90% of coal-fired power plants under 20 years of age

As matters stand, the existing global fleet of coal-fired power plants would continue to produce electricity and release CO_2 emissions for many years to come. The design lifetime of coal plants is typically 50 years. On the basis of continued operations and economics in line with stated policies, CO_2 emissions from the existing coal fleet would steadily decline to 2040, falling by about 40% over the next two decades (Figure 6.17).

Subcritical coal-fired power plants, the least efficient designs when producing only electricity, have been built around the world and account for over 40% of global CO_2 emissions from the coal fleet today. Subcritical plants produce significantly more emissions per unit of power generated than more modern coal plants. Close to half of subcritical plants in operation are under 20 years of age, and could emit more than 60 gigatonnes (Gt) of CO_2 over the next two decades. More efficient supercritical and advanced designs account for about 30% of coal plant CO_2 emissions today, with the remainder coming from combined heat and power plants. To 2040, CO_2 emissions from efficient coal-fired power plants are set to remain largely unchanged under stated policies.

Figure 6.17 ▷ Global CO_2 emissions from existing coal-fired power plants by technology with a 50-year lifetime in the Stated Policies Scenario

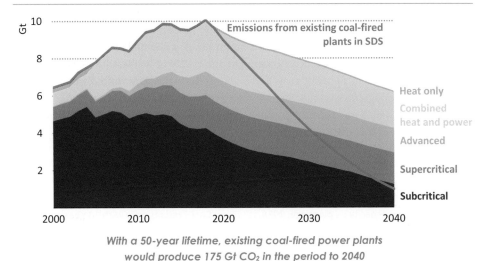

With a 50-year lifetime, existing coal-fired power plants would produce 175 Gt CO_2 in the period to 2040

Note: SDS = Sustainable Development Scenario.

Recent developments around the world

Government policies towards coal differ significantly from country to country. This section provides an overview of recent major developments in key regions.

Excess power capacity, cheaper renewables and concerns about local pollution have prompted China's central government to look closely at the country's fleet of coal-fired power plants in recent years. Efforts to fight pollution and reduce emissions have led to ultra-low emissions retrofits, efficiency improvements and flexibility retrofits. Over the past decade, more than 700 GW of coal-fired power units – nearly 70% of its total installed coal capacity – have been retrofitted to reduce their pollutant emissions to around the level of gas-fired units (for SO_2, NO_X and $PM_{2.5}$). In addition over the past decade, 650 GW of coal-fired capacity completed efficiency upgrades, and at least 54 GW was retired, all of it small scale and of subcritical design.

In the United States, the total capacity of the ageing coal fleet has declined by 20% over the past decade, due to challenging market conditions and stricter pollution standards. On average, the 75 GW of retired capacity had operated for 53 years, and three-quarters of it used subcritical designs. Low natural gas prices have driven down wholesale electricity prices and tightened margins for participants in competitive markets, including coal-fired capacity, while stricter pollution controls have raised costs. For example, the Mercury and Air Toxics Standards rule was established in 2011, setting limits on pollutants such as mercury, arsenic and heavy metals. As a result, nearly 90 GW of coal capacity added pollution control equipment by mid-2016, at a cost of less than $10 million for most facilities (US EIA, 2016). In 2018, the United States raised the 45Q tax credits that provide a financial boost for CCUS development in power and industry, tied to the amount of CO_2 captured and the type of storage employed.

Germany offers a real-world example of the challenges and opportunities of looking to address coal plant retirements. It announced plans in January 2019 to end its use of coal-fired power plants by 2038, with the possibility of bringing the deadline forward to 2035. Germany is Europe's largest coal consumer, and about 35% of its electricity was from coal plants in 2018. In recent decades, coal-fired plants in Germany were retired when reaching 35 or 40 years of age on average. The average age of the coal fleet in Germany today is about 30 years and a quarter of it is more than 40 years old (Figure 6.18). The installed coal-fired power capacity was 46 GW in 2018.

Figure 6.18 ▷ **Germany's existing coal-fired power capacity by age and phase-out plan**

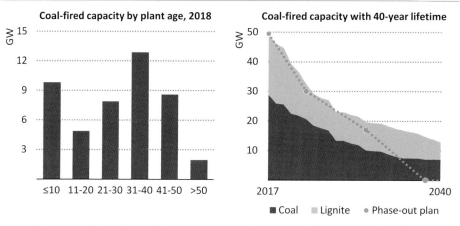

Germany's fleet of coal-fired power plants is a mix of old and young facilities, and the announced phase-out plan will accelerate retirements

Launched by the United Kingdom and Canada in 2017, the Powering Past Coal Alliance includes Germany along with 31 other countries, 25 states and cities, and 34 private sector members that have committed to phase out existing coal-fired power plants and to introduce a moratorium on new coal power stations not equipped with CCUS. Alliance members have over 110 GW of coal plants currently in operation, accounting for 5% of the total global coal fleet, of which about 60% is more than 30 years old. The move to phase out coal is driven by concerns about the associated emissions. In some cases there are emissions performance standards and air quality regulations that reflect those concerns. For example, half of the Alliance countries are European Union members, accounting for some 60% of the EU's coal-fired capacity. By 2021, all coal-fired power plants in the European Union need to meet tighter environmental standards and about 82% of plants currently do not meet the new standards (Climate Analytics, 2019). The cost of upgrading them to meet the new standards could be around EUR 11 billion, plus operating costs would rise if they are not shuttered. Plant owners may therefore be better off decommissioning the plants rather than retrofitting them.

What are the options?

A crucial question for tackling CO_2 emissions is what average lifetime for the world's coal-fired power plants would be compatible with climate goals. To illustrate the point, capping the lifetime of all existing coal-fired power plants at 25 years would bring their emissions in line with the Sustainable Development Scenario. Applying a strict 40-year limit would result in emissions exceeding climate goals by 50% (Figure 6.19).

Restricting the lifespan of operating coal-fired power plants is appealing in its simplicity, but it would not be practical for reasons of cost and energy security. Almost 700 GW of coal-fired capacity in operation today is more than 25 years of age. If these plants were to shut down immediately, it would leave many countries short of sufficient supply and significantly compromise electricity security. Applying a strict lifespan limit to existing coal-fired power plants would also entail immediate financial challenges to the owners, which are a mix of state-owned and privately held companies.

Figure 6.19 ▷ Cumulative CO₂ emissions from existing coal-fired power plants by assumed lifespan, 2019-2040

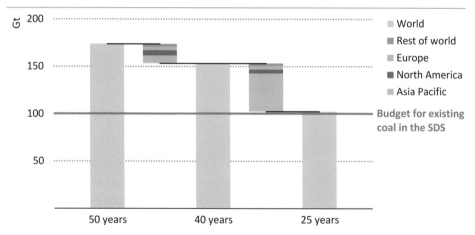

Limiting coal plants to 25-year lifetimes would bring emissions in line with the Sustainable Development Scenario, but would have consequences for costs and energy security

Note: SDS = Sustainable Development Scenario.

A multifaceted approach needs to be considered, analysing each plant and the market in which it operates as well as the economic implications of various options. There are several options that could move the dial on emissions from existing coal-fired power plants besides limiting their lifetimes. These include: *retrofitting* with CCUS or biomass co-firing equipment; *repurposing* facilities to focus on providing power system adequacy and flexibility while reducing operations (i.e. electricity produced); and *retiring* early if the above options are not viable. Each option would offer CO_2 emission reductions ranging from only a few percentage points with limited changes in operations to close to 100% for high levels of CCUS or CCUS in combination with biomass co-firing.

Further investment is likely to be needed to make existing coal-fired power plants more compatible with a sustainable energy pathway. Increasing flexibility or enabling biomass co-firing would require millions of dollars per gigawatt of capacity, and equipping with CCUS would cost $1 billion or more per gigawatt at current technology costs. Allowing coal plants to continue operating would help to maintain system adequacy and reliability, and would delay or soften the financial impact of closures (Table 6.6). Other considerations, such as the potential costs of remediating the site of a decommissioned facility or a plant's suitability for CCUS retrofits, also influence the attractiveness of each option.

Table 6.6 ▷ Measures to reduce CO_2 emissions from coal-fired power plants

	Retrofit		Repurpose operations, focus on flexibility	Retire
	CCUS	Biomass co-firing		
Impact at facility				
CO_2 emission reductions	Up to 99.7%	5-20%	1-80%	100%
Considerations				
Certainty of emissions reductions	High	Medium	Low	High
Additional investment scale per GW	Billions	Millions	Millions	None
Provides system adequacy & flexibility	◐	◐	◐	●
Delays financial impact	◐	◐	◐	●
Other	Site suitability	Biomass availability		Costs of site remediation, employment impacts

Meeting global climate goals in a cost-effective manner requires a combination of the available options in order to reduce CO_2 emissions from the existing fleet of coal-fired power plants by 70 Gt over the next two decades. We assessed the lowest cost option for each plant worldwide in a scenario compatible with meeting global climate goals and other sustainable development targets. Retrofitting capacity to capture carbon or co-fire with biomass plays an important role, particularly in China. In this analysis, about 110 GW of coal plants are retrofitted with CCUS by 2040 in China, offering an asset protection strategy and supporting energy security with capital investment of around $160 billion. In 2016, an assessment of the existing fleet in China concluded that over 300 GW of existing coal-fired power plants met the basic criteria to be suitable for retrofit (IEA, 2016). CCUS retrofits are most attractive for young and efficient coal-fired power plants that are located near places with opportunities to use or store CO_2, including for enhanced oil recovery and where alternative generation options are limited. Although significant cost reductions have been identified through early operational experience, widespread adoption of retrofitting would require substantially enhanced policy support, including preferential dispatch to ensure high utilisation rates, together with some combination of feed-in tariffs, capital grant funding and tax credits. In this analysis, some 10 GW of US coal-fired power plants are retrofitted with CCUS, as the technology benefits from the recently increased tax credit. Other countries such as Canada, Norway and United Kingdom are also looking to scale up efforts on CCUS.

Repurposing coal-fired power plants to provide system flexibility while reducing electricity generation plays the largest role in our assessment, accounting for about 60% of the emissions reduction. Several countries, including India, are looking at ways to make coal an integral part of system flexibility while transitioning to a low-carbon power sector, providing electricity access to all and maintaining system reliability. The option of retiring coal-fired capacity before 50 years of operations accounts for about one-quarter of the emissions reduction in the assessment (Figure 6.20).

Figure 6.20 ▷ Reducing CO_2 emissions from existing coal-fired power capacity by measure

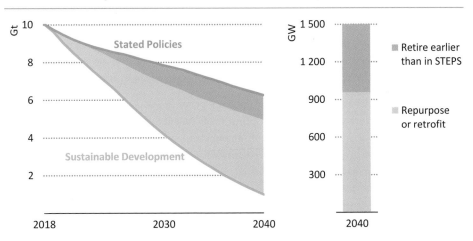

Curbing CO_2 emissions from coal-fired power plants can be done cost effectively by retrofitting, repurposing and retiring the existing fleet

Most of the 2 080 GW of existing coal-fired capacity worldwide would be affected by a shift in policies to meet global sustainability goals. About 600 GW reaches 50 years of age and retires (as in the Stated Policies Scenario). In addition, around 240 GW of existing coal-fired capacity would be retrofitted with CCUS or biomass co-firing equipment. Some 720 GW of would reduce operations to cut emissions, limiting electricity production but still providing system adequacy and flexibility. About one-quarter of the existing fleet would be retired in the Sustainable Development Scenario before reaching the typical 50-year lifespan.

What would be the financial impact?

Implementing the retrofit, repurpose, retire strategy to make the world's existing coal-fired power plant fleet compliant with climate goals would entail CCUS investment of $225 billion over the period to 2040 and lead to balance sheet write-downs for some owners of existing facilities. Coal plant retirements would also call for greater investment in lower-carbon sources of electricity and the associated network infrastructure. While immediate action is needed, most of the financial impact would not occur until the 2030s.

More than $1 trillion of capital invested in the existing fleet of coal plants has yet to be recovered, most of it in Asia. Asset owners would normally look to recover this capital through revenues that exceed short-term fuel and operating costs. Continued operations while lower carbon sources of electricity were scaled up would provide an opportunity to most facilities to recover some of the remaining capital at risk. Some facilities would need to shut down before fully recovering their upfront capital investment, whether because of policy changes or changes in market conditions.

Implications

Action must be taken to curb emissions from the existing fleet of coal-fired power plants around the world in order to keep the door open to limiting global average temperature increases to 2 °C or less, and to address air pollution. There are a number of options available, providing some flexibility for governments to adopt policies that reflect the opportunities and constraints of their respective energy systems. The options chosen have financial and social implications, as well as impacts on electricity security.

Coal-fired power plants provided 38% of global electricity supply in 2018. Filling the gap created by an accelerated reduction in output from existing coal capacity would involve a range of sources. Varying combinations of renewables, natural gas and nuclear power – all of which have lower CO_2 intensities than existing coal plants – could be used, depending on conditions and resources. Existing coal-fired power plants can have very low operating costs of $20 per megawatt-hour (MWh) or less, making it challenging to replace existing facilities with new sources of electricity without raising electricity supply costs. However, the falling costs of renewables are closing the gap with the operating costs of coal plants.

Tackling emissions from coal-fired power plants clearly has potential employment and social implications. Policies that support the development and deployment of CCUS technologies would spur investment in the technology and support related research, engineering and construction jobs. CCUS deployment would also boost employment and local economies in coal producing regions. Putting the power sector on a path to meet sustainable development goals could mean the potential closure of more than 500 GW of coal-fired power capacity, which would have significant negative consequences for local employment and local economies. Clear plans for retraining and other support measures would be vital to mitigate the impact in affected areas. Addressing emissions from coal power generation worldwide also raises the challenge for policy makers of how to finance the transition, which is global in scale and unevenly distributed geographically. Developing countries may also face a significant share of the burden.

6.12 Exploring the new frontiers of flexibility

Power system flexibility requirements will increase

The Stated Policies Scenario sees a significant increase in the need for flexibility – the ability of power systems to respond in a timely way to changes in electricity supply and demand. All regions will need more flexibility. Expressed as peak ramping requirements, flexibility needs[8] will increase much faster than electricity demand. They increase fastest in developing economies where almost 90% of the electricity demand growth in this scenario takes place, and particularly in India (Figure 6.21).

The speed of that increase depends mainly on how fast the share of variable renewable energy expands. The share of variable renewables in the power generation mix is set to more than triple in China and the United States in the Stated Policies Scenario, as well as at the global level. In India, it increases fivefold, and in Southeast Asia sevenfold.

Flexibility needs are also affected by the changing demand profile, how well the rising variable renewables supply matches the demand profile of a particular power system and its size. Increasing use of air conditioners is adding to loads during the summer, particularly during peak periods. Electric vehicles potentially may strongly affect peak demand, especially if smart charging is not fully developed (See Chapter 7).

Figure 6.21 ▷ Growth in electricity demand and flexibility needs by selected region and scenario, 2018-2040

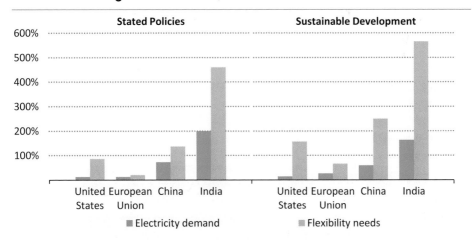

Flexibility needs increase much faster than electricity demand, driven by rising shares of variable renewables, more electric vehicles and higher demand for cooling

---

[8] Flexibility is a multifaceted concept that refers to the ability of power systems to balance demand and supply, and can be provided by different services (e.g. frequency regulation, operational reserves, load balancing). The change in the net load from one hour to the next (hourly ramping requirements) provides a useful indicator of flexibility and is used in this analysis. For more information on the drivers of increasing demand for flexibility and flexibility sources, see the *WEO-2018 Special Focus on Electricity* (IEA, 2018a).

In the Sustainable Development Scenario, as the power sector moves towards decarbonisation and as electric mobility spreads, flexibility needs increase even more strongly. In this scenario, flexibility requirements in India's power system are six times today's level, in China they more than triple and in the United States they are 150% higher.

A diverse portfolio of flexibility options will be required

Flexibility needs in the scenarios are based on analysis in which hourly demand profiles for projected years in different regions are assessed and fluctuations in net load are calculated in our World Energy Model. Based on the capacity mix of the specific region, the capability of the power system assets to change their output by the hour is simulated to identify which technologies can provide the flexibility required.

Conventional sources of flexibility in the form of power plants and interconnections have long maintained the reliability of power systems around the world. Today thermal power plants provide the bulk of the flexibility required by many electricity systems and this remains the case to 2040 in the Stated Policies Scenario (Figure 6.22). This is made possible by the retrofitting of existing thermal power plants, which helps increase ramp rates (IEA, 2018d), and by the construction of more flexible power plants such as gas turbines. Hydropower also remains an important source of flexibility in many regions. Interconnections between power systems and regions continue to alleviate network congestion by taking advantage of varying supply and demand patterns and pooling available flexibility resources.

Figure 6.22 ▷ Sources of flexibility by region in the Stated Policies Scenario

■ Hydro ■ Gas ■ Coal ■ Oil ■ Nuclear ■ Other ■ Interconnections ■ Batteries ■ Demand response

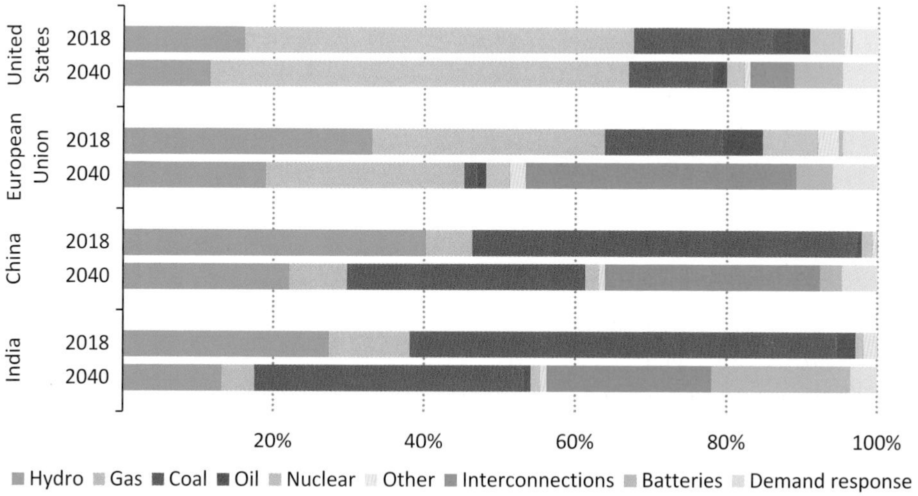

Thermal power plants continue to provide the bulk of flexibility needs, along with interconnections, but batteries and demand-side response are rising fast

Nonetheless, new flexibility sources will be needed. Batteries, demand response and sector coupling are poised to play pivotal roles in making sure future power systems are secure and reliable. Demand-side response also has a large part to play in meeting rising flexibility needs, for example by shaving peak demand and redistributing electricity to time periods when the load is smaller and electricity is cheaper. Distributed resources including variable renewables themselves, storage and demand response can also become key flexibility sources with appropriate market designs, as is happening in several countries (IEA, 2019f). Digitalisation is likely to have a major role to capitalise on the flexibility options.

Regional trends to 2040 show there is no one-size-fits-all approach to flexibility. The European Union is expected to source a significant portion of its flexibility needs from the large-scale deployment of interconnections. China is set to rely on more flexible coal-fired power plants and large-scale interconnections. In the United States gas-fired power plants are set to remain a cheap source of power system flexibility through to 2040. Most of India's additional flexibility needs are to be met by flexible coal-fired power plants, batteries and interconnections.

Changes in policy and regulatory frameworks, as well as economic incentives, are essential to ensure timely investment in flexibility assets and to make the most of the flexibility potential of existing power plants. Competitive electricity markets were originally designed with dispatchable power plants in mind. The rise of variable renewables is now challenging the suitability of those market designs to deliver efficient and timely investment. For example, there is a widening gap between electricity supply costs and revenues from energy sales, particularly in the European Union and the United States (IEA, 2018a). These markets may require reforms to spur investment and to establish a cost-effective set of flexibility measures.

The transformation of the power generation fleet is even more pronounced in the Sustainable Development Scenario, with variable renewables making up 40% of electricity generation by 2040. The increased reliance on variable renewables often translates into higher hourly ramps, which requires more flexibility including from batteries and demand-response measures.

Focus on battery storage

The use of battery storage in the power sector is accelerating worldwide. In 2018, the global power sector added 3 GW of batteries boosting the total installed capacity to 8 GW. Major cost reductions – down 45% from 2012 to 2018 – and increased policy support in many regions underpin this development. For example, several US states, including New York and California, have introduced specific targets for batteries. Australia has subsidy schemes for residential storage and funding for utility-scale batteries.

Battery storage is the fastest growing source of power system flexibility today and over the next 20 years. The modularity of batteries and their simple design allow them to be installed almost anywhere. They can provide fast response flexibility and help to balance the network through the provision of various remunerated services. They can also increase the value of variable renewables by enabling the electricity produced from solar and wind to be stored and injected into the grid at another time when system needs are higher.

The outlook for batteries is improved further by the prospect of economies of scale, learning from experience and advances in chemistry. Battery storage system costs are projected to halve in the next two decades, e.g. four-hour storage systems falling from $400 per kilowatt-hour (kWh) to less than $200/kWh. By 2040, batteries provide 330 GW of flexibility in the Stated Policies Scenario and 550 GW in the Sustainable Development Scenario (Figure 6.23). India is projected to lead the way in battery storage deployment, reaching 120 GW of installed capacity by 2040 in the Stated Policies Scenario, with most of this capacity being coupled with variable renewables such as solar PV and wind. China and the United States both install around 50 GW of utility-scale batteries in this scenario, while the European Union installs 35 GW.

Batteries are projected to become more and more competitive with other flexibility options as a result of cost reductions and the value of the energy arbitrage opportunities. Batteries can charge for a number of hours when electricity prices are low and then discharge during periods of higher prices.

The widespread deployment of battery storage requires electricity market reforms to incentivise sufficient investment in flexibility, together with other supporting conditions. Electricity market reforms ranging from scarcity pricing, operating reserve prices (as in US markets), frequency control ancillary services (as in Australia) and capacity mechanisms already appear in various places. Many of these market measures would help the business case for investment in battery storage, rewarding their dispatchability, fast response and contributions to system adequacy and reliability.

Figure 6.23 ▷ **Battery storage capital costs and installed capacity by scenario**

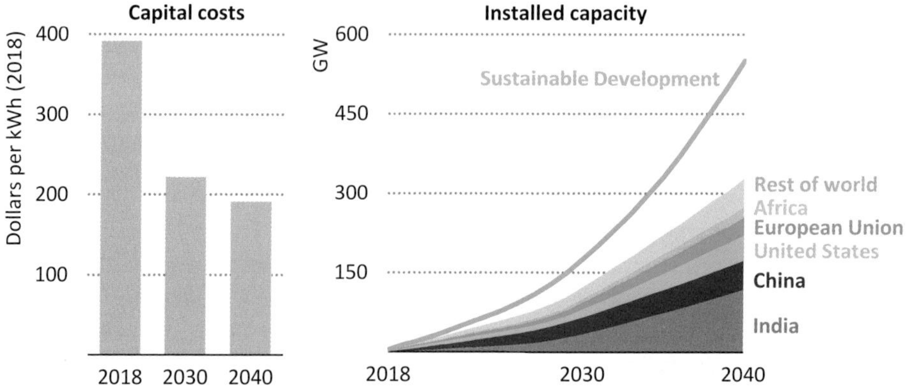

In the next two decades, batteries are projected to be the fastest growing technology in the power sector, thanks to reduced costs and policies supporting deployment

Note: the figure refers to four-hour battery storage.

Battery storage in India

With ambitious plans for renewables, in particular solar PV, to satisfy rapidly increasing electricity demand, particularly at peak times, India has the most significant need for additional power sector flexibility among the countries in the Stated Policies Scenario. How India tackles its rapidly rising flexibility needs is critical for its energy transition, and also has global implications. Current flexibility needs are met by coal, hydro and gas-fired power plants, and by interconnections. But new and cheaper sources of flexibility are becoming available. India's significant uptake of variable renewables makes it a potential large-scale launch pad for battery storage technology. This section explores how that could happen.

Demand soars: In the Stated Policies Scenario, electricity demand is set to rise faster in India than in any other region to 2040 as a result of population growth and economic development. Increasing wealth drives demand for cooling, which expands almost eightfold and accounts for one-fifth of the country's total electricity demand growth.

Renewables expand: Massive deployment of solar PV is projected in India thanks to a strong policy push. Its solar PV capacity is projected to reach 620 GW by 2040. Flexibility needs increase dramatically as demand peaks get higher, and the share of solar PV and wind in total electricity generation increases from 6% in 2018 to 34% in 2040.

Coal continues to grow: Total coal-fired power capacity is set to increase with 40 GW under construction in India at the start of 2019 and more in the pipeline. Uncertainty remains over whether new coal-fired power plants will continue to be built after 2030.

Battery opportunity: Battery storage is one of the technologies best suited to provide the flexibility that power systems in India need. The combination of solar PV and battery storage is emerging as a potentially potent combination that could drive significant new growth in storage capacity. Moreover, electric vehicle use is poised to increase significantly encouraged by policies to combat air pollution, thereby expanding demand for batteries exponentially. Boosting deployment could cultivate domestic battery manufacturing, potentially driving down their costs.

Rewarding flexibility: India needs flexible sources of electricity to ensure grid stability in the face of very large increases in variable renewable generation. Battery storage capacity could thrive in an ancillary services market that remunerates flexible options depending on their response speed (Central Electricity Authority, 2017). There is already an increasing need for energy storage to facilitate the integration of solar PV and wind generation as these technologies begin to affect the economics and technical operation power systems.

Battery storage coupled with variable renewables can both provide flexibility for power system operation as well as improve the competitiveness of variable renewables, especially solar PV. The rapidly falling costs of both battery storage and solar PV help to make combined facilities competitive with coal-fired power plants in India by 2030. As such, coupling battery storage with solar PV would offer an affordable option for displacing the need for some coal-fired capacity.

Cheap Battery case: India

Figure 6.24 ▷ Value-adjusted LCOE for select power technologies in India in the Stated Policies Scenario

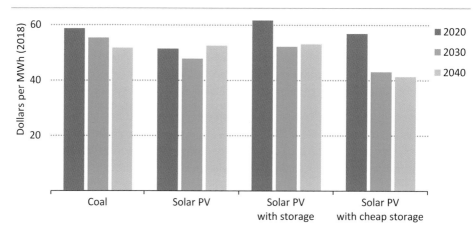

Compared with new coal-fired power plants, solar PV alone is already more competitive and solar PV with storage is rapidly closing the gap, especially if batteries are cheap

Notes: LCOE = levelised cost of electricity. Assumed CO_2 prices in India are zero.

The cost of battery storage in the future is uncertain. Here we consider a case in which further cost reductions are achieved through improved chemistry; economies of scale in manufacturing and learning from experience. Battery system costs, which include the battery pack and the remaining balance-of-system costs, for a four-hour storage system go below $200/kWh by 2040 in the Stated Policies Scenario. In the Cheap Battery case, they are further reduced by 40% reaching $120 /kWh, making solar PV paired with batteries far more competitive with coal-fired power plants (Figure 6.24).

The Cheap Battery case projects 200 GW of installed battery storage capacity in India by 2040, around 80 GW higher than in the Stated Policies Scenario. This would have a direct impact on the need for new coal-fired power plants: once the existing coal project pipeline is completed, additions after 2030 would be cut by almost three-quarters, avoiding the construction of almost 80 GW of new coal-fired capacity. Adding battery capacity could bring additional benefits:

- **Tackle local pollution:** With a larger amount of power system flexibility from battery storage, higher shares of variable renewables can be accommodated, helping to reduce local air pollution. In the Cheap Battery case, the share of renewables in India's power sector would exceed 50% by 2040 without additional curtailment. Solar PV capacity would increase by almost 30%, reaching 800 GW, and an additional 25 GW of wind capacity would be added, bringing the total close to 250 GW.

- **Reduce CO_2 emissions**: In the Cheap Battery case, CO_2 emissions from electricity generation in India would peak just after 2030 (Figure 6.25). This would help the transition from the Stated Policies Scenario to a more sustainable path in which more renewables are deployed, although further changes would be needed to follow the pathway in the Sustainable Development Scenario.

Figure 6.25 ▷ Installed capacity by source and CO_2 emissions from electricity generation in India in the Stated Policies Scenario and Cheap Battery Case

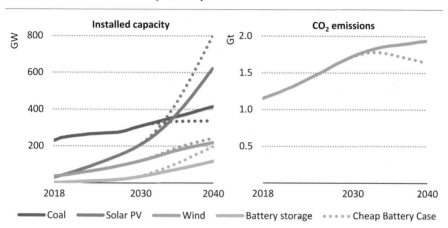

Coal — Solar PV — Wind — Battery storage ····· Cheap Battery Case

Increased battery use in the Cheap Battery Case enables more solar PV and wind installations and cuts coal use so that CO_2 emissions peak just after 2030

- **Provide alternatives to grid upgrades:** Total power systems costs would be unchanged from the Stated Policies Scenario as battery storage could reduce network requirements and displace some investment in transmission and distribution grids.
- **Improved operation of conventional plants:** Enhanced use of battery storage in India could improve the economic performance of the thermal power plant fleet because they would be able to operate at higher annual capacity factors. This could also provide cost savings for distribution companies by reducing overcapacity needs.

Chapter 7

Outlook for energy efficiency and renewables
Untapped potential for sustainable growth?

SUMMARY

- Total global final energy consumption was almost 10 000 million tonnes of oil equivalent (Mtoe) in 2018, an increase of 2.2% compared with 2017. In the Stated Policies Scenario, it rises to almost 12 700 Mtoe by 2040, an increase of around 1.1% per year on average, while global energy intensity improves by 2.3% per year. Implementation of the Stated Policies Scenario results in energy intensity improvements and an expansion of renewable energy, but the rate of improvement is not sufficient to achieve the energy-related Sustainable Development Goals.

Figure 7.1 ▷ Energy intensity improvement and renewables share of total final consumption by scenario

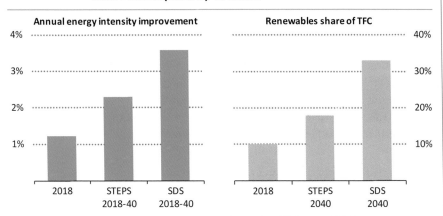

A significant step up in effort is required if the world is to achieve the targets set out in the Sustainable Development Goals

Note: TFC = total final consumption; STEPS = Stated Policies Scenario; SDS = Sustainable Development Scenario.

- Annual investment in energy efficiency increases in the Stated Policies Scenario from $240 billion in 2018 to around $445 billion a year by 2030 and to $635 billion a year thereafter to reach a cumulative total of $11.7 trillion over the period to 2040. In the Sustainable Development Scenario, there is a much larger increase in energy efficiency investment, with spending rising to a cumulative total of $16.7 trillion by 2040 (around $625 billion a year to 2030 and almost $920 billion thereafter).
- The story for renewables is similar. While investment in the Stated Policies Scenario is expected to increase from around $390 billion in 2018 to an average of almost $440 billion a year until 2030, it falls short of the annual $650 billion that would be required on average to 2030 to meet the Sustainable Development Goals.

- Energy transitions worldwide imply changes in how we supply and consume energy. Demand for materials and industrial products plays a central role in shaping energy consumption and carbon dioxide (CO_2) emissions in industry. In the face of demand growth, the policies so far implemented or announced will not halt a future increase in industry emissions: which grow by 16% in 2040 the Stated Policies Scenario.

- The energy transformation in the Sustainable Development Scenario implies changes in demand for materials, some of them counter-intuitive, including more steel for renewable energy infrastructure and more aluminium for electric vehicles. Greater emphasis on materials efficiency and materials recycling, reuse and substitution succeeds in reversing the historic trend of growing emissions for steel and cement, leading to a stronger decline in industrial CO_2 emissions.

- As the electrification of economies progresses and the share of variable renewables in generation increases, the carbon footprint of electricity use increasingly fluctuates depending on the time of day or night. In India, our analysis indicates that, when the share of variable renewables reaches 50%, average CO_2 emissions from using electricity at midday or 23:00 differ by a factor of seven. In Europe, the difference is a factor of three.

- Targeting efficiency measures on appliances and equipment used when power sector emissions are high could bring major reductions in emissions of CO_2 and other pollutants. In the case of India and the European Union, enhanced energy efficiency sees emissions from electricity generation decrease by around 30% and 20% respectively in the Sustainable Development Scenario relative to the Stated Policies Scenario.

- Facilitated by digitalisation and new business models, demand-side response has a role to play in shifting consumer demand to low CO_2 and energy price periods and selling flexibility services to the grid. In doing so, it can both increase the share of variable renewables used and also cut CO_2 emissions from power generation. IEA analysis shows potential for further power sector emissions reductions of 25% in the United States and 16% in China, relative to the Sustainable Development Scenario.

- While the technologies to produce biogas are not new, there has been a resurgence of interest in their potential in recent years. IEA analysis indicates that over 570 Mtoe per year of biogas could be produced sustainably today, equivalent to almost 20% of global natural gas demand. Emerging economies account for two-thirds of the global biogas potential. However, biogas everywhere requires supportive policies if it is to be fully utilised.

- The benefits of using biogas are many and the economic case improves considerably if these non-economic benefits are fully taken into account. In the Stated Policies Scenario, close to 150 Mtoe of biogas is produced globally by 2040, over 40% of which is in China and India. In the Sustainable Development Scenario, there is a more pronounced increase in biogas production: it reaches around 330 Mtoe in 2040, utilising around 40% of the total sustainable technical potential.

Introduction

This chapter examines current trends in energy efficiency and renewable energy. The past decade has been characterised by continuous growth in the deployment of renewable energy technologies. The power sector has been leading the way but the uptake of renewables has been slower in other sectors such as industry and buildings[1]. Energy efficiency has also seen significant investment in recent years, although growth stalled in 2018 owing to lacklustre progress in implementing new efficiency policies and increasing the stringency of existing ones. Efficiency gains and increased use of renewables remain among the most cost-effective ways to enhance security of energy supply, reduce carbon dioxide (CO_2) emissions and improve local air quality. The first part of this chapter presents the key findings on energy efficiency and renewable energy from the Stated Policies and Sustainable Development scenarios. It explores three important topics in detail:

- The production of steel, cement and aluminium accounts for 10% of global energy consumption, and is generally carbon intensive. This means that curbing material demand has an important role in achieving the goals of the Sustainable Development Scenario, in which improved material efficiency contributes over 20% of the savings in CO_2 emissions from steel, cement and aluminium production by 2040. Material efficiency improvements can be realised along the entire product lifecycle, covering design, fabrication, use and recycling.

- Smarter electricity use can significantly reduce the environmental footprint of electricity consumption. As the share of electricity generation from variable renewables increases, so does the hourly variability of power sector emissions: in other words, the level of emissions from electricity consumption will vary depending on whether power is used at a time of day when most electricity comes from renewable sources or not. Targeted energy efficiency measures and demand-side response measures can reduce electricity demand in periods of high CO_2 intensity of electricity supply, and thus reduce emissions. Additional benefits include lower bills, improved power system operation and a reduction in air pollutant emissions.

- Biogas is emerging as having significant potential to reduce greenhouse gas (GHG) emissions, improve waste management, enhance security of supply and improve access to clean cooking in rural areas. We provide a complete overview of the key technology and production pathways for biogas, exploring the potential volumes and production costs, and discussing various biogas applications. While there is significant global potential for biogas, most of this requires supportive policies if it is to be exploited.

Figures and tables from this chapter may be downloaded from www.iea.org/weo2019/secure/.

---

[1] Industry is defined in this chapter as including transformation in blast furnaces and coke ovens and petrochemical feedstock, unless where reference is made to final consumption of industry.

Scenarios

7.1 Energy efficiency overview

Table 7.1 ▷ **Key indicators by scenario**

	2018	Stated Policies		Sustainable Development		Current Policies	
		2030	2040	2030	2040	2030	2040
TPED (Mtoe)	14 314	16 311	17 723	13 750	13 279	16 960	19 177
Share of fossil fuels (%)	81%	77%	74%	72%	58%	79%	78%
TFC (Mtoe)	9 954	11 607	12 672	9 904	9 500	11 996	13 540
Energy intensity of GDP (2018=100)	100	75	60	63	45	78	65

Notes: TPED = total primary energy demand; Mtoe = million tonnes of oil equivalent; TFC = total final consumption of energy; TPED indicators exclude fossil fuel statistical differences.

Global primary energy demand grew by almost 2.3% in 2018, its largest annual increase since 2010, with China, United States and India accounting for over two-thirds of total energy demand growth. Global energy intensity reached 106 tonnes of oil equivalent (toe) per $1 million of gross domestic product (GDP), a 1.2% improvement compared to 2017, but around half the rate of improvement in the 2010-17 period.[2] This third consecutive year of slowdown was the result of weaker energy efficiency policy implementation and strong demand growth in energy-intensive economies.

As always, worldwide trends conceal considerable regional variations. In India, energy intensity improved by around 3%. Energy intensity in China improved by 2.8% while energy demand reached an all-time high. In the European Union, energy intensity improved by 2.3% as efforts to meet the 2020 energy efficiency targets continued. The United States, in contrast, saw energy intensity deteriorate, by close to 1%, driven by heating demand in winter and cooling demand in summer as well as growing demand in the transport sector.

In the **Stated Policies Scenario**, primary energy demand expands by around a quarter between 2018 and 2040 at an average annual growth rate of 1%. This rate is slower than in the past (it was 1.4% between 2010 and 2018, and 2.7% in the preceding decade) as energy consumption and economic growth continue to decouple. In the Current Policies Scenario, energy demand rises at an average annual rate of 1.3% to reach over 19 000 million tonnes of oil equivalent (Mtoe) by 2040.

In the **Sustainable Development Scenario**, energy demand in 2040 is 25% lower than in the Stated Policies Scenario. Further improvements in energy efficiency in final energy consumption in the Sustainable Development Scenario save over 1 900 Mtoe of energy

---

[2] In the *WEO-2019*, energy intensity is calculated using GDP in purchasing power parity (PPP) terms to enable differences in price levels among countries to be taken into account. In our scenarios, PPP factors are adjusted as developing countries become richer.

demand in end-use sectors by 2040 relative to the Stated Policies Scenario. Energy efficiency is the single biggest contributor to reducing final energy consumption in end-use sectors in the Sustainable Development Scenario relative to the Stated Policies Scenario, responsible for 60% of the savings. Almost half of the energy efficiency savings come from industry, with major contributions also from transport and buildings.

Table 7.2 ▷ **Energy intensity of GDP by scenario** (toe/$1 000 PPP)

		Stated Policies		Sustainable Development		Current Policies	
	2018	2030	2040	2030	2040	2030	2040
North America	0.11	0.09	0.07	0.08	0.05	0.09	0.07
United States	0.11	0.09	0.07	0.08	0.05	0.09	0.07
Central and South America	0.09	0.07	0.06	0.06	0.05	0.08	0.07
Brazil	0.08	0.08	0.06	0.07	0.05	0.08	0.07
Europe	0.08	0.06	0.05	0.05	0.04	0.06	0.05
European Union	0.07	0.05	0.04	0.05	0.04	0.06	0.05
Africa	0.12	0.10	0.08	0.06	0.05	0.10	0.08
South Africa	0.17	0.13	0.10	0.11	0.08	0.14	0.12
Middle East	0.12	0.11	0.10	0.09	0.07	0.11	0.10
Eurasia	0.17	0.14	0.12	0.12	0.09	0.14	0.12
Russia	0.18	0.15	0.12	0.13	0.10	0.15	0.13
Asia Pacific	0.10	0.07	0.05	0.06	0.04	0.07	0.06
China	0.12	0.08	0.06	0.07	0.04	0.08	0.07
India	0.09	0.06	0.05	0.05	0.03	0.06	0.05
Japan	0.08	0.06	0.05	0.06	0.05	0.07	0.06
Southeast Asia	0.08	0.06	0.05	0.05	0.04	0.06	0.05
World	0.11	0.08	0.06	0.07	0.05	0.08	0.07

Note: toe = tonne of oil equivalent; PPP = purchasing power parity.

Energy intensity improves in the Stated Policies Scenario by 2.3% annually to 2040. This is a slightly higher rate than in the period since 2010 and about twice the rate in 2000-10. The link between GDP growth and energy demand growth continues to weaken by 2040. The largest improvements are projected in China, India and European Union, with the implementation of more stringent policies such as the 13th Five-Year Plan for National Energy Conservation Action in China, the Perform, Achieve, Trade scheme in India, and emission performance standards and directives on the energy performance of buildings and Ecodesign in the European Union. Energy intensity in the Sustainable Development Scenario, in contrast, improves much faster at a rate of 3.6% a year on average as all economically viable energy efficiency opportunities are pursued. The biggest improvements are in China, India and Africa.

7.2 Renewables overview

Table 7.3 ▷ Share of renewables in energy demand by scenario

	2018	Stated Policies 2030	Stated Policies 2040	Sustainable Development 2030	Sustainable Development 2040	Current Policies 2030	Current Policies 2040
Primary demand (Mtoe)	1 391	2 287	3 127	2 776	4 381	2 138	2 741
Solid biomass (Mtoe)	620	613	546	140	75	613	546
Share of total bioenergy	*46%*	*37%*	*30%*	*11%*	*5%*	*38%*	*31%*
Electricity generation (TWh)	6 799	12 479	18 049	15 434	26 065	11 627	15 485
Bioenergy	636	1 085	1 459	1 335	2 196	1 022	1 256
Hydro	4 203	5 255	6 098	5 685	6 934	5 171	5 923
Wind	1 265	3 317	5 226	4 453	8 295	2 955	4 258
Of which offshore wind	67	567	1 281	764	2 072	416	860
Solar PV	592	2 562	4 705	3 513	7 208	2 265	3 658
Concentrating solar power	12	67	196	153	805	46	104
Geothermal	90	182	316	282	552	161	258
Share of total generation	*26%*	*37%*	*44%*	*49%*	*67%*	*33%*	*36%*
Final consumption* (Mtoe)	992	1 748	2 259	1 961	3 137	1 531	1 969
Modern bioenergy	430	592	718	729	873	555	664
Of which biogas/biomethane	12	47	72	106	177	29	43
Electricity	203	409	600	494	1 023	354	499
*Share of global TFC***	*10%*	*15%*	*18%*	*20%*	*33%*	*13%*	*15%*
Heat consumption* (Mtoe)	506	731	936	830	1 176	704	869
Modern bioenergy	342	423	492	428	495	419	493
Of which biogas/biomethane	12	27	45	81	132	18	30
Electricity	91	167	240	215	368	156	206
District heating	22	37	46	30	44	42	46
*Share of total heat demand**	*10%*	*13%*	*16%*	*19%*	*30%*	*13%*	*14%*
Transport* (Mtoe)	96	199	302	354	606	153	203
Biofuels (road)	88	149	186	248	243	130	162
Biofuels (aviation and shipping)	0	17	36	50	129	3	6
Electricity	8	33	79	56	234	21	35
*Share of total transport demand**	*3%*	*6%*	*8%*	*12%*	*23%*	*4%*	*5%*

* Includes indirect renewables contribution, but excludes environmental heat contribution. Notes: Mtoe = million tonnes of oil equivalent; TWh = terawatt-hours; PV = photovoltaics; TFC = total final consumption. Solid biomass includes its traditional use in three-stone fires and in improved cookstoves.

Electricity generation from renewables continued to grow in 2018 with output up by 450 terawatt-hours (TWh) (or 7%) compared to the previous year and accounting for more than a quarter of total power generation output (this compares to around a fifth in 2010). Growth in output from solar photovoltaics (PV), wind and hydro accounted for 90% of the increase. Around 180 gigawatts (GW) of new renewable power capacity was added in 2018, which is the same level as the previous year, although the IEA's provisional estimate for 2019 suggests a resumption of robust growth in annual renewable additions.

Hydropower remains the largest source of renewables-based power generation today, although its rate of deployment slowed again in 2018: 20 GW of new capacity was added compared with 25 GW in 2017 and 36 GW in 2016. Wind power, the second-largest renewable energy power source, added 50 GW: this represents a slight increase compared with 2017. Offshore wind continues to pick up with 4.3 GW of new capacity, led by China with 1.6 GW, United Kingdom with 1.3 GW and Germany with 1.0 GW (See Chapter 14). Solar PV capacity additions expanded by almost 100 GW in 2018: China accounted for 44 GW or almost half of the increase, though this is a lower figure than in 2017. Electricity output from wind and solar PV combined was almost 20% higher in 2018 than in 2017.

The use of renewables in the transport sector also continued to expand in 2018. Biofuels are the only renewable energy source used directly in the transport sector, and they contributed around 90 Mtoe or almost 2 million barrels of oil equivalent (mboe) per day. About 70% of this was accounted for by ethanol, with biodiesel the next biggest contributor.

The use of renewables also increased in the heating sector, with renewable sources of heat excluding traditional use of solid biomass (over 500 Mtoe) accounting for 10% of heat supply. Direct use of biomass (67%) was the main source of renewable-based heat followed by electricity from renewable energy sources (18%) and solar thermal (7%).

In the **Stated Policies Scenario**, the amount of renewables excluding traditional biomass in final energy consumption increases from more than 990 Mtoe today to almost 2 260 Mtoe in 2040 and its share of consumption nearly doubles to 20%. The share of renewables in global heat increases by 60% and reaches nearly 940 Mtoe in 2040, thanks to substantial increases in the modern use of biomass (pellets in boilers and stoves, biogas and biomethane, biofuels), renewable electricity and also of solar thermal. China, the European Union, India and the United States supply more than half of the of the world's renewable heat today and are expected to account for around 65% of the increase by 2040. Meanwhile the amount of renewables consumed in the transport sector triples to around 300 Mtoe, three-quarters of which comes from biofuels, with the remainder taking the form of renewable electricity consumed by electric vehicles and rail transport.

In the **Sustainable Development Scenario**, additional measures to incentivise investment in renewables-based electricity, biofuels, solar heat, geothermal heat and electrification push the share of renewables to two-thirds of electricity generation output and 37% of final energy consumption. By 2040, expected output from wind (8 300 TWh) and solar PV (7 200 TWh) are expected to exceed hydropower (6 950 TWh), while the share of heat coming from renewables in 2040 increases to 30% of the total or to 1 200 Mtoe. In the transport sector, consumption of energy from renewable sources is projected to increase to 600 Mtoe, with biofuels accounting for around 60% and electricity from renewable sources consumed by electric vehicles and rail accounting for the remainder.

7.3 Efficiency by sector and investments

The coverage and strength of mandatory energy efficiency policies continued to increase with policy coverage reaching 35% in 2018, a 1.3 percentage point change improvement since 2017 (Figure 7.2). Despite the cost effectiveness of efficiency investments, in many areas there has been limited progress in expanding policy coverage despite a small increase in industry and in residential buildings. Once again, almost all of the increase in coverage is attributable to more end-uses being covered by existing standards, rather than new standards (IEA, 2019a).

Figure 7.2 ▷ Year-on-year changes in the share of global energy consumption covered by mandatory efficiency standards by selected end-uses

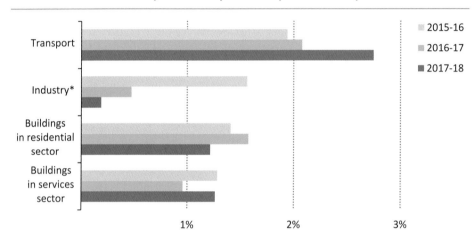

Efficiency standards expanded for transport in 2018 while there was very little coverage increase in efficiency policies in the industry and buildings sectors

* In this case industry includes only final energy consumption of industry and not feedstock demand for chemicals or fuel transformation in blast furnaces and coke ovens.

Trends for passenger cars were not encouraging with a rise in market share by less-efficient sports utility vehicles which on average consume around 25% more oil than medium-size cars, leading to a stagnation of fuel economy for road transport (see Chapter 3). For the first time since the 2008 recession, key markets saw a slowdown or drop in sales of internal combustion engine (ICE) cars in 2018. New vehicle standards in India (2022) and the European Union (2022) are expected to support more efficiency-related spending in the future.

In the **Stated Policies Scenario**, energy demand in buildings increases by about 1% per year on average between today and 2040: this reflects a large increase in the number of electric appliances and connected devices and in demand for space cooling. In industry, efficiency improvements slow the average annual increase in demand between today and 2040 to

1.3%, an increase from the 0.9% recorded over the 2010-18 period. In the transport sector, efficiency improvements help to constrain growth in demand to around 25% despite the number of light-duty passenger vehicles increases by 70%.

In the **Sustainable Development Scenario**, stronger efficiency policies and greater access to clean cooking technologies such as more efficient stoves and LPG see energy consumption in the buildings sector fall by around 400 Mtoe or 13% over the period to 2040. Energy consumption in industry remains at a similar level as today, while oil demand in the transport sector decreases by as much as 40%, despite an increase in the global vehicle stock of 45%, reflecting more stringent efficiency standards and growth in electric vehicle sales, which reach 54% of the total vehicle stock by 2040.

Table 7.4 ▷ Annual average investment in energy efficiency in selected regions by scenario ($2018 billion)

		Stated Policies		Sustainable Development		Change 2031-40 versus 2018	
	2018	2019-30	2031-40	2019-30	2031-40	STEPS	SDS
United States	42	74	94	127	186	52	144
European Union	70	136	167	149	141	97	71
China	61	68	98	87	111	37	50
India	10	22	40	30	58	30	48
World	240	445	635	625	916	395	676
Cumulative		5 338	6 345	7 498	9 156		

In 2018, investment in energy efficiency across the buildings, transport and industry sectors totalled around $240 billion, the same level as the previous year. The buildings sector accounted for most energy efficiency spending despite declining by 2% to under $140 billion (IEA, 2019b).

In the **Stated Policies Scenario**, investments in energy efficiency are expected to rise from current levels, averaging around $445 billion a year in the period from 2019 to 2030 and to around $635 billion a year from 2031 to 2040 (Table 7.4). The transport sector accounts for more than half of cumulative investment spending to 2040, followed by buildings and industry. The European Union, China and United States together account for more than half of all energy efficiency investments in 2040.

In the **Sustainable Development Scenario**, expected cumulative spending on efficiency reaches $16.7 trillion between now and 2040, with an annual average of around $625 billion between 2019 and 2030 rising to almost $920 billion a year from 2031 to 2040.

7.4 Renewables policies and investments

Renewables have grown rapidly in recent years, accompanied by sharp cost reductions for solar PV and wind power in particular. By 2018, most countries in the world had some form of renewable energy targets in place, and more than 150 countries had renewable energy policies in place in the power sector. More than 45 countries also had policies in place to support the use of biofuels in the transport sector: no new countries added regulatory incentives or mandates for renewable transport in 2018, but some countries strengthened existing ones. In addition, around 45 countries had renewable energy policies specifically for heating and cooling (REN21, 2019).

Figure 7.3 ▷ Renewable energy in total primary energy demand by category and region in the Stated Policies Scenario, 2018 and 2040

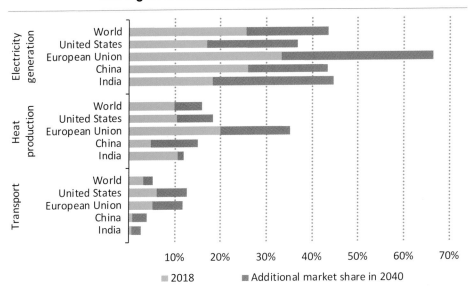

Strong support policies for renewable-based electricity are boosting their penetration, but more policy action is needed in other sectors

Several new or revised renewable energy targets were established in 2018. China's National Energy Administration raised the country's renewable portfolio standard (RPS) in 2019 to 35% of electricity consumption by 2030. The amended RPS established local targets for the provinces in the form of minimum mandated levels of renewable electricity supply. In December 2018, the European Union's revised Renewables Energy Directive (2018/2001) established a new binding 2030 renewable energy target of at least 32% of gross final consumption of energy from renewable sources, with scope for possible upwards revision from 2023.

In the United States, the increase in renewable capacity in 2018 came mainly as a result of onshore wind expansion, although uncertainty about PV module import tariffs resulted in delays to some solar projects (IEA, 2019c). Elsewhere, renewable capacity expansion accelerated in many countries in the Middle East, North Africa and parts of Asia, reflecting declining wind and solar PV costs, strong policy support and commitments made under the Paris agreement (IEA, 2019c).

Table 7.5 ▷ **Global annual average investment in renewables by scenario ($2018 billion)**

	2018	Stated Policies		Sustainable Development		Change 2031-40 versus 2018	
		2019-30	2031-40	2019-30	2031-40	STEPS	SDS
Power	304	347	382	524	661	78	357
Wind	89	119	138	185	235	49	146
Solar PV	135	116	117	169	189	-18	54
End-use sectors	84	89	95	125	146	11	62
World	388	436	477	649	807	89	419
Cumulative		5 233	4 772	7 793	8 075		

Note: Renewables for end-use include solar thermal, bioenergy and geothermal applications for heating.

Renewables-based power investment declined modestly in 2018 to around $390 billion. A dollar of renewables spending continued to buy more generation capacity than in the past, putting downward pressure on spending levels (IEA, 2019b), and meaning that net additions to capacity remained at the same level as in 2017 despite the decline in investment. Solar PV spending declined by around 4% to $135 billion, mostly as a result of policy changes in China. Wind investment remained relatively flat compared to 2017 at around $90 billion, dominated by China and Europe, which together accounted for almost 60% of spending.

In the **Stated Policies Scenario**, investment in renewables is expected to continue to grow, reaching a cumulative total between now and 2040 of around $10 trillion (Table 7.5). Wind and solar PV account for around two-thirds of the expected spending on renewable electricity generation over the *Outlook* period. China and Europe see more than half of the expected investments in wind power while China, United States and India dominate the projected spend on solar PV with over 55%.

In the **Sustainable Development Scenario**, renewable investments grow at a faster rate, reflecting stronger policy support, and reach almost $840 billion in 2040. The cumulative total amounts to nearly $16 trillion. Wind and solar PV make up two-thirds of the investment in renewable electricity generation over the period, with China and Europe combined responsible for around half of the expected spend. China (27%) and Europe (27%) also lead investment in wind.

Key themes

7.5 Material efficiency in heavy industries

In 2018, the industry sector accounted for 29% of final consumption and 42% of direct CO_2 energy-related and process emissions. In the Stated Policies Scenario, this contribution will grow to 30% of final consumption and 43% of CO_2 emissions in 2040, driven by increasing demand for materials and other industrial products.

Emissions from industry are more difficult to abate than those from other sectors for a number of reasons. Some industrial processes require high-temperature heat and some involve chemical reactions that produce emissions. Furthermore, industrial products tend to have low profit margins and to be traded in highly competitive markets, and this makes it difficult for producers to adopt costly low-emission technologies while remaining competitive. A strong push on energy efficiency and on fuel and feedstock switching contributes to significant reductions in the emissions intensity of production in the Sustainable Development Scenario. Electrification, the development and deployment of carbon capture, utilisation and storage (CCUS) and the introduction of hydrogen and other innovative technologies also have roles to play.

Material efficiency – an often under-appreciated strategy – can assist with pushing reductions even further. Alongside the closely related topics of the circular economy and resource efficiency, it has been gaining increased prominence in public debate. Several recent policy developments have placed a new emphasis on material efficiency.[3] These include:

- The G20 Osaka Leaders Declaration of 28-29 June 2019, acknowledged that resource efficiency, including sustainable materials management, contributes to the Sustainable Development Goals and enhances competitiveness. Leaders also looked forward to the G20 Resource Efficiency Dialogue resulting in the development of a roadmap.

- In March 2019, the European Union approved a ban on single-use plastic cutlery, cotton buds, straws and stirrers by 2021 and introduced a requirement for plastic bottles to contain 30% recycled content by 2030. In June 2019, Canada announced a similar ban that would come into effect in 2021 or afterwards. In Africa, 34 nations have banned the use of plastic bags.

- The European Union (EU) Ecodesign Directive, adopted in 2009, established a requirement to improve the environmental performance of products, focusing on lifecycle impacts during product design, as well as material efficiency. The European Union's 2018 Circular Economy Action Plan includes strategies to reduce further the lifecycle impact of plastic products.

---

[3] Material efficiency refers to actions across value chains such as lightweighting, using products for longer and direct material reuse, which result in lower demand for materials. It also includes other changes in materials use and management that enable emissions reductions, such as switching to lower emissions construction materials and increasing metals recycling to enable more energy-efficient secondary production.

- In April 2019, the European Union adopted a new regulation for passenger cars and light commercial vehicles that requires an evaluation by 2023 of the potential to develop a common methodology for assessing and reporting full lifecycle CO_2 emissions of cars and vans.

- In 2017, the Netherlands adopted a mandatory carbon cap for the environmental profile of new homes and offices that includes taking into account emissions from buildings materials production.

In this section, we quantify the impact on material demand, energy needs and CO_2 emissions in industry arising from economy-wide resource efficiency strategies including those applied in other sectors, such as transport, buildings and power generation. Our analysis focuses on widely used emissions-intensive materials such as steel, cement and aluminium and, to a lesser extent, petrochemicals.

Demand for key materials

Material demand growth tends broadly to reflect economic and social development. At lower levels of economic development, per capita demand for materials is generally relatively low. As economies develop and urbanise, they consume more goods and build up their infrastructure (for example high-rise buildings, roads, railways and electricity generation infrastructure), and their material demand per capita tends to increase significantly. Demand for cement and steel for infrastructure generally rises sharply at this stage. Once an economy is industrialised, material demand per capita tends to level off or even decline: materials like cement and steel tend to be used primarily for replenishing and renovating stocks, though demand for other materials such as aluminium and plastics may increase, depending among other things on product demand, product lifetimes and behavioural patterns.

Global demand for materials has grown considerably in recent decades (Figure 7.4). Since 1990, global demand for steel has grown almost 2.5-times, for plastics by around three-times, and for aluminium and cement by around 3.5-times. This growth in demand has coincided with a 45% increase in global population and a doubling of industry's contribution to GDP growth.

There are differences in regional trends both in terms of the rate and magnitude of this growth in demand (Figure 7.5). Since 2000, the rapid development of China has fuelled global material demand growth, with per capita demand for some materials increasing to levels higher than in most developed economies. In recent years, however, demand in China for some materials appears to have levelled off: for example, cement production has declined from its 2014 peak and this decline is expected to continue in line with a slower pace of new infrastructure development. Even as demand growth slows in China, however, other developing economies including India (TERI, 2019) and economies in Asia, Latin America and Africa are set to drive growing global demand for materials.

Figure 7.4 ▷ Demand for key materials in the Stated Policies Scenario

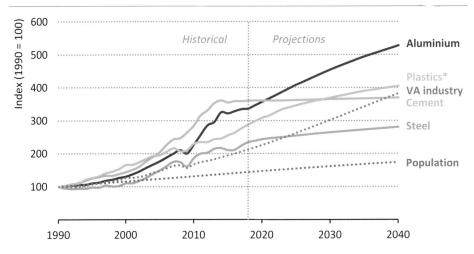

Demand for materials has increased considerably in recent decades and is set to continue for aluminium, plastics and steel

* Demand for high-value chemicals (ethylene, propylene, benzene, toluene and mixed xylenes), which are the key precursors to plastics, are used as a proxy for plastics demand. High-value chemicals are also precursors to synthetic fibre and rubber production. Note: VA = value added.

Sources: Platts (2019); USGS (2018a); USGS (2018b).

Figure 7.5 ▷ Per capita material consumption and GDP for selected countries, 2000-2017

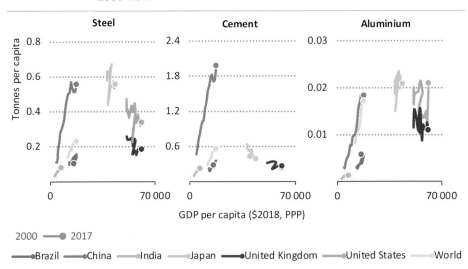

Per capita demand for materials is increasing but varies considerably by stage of economic development

The Stated Policies Scenario sees continued growth in material demand, including increases in demand by 2040 of 20% for steel, 57% for aluminium, 41% for high-value chemicals and 2% for cement. Industry emissions are projected to be 16% higher in 2040 than they are today. However, material efficiency gains have the potential to reduce this level of emissions by helping reduce demand for materials, complementing efforts to change the means by which materials are produced through energy efficiency improvements, fuel switching and the adoption of new low-carbon process routes that incorporate hydrogen or CCUS.

Material demand and interlinkages with CO_2 emissions abatement in other sectors

Energy transitions worldwide imply considerable changes in how we produce and use energy. Changes in activity in turn lead to changes in material demand and the share of materials going to various end-uses (Figure 7.6).

Figure 7.6 ▷ Material demand worldwide by scenario and end-use

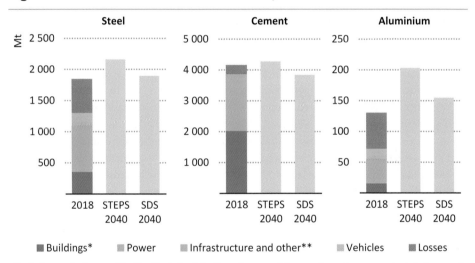

Emissions abatement in the Sustainable Development Scenario reduces material demand. Buildings, infrastructure and other are the major end-use sectors to benefit.

* Buildings: For aluminium, this category includes all buildings and infrastructure, except for power infrastructure. ** Infrastructure and other: Includes transport infrastructure (e.g. roads, rail, bridges), other infrastructure (e.g. below-ground water supply networks, pipelines), ships, airplanes, mechanical and electrical equipment, consumer goods, domestic appliances and food packaging.

Notes: Mt = million tonnes; STEPS = Stated Policies Scenario; SDS = Sustainable Development Scenario. Material demand changes include both activity levels and material efficiency. Losses are materials lost during manufacturing and semi-manufacturing stages, which do not end up in a specific end-use. Aluminium is liquid aluminium (includes internal scrap).

Road vehicles currently account for 12% of global demand for steel and for 14% of global demand for aluminium. In the Sustainable Development Scenario, there is a reduction of 12% in passenger light-duty vehicle sales and a 12% reduction in truck sales by

2040 compared with the Stated Policies Scenario, driven by a shift towards public transport, car sharing, improved fleet management aided by digitalisation, and increased average mileage per vehicle. This leads to fewer materials being needed for the manufacture of passenger vehicles and trucks, the effect of which is partially offset by increases in materials needed to produce more buses and railway stock. Considerable efforts go into vehicle lightweighting, i.e. making vehicles weigh less in order to increase their efficiency. Lightweighting reduces demand for steel, while increasing demand for aluminium, plastics and composites. It delivers overall lifecycle energy and CO_2 emissions savings: for example, it reduces the net lifecycle emissions in 2030 of the average internal combustion engine passenger car in the United States by around 4%.

The increasing shift from internal combustion engine vehicles towards electric and (to a lesser extent) hydrogen fuel-cell vehicles also changes the nature of material requirements for manufacturing. Electric vehicles for example, contain less steel but more aluminium than internal combustion vehicles. In the Sustainable Development Scenario, material demand for vehicles in 2040 falls by 30% for steel and by around 20% for aluminium compared to the Stated Policies Scenario.

Requirements for transport infrastructure also change. By 2040, additional rail infrastructure leads to an increase in demand of about 35% for both steel and cement in the Sustainable Development Scenario relative to the Stated Policies Scenario. On the other hand, reduced vehicle travel requires less use of materials for the construction and maintenance of roads, leading to a decrease of about 15% in demand for these uses for both steel and cement. In the Sustainable Development Scenario, the combined material demand for rail and road transport infrastructure is 25% higher for steel and 5% higher for cement than in the Stated Policies Scenario, as the increase in demand for both materials for rail is higher than the decrease in demand from roads. The increase for cement is less than that for steel because roads require considerably more cement than railways do, and considerably more cement than steel.

Buildings account for an estimated 20% of global demand for steel and for about half of cement demand. Currently, many buildings, especially non-residential buildings, are demolished before the end of their useful lifetime. In the Sustainable Development Scenario, a large portion of the buildings stock undergoes deep retrofitting in an effort to improve the energy efficiency of building envelopes, thereby creating an incentive to continue using the building for longer in order to realise savings and reducing the need to construct new buildings. In 2040, demand for steel and cement for buildings is lower by around 15% in the Sustainable Development Scenario than it is in the Stated Policies Scenario.

Power capacity additions account for a small share of global demand for steel, cement and aluminium today. Demand from the power sector is likely to increase in the future. In the Sustainable Development Scenario, the power sector accounts for around 4% of steel demand, over 2% of aluminium and almost 1% of cement demand in 2040. Renewables make up a large share of power sector demand, accounting for up to 95% of some materials.

Figure 7.7 ▷ Power sector demand for steel, cement and aluminium by scenario

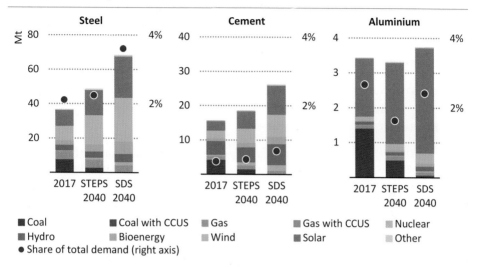

Energy transitions lead to increasing demand for materials for the power sector, but it accounts for a relatively small portion for steel, cement and aluminium demand

Notes: STEPS = Stated Policies Scenario; SDS = Sustainable Development Scenario. Material intensity estimates for power were based on the work of Gibon et al., (2017). Percent of total demand accounts for material flows into final products; semi-manufacturing and manufacturing losses are not included.

The combined emissions from steel, cement and aluminium produced for use in power sector infrastructure in 2040 are about 4% higher in the Sustainable Development Scenario than they are in the Stated Policies Scenario, despite the much larger increases in material demand in the former scenario (Figure 7.7). This is the result of considerable reductions in the emissions intensity of materials production. Although materials production emissions account for a larger proportion of total power sector emissions in the Sustainable Development Scenario, this has to be seen in context: the combined overall emissions in 2040 from power generation and from steel, cement and aluminium production for power capacity additions in the Sustainable Development Scenario are less than 30% of those in the Stated Policies Scenario.

Material efficiency potential in the Sustainable Development Scenario

In addition to changes in material demand related to activity level changes, the Sustainable Development Scenario sees major efforts being undertaken to improve material efficiency – that is, to reduce material demand and to manage materials use and production in ways that reduce waste and increase recycling and reuse, in turn leading to lower energy use and emissions across value chains. These efforts are focused on steel, cement and aluminium, with a particular eye to material efficiency potential in the buildings and vehicles value chains.

Material efficiency improvements in the Sustainable Development Scenario stem from a diverse range of strategies applied to different materials and value chains (Table 7.6). Product or building design is a key determinant of lifecycle material demand, energy consumption and CO_2 emissions. Maximising the efficiency of fabrication and construction, optimising the useful lifetime of a product, and ensuring the appropriate treatment of its constituent materials at the end of its useful life are all also important.

Table 7.6 ▷ Summary of material efficiency strategies in the Sustainable Development Scenario

Design	Manufacturing	Use	End-of-life
Overview			
• Lightweighting. • Reduce over-design, optimise design. • Design for use, long life and reuse.	• Reduce material losses. • Reduce material overuse.	• Lifetime extension and repair. • More intensive use.	• Remanufacture and repurpose. • Direct material reuse. • Recycle.
Steel			
☑ Vehicle lightweighting. ☑ Building design, reduce over-specification and concrete-steel composite construction; modular design for future materials reuse.	☑ Improve steel semi-manufacturing and end-use product manufacturing yields.	☑ Use buildings for longer through refurbishment. ☑ Mode shift to reduce the number of vehicles being produced.	☑ Direct reuse of steel (with highest potential in specific end-uses such as ships). ☑ Recycle.
Cement			
☑ Building design, reduce over-specification and concrete-steel composite construction; modular design for future materials reuse.	☑ Improved construction, including reducing onsite construction waste, reducing cement content in concrete and pre-cast fabrication.	☑ Using buildings for longer through refurbishment.	☑ Concrete reuse.
Aluminium			
☑ Vehicle lightweighting (steel-aluminium substitution) offsets some reductions from other strategies.	☑ Improve aluminium semi-manufacturing and end-use product manufacturing yields.	☑ Mode shift to reduce the number of vehicles being produced.	☑ Direct reuse of aluminium. ☑ Recycle.

☑ High potential   ☑ Medium potential   ☑ Low potential   ☑ Increase in demand

By 2040, overall net demand in the Sustainable Development Scenario is reduced relative to the Stated Policies Scenario by 12% for steel, 10% for cement and 24% for aluminium. These net reductions occur despite some upward pressure from increased aluminium demand owing to vehicle lightweighting and increased material demand for power and transport infrastructure. They are achieved through a combination of material efficiency strategies and changing patterns of demand.

Some material efficiency strategies are focused on the nature of materials production. For example, increasing metal scrap collection rates, improving sorting and reducing contamination result in more scrap availability, thus enabling higher shares of secondary production, which is less energy and emissions intensive than production using virgin materials. However, material efficiency can also affect the total amount of scrap available. While improved end-of-life collection rates help increase the amount of scrap metal available, improved manufacturing yields and lifetime extension and reuse place downward pressure on scrap availability. In the Sustainable Development Scenario, the net impact of these pressures is a reduction in both steel and aluminium scrap availability. This underlines the importance of a systems approach: savings from material efficiency cannot simply be added on top of energy efficiency savings.

In the case of cement, reducing the clinker-to-cement ratio is an important way of improving materials efficiency during manufacturing. Clinker is the main active ingredient in cement and its production is the most emissions intensive part of cement production. While common Ordinary Portland Cement typically contains more than 90% clinker, it is possible to replace a considerable portion of clinker with alternative cement constituents, such as fly ash, granulated blast furnace slag, ground limestone, calcined clay, volcanic ash, rice husk ash and silica fume. This makes more efficient use of clinker and reduces the carbon footprint of cement substantially.

Improvements for other materials also play an important role. Improving recycling rates of plastics, for example, can displace considerable chemical sector demand (Box 7.1).

Box 7.1 ▷ Plastics recycling is a key to reduce chemical material demand

Petrochemical products are found everywhere and are an integral part of modern society. They include plastics, fertilisers, packaging, clothing, digital devices, medical equipment, detergents, tyres and many other products. Led by developing economies, demand for petrochemicals is surging and will continue to grow. As a result, petrochemicals are rapidly becoming the second-largest source of oil demand growth.

In 2018, the IEA examined the consequences of increased demand for these products, and what can be done to accelerate a clean energy transition for the petrochemical industry (IEA, 2018a). It found that the reduction in direct CO_2 emissions in the Sustainable Development Scenario would result in a decrease in environmental impacts associated with petrochemicals, but that a broad range of efforts will be required to tackle plastic pollution in oceans and other forms of environmental damage.

Plastic recycling is a key underlying source of emissions reduction in the Sustainable Development Scenario, which sees a substantial increase in waste plastic collection rates, recycling yield rates and displacement rates (the extent to which recycled plastics displace demand for their virgin counterparts). Recycling yield rates increase from an average of 75% in 2017 to nearly 85% in 2050. Displacement rates double, from about one-third today to two-thirds in 2050. However, achieving the improvements is dependent on significant technical advances in recycling processes.

Contribution of material demand changes to energy and CO_2 savings

The policy changes in buildings, transport and power generation, and the industrial material efficiency strategies included in the Sustainable Development Scenario together have the potential to bring about significant changes in energy demand and CO_2 emissions trends for key energy-intensive industries. Thanks to a combination of these and other policies to reduce the emissions intensity of material production, CO_2 emissions from steel, cement and aluminium stabilise after 2019 in the Sustainable Development Scenario and enter a steady decline thereafter. Without those policies, emissions trends instead plateau under the Stated Policies Scenario (Figure 7.8). Material demand reductions alone contribute to over 20% of the Sustainable Development Scenario reductions in 2040. Additional material efficiency strategies such as reducing the clinker-to-cement ratio and increases in secondary metals production also contribute to the overall emissions savings, while also bringing air quality and health benefits.

Figure 7.8 ▷ **CO_2 emissions reductions resulting from material demand reductions for steel, cement and aluminium**

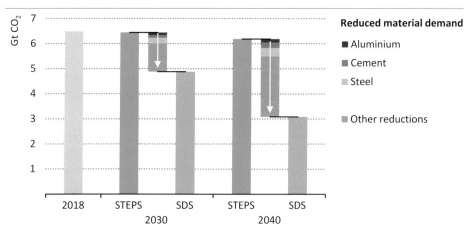

Policies in transport, buildings and power sectors can change material demand and contribute to a reduction in heavy industry emissions

Notes: STEPS = Stated Policies Scenario; SDS = Sustainable Development Scenario. Emissions include direct energy-related, process and indirect electricity emissions. Other reductions include factors that reduce the emission intensity of material production, i.e. energy efficiency, fuel switching, CCUS, hydrogen-based production, other innovative technologies, and other aspects of material efficiency not directly related to demand reduction (e.g. increased secondary metals production).

Challenges, costs and enabling policies and actions

Various barriers and challenges prevent material efficiency from maximising its full potential even in cases where it would be economically beneficial. These include real and perceived risks; financial costs and time constraints; limited awareness of and training in material-efficient design and construction methods; fragmented supply chains that pose challenges for direct materials reuse; and restrictive design standards that may hamper the uptake of new materials or design methods.

As far as costs are concerned, comprehensive analysis of material efficiency costs found that some strategies may result in savings (car sharing, reducing waste in buildings construction and increasing collection rates of aluminium), while other strategies have moderate abatement costs (Material Economics, 2018). All the strategies in the analysis had costs of below EUR 100/tonne CO_2, which is below the level of most carbon prices in 2040 in the Sustainable Development Scenario.

Further efforts will be required to overcome non-financial barriers related to perceived risks, co-ordination and behaviour. This is a matter for governments to consider in the light of their own goals and circumstances, but also for industry, researchers and consumers. In pondering what might be useful, stakeholders might want to consider the scope to:

- **Improve data collection, lifecycle assessment, benchmarking and material flow analysis**: More robust data and better modelling tools would help to support the development of benchmarks. They would also support the development of regulations if governments adopt such measures.

- **Pay greater attention to the lifecycle impact at the design stage and in climate regulations**: Designers and manufacturers should take into account the trade-offs between production and use-phase emissions, and design for long lifespans, repurposing, reuse and recycling. Governments may wish to consider the case for underpinning this by moving towards lifecycle based requirements in emissions regulations.

- **Increase product lifespans and promote repurposing, reuse and recycling**: Government and industry should promote durability and long lifetimes for building and products, except in cases where doing so would lock in CO_2 emissions and lead to higher lifecycle emissions. Increased materials reuse and recycling could be facilitated by better supply chain integration, mandating the use of a share of recycled materials in products, adopting landfill disposal fees and expanding the coverage of recycling requirements.

- **Develop regulatory frameworks and fiscal incentives to support material efficiency**: Moving from prescriptive to performance-based design standards would facilitate efficient use of materials. Carbon pricing would incentivise material efficiency, green certification programmes and government procurement could also help to support it.

- **Adopt business models and practices that advance circular economy objectives**: Business models that promote a sharing economy and incorporate digitalisation may help to reduce materials use. Company monitoring and reporting requirements could deter practices that may increase material use.

- **Train, build capacity and share best practices**: Material efficiency has an important part to play in education programmes for designers, engineers, construction workers, and manufacturing and demolition companies. Government-supported capacity building could help, as could best practice sharing among companies.

- **Shift behaviour towards material efficiency**: Consumers can seek to make a difference by opting for products with materially efficient design, particularly if eco-labels give them relevant information, and can also play their part in improving recycling rates.

7.6 Smart electricity use: the power of the hour in reducing emissions

Emissions linked to the use of electricity are often calculated using average annual CO_2 emissions intensity of electricity supply. In reality, the CO_2 emissions intensity of electricity supply, and therefore electricity use, varies significantly depending on when the electricity is used and the electricity generation mix of the region. This variability is projected to increase in both the Stated Policies and Sustainable Development scenarios as shares of variable renewables increase and demand becomes more concentrated in peak hours.

This presents opportunities for reducing CO_2 emissions. Targeted energy efficiency and demand-side response can significantly reduce CO_2 emissions linked to electricity use, as well as contributing to power system operation and flexibility needs, lower consumer electricity bills and improved air quality. Tapping these potential gains however depends on adequate and accessible markets for flexibility, and on electricity pricing models which incentivise consumers to reduce demand at peak periods.

7.6.1 Hourly electricity supply CO_2 emissions intensity: status and outlook

The time-based variation of the emissions intensity of electricity generation is a result of the changing contribution of different technologies to electricity generation at various times of the day. The variation in the hourly average carbon emissions intensity of electricity supply is especially affected by the increases in generation from variable renewables (VRE) such as solar PV and wind. The share of VRE in global electricity generation increased from under 2% in 2010 to 7% in 2018. By 2040, this share reaches 24% in the Stated Policies Scenario and rises to 40% in the Sustainable Development Scenario. Using the World Energy Model (WEM) hourly electricity supply model, we are able to determine the implications of increasing shares of VRE on the average hourly variation of electricity supply CO_2 intensity.

In some countries the carbon intensity of electricity supply in 2040 is generally lowest during the middle of the day when generation from solar PV is highest, with CO_2 emissions intensity becoming higher during evening peaks when the sun sets and more fossil fuel-fired generation is used to meet demand. This is especially the case in systems where solar PV and gas (or coal) represent the largest sources of generation. In both the Stated Policies and the Sustainable Development scenarios, India moves into this position in the years ahead. The share of solar PV in generation rises in India from 3% today to 31% by 2040 in the Sustainable Development Scenario, significantly reducing average electricity supply CO_2 intensity during the middle of the day. As a result, the CO_2 intensity of generation in 2040 varies by a factor of seven across an average day in this scenario.

Figure 7.9 ▷ Average hourly CO_2 emissions intensity, electricity demand and wholesale electricity prices in India and the European Union

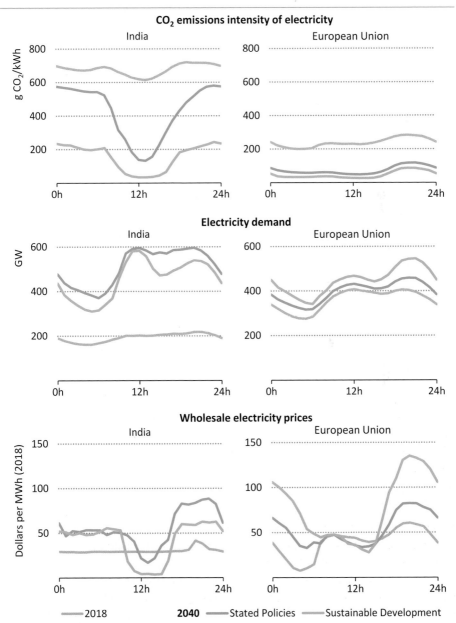

─── 2018   2040 ─── Stated Policies   ─── Sustainable Development

Daily variation in the average CO_2 intensity of electricity increases to a factor of seven in India, with generation oscillating between solar PV and gas or coal. The European Union, with higher shares of wind generation, reaches a factor of three.

Note: g CO_2/kWh = grammes of CO_2 per kilowatt-hour; MWh = megawatt-hour.

Power systems with a varied mix of generation technologies will have a different profile of average CO_2 emissions intensity of electricity supply. For example, in systems with large shares of generation from wind, the CO_2 intensity can be lowest during the night when demand is low and the wind is blowing. The European Union is an example of a system where wind emerges as the largest source of generation in both the Stated Policies and Sustainable Development scenarios. Hourly average CO_2 emissions intensity is lowest during the night and middle of the day, and peaks during winter evenings when demand is highest. The result is a difference of up to a factor of three in the average CO_2 intensity of electricity between certain hours of an average day (Figure 7.9).

7.6.2 Implications for policy makers and consumers

Two pathways offer policy makers and consumers opportunities to translate time-based electricity supply CO_2 intensity information into a reduction in the CO_2 emissions associated with electricity demand.

Targeted energy efficiency to deliver the Sustainable Development Scenario

Decarbonisation of electricity supply accounts for the bulk of the difference in CO_2 emissions from electricity generation between the Stated Policies and Sustainable Development scenarios. Nevertheless, energy efficiency has a major role in reducing electricity sector emissions in the Sustainable Development Scenario, and energy efficiency policies could support the reduction of emissions by targeting demand reductions when the average emissions intensity of generation is high (for example evening peaks, or nights in solar PV dominated systems).

For the first time, this *World Energy Outlook* computes average CO_2 emissions intensity of different uses of electricity and provides an indication of where additional energy efficiency efforts could maximise emissions reductions. Our assessment of hourly electricity usage and the CO_2 intensity of electricity supply illustrates that the average annual CO_2 intensities of different uses of electricity can vary by over 30% in the Stated Policies Scenario. The largest variation is in India, where the majority coal and solar PV based power system sees the average CO_2 emissions intensity of end-uses vary by over 120 grammes of CO_2 per kilowatt-hour (g CO_2/kWh) in the Stated Policies Scenario in 2040. Space cooling in commercial buildings has the lowest average emissions intensity as a result of its strong correlation with generation from solar PV. Lighting in homes and electric vehicle (EV) charging have the highest CO_2 intensities as a result of the concentration of these uses during evening peak-load periods when solar PV is not available. By way of comparison, the CO_2 intensities of the lowest and highest end-uses in the Stated Policies Scenario vary by almost 30% in the European Union, 20% in China, and 15% in the United States in 2040.

Energy efficiency efforts in the Sustainable Development Scenario reduce electricity demand by 23% in India relative to the Stated Policies Scenario. Thanks to targeting of additional energy efficiency efforts towards the most carbon-intensive end-uses, this leads to a larger reduction in CO_2 emissions, with a decrease of almost 30% (Figure 7.10). In the

European Union and the United States, achieving energy efficiency gains in line with the Sustainable Development Scenario, including by targeting energy efficiency improvements on CO_2 intensive end-uses such as home appliances and air conditioners, also has more of an impact on CO_2 emissions than electricity demand. These reductions are additional to the impact of decarbonising electricity supply in the Sustainable Development Scenario.

Figure 7.10 ▷ **Impact of energy efficiency on CO_2 emissions from electricity supply in the Sustainable Development Scenario, 2040**

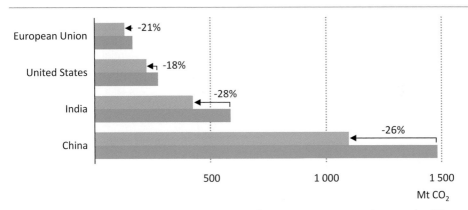

Energy efficiency reduces global emissions from electricity by one-quarter, on top of the benefits of decarbonising electricity supply in the Sustainable Development Scenario

Note: SDS = Sustainable Development Scenario; STEPS = Stated Policies Scenario; Mt CO_2 = million tonnes of carbon dioxide.

The emission reductions in Figure 7.10 depend on measures being taken to incentivise them. Equipment with a long lifetime can in effect lock in electricity demand at times when variable renewables are rarely available to meet demand, and this has the potential to prolong reliance on CO_2 emissions intensive sources of generation if storage or other dispatchable low-carbon sources of supply are not available. Governments can take advantage of the opportunity to multiply the CO_2 emissions abatement impact of energy efficiency measures by improving minimum energy performance standards for appliances and equipment (especially in the residential sector); providing financial incentives for building retrofits and replacement of inefficient equipment, and supporting the further digitalisation of electricity use and associated opportunities.

Demand-side response and consumer awareness: unlocking further reductions

Demand-side response (DSR) draws on the willingness and ability of certain users of electricity to shift electricity demand in time in response to signals from the market or system operator. Typically DSR is used to provide flexibility to power systems by reducing demand during periods of high relative electricity prices, drawing principally on the flexibility of large-scale industrial and commercial customers.

DSR can also be utilised as a tool for CO_2 emissions reduction, shifting demand from periods of CO_2-intensive electricity supply to less CO_2-intensive periods, reducing the curtailment of variable renewables and lowering CO_2 emissions. The projected increase in the average hourly variation of the CO_2 intensity of electricity offers opportunities for exactly this.

If enough consumers are willing to participate in demand-side response and be flexible in their electricity consumption behaviours, the potential of DSR could be considerably increased beyond what is seen in the Sustainable Development Scenario. The impacts of such enhanced DSR were assessed using the WEM hourly electricity supply model, with economic dispatch that maintains systems reliability. Analysis showed that up to 20% of peak load could potentially be shifted to periods of lower CO_2 intensity in electricity supply, thus reducing overall CO_2 emissions. In addition to tapping DSR potential from uses such as water heating, refrigeration, heating and cooling, drawing on low-cost demand-side flexibility resources such as optimised EV charging is central to enhancing DSR (see Spotlight).

SPOTLIGHT

When to plug in: the system impact of EV charging times

With stronger policy support and battery costs falling, the global electric vehicle (EV) market is growing rapidly. EVs account for 75% of new vehicles sales by 2040 in the Sustainable Development Scenario. Rapid growth in EV deployment could have major implications for power systems. Currently the majority of EV charging occurs at home (IEA, 2019d). If charging is concentrated during evening peaks when people return from work, and storage is not deployed at scale, the average CO_2 emissions intensity of road transport in 2040 would be the highest of all electricity end-uses; up to 25% higher than the average CO_2 emissions intensity of electricity use in India and the European Union in the Sustainable Development Scenario, and close to 20% higher than in China and the United States.

The flexibility of EV charging times and locations creates an opportunity to further improve the CO_2 emissions benefit of switching to EVs. "Smart charging" at home could use overnight off-peak electricity (Figure 7.11). If widely adopted, this type of charging would reduce peak electricity demand and also reduce overall CO_2 emissions. In Europe for example, the average CO_2 intensity of EV charging in 2040 using smart charging would be around 20% less than if charging was concentrated during evening peaks.

An increase in the share of vehicles charging at workplaces or communal charging locations during the day, when solar PV output is at its highest, offers the greatest potential for CO_2 emissions reductions. A balanced approach combining daytime workplace charging and night time home charging could reduce the average CO_2 emissions intensity of EV charging by 45% in the European Union and more than double CO_2 emissions reductions relative to majority smart charging at home. Vehicles could also be used to supply electricity and other services to the grid or buildings during peak times, known as "Vehicle2X". With the right regulatory framework and incentives, this approach could unlock further emissions reductions and system benefits.

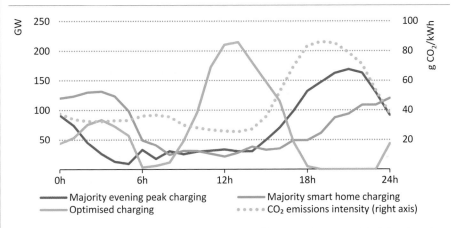

Figure 7.11 ▷ EV charging patterns and average hourly CO_2 emissions intensity of electricity in the European Union in the Sustainable Development Scenario, 2040

Changing behaviour to increase EV charging during times of low CO_2 intensity of electricity supply could reduce EV CO_2 emissions by over 45%

In an enhanced DSR case in China, an average of around 250 GW of electricity demand are shifted by 2040 from hours of high average CO_2 intensity of electricity supply to periods of near-zero average CO_2 intensity (Figure 7.12). The additional demand during the middle of the day allows for the integration of a further 330 GW of solar PV, representing an increase of 20%, without increasing curtailment rates. Enhanced use of DSR could see CO_2 emissions from electricity supply reduced by 16% relative to levels of DSR in the Sustainable Development Scenario.

The United States would also see important gains from significant increases in the use of DSR, with the possibility of shifting an average of over 100 GW of load to the middle of the day and integrating an additional 130 GW of solar PV, potentially reducing emissions from electricity supply by up to 65 Mt CO_2 (-25%). The exact emissions savings from enhanced DSR would depend on what technologies are used to meet the increase in demand during the middle of the day (Box 7.2).

Achieving the potential for CO_2 emissions abatement from DSR requires the existence of market opportunities for DSR and incentives for consumer participation. In some markets today, regulatory and market barriers either prohibit the participation of aggregators[4] or other DSR providers in flexibility and energy markets, or render participation expensive (IEA, 2018b). Policy makers have a role to play in creating adequate markets for flexibility

---

[4] A typical DSR aggregator is a third-party company that contracts with individual demand sites (industrial, commercial or residential consumers) and aggregates them to operate as a single DSR provider.

with access for a range of actors including aggregators, and also in legislating connectivity as part of energy efficiency standards, while ensuring that cybersecurity risks and other consumer concerns are adequately addressed. Standardising market and regulatory frameworks across jurisdictions would help to accelerate deployment. Progress is already being made in many major markets and DSR is beginning to spread beyond the traditional centres of the United States and Europe (IEA, 2018b).

Figure 7.12 ▷ Average hourly electricity supply and demand in China in the Sustainable Development Scenario, 2040

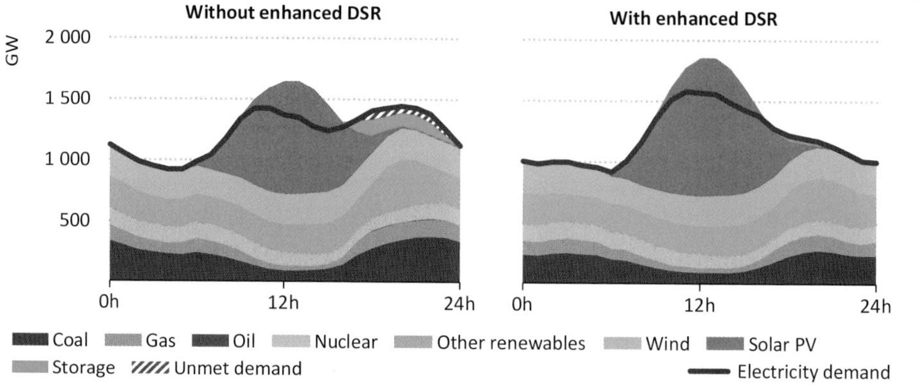

Tapping additional demand-side flexibility can shift electricity demand to periods of lower CO_2 intensity of electricity supply, with the potential to reduce emissions by 16% in China

Box 7.2 ▷ Measuring the benefits: average or marginal CO_2 intensity?

The World Energy Model includes a module simulating hourly electricity demand and supply through economic dispatch. Dispatch modelling captures the effects of generator availability and changing demand on the mix of generators dispatched and therefore the CO_2 emissions associated with electricity supply. This effectively captures the marginal effect of higher utilisation of demand-side flexibility on CO_2 emissions.

Without the use of an economic dispatch model, assessing the impact of shifting demand on power sector CO_2 emissions could be estimated using the average CO_2 intensity of electricity supply or the marginal CO_2 intensity. However, either approach would miss parts of the complex dynamics of power system operation, limiting the accuracy of estimates.

Applying the average CO_2 intensity of electricity supply overlooks the fact that the marginal unit responding to increases (or decreases) in demand often has a higher CO_2 intensity than the average, and this could lead to overestimation of the CO_2 emissions reductions. However, in the case of structural changes to electricity demand, such as energy efficiency, using the average CO_2 intensity provides a reasonable estimate of the minimum impact on emissions.

Applying the CO_2 intensity of the marginal unit approach may not pick up all the changes in the generation mix and overall CO_2 emissions that may result from an economic dispatch modelling based on a new demand profile incorporating enhanced DSR. For example, increasing day-ahead demand may lead nuclear power plants to increase operation during all hours of the day, not just during periods of increased demand, and thus lower CO_2 emissions.

Small dispersed loads in the buildings sector such as residential air conditioners and water heaters represent the largest opportunity for DSR globally. For this potential to be realised, digital infrastructure such as smart meters and price incentives for consumers are necessary, as are aggregators or electricity retailers able to connect consumers to markets that value their flexibility.

How and what share of the value of DSR is channelled to consumers will depend on business models, electricity tariff design, taxation and other charges. Real-time pricing exposes consumers to variations in wholesale prices, creating direct incentivises to shift flexible demand to times of low prices, but they can also expose consumers to unexpected price hikes, since consumers are unlikely to be able and willing to monitor and manage their domestic electricity consumption more or less continuously. Fixed time of use tariffs provide more predictable incentives, but have less ability to reflect the variability of the intensity of CO_2 emissions. Business models which rely on the automation of DSR from household appliances and equipment may be of wider appeal to consumers, and also be capable of generating significant savings in CO_2 emissions.

7.6.3 System benefits

Optimising electricity demand for CO_2 emissions reductions also has wider system benefits: using targeted energy efficiency and demand-side response has the potential to assist power system operation, reduce electricity bills and moderate air pollution.

Implications for flexibility

Flexibility needs globally are expected to increase much faster than electricity demand in the Stated Policies Scenario, owing to a combination of a rising share of variable renewables in electricity supply and a rapid rise in demand for uses such as cooling and EV charging which tend to add to peak demand (see Chapter 6).

Focusing energy efficiency efforts on the most CO_2 intensive end-uses will reduce not only CO_2 emissions but also peak electricity demand. Lowering peak demand reduces hourly ramping requirements for power systems, and this reduces the need for flexibility.

The deployment of demand-side response to shift electricity usage from times of high CO_2 intensity of supply to times of high shares of generation from variable renewables will also tend to reduce hourly ramping requirements, as, in general, it implies moving demand away from peak hours when the system is constrained and flexibility needs are highest. The ability of electricity demand to be a source of flexibility for the system, with the potential to shift demand in time, is a subject explored in detail in the Special Focus on Electricity in the *World Energy Outlook-2018* (IEA, 2018b).

Implications for consumer bills

Energy efficiency lowers consumer bills directly by reducing overall electricity demand. DSR can also lower consumer bills, but its effect is less direct. Variable renewables such as wind and solar PV have a short-run marginal cost of near-zero, so using DSR to shift electricity demand to periods when a very high share of electricity comes from solar PV or wind can result in a considerable reduction in average wholesale prices. Enhanced deployment of DSR reduces average annual wholesale prices by 5% in the United States and almost 10% in China. The share of this price reduction passed on to consumers will depend on a number of factors, including the share of energy costs in consumer bills in each jurisdiction.

Optimising electricity demand for CO_2 emissions reductions will also have an impact on the fixed component of consumer bills, albeit over a longer timeframe. Reducing peak demand and using DSR to provide low-cost flexibility to power systems will reduce the need for other flexibility solutions such as transmission capacity and distribution network upgrades. This is especially relevant in the case of EV charging, where uncoordinated charging has the potential to place considerable strain on distribution networks.

7.7 Biogas: turning organic matter into renewable energy

Over one billion tonnes of organic by-products and waste are thrown away or abandoned every year. Their decomposition can lead to emissions of methane, which has a significantly higher global warming potential than CO_2: the waste, if left unmanaged, can cause land and groundwater contamination. If these waste products were collected and processed in an appropriate way, they could provide a valuable source of renewable energy in the form of biogas.

The technologies to produce biogas are not new, yet only a handful of countries produce biogas today. Nonetheless recent years have seen a resurgence of interest in how biogas technologies might aid progress towards the energy-related Sustainable Development Goals while improving sanitation and waste management. Biogas can provide a local source of power and heat for rural communities, as well as a source of clean energy for cooking and heating in communities that lack access to energy and sanitation, and that suffer from indoor air pollution. If biogas is upgraded to pipeline-quality gas, it is typically then known as biomethane and it could help to reduce the emissions intensity of gas supply in gas-consuming economies. In addition, when it results in the productive use of methane that would otherwise be released into the atmosphere, the use of biogas can lead to an additional reduction in GHG emissions.

This section provides an overview of the key technologies and production pathways for biogas (Box 7.3), explores potential volumes and production costs, and discusses various biogas applications. There is further discussion about the potential role of biogas in helping to achieve clean cooking access in African countries in Chapter 9 and about its role in decarbonising gas supply in Chapter 13.

Box 7.3 ▷   **Glossary: biogas and biomethane**

> **Biogas**: A mixture of methane, CO_2 and small quantities of other gases produced by anaerobic digestion of organic matter in an oxygen-free environment. The composition of biogas depends on the type of feedstock and the production pathway. The methane content of biogas typically ranges from 45% to 75% and the CO_2 content from 25% to 55% by volume.
>
> **Biomethane** (also called "renewable natural gas"): A near-pure source of methane produced either by "upgrading" biogas (a process that removes any CO_2 and other contaminants present in the biogas), or through the gasification of solid biomass followed by methanation.
>
> **Energy content**: The net calorific value of biogas is less than that of methane due to its higher CO_2 content. The lower heating value (LHV) of biogas varies by feedstock between 16 and 28 megajoules per cubic metre (MJ/m^3). Biomethane is assumed here to have an LHV of 36 MJ/m^3 (natural gas is around 35 MJ/m^3). Biogas and biomethane are reported here in million tonnes of oil equivalent (Mtoe); biomethane can also be given in volumes that are equivalent to natural gas (i.e. billion cubic metres).

State of play: main technologies, uses and production today

Around 30 Mtoe of biogas was produced worldwide in 2017, of which 90% was in Europe, China and United States (Figure 7.13). Around 60% of production is used for electricity and heat generation: in terms of electricity, there is some 18 GW of biogas electricity capacity globally, half of which is in Germany and the United States. Around 30% of global biogas production is consumed in buildings, mainly in the residential sector. A further 8% production is upgraded to biomethane and is blended into the gas networks, and a small proportion is used directly as transport fuel.

A wide variety of feedstocks can be used to produce biogas. In a detailed bottom-up analysis of global supply potential conducted for this *Outlook*, we considered 17 individual types of residue or waste, grouped into four broad feedstock categories: crop residues, animal manure, wastewater sludge, and municipal solid waste (MSW) which includes landfill gas.[5] Biogas production pathways vary by feedstock and region and rely on the following main technologies:

- Biodigesters: Naturally occurring micro-organisms break down crop, manure or MSW feedstock to produce biogas through anaerobic digestion. Contaminants and moisture are usually removed prior to use of the biogas.
- Landfill gas recovery systems: The decomposition of MSW under anaerobic conditions at landfill sites produces biogas. Capturing this gas requires pipes and extraction wells within the landfill site along with compressors to induce flow to a central collection point.
- Wastewater treatment plants (WWTP): These are designed to treat and recover wastewater, but biogas can be produced as a by-product. Anaerobic digestion is used, but with an additional filtration and drying step because of the high moisture content of sewage sludge.

Figure 7.13 ▷ **Biogas production by region and feedstock, 2017**

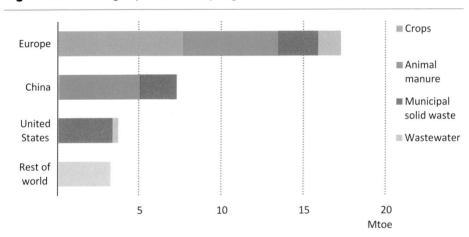

Agricultural feedstock is the main source of biogas production in Europe today, China relies mostly on animal manure, while municipal solid waste is the primary feedstock in the United States

Notes: Crops include energy crops, crop residues and sequential crops. Sequential crops are grown between two harvested crops as a soil management solution that helps to preserve the fertility of soil and avoid erosion; they do not compete with food for agricultural land.

---

[5] Another possible feedstock source is industrial waste, especially from the food processing sector (sometimes called "agro-industrial" feedstock). This could have a strong economic case since the feedstock is produced on-site and a co-located biodigester plant can offset on-site energy demand; however there is limited data available on its potential. In our analysis, we include only distiller dry grain, a by-product from ethanol production, for which comprehensive and reliable data are available. This source is included in the MSW category.

The growth of biogas across the three main production regions to date – Europe, China and United States – has been shaped by three main factors: local conditions, policy support and feedstock availability.

- In Europe, biodigesters processing agricultural and municipal wastes, most notably in Germany, have provided the majority of the two-thirds growth in biogas production since 2009 (EBA, 2017). Energy crops have played a key role as biogas feedstock in the growth of Germany's biogas industry, but there is a vigorous debate about their land-use impacts, and so policy has shifted towards the use of crop residues, sequential crops, livestock waste and the capture of methane from landfill sites (Theuerl et al., 2019).

- In China, policies have supported the installation of household-scale digesters in rural areas as part of a drive to increase access to modern energy and clean cooking fuels. This type of deployment accounts for around 70% of installed biogas capacity in the country, but it has increasingly been accompanied by larger scale combined heat and power plants, which currently provide around 600 MW capacity, and by the production of biomethane.

- In the United States, policies and regulations have traditionally focused on landfill gas collection, which today accounts for nearly 90% of total biogas production in the United States. There is also growing interest in biogas production from agricultural waste, since domestic livestock markets are responsible for almost one-third of methane emissions in the United States part of which could be reduced producing biogas (USDA, 2016). The United States is also leading the way globally on the use of biomethane in the transport sector, primarily because of policy support in the form of the federal Renewable Fuel Standard (US EPA, 2018) and California's Low Carbon Fuel Standard (California Air Resources Board, 2019).

- Around half of the remaining production comes from developing countries in Asia. Access to the Clean Development Mechanism[6] (CDM) was a key factor underpinning this growth, particularly between 2007 and 2011: the development of new projects fell sharply after 2011 when the value of emission reduction credits awarded under the CDM fell. Today, most production in the region comes from Thailand and India. Thailand produces biogas from the waste streams of its cassava starch sector, its biofuel industry and pig farms (Suwanasri et al., 2015). Despite the stalling of projects incentivised through the CDM, India expects to see the development of 5 000 compressed biogas plants over the next five years (GMI, 2019).

Biogas cost supply curve

For this *Outlook*, we have developed a new global assessment of the sustainable technical potential and costs of biogas supply, and how this might evolve in the future. This is based on a detailed assessment of the availability of 17 individual residues and wastes across the

---

[6] Under the Clean Development Mechanism of the Kyoto Protocol emission-reduction projects in developing countries can earn certified emission reduction credits.

25 regions modelled in the World Energy Model. Biogas production pathways were generated for each feedstock based on the range of technologies available, including different sizes of biodigesters and the availability of sustainable feedstocks. On this point, we have assumed that most of the organic content of MSW (when not composted or recycled) could be recovered sustainably even though a considerable amount is at present incinerated or disposed of in landfills. We have also assumed that the use of crop residues for biogas is limited to half of annual production in order to avoid soil erosion and soil nutrient depletion. Our estimates of biogas supply potential and production costs today were then projected forward taking account of GDP and population growth, urbanisation trends, changes in waste management processes and rates of technology evolution.

We estimate that over 570 Mtoe of biogas could be produced sustainably today, which is the equivalent of almost 20% of global natural gas demand. By 2040, this potential increases to over 880 Mtoe (Figure 7.14). Of the different feedstocks:

- Crop residues account for almost half of the global biogas potential today and over 40% in 2040. Maize residues alone account for over 40% of the crop residues global potential.

- Among wastes, animal manure is the largest potential source of biogas; its availability is projected to increase by around 2.5% on average each year, double the rate of increase for crop residues.

- Municipal solid waste provides a smaller fraction of total potential, but landfill gas is the lowest cost source of supply. A global shift over time towards more sophisticated and sustainable waste management practices is likely to limit the long-term potential for landfill gas, but there is still scope for more than 80 Mtoe to be produced in 2040.

- The level of wastewater feedstock increases by around 6% annually to 2040. However, the primary purpose of wastewater facilities is water treatment and recovery and not the production of biogas: their contribution to global supply potential is about 5%, and they tend to be a relatively expensive way to produce biogas.

The potential for biogas supply varies widely between different countries and regions. Developing economies currently account for two-thirds of all potential supply, and their potential supply grows at twice the rate of advanced economies. In India and Southeast Asia, the rationalisation of waste management programmes leads to huge growth in the potential from MSW (reaching 36 Mtoe in 2040, three-times the current potential). In Brazil, crop residues are the most promising feedstock: maize and sugar cane residue traditionally used by the sugar and ethanol industries, each represent a quarter of the country's total biogas potential in 2040. There is a more homogenous picture among the advanced economies. In both Europe and the United States, over 70% of the total potential comes from manure and crop residues, but the cheapest production options are generally from MSW.

Only a fraction of the potential for biogas is exploited today. The main reason is the relatively high cost of biogas production: the average cost is currently around $12 per million British thermal units (MBtu). While constructing larger and more industrialised facilities could provide some economies of scale, there is only modest scope for costs to fall because the technology is generally mature. In our new assessment, the average cost of production falls slightly to around $10/MBtu in 2040, but this is still considerably higher than average projected natural gas prices around the world at that time.

Figure 7.14 ▷ Cost curves of potential global biogas supply by feedstock

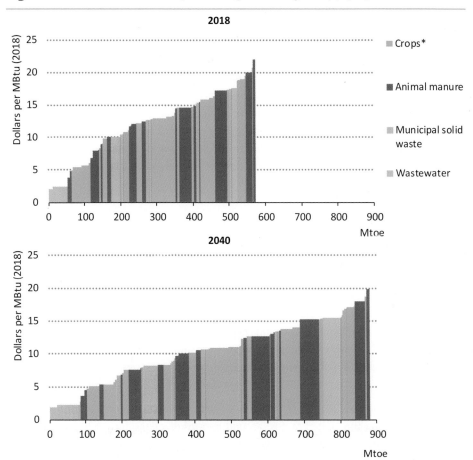

Landfill gas (MSW) is the lowest cost source of biogas. By 2040, 880 Mtoe of biogas could be produced globally, around half of which would cost of less than $10/MBtu.

* Includes crop residues only, energy crops are excluded given concerns over their sustainability.
Notes: MBtu = million British thermal units; Mtoe = million tonnes of oil equivalent.

Biodigesters provide the largest proportion of overall biogas potential, but their use is considerably more expensive than the cheap landfill gas. The cost of purchasing and operating a biodigester accounts for between 70% and 90% of the final biogas cost, with the variation stemming from differences in the size of the biodigester used, and the quality and weight of various feedstocks across different regions. For example, in Southeast Asia the cost of collecting and processing feedstocks is around 10% of the final biogas cost, while in Europe and Australia, the high cost of collection means that feedstocks can represent over 20% of final biogas cost.

There are however cheaper biogas options that could be cost competitive with natural gas. In all regions, landfills equipped with a gas recovery system could provide biogas for less than $3/MBtu. Around 8% of the global potential could be produced at this cost, a proportion that slightly increases over time.

In total, we estimate that around 130 Mtoe of today's biogas potential could be exploited at a cost equal to or lower than regional gas prices. This is three-and-a-half times the current level of production. China and Southeast Asia are well positioned to increase their levels of biogas production, largely due to relatively high natural gas prices. Biogas production costs are projected to decrease slightly over time while natural gas prices tend to increase. As a result, we estimate that around 300 Mtoe of biogas could be produced globally for less than the respective regional price of natural gas by 2040.

Outlook for biogas

In the Stated Policies Scenario, biogas production grows globally by around 7% per year over the *Outlook* period. Most of the production increase comes from developing economies in Asia where there is widespread availability of relatively low-cost biogas feedstock and where natural gas prices are rising. By 2040, over 70 Mtoe of biogas is consumed globally and a further 80 Mtoe of biogas is upgraded to biomethane. Over 40% of this is in China and India, reflecting their ambitious biogas output goals.

In the Sustainable Development Scenario, there is a more pronounced increase in biogas production: globally nearly 120 Mtoe of biogas is consumed directly in 2040 and a further 200 Mtoe is upgraded to biomethane (Figure 7.15). Nevertheless, less than 40% of the total sustainable technical potential is used globally in 2040.

In advanced economies, especially in countries with well-established gas infrastructure, such as Europe and the United States, around two-thirds of the biogas produced in the Sustainable Development Scenario is upgraded to biomethane. Biomethane can be injected into existing networks to reduce emissions across all end-use sectors that rely on natural gas (see Chapter 13). Most of the biogas that is consumed directly is used in the power sector.

In developing economies, over 210 Mtoe of biogas is produced in 2040 in the Sustainable Development Scenario. It plays an important role as a household fuel, helping to reduce indoor air pollution by displacing the traditional use of solid biomass and contributing significantly to improving access to clean cooking in poor rural areas. By 2030, around 200 million people move away from the traditional use of biomass through biogas; over 100 million people gain access to clean cooking in this way in Africa alone (see Chapter 9).

Using biogas for clean cooking requires around 20 toe gas per 1 000 people each year, and this improvement in welfare for such a large number of people requires around 4 Mtoe of biogas in 2030.

Figure 7.15 ▷ Global biogas demand in the Sustainable Development Scenario

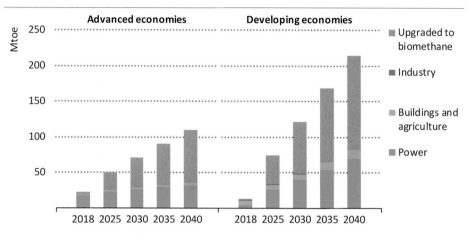

In 2040, biogas is still mostly used as fuel for electricity and heat production, but it is also increasingly used for clean cooking and for the production of biomethane

Note: Biomethane here does not include production through thermal gasification.

In summary, our analysis suggests that there is significant potential for biogas globally, but that most of this requires supportive policies if it is to be exploited. Biogas is relatively expensive to produce, but it offers benefits including GHG emissions savings, improved waste management, enhanced security of supply, improved access to clean cooking facilities, local and rural job creation, and the production of sustainable fertiliser. The case for biogas improves considerably if these non-economic benefits are fully taken into account.

PART B
SPECIAL FOCUS ON AFRICA

The conclusion in 2018 of a new strategic partnership on energy with the African Union marked the start of a new era in IEA-Africa relations and presented the IEA with an opportunity to update and expand its analysis of the energy outlook for the continent.

Energy is a crucial element of Africa's development agenda, and changing global energy dynamics are affecting the way that Africa might provide for its growing energy needs in the coming decades. The falling cost of key renewable technologies, notably solar PV, and the shale revolution in the United States provide a new backdrop for African energy strategies and ambitions. The rise of Africa's cities and its industries are set to have profound implications for energy. Electrification and digitalisation are also opening up new avenues for innovation and growth. The stage is ready for a new wave of dynamism among African policymakers and the business community.

This renewed focus on Africa is a reflection of the continent's increasing importance in global energy affairs and its deepening relationship with the IEA. At the heart of this special focus are analyses at country level that break new ground and give important policy insights for global and Africa energy stakeholders. This analysis has enabled us to produce detailed, comprehensive, data-rich profiles for these countries and draw important implications for Africa and the world.

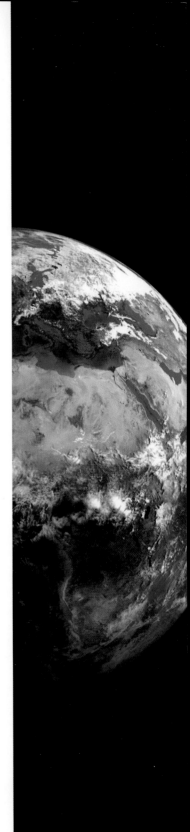

Part B

Special Focus on Africa
Introduction

Five years after a first special report on Africa in the *World Energy Outlook-2014 (WEO-2014)* series, the IEA has updated and expanded its analysis of the energy outlook for the continent in this new report. The renewed focus on Africa in this year's *World Energy Outlook* reflects Africa's increasing importance in global energy affairs and the deepening relationships between African energy decision makers and the IEA. Institutionally, South Africa and Morocco have become IEA Association countries, and the IEA concluded a new strategic partnership on energy with the African Union in 2018.

At the heart of this *Special Focus* report are analyses at country level that break new ground and give important policy insights for global and African energy stakeholders. We carried out quantitative analysis of a number of important factors and used the results to develop modelling for eleven sub-Saharan countries for the first time. This enabled us to produce detailed, comprehensive, data-rich profiles for these countries and draw implications for Africa and the world. The profiles provide much greater granularity than the regional analysis carried out in the *WEO-2014 Special Report*. We draw on these profiles in the analysis and discussion in the chapters that follow, and present the detailed output of our analysis for each country in an easy-to-read graphic format in Chapter 12.

This introduction covers:

- **Context**: Africa's population is growing rapidly. Its economies have the potential to do the same, and their demand for energy is set to grow very fast. Africa is also an important supplier of energy and has seen major recent gas discoveries. At the same time it still faces significant challenges in terms of access to electricity and clean cooking for all.

- **Structure**: This *Special Focus* begins by looking at the energy landscape of Africa today and then moves on to look at a variety of future energy challenges. It concludes with country profiles for eleven key sub-Saharan countries.

- **Focus countries and country profiles**: The report looks at Africa as a whole, but focuses on sub-Saharan Africa, and on the eleven countries selected for detailed country profiles. These profiles reflect modelling capabilities developed for the first time for this report.

- **Scenarios**: We use the Stated Policies Scenario, as elsewhere in the *WEO*, and a new Africa Case developed for this *Special Focus* that reflects Agenda 2063, in which African leaders set out their vision for the future growth and development of the continent.

Context

In 2000, Africa's population of around 820 million accounted for about 13% of the world's 6.1 billion people. In 2018, this share had increased to around 17%, as its population expanded at more than twice the global rate. Africa has the world's fastest growing and youngest population. The last two decades have seen the number of people living in cities increase by 90% and this trend continues over the next two decades. By 2040, an additional 580 million Africans will be living in cities, an amount greater than the entire population of the European Union today, and a pace of urbanisation that is unprecedented.

Despite its large and growing population, Africa accounts for a very small share of global energy sector investment. In 2018, around $100 billion was invested in the energy sector in Africa, or about 5.5% of the global total. Of this, $70 billion was invested in fossil fuels and $13 billion in renewables. Another $13 billion was spent on electricity networks.

Africa accounts for a relatively small, but nonetheless growing share of the world's carbon dioxide (CO_2) emissions. In 2010, the continent accounted for 3.3% of global energy-related CO_2 emissions; by 2018 this share had increased to 3.7% or around 1.2 gigatonnes (Gt) CO_2. North Africa accounted for the largest share of the continent's energy-related CO_2 emissions, with 40% or around 490 million tonnes (Mt) CO_2 and South Africa accounted for 35% (420 Mt CO_2).

Figure B.1 ▷ Africa's share of selected global indicators, 2018

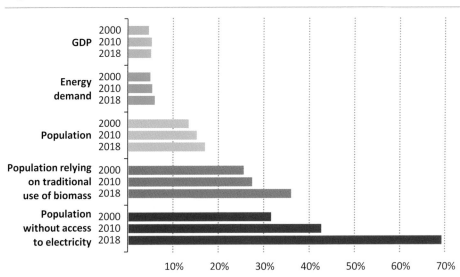

Africa accounts for a low share of the world's energy demand and a high share of the population without access to modern energy services

Africa is among the regions most exposed to the effects of climate change. Its ecosystems already suffer disproportionately from global climate change and future impacts are expected to be substantial. This will have implications for food security, migration and ultimately development. All African countries signed the Paris Agreement and, through their Nationally Determined Contributions (NDCs), they committed to contribute to the global effort to mitigate GHG emissions (IPCC, 2019). The majority of countries that have already enhanced the ambition stated in their NDCs in 2019 are in Africa.

The continent has made progress on metrics such as economic growth, income per capita, educational attainment, access to clean drinking water, child mortality and life expectancy. While the region is also home to a growing share of the world's energy-poor, new renewable energy technologies, innovative digital technologies and finance tools halted the growth in the number of people without access to electricity in 2013, giving new impetus towards achieving the UN Sustainable Development Goal 7 (access to affordable, reliable and sustainable modern energy for all by 2030). In East Africa, countries such as Kenya, Ethiopia and Uganda have made remarkable progress in providing modern energy services to millions over the past five years. Nonetheless, much remains to be done to deliver universal access to electricity, and in particular to expand access to clean cooking, where progress is being outpaced by population growth.

Africa's vast natural resources mean that low-cost clean energy technologies have plenty of potential. Solar is rightly seen as a huge opportunity for Africa, both at utility-scale and off-grid. Deployment is low today, but it is expanding fast. Recent new utility-scale capacity additions include the first phases of the 1.6 gigawatt (GW) solar photovoltaics (PV) park at Benban in Egypt and the 510 MW Noor solar concentrating solar power development in Morocco, both of which are among the world's largest of their kind. In East Africa, Kenya commissioned the 310 MW Lake Turkana Wind Power plant and the 185 MW Olkaria Geothermal Power Plant, both among Africa's largest in their respective technologies. Large-scale projects have successfully been confirmed in other countries and many more are under way in Angola, Ethiopia, South Africa, Senegal, Uganda and Zambia, among others.

There have been a number of natural gas discoveries, most notably in Mozambique and Tanzania but also in Egypt, Mauritania, Senegal and South Africa. These finds accounted for over 40% of global gas discoveries between 2011 and 2018 and the majority have occurred since our last focus on Africa in 2014 (Box B.1). The possible benefits are many: natural gas can provide the continent with an additional source of electricity for baseload and flexibility needs, it can supply feedstock for industrial growth and bring new export opportunities and revenues; and, when combined with the continent's huge renewable energy potential, gas can offer a new pathway towards the provision of modern energy services for all. In addition, many of the minerals essential to modern energy transitions are found in large quantities in Africa, including uranium and strategic metals and minerals such as cobalt, copper and platinum.

Box B.1 ▷ A retrospective assessment: Africa Energy Outlook 2014

The New Policies Scenario in the *Africa Energy Outlook* (IEA, 2014) provided a set of projections that were based on the policies firmly in place or planned at the time (the same principles that apply to this edition's Stated Policies Scenario). When we look back at those we find:

- Electricity access has progressed more rapidly than projected in 2014, as countries have stepped up their policy efforts in the meantime. A combination of lower cost renewable technologies and new business models helped to facilitate progress. In the case of clean cooking, progress was not as strong as projected in 2014.

- Since 2014, accelerated global deployment has made renewable technologies a much more cost-effective option to support Africa's energy and development objectives. The outlook for natural gas is also affected by new discoveries not just in Mozambique and Tanzania (covered in the 2014 report) but also in Egypt, Mauritania and Senegal and South Africa.

- Another consideration is energy prices. The 2014 report was written at a time when oil prices were relatively high and the focus of the analysis was on responsible and transparent management of revenues. Since then, the net income from oil and gas production has been very volatile, with the impacts of low fossil fuel prices amplified by domestic challenges in major producer economies. The combined net income from oil and gas production of the top-ten producing countries in Africa declined by 70% between 2014 and 2016.

Structure

This first part of this *Special Focus on Africa*, Chapter 8, analyses the changing dynamics and energy landscape of Africa today. It sets out the existing economic and demographic architecture, the continent's key energy demand and supply trends, the scale of its energy resources and assesses the factors that are likely to influence the future energy landscape.

The second part, consisting of three chapters, seeks to capture a sense of the continent's diversity by exploring in detail a number of themes, which touch on some of the most fundamental energy challenges facing sub-Saharan Africa. These include the outlook for energy demand, with a focus on urbanisation, industrialisation and access to clean cooking (Chapter 9); access to electricity and reliable power together with investment and financing for power supply (Chapter 10); and natural gas and resource management (Chapter 11). Woven through these themes are insights on gender and other forms of social inequality, which present major impediments for optimising economic and human development in African countries.

The final part of this Africa focus, in Chapter 12, serves two purposes. First, it sets out the implications for Africa and the global energy sector of the analysis contained in the preceding chapters. Second, it presents a comprehensive profile of the energy sector not only in sub-Saharan Africa but in each of our eleven focus countries. These profiles provide a snapshot of the energy sector in each and help identify not only the challenges ahead but also highlight the huge potential of the continent's plentiful natural resources.

Focus countries and country profiles

This special report covers Africa as a whole, but much of the detailed analysis and discussion focusses on sub-Saharan Africa and on the eleven countries highlighted in our country profiles: Angola, Côte d'Ivoire, Democratic Republic of the Congo (DR Congo), Ethiopia, Ghana, Kenya, Mozambique, Nigeria, Senegal, South Africa and Tanzania. These eleven sub-Saharan countries are a diverse group: some are among the world's fastest growing economies while others are among the poorest in the world. Together, they represent three-quarters of sub-Saharan Africa's 2018 gross domestic product (GDP) and energy demand, and two-thirds of population. They also account for the majority of Africans without access to modern energy services. They are also home to a large share of the continent's abundant energy and mineral resources, with the potential not only to transform Africa, but also to make a substantial contribution to the world's energy transitions.

For the purpose of this *Special Focus*, North Africa, South Africa, ten additional sub-Saharan countries and the rest of sub-Saharan Africa have been modelled separately. Each of these countries and groupings was assessed on the basis of a number of factors including: demographics (population, urban population, age profile); economy (size and structure of the economy, GDP per capita, level of foreign direct investment, transport infrastructure); key energy indicators (total final energy consumption, energy consumed by end-use, consumption of energy per capita), scale of power sector (installed generation capacity, renewable energy capacity and potential, level of electricity exports and trade); the level of access to both electricity and clean cooking; fossil fuel resources; and the full range of energy and environment policies and strategies. For ten African countries (except South Africa), this led to the development of detailed modelling capabilities for these countries for the first time.

Understanding the scenarios

This *Special Focus* provides a framework for analysing the outlook for Africa's energy sector, particularly sub-Saharan Africa. It sets out what the future could look like based on different scenarios or pathways, with the aim of providing insights to inform decision making by governments, companies and others concerned with energy. The two scenarios assessed in this *Special Focus* are:

The **Stated Policies Scenario** provides a measured assessment of where today's policy frameworks and announced policies, together with the continued evolution of known technologies, might take the energy sector in sub-Saharan Africa in the coming decades. Given that announced policies are by definition not yet fully reflected in legislation or regulation, the prospects and timing for their full realisation are based upon our assessment of the relevant political, regulatory, market, infrastructural and financial constraints. This scenario does not focus on achieving any particular outcome: it simply looks forward on the basis of announced policy ambitions in various sectors.

The **Africa Case** is built on the premise of Agenda 2063 (Box B.2), and takes into account each country's own vision for economic growth, based on regional economic blueprints and typically incorporating an accelerated industrial expansion. Enhanced economic growth also increases the means to achieve the ambitions included in energy master plans and other policy announcements, so these are more fully realised than in the Stated Policies Scenario (Table B.1). The Africa Case also incorporates key sustainable development goals by 2030, including achieving full electricity and clean cooking access as well as significant reductions in pollution-related premature deaths. As a tangible representation of the Agenda 2063 vision, it presents a pathway to attain inclusive and sustainable economic growth and development.

Box B.2 ▷   Agenda 2063: the future that Africa wants

In 2015, the Heads of State and Governments of the African Union adopted *Agenda 2063*. It sets out a vision for "an integrated, prosperous and peaceful Africa, driven by its own citizens and representing a dynamic force in the international arena". Closely linked to the United Nations Sustainable Development Goals, it is an ambitious vision and one which will require significant political will if its goals are to be realised.

Agenda 2063 builds on previous Pan-African initiatives, but is distinct in many respects: it sets out clear goals, implementation plans and targets alongside elements of accountability; it identifies key flagship programmes as well as monitoring and review mechanisms; and it proposes a clear resource mobilisation strategy. Successful delivery of Agenda 2063 is likely to depend on good governance, transparency and effective intra-African co-ordination, among other things. It will also depend on resources being available to implement it and in particular on the mobilisation of private sector resources.

Agenda 2063 emphasises the implementation of the Grand Inga Dam Project as a key development priority and a means to support regional power pools and help transform the continent from traditional to modern sources of energy. Energy-related targets contained in the framework for the first ten years include increasing access to electricity by at least 50% compared to 2013 levels and increasing the efficiency of household energy use by at least 30% before 2023.

The Africa Case sees an annual average rate of growth in gross domestic product (GDP) of 6% across sub-Saharan Africa, significantly higher than the 4.3% assumed in the Stated Policies Scenario. The population assumptions are held constant across the two scenarios (Table B.2); these are based on the medium variant of the United Nations population projections.

Table B.1 ▷ Real GDP growth assumptions by scenario
(compound average annual growth rate)

Country/Region	2000-2018	2018-2040	
		Stated Policies Scenario	Africa Case
North Africa	3.5%	3.6%	4.8%
South Africa	2.7%	2.5%	3.3%
Other sub-Saharan Africa	5.4%	5.0%	7.3%
Angola	5.8%	3.3%	5.3%
DR Congo	5.3%	5.5%	8.7%
Côte d'Ivoire	3.6%	5.3%	7.9%
Ethiopia	9.0%	6.5%	8.9%
Ghana	6.3%	3.9%	6.3%
Kenya	4.8%	5.9%	9.0%
Mozambique	7.1%	6.0%	8.1%
Nigeria	6.3%	3.4%	5.3%
Senegal	4.6%	6.5%	8.7%
Tanzania	6.5%	5.6%	9.3%
Africa	4.3%	4.3%	6.1%

Note: GDP In PPP terms, $2018.

Table B.2 ▷ Population assumptions

	Total population (million)		2018-2040	
	2018	2040	Delta (million)	CAAGR*
North Africa	196	263	67	1.3%
South Africa	57	71	14	1.0%
Other sub-Saharan Africa	1 034	1 761	728	2.5%
Nigeria	196	329	133	2.4%
Ethiopia	108	173	65	2.2%
DR Congo	84	156	72	2.8%
Tanzania	59	108	49	2.8%
Kenya	51	79	28	2.0%
Angola	31	60	29	3.1%
Mozambique	31	55	24	2.7%
Ghana	29	44	15	1.9%
Côte d'Ivoire	25	42	17	2.4%
Senegal	16	28	12	2.5%
Africa	1 287	2 095	808	2.2%

* CAAGR = compound average annual growth rate.

Chapter 8

Africa today
Viewing Africa through a new lens?

SUMMARY

- The pace of change in Africa's energy sector has quickened, imparting to the continent a growing sense of confidence despite many setbacks. Africa's economy is also on an upward trajectory, with gross domestic product (GDP) likely to rise by around 4% this year. East Africa looks to be the fastest-expanding region today, led by Rwanda, Ethiopia, Kenya, and Tanzania. The way in which the energy sector develops will have a crucial influence on Africa's future.

Figure 8.1 ▷ Selected indicators for Africa, 2000, 2010, 2018

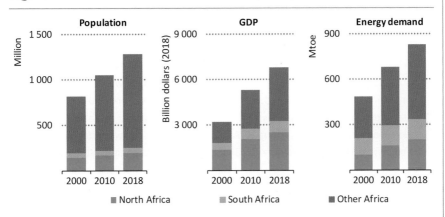

■ North Africa   ■ South Africa   ■ Other Africa

Africa's urban population is expanding fast while energy services and GDP struggle to keep pace

Note: GDP = Gross domestic product in PPP terms, $2018.

- The number of people gaining access to electricity in Africa doubled from 9 million a year between 2000 and 2013 to 20 million people between 2014 and 2018, outpacing population growth. As a result, the number of people without access to electricity, which peaked at 610 million in 2013, declined slowly to around 595 million in 2018. Recent progress has been led by Kenya, Ethiopia and Tanzania, which accounted for more than 50% of those gaining access. However, sub-Saharan Africa's electrification rate of 45% in 2018 remains very low compared with other parts of the world.

- Since 2015, only seven million people have gained access to clean cooking in sub-Saharan Africa, meaning that the number of people without access increased to over 900 million in 2018 as population growth outpaced provision efforts. Progress has been strongest in parts of West Africa such as Côte d'Ivoire and Ghana which

- have promoted liquefied petroleum gas (LPG). However, the problem remains acute and Sub-Saharan Africa is one of the only regions worldwide where the number of people without access to clean cooking continues to increase.

- Energy demand in Africa has been driven by the growing needs of North Africa, Nigeria and South Africa. There are also very strong regional variations. Countries such as Democratic Republic of the Congo (DR Congo), Africa's fourth most populous country, and Mozambique have seen their primary energy demand increase by over 50% between 2010 to 2018, whereas others such as Côte d'Ivoire and Ghana have witnessed only a gradual increase in energy demand (or even a decline).

- With a fifth of the world's population, Africa accounts for only 6% of global energy demand and little more than 3% of electricity demand. Average energy consumption per capita in most African countries is well below the world average and largely comparable to that of India. Bioenergy is the largest source of energy in Africa today, meeting 45% of primary energy demand and over half of final energy consumption.

- Africa has plentiful renewable energy resources and its economic potential is substantially larger than the current and projected power consumption of the continent. Bioenergy, hydropower, solar and wind power account for the bulk of the resources. East Africa also has rich geothermal resources. To date, limited use has been made of this vast potential: Africa has only 50 gigawatts (GW) of renewable capacity, mostly hydropower (36 GW). But this is changing: utility-scale projects have entered service in Egypt, Ethiopia, Kenya, Morocco and South Africa. Meanwhile, mini-grids, micro-grids and solar home systems are anchoring efforts to bring modern energy services and new sources of productive employment to remote populations, facilitated by digital technologies and payment tools.

- The future of natural gas in Africa is at an important juncture. Since 2010, there have been major gas discoveries in every part of the continent: immense finds in East Africa (Mozambique and Tanzania) were followed by more in Egypt, West Africa (Mauritania and Senegal) and South Africa. While Africa accounts for 6% of global gas production today, over 40% of global gas discoveries between 2011 and 2018 were in Africa. These resources offer new opportunities for Africa's energy and industrial development. The prospects for gas, however, hinge upon well-articulated strategies to bring the discoveries into production and build infrastructure to deliver gas to consumers at competitive prices.

- Africa is home to many of the minerals essential to the energy industry, for example, DR Congo accounts for almost two-thirds of global cobalt production. The continent also produces a large share of key minerals such as platinum (cars and fuel cells), chromium (wind turbines) and manganese (batteries), which will play a major role in powering the global energy transitions.

8.1 Context

8.1.1 Economic growth and industrialisation

Africa has experienced relatively low gross domestic product (GDP) growth since 2010, an average of 3.1% per year compared with a global average of 3.5% per year. GDP per capita in the continent is less than a third of the global average and in sub-Saharan Africa as low as a fifth of the global average. Two countries, Nigeria (17%) and South Africa (12%) account for a large portion of Africa's economic activity. Recent growth in some countries has been significantly influenced by their dependence on commodities. The greater this reliance, the more severe the impact of the 2014 commodity price decline; and the greater the fall, the more challenging the recovery (Figure 8.2). While countries such as Ethiopia, Kenya and Rwanda have successfully boosted growth through public investment and a growing services sector (AUC/OECD, 2018), Nigeria is only slowly pulling out of the recession that was triggered by a combination of lower oil prices and production outages associated with conflict (IEA, 2018a).

Figure 8.2 ▷ Annual GDP growth rates in selected African economies

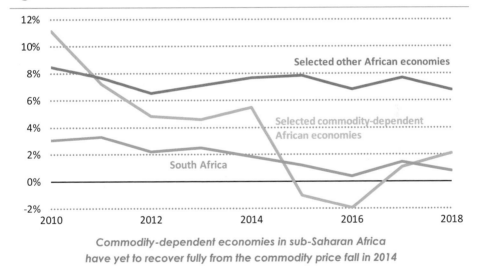

Commodity-dependent economies in sub-Saharan Africa
have yet to recover fully from the commodity price fall in 2014

Note: Selected commodity-dependent economies are Algeria, Angola and Nigeria; Selected other African economies are Ethiopia, Kenya, Rwanda and Senegal.

Despite this, the overall sub-Saharan economy has expanded by more than one-third since 2010, reaching more than $4.3 trillion in 2018. Growth in Nigeria and South Africa has slowed, but GDP elsewhere is now growing at the fastest pace since 2013. GDP growth for the continent as a whole is forecast to accelerate to 4% in 2019, up from an estimated 3.3% in 2018, making it the second fastest-growing region in the world, after Asia. Some countries are growing much faster than this average. In Ghana for example, the International Monetary Fund estimates that GDP will rise by almost 9% in 2019, double the pace of emerging economies as a whole, and well ahead of world growth (IMF, 2019). Ethiopia, Côte d'Ivoire, Rwanda, Senegal and Tanzania all feature on the African Development Bank Group's list of the ten fastest-growing economies for 2018 (AfDB, 2019). Foreign direct investment (FDI) into Africa rose by 11% to $46 billion in 2018, reversing declines in 2016 and 2017 while FDI into sub-Saharan Africa increased by 13% to $32 billion (UNCTAD, 2019).

North Africa and South Africa are relatively industrialised, but other sub-Saharan African countries represent only a small share of global industrial production, and industry is hampered in many countries by unreliable electricity supply and high energy costs. The contribution of different sectors to GDP and employment varies significantly between individual economies, but for sub-Saharan Africa as a whole the relatively low share of employment in industrial sectors stands out. By contrast, services contribute 55% to the economy and a third of employment, while agriculture accounts for only 18% of GDP but well over half of employment (Figure 8.3).

Figure 8.3 ▷ Shares of value added and employment by sector in North Africa and sub-Saharan Africa, 2018

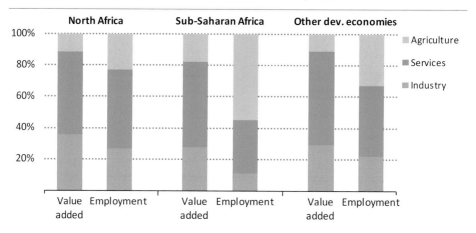

Agriculture accounts for a very large share of employment in sub-Saharan Africa even when compared to other developing economies

Incomes and personal wealth vary greatly across Africa, but sub-Saharan Africa is home to some of the world's poorest people. While global poverty rates have been reduced by more than half since 2000, and good progress is being made in many African countries such as DR Congo, Ethiopia and Nigeria, more than 40% of the population in sub-Saharan Africa continues to live below the poverty line on an income of less than $1.90 a day (UN, 2019a). Sub-Saharan Africa also remains one of the most unequal regions in the world. Half of the twenty most unequal countries in the world (measured by the Gini co-efficient) are in sub-Saharan Africa (UNDP, 2017). Average per capita incomes across sub-Saharan Africa range from over $10 000 in Mauritius and South Africa to less than $500 in the Central

African Republic, Madagascar and Niger. In 2018, average per capita GDP in sub-Saharan Africa was around $4 000 ($2018 at power purchasing parity) compared to about $3 700 in 2010. This represents an average annual increase of around 1% a year over the period. This compares with an average annual GDP per capita increase of 2.3% worldwide and 5.5% in developing Asia over the same period. The small size of the annual increase of per capita GDP in sub-Saharan Africa may seem hard to square with its faster GDP growth rate: the reason for the difference is that the population of sub-Saharan Africa grew rapidly over the period.

8.1.2 Demographics and urbanisation

In 2000, Africa's population of 817 million accounted for just over 13% of the world's 6.1 billion people. By 2018, this share had increased to around 17%, as Africa's population expanded at more than twice the global rate to reach almost 1.3 billion, of which almost 85% or 1.1 billion live in sub-Saharan Africa. Eleven countries account for almost two-thirds of sub-Saharan Africa's population today (Figure 8.4). The average age of the population of Africa is very young: in 2017, the median age was 17 while children under age 15 accounted for 41% of the population and 42% of the population of sub-Saharan Africa (UNDESA, 2019).

Almost 60% of Africa's population lives in rural areas although an increasing share is moving to the expanding urban areas. Africa already has two megacities in sub-Saharan Africa (Kinshasa and Lagos) and another in North Africa (Cairo). There are another five large cities on the continent with a population of between five and ten million each: Alexandria, Dar es Salaam, Johannesburg, Khartoum and Luanda (UNDESA, 2018).[1] Of these, Dar es Salaam and Luanda are likely to become sub-Saharan Africa's next megacities. The implications of an increasingly urban population for the energy sector are profound. In general, urban residents tend to consume more energy than those in rural areas, in large part because of differences in income levels. Smart urban planning and sustainable development offer a huge opportunity to shape patterns of future energy use. However, there are also likely to be major challenges arising from further strains on air quality, housing, transport, public utilities and sanitation (see Chapter 10).

The gender equality landscape in African countries is complex and sometimes contradictory, presenting both challenges and opportunities for the future. Nearly one-in-four households in Africa are headed by a woman: those in Southern Africa are most likely to be headed by a woman, while households in West African countries are least likely to be headed by a woman (Van de Valle, 2015). The percentage of women elected to parliament in many African countries, (e.g. Rwanda, Namibia, South Africa and Senegal) is among the highest in the world (IPU, 2019). Countries such as Mauritius, Seychelles and South Africa have female literacy rates on par with or exceeding those in many developing and emerging

---

[1] Large cities are generally defined as having between five and ten million inhabitants and megacities as having ten million or more inhabitants.

economies. Conversely, sub-Saharan Africa contains nine of the ten countries with the lowest levels of female literacy in the world. The complicated gender terrain in sub-Saharan Africa, evidenced by the existence of high levels of disparity between different groups of women as well as between regions and countries, has important implications for access to energy and for socio-economic development.

Figure 8.4 ▷ Share of urban/rural population in sub-Saharan Africa, 2018

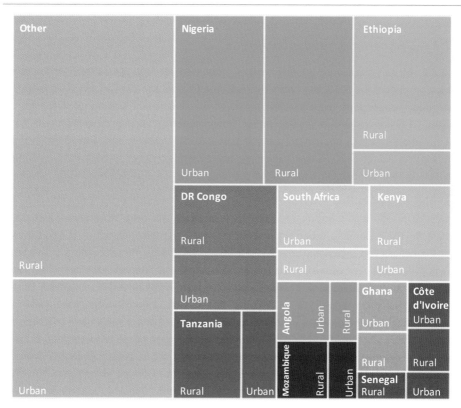

Eleven countries account for almost two-thirds of sub-Saharan Africa's 1.1 billion people today

8.1.3 Infrastructure and investment

Infrastructure is an essential building block for economic development and quality of life, but Africa, especially sub-Saharan Africa, lags behind other developing economies in virtually all aspects of infrastructure quality. Over the past three decades, the level of per capita power generation capacity in sub-Saharan Africa has remained flat, whereas in India and Southeast Asia (which had less generation capacity per capita than sub-Saharan Africa in 1990) it has grown fourfold. Sub-Saharan Africa has made relatively good progress on telecommunications infrastructure, but still compares unfavourably with other developing economies. Paved road network density has remained unchanged over the past three decades despite the growth in population and trade, owing mainly to continued under-investment in road expansion, maintenance and rehabilitation (Figure 8.5).

Figure 8.5 ▷   Infrastructure quality developments in selected regions

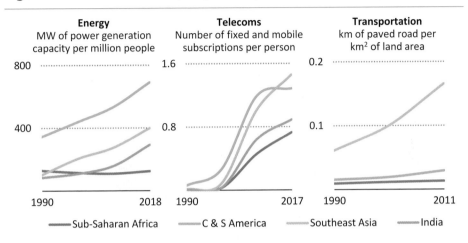

Note: C&S America = Central and South America; MW = megawatt.
Sources: International Telecommunications Union Statistics; World Bank World Development Indicators Database.

Making up the deficit of energy infrastructure in Africa will require a massive ramp-up in investment, but actual spending trends have been moving in the opposite direction. Energy supply investment in sub-Saharan Africa has dropped by over 30% since 2011[2], and oil and gas investments have more than halved because of low oil prices and investor concerns about regulation and security in major producing countries. Power supply investment registered strong growth until 2014, but has since stalled. The one bright spot has been rising investment in solar photovoltaics (PV), which is set to surpass that in hydropower for the first time in 2019, according to early data. Nonetheless, levels of spending still fall significantly short of what would be needed in the Stated Policies Scenario, particularly in the power sector (Figure 8.6).

Attracting capital for oil and gas projects in sub-Saharan Africa has generally been hampered by uncertainties around fiscal and regulatory frameworks - the design of local content rules has been a particular source of contention (Box 8.1). Moreover, difficulties in reaching agreement on contractual terms have often led to reliance in practice on a

---

[2] Energy supply investment includes capital spent on building infrastructure for fuel supply (extraction, processing and transportation of oil, gas, coal and biofuel) and power supply (generation, networks and storage).

handful of large companies that have the capacity to bear the risks. As a result, investment in oil and gas in sub-Saharan Africa has largely been driven by international oil companies. This contrasts with the prevailing trend in many other resource-rich countries where domestic companies, and in particular national oil companies (NOCs) take the lead. In those cases where sub-Saharan countries have established a NOC, they have generally not been effective in accelerating resource developments in the country due to their limited financial capacity and lack of technical expertise in handling complex projects (see Chapter 11). The lack of a competitive service industry is another constraint that has weighed on development costs. The limited attractiveness of the domestic market also means that most spending in oil and gas has been directed at export-oriented projects (e.g. upstream and liquefied natural gas) rather than projects geared towards domestic markets (e.g. gas pipeline, refineries).

Figure 8.6 ▷ Historical energy supply investment and average by sector in sub-Saharan Africa in the Stated Policies Scenario

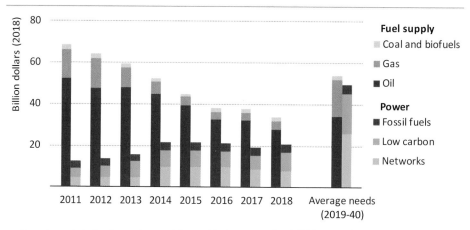

Investments in fuel supply would need to increase by 60% from today's levels to meet needs; investments in power would need to grow by two-and-a-half times through to 2040

Investment in power infrastructure in sub-Saharan Africa has mainly been financed by state budgets with substantial contributions from international donors. Public and international development finance collectively accounted for over 90% of the capital committed to power infrastructure in 2017 (ICA, 2017). While public sources of finance have an important role to play, they are unlikely to be sufficient to address the significant investment gaps that exist, and need to be supplemented by private sector financing. Africa has so far had limited success in mobilising private finance. Between 2013 and the first-half of 2018, power sector investment based on private participation in infrastructure (PPI) models in sub-Saharan Africa amounted to around $4.5 billion per year on average, less than 10% of the annual needs between today and 2040. Most of the region's PPI investment has gone to a handful of countries with South Africa alone accounting for more than half.

Ultimately whether projects can attract financing depends on whether developers and investors believe that they will deliver adequate returns and the timely repayment of debt to lenders. This requires a sound investment framework (e.g. tariff schemes, institutional and regulatory structures) as well as a robust contracting framework (e.g. offtake agreements and financing structures) to manage risks around future cash flows (see Chapter 10). The role of public and development finance is also important: these sources can not only provide necessary capital, but also encourage private sector investment through targeted interventions such as risk sharing, liquidity support and take-out financing. Outside South Africa, public and development finance has not been very effective in catalysing private capital, suggesting that much more needs to be done to plug the investment gaps in power and energy infrastructure (see Chapter 12).

Box 8.1 ▷   Governance and policy frameworks

Good governance is closely correlated with faster growth, higher investment and faster poverty reduction. The World Bank Governance Indicators show that there was little to no progress in institutional quality across sub-Saharan Africa from 2000 to 2015 (World Bank, 2018). Progress was however recorded in many countries on perceived corruption. Côte d'Ivoire and Senegal, for example, are among the countries that improved their position on the Corruption Perceptions Index while, Angola, Nigeria, Botswana, South Africa and Kenya all displayed some promising developments (Transparency International, 2019).

Stable and effective governance and regulatory frameworks are crucial for increasing competition and attracting investments in the energy sector, and weak governance and regulatory frameworks at national and sub-national levels continue to impede performance in the energy sector. A key issue is the need for transparent and responsible management of hydrocarbon revenues (discussed in more detail in Chapter 11).

8.2   Access to modern energy

Access to modern energy is a central pillar of efforts to reduce poverty and support economic growth in sub-Saharan Africa. Modern household energy services have two components: first, access to clean cooking facilities, where progress remains slow, with around 900 million people without access today; second, access to electricity, where there has been strong progress in several countries over the past decade but almost 600 million people in sub-Saharan Africa remain without access today (Box 8.2). Beyond households, gaining access to modern energy services is also essential for businesses, farmers and community buildings.

Box 8.2 ▷ **Defining and tracking household energy access**

The IEA defines a household as having energy access when it has reliable and affordable access to both clean cooking facilities and electricity, which is enough to supply a basic bundle of energy services initially, and with the level of service capable of growing over time (IEA, 2019a).[3] We consider that this basic bundle of electricity services should encompass, at a minimum, several lightbulbs, phone charging, a radio and potentially a fan or television. Access to clean cooking facilities means access to (and primary use of) modern fuels and technologies, including natural gas, liquefied petroleum gas (LPG), electricity, bioethanol and biogas, or improved biomass cookstoves which deliver significant improvements[4] compared with basic biomass cookstoves and three-stone fires traditionally used in some developing countries. This definition of energy access serves as a benchmark to measure progress towards Sustainable Development Goal (SDG) 7.1 and as a benchmark for our forward-looking analysis.

The *World Energy Outlook (WEO)* electricity and clean cooking access databases are updated annually. They contain the most recent country-level data on the share of national, urban and rural households with electricity and clean cooking access for the 2000-18 period. The Access to Electricity Database sources data, where possible, from government-reported values for household electrification.[5] It takes into account connections to the main grid, and where available access through decentralised systems able to supply the basic energy services mentioned above. Despite their development benefits, "pico solar" products, mainly solar lanterns which may include mobile phone chargers, are considered to be below the minimum threshold to count as having access.

Access to electricity is considered to be binary (a household either has or does not have access) as the availability and quality of reported data limits the capacity to describe the level of service, reliability and affordability.[6] Within these limits, this special focus

---

[3] A full description of the *World Energy Outlook* energy access methodology can be found at www.iea.org/energyaccess/methodology.

[4] Most improved cookstoves currently in use have not been found to significantly improve household air pollution and thus are not considered as access to clean cooking. For our projections, only the most improved biomass cookstoves that deliver significant improvements are considered as contributing to energy access.

[5] The IEA Electricity Access Database, based on administrative data reported by ministries, differs from the World Bank Global Electrification Database, which derives estimates from household surveys. The IEA administrative data on electrification provides information from the perspective of supply-side data on utility connections and decentralised systems distributions, which in particular have the advantage of being updated annually. More information on the differences between databases can be found in the *Tracking SDG 7 Report*, which the IEA chaired in 2019, as one of the United Nations-appointed co-custodians of SDG 7 (IEA, IRENA, UNSD, WB, WHO, 2019).

[6] The World Bank's Energy Sector Management Assistance Program is undertaking surveys in several countries to measure energy access according to the "Multi-Tier Framework", a methodology that measures multiple attributes of the supply and use of electricity and cooking fuels; this initiative is helpful to map the current levels of access to affordable, reliable and modern energy, though assessment difficulties are likely to remain on a large scale.

integrates comprehensive analyses on the reliability of power systems, the level of energy demand from households, and the affordability of energy services, which are all important in ensuring that access to modern energy services delivers social and economic benefits (see section 8.3.3, and Chapters 9 and 10).

The Access to Clean Cooking Database comes from IEA analysis based on the Household Energy Database 2018, collected by the World Health Organization, which compiles data on reliance on primary cooking fuels at urban and rural level using national surveys. This is complemented by IEA's *World Energy Balances*, which contain data on residential energy consumption, as well as government sources of data.

8.2.1   Clean cooking

Lack of access to clean cooking remains very acute in sub-Saharan Africa with access increasing only slightly from 15% in 2015 to 17% in 2018 (Figure 8.7). Progress has been registered in a handful of countries: West Africa has made the fastest progress since 2010, with almost 3 million people gaining access each year, followed by East Africa with nearly 1.5 million people per year. The number of people without access exceeded 900 million in 2018 as population growth outpaced efforts to provide access. Sub-Saharan Africa is the only region where the number of those without access continues to rise significantly, highlighting the urgent need for action. Almost 500 000 premature deaths per year are related to household air pollution from the lack of access to clean cooking facilities, with women and children the worst affected. Lack of access to clean fuels is also one of the most significant contributors in low-income countries to women's workloads, and poses a barrier to the economic advancement of women. It leads to women collecting and carrying loads of wood that weigh as much as 25-50 kilogrammes, which can also damage their health (UNEP, 2017).

Forest degradation, sometimes leading to deforestation, is another serious consequence of the unsustainable harvesting of fuelwood, mainly driven by inefficient charcoal production for cities (see Chapter 9). The forest area per capita in the sub-Saharan African region, a rough proxy for biomass potential available for consumption, is declining at an annual average rate of about 3%, almost double the rate seen in other developing regions. Biomass consumption outstrips the sustainable potential by the largest margin in Nigeria and Kenya. It is estimated that 27-34% of wood-fuel harvesting in tropical regions is unsustainable (FAO, 2018). Deforestation and the resultant shortage of fuel affect millions of people, especially women and children, who bear most of the responsibility for gathering firewood and cooking.

The shares of cooking fuels in sub-Saharan Africa (excluding South Africa) have remained relatively stable in recent years (Figure 8.8). Solid biomass – including fuelwood, charcoal and dung – is the most widely used fuel across the region (Box 8.3). Several governments, including Ghana, Cameroon and Kenya, are promoting LPG as a better alternative, largely in

Figure 8.7 ▷ Population without access to clean cooking in Africa, 2018

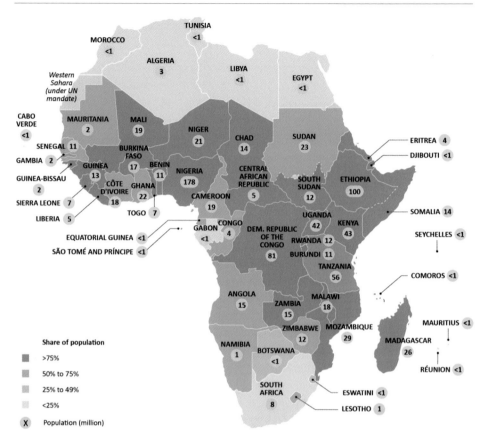

Around 900 million people are without access to clean cooking in Africa; in 32 countries more than 75% of the population are without access to clean cooking

This map is without prejudice to the status of or sovereignty over any territory, to the delimitation of international frontiers and boundaries and to the name of any territory, city or area.

Sources: IEA analysis; World Health Organization (WHO) Household Energy Database.

urban areas. Ghana has been promoting LPG since 1989 and 24% of the population relied on LPG in 2018; as of December 2017, the government had distributed LPG cookstoves to 150 000 households in 108 districts under the LPG Promotion Programme launched in 2017. It intends to distribute them in all 217 districts of Ghana by 2020 (Asante, et al., 2018). In other countries, for example Nigeria, LPG uptake primarily displaces kerosene. Clean cooking has only increased by 0.7 percentage point since 2013 in rural sub-Saharan Africa, in part because supply chains for cleaner fuels lack the necessary scale to reach many rural communities.

Figure 8.8 ▷ **Main fuels used by households for cooking, 2018**

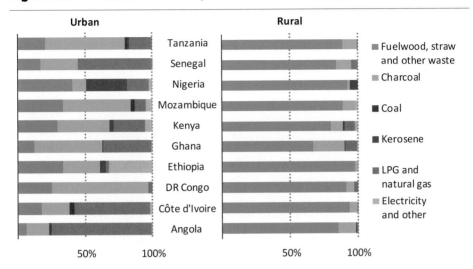

Use of clean cooking fuels such as LPG continues to increase in urban areas, but reliance on traditional use of biomass still dominates in rural areas

Sources: IFA analysis; WHO Household Energy Database.

Box 8.3 ▷ **Traditional and modern uses of biomass**

The various forms of bioenergy differ in terms of sustainability. Bioenergy feedstocks include different products and by-products from the agriculture, forestry and waste sectors (e.g. wood, charcoal, sugarcane, palm oil, animal waste) and there are many ways to use them to produce energy (heat, electricity and fuels).

In Africa, as well as developing Asia, solid biomass remains the largest source of energy used by households (in energy-equivalent terms) and is often burned as fuel in a traditional manner in inefficient and polluting cookstoves, using very basic technologies often with no chimney or one that operates poorly. This so-called "traditional use" of solid biomass is not sustainable and is associated with a range of damaging impacts to health and well-being. The volumes concerned are generally excluded when presenting shares of energy from renewable sources.

Solid biomass can also be used for cooking and heating in more advanced, efficient and less polluting stoves. It can likewise be used as a fuel in combined heat and power plants or transformed into processed solid biomass (pellets), liquid biofuels or biogas. These are classified as modern uses of bioenergy.

Bioenergy has the potential to contribute to the decarbonisation of the power, heat and transport sectors, bringing wider benefits in terms of rural development and diversification of energy supply. There are a number of potential concerns regarding sustainability that have to be considered when planning to use biomass, however, including deforestation, loss of biodiversity, lifecycle greenhouse-gas (GHG) emissions, land-use changes and air pollution linked to combustion. It is thus important that the potential benefits of using bioenergy are balanced against the sustainability considerations that are unique to each bioenergy supply chain application (see Chapter 9).

Several programmes support the diffusion of improved biomass cookstoves. Nonetheless, extensive analysis conducted by the IEA for *Energy Access Outlook 2017*, in collaboration with the International Institute for Applied Systems Analysis, showed that such programmes had limited success (IEA, 2017a). Improvements in pollutant levels from improved biomass cookstoves were often overstated, with virtually no biomass cookstoves on the market meeting WHO standards for exposure to household air pollution.

Conversely, while their reach has been limited to date, alternative biomass-based cooking fuels (such as bioethanol, biomass pellets, briquettes and biogas) are increasingly considered as viable alternatives to the unsustainable use of biomass. Where infrastructure and production can be efficiently developed, bioethanol in particular could prove to be not only safer but also cheaper than charcoal or kerosene. KOKO Networks, a company that focuses on promoting liquid bioethanol as a clean cooking fuel, recently launched 700 distribution points across Nairobi following a successful pilot project. Government support however will be essential to support production and distribution in many areas, especially in rural areas. In Ethiopia, for example, following initiatives such as Project Gaia, the government developed a National Biofuels Policy; promoting ethanol both for stoves and for blending with gasoline as a transport fuel, and production of ethanol now stands at around 40 million litres per year.

Very efficient electric cooking solutions are meanwhile increasing the attractiveness of electric cooking options (Couture and Jacobs, 2019). Increasing the efficiency of electric cooking could help merge initiatives on access to electricity and clean cooking by facilitating the integration of very efficient cooking appliances such as pressure- and slow-cookers in decentralised electric systems.

While clean cooking fuels and technologies are now more available, consumer awareness, accessibility and affordability remain significant challenges. The provision of clean cooking solutions does not guarantee that rural and urban communities will stop using traditional cooking methods. In Kenya, while only 26% of households said that charcoal was their primary cooking fuel, almost 70% were using it some of the time (Dalberg, 2018); and in an experiment testing several improved biomass cookstoves solutions some rural Kenyan households said that, although many of the proposed cookstoves allowed faster and more efficient cooking, they were much less flexible and adapted to their needs than traditional three-stone fireplaces (Pilishvili et al., 2016). Many poor and rural recipients of clean cookstove programmes thus continue to use traditional fuels and solutions for socio-cultural, economic and pragmatic reasons. Programmes to replace traditional but unsustainable fuel use are likely to succeed only if they are able to take account of these barriers to adoption in their design.

8.2.2 Electricity

More than two-thirds of people without access to electricity in the world today live in sub-Saharan Africa. North Africa reached almost universal access to electricity by 2018, but the electrification rate in sub-Saharan Africa was 45% in that year. Electrification levels in sub-Saharan Africa remain very low compared to the levels in other developing parts of the world, most notably the 94% rate reached on average across developing countries in Asia. Lack of electricity often obliges households, small businesses and community services that can afford it to use inefficient, polluting and expensive alternative solutions for essential services.

Despite the comparatively low access rate, sub-Saharan Africa has made progress with the pace of electrification accelerating over the past five years. The number of people gaining access to electricity for the first time more than doubled from 9 million a year between 2000 and 2013 to more than 20 million a year between 2014 and 2018, outpacing population growth for the first time. As a result, the number of people without access to electricity in sub-Saharan Africa peaked at 610 million in 2013, before slowly declining to around 595 million in 2018 (Figure 8.9). The region now faces a dual challenge: how to provide access to the 600 million currently deprived while at the same time reaching the millions born every year in areas without access to electricity.

Figure 8.9 ▷ Electricity access by country in sub-Saharan Africa, 2000-2018

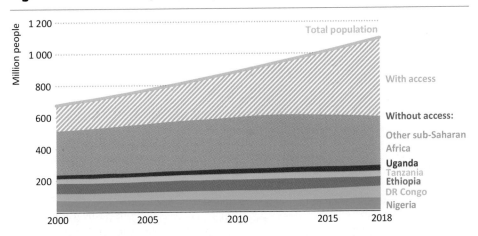

Population without electricity access has plateaued since 2013 thanks to the acceleration of connections; almost 50% of those without access live in five countries

About half of the sub-Saharan African population without access to electricity live in five countries: Nigeria, DR Congo, Ethiopia, Tanzania and Uganda (Figure 8.10). Conversely, Ethiopia, Tanzania and Kenya connected the highest number of people between 2014 and 2018, with these three countries accounting for more than 50% of those gaining access.

Figure 8.10 ▷ Population without access to electricity by country in Africa, 2018

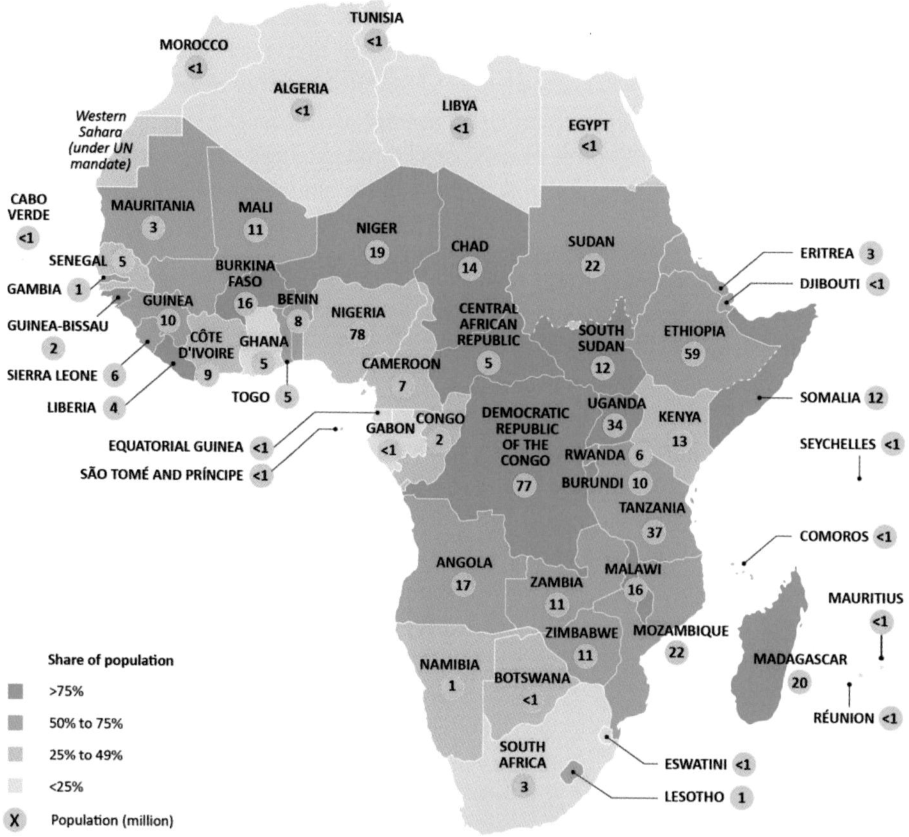

In sub-Saharan Africa 55% of people lack access to electricity; in thirteen countries, more than three-quarters of the population do not have access to electricity

This map is without prejudice to the status of or sovereignty over any territory, to the delimitation of international frontiers and boundaries and to the name of any territory, city or area.

The energy challenges facing households vary significantly across Africa. In urban areas, on average, almost three-quarters of households have access to electricity, whereas in rural areas this figure falls to one-quarter. In remote rural areas and small cities not connected to a grid, finding affordable off-grid solutions and business models is key. But there are also many people living in informal settlements, with grid infrastructure nearby, that are not connected at all, or are connected illegally to the distribution grid, resulting in a revenue loss for utilities as well as exposure to hazards such as fires and risk of electrocution for those with illegal connections (see Spotlight on communities that live "under-the-grid" but without electricity). Getting or formalising grid access is often complicated not only by the high upfront connection costs for poor households, but also by the quality or absence of local distribution infrastructure. For households already connected, located mainly in urban areas, strengthening the reliability of supply from the grid and the affordability of electricity remains the priority (see section 8.3.3).

East Africa stands out as a beacon of progress. It has more than quadrupled the increase in its electrification rate, going from an increase of around one percentage point per year between 2000 and 2013 to more than four percentage points per year from 2014 to 2018. It contains three strong performing countries in terms of electricity access rate progression: Kenya, Rwanda and Ethiopia (Figure 8.11). Kenya has performed best in recent years, with its access rate going from 25% in 2013 to 75% in 2018. Progress in Kenya is attributable to a combination of factors: a strong grid connection push through the Last Mile Connectivity Project; continuous support by government for decentralised systems expressed through exemption from import and value-added taxes for solar products and the adoption of international standards; and the development of a mature mobile payment infrastructure that enabled innovative business models and payment mechanisms to emerge. These factors allowed the country to increase grid connections by almost one million households per year (or more than five million people), and to provide more than 700 000 households with access to electricity through decentralised systems by 2018.

Figure 8.11 ▷ Electricity access progress in sub-Saharan Africa

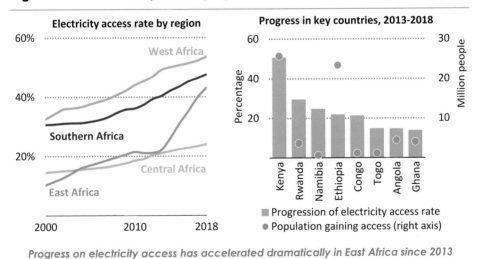

Progress on electricity access has accelerated dramatically in East Africa since 2013

Progress has been much slower in Central Africa, but there are brighter signs in both West and Southern Africa, which achieved 53% and 48% access rates respectively in 2018. Early progress in Ghana demonstrated the need for an integrated approach that takes into account all possible solutions to achieve universal access. In 1991, the Ghanaian government developed the National Electrification Scheme, a master plan containing six five-year implementation phases to reach universal access to electricity by 2030 which was drawn up after a two-year National Electrification Planning Study. This scheme supported the use of main grid, mini-grid and renewables stand-alone systems according to which was most appropriate for a given area. Under its Self-Help Electrification Scheme, clearly stated contributions or commitments were expected from both the government and communities seeking to get connected to grid electricity. By 2018, 84% of the population had access to electricity, up from 45% in 2000.

Lessons from other countries confirm the need for strong government leadership; for adequate planning based on precise analyses of the situation; and for clear allocation of responsibilities at the national and local levels in order to be able to achieve steady and effective progress. In Morocco, for example, the national utility (ONEE) increased the rural electrification rate from 18% in 1995 to 97% in 2009. It implemented a utility-led model that focused on grid extension for 95% of households and that used solar home systems to provide access to electricity for those in isolated or dispersed areas on the basis of a fee-for-service model. In India, a strong push from the government resulted in almost 100 million persons gaining access to electricity in 2018. The Saubhagya Scheme, which ran from October 2017 to March 2019, connected 26 million households to electricity for free; almost 99% of these connections were realised through the main grid, while mini-grids and solar home systems helped reach the remaining 1%, usually in remote areas.

While Morocco and India have provided access to electricity primarily through grid connections, the geography, demography and level of infrastructure provision in sub-Saharan Africa points towards the need for strategies specific to each country that integrate centralised and decentralised solutions to reach universal access. As such, governments in sub-Saharan Africa are increasingly allowing for flexibility in their policy design. In Nigeria, the energy ministry has developed a set of policies that cover a wide range of renewable systems as part of their efforts to reach remote populations across the country. The Rural Electrification Agency is currently implementing a large-scale strategy including energy service company-led and utility-led models to accelerate the rate of electrification through grid extension and green mini-grids, and is targeting market clusters, manufacturing centres, schools, universities and hospitals, for electrification using solar PV and hybrid solar PV-diesel systems.

The majority of progress over the past decade has been made as a result of grid connections[7], but the balance has been shifting. Provision of access by means of decentralised solutions has increased considerably since the IEA's *Energy Access Outlook* was published in 2017 (IEA, 2017a). Around 15 million people are now connected to mini-grids in Africa (ESMAP, 2019), while the number of people gaining access through solar

---

[7] Connections to the grid have been both formal and informal: in Côte d'Ivoire and Ghana, for example, one single formal metered connection can legally serve more than one household.

home systems in sub-Saharan Africa increased from two million in 2016 (IEA, 2017a) to almost five million in 2018 (IEA analysis based on sales data provided by the Global Off-grid Lighting Association). This route to energy access has been concentrated in a few countries: Kenya, Tanzania and Ethiopia accounted for almost 50% of new connections in 2018. In Ethiopia, 32% of rural households are connected through solar home systems (Padam et al., 2018) and in Rwanda around 15%. The market for solar home systems is largely made up of systems below 50 watts which provide access to energy services such as energy-efficient televisions and cooling fans.

The digitalisation of communication and financial services has been critical to the development of mini-grids and solar home systems markets. In some countries in sub-Saharan Africa, the widespread availability of mobile phones, mobile money accounts, and associated telecommunication and payment infrastructures have helped the development of a wide array of energy services (IEA, 2017b). Solar home system providers are offering customers affordable payment plans over several months or years, often with an initial deposit followed by daily payments that cost less than customers currently spend on kerosene (see Chapter 10). Mobile networks enable direct communication with customers and remote control of devices, enabling solar home systems to be disabled when the customer fails to pay. By means of such a scheme, the company Fenix International Inc. has brought affordable solar power to over 500 000 homes in several regions of sub-Saharan Africa with its ReadyPaySolar Systems (Fenix, 2019).

The relationship between electricity access and priorities such as local development and human capital is an important element of the United Nation's 2030 Agenda for Sustainable Development. The vast majority of rural households in Africa rely on agriculture, and integrating agricultural needs such as irrigation, agro-processing and storage into the design of electricity access business models and technologies can have a very positive impact. Cold storage powered by renewable energy supply, for example, could help reduce post-harvest losses, which are estimated at between 20% and 50% of food produced (depending on the food) in sub-Saharan Africa. Electricity can also play an important role in improving agricultural productivity through irrigation, as several successful examples of stand-alone solar water pumps show, provided that policy makers also tackle wasteful irrigation practices.

The absence of electricity access, or access only to intermittent supply, also deeply impacts the quality of services available to the population. In 2016 in sub-Saharan Africa, around half of lower secondary schools and 57% of upper secondary schools had no access to electricity (UNICEF Institute for Statistics, 2019). Moreover, in 27 sub-Saharan African countries, close to 60% of health centre facilities have no access to reliable electricity (Cronk and Bartram, 2018). Access to electricity is essential to a proper provision of essential services: in health centres, for instance, it supports the use of efficient modern equipment, the preservation of vaccines and medicines, and the ability conduct emergency medical procedures, for instance during childbirth.

SPOTLIGHT

What approaches can help communities that live "under-the-grid" but without electricity?

More than half of the urban population in African countries lives in informal settlements often lacking access to formal electrification services (Tusting et al., 2019). Furthermore, at least 110 million of Africa's 600 million people without electricity access live in informal urban settlements close to or directly under a grid (GTM Research, 2017). A 2017 World Bank study on infrastructure in Africa discovered that connection rates for populations living under-the-grid is lower than 50% in most countries in sub-Saharan Africa, with a few exceptions such as South Africa, Nigeria, Gabon and Cameroon. Depending on the data source, estimates for under-the-grid populations without access to legal electricity in other African countries range from 61% to 78% (World Bank, 2017a).

A few studies provide insights into the consumption of poor urban customers and the reasons why utilities are unwilling or unable to serve them despite the immense commercial opportunity they represent (Baruah B. , Energy Services for the Urban Poor: NGO Participation in Slum Electrification in India, 2010). The price of grid connection is one of the major barriers to connection. Typically, people who live in urban informal settlements are poor, with low incomes and low power demands, and connection prices are frequently unaffordable. Before the Last Mile Connectivity Project, the price of a connection in Kenya was around $400 per household (Lee, Miguel and Wolfram, 2016), nearly one-third the annual average per capita income and over three-times the mean of the willingness to pay of surveyed Kenyans. In Nigeria, 62% of under-the-grid households cite high connection costs as the major reason for not being connected to the grid (GTM, 2017). The Centre for Global Development estimates that there may be up to 95 million people living in under-the-grid areas in Nigeria, Kenya, Tanzania, Ghana and Liberia.

There is an urgent need for innovative approaches to provide affordable legal electricity to those living under-the-grid in African countries. A look at promising practices for providing legal access to electricity to urban poor consumers offers some guidance (Shrivastava, 2017). In 2009, Tata Power Delhi Distribution Limited (TPDDL) in India focused on connecting a particular segment of its customer base living in impoverished neighbourhoods. The most significant interventions introduced by TPDDL included reducing new connection charges to INR 350 (about $7 at the time), offering an affordable 24-month payment plan, waiving outstanding dues, reducing requirements for proof of identification and residency, relaxing commercial formalities on land rights and promoting insurance offerings for those with metered connections. Between 2010 and 2015, TPDDL's unique model connected 175 000 new rate-paying customers in 217 impoverished neighbourhoods near New Delhi. In the process, the utility doubled its customer base and increased its revenues fourfold.

> Many of the features TPDDL adopted were replicated in Jamaica, Brazil, South Africa, Philippines and Kenya (in Kenya it was estimated in 2013 that more than three-quarters of off-grid homes are located within 1.2 kilometres of a power line) (Wolfram, 2013). In 2015, Kenya Power & Lighting Co. (KPLC) discovered through a survey that customer willingness to pay for an electricity connection upfront was only 57%. KPLC responded by instituting a subsidised connection fee of around $170 through the Last Mile Connectivity Project and an instalment-based payment plan that led to a thirty-fold increase in legal electricity connections in impoverished neighbourhoods in one year.

8.2.3   Affordability: energy prices and fossil fuel subsidies

Affordability remains a challenge when it comes to providing households with access and with modern energy services that they can use. The two central elements here are the cost of being connected and equipped (connection to the grid, acquisition of the decentralised system or the stove, appliances); and the cost of the energy used (i.e. electricity supply and cooking fuel refills). Making access to electricity and clean cooking affordable requires an understanding of the current payments made by households for similar services and of the structure of their income. For example, many households in rural areas relying on traditional use of solid biomass use a stove and a fuel which do not require significant capital outlay. As noted above, programmes which have proved to work on a large scale have had to respond to household concerns about capital costs either by subsidising the costs or by spreading payments over time to reduce the required capital outlay.

On the basis of electricity prices in several countries, it appears that paying for electricity to power a few basic appliances (four lightbulbs, a fan, a mobile phone charger and a television) would represent around 10% of the average income of the bottom 40% poorest households (Figure 8.12), while in Mozambique and Togo it would represent more than 15%. As a result of the high cost of power relative to income, electricity consumption rates per household in many sub-Saharan African countries are among the lowest in the world. The average household in most sub-Saharan Africa consumes less than 1 000 kilowatt-hours (kWh) of electricity each year, less than one-seventh of the average consumption of households in advanced economies. Other than South Africa, it is only in Angola that average household consumption exceeds 2 500 kWh, partly because the government subsidises tariffs.

There are actions that governments can take to improve affordability. Cross-subsidy schemes could help lower the costs of electricity for the poorest households. So could the promotion of highly efficient appliances. While such appliances tend to have higher upfront costs, we estimate that they could cut the proportion of income spent on electricity bills. Incentives could be provided to poorer households to encourage the purchase of more efficient appliances instead of or in conjunction with subsidies or cross-subsidies to reduce the price of their electricity. Digital payments could facilitate such schemes: around 90% of the 147 000 televisions sold by in the second-half of 2018 sold by solar home systems companies were through pay-as-you-go mechanisms, with more than 100 000 sold in East Africa (GOGLA, 2019).

Figure 8.12 ▷ Electricity expenses required to power basic appliances as a percentage of household revenues for the poorest 40%

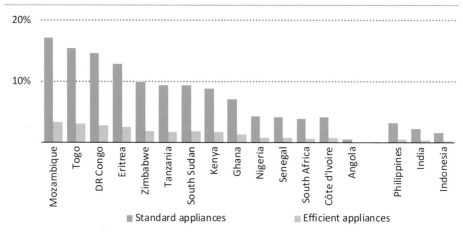

Affordability of electricity remains an issue for many people in African countries. Efficient appliances can help keep costs down

Notes: Electricity consumption is based on a basic bundle of energy services, equating to around 500 kWh per household annually with standard appliances and 100 kWh with highly efficient appliances. This delivers four lightbulbs operating four hours per day, a mobile phone charger, a fan operating three hours per day and a television operating two hours per day. The household revenue is the average gross national income per household for the bottom 40% of households and is computed using the World Development Indicators.

Sources: IEA analysis using World Bank World Development Indicators in some cases.

End-user energy prices vary significantly across countries in Africa, and reflect differences in domestic energy resources, levels of energy access, subsidies and taxes. Retail prices for road transport (gasoline and diesel), for example, are often higher than the world average (Figure 8.13) There are however exceptions: some major hydrocarbon exporting countries supply fuels to their domestic markets at prices lower than those in international markets. Some countries abstain from energy consumption subsidies in order to focus on other policy priorities. Instead of supporting gasoline and diesel prices, for example, Ghana prioritises subsidising kerosene and LPG as part of a strategy to promote switching away from the harmful and unsustainable use of solid biomass.

The interaction of subsidy policies with energy access is a challenge for many sub-Saharan African countries, raising questions about fiscal priorities and about how best to improve access to electricity and achieve sustainable development goals. Consumption subsidies for fossil fuels may once have seemed necessary for development goals, but renewables are increasingly cost competitive with other forms of generation, and many countries are now looking instead at an expansion of low-carbon electricity provision, both via centralised grids and on a decentralised basis (which avoid the costs of transmission and associated losses as well as incurring lower costs for distribution). The situation is different for clean cooking, where some of the viable alternatives to solid biomass are fossil fuels, in particular

LPG, but examples in India and in some African countries show how subsidies can quite effectively be targeted at specific sectors of the population, or limited to a certain number of LPG cylinders per month.

Figure 8.13 ▷ Gasoline end-user prices in selected African countries, 2018

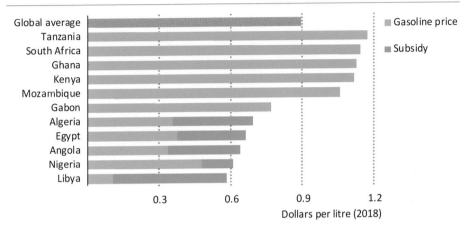

Gasoline end-user prices vary substantially in African countries

For oil-exporting countries in Africa, many of which subsidise fossil fuel consumption, lower prices since 2014 have created strong pressure for pricing reforms to improve fiscal balances and to diversify economies which are highly dependent on hydrocarbons. However, the reform process remains unfinished business in a number of countries. Despite being a net exporter of oil, Nigeria imports most of the oil products consumed in the country (see section 8.3.3) and continues to supply them at subsidised prices; we estimated the value of these subsidies in 2018 at $2.9 billion section. Fossil fuel consumption subsidies are much more prevalent in North Africa, in particular in Egypt (with an estimated consumption subsidy bill of $27 billion in 2018), Algeria ($17 billion) and Libya ($5 billion).

8.3   Energy trends in Africa today

8.3.1   Energy demand

Primary energy demand

In recent decades, African energy demand has been driven by the growing needs of North Africa, Nigeria and South Africa. In 2018, primary energy demand in Africa was more than 830 million tonnes of oil equivalent (Mtoe): North Africa (24%), Nigeria (19%), and South Africa (16%) together accounted for almost 60% of this despite making up only 35% of the population. Average energy consumption per person in most African countries is well below the world average of around 2 tonnes of oil equivalent (toe) per capita and is broadly comparable to India's average of 0.7 toe/capita. In 2018, per capita consumption in sub-Saharan Africa was highest in South Africa at 2.3 toe/capita and in Nigeria at 0.8 toe/capita (Figure 8.14). Most other sub-Saharan African countries have per capita consumption of around 0.4 toe/capita and in most a large part of it consists of the relatively inefficient use of solid biomass.

Figure 8.14 ▷ Energy consumption per capita and population in selected sub-Saharan African countries, 2018

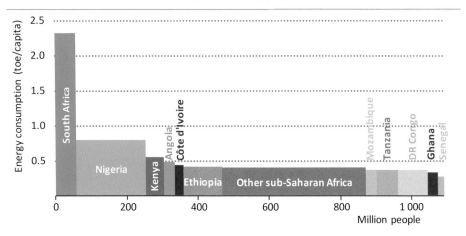

Excluding South Africa, per capita consumption in sub-Saharan Africa is 65% below the average for developing economies

Figure 8.15 ▷ Primary energy demand and GDP annual growth in selected sub-Saharan African countries, 2010-2018

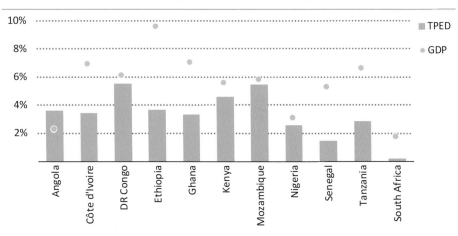

There is wide disparity between annual GDP growth and primary energy demand in many sub-Saharan African countries

Note: TPED = total primary energy demand.

The rate of growth in energy demand in sub-Saharan Africa has slightly slowed in recent years and remains lower compared to GDP growth (Figure 8.15). Between 2000 and 2010, energy demand increased at an annual average rate of 3%, but this slowed to 2.5% from 2010 to 2018, with very marked variations. Countries such as the DR Congo (Africa's fourth most populous country) saw their primary energy demand more than double between 2000 and 2018, whereas others such as Côte d'Ivoire, Ghana and Mozambique have witnessed an increase in demand of around half. The smaller increase in demand does not mean energy services didn't grow at the same rate: in the case of Côte d'Ivoire, the push towards LPG for cooking has resulted in a decline in solid biomass use, and this has produced large efficiency gains.

Traditional biomass is used mostly for cooking in Africa, but is also used in industry. It is by far the most widely used energy source across Africa, with the exception of North Africa, where oil and gas dominate, and South Africa, where the energy mix is coal-heavy (Figure 8.16). In sub-Saharan Africa, bioenergy's share in the overall energy mix has barely changed over the last 25 years, and it continues to dominate the primary energy mix, accounting for 60% of total energy use in the region (if South Africa is excluded, this share increases to almost three-quarters). There is no other region in the world that relies so heavily on bioenergy.

Figure 8.16 ▷ Total primary energy demand by fuel for selected African regions, 2018

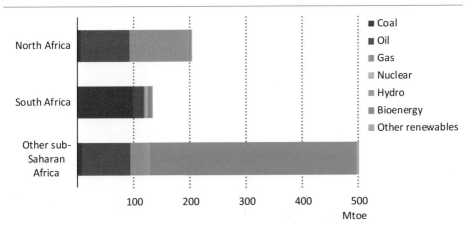

With the exception of South Africa, the sub-Saharan Africa energy mix is dominated by solid biomass and oil

Fossil fuels represent almost 40% of the overall energy mix in sub-Saharan Africa and more than half of the African energy mix. Oil demand stands at almost four million barrels per day (mb/d). The transport sector accounts for most oil use (60%), but diesel is also consumed for back-up generators, kerosene or LPG within households for lighting and cooking, and a variety of oil products are used by industry (Table 8.1). Natural gas overtook coal as the third fuel in the African energy mix in 2015. Today, natural gas accounts for 16% of that mix, with nearly 160 billion cubic metres (bcm) consumed each year: almost 80% of this is consumed in North Africa and over 10% in Nigeria. Coal now accounts for 13% of the primary energy mix (compared with around a quarter globally), with consumption of almost 160 Mtce. South Africa accounts for the overwhelming majority of the continent's coal consumption, where it is used for power generation, industrial processes, transport (after coal-to-liquid conversion), and household heating.

Table 8.1 ▷ Total final consumption by fuel and sector in sub-Saharan Africa, 2018 (Mtoe)

	Industry	Transport	Residential	Other
Coal	12	0	5	5
Oil	9	69	5	13
Gas	9	0	0	0
Electricity	17	0	10	7
Bioenergy	18	0	281	13
Other renewables	-	-	0	0
Total	65	69	301	38
Share of total final consumption	*14%*	*15%*	*64%*	*8%*

Households

The affordability of basic services is an important concern in many African countries, where there are very low levels of appliance and vehicle ownership (Table 8.2). There are important disparities between urban and rural areas, reflecting their different levels of income (see section 8.2.3). There are also major inequalities within urban and peri-urban areas, where many people live in informal settlements under poor conditions as measured in terms of access to energy, sanitation and water services. While the situation has improved slowly over the last ten years, many households still lack appliances that could improve their quality of life, such as a fan (or even air conditioning) and a refrigerator: ownership levels of these appliances are far below the average of developing countries.

In many parts of Africa, ownership of a car remains a luxury, while ownership of two/three-wheelers is comparatively more common. The number of passenger light-duty vehicles is increasing in many countries as incomes rise, but the efficiency of the fleet is low as many are older vehicles imported second-hand from Europe and Asia (although some countries, for example Angola and Mozambique, restrict the importation of older vehicles). Public transport is also less developed in many places though it could play a pivotal role in boosting economic and social welfare in one of the world's most rapidly urbanising region. Rail networks are scarce, and many were originally built to meet the needs of extractive industries rather than to provide passenger services. In many parts of the continent, households rely on buses and minibuses to travel within or between cities. Providing safer and faster alternatives for transporting large numbers of people would be the first among the many benefits of investing in mass transport. There are a number of success stories in the region to inspire the further development of public transport (Box 8.4).

Table 8.2 ▷ Household size and average household ownership of key items in rural and urban areas in Africa

	Household occupancy (number of people)		Air conditioner (ownership per 1 000 households)		Car (ownership per 1 000 households)		Two/three-wheeler (ownership per 1 000 households)	
	Urban	Rural	Urban	Rural	Urban	Rural	Urban	Rural
Sub-Saharan Africa	3.8	5.5	68	11	125	21	129	96
Angola	3.7	5.0	164	22	171	12	162	202
Côte d'Ivoire	4.0	5.6	25	10	64	8	161	220
DR Congo	4.3	5.9	9	8	43	1	79	38
Ethiopia	3.2	4.8	13	9	33	2	25	8
Ghana	3.1	4.1	119	33	126	44	91	132
Kenya	3.4	4.4	20	3	109	29	78	107
Mozambique	3.6	4.9	9	8	108	12	111	74
Nigeria	3.9	5.9	92	14	205	64	304	356
Senegal	6.4	11.3	19	9	25	14	112	104
Tanzania	3.8	5.5	16	9	59	14	115	111
South Africa	3.1	3.9	107	34	343	168	31	9
Other	4.1	6.0	62	8	42	1	77	37
North Africa	3.7	5.5	235	58	147	57	60	106

Box 8.4 ▷ A transport success story: Bus Rapid Transit in Dar es Salaam

Dar es Salaam, Tanzania is growing at an unprecedented rate and is projected to become a megacity by 2035.[8] Most of the expansion has emerged on the edges of the city, compelling many commuters to travel very long distances to work. After years of struggling with an outdated and overcrowded transportation system that was extremely time-consuming to use, the city has introduced a new bus rapid transit (BRT) system which carries 200 000 passengers a day and has cut average travel times from the city centre to the terminus from two hours to 45 minutes each way. Recognised by the Institute for Transportation and Development (ITDP) as Africa's only "gold standard" BRT rating, this high-quality transit system of dedicated bus lanes offering fast, comfortable and cost-effective services at metro-level capacities is a potential model for other cities.

A similar initiative in Johannesburg in South Africa is called the Rea Vaya or "We are going" bus system. It runs on low-sulfur diesel and operates on predetermined routes in dedicated bus lanes, significantly cutting the time spent travelling through the congested streets. The system is reported by the World Resources Institute to have saved the country around $900 million so far by reducing travel time, improving road safety and reducing carbon emissions.

---

[8] A megacity is defined by the United Nations as a metropolitan area with a total population of more than 10 million people.

Households in Africa generally have a low level of energy consumption (Figure 8.17). Nonetheless, their overall energy use (including passenger transport needs) accounts for around two-thirds of final energy consumption. The share of households in final consumption is even higher in sub-Saharan Africa, where it reaches more than 70%, largely because of the extent of their reliance on inefficient solid biomass for cooking and poor quality cookstoves.

Figure 8.17 ▷ Urban and rural household energy consumption per capita and by fuel for selected African countries, 2018

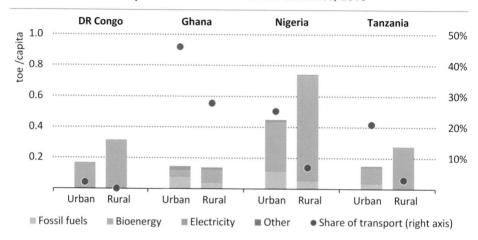

■ Fossil fuels   ■ Bioenergy   ■ Electricity   ■ Other   ● Share of transport (right axis)

Average household energy consumption varies between urban and rural and across countries, as does the fuel mix, though generally with a high share of bioenergy

Productive uses

Productive uses, including industry, agriculture and services, account for around a quarter of total final consumption of energy in Africa. Industry employs only 13% of the workforce and generates only a third of GDP but it uses almost 70% of the energy that goes into productive uses. The services sector uses only limited amounts of energy even though it generates half of GDP. Agriculture employs half of the African workforce, but accounts for only 16% of GDP and less than 10% of energy for productive uses.

Agricultural productivity per hectare in sub-Saharan Africa is well below that of other regions in the world: this reflects low energy inputs, but also a lack of modernisation, limited use of irrigation to raise crops yields and unpredictable weather (IEA, 2017a). As a result, food production per capita has not changed significantly since 2000.

The lack of transport infrastructure acts as a brake on the development of the African economy. It hinders the development of trade within the continent as well as export (and import) of finished goods. Tackling this would help Africa to take advantage of opportunities arising from the new African Continental Free Trade Agreement (AfCFTA), which entered into force in May 2019. With the right enabling conditions, the AfCFTA has the potential to have a significant impact on the continent's development. The United Nations Economic Commission for Africa UNECA has predicted that by 2040 it will raise intra-African trade by 15-25%, or from $50 billion to $70 billion, compared to an Africa without the AfCFTA (UNECA, 2018a).

Figure 8.18 ▷ Per capita consumption of key materials, 2017

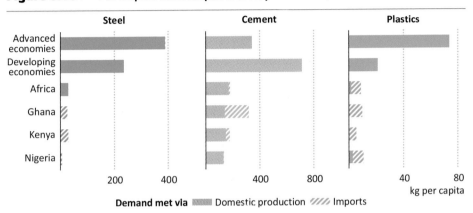

Per capita consumption of key materials such as cement, plastics and steel in Africa is low compared to developing economies elsewhere

Notes: Plastics consumption is based on 2015 data and includes key thermoplastic resins; developing economies elsewhere refers to Developing Economies less African economies in the group.

As things stand, sub-Saharan African countries represent a very small share of global industrial production: they are responsible for around 2% of global cement and aluminium production and less than 1% of steel production. As a result, Africa continues to rely on imports of many energy-intensive materials and manufactured goods (Figure 8.18).

8.3.2   Power sector

Electricity demand

Despite being home to almost a fifth of the world's population, Africa accounts for little more than 3% of global electricity demand and North African countries (42%) and South Africa (30%) represent nearly three-quarters of this. Africa's electricity demand is growing, but only at half the rate of developing Asian countries: it rose to 3% a year on average between 2010 and 2018, increasing from 560 terawatt-hours (TWh) in 2010 to around 705 TWh. The latter figure is equivalent to a fifth of electricity demand in Europe in 2018.

Electricity accounts for around 10% of Africa's total final energy consumption, but per capita electricity demand in Africa remains very low at around 550 kWh (370 kWh in sub-Saharan Africa) compared with 920 kWh in India and 2 300 kWh in Developing Asia.

Higher demand from the buildings sector accounted for almost 70% of the increase since 2010, largely, as a result of increased demand (more than 60 TWh) in residential buildings from appliances, water heating and cooling. Demand in heavy industry was largely stagnant over 2010-18 with lighter industries making up 90% of the almost 40 TWh demand increase across industry. South Africa alone accounted for more than 40% of African electricity demand from industry in 2018 although demand in the sector has been largely flat since 2010. Electricity use in transport remains very low across Africa, but is highest in South Africa, where parts of the rail network are electrified.

Electricity supply from centralised grids

Electricity generation in Africa increased to 870 TWh in 2018 from 670 TWh in 2010. Natural gas and coal (the latter largely in South Africa) accounted for 40% and 30% of generation output in 2018 respectively. Hydropower accounted for a further 16% and oil for 9%. However, there are large regional differences. In North Africa, for example, natural gas contributed more than three-quarters to power generation in 2018. South Africa in contrast is hugely reliant on coal and to a modest extent on nuclear power while in the remainder of sub-Saharan Africa, hydropower provides over half of generation output with oil and gas accounting for most of the balance. Although non-hydro renewables in sub-Saharan Africa (excluding South Africa) increased by 250% over the 2010-18 period, accounting for slightly more than 7% of all renewables and 4% of total generation output (Figure 8.19) in 2018.

Figure 8.19 ▷ **Electricity generation by fuel in Africa, 2010-2018** (left) **and in key regions in 2018** (right)

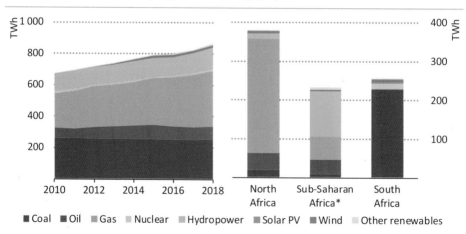

Natural gas fuelled most of the increase in electricity supply for the continent on the whole, but fuel shares varied by region and coal dominated in South Africa

* Excluding South Africa.

Figure 8.20 ▷ Installed power capacity by fuel in selected regions/countries

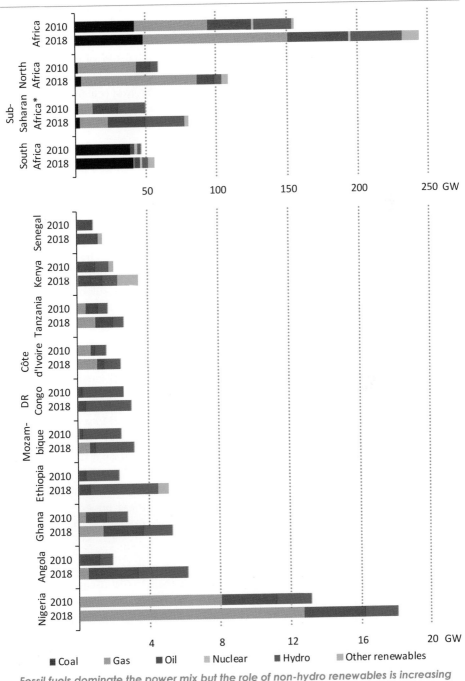

■ Coal ■ Gas ■ Oil ■ Nuclear ■ Hydro ■ Other renewables

Fossil fuels dominate the power mix but the role of non-hydro renewables is increasing

* Excluding South Africa.

Chapter 8 | Africa today

Between 2010 and 2018, total installed generation capacity in Africa increased from around 155 gigawatts (GW) to almost 245 GW, or about a quarter of the capacity in European Union countries. South Africa and North African countries account for around 165 GW of this installed capacity. The capacity mix by fuel varies across the continent by country and region. North Africa accounts for almost 85 GW of Africa's 100 GW of gas-fired power plants, while the remainder is concentrated in Nigeria, Ghana, Côte d'Ivoire, Tanzania and Mozambique. South Africa accounts for 85% of the almost 50 GW of coal-fired capacity on the continent. Oil-fired capacity totals just over 40 GW; its relative importance varies greatly by country.

Renewable power capacity increased from 28 GW in 2010 to almost 50 GW in 2018. Hydropower is the largest source of renewable power by far and its capacity increased from 26 GW in 2010 to 35 GW in 2018, although its share in the overall generation mix has remained relatively constant at around 15%. Other renewable sources have started to develop but, for the moment, their share in generation and capacity is low. Although it has expanded in recent years, wind power development in Africa has been limited in scale compared to hydro with close to 5.5 GW of installed capacity in 2018, up from almost 1 GW in 2010. North Africa accounts for around 2.6 GW of this capacity and South Africa for around 2 GW. The growth of wind power in South Africa is in part a result of its Renewable Energy Independent Power Producer Procurement Programme that was launched in 2011 and has delivered close to 3 GW of new capacity over the past five years: notable projects include the Loeiresfotein and Khobab Wind Farms (140 megawatts each) which were commissioned in 2017. Countries such as Ethiopia, Ghana, Tunisia, Kenya and Morocco are making efforts to increase their wind deployment by adopting the independent power producers (IPPs) model (Greentech Media, 2019).

Solar PV installed capacity is around 4.5 GW. Capacity increased in 2019 when the 1.6 GW Benban Solar project, the largest utility-scale solar PV project on the continent to date, recently started service in Egypt. South Africa currently has close to 2 GW of installed solar PV capacity and a number of concentrating solar power (CSP) projects including the 100 megawatt (MW) Xina Solar One project and the 100 MW Ilanga-1 plant, which were commissioned in 2017 and 2018 respectively (IEA, 2018b). These projects brought the country's total installed CSP capacity to 0.4 GW, close to 40% of Africa's installed capacity of CSP.

Geothermal resources are generally concentrated in the eastern part of Africa where tectonic regimes indicate potential equivalent to more than 15 GW (Geothermal Energy Association, 2019). With excellent geothermal resources, Kenya has installed more than 600 MW of capacity: plans are underway to develop an additional 1 000 MW from three geothermal projects (Geothermal Development Company, 2019). Other countries in East Africa, including Ethiopia, Djibouti, Eritrea, Tanzania and Uganda are also looking to tap their geothermal resources.

Electricity trade

Africa is home to five regional power pools: Eastern Africa Power Pool (EAPP); Central African Power Pool (CAPP); Southern African Power Pool (SAPP); West African Power Pool (WAPP); and Maghreb Electricity Committee (COMELEC) (*Comité Maghrébin de l'Electricité*) (Figure 8.21). These pools vary greatly in terms of scale, governance and effectiveness.

Figure 8.21 ▷ Electricity trade between power pools in sub-Saharan Africa, 2018

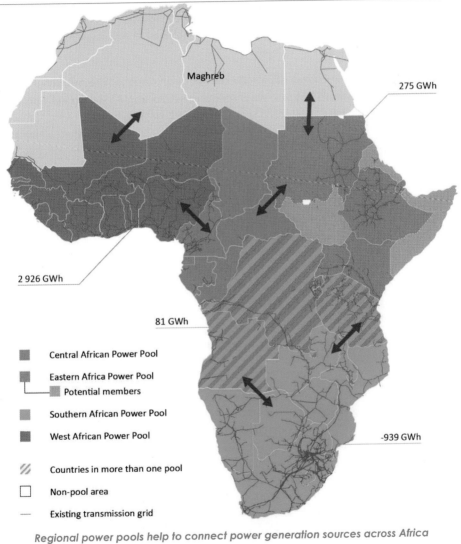

Regional power pools help to connect power generation sources across Africa

This map is without prejudice to the status of or sovereignty over any territory, to the delimitation of international frontiers and boundaries and to the name of any territory, city or area.

Sources: World Bank (2017b) and AfDB's Africa Energy Portal (AfDB, 2018).

Chapter 8 | Africa today

The SAPP has 17 members from 12 countries and operates four competitive electricity markets. Annual power trade amounts to around 3% of the member countries' demand, of which a growing portion is traded in the market. The COMELEC is relatively advanced while the WAPP and EAPP, both comparatively new, are making progress. The WAPP was established in 1999 and currently has 29 members, utilities and IPPs but remains underdeveloped compared with SAPP and COMELEC. The EAPP, established in 2005, comprises 11 state-owned utilities in Eastern Africa. The WAPP and EAPP have been working to harmonise their regulatory systems and develop market rules (e.g. standardised contracts and regional wheeling tariff methodology). Various transmission interconnections are under construction in EAPP, including an interconnection with SAPP (through a 400 kilovolt Tanzania-Zambia line). The CAPP was established in 2003 by utilities of eleven central African countries and is the least advanced.

Figure 8.22 ▷ Power traded bilaterally and through competitive markets in the Southern African, West African and Eastern Africa power pools

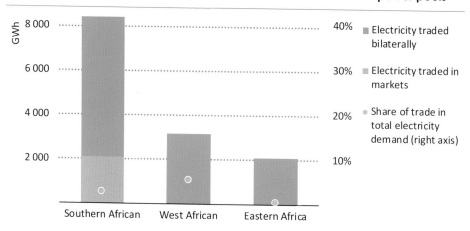

Power trade is low across the power pools and, except for the Southern African Power Pool, only bilateral trade

Note: GWh = gigawatt-hours.

Trade across the region remains low and is mostly realised through bilateral contracts (Figure 8.22). At present, in sub-Saharan Africa only SAPP has a functioning market. Some countries remain isolated from regional grids. Even where transmission interconnections exist, these are sometimes congested and need to be upgraded to facilitate trading. Around 1.8 TWh of electricity were matched in the competitive markets in SAPP in 2016/17, but were not traded because of transmission constraints (Figure 8.23).[9]

---

[9] Market players submit bids to buy and sell electricity in the wholesale market. When the power offered coincides with that requested, or vice versa, the electricity is said to be 'matched'. Yet, electricity can be matched but not traded, as technical constraints can come into play, like lack of transmission capacity to transport the matched electricity.

Figure 8.23 ▷ Electricity matched and traded in Southern African Power Pool and share of power traded in competitive markets, 2011-2018

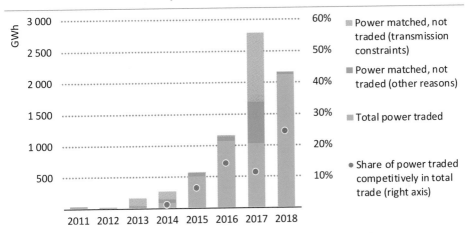

Power traded in the competitive market has increased, though transmission constraints risk impeding further exchanges

Note: GWh = gigawatt-hours.
Source: SAPP annual reports (SAPP, 2019).

Most interconnections are state- or utility-owned and publicly funded. Transmission interconnections are capital-intense investments that require strong co-ordination among two or more governments and have high perceived risks, including risks related to transmission pricing. This makes it difficult for them to attract private sector finance, while domestic utilities are loss-making and have a low ability to raise funds themselves to finance these assets. This results in low investment and substantial delays. For example, the Zimbabwe-Zambia-Botswana-Namibia Interconnector (ZiZaBoNa) was initiated with the signing of an inter-utility memorandum in 2007 but has yet to reach financial closure. The situation looks set to improve with foreign donors such as the World Bank, the European Union, the European Investment Bank and US Agency for International Development Power Africa committed to providing financial and technical assistance for transmission interconnection projects, including the Cameroon-Chad Electricity Interconnection Project (PIRECT) and the 225 kilovolt (kV) interconnection of the electricity grids between Guinea and Mali.

Regional power integration has the potential to reduce the level of infrastructure investment needed to meet demand by opening up a wider range of sources of supply (SAPP estimates savings in generation and transmission infrastructure of $37 billion over the 2017-40 period). It would also help to improve the resilience of countries energy systems by providing access to diverse and complementary markets, and help countries take advantage of economies of scale and realise large, low-cost projects that would not be justified based on domestic electricity demand alone. The potential benefits of the power pools mean that it makes sense to continue efforts to reduce investment risks and increase the bankability of projects, despite the challenges. This means improving transmission infrastructure and finding the right business model to scale up cross-border investments. Setting up a market in the EAPP and WAPP would allow countries to benefit from price differences within power pools, but will need improvements in trading regulations, wheeling methodology and regional grid codes.

Power system performance

According to World Bank Enterprise Surveys, unreliable electricity is perceived as a major constraint by almost 40% of firms in sub-Saharan Africa (World Bank, 2019). The vast majority of firms in sub-Saharan Africa experience electrical outages on a regular basis. In many countries, outages average 200 to 700 hours each year. In some countries they can be much worse than the average: A typical Nigeria firm experienced more than 32 electrical outages in 2018. These outages can vary in duration from less than one hour to more than a day, and in some countries they cost firms as much as a quarter of potential annual turnover and up to 2% of annual GDP.

Technical electricity losses are also high in Africa: in 2018, average losses amounted to 16%, which is almost seven percentage points higher than the average losses observed in other developing countries (Figure 8.24). The scale of these losses also varies significantly by region. In South Africa, average electricity losses were 9%. This is markedly lower than in other sub-Saharan Africa and North Africa countries, where electricity losses hovered between 17% and 19% respectively. Higher losses in these regions are a combination of a number of factors including poor operational performance on the part of utilities and theft of electricity from utilities. Reducing the level of losses would bring large efficiency gains.

Figure 8.24 ▷ Average electricity losses in selected African power systems today

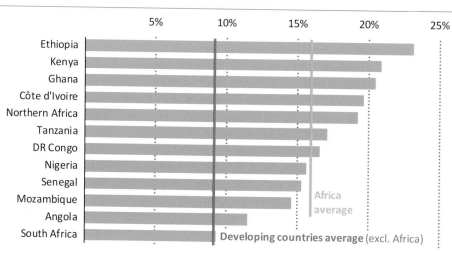

Potential efficiency gains are possible by addressing the high levels of losses

The frequent power outages in sub-Saharan Africa have resulted in a proliferation of stand-alone solar capacity and increased use of back-up generators to supplement the power needs of industry and households (Box 8.5). In Ethiopia, which depends on hydropower for much of its electricity generation, power rationing was in force during part of 2019 as a result of mechanical difficulties at dams and low water levels resulting from diminished rainfall (Africa Report, 2019). Kenya, South Africa, Zambia and Zimbabwe also experienced electricity price hikes or outages during the dry season in 2019 as a result of low water levels in hydropower systems. This may be a recurring problem if climate change results in lower rainfall.

Box 8.5 ▷ Use of back-up generators in Africa

The poor maintenance regime and ageing infrastructure in certain countries means that electricity outages are an everyday affair for many Africans. On average, many parts of countries such as Ghana and Mozambique experience outages once or twice a week. This has a direct impact on the daily lives of citizens and on the ability of the country to attract business, while the use of diesel fuel in back-up generators contributes significantly to emissions of CO_2 and air pollutant emissions.

IEA analysis estimates that 40 TWh of power was generated from 40 GW of back-up generating capacity in sub-Saharan Africa in 2018, which is equivalent to 8% of electricity generation. Nigeria accounted for almost half, generating 18 TWh of power from about 9 GW of back-up generation (Figure 8.25). Most back-up generation is used by businesses; households are often unable to afford the extra costs.

Figure 8.25 ▷ Electricity demand served by back-up generators and share of hours of electricity supply lost to outages in 2018

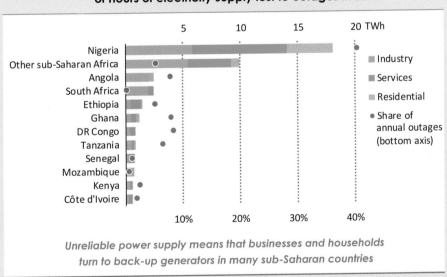

Unreliable power supply means that businesses and households turn to back-up generators in many sub-Saharan countries

Mini-grid and stand-alone systems

Grid extension is generally most cost effective when built to serve an area with a high density of demand. In more isolated (and often rural) areas, decentralised systems may be more cost effective. There are two main kinds of decentralised system: mini-grids and stand-alone systems.

Mini-grids are localised power networks, with infrastructure to transmit electricity within a defined service area. Generally, mini-grids provide electricity at a higher levelised cost than the main grid. Like any grid, mini-grids need a stable flow of power to function properly and often use either a small diesel generator or (increasingly) renewable-based power and battery systems for back-up. Mini-grids also require a certain demand threshold to justify the initial investment in the network, and therefore benefit from sizeable anchor loads provided by industrial and commercial customers or public buildings such as hospitals and schools. Mini-grids can be scaled up in line with rising demand, and eventually be connected to the main grid.

Electricity access can also be provided through stand-alone systems. These are systems not connected to a grid and typically power single households. Today, this market is dominated by diesel generators and solar PV systems (solar home systems). Stand-alone systems may be the most cost-effective option (from a system cost perspective) in sparsely populated and remote areas. Both solar PV systems and batteries can be built at different scales to match a need, which has led to innovative products coupling stand-alone generation with appliances. These products can be scaled up as power demand grows, and can supply a range of needs, from lighting and mobile phone charging to fans, televisions and sometimes refrigerators. Table 8.3 describes the main features of the mini-grids and solar home system markets and business models in sub-Saharan Africa.

The least-cost option to provide electricity access for a given area depends on its distance from the main grid, current and expected demand, and the lifetime of the assets needed for service delivery.[10] Least-cost options are generally identified in National Electrification Plans. Countries that have been successful in rolling these out tend to have plans that specify the technical, institutional and financial aspects of implementation.

There are around 1 500 mini-grids installed across Africa today and 4 000 more are planned; over half of them in Senegal and Nigeria, according to a 2019 World Bank market report (ESMAP, 2019). Demand has increased reflecting reductions in the capital costs of renewable-based generation. Policies and regulations have helped to promote investment and increase private sector participation.

Access through stand-alone systems has also been on the rise in sub-Saharan Africa, where East African countries are taking the lead. Almost five million people gained access to electricity through solar home systems in sub-Saharan Africa in 2018.[11] Many of these

---

[10] The IEA and the KTH Royal Institute of Technology Stockholm have developed a detailed geospatial model determining least-cost technologies to achieve universal access to electricity in sub-Saharan Africa, which is presented in detail in Chapter 10.

[11] This figure is based on solar home systems of at least 8 W and does not include multi-lighting systems.

bought through the pay-as-you-go (PAYG) method, where private companies lease the solar products to customers who pay periodic instalment costs and can become owners once the loan is repaid. A market report prepared by a global association for the off-grid solar energy industry (GOGLA) and others shows that PAYG accounted for more than 80% of the value of sales (mostly solar home systems) made by the main private companies in the second-half of 2018.

Table 8.3 ▷ Features of mini-grids and solar home systems in sub-Saharan Africa

		Mini-grids	Solar home systems
Systems installed		1 500*	n.a.**
Capacity of systems		Installed capacity varies substantially by system and can range from a few kW to above a MW, depending on number of people supplied, demand and uses.	• Entry level: 11.0-20.9 watts (W) • Basic capacity: 21.0 - 49.9 W • Medium capacity: 50.0 - 99.9 W • Higher capacity: > 100 W***
Business models applied		• Private models, led mainly by private developers but also demand-driven (e.g. small industries). • Public models, led by utilities. • Co-operative models.	• Private models, led by private developers. • Some public programmes, led generally by rural electrification agencies.
Private business models	Description	• Mainly pay-as-you-go (PAYG). Private companies finance upfront investment costs and customers prepay for electricity. • Revenues depend on customer demand and tariffs levels.	• Mainly PAYG (e.g. lease-to-own: customer makes small deposit upfront and pays periodic instalments). • Revenues depend mainly on loan repayment rates.
	Tariff setting	Depends on regulatory framework and/or system capacity, including: • Set as per national uniform tariff. • Set according to system cost-recovery level and return on investment, often with approval from regulator.	n.a.**
	Ownership of assets	Mini-grids owned by developers or community.	Customer acquires solar home system by the end of lease period (lease-to-own model). Can also own it from the beginning (in cash-based models), though this is less common.
	Main risks	Regulatory risk (mainly around tariff setting and what happens when main grid arrives) and revenue risk.	Revenue risk (given customers have low and unpredictable income).

* ESMAP (2019). ** Data on number or cumulative capacity of solar home systems in the region is not available (n.a.). *** Categories as defined by GOGLA (2019).

Countries with clear targets, well-designed electrification strategies, and predictable policy and regulatory frameworks show a consistent growth in sales and uptake of stand-alone systems. Kenya and Ethiopia are cases in point. Tanzania was one of the first countries to promote stand-alone systems, but recent policy uncertainty has led to stagnation in sales. The commercialisation of these products has also been linked to business models that promote local job creation and training. "Solar Sisters" is an initiative that trains local women entrepreneurs to supply clean energy products to rural households and it is impacting the lives of women and local communities (Box 8.6).

Box 8.6 ▷   Impacts of clean energy entrepreneurship by women

Much of the existing research on access by women to renewable and clean energy technologies has focused on decentralised solar and bioenergy projects that disseminate technologies such as solar lanterns, improved cook stoves and solar home systems. Initiatives such as Solar Sisters, Envirofit, Barefoot College, Kopernik and Grameen Shakti have reached millions of low-income people in African and Asian countries using similar strategies.

In 2015, the International Center for Research on Women conducted a qualitative assessment in Tanzania to better understand whether and how being a Solar Sister clean energy entrepreneur impacts women's and men's lives at the individual, family and community levels. A secondary focus of the study was to reveal initial insights about the benefits experienced by customers as a result of using Solar Sister's clean energy products. Solar Sister's unique model of recruiting, training and supporting female clean energy entrepreneurs was found to create benefits for individual women and their households and communities by enhancing income and autonomy; business skills and leadership; equality and communication; household health and stability; child education; mobility and status; and community safety (Solar Sisters, 2019).

8.3.3   Fossil fuel resources and supply

Africa has large fossil fuel resources, with sub-Saharan Africa holding around half of the continent's oil and gas resources and nearly all of the coal resources. Remaining technically recoverable oil resources in Africa amount to some 450 billion barrels or around 7% of the world's oil resources. The 100 trillion cubic metres of remaining recoverable gas resources in Africa represent 13% of the world's gas resources. Coal resources are relatively small and concentrated in South Africa and, to a lesser extent, Mozambique (Figure 8.26).

Africa is also home to many of the minerals essential to the energy industry. It has around 20% of the world's uranium resources and 40% of the manganese reserves. It also produces a large share of key precious and base metals – for example, two-thirds of global cobalt production comes from DR Congo.

The continent's resource wealth has attracted interest from international companies. Between 2011 and 2014, Africa accounted for around 20% of global oil discoveries with six countries – Angola, Nigeria, Republic of the Congo (Congo), Ghana, Mozambique and Senegal – adding around 5 billion barrels of offshore resources. With major discoveries in Mozambique and Tanzania, Africa also accounted for around 45% of global gas discoveries during this period.

Figure 8.26 ▷ Remaining recoverable fossil fuel resources in Africa, 2018

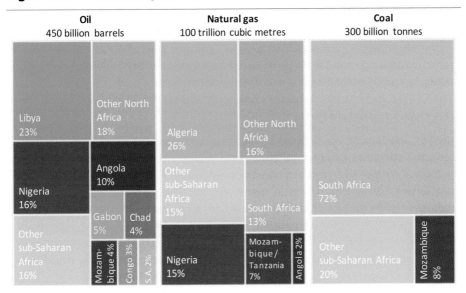

Africa has abundant fossil fuel resources; sub-Saharan Africa accounts for around half of the continent's oil and gas resources

Since the fall in oil prices in 2014, oil exploration has fallen sharply, and Africa accounted for less than 10% of global oil discoveries between 2015 and 2018. There has however been a series of major offshore gas discoveries in Egypt (2015), Mauritania and Senegal (2015-17) and South Africa (2019) (Figure 8.27).

Figure 8.27 ▷ Global discoveries of oil and natural gas by region

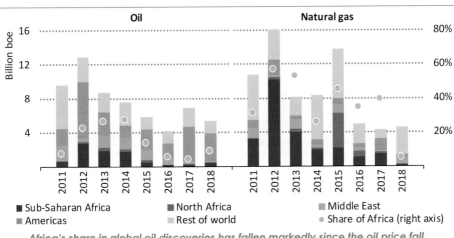

Africa's share in global oil discoveries has fallen markedly since the oil price fall, but the region has seen significant gas discoveries

Note: boe = barrels of oil equivalent.

Oil

Oil production in Africa has seen major swings since 2000. In the early years after 2000, sub-Saharan Africa showed strong production growth as the expansion in Nigeria and Angola was joined by new producers such as Chad and Equatorial Guinea. The pace of production growth in sub-Saharan Africa was four-times faster than the global average and the region accounted for almost a quarter of global production growth between 2000 and 2010. This resulted in a 50% increase in net export volumes and, thanks to rising oil prices, a threefold increase in oil revenue.

However, sub-Saharan Africa faced a sharp reversal of fortune after 2010. Nigerian oil production started to decline from 2010 as regulatory uncertainties, militant attacks and the theft of oil took their toll, and Nigerian sweet crude oil also faced fierce competition from surging US tight oil output in export markets. Angola too struggled to keep up production levels as new investments failed to compensate for the rapid decline in maturing fields. Other producers such as Equatorial Guinea and Gabon also registered a gradual output decline. As a result, oil production in sub-Saharan Africa decreased by 15% from its peak in 2010 to 5 mb/d in 2018. Coupled with a 35% increase in domestic demand, this led to a 35% decline in net exports and associated revenue (Figure 8.28).

Figure 8.28 ▷ Changes in oil demand, production and net exports in sub-Saharan Africa

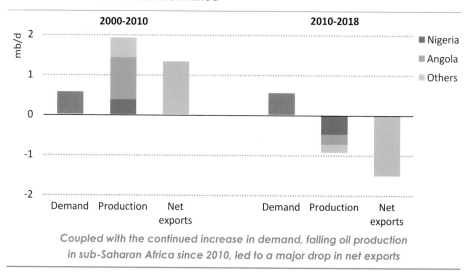

Coupled with the continued increase in demand, falling oil production in sub-Saharan Africa since 2010, led to a major drop in net exports

Major producers have recently managed to halt output declines. Nigeria's production has risen since 2016 as militant attacks in the Niger Delta have eased, but remains below the peak level reached in 2010. Long-standing issues holding back upstream investment, notably the uncertainties around the Petroleum Industry Bill, remain unresolved. In 2019, Angola succeeded in mitigating output declines due in part to the start-up of the Kaombo project, and the new government has initiated an overhaul of its oil and gas sector to stimulate investment, creating a new regulator, reorganising the role of Sonangol and streamlining investment procedures. In North Africa, output in Libya has been highly volatile given the instability and unrest there. Production registered a major drop in 2011 and remains well below 2010 levels. Output in Egypt has been in gradual decline since 2014 as investments have been directed towards natural gas developments.

Although Africa remains a major crude oil exporter, its growing oil demand and under-performing refining sector have combined to make it the world's largest importer of refined products. Africa holds 3.5 mb/d of refining capacity, which could theoretically serve three-quarters of its oil product demand, but it runs at low utilisation rates. On average, the utilisation rate was 58% in 2018; the rate was 25% in West Africa and just 9% in Nigeria.[12] African refineries tend to have fairly simple configurations, with low upgrading capabilities, and often suffer from poor maintenance. They also face growing competition from Asian and Middle Eastern refiners who are keen to export surplus diesel, and from European refiners who want to export excess gasoline. Poor refining performance has led to a growing deficit of refined products (Figure 8.29). Africa has now overtaken North America as the world's largest gasoline importer and is also the second-largest diesel importer after the European Union (IEA, 2019b).

Figure 8.29 ▷ Net imports of key refined products in selected regions

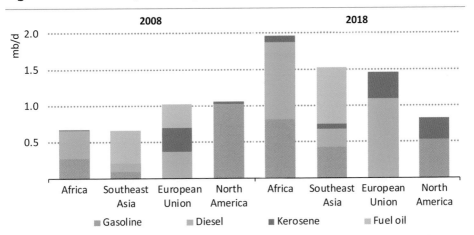

Growing demand for transport fuels and under-performing refineries have made Africa the world's largest importer of refined products

Natural gas

Africa's gas production increased rapidly in the 2000s, led by strong growth in Nigeria where the rise in oil production was accompanied by a large amount of associated gas, and in Egypt where shifting attention to gas use brought about threefold production growth. However Africa's gas production stagnated from about 2010. Egyptian production started to trend downwards until 2016 as unfavourable energy price schemes, mounting arrears to

---

[12] Early data suggest that the utilisation rate dropped further to below 6% in the first-half of 2019.

international companies and social unrest caused a significant reduction in investment. Nigeria's rapid production growth also came to a halt as fiscal and legislative uncertainties weighed on investment. Algeria managed to maintain output levels, although its largest gas field, Hassi R'Mel, is already mature.

A series of major new gas discoveries seem likely to boost future gas production in Africa (Figure 8.30). The start of production at the large Zohr offshore field has already led to a turnaround in Egypt. A gas discovery on the maritime border of Mauritania and Senegal has been followed by a final investment decision (FID) on the Tortue liquefied natural gas (LNG) project. FIDs on new onshore liquefaction plants are coming to fruition in Mozambique to exploit the huge offshore resources in the Rovuma basin.[13] Total has also recently made a significant gas condensate discovery off the southern coast of South Africa, and the estimated volume of resources is over 20% of the world's entire gas discoveries in 2018.

Gas in Africa is at a critical juncture (see Chapter 11). Where resources are plentiful, it could provide the continent with additional electricity for baseload and flexibility needs, energy for industrial growth and a sizeable source of revenue. But whether that happens depends on countries with gas putting in place well-articulated strategies to turn the discoveries into production and to build infrastructure to deliver gas to consumers cost-effectively in a competitive global LNG market.

Figure 8.30 ▷ Share of Africa in global gas demand and production, 2018, and new discoveries, 2011-2018

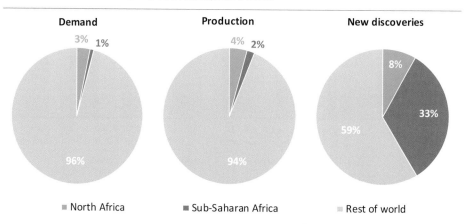

Recent discoveries offer the potential to fundamentally change the role of gas in Africa

Coal

Coal production in Africa is dominated by South Africa, which accounted for 93% of the continent's output in 2018. Production in the main current producing region in

---

[13] Coral LNG reached a FID in 2017 and started construction in 2018. Mozambique LNG reached a FID in 2019 and Rovuma LNG by ExxonMobil is approaching a FID at the time of writing.

Mpumalanga province is starting to fall, and mining activities are now shifting to the northern Limpopo province on the border with Mozambique (IEA, 2018c). Around two-thirds of the country's output is consumed in domestic markets and most of the rest is exported via the Richards Bay Coal Terminal. Mozambique started coal production in 2010 and is the second-largest coal producer in Africa. Other countries such as Botswana and Zimbabwe are aiming to ramp up coal output, although building infrastructure – rail or roads – to connect production sites to demand or export centres remains a challenge.

Imperatives for resource producers

Energy resources have long been a crucial element in the economic outlook for Africa. There is a chance that recent discoveries will lead more countries to join the ranks of resource exporters. However, translating resource endowments into economic prosperity can be a daunting task. The net income[14] from resource production has been volatile over the past few decades and has tended to lead to high levels of public spending during boom times followed by periods of fiscal strain during downturns (Figure 8.31). This pro-cyclical spending has often undermined the effectiveness of government expenditure in promoting economic growth and structural transformation. Moreover, the rise of shale and the shift to clean energy also pose serious questions for development models that rely heavily on fossil fuels. It therefore is critical for both existing and potential producers to assess the resilience of their resource production and associated revenues and devise productive ways to manage and utilise resource income, a topic explored in more detail in Chapter 11.

Figure 8.31 ▷ Net income from oil and gas production and government expenditure in top-ten producers in Africa

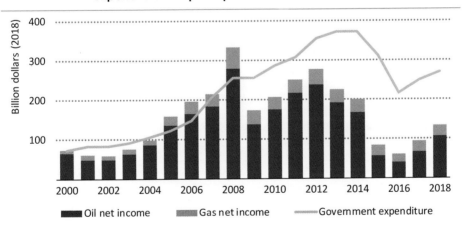

Net income from oil and gas production in Africa has been volatile as the impacts of fluctuating commodity prices have been amplified by domestic circumstances

---

[14] Net income from oil and gas production is defined as the difference between the costs of various types of oil and gas production and the value realised from their sale on either domestic or international markets.

8.3.4 Renewable resources and supply

Africa is home to abundant renewable energy resources and its renewable energy power potential is substantially larger than the current and projected power consumption of the continent. Growth has been constrained, so far, by limited access to financing, underdeveloped grids and infrastructure, unstable off-taker financial arrangements and, in many countries, an uncertain policy environment (IEA, 2018b). Despite this, recent advances in renewable energy technologies and accompanying cost reductions mean that the large-scale deployment of renewable energy now offers Africa a cost-effective path to sustainable and equitable growth. In many parts of Africa, decentralised renewable energy technologies offer an economical solution for electrification in remote areas as well as for grid extension.

Solar

A study undertaken by the International Renewable Energy Agency (IRENA, 2014) assessed the theoretical potential of a range of renewable energy technologies in Africa (Figure 8.32). It estimated that Africa's solar PV theoretical potential could provide the continent with more than 660 000 TWh of electricity a year, far above its projected needs. East Africa was identified as having the highest theoretical potential (more than 200 000 TWh/year), followed by Southern Africa (more than 160 000 TWh/year).[15]

Figure 8.32 ▷ Solar energy resource potential per year in Africa

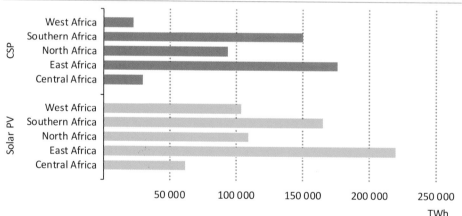

East Africa and Southern Africa contain the highest solar resource potential

Note: CSP = concentrating solar power.
Source: IRENA (2014).

---

[15] These potentials are purely theoretical potentials, with no techno-economic evaluation undertaken. These resource potentials, therefore, are subject to a significant reduction when economic parameters are applied.

Development of solar in Africa has been slow, with only around 4 GW of new solar PV capacity added between 2010 and 2018, more than two-thirds of it in sub-Saharan Africa. The main challenges and barriers that countries face include limited institutional capacity within government, lack of scale and competition, high transaction costs and the perceived high risk of such projects (World Bank, 2018). This has prompted the World Bank to start the Scaling Solar initiative to address these challenges by providing a "one-stop shop" to help governments mobilise privately funded grid connected solar projects at competitive costs.

IRENA also assessed the potential of CSP on the continent and estimated the likely potential as being around 470 000 TWh a year. Again, East Africa has the highest potential, followed by Southern Africa. Here too development has been slow with the exception of large solar CSP projects in Morocco and South Africa.

Hydropower

Hydropower has been the main renewable energy resource developed to date with around 35 GW of hydro capacity across Africa, with Angola, Ethiopia, DR Congo, Zambia, South Africa, Sudan, Mozambique and Nigeria each having 2 GW or more. Ethiopia has hydropower capacity of nearly 4 GW and more developments are planned, most notably the 6 GW Grand Ethiopian Renaissance Dam, which will be the largest in Africa when it comes into service in 2022. South Africa has installed hydropower capacity of close to 4 GW including the recent 1.3 GW Ingula plant.

Central Africa has very rich hydropower resources thanks mostly to the Congo River, the deepest river in the world and the second-longest in Africa after the Nile. There is a large mismatch between the significant hydropower potential in this region and the much more limited local electricity demand. This means that large-scale regional interconnections will be essential to promote its development. The DR Congo in particular has enormous hydropower potential that has been estimated at 100 GW, which could generate about 774 TWh of electricity per year. Plans in DR Congo to develop the Grand Inga Dam further have been beset with difficulties, but projects have been moving forward elsewhere.

While state-owned enterprises remain the largest developers of hydropower projects, many have been built by Chinese developers and backed by concessional financing. Chinese investors accounted for 60% of investment in sub-Saharan hydropower projects between 2010 and 2015.

Small-scale hydropower (1-10 MW) and mini-hydro power (0.1-1 MW) could provide power for rural electrification in some areas of sub-Saharan Africa, and there is particular potential in the central and south-eastern parts of the continent. A recent study estimated around 21 800 MW of small-scale hydropower technical potential (Korkovelos et al., 2018), with the central corridor of the sub-continent and especially South Africa, DR Congo and Sudan having the most potential. The same study also estimated that total mini-hydropower technical potential in sub-Saharan Africa was around 3 400 MW.

Wind

With close to 5.5 GW of installed wind power capacity in 2018, there is plenty of room for expansion given its theoretical potential to produce as much as 460 000 TWh of electricity a year (IRENA, 2014). Most wind resources are found close to coastal locations, mountain ranges and other natural channels in the eastern and northern regions of the continent. Algeria, Egypt, Somalia, South Africa and Sudan are among the countries with the highest wind energy potentials (IRENA, 2014). The best offshore wind energy potential is found off the coasts of Madagascar, Mozambique, Tanzania, Angola and South Africa.

Wind can be cost competitive with other technologies where the resources are good, but other factors could limit its deployment. For example, in East and North Africa, where the best resource potential is estimated, domestic markets are small and power grids are not well developed, meaning that significant variable generation from wind could be challenging to manage without additional grid investment.

Geothermal

Geothermal resources can be found throughout Africa but the bulk of the potential is concentrated in the East Africa Rift System, where total potential could be as much as 15 GW (BGR, 2016). This potential is largely untapped at present. Only Kenya has tapped its geothermal potential and installed capacity of almost 700 MW. Other countries in East Africa are now taking steps to make use of geothermal energy: Ethiopia is operating a 7 MW pilot plant and new developments totalling more than 1 GW are planned in Djibouti, Eritrea, Tanzania and Uganda. The expansion of geothermal power in the East Africa region faces a number of barriers, but technical and financial support is available (notably from Japan) to help countries formulate geothermal master plans and to promote private sector funding and local capabilities.

Bioenergy

Bioenergy continues to dominate the sub-Saharan energy mix and made up almost 60% of primary energy use in 2018. Almost three-quarters of bioenergy demand are accounted for by the traditional use of solid biomass in the residential sector, although there is also some use of solid biomass and biogas for modern power generation and heat.

Bioenergy can generate around 800 MW of electricity from current installed capacity, mainly in East and South Africa. However, large-scale deployment will be challenging, as the levelised costs of power generation from bioenergy are often higher than gas-fired generation and hydropower, due in part to the cost of collecting the biomass. Biogas has emerged as a substitute for firewood for cooking purposes in some areas, primarily in rural East Africa. Recently bio-slurry obtained from biogas production has started to be collected and utilised as fertiliser to increase agricultural production. Although at present there are technical and financial barriers limiting its application, biogas has a potentially important role to play in reducing indoor air pollution and related premature deaths (especially of women and children), limiting deforestation, and improving sanitation and the quality of life (especially for women) in rural and agricultural areas.

Advanced biofuels for transport have significant potential in many African countries. West Africa alone is estimated to have the potential to produce over 100 Mt per year of agriculture residues that could be converted into electricity or advanced biofuels such as ethanol and bio-butanol.

8.3.5   Environment

Water

Africa has less than 9% of the world's renewable freshwater water resources, and more than 50% of it is held in just six countries in Central and West Africa (UNESCO, 2019). The continent is home to roughly 80 transboundary lakes and rivers with most large river basins shared by five or more countries (UNECA, 2018b). Today, sub-Saharan Africa is in the midst of its worst drought in 35 years. In 2018, Cape Town almost ran out of water. In 2019, several cities in Mozambique, Zimbabwe, Ghana and Côte d'Ivoire experienced water shortages. Alongside changing and uncertain precipitation patterns brought on by climate change, Africa's water scarcity is compounded by a lack of water storage, supply and management infrastructure. By 2025, it is estimated that over 450 million people, mostly in West Africa, could be at risk of water stress (UNECA, 2018b).

Agriculture is the largest water user in sub-Saharan Africa today, despite the fact that just 3% of its total cultivated land is irrigated. Groundwater is estimated to be plentiful — the region withdraws less than 5% of its renewable groundwater whereas India, for example, withdraws almost 60% — but there is increasing evidence that some aquifers are being depleted (WWAP, 2019). Water use by the energy sector is low today, with coal and oil accounting for most of it. However, water availability could increasingly become a critical issue for energy sources, in particular for hydropower (see Spotlight in Chapter 10).

Household water use is also low and the World Health Organization (WHO) estimates that on average a person in Africa uses just 20 litres of water per day, well below the recommended minimum of 50 litres. Sub-Saharan Africa is also home to 745 million people that have no access to safely managed drinking water (over 70% of which reside in rural areas) and almost 840 million people that lack access to safely managed sanitation (roughly 60% of which are in rural areas) (UNICEF and WHO, 2019). Low rates of wastewater collection and treatment mean that a significant amount of untreated wastewater is released into the open. Contaminated drinking water has a significant health impact: diarrhoea is a major cause of mortality for children under five in sub-Saharan Africa.

Significant progress will be required to reach the targets set under the Sustainable Development Goal 6 (clean water and sanitation) by 2030 (Box 8.7). Water demand is projected to increase more in Africa than in any other part of the world rising by almost 300% from 2005 to 2030, and a large share of it is projected to occur in municipalities reflecting rapid urbanisation and more people gaining access to clean water (Wijnen et al., 2018). This will also result in larger amounts of wastewater to be collected and treated (which can increase energy demand depending on the level of treatment).

Box 8.7 ▷ How does energy provision look different if viewed alongside water and food?

Millions of people in sub-Saharan Africa do not have adequate access to the basic building blocks of economic and societal development: energy, water and food. While providing electricity access at a household level is critical, it is not enough on its own to ensure economic development. Approaching development from an integrated water-energy-food perspective allows for a broader and more durable view of local economic development, and could also change the scale and type of the energy technologies deployed. The value of this perspective for sub-Saharan Africa is visible at both a household and a broad economic level. At a household level, it is clear that the technologies being deployed to provide access to electricity can also be used to provide access to clean drinking water (IEA, 2019c). At an economic level, viewing energy development though this prism can advance the prospects for sustainable productive uses, such as agriculture. Moreover, such an approach can help address the central role of women in providing these resources.

The potential for such an approach is highlighted by a recent micro and macroeconomic modelling exercise that looked at the Ikondo-Matembwe project in rural Tanzania (RES4Africa Foundation, 2019). This project, which serves eight villages, consists of two community-scale hydro-powered mini-grids that power an anchor client, an agribusiness focused on producing animal feed, hatching poultry, and providing electricity and water to the surrounding households. The preliminary results of this study, which used a cost-benefit analysis based on the project's investment data to assess the benefits of single versus multi-service scenarios, indicate that over its 20-year lifespan a renewable energy-based integrated project has more than twice as much economic impact[16] on a local community as a project geared to the provision of energy alone. Investing in an integrated approach also had a multiplier effect, significantly increasing local purchasing power which translated into improvements in other areas. Some of the biggest benefits came in the form of increased access to better education, improved agricultural productivity and time saved from having water and energy access on site.

More research and examples are needed of these kinds of projects to understand the scope for replicating them at scale, but it is evident that looking at energy, food and water together in an integrated way has clear potential to trigger captive energy demand, increase economic productive capacity, and set African communities and economies on a path that looks beyond the immediate imperatives to meet the 2030 sustainable development goals.

---

[16] Measured in net-present value.

GHG emissions

In 2018, Africa accounted for around 4% of the world's energy-related carbon dioxide (CO_2) emissions despite being home to around 17% of the population. The power sector was the largest emitting sector (480 Mt CO_2) followed by transport (355 Mt CO_2) and industry (150 Mt CO_2).

Total energy related CO_2 emissions in North African countries in 2018 were around 490 Mt or 40% of Africa's energy-related CO_2 emissions (1 215 Mt CO_2). South Africa's energy sector emitted 420 Mt CO_2 with its coal-fired power fleet responsible for more than half of the country's energy-related CO_2 emissions and more than three-quarters of the sub-Saharan region's power sector emissions.

While the sub-Saharan African energy sector makes a very small contribution to global CO_2 emissions, the region is among those most exposed to the effects of climate change. For sub-Saharan Africa, which has experienced more frequent and more intense climate extremes over the past decades, the consequences of the world's warming by more than 1.5 degrees Celsius (°C) would be severe. Temperature increases in the region are projected to be higher than average global temperature increase; regions in Africa within 15 degrees latitude of the equator are projected to experience an increase in warmer nights and longer and more frequent heat waves (UN, 2019b). A recent report published by the Intergovernmental Panel on Climate Change found that the potential impacts of global warming levels on key sectors at local to regional scales, such as agriculture, energy and health, remain uncertain in most regions and countries of Africa (IPCC, 2018).

Local air pollution

Around 6.8 Mt of fine particulate matter ($PM_{2.5}$) were emitted in Africa in 2018, of which almost 85% was from the burning of biomass indoors. Damage to air quality from these sources disproportionately affects the poorest in Africa. Nigeria, with its large population and low levels of access to modern energy services, accounted for around a third of Africa's $PM_{2.5}$ emissions, emitting 2.1 Mt in 2018. This compares to around $PM_{2.5}$ emissions of 0.4 Mt in South Africa, 0.3 Mt in Tanzania and 0.2 Mt in Kenya in 2018. Efforts have been made across the continent to reduce $PM_{2.5}$ emissions mainly through incentivising the use of modern cooking fuels, such as LPG and natural gas. South Africa's National Environmental Management Air Quality Act of 2004 is one example of an African country regulating air quality and setting emissions standards, imposing limits on new and existing power plants and industrial installations.

Nitrogen oxides (NO_X) emissions in Africa were around 7.5 Mt in 2018, of which nearly 50% came from vehicle tailpipe emissions, a further 18% from industry, 16% from the power sector and 11% from buildings. In North Africa, vehicle tailpipe emissions made up almost two-thirds of NO_X emissions, reflecting the relatively high number of cars on the road. In South Africa, which has sub-Saharan Africa's largest car fleet, vehicle tailpipe emissions accounted for about a third of NO_X emissions: a further third came from the power sector (33%), largely as a result of the large share of coal in South Africa's power mix.

Figure 8.33 ▷ Emissions of PM₂.₅, NOₓ and SO₂ by sector in Nigeria, South Africa and North Africa, 2018

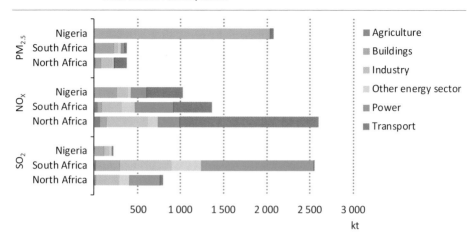

Nigeria accounts for a very large share of Africa's PM₂.₅ emissions, North Africa for a large share of NOₓ emissions and South Africa for its SO₂ emissions

Note: kt = kilotonnes.
Source: International Institute for Applied Systems Analysis.

Sulfur dioxide (SO₂) emissions were almost 5 Mt in 2018, almost 40% of which came from power generation, largely as a result of coal combustion in South Africa, and nearly 40% from the industry and transportation sectors (Figure 8.33).

In Africa, almost 500 000 premature deaths each year can be attributed to household air pollution, a health problem which is closely related to the lack of access to modern forms of energy. Fewer deaths in Africa are attributable to outdoor pollution than to household air pollution, but the number still stands at more than 300 000 per year, with most occurring in sub-Saharan Africa (excluding South Africa).

Chapter 9

Urbanisation, industrialisation and clean cooking
How fast will Africa grow?

> **SUMMARY**
>
> - Africa has the world's fastest growing population: one-in-two people added to the world population in the period to 2040 are African. With over 40% of the continent's population under the age of 15, it also has the world's youngest population. This young, expanding population is rapidly becoming more urban. The last two decades have seen the number of people living in cities increase by 90%, and this trend is set to accelerate over the next two decades. By 2040, an additional 580 million Africans are living in cities – a pace of urbanisation that is unprecedented.
>
> - The energy sector has a vital role to play in Africa's future. Growing urban populations imply rapid growth in material demand to build infrastructure, expansion of industrial and agricultural production, and increased mobility of people and goods, boosting energy demand. In the Stated Policies Scenario, the continent's economy grows to two-and-a-half times its current size by 2040.
>
> - Our Africa Case outlines the implications for the energy sector of countries across the continent realising their ambitions for accelerated economic growth and universal access to energy. If the energy needs of the future set out in the Africa Case are to be met in a sustainable way, a strong emphasis on improvements in energy efficiency will be vital. Focus areas include fuel economy standards for cars and two/three-wheelers (largely absent today), highly efficient industrial processes, building codes and efficiency standards for appliances and cooling systems.
>
> - The growth in population in areas with high average temperatures means that by 2040, if the global average temperature increase is kept to the limits of the Paris Agreement, over 1 billion people in sub-Saharan Africa will be living in areas that need space cooling (a number that increases to 1.2 billion if the world continues on its current trajectory of warming). This means that cooling is set to be one of the most important factors in determining the extent of future energy demand.
>
> - Bioenergy is the largest source of energy in sub-Saharan Africa today and accounts for two-thirds of final energy consumption. Around 850 million people in sub-Saharan Africa rely on the traditional use of biomass, cooking with inefficient stoves, while another 60 million rely on kerosene or coal to meet their daily energy needs. Cooking with polluting fuels and stoves has major health and environmental consequences, and was linked to almost 500 000 premature deaths in 2018. Less than 200 million people in sub-Saharan Africa currently have access to cleaner options such as liquefied petroleum gas (LPG), natural gas, electricity or improved biomass stoves.

Figure 9.1 ▷ Final consumption by fuel, sector and scenario in sub-Saharan Africa (excluding South Africa)

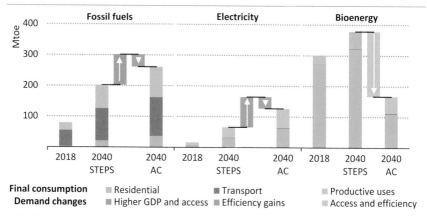

In the Africa Case, efficiency gains reduce potential demand growth to 2040 by one-third, while achieving SDG 7 reduces bioenergy use for cooking

Note: STEPS = Stated Policies Scenario; AC = Africa Case.

- Some progress is made in the Stated Policies Scenario in reducing reliance on the inefficient use of bioenergy for cooking, but the number of people gaining access to clean cooking barely exceeds population growth. Switching from the traditional use of biomass to cleaner options faces both economic and non-economic barriers, but the Africa Case sees all households across the continent gain access to clean cooking by 2030, with benefits in particular for women and girls. In the Stated Policies Scenario, the number of premature deaths linked to indoor air pollution continues to increase to 500 000 by 2030; in the Africa Case, clean cooking for all by 2030 reduces that number by more than two-thirds.

- Although nearly 600 million people in sub-Saharan Africa still use solid biomass in improved cookstoves in the Africa Case in 2040, improved efficiency is enough to cut bioenergy demand in half from today's level. Charcoal use for cooking is increasing, especially in urban areas, it remains an important source of energy in both cases in 2040. This makes improving efficiency along the charcoal value chain an important priority.

- Biogas has the potential to provide 50 million tonnes of oil equivalent (Mtoe) of locally produced low-carbon energy in Africa, largely via household-scale biodigesters; this potential doubles to almost 100 Mtoe by 2040, at an average cost of around $10 per million British thermal units (MBtu). The main economic barrier to increased uptake is the upfront cost of installing a biodigester. Other barriers such as maintenance requirements and feedstock availability can be overcome with well-targeted policies and programmes.

9.1 Introduction

Africa's population has grown by 470 million since 2000, making it the fastest growing region in the world. It is also by some distance the youngest, with a median age of 20 years, compared to a global average of 30 years (United Nations, 2019). Changing demographics are at the heart of any discussion of economic development. Beyond the increase in the number of people in the region, the unprecedented magnitude of the shift from rural to urban living and the extent to which industrial development contributes a greater share of economic output will have a profound impact on energy demand growth in the future. Africa's projected population growth could act as a catalyst for economic growth. However, this will only be the case if there are enough job opportunities in productive enterprises to fully capture the potential dividends of the demographic change.

As an ever-higher proportion of the African population moves to cities in the period to 2040, demand for energy services will inevitably grow. The average household in a city consumes three-times more oil and electricity than the average rural household in sub-Saharan Africa (excluding South Africa), although there are large disparities within cities. A large part of the migration to cities is comprised of low or unskilled labourers — often moving to peri-urban areas which are home to 55% of the urban population in sub-Saharan Africa today (Odamo, 2019). In these areas, electricity access rates are lower than in the heart of the city, although still much higher than in rural regions. Charcoal and firewood remain the predominant cooking fuels in many peri-urban households, as traditional practices tend to endure even when cleaner options are available.

Most energy in sub-Saharan Africa (excluding South Africa) is used for cooking, which accounts for around 70% of total final consumption, compared to less than 10% globally. Low rates of access to electricity, low levels of appliance and vehicle ownership, limited transport infrastructure and low levels of industrial production explain why this is so. Bioenergy is the major source of energy used to meet cooking energy needs in sub-Saharan Africa (excluding South Africa). This cheap or free resource accounts for three-quarters of total final consumption. Traditional use of biomass, often in poorly ventilated spaces, has severe impacts on health and the environment. Efforts to improve access to clean cooking lag far behind efforts to secure access to electricity. Globally, nearly 2.4 billion people continue to use inefficient open fires or simple cook stoves today, around 840 million of whom live in sub-Saharan Africa (excluding South Africa). To date, cleaner processed forms of bioenergy like biogas and biofuels have made limited progress in the region.

In both the Stated Policies Scenario and the Africa Case, higher economic outputs and higher household incomes lead to increased energy service demand in every sector. Total final consumption in sub-Saharan Africa (excluding South Africa) grows by around 65% from today to 2040 in the Stated Policies Scenario, with the share of all productive uses within total final consumption rising as the scaling up of production leads to industry sector energy demand more than doubling by 2040 (Figure 9.2). In the Africa Case, industry energy demand more than triples on the assumption of accelerated industrial and economic growth. Increased economic prosperity is reflected in terms of higher household appliance ownership and expanding vehicle stocks. In the Africa Case, electricity demand within households for cooling and appliances in sub-Saharan Africa (excluding South Africa) rises 14% per year to reach more than 500 terawatt-hours (TWh) by 2040 (compared to 30 TWh today), with cities being responsible for more than 60% of this growth. Similar trends are observed for passenger vehicles, with car stocks more than tripling to reach 27 million within two decades in the Stated Policies Scenario and more than 35 million in the Africa Case. In 2040, over 80% of this passenger car fleet is operating in cities.

Figure 9.2 ▷ Final consumption by sector and fuel in sub-Saharan Africa (excluding South Africa) **in the Stated Policies Scenario and Africa Case, 2018 and 2040**

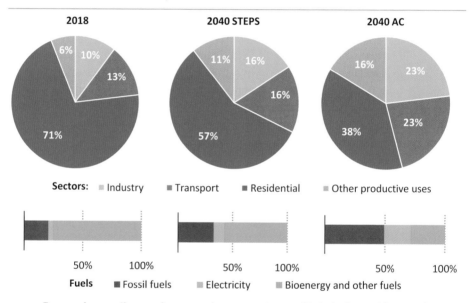

Economic growth spurs increases in energy demand in industry and transport. Bioenergy's share declines as households increasingly switch to cleaner fuels

Note: STEPS = Stated Policies Scenario; AC = Africa Case.

The share of fuels in the energy mix of every sector sees some significant changes in both scenarios. In the Stated Policies Scenario, bioenergy remains the largest source of energy, with a share of just under 60% in 2040 in sub-Saharan Africa (excluding South Africa), and total bioenergy consumption increases from 300 million tonnes of oil equivalent (Mtoe) today to 380 Mtoe in 2040. In the Africa Case, extending energy access to the entire population by 2030 (UN Sustainable Development Goal [SDG] 7) is achieved via a combination of improved cook stoves, liquefied petroleum gas (LPG), electrification, ethanol and other solutions. Bioenergy demand is cut in half, as the number of people relying on the traditional use of biomass drops to zero and kilns to produce charcoal become more efficient.

We analyse how the electricity sector can be expanded to supply the increased demand caused by urbanisation and electrification in Chapter 10, and carry out the corresponding analysis for fuel supply in Chapter 11. In this chapter, we assess the role of demand-side policies and aim to answer three crucial questions. What are the implications of rapid urbanisation and industrialisation for energy use in Africa? What role could a more efficient use of energy play in accelerating Africa's economic development? What is needed to reach full access to clean cooking?

9.2 Urbanisation and industrialisation, drivers of growth

The urban population in sub-Saharan Africa is growing rapidly, more than doubling since 2000 to reach 440 million today. The share of the population living in cities is now 40%, up from 32% in 2000. There are 520 million more people in cities in sub-Saharan Africa in 2040 than there are today (along with an extra 60 million in North Africa). Urbanisation of this scale and at this speed has never been seen before, and is expected to be twice as large as the projected growth of urban population in India over the next two decades.

Figure 9.3 ▷ Urban population and cement demand growth in China (historical), India (Stated Policies Scenario) and Africa (Africa Case)

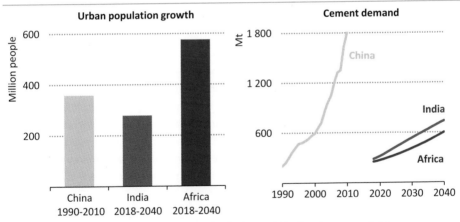

Africa experiences unprecedented growth in its urban population, but follows its own trajectory for cement demand

Notes: Urban population in Africa is assumed to be the same in both the Africa Case and Stated Policies Scenario.

Africa's rapidly increasing population and growing urbanisation bolsters demand for materials. This could underpin industries that in turn stimulate further growth and income generation on the continent. As a point of comparison, China's economic boom between 1990 and 2010 saw the population of cities increase by 360 million: cement production grew ninefold during this period, while iron and steel production increased ten-fold. Africa's future infrastructure is however unlikely to follow the same path as China's. Traditional materials and designs look set to play an important role in Africa, and so do local innovations.[1] Therefore, Africa is likely to follow its own growth trajectory (Figure 9.3). As well as mitigating the growth of cement and steel demand, an approach tailored to African conditions would help promote a housing stock suited to the climate, which in turn would act to offset some of the growth in demand for cooling. New initiatives could bring about urbanisation with reduced resource needs. Adobe houses, and bamboo- or wood-based low-storey buildings are examples of what may be possible in this context

Per capita energy consumption across sub-Saharan Africa is extremely low, at less than a third of the global average. This is, in part, a reflection of the fact that a higher percentage of people live in rural areas in sub-Saharan Africa than in any other region in the world, which makes it more difficult to achieve access to electricity and clean cooking for all (see Chapter 8). But it also reflects lower than average consumption across end-use sectors. There are, of course, large disparities both between and within countries, but the level of car ownership illustrates the size of the gap in energy services: at 115 cars per 1 000 people, South Africa's ownership rate is eight-times the sub-Saharan African average, but stands at one-quarter the average for advanced economies globally, while in Ethiopia the ownership of vehicles is less than 2 cars per 1 000 people. The potential for growth in energy consumption in the transport sector is very large, and an increase in the number of two/three-wheelers is already taking place in a number of sub-Saharan African countries.

There are other examples as well of low levels of energy consumption and potential for strong future growth. About 680 million people are located in hot areas in sub-Saharan Africa (where the average perceived daily temperature[2] exceeds 25 degrees Celsius (°C) over the whole year) which would typically require the use of cooling devices such a fan or air conditioner. However, only around 10 million households own an air conditioner across sub-Saharan Africa, while 100 million own an electric fan. The inevitable increase in future demand for cooling (spurred by population and income growth, the move to cities and climate change), requires a fundamental rethink not only of the way energy is produced and consumed across the continent, but the way the urban environment is constructed (Box 9.1).

Large infrastructure projects, from maritime port expansions to railway modernisation, are one of the main drivers of growth in materials demand (Table 9.1). The choice of whether industrial products will be imported (as they mostly are now, with the exception of cement), or produced domestically (underpinning a potentially transformative industrial growth story), will have major implications for Africa's energy and economic sectors.

---

[1] For example, a Colombian company has exported its manufacturing experience of producing bricks from used plastics to Côte d'Ivoire, where the aim is to build 500 classrooms using this technology by 2020.

[2] The combined effects of air temperature, relative humidity and wind speed.

Table 9.1 ▷ Selected large infrastructure projects in Africa

Project	Countries involved	Proposed construction start	Cost estimate ($ million)
Abidjan - Ouagadougou transport corridor	Burkina Faso and Côte d'Ivoire	Unknown	600
Batoka Gorge Hydropower Project	Zambia and Zimbabwe	2020	4 500
Brazzaville - Kinshasa Road/Rail Bridge	DR Congo and Republic of Congo	2020	550
Dar es Salaam Port Expansion	Tanzania and neighbouring countries	2020	420
Grand Ethiopian Renaissance Dam	Ethiopia	In progress	5 000
Inga 3 Hydropower Plant	DR Congo and regional partners	2024-25	12 000-14 000
Juba - Torit - Kapoeta - Nadapal - Eldoret Road Project	Kenya and South Sudan	Unknown	420
Lamu Port	Kenya	2020	3 100
Modernisation of Dakar-Bamako Rail Line	Senegal and Mali	Unknown	3 000
New Administrative Capital (Egypt)	Egypt	Unknown	20 000
Ruzizi III Hydropower Project	Burundi, DR Congo and Rwanda	2020	600
Sambangalou Hydropower Project	Guinea, Senegal and other West African Power Pool member countries	Unknown	455
Serenje - Nakonde Road Project	Zambia and regional partners	In progress	674
Trans-Saharan Gas Pipeline	Nigeria, Niger and Algeria	Unknown	10 000-13 700
Zambia Tanzania Kenya Transmission Line	Kenya, Tanzania, Zambia	Unknown	1 200

Box 9.1 ▷ Design, build and strengthen sustainability for African cities

Among the 534 cities of more than one million inhabitants across the world today, 66 are in Africa (Demographia, 2019). Uncontrolled urban development is making living conditions more precarious: some cities in Africa already suffer from population displacement and serious congestion.

About 85% of the inhabitants of today's African cities live in areas denser than metropolitan Paris (Figure 9.4). Although densely populated cities can ease the provision of services, they can exacerbate air and noise pollution and increase demand for space cooling. Temperatures in densely populated cities can be 3-5 degrees higher than in low-density neighbourhoods due to human activity, heat radiation from concrete, asphalt and other materials as well as an increasing number of air conditioners that move heat from inside to outside of buildings (Tremeac et al., 2012).

In 2040, there are twice as many people in cities in Africa as there are today. Strong policy ambitions, long-term investment and sound urban planning are needed to harness the opportunities and overcome the challenges of urbanisation and the rise of medium-size cities. Training, capacity building and engagement at the national, regional and community levels are critical to ensure that solutions are tailored to local needs. In order to balance economic and environmental aspirations, policy makers and urban planners need to plan for growth in energy demand related to the manufacturing, transport and use of building construction materials. From now to 2040, the equivalent of all built surfaces in Ethiopia are added every year to Africa's built environment. Material efficiency strategies including building design, lifetime extension and waste reduction are an essential part of this planning.

Figure 9.4 ▷ Urban population by density in cities of more than one million inhabitants in Africa and density of selected cities, 2018

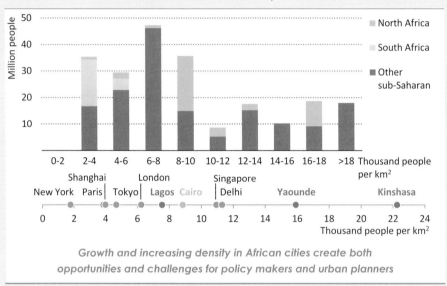

Growth and increasing density in African cities create both opportunities and challenges for policy makers and urban planners

High-density urban areas could provide opportunities to scale up the high energy performance construction market. Passive design allowing for natural ventilation and solar heat gain reductions could slash air conditioning demand by 65-70% in climates similar to coastal Senegal (Harkouss et al., 2018). They also fit well with other energy-efficient design strategies such as increasing natural daylight intake.

Providing infrastructure and promoting rules that encourage public or non-motorised transit could reduce the impact of urban growth on energy demand for transport in Africa. Bus rapid transit systems generally require low shares of public subsidies since moderate capital investment can sustain a large passenger throughput. Urban planning and city structure also matter. Planning should ideally lead to a mix of land uses – residential, commercial and industrial – while promoting walking, cycling and mass transit. Urban planners equally need to address current practices leading to sub-optimal choices such as importing old second-hand vehicles that pollute, the provision of free/cheap parking and the remuneration of public transport operators on the basis of number of passengers transported per day.

A number of countries, including Nigeria, Mozambique and Angola, have included steel and chemicals in their industrial development strategies. The Action Plan for the Accelerated Industrial Development of Africa (AIDA) initiative counts among its ambitions the development and implementation of industrial policy that promotes local production and upgrading of existing industrial technologies. At regional level, the initiative encompasses programmes such as the development of regional industrial strategies, the creation of industrial development funds for infrastructure and heavy industry, and support for incubators. At continental level, the AIDA initiative aims to harmonise industrial policies and promote development partnerships. The realisation of these ambitions is closely bound up with the evolution of the energy sector. Industrial expansion requires more energy than is produced today. It is also likely to accelerate urbanisation, leading to increased energy demand in households and for transport. At the same time it could generate higher incomes, further increasing energy demand. The upside for industrial output is explored in detail in the Africa Case.

Figure 9.5 ▷ Change in total final consumption by sector and energy type in sub-Saharan Africa (excluding South Africa) in the Stated Policies Scenario and Africa Case, 2018-2040

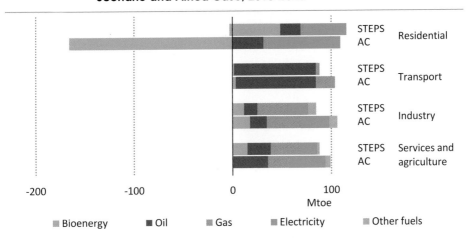

Higher economic growth in the Africa Case sees increasing demand for all fuels, except bioenergy, with access to clean cooking for all and efficiency gains driving its reduction

Note: STEPS = Stated Policies Scenario; AC = Africa Case.

Greater economic output, higher household incomes and improved living standards also mean that car, appliances and cooling systems ownership rates in sub-Saharan Africa are higher in the Africa Case than the Stated Policies Scenario. The impact on energy consumption is mitigated to some extent by the implementation of more stringent vehicle efficiency policies and regulations, notably on the second-hand market. In the residential sector, building energy efficiency codes and higher minimum energy performance standards (MEPS) for appliances and cooling systems help to slow electricity demand growth in the Africa Case. Building codes also play an important role in the construction industry (only 30% of the building stock anticipated for 2040 currently exists). These codes require countries to establish regulatory frameworks to set performance standards and ensure compliance with them.

Overall, the reduction in bioenergy use in the residential sector in the Africa Case and the increased focus on energy efficiency, mean that total final consumption in 2040 in sub-Saharan Africa (excluding South Africa) is lower in the Africa Case than in the Stated Policies Scenario, despite an economy that is 60% larger (Figure 9.5).

9.2.1 Residential sector

The residential sector accounts for around 65% of total final consumption of energy in sub-Saharan Africa (compared to 22% globally and less than 20% in advanced economies), making it the largest end-use sector across sub-Saharan Africa. In the Stated Policies Scenario, increasing incomes and living standards, as well as improved access to affordable and reliable electricity, leads to a steady increase in energy consumption in the residential sector. However, its share in overall consumption diminishes to around half in 2040 as energy demand in industry grows at a rate that is nearly triple that of the residential sector. In the Africa Case, overall demand in residential buildings actually decreases by around one-quarter as a result of efficiency improvements and the rapid displacement of biomass in cooking (see section 9.3). Together with the accelerated industrialisation included in the Africa Case, this fall in demand reduces the share of the residential sector to around one-third of overall consumption by 2040.

Figure 9.6 ▷ Final consumption of selected fuels in the residential sector in sub-Saharan Africa (excluding South Africa) in the Stated Policies Scenario and Africa Case, 2018 and 2040

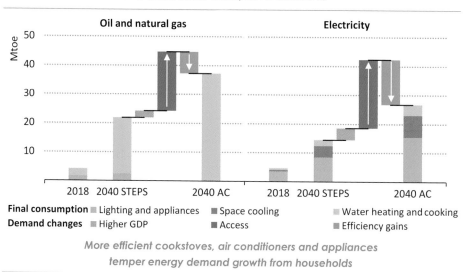

More efficient cookstoves, air conditioners and appliances temper energy demand growth from households

In the residential sector, oil and electricity are increasingly used for cooking and water heating in both scenarios, with LPG making big inroads (Figure 9.6). There is a major shift away from the traditional use of biomass, and universal access to clean cooking is achieved by 2030 in the Africa Case, fulfilling SDG target 7.1. Use of natural gas expands in cities that are able to make use of nearby gas resources.

In the Africa Case, rising incomes and access to basic energy services mean that demand for electricity grows more than for any other energy source (see Chapter 10). Demand for lighting and other appliances grows sixfold to 2040, while demand for cooling increases by more than 13% per year to 2040 in the Africa Case, making the issue of cooling one of the most consequential in determining the extent of future residential energy demand (see Spotlight). The consequences for energy demand of meeting this challenge can be mitigated by ensuring that sustainability is central to the choices made when designing Africa's future built environment (Box 9.1). Rigorous efficiency standards for new cooling systems (fans and air conditioners) and better design of new buildings are both central to the Africa Case and help to avoid 110 TWh of additional electricity demand (around 15% of residential electricity demand in 2040). Efficiency standards for major appliances avoid an additional 180 TWh.

SPOTLIGHT

Is cooling comfort achievable for Africa in a warming world?

Roughly 680 million people in Africa (more than half of the population) currently live in areas that may need cooling systems.[3] This share varies by country: in Egypt and Tanzania, less than 40% of the population live in places that have daily average temperatures above 25 °C, while in countries such as Niger, Senegal and Sudan, nearly the entire population does.

Overall, around one-quarter of the global population that needs cooling today lives in Africa. Yet ownership of cooling devices is rare; air conditioner ownership across Africa averages only 0.06 units per household, while fans are somewhat more common averaging 0.6 units per household. Ownership rates reflect differences in income levels and climate. Wealthier countries such as Morocco, Algeria and Tunisia have air conditioner ownership rates that are three-times the African average, despite a lower than average number of cooling degree days (Figure 9.7).[4] In contrast, less affluent

---

[3] Cooling needs exist across multiple sectors, *inter alia*, providing health services, air conditioning in commercial buildings, cold storage for agricultural products, transport cold chains; this Spotlight highlights the biggest growth projected which is for residential buildings.

[4] Thermal comfort is measured in cooling degree days (CDDs), a universally recognised metric that allows comparison of cooling needs between regions. A CDD measures how warm a given location is by comparing actual temperatures with a standard base temperature (usually 18 °C). Calculating annual CDDs for a location gives an indication of cooling needs. CDDs can also include a heat index correction to account for the influence of humidity on perceptions of temperature and cooling needs. Many locations across Africa experience 4 000 – 5 000 CDDs annually, an order of magnitude higher than in countries such as France (230) and Italy (630). The number of CDDs in Africa dwarfs the major cooling demand centres of the United States (3 150) and China (1 100).

Figure 9.7 ⊳ Cooling degree days in the Stated Policies Scenario and cooling electricity demand, 2018 and 2040

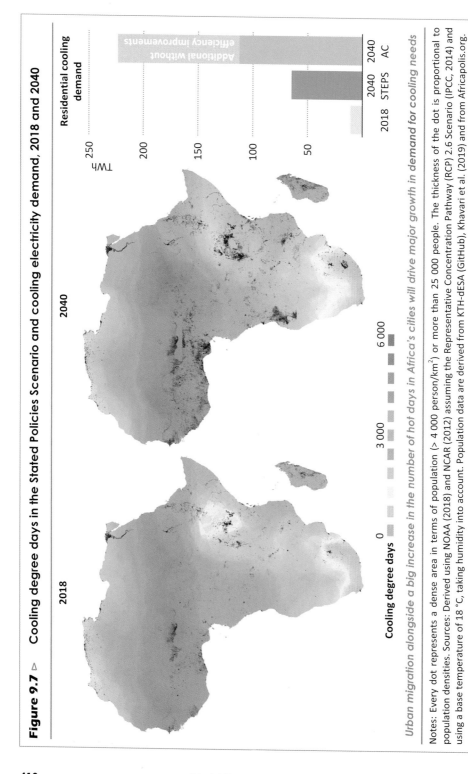

Urban migration alongside a big increase in the number of hot days in Africa's cities will drive major growth in demand for cooling needs

Notes: Every dot represents a dense area in terms of population (> 4 000 person/km²) or more than 25 000 people. The thickness of the dot is proportional to population densities. Sources: Derived using NOAA (2018) and NCAR (2012) assuming the Representative Concentration Pathway (RCP) 2.6 Scenario (IPCC, 2014) and using a base temperature of 18 °C, taking humidity into account. Population data are derived from KTH-dESA (GitHub), Khavari et al. (2019) and from Africapolis.org.

countries with much higher cooling needs, such as Togo, Senegal and Niger have ownership levels that are half the African average or less.

The costs associated with operating an air conditioner are a big barrier for many households, an issue that is compounded by comparatively inefficient equipment. The average seasonal energy efficiency ratio (SEER), the level of cooling for any given unit of energy consumed, is on average almost 30% lower for a typical air conditioner in Africa than the world average. The average air conditioner sold in Africa is also typically less than half as efficient as the best available units on the market, reflecting Africa's currently weak air conditioner energy performance standards: most countries in Africa lack any mandatory standards for air conditioners, while standards are also weak for fans. Recent policy progress and proposed minimum energy performance standards in countries such as Kenya, Rwanda and Morocco point the way forward.

Population growth, urbanisation and climate change significantly increase the need for cooling in Africa. By 2040 more than one billion people need access to cooling in the Africa Case (Figure 9.8), a scenario that matches the ambitions of the Paris Agreement (this number increases to 1.2 billion if the world continues on its current trajectory). The anticipated increase in temperatures across the continent is higher than the global average, with Somalia, Ethiopia and Gabon being the countries most exposed.

Figure 9.8 ▷ Population with cooling needs and ownership in the Stated Policies Scenario and Africa Case, 2018 and 2040

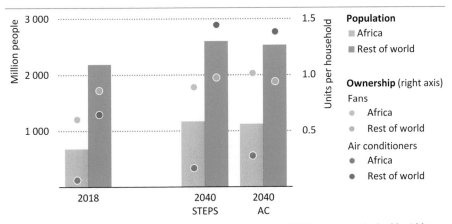

Growth in the global population with cooling needs by 2040 is concentrated in Africa, but ownership of cooling devices remains lower than the global average

Note: STEPS = Stated Policies Scenario; AC = Africa Case.

While the African continent experiences the largest increase in cooling needs by 2040, the Stated Policies Scenario points to air conditioner ownership in Africa growing to an average of just over 0.15 units per household (compared to a world average of 1.15 in 2040), ownership of fans increases to almost 0.8 units per household. While this means tens of millions of air conditioners and fans are sold in Africa in the coming decades, it still falls short of providing access to thermal comfort to millions of Africans.

In the Africa Case, cooling is more widely available: universal electricity access and higher levels of household income lead to demand for cooling increasing to over 110 TWh in 2040 in Africa, almost double the level of demand seen in the Stated Policies Scenario. More stringent policies for cooling equipment efficiency, and passive cooling through better design of buildings and use of vegetation in the Africa Case mean that around 110 TWh of additional demand are avoided.

9.2.2 Transport sector

Sub-Saharan Africa (excluding South Africa) has the world's lowest per capita car ownership level. It has a smaller passenger car stock than Australia, whose population is 95% smaller. In all scenarios and cases presented in this *World Energy Outlook*, there is a large expansion of the passenger car stock in the period to 2040 (Figure 9.9). In the Stated Policies Scenario, the car fleet in 2040 in sub-Saharan Africa (excluding South Africa) triples to reach 27 million, but this still means average ownership levels of only 15 cars per 1 000 people, equivalent to 60% of the rate in India today. Factoring in an accelerated rate of economic growth, as in the Africa Case, the car fleet reaches more than 35 million.

Figure 9.9 ▷ Vehicle ownership by country in the Stated Policies Scenario and Africa Case, 2018 and 2040

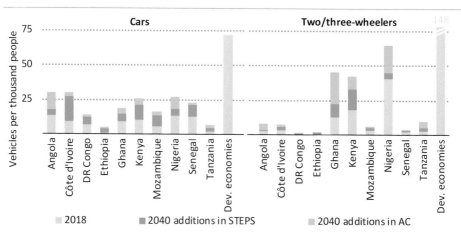

Vehicle ownership doubles in sub-Saharan Africa (excluding South Africa) in the Stated Policies Scenario; the Africa Case provides a further boost, yet rates remain low

Note: Dev. economies = developing economies; DR Congo = Democratic Republic of the Congo; STEPS = Stated Policies Scenario; AC = Africa Case.

The increase in road transport demand fuels a significant rise in oil use in Africa. Under the assumptions of the Stated Policies Scenario, oil demand for transport grows from just over 1 million barrels per day (mb/d) today to reach 2.2 mb/d by 2040. Demand is slightly higher in the Africa Case, at 2.5 mb/d, but this assumes 8 million additional cars, with the increased efficiency of the fleet avoiding the consumption of nearly 320 thousand barrels per day (kb/d) in 2040 (Figure 9.10).

As is the case with residential energy demand, there is a large discrepancy between demand in cities and rural areas. The 6.6 million passenger cars in cities in sub-Saharan Africa (excluding South Africa), represent almost 80% of the total car stock of the region today. This number rises by almost 6% per year in the Stated Policies Scenario, raising the total to close to 23 million cars in cities by 2040 and bringing the share of cars in urban areas to 85% of the total stock. In the Africa Case, the number of cars grows by more than 7% a year increasing the urban car stock to nearly 30 million by 2040. Rural areas account for over half of the current stock of 16 million two/three-wheelers in the region: in both the Stated Policies Scenario and the Africa Case, they continue to account for a sizeable 40% share of the stock of two/three-wheelers through to 2040.

Future demand for energy in the transport sector is determined not just by the number of vehicles on the road but also by their condition. Up to 80% of cars for personal transport are used cars imported from Japan and Europe which no longer meet emissions standards in those countries. The situation is not much better for new cars: only Nigeria (Euro 3) and South Africa (below Euro 3) have any emissions standards in place for new car sales (UNEP, 2017), and the Euro 3 standard was superseded in Europe almost two decades ago. Members of the East African Community recently agreed to adopt Euro IV/4 equivalent standards for new vehicles (UN Environment, 2019). The average fuel economy of cars on the road in sub-Saharan Africa is 8.4 litres per 100 kilometres (L/100 km), less efficient than the average 7.4 L/100 km in North African countries. Among several possible solutions, the most practical would be to ensure a uniform age limit on imported cars across all African countries and to ban the import of cars that do not meet minimum emissions standards. Currently more than half of African countries do not impose any restrictions on second-hand vehicle imports, while for those that have set restrictions, half of them apply age limits of between 8-15 years (UN Environment, 2019). Lower age limits on imported vehicles would significantly improve average fleet efficiency. Angola was a first mover among sub-Saharan countries, with an age limit of three years. Fuel quality specifications could also bring significant benefits, not least in reducing pollution. Here the East African Community is leading the way: it harmonised fuel quality standards in 2016, setting sulfur limits for gasoline of 150 parts per million (ppm) and for diesel of 50 ppm.

Africa is one of the world's fastest growing markets for two/three-wheelers, with momentum partially driven by the increasing number of cheap second-hand motorcycles imported from Asia. Market growth means that the number of motorcycles in Africa is set to surpass the number of private cars by 2040. However, the market for two/three-wheelers in Africa is largely informal and unregulated, and market growth has not been accompanied by a shift to electric options as part of a transition to low-carbon transport and emissions reductions. Electric two/three-wheelers have a lot of potential value given the expected increases in demand for mobility and the current low levels of mobility in most parts of rural and urban areas in Africa, especially since travel in rural areas often involves long distances and fuel quality is less reliable. The higher upfront costs of an electric two/three-wheeler as compared with a conventional one would be offset by lower maintenance and operating costs, depending on the distances travelled and fossil fuel prices; the payback period could be two to three years. The lack of reliable access to electricity however is a major barrier to electrification of these vehicles, especially in rural areas where two/three-wheelers play a prominent role. Even when full access is achieved, decisions will be required on the extent to which transport should be a priority for electrification over other end-use sectors.

Figure 9.10 ▷ **Oil demand for transport in sub-Saharan Africa** (excluding South Africa) **in the Stated Policies Scenario and Africa Case**

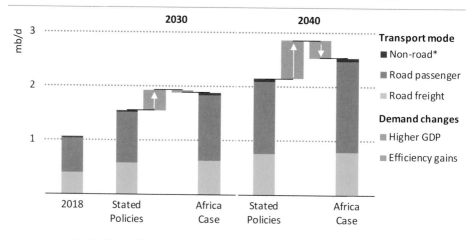

Introduction of fuel economy standards could avoid 0.32 mb/d of oil demand growth driven by the fleet expansion

* Non-road includes aviation, shipping, rail and other transport.

Increased industrialisation means more transport of raw and finished goods, leading to higher demand for freight vehicles, rail, navigation and aviation. The new African Continental Free Trade Agreement should also stimulate improved connectivity between countries to facilitate goods transport from regions of production to major commercial centres and ports. A transcontinental railway financed by China is already in the works: it plans to connect shipping ports in West and East Africa and it is possible that it may link more than ten countries including Angola, DR Congo, Zambia, Tanzania and Kenya. Such projects should help to support the economic development of landlocked African nations and enhance trading and mining activities in the region significantly.

Box 9.2 ▷   Gender and mobility in Africa

> Improving transportation is a big challenge for sub-Saharan Africa countries, and discussions on how best to deliver improvements must take account of the needs of women in both urban and rural communities. In cities, factors such as financial constraints, cultural norms prioritising asset allocation to men, and security concerns limit access to transport for women. In rural areas, women and children are more vulnerable than men to the negative impacts of a lack of transport options. Long distance walks to fetch water or wood for fuel, and to reach healthcare facilities and schools are a major obstacle in the daily lives of women. The effects range in severity from loss of productive time to reduced literacy rates and risks to personal safety. Proper transport planning needs to consider the different mobility needs of women and men throughout the day and to seek to ensure that women remain safe when travelling.
>
> In sub-Saharan Africa, maternal mortality and morbidity rates continue to be high (Hofman et al., 2008; AMANHI, 2018). Although there are many factors in addition to obstetric causes that contribute to high mortality in sub-Saharan Africa, one is the delay in reaching emergency care. Emergency and routine medical check-ups for the elderly, children and women are critical to achieving most of the sustainable development goals, specifically SDG 3 on healthy lives and wellbeing, but rural populations in sub-Saharan Africa often still find it difficult to access healthcare and other essential services due to lack of mobility. Light two/three-wheelers could make a big difference to rural mobility in Africa. They are extremely popular in large, developing countries such as China and India and have great potential to improve women's lives in Africa. Electric two/three-wheelers would avoid air pollution issues and, provided that electricity is available, may be cheaper than fossil fuel powered alternatives.

Is there a role for biofuels for transport?

Biofuels account for less than 0.1% of transport energy use in Africa today, but there is strong potential for growth. The market grew 5% in 2018, mainly led by South Africa and Nigeria.

The potential for production of advanced biofuels in many African countries is enormous, thanks to the size of the continent's agricultural sector. Increasing biofuel production from the transformation of agricultural waste can be sustainable if based on the intensification of crop production and livestock grazing on existing agricultural lands, rather than the extension of crop and grazing lands (IRENA, 2017). Relying on intensification rather than extension reduces negative environmental impacts such as deforestation, land-use change and associated greenhouse-gas emissions. East Africa alone is estimated to have the potential to produce over 100 million tonnes (Mt) of agricultural residues per year, which could be converted into advanced biofuels like ethanol and bio-butanol (Bentsen et al., 2014). Realisation of this potential will depend on whether using this resource for biofuels production can compete in terms of cost-effectiveness with other potential uses of these residues such as direct combustion and electricity generation.

Many countries have announced mandates for boosting the use of biofuels in the transport sector, with the most popular mandates being ethanol blending rates of 5% or 10% (E5 and E10). Table 9.2 lists the intended biofuel blending targets and mandates of major African economies. In both the Stated Policies Scenario and the Africa Case, biofuels represent only a small share of total fuel consumption for the road transport sector, accounting for 1.5% and 2.5% respectively of the overall fuel consumption in road transport. The evolution of the biofuels market in the Africa Case takes into account the potential supply of agricultural residues in key African countries.

Table 9.2 ▷ Biofuel blending mandates for transport for selected countries in sub-Saharan Africa

Country	Mandate	Target	Potential from agricultural residues
Angola	Ethanol 10		11%
Ethiopia	Ethanol 5	Ethanol 10	n.a.
Ghana		Replace 10% of fossil fuels by 2020 and 20% by 2030	n.a.
Kenya	Ethanol 10	Ethanol 5, Ethanol 10	5%
Malawi	Ethanol 10		n.a.
Mozambique	Ethanol 10		1%
Nigeria	Biodiesel/Ethanol 2	Ethanol 10	n.a.
South Africa	Ethanol 2, Ethanol 10		n.a.
Sudan	Ethanol 5		n.a.
Uganda		Ethanol 20	n.a.
Zimbabwe	Ethanol 10	Ethanol 15, Ethanol 20	n.a.
Zambia		Ethanol 10	n.a.

Notes: n.a. = not available. Mandate or target numbers refer to the blending ratios of each type of biofuel. The distinction between mandate and target is related to the policy framework strictness and the mechanisms in place to enforce these shares. Enforcement of these mandates and targets vary by country: with some having announced them but not yet enforced or are not strict. For example, in Zimbabwe, when ethanol production is low, the E10 mandate is suspended. Potential from agricultural residues refers to the maximum share of oil use in transport that could be replaced with biofuels from agricultural residues.

Sources: REN21 (2017); BiofuturePlatform (2018); Lane (2019); Sekoai and Yoro (2016); Ministry of Energy and Petroleum (Kenya) (2014); National Environment Management Authority (Uganda) (2010); Fundira and Henley (2017) ; UNCTAD (2014).

9.2.3 Productive uses

Much of the industrial production that does exist in Africa relies on the extraction and sale of natural resources without further transformation into higher value products. This is reflected in energy use in industry, which is low by international standards. Other productive activities, in the services and agriculture sectors, contribute higher shares to GDP in many African economies, with an average of 54% and 19% respectively across sub-Saharan Africa.

Future GDP growth is largely driven by the industry and services sectors, especially in the Africa Case. Industry and services value-added in some countries (Ethiopia, Kenya and Tanzania) in the Africa Case increase almost by an eightfold (Figure 9.11). In global average terms, generating $1 of value added within the industry sector consumes twice as much energy as $1 of value added from the agriculture sector, and up to eight-times the amount consumed to produce $1 of value added from the services sector. In sub-Saharan Africa the situation is different because the agriculture sector is less motorised than anywhere else, and the services sector includes many small-scale activities with limited contribution to GDP. This leads to the industry sector in sub-Saharan Africa consuming six-times more energy to generate $1 of value added than in either the agriculture or services sectors.

Figure 9.11 ▷ Increase in value added by sector in selected countries in the Stated Policies Scenario and Africa Case, 2018 to 2040

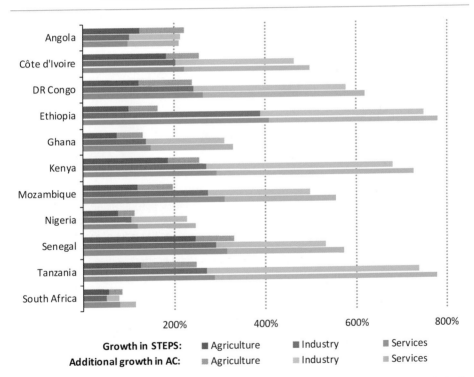

Industry and services sectors are major drivers of GDP growth in sub-Saharan Africa in both projections, while the share of agriculture in GDP drops to 13% in the Africa Case

Note: DR Congo = Democratic Republic of the Congo; STEPS = Stated Policies Scenario; AC = Africa Case.

The agriculture, industry and services sectors are all set to increase their energy demand in sub-Saharan Africa, with most of the demand growth being driven by light industry and services. Energy demand across these productive uses increases by 110% in the period to 2040 in the Stated Policies Scenario. In the Africa Case, GDP is more than 50% larger by 2040 than in the Stated Policies Scenario, but energy consumption for productive sectors is only 20% larger as a result of efficiency gains. The Africa Case also projects a different energy consumption mix by 2040 as electricity is increasingly used in the services sector and for motors in industry, causing electricity consumption to grow 50% faster than demand for oil and gas together. The share of consumption by sector also varies over time and by case. By 2040, as more industrial products are produced domestically (cement, steel, aluminium and in some countries, chemicals), the energy consumed by heavy industry doubles in the Stated Policies Scenario and triples in the Africa Case.

Figure 9.12 ▷ Final consumption in productive uses in sub-Saharan Africa (excluding South Africa) **by scenario**

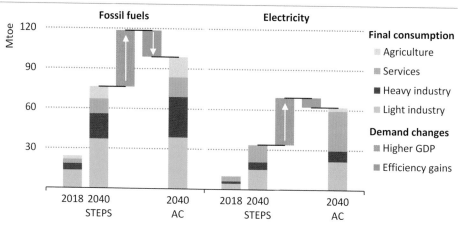

Energy efficiency standards and material efficiency temper oil and gas demand growth by around 25% and electricity by 15% for productive uses

In the Africa Case, around 35 Mtoe of energy consumed by productive uses in sub-Saharan Africa (excluding South Africa) is avoided thanks to energy efficiency (Figure 9.12). There is an important opportunity to put in place robust efficiency policies today in anticipation of expected energy demand growth. In many countries around the world there are now standards for electric motors used within industry and the use of variable speed drives and other measures to increase the efficiency of motor systems is becoming increasingly common. In Africa, some countries have recognised the potential efficiency benefits from the implementation of MEPS for electric motors, but no countries have yet enforced them.[5]

---

[5] In Uganda, standards have been developed, but are not yet mandatory. In South Africa, it has been listed as a planned policy in its post-2015 National Energy Efficiency Strategy. In Egypt, market assessments are seeking to pave the way for development of standards. In Ghana, planned implementation was announced in its National Energy Efficiency Action Plan, but not yet implemented.

Efficiency improvements in the Africa Case are not confined to industrial activities: efficient pumps for irrigation, and efficient appliances and cooling devices also help to reduce the energy intensity of other sectors compared to the Stated Policies Scenario.

Heavy industry

Currently, Africa's per capita use of construction material is a small fraction of the global average, and most of what is consumed is imported (Figure 9.13). However, Africa's rapid urbanisation presents considerable potential for industrial growth across the continent: domestic industries have an opportunity to compete effectively to provide the materials needed to build new cities and expand current ones, and in so doing to underpin wider industrial development.

Figure 9.13 ▷ Steel and cement demand per capita in selected sub-Saharan countries in the Africa Case compared with 2018 levels in India

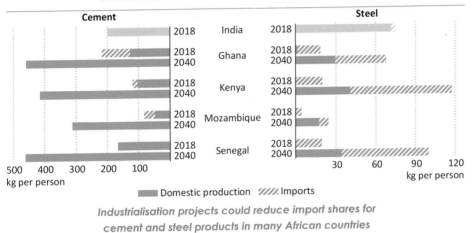

Industrialisation projects could reduce import shares for cement and steel products in many African countries

Note: kg = kilogrammes.

Cement production is, by a large margin, the most significant energy-intensive industry in sub-Saharan Africa. It currently accounts for around 2% of global cement production, and the growth of cities and infrastructure across the continent provides a significant opportunity for production to expand. Current production ranges from traditional small-scale facilities to large-scale projects, notably in West Africa. There are ambitious plans in Nigeria to further expand cement production facilities such as Unicem. Cement is an energy-intensive industry, and relies heavily on coal for the clinker/limestone calcination process, particularly in South Africa, but the share of gas is projected to grow. A cement-based industrialisation pathway implies a significant increase in energy consumption. In the Africa Case, energy demand for cement almost quadruples by 2040 and cement production increases to around 430 Mt per year in sub-Saharan Africa (excluding South Africa).

Box 9.3 ▷ Potential for hydrogen in Africa

Hydrogen could play a number of important roles in a sustainable energy future, including clean energy trade and displacing fossil fuel use in industry. Technologies for the production and use of low-carbon hydrogen are developing and projects around the world are scaling up. This promises to reduce costs, including the costs for electrolysis, the process by which water and electricity, including from renewables, can be converted to hydrogen fuel.

Globally, Africa has some of the highest potential for producing hydrogen from low-cost renewable electricity, especially from solar power. Production costs in North Africa are expected to be two-to-three times lower than in most of Europe or Japan. Energy storage for off-grid and back-up power could be an attractive application for hydrogen fuel cells; the telecommunications industry is already deploying fuels cells running on methanol to replace some diesel engines in South Africa, Namibia and other countries.

Today, hydrogen is produced and used at industrial scale in Africa to make ammonia-based fertilisers and to refine oil. Among the larger suppliers, Algeria, Egypt and Nigeria use natural gas to produce the vast majority of Africa's hydrogen for ammonia, while South Africa produces ammonia from coal at a smaller scale (all with accompanying greenhouse gas emissions). African countries also import considerable amounts of ammonia and ammonia-based fertilisers produced from fossil fuels without carbon capture, utilisation and storage. As technology for ammonia production from water electrolysis gets cheaper and smaller, hydrogen from solar power could help to avoid greenhouse gas emissions while meeting latent fertiliser demand.

Hydrogen also presents a potential export opportunity, whether as hydrogen itself or in the form of ammonia or other synthetic fuels. African countries endowed with natural gas resources and CO_2 storage options could also export low-carbon hydrogen produced using CCUS. A similar logic could apply to goods produced with low-carbon hydrogen, such as low CO_2 intensity steel. The first commercial plants using hydrogen as an alternative to fossil fuels for steel production are planned for the 2030s, a period when Africa's steel capacity is expected to grow. By looking at synergies with local steel capacity expansions in this timeframe, African countries could be good places to test this new technology, if governments and financers are supportive.

At present, more than half of steel demand in sub-Saharan Africa is met through imports but projections indicate expanding domestic steel production to meet local demand with the share of imports reducing gradually over time. Only a handful of sub-Saharan African countries currently produce steel, with South Africa accounting for the vast majority of the regions production. In the Africa Case, primary steel production in sub-Saharan Africa, excluding South Africa, increases from less than 1 Mt today to almost 20 Mt in 2040, pushing up energy demand by 12 Mtoe. This increase in production is not sufficient to meet all growth in domestic demand, and the share of imports remains high. New developments in the steel sector in Africa include investment in electric steel-making furnaces which enable scrap reuse. Another possibility, given Africa's endowment of low-cost renewable power resources, is the use of hydrogen through renewables-based electrolytic hydrogen production (Box 9.3). This, however, would require levels of capital investment that do not appear to be on the horizon for the moment. Overall, less than 1% of global steel is produced in sub-Saharan Africa and its global market share remains low in both scenarios.

Africa is home to one-third of the world's proven reserves of bauxite, but only a handful of countries, including South Africa, Mozambique, Cameroon, Ghana and Kenya, process it to produce aluminium. Together, sub-Saharan Africa (excluding South Africa) countries produce less than 1% of global aluminium, less than the region's demand. Aluminium production is electricity-intensive and plants are often located close to hydropower dams where cheap and reliable power can be delivered. The Inga Dam projects in DR Congo raises the prospect of attracting investment in new plants to produce aluminium, but this would require concurrent efforts to build local expertise. Aluminium production in the Africa Case increases at a rate that keeps pace with strong domestic demand growth, and energy consumption for aluminium production nearly quadruples by 2040 as a result.

The chemicals industry in Africa is heavily concentrated in South Africa and Nigeria, and demand for chemicals and energy inputs into the sub-sector is set to grow through to 2040. Chemical industries are often linked to the availability of oil and gas infrastructure, due to the need for petrochemical feedstock. Nigeria and Tanzania are examples of countries where the emergence and development of upstream resource extraction is attracting chemical industries, especially for methanol, with production expected to begin in coming years. By 2040, Gabon, South Africa and Nigeria use hydrocarbons in the fertiliser industry for large-scale ammonia production in all scenarios. As agriculture continues to account for an important share of African GDP in 2040, the case for local fertiliser industries is strong, and production is projected to increase. Clean hydrogen-based ammonia production, as tested in Morocco, could offer a sustainable way to boost ammonia production.

Light industry, agriculture and services

Light industries such as food processing and manufacturing are often located close to urban areas with existing electric grid infrastructure. They provide significant employment and are less energy intensive than heavy industries. The sub-sector is characterised by a strong degree of electrification and need for low-temperature heat. Security of energy supply is of paramount importance. If reliable energy supply can be guaranteed, the development of light industries including manufacturing is an attractive option for many sub-Saharan African countries. The Africa Case sees the sub-sector growing strongly throughout the projection horizon, and energy demand follows suit.

The agricultural sector accounts for more than half of employment in sub-Saharan Africa today. Agricultural productivity per hectare in sub-Saharan Africa is well below that of other regions, largely due to the limited use of irrigation to raise crop yields and lack of mechanisation (IEA, 2017). As a result, food production per person has not changed significantly since 2010. In the Africa Case, energy consumption in agriculture in sub-Saharan Africa increases threefold: most of the additional demand is for oil and electricity as the sector becomes increasingly mechanised. Agricultural value added represents around $830 billion today (18% of total GDP in sub-Saharan Africa). In the Stated Policies Scenario, it more than doubles, while in the Africa Case it almost triples.

The services sector consumes only a quarter of energy consumed by productive uses in sub-Saharan Africa today, but generates 55% of GDP. It depends principally on bioenergy and electricity; planned increases in electricity generation would support growth. Alongside increased demand for office space and associated cooling demand, new data centres and information technology infrastructure also propel growth in electricity demand. However, a shift from relatively informal services activities to high-tech services largely delivered in offices lowers the use of bioenergy and, as a result, leads to a reduction of the sector's energy intensity (measured in terms of energy consumed per unit of value added).

9.3 Clean cooking: the role of cities and higher incomes

While the world has seen considerable progress towards achieving universal access to electricity in recent years, increasing access to clean cooking facilities remains challenging. In sub-Saharan Africa, around 900 million people lack access to clean cooking (five-out-of-six people), accounting for a third of the global total. Almost 95% of them use solid biomass in the form of fuelwood, charcoal or dung in open fires, while the remainder use kerosene (especially in Nigeria) or coal (mostly in Southern Africa). At the global level, 80% of those without access to clean cooking are located in rural areas, and they make up 60% of the world's rural population. Less than 15% of the urban population globally lacks access to clean cooking, thanks to wider access to clean options, such as LPG, and higher average incomes. In sub-Saharan Africa, the lack of access to clean cooking in cities remains much higher than in other regions, with just one-third of the urban population having access to clean cooking solutions, but the problem is much worse in rural areas, where only 6% do. While fuelwood remains the dominant fuel in rural areas, for many urban households, charcoal has become the fuel of choice, as distances to sources of fuelwood increase.

The household air pollution resulting from reliance on inefficient and polluting cookstoves is directly linked to nearly 500 000 premature deaths in sub-Saharan Africa in 2018, and 2.5 million globally – a figure that equals the combined death toll of malaria, tuberculosis and HIV/AIDS. Despite the size of the gap between the current level of access to clean cooking and the goal of providing clean, reliable and affordable energy for all by 2030, the issue is often not given importance commensurate with its impacts. In addition to severe health effects, the reliance on traditional use of solid biomass for cooking contributes to forest depletion through unsustainable harvesting of fuelwood, as well as climate change (see Chapter 12). Moreover, cooking with solid biomass incurs a considerable cost in terms of time and income: on average, households dedicate 1.4 hours a day to collecting fuel, a burden borne primarily by women and children (IEA, 2017). The time spent gathering fuelwood exposes people to various dangers and reduces the time available for educational or productive ventures. Progress in switching to cleaner cooking fuels has so far been slow, despite growing awareness of the associated health, environmental and socioeconomic impacts, as well as decades of programmes targeting access to modern cooking.

In the Stated Policies Scenario, the population lacking access to clean cooking in sub-Saharan Africa slowly increases from around 900 million to 970 million in 2030 before declining to 870 million in 2040. The share of the population with access to clean cooking increases from one-third today to 65% in 2040 in urban areas, and from 6% today to around 40% in 2040 in rural areas. But progress across the continent differs by country depending on the existence, ambition and implementation of relevant policy frameworks (Figure 9.14). In this scenario, 80 million people in Nigeria and 70 million people in Ethiopia are expected to gain access by 2040. On the other hand, in several countries population growth outpaces the number of additional people gaining access. In DR Congo, the number of people without access to clean cooking facilities almost doubles to 150 million by 2040.

Figure 9.14 ▷ Access to clean cooking in selected sub-Saharan Africa countries in the Stated Policies Scenario, 2018 and 2040

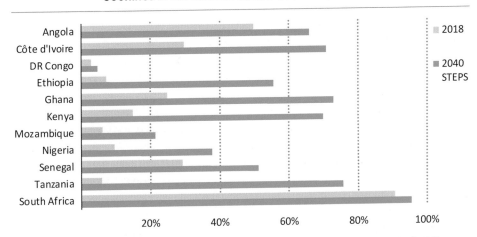

Progress towards universal clean cooking access is slow, with just under half of the sub-Saharan African population remaining without access in 2040

Note: STEPS = Stated Policies Scenario.
Sources: IEA analysis; WHO Household Energy Database.

In the Africa Case, full access to clean cooking by 2030, in line with SDG 7, means that more than 1.1 billion people in sub-Saharan Africa move away from the traditional use of solid biomass by 2030. Improvements in both scenarios have important impacts on household air pollution. In the Stated Policies Scenario, premature deaths related to household air pollution increase in the short term before declining to 2% below today's level by 2040, while in the Africa Case, the number of premature deaths linked to household air pollution falls by two-thirds.

9.3.1 Increasing access to clean cooking options

Future trajectories among the many countries in sub-Saharan Africa are very diverse and largely dependent upon policy choices and domestic circumstances. For example, 30% of urban households in Nigeria today rely on kerosene as a cooking fuel, but this is reduced through the increased uptake of LPG and gas in the Stated Policies Scenario, and completely replaced in the Africa Case by 2030 (Figure 9.15). Progress in Nigeria is facilitated by the National Cookstove Programme which provides funding for the distribution of clean cookstoves and encourages state and non-state actors to build on the national scheme via a market-based approach. In Côte d'Ivoire, the national plan seeks to increase efficiency of charcoal production and promote LPG use. In Ethiopia, recent gains in electricity access are expected to continue and result in 50% of urban households cooking with electricity in the Stated Policies Scenario (and 80% in the Africa Case in 2040), compared with 32% today. Examples of regional level frameworks to expand clean cooking access include the West Africa Clean Cooking Alliance which aims to disseminate clean, efficient and affordable cooking fuels and devices to all Economic Community of West African States citizens by 2030.

Households seeking to switch to cleaner cooking solutions, such as LPG, ethanol, natural gas, electricity and improved cookstoves, face economic and non-economic barriers. Evidence has shown that fuel and technology choices do not follow an energy ladder; higher incomes do not necessarily result in households switching from the traditional use of solid biomass to clean cooking options. Instead, a phenomenon called "fuel stacking" is increasingly prevalent in Africa, with many households using a number of different cooking solutions depending on needs and economic circumstances. The relatively high price of the technologies (and the lack of adequate and accessible financing) is an important impediment to the dissemination of clean cooking options. Even with declining prices for clean cooking technologies, and financing through loans and microcredit, millions of poor rural and urban households may not be able to afford the cost of these technologies. The variability of the fuel price is another barrier that impedes households considering switching to other options or prevents them from fully relying on cleaner cooking options.

Cultural habits, traditional cooking practices, low levels of empowerment of women and lack of awareness of the health, social, economic and environmental benefits of using cleaner options also remain persistent obstacles to widespread diffusion of clean cooking technologies to poor households. For example, wood smoke can be regarded as beneficial for avoiding bad odours, for the taste of food, or for repelling insects, exemplifying the importance of cultural habits. Health education has an important part to play in this context.

In terms of accessing clean cooking technologies, women may be disadvantaged by the fact that men often make purchasing decisions within the household. This can result in decisions to purchase or finance technologies such as solar lighting systems that are perceived as beneficial for the entire family rather than technologies such as clean cookstoves that can be perceived as having more limited benefits. Understanding how intra-household gender hierarchies influence technology acquisition is crucial for designing responses to address them. Simple design adaptations such as adding cell phone chargers to cook stoves rather than solar lanterns can sometimes give a competitive advantage to technologies that women tend to use more often than men.

Figure 9.15 ▷ **Primary fuels used for cooking in selected sub-Saharan countries in the Stated Policies Scenario and Africa Case, 2018 and 2040**

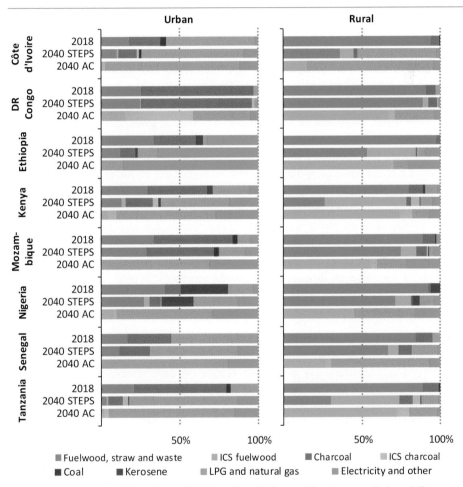

Displacing the traditional use of fuelwood and charcoal involves multiple solutions, depending on the availability and affordability of cleaner options

Note: STEPS = Stated Policies Scenario; AC = Africa Case; DR Congo = Democratic Republic of the Congo; ICS = improved cookstoves; LPG = liquid petroleum gas.

Sources: IEA analysis; WHO Household Energy Database.

Chapter 9 | Urbanisation, industrialisation and clean cooking

In urban areas in particular, the large-scale uptake of LPG, electricity, ethanol and gas in the Stated Policies Scenario enables a switch away from the use of inefficient and polluting stoves, and fuels such as lignite and kerosene. Charcoal continues to play a key role in urban areas due to its light weight, low price and high energy content compared to fuelwood (30 megajoules per kilogramme [MJ/kg] for charcoal compared to around 16 MJ/kg for fuelwood). In rural areas, reliance on fuelwood, straw and waste is set to remain relatively high, although they are increasingly burned in more efficient and improved cookstoves. Continued reliance on bioenergy in rural areas is often compounded by lack of infrastructure to supply clean cooking fuels like LPG, and by fragile supply chains.

Cooking practices are very context-specific. There is no one-size-fits-all option and each clean cooking solution comes with its disadvantages. Cooking with LPG on easy to connect burners offers clean indoor air as well as a very comfortable user experience. Meals can be cooked or reheated quickly with no difficulty in igniting or maintaining the burning flame. However, LPG can raise problems of affordability, given the high upfront cost for the burners and hoses, initial deposit for a cylinder, plus the gas content of the cylinder (users might also need to save money to be able to pay upfront to refill their cylinder). Although LPG is relatively safe compared to kerosene or biomass, illegal and unsafe refilling of the pressurised cylinders does carry dangers. Some innovative pay-as-you-go enterprises aim to tackle these challenges by supplying LPG bottles to customers that are equipped with smart meters and release small quantities of gas instantaneously when payment is received via mobile money services. Other innovative service models are also gaining traction: for example some companies are piloting the distribution of biomass pellets or briquettes, with sale costs designed to cover the cost of a subsidised or loaned gasifier stove (see Chapter 8, section 8.2.1 for more on innovative bioethanol solutions). Several biogas programmes offer support at a bigger scale, for example by providing village-scale biodigester installations including training and assistance to the community.

The economic and social barriers to the use of charcoal are much lower than for most other fuels. Using charcoal for cooking and heating comes at a lower upfront cost than using electricity, biogas or natural gas, all of which require the development of capital-intensive, durable infrastructure for fuel supply. The limited capacity for a typical solar panel to produce enough electricity for cooking may also weigh in favour of using solid biomass, as may the ability to continue traditional cooking practices with charcoal. All of the above are factors in the attractiveness of charcoal relative to alternative fuels, driving growth in its demand. Burning charcoal in improved cookstoves however can significantly reduce air pollutant emissions and fuel requirements. In the Stated Policies Scenario more than 10% of households still cook with charcoal by 2040 in sub-Saharan Africa (excluding South Africa), but 15% of them do so with improved cookstoves: this increases to 100% in the Africa Case.

Over recent years, several international organisations and initiatives have promoted access to clean cooking, including Sustainable Energy for All and the Clean Cooking Alliance, both of which have been instrumental in researching, building evidence and raising the profile of issues linked to lack of access to clean cooking. Acknowledging that progress has been slow, despite all efforts, the international community is seeking to create a renewed and stronger momentum for tackling the lack of access to clean cooking. To accelerate progress towards universal access, several international development organisations are calling for joint efforts from the energy and the health communities, on the basis that addressing the lack of clean cooking solutions and its consequent indoor air pollution should be a key priority for both. Combining strategies and actions to reach the SDG 7 on energy and SDG 3 on health will require more policy attention as well as more funding to support technology and business model innovation.

In 2017 only $2 million of equity were invested in companies in the clean cooking sector, a mere 0.1% of what is estimated to be necessary to bring clean cooking solutions to scale (ACUMEN, 2018). Private sector involvement in clean cooking remains highly fragmented, not least because a majority of clean cooking companies are small scale and face difficulties in accessing adequate funding (Clean Cooking Alliance, 2019). The recently launched Clean Cooking Fund, backed by the World Bank, aims to help address these problems by providing result-based finance, grants and technical assistance to organisations that offer innovative solutions to accelerating deployment of clean cooking solutions.

9.3.2 *Rapid urbanisation requires better use of charcoal*

More than 750 million cubic metres of wood was harvested on the African continent in 2017, accounting for roughly 20% of the global total. The share of harvested wood used for energy is around 90% in Africa, considerably higher than the global average of around 50% (FAO, 2017). With Africa producing over 60% of the world's charcoal, much of this wood is converted to charcoal to enhance its calorific value and make it easier to transport to urban areas (FAO, 2017). Charcoal use across sub-Saharan Africa (excluding South Africa) is thriving, with demand growing at an annual average rate of around 4% since the year 2000. The use of charcoal is rising most rapidly in urban areas where population growth and the often unsustainable use of forest resources limits the availability of fuelwood (Figure 9.16). As a source of income and employment creation along the value chain (for production, transportation, sales and distribution), charcoal manufacturing and trade have shaped patterns of economic development in many areas.

As urbanisation increases across Africa, the growing demand for both land and charcoal is likely to put additional pressure on traditional forest management and extraction practices. While sustainable land and biomass use is part of the climate mitigation strategy of most African countries, few countries include commitments on charcoal production or use in their Nationally Determined Contributions pledged under the Paris Agreement. Some countries, for example Kenya, however have decided to ban the use of charcoal to reduce stress on forests. Given the importance of charcoal use for cooking in sub-Saharan Africa (excluding South Africa), and the implications of its use, improving the sustainability of the charcoal value chain could bring significant benefits.

In the Stated Policies Scenario, charcoal consumption per capita continues to increase over time, with overall charcoal consumption in sub-Saharan Africa (excluding South Africa) increasing by 50% by 2040 compared to today's level. By contrast, the increasing number of improved cookstoves in the Africa Case along with other clean cooking options allows for a reduction in charcoal consumption in sub-Saharan Africa by over 25%. With urbanisation rates increasing to 51% in 2040, reducing total charcoal use requires dramatic changes in consumption patterns to decouple urban migration and increased charcoal use.

Addressing the inefficiencies of charcoal stoves is a first step towards reducing demand for charcoal, and would also reduce indoor air pollution. Twenty-three countries in sub-Saharan Africa have already committed to promote efficient or improved cookstoves as part of their updated Nationally Determined Contributions. Further opportunities exist to improve the upstream efficiency of charcoal production and reduce related greenhouse gas emissions (Box 9.4).

Figure 9.16 ▷ Charcoal consumption per capita and urbanisation rates in sub-Saharan Africa (excluding South Africa) in the Stated Policies Scenario and Africa Case

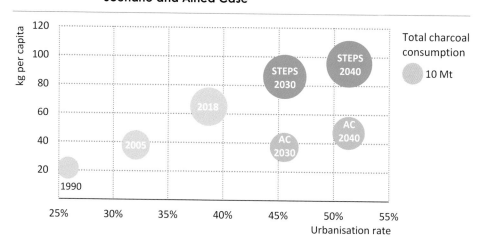

Close ties between urbanisation and charcoal use are loosened in the Africa Case, reducing total charcoal consumption and subsequent pressure on forest resources

Notes: STEPS = Stated Policies Scenario; AC = Africa Case.

It is unlikely to be possible to achieve a sustainable charcoal value chain and access to affordable, efficient cook stoves for end-users without changes in policies and regulations. A comprehensive policy framework could lead to a number of benefits including improved efficiency of charcoal production, reduced stress on forest resources and improved health outcomes. The costs involved in promoting efficient stoves and kilns, and improving forest management would need to be set against wider financial and societal benefits, not least in advancing progress towards multiple sustainable development goals.

Box 9.4 ▷   Can charcoal production be made more efficient?

Charcoal manufacturing, transportation and distribution are responsible for two-thirds of the efficiency losses in the overall charcoal value chain: the other third comes from the energy performance of stoves. There is sizeable potential to reduce upstream efficiency losses: conventional production practices rely on earth-mound kilns that operate at 8-20% efficiency, but more efficient kilns can reach efficiency levels of at 35-40%. There is similarly sizeable potential to improve the efficiency of stoves: energy-efficient charcoal equipment for cooking can achieve a thermal efficiency of 30-40%, compared with 10-20% for most traditional cookstoves. Shifting to efficient stoves for both fuelwood and charcoal burning in the Africa Case would cut biomass consumption for cooking in sub-Saharan Africa (excluding South Africa) by 60% by 2040, and would reduce charcoal consumption for cooking by 35%. These types of gains are instrumental in achieving the reductions in wood demand for charcoal production in the Africa Case (Figure 9.17).

Figure 9.17 ▷   Wood demand for charcoal production in the Stated Policies Scenario and Africa Case

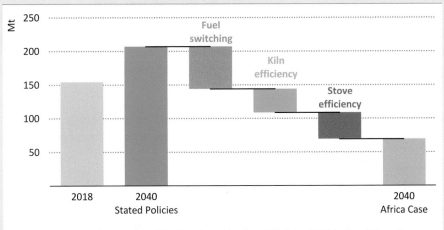

Wood use for charcoal production is cut by two-thirds in 2040 in the Africa Case with more efficient kilns and stoves accounting for one-third of the gains

Putting a price on wood resources and reinvesting revenues (for example from wood cutting taxation, licensing fees, certifications) to help ensure sustainable forest management and wider use of efficient stoves is a key step in improving the sustainability of the charcoal value chain (FAO, 2017). The diversification of bio-based fuels (using agricultural waste and wood residues) would also reduce the need for wood extraction from forests, and consequently the time spent by women and children in gathering wood and cooking.

9.3.3 Rural areas – how to unleash the potential of biogas?

Rural areas face a unique array of challenges in transitioning towards clean cooking, with the lack of availability of modern fuels being one of the principal barriers to change. LPG is not always available due to long distances and poor transport links between distribution centres and households, and to demand competition from urban areas. A move to electric cooking is impeded by very low rates of electricity access in rural areas in Africa, the unreliability of electricity supply in many places where it does exist, and the prioritisation of electricity for uses such as lighting and appliances. Other modern fuels such as ethanol and processed biomass pellets or briquettes often face similar barriers to access.

Biogas provides an alternative clean cooking solution that is ideally suited to many rural areas and can also be used as a local source of clean energy for heating. A mixture of methane and carbon dioxide (CO_2), biogas can be produced from organic by-products and waste otherwise thrown away or abandoned (see Chapter 7). Biogas is ideally suited to communities where agricultural residues and animal manure are available as a feedstock. In addition to providing a source of clean energy, anaerobic digestion produces as a by-product a valuable fertiliser that can enhance agricultural production.

Based on our new bottom-up assessment, we estimate that today in Africa there is sustainable technical potential available to produce around 50 Mtoe of biogas. The potential doubles by 2040 to almost 100 Mtoe at an average cost of just over $10 per million British thermal units (MBtu), which would represent around one-third of the projected natural gas demand in the region in the Stated Policies Scenario. About 80% of this potential is in sub-Saharan Africa (Figure 9.18).

Figure 9.18 ▷ Cost curve of potential biogas supply by feedstock in Africa, 2040

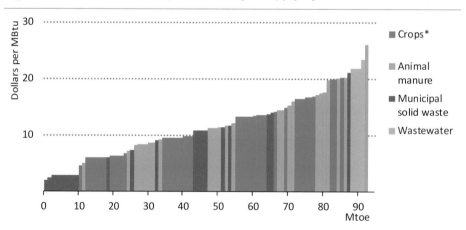

By 2040, over 90 Mtoe of biogas could be produced in Africa, around half of which would cost less than $10/MBtu

* Includes crop residues only, energy crops are excluded given concerns about their sustainability.

Note: MBtu = million British thermal units.

The biggest contribution to the potential comes from rural areas with strong agricultural sectors. Crop residues, especially cereals, account for almost 60% of the total potential, animal manure for close to 25%, and municipal solid waste (MSW) for most of the remainder. At the end of the outlook period, the picture changes slightly as further urbanisation increases the availability of MSW and as anticipated changes in diet bring an increase of livestock and therefore of animal manure.

The main route to biogas production (over 80% of our estimated 2040 potential) is via a biodigester, an airtight system where anaerobic digestion occurs and biogas is produced. These can be either decentralised (household-scale) or centralised. A decentralised biodigester has a basic design and is built to provide enough biogas to fulfil the energy needs for cooking and water heating of a single family. It produces biogas at an average cost of $6/MBtu, with variations depending on the capital cost of the biodigester installed. Centralised biodigesters can be of small, medium or large scale and the feedstock is usually provided from a single farm or a group of farmers with the involvement of local entrepreneurs. In this case costs are higher – in the range of $12-20/MBtu – with limited economies of scale and additional labour and feedstock costs.

A clear picture of today's consumption of biogas in Africa is not available due to lack of data. We estimate that current biogas use is around 5 000 tonnes of oil equivalent (toe) (6 million cubic metres of natural gas equivalent), and its use is concentrated in countries with specific support programmes for this fuel. Some governments, such as Benin, Burkina Faso and Ethiopia, provide subsidies that can cover from half to all of the investment, while numerous projects promoted by non-governmental organisations provide practical know-how and subsidies to lower the net investment cost. In addition to these subsidies, credit facilities have made progress in a few countries. A new lease-to-own[6] (LtO) arrangement has recently been developed by a limited number of companies in Kenya, and around 45% of the households in Kenya that installed a digester in 2018 financed their unit through an LtO arrangement (ter Heegde, 2019).

Research in East Africa shows that families with access to biogas see benefits in terms of ease of cooking and a reduction in the time spent collecting fuelwood, as well as a lower incidence of health and respiratory problems. There are also co-benefits in terms of agricultural productivity (as a result of using the bio-slurry as fertiliser) and reducing deforestation (Clemens et al., 2018).

In the Stated Policies Scenario, consumption rises to over 3 Mtoe of biogas in Africa by 2040. However, there is much larger uptake of biogas in the Africa Case, spurred by the drive in this scenario to provide universal access to clean cooking by 2030. Biogas demand rises to 9 Mtoe in 2040, over half of which is used for providing access to clean cooking in sub-Saharan Africa, the remainder is used for power generation. An additional 2 Mtoe is used for biomethane production in South Africa. In this scenario, over 135 million people in Africa use biogas to move away from reliance on traditional use of solid biomass.

---

[6] Lease-to-own credit mechanism allows the user to purchase the biodigester when the lease period expires.

The main economic challenge is the relatively high upfront cost of the biodigester. In Africa, upfront costs for an average sized household biodigester with a technical lifetime of over 20 years can range between $500-800 (ter Heegde, 2019). A part of the capital cost can be reduced by using traditional and locally available construction materials such as sand and gravel, and by relying on local labour. For the remainder, financing help is often needed.

There are also significant non-economic barriers, notably biodigester maintenance and availability of gathered feedstock. These barriers can be even more pronounced for a biodigester at the community scale or larger. In a survey in East Africa, more than a quarter of biodigesters installed between 2009 and 2013 were out of operation by 2016 because of a lack of readily available maintenance expertise (Clemens et al., 2018). Feeding a household biodigester regularly with animal manure requires at least two mature cattle, so any deterioration in household circumstances quickly affects biogas production, while local communities need to develop and maintain a system to collect waste and residues for centralised biodigesters. Local entrepreneurs and government partnerships with the private sector have a crucial role to play in overcoming these barriers, with governments promoting biogas utilisation through a range of programmes and facilities while the private sector ensures a proper and sustainable development of the sector along all the supply chain.

Chapter 10

Access to electricity and reliable power
Generating a brighter future for Africa

SUMMARY

- Today, 600 million people in sub-Saharan Africa (one-out-of-two people) do not have access to electricity, according to our latest country-by-country assessment. A number of countries make important headway in the Stated Policies Scenario, with South Africa, Ethiopia, Ghana, Kenya, Rwanda and Senegal reaching full access by 2030. This allows around 20 million people to gain access every year. Yet progress is uneven: 530 million people (one-out-of-three people) remain without electricity in 2030. Annual gains in access would need to triple to reach universal access by 2030.

- South Africa differs from its sub-Saharan African peers with its mature economy, successful access programmes and integrated policy making. Competitive auctions for renewables are stimulating private investment. The financial health of the state-owned utility remains vulnerable, strengthening its commercial and operational performance is essential to the future well-being of the power sector.

Figure 10.1 ▷ Electricity demand and generation in sub-Saharan Africa (excluding South Africa) by scenario, 2018 and 2040

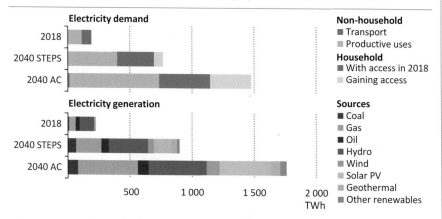

Demand quadruples by 2040 in the Stated Policies Scenario and increases almost eightfold in the Africa Case, renewables and gas rise to meet demand growth

Note: STEPS = Stated Policies Scenario; AC = Africa Case.

- Electricity demand in the rest of sub-Saharan Africa is set to quadruple by 2040, driven by rising incomes and industrialisation. However, per capita demand remains low, at less than 15% of today's global average. Accelerated economic development and universal access to electricity in the Africa Case push demand to almost 1 500 terawatt-hours (TWh) by 2040, with households in urban areas approaching the ownership and consumption levels of middle-income countries.

- Keeping pace with soaring needs, electricity supply in the rest of sub-Saharan Africa increases fourfold in the Stated Policies Scenario. Generation capacity triples to 270 gigawatts (GW) by 2040, but this is far short of the 600 GW reached in the Africa Case. The expansion is achieved through a combination of renewables and natural gas. Solar photovoltaics (PV) plays a key role in delivering access and becomes the largest source in terms of installed capacity in the mid-2030s in the Stated Policies Scenario, and in the mid-2020's in the Africa Case, overtaking hydropower.

- While hydropower generation grows in both scenarios, new detailed analysis on its vulnerability to climate change shows increased variability of outputs and the need to plan for long-term resilience with a diverse power mix and regional co-operation.

- The expansion of generation capacity is not sufficient to provide reliable and affordable electricity. Around 80% of sub-Saharan African businesses recently suffered from electricity disruptions, leading to average annual losses of around 8% of sales. Sustaining economic growth of 7.3% per year in the Africa Case requires continuous focus on transmission and distribution assets to reduce the incidence of power outages by over three-quarters and take network losses to below 10% (from 18% today) improving economic outcomes for companies, including power utilities.

- Focus on improved network management, densification and extensions see the grid provide about 70% of the 230 million new connections expected by 2030 in the Stated Policies Scenario. In the Africa Case, mini-grids and stand-alone systems, mostly based on renewables, are essential to bridge the gap to achieve universal access; they are the least-cost solutions for over two-thirds of the additional people that attain access, connecting almost 450 million people by 2030 in the Africa Case.

- Investment in the power sector in sub-Saharan Africa averages more than $45 billion per year over the outlook period in the Stated Policies Scenario (compared to $21 billion today). The Africa Case requires a fivefold increase to ensure reliable and affordable power for all (over $100 billion per year). In both scenarios, half of the total investment goes to expansion, reinforcement and maintenance of grids, increasingly for mini-grids and cross-border infrastructure. Investment in low-carbon generation accelerates, driven by a rise in spending on solar PV projects, which reaches almost $25 billion per year on average in the Africa Case.

- Most power sector investment in Africa today is underpinned by public funds, with heavy reliance on international development finance. Given the financial constraints of utilities and limited fiscal capacity of governments, private sources of finance will be essential to bridge investment gaps. Policy and regulatory improvements are needed to address investment risks, facilitate a more effective use of public funds and help reduce the cost of capital. Four areas are crucial to foster a more self-sustaining environment for investment: better financial performance of utilities; improvements in procurement frameworks; more sustainable business models for the decentralised sector; and strengthened provision of long-term finance.

10.1 Introduction

The achievement of universal access to reliable electricity is vital if Africa is to thrive, and that means providing access for the first time to the 600 million people currently deprived of electricity, electrifying schools and hospitals, and ensuring that electricity is available for companies and entrepreneurs. China and India faced similar challenges and it took them 35 and 16 years respectively to reach a 95% access level and to connect as many people as now need to be connected in sub-Saharan Africa (Figure 10.2). More decentralised and modular technologies are now available and they are reducing the length of time it takes to provide access and the costs of doing so.

The regional and country-by-country projections for the Stated Policies Scenario and the Africa Case are described in this chapter. We analyse the electricity demand and supply outlook and their drivers, the changing electricity generation mix by technology and access solution, and the implications for affordability, reliability and investment needs.

Figure 10.2 ▷ Reaching universal access to electricity in sub-Saharan Africa compared with achievements in China and India

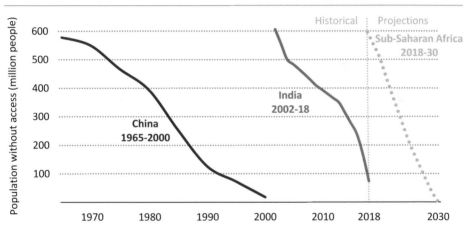

Achieving access for all in sub-Saharan Africa in only twelve years will require an unprecedented effort

An increasing number of countries are implementing policies with a view to meet the United Nations Sustainable Development Goal 7[1] by 2030, resulting in substantial progress in the Stated Policies Scenario. New technologies and business models are attracting investment from donors, development banks and increasingly the private sector. Nonetheless, without a significant step up in efforts, the population without access to electricity will remain as high as 530 million in 2030 in sub-Saharan Africa.

---

[1] Sustainable Development Goal 7: Ensure access to affordable, reliable, sustainable and modern energy for all.

In the Africa Case, we examine by country, and for sub-Saharan Africa as a whole, the range of technologies, policies and investment frameworks needed to achieve the universal electricity access target. Our least-cost analysis points to the best way forward as being comprehensive policies that make use of all solutions, centralised and decentralised, with mini-grids and stand-alone systems providing power to more than half of the population gaining access by 2030.

We go beyond looking at the achievement of universal access and also examine what it would take to develop a reliable, affordable and sustainable power system capable of making the African Union's Agenda 2063 a reality (see Part B introduction, Box B.2). Reliable electricity is an essential element of a thriving economy, and Africa has the opportunity to be the first continent to industrialise and build resilient and reliable power systems based on cleaner sources, with a combination of readily available renewables and natural gas now looking like the most competitive way to provide electricity.

10.2 Outlook for electricity access

While more than 99% of the population in North Africa has access to electricity, the situation is very different in the rest of the continent. In the Stated Policies Scenario, the number of people without access to electricity across sub-Saharan Africa declines slightly to 530 million by 2030, but increases after that to 600 million as rapid population growth outruns efforts to increase access (Figure 10.3). Many countries on the continent are putting in place policies which, if effectively implemented, will allow around 20 million people to gain access to electricity each year by 2030, a rate similar to what the region has witnessed since 2013 (see Chapter 8).[2] However, the rate is less than a third of what would be needed to reach full access by 2030. The share of the population with access to electricity in sub-Saharan Africa rises from 45% today to nearly 65% in 2030 in the Stated Policies Scenario, with over 230 million people gaining access.

Projected progress in the Stated Policies Scenario is most rapid in East Africa, as it moves from a regional access rate of 43% today to more than 70% by 2030. Kenya, Ethiopia and Rwanda are all set to achieve universal access before 2030 (Table 10.1). Ethiopia brings access to the highest number of people in the region by 2030 (more than 70 million). Tanzania also sees rapid progress, with its electrification rate climbing to around 70% in 2030 from less than 40% in 2018. Progress is also made in West Africa and Southern Africa, where the regional access rates reach over 60% by 2030. South Africa and Ghana, which achieved two of the highest access rates on the continent in 2018 after two decades of effective government leadership, are expected to reach full electrification by 2030. Senegal is expected to achieve universal access in 2025. Strong efforts in Nigeria and Côte d'Ivoire result in their rates of access increasing to 80% and more than 90% respectively by 2030. Countries across Central Africa see limited progress under the Stated Policies Scenario, but there are some bright spots: Gabon and Cameroon both reach more than 90% by 2030.

---

[2] The investment and financing implications are discussed in sections 10.7 to 10.8.

Figure 10.3 ▷ Electricity access progress in sub-Saharan Africa in the Stated Policies Scenario

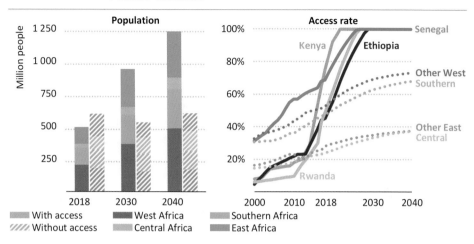

Strong policy support is instrumental to drive the rapid increase in access rates observed in several countries, but many struggle to provide access to increasing populations

Table 10.1 ▷ Electricity access policies and targets in selected countries

Country	Target	Implementation measures
Kenya	Full access by 2022	Kenya National Electrification Strategy (2018): investment of $2.8 billion from 2018-22. Kenya Off-grid Solar Access Project: distribute 250 000 solar home systems to power households, schools, health facilities and agriculture by 2030.
Ethiopia	Full access by 2025	Electrification Program (2017): geospatial least-cost roll-out plans, fast-paced extension of the grid to reach 65% of the population with the grid and 35% with decentralised systems by 2025; public-private off-grid programme for 6 million households.
Rwanda	Full access by 2024	Energy Sector Strategic Plan and Rural Electrification Strategy: connect 52% households to the grid and 48% to decentralised systems by 2024; connect all productive users; cut by half the duration and number of interruptions; introduction of appliance efficiency standards.
Senegal	Full access by 2025	National Rural Electrification Program (PNER), aiming to electrify 95% of rural clients through grid extension, 4% through solar only or solar-diesel hybrid mini-grids, and the rest through solar home systems.
Côte d'Ivoire	Connect all areas by 2025	Programme Electricité pour Tous: electrify 1 million households. Programme National d'Electrification Rurale: connect all towns above 500 inhabitants by 2020, and all areas by 2025. Tariff reductions for poor households.

Despite this impressive projected progress in a number of countries, around 20 countries, accounting for 30% of the population of sub-Saharan Africa in 2030, still have less than half of their population with access to electricity in 2030 on the basis of current and stated policies. Across sub-Saharan Africa, 36% of the population have no access to electricity in 2030, of which three-quarters, or more than 400 million people, live in rural areas.

Achieving full access by 2030 would require tripling the current rate of annual connections to reach over 60 million people on average each year. This would mean finding ways to connect people living "under the grid" but lacking access (see Spotlight in Chapter 8, section 8.2.2). It would also mean accelerating the deployment of mini-grids and stand-alone systems, which are the least-cost way to provide power to more than half of the population gaining access by 2030 (see section 10.4.2). In 2030, around 50% of the population without access in the Stated Policies Scenario live in the Democratic Republic of the Congo (DR Congo), Nigeria, Uganda, Niger, and Sudan: scaling up efforts in these countries is particularly important.

Delivering access to electricity in an integrated way would support economic growth and overall development. Access could bring new sources of productive employment to remote populations, in particular for women. Less time to complete domestic chores provides more time for paid jobs. Access to electricity also benefits women-owned businesses, helping women to move from extreme poverty to near middle-class status, as shown within areas connected by a mini-grid company in Ghana (Power Africa, 2019). A recent study shows that the decentralised renewables sector is beginning to support employment at a similar scale to the traditional utility sector, with strong potential for future growth (Power for All, 2019).

10.3 Outlook for electricity demand

Electricity demand[3] in Africa today is 700 terawatt-hours (TWh), with South Africa and the North African countries accounting for over 500 TWh of this total. Yet it is the sub-Saharan Africa countries (excluding South Africa) that see the fastest growth in electricity demand in the Stated Policies Scenario, with demand increasing at an average annual rate of 6.5%, the highest rate of any region worldwide. By 2040, electricity demand in sub-Saharan Africa, excluding South Africa, reaches 770 TWh under current and stated policies, four-times today's level. Electricity demand per capita increases from an average of 185 kilowatt-hours (kWh) today to over 430 kWh in 2040, but still represents less than 15% of today's global average of over 3 000 kWh.

Low per capita electricity demand masks large inequalities that seem likely to persist. About 440 million people across sub-Saharan Africa (excluding South Africa) live in households that have access to electricity today, predominantly in urban areas. In the Stated Policies Scenario, households with access today and new ones in areas with existing

---

[3] Electricity demand is defined as total gross electricity generated less own use generation, plus net trade (imports less exports), less transmission and distribution losses.

access to electricity consume an additional 220 TWh by 2040. This compares to an increase of only 70 TWh in order to provide access to electricity for the first time to 320 million people in this scenario. Growth in electricity demand from an emerging middle-class and newly connected households goes hand-in-hand with growth in demand from the productive sectors of the economies (Figure 10.4). Demand from industry and the services sectors more than triples to 390 TWh by 2040 in the Stated Policies Scenario, fuelled by domestic consumption and economic growth. Electricity demand from agriculture increases by 150%, but still accounts for only 6 TWh in 2040.

Figure 10.4 ▷ Electricity demand by scenario in sub-Saharan Africa (excluding South Africa)

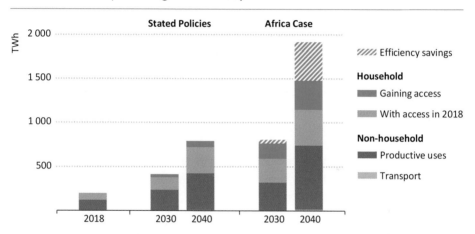

Demand quadruples by 2040 in the Stated Policies Scenario and increases almost eightfold in the Africa Case. Demand would be even higher without efficiency savings.

The Africa Case sees national strategic plans and the Agenda 2063 ambitions realised in full, with important implications for electricity demand. A virtuous cycle emerges in which electricity demand growth is fuelled by the development of local industries and services, increasing employment and incomes, and this in turn increases the consumption of locally produced goods and services. Electricity demand in the Africa Case grows at close to 10% per year to reach almost 1 500 TWh in 2040. Extension of electricity access to all households in the Africa Case adds 260 TWh of demand relative to the Stated Policies Scenario by 2040. Nonetheless, achieving universal electricity access still accounts for only a quarter of demand growth to 2040.

The electricity demand growth rate in sub-Saharan Africa (excluding South Africa) would reach 11% per year in the Africa Case without efficiency improvements in appliances and equipment. Energy efficiency measures are essential to achieve the vision of the Africa Case, helping to improve the competitiveness of local industries and reduce the impact of increases in energy services on electricity bills (see section 10.6). A handful of countries in sub-Saharan Africa are already leaders in energy efficiency when it comes to the residential appliances that accompany decentralised models of electricity access. Extending innovations of this kind to the wider appliance market is central to the savings achieved in the Africa Case.

10.3.1 Electricity demand growth by sector

The residential sector is the largest contributor to electricity demand growth, accounting for some 50% of the growth to 2040 in sub-Saharan Africa (excluding South Africa) in both the Stated Policies Scenario and Africa Case (Figure 10.5). As income levels increase across Africa, households increasingly own appliances such as refrigerators, washing machines and phones: the wealthier ones also own cooling devices. Domestic appliances are the biggest contributor to growth in residential electricity demand in the Stated Policies Scenario, adding 175 TWh to demand by 2040. Space cooling adds another 40 TWh, with the number of air conditioners across the region expected to increase almost sixfold to about 45 million by 2040. In the Africa Case, universal access to electricity is achieved by 2030 and household incomes rise more rapidly. As a result, residential electricity demand is 370 TWh higher in 2040 relative to the Stated Policies Scenario, with the increase equally split between rural and urban areas.

Figure 10.5 ▷ Residential electricity demand and household appliance ownership by scenario in sub-Saharan Africa
(excluding South Africa)

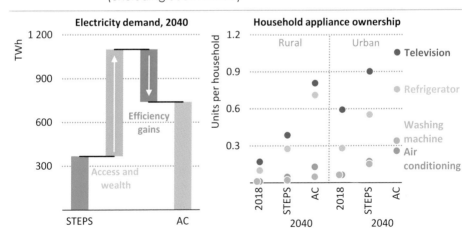

In the Africa Case, an increase in residential demand stemming from better access to electricity and increased ownership of appliances is partially offset by efficiency gains

Notes: STEPS = Stated Policies Scenario; AC = Africa Case. Access and wealth refers to the increase in demand associated with higher electricity access and higher average household incomes in the Africa Case relative to the Stated Policies Scenario. Efficiency gains refers to the reduction in electricity demand due to efficiency gains in the Africa Case relative to the Stated Policies Scenario.

By 2040, the urban population in sub-Saharan Africa (excluding South Africa) more than doubles to over 900 million, and average incomes in urban households increase by close to 40%, driving up appliance ownership rates. In the Stated Policies Scenario, on average 90% of urban households own a television and 55% a refrigerator by 2040. Ownership of air conditioners triples by 2040, but even in urban areas air conditioner ownership rates remain among the lowest in the world, despite the hot climate in many areas (see Chapter 9). Average urban household incomes are more than twice as high in the Africa Case relative to the Stated Policies Scenario, which leads to urban households purchasing more appliances, and electricity demand in urban areas increases at 10% per year, compared with 7.4% in the Stated Policies Scenario. More people own air conditioners in the Africa Case, but two-thirds of urban households remain without in 2040.

In the Africa Case, universal access to electricity in rural areas of sub-Saharan Africa (excluding South Africa) results in an additional 210 TWh of electricity demand by 2040. Rural households also benefit from higher levels of appliance ownership, roughly doubling the average number of televisions, refrigerators and washing machines in the Africa Case relative to the Stated Policies Scenario. Rural ownership of air conditioners remains uncommon. The impact of universal access and higher incomes is enough to see average per capita consumption in rural areas increase ten-fold to 320 kWh in the Africa Case, compared to only 100 kWh in the Stated Policies Scenario.

The services sector benefits from increasing electrification which contributes to economic growth. Electrification of the services sector is often a by-product of household electrification efforts, but it can also be an objective in itself: in Rwanda, the Energy Sector Strategic Plan announced in 2018 aims to bring electricity access by 2024 to all public infrastructure, schools, health facilities, small businesses and administrative offices, in addition to households. In the Stated Policies Scenario, electricity demand from the services sector reaches 170 TWh in sub-Saharan Africa (excluding South Africa) by 2040, with the majority of growth stemming from demand for cooling and appliances. Achievement of the electrification and economic growth targets in the Africa Case sees demand from the sector increase by a further 170 TWh.

Industry contributes to around 30% of the growth in electricity demand to 2040 in the Stated Policies Scenario. Electricity demand from industry increases at an annual average of 6%, which is a third faster than the rate of growth of total industry sector energy demand. Much of the growth comes from the use of electric motors in processing, manufacturing and other light industries. The Africa Case sees a step up in the rate of electricity demand growth in industry to 7.5%, driven by the modernisation of industry and increasing domestic demand for locally produced goods as well as expanding exports. Improvements in industrial energy efficiency in the Africa Case temper demand growth as well as helping to improve the competitiveness of industry. By 2040, electricity demand from industry exceeds 340 TWh, 100 TWh higher than the Stated Policies Scenario.

The agricultural sector sees increasing electricity demand for irrigation (some of it met through the use of stand-alone solar photovoltaic [PV] powered pumping systems) and for cooling (to support refrigerated storage of produce). The expansion of irrigation and cooling leads to important productivity gains,[4] and these bring further increases in electricity demand. Electricity demand for agriculture increases from about 3 TWh today to over 6 TWh in the Stated Policies Scenario, while in the Africa Case it rises to 36 TWh as a result of a larger increase in value added in the sector and a bigger shift from other sources of energy to electricity.

The electrification of transport struggles to get started in the Stated Policies Scenario: there are very few policies that support electric vehicles (EVs) (cars, buses, trucks and two/three-wheelers) and electricity accounts for only 0.5% of transport energy demand by 2040. Progress is faster in the Africa Case, but electricity still powers less than 1% of cars by 2040, together with around 18% of two/three-wheelers. By 2040, electricity demand for transport reaches 15 TWh in the Africa Case: this is almost triple the level in the Stated Policies Scenario.

The limited electrification of transport even in the Africa Case is a result of the size of the power requirements for EV charging, relative to other uses. Designing the extension of electricity access with the electrification of transport in mind would significantly increase the costs of achieving universal access. Concerns over the reliability of electricity supply and the costs of EVs also hinder their uptake. Conditions for the electrification of transport are more favourable in urban areas with existing grid connections: as a result, the majority of EV uptake in Africa is concentrated in cities, with almost 30% of the urban two/three-wheeler fleet electrified by 2040.

10.3.2 Electricity demand growth by region

The evolution of electricity demand is far from homogenous across African countries; it ranges from 4.5% to 8.5% per year through to 2040 in the Stated Policies Scenario, reflecting disparities in economic developments and the rate of progress in improving electricity access. Kenya, Ghana and Ethiopia all reach universal access to electricity before 2030, raising average per capita electricity demand to 450 kWh, 750 kWh and 375 kWh respectively in 2040. Per capita electricity demand growth is more limited in countries with lower average incomes today, and in those that experience the fastest population growth or make slower progress on electricity access, such as DR Congo and Nigeria (Figure 10.6).

While higher than average incomes contribute to higher per capita electricity demand by 2040 in Angola, Ghana and Côte d'Ivoire relative to the regional average, it is the larger economies that lead total electricity demand growth. About 15% of the increase in total electricity demand in the Stated Policies Scenario across sub-Saharan Africa (excluding South Africa) comes from Nigeria. Ethiopia accounts for a further 9%, with Tanzania, DR Congo and Angola not far behind. Overall, just seven sub-Saharan African countries account for over half of demand growth over the outlook period in the Stated Policies Scenario.

---

[4] Chapter 4 of *Energy Access Outlook 2017: From Poverty to Prosperity* takes an in-depth look at how energy can improve agricultural productivity in Africa (IEA, 2017).

Figure 10.6 ▷ Electricity demand per capita and share of regional electricity demand by scenario, 2018 and 2040

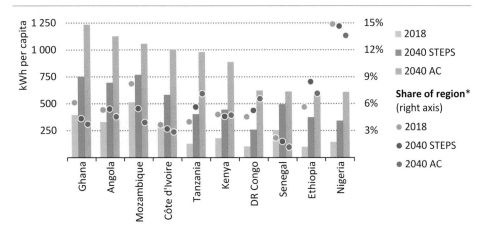

Electricity demand per capita rises fastest in the Africa Case, but in 2040 it is still only around a third of today's average in other developing countries

* Region = sub-Saharan Africa excluding South Africa.
Note: STEPS = Stated Policies Scenario; AC = Africa Case.

In the Africa Case, even those countries that have lower average incomes and currently are making slow progress on access to electricity see a jump in per capita electricity demand, thanks to the combined impacts of universal access to electricity and accelerated economic growth across all sectors of the economy. In DR Congo, for example, electricity demand is 625 kWh per capita in 2040 in the Africa Case, more than double the level in the Stated Policies Scenario as a result of a near 9% gross domestic product (GDP) growth rate and the electrification of an additional 120 million people. In absolute terms, this translates to an additional 55 TWh of electricity demand in 2040, enough to see DR Congo's share of total electricity demand in sub-Saharan Africa (excluding South Africa) increase from 5% in the Stated Policies Scenario to 7% in the Africa Case.

Box 10.1 ▷ Electricity in South Africa – deep transformation ahead?

South Africa's energy landscape looks different from that of other sub-Saharan countries. The country has a more mature economy than its neighbours and a history of relatively low energy prices, in particular for coal and electricity. The competitiveness of electricity relative to other fuels results in a share of electricity in final energy consumption of 25% today, which is high by international standards. This has favoured the development of energy-intensive industries and the extensive electrification of energy use in buildings. South Africa currently accounts for over half of all electricity consumed in sub-Saharan Africa.

Competitive electricity prices and the coupling of clean cooking efforts with electrification mean that around 85% of South African households cook with electricity today. Electrifying cooking increased residential electricity demand, but the largest consumer of electricity in South Africa remains the industry sector, which accounts for 60% of demand.

In the Stated Policies Scenario, South Africa sees further electrification of the economy with electricity demand growing at nearly 2% per year to reach 320 TWh by 2040, equivalent to demand in the United Kingdom today. Average residential electricity consumption reaches nearly 1 400 kWh per capita in 2040, the same level as Korea today and seven-times higher than the average for the rest of sub-Saharan Africa in 2040. Thanks to very proactive government programmes, only 5% of the population does not have access to electricity today, mainly in remote areas. South Africa is on track to achieve universal access well before 2030.

In the Africa Case, the impact of increased electrification across the economy (and more rapid economic growth) is offset by significant improvements in efficiency. The scope for further energy efficiency improvements remains large for motors in industry, heating in buildings and air conditioning. Maximising this potential moderates demand growth. By 2040, electricity demand in the Africa Case is 6% (20 TWh) lower than in the Stated Policies Scenario. Pulling the efficiency lever is central to ensuring reliable, secure and affordable electricity supply.

Although access to electricity has improved, the reliability of electricity supply has deteriorated over recent years with severe power disruptions. A shortage of generating capacity, mainly caused by disruptions and maintenance needs of old coal-fired power plants and delays in the construction of new thermal plants, has caused the vertically integrated state-owned utility, Eskom, to regularly resort to rotational load shedding.

South Africa's latest draft Integrated Resource Plan (IRP 2018) points to a new direction for the power sector, and opens the door for alternatives to coal-fired generation based on a market-based model. The government seeks to procure over 30 GW from independent power producers, half of which will come from the Renewable Energy Independent Power Producer Procurement Programme (REIPPPP). Falling costs are indeed making these solutions more competitive. The average levelised cost of electricity (LCOE) of renewable energy technologies in South Africa has declined substantially over the last five years – by an estimated 55% for utility-scale solar PV (about $90 per megawatt-hour [MWh] on average in 2018) and by more than 20% for onshore wind ($70/MWh). To date, the REIPPPP has attracted about $15 billion in investment in the power sector (20% foreign) and it is one of the most advanced private procurement programme for the power sector in Africa (see sections 10.7 and 10.8). Despite its early success, however, many projects found it difficult to get to the stage of triggering the release of funding, in part because of political uncertainty and the deteriorating performance of Eskom.

This shift away from coal reflects South Africa's goal of lowering carbon dioxide (CO_2) emissions and its commitment to the "peak, plateau and decline" emissions trajectory that led to the adoption of a Carbon Tax Act in 2019. This strategy implies a significant decline in the use of coal for electricity generation: its contribution shrinks in the Stated Policies Scenario from 90% today to just over 40% of in 2040 (Figure 10.7).

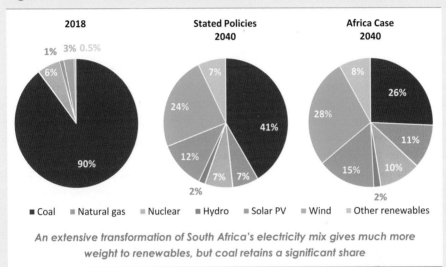

Figure 10.7 ▷ **Electricity generation mix by scenario in South Africa**

An extensive transformation of South Africa's electricity mix gives much more weight to renewables, but coal retains a significant share

About 85 GW of additional capacity is required by 2040 to meet growing demand and compensate for ageing coal plant retirements in the Stated Policies Scenario. Most additions are renewables units, led by wind and solar (about 25% each), making up close to 45% of electricity supply by 2040. South Africa is building two coal-fired power plants (Medupi expected to be completed by 2020 and Kusile by 2024) and despite the retirement of 30 GW of existing capacity over the period to 2040, coal remains the dominant fuel in terms of both capacity and generation.

In the Africa Case, the power sector in South Africa proceeds further and faster with diversification of the generation mix, driven by improved maintenance and management of the power system as well as the increased effectiveness of the procurement programme. The contribution of renewables to electricity supply grows at a much faster rate to provide over half of generation. By 2040, wind and solar PV become some of the most attractive options while generation costs from fossil fuel plants increase and wind overtakes coal as the primary source of electricity generation. Deeper regional co-operation and integration also sees South Africa benefit from competitive electricity imports, as large hydro projects such as Grand Inga in DR Congo move ahead more quickly.

Achieving the Stated Policies Scenario would require multiplying current investment levels by almost three (from $3.7 billion in 2018 to an average of almost $10 billion per year over the period to 2040). The Africa Case would require 8% less ($9 billion per year), aided by the impact of additional energy efficiency pushing down electricity demand. In both the Stated Policies Scenario and the Africa Case investment would need to shift away from coal (around 35% of total power sector investment in 2018) and focus more on low-carbon generation and on-grid extension and strengthening.

10.4 Outlook for electricity supply

Efforts to meet rapidly growing electricity demand lead to a significant expansion of the power system over the period to 2040 in the Stated Policies Scenario. Total power generation capacity in Africa (which includes on-grid, mini-grid, stand-alone systems and back-up generation capacity) more than doubles to reach 615 gigawatts (GW) in 2040. Natural gas remains the primary source of electricity generation, in particular in North Africa, while the contribution of coal gradually decreases as new projects are offset by ageing plants retirements in South Africa (Box 10.1). Many countries are actively developing their considerable renewable energy resource and over two-thirds of the additional power needs are met by renewables.

Excluding South Africa, the sub-Saharan Africa power sector is already relatively low-carbon and it remains so in the future in both scenarios (Figure 10.8). In the Stated Policies Scenario, electricity output increases fourfold, from around 225 TWh in 2018 to just over 900 TWh in 2040. On-grid supply continues to serve as the primary means of delivering electricity, but decentralised solutions for access play a larger role than anywhere else in the world, especially in the Africa Case. Although on-grid solutions have traditionally served as the most cost-effective option to supply electricity in areas close to an existing grid, the falling costs of stand-alone solar PV and battery storage technologies as well as new business models using digital and appliance innovations are making these solutions more competitive. In the Africa Case, mini-grids and stand-alone systems offer the least-cost solution to deliver over 160 TWh, or nearly 10% of electricity supply, enabling access to new or improved energy services to more than half of people gaining access by 2040.

In the **Stated Policies Scenario**, hydropower output almost triples over the period to 2040 and remains the largest source of electricity, although its share of supply declines from a half today to 35%. Natural gas provides more than a fifth of the additional generation to 2040, and retain a market share above 20%. Falling cost drives fast deployment of utility-scale and distributed solar PV, and also geothermal and wind: the combined contribution of these non-hydro renewable resources increases to over a quarter of overall supply. Coal-fired generation increases from a low base, providing cheap baseload power to meet fast-growing demand. Generation from oil increases in absolute terms, but its share in generation declines markedly to 7% in 2040, half its share in 2018.

Figure 10.8 ▷ Electricity supply by type, source and scenario in sub-Saharan Africa (excluding South Africa), **2018 and 2040**

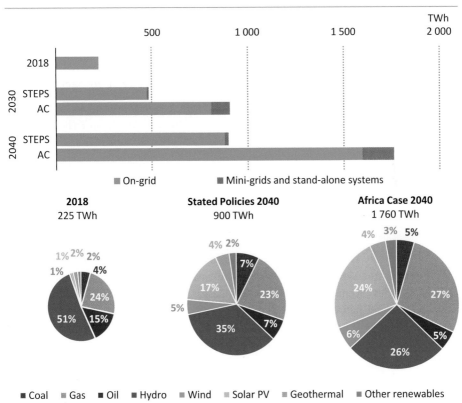

Most of the soaring electricity needs are met through new grid connections; renewable sources make the largest contribution, followed by gas

In the **Africa Case**, electricity output in sub-Saharan Africa (excluding South Africa) soars to 1 760 TWh by 2040, nearly twice the level in the Stated Policies Scenario and about eight-times 2018 levels. Renewables-based generation accounts for the largest share of the additional 860 TWh needed, bringing the total share of renewable-based generation to over 60%. On-grid hydropower and solar PV account for over 40% of the overall generation mix by 2040, but decentralised renewable solutions also play a much bigger role in delivering power, providing electricity access to 400 million people across sub-Saharan Africa (excluding South Africa) by 2040. The substantial increase in electricity demand also requires major new contributions from gas-fired generation, which account for over a third of the extra needs relative to the Stated Policies Scenario. With these additions, the share of gas in the electricity mix increases to nearly 30% in 2040 and it becomes the largest source of generation in the region. The share of coal declines compared to the Stated Policies Scenario, as does that of oil. Other renewable power sources such as geothermal and wind expand to significant levels in several countries benefiting from high quality sites.

Box 10.2 ▷ Geospatial estimation of the least-cost pathway to universal access to electricity

Over the years, our World Energy Model (WEM) has been expanded and coupled with other tools to provide a detailed outlook for electricity access in the next decades.[5] As part of this work, the IEA has been working closely with several leading universities, including the KTH Royal Institute of Technology (KTH), to analyse the least-cost route to achieve full access to electricity, using the most recent tools available. Analysis was done for a few individual countries in 2014 for our first focus report on Africa (Nigeria and Ethiopia) (IEA, 2014), and for all sub-Saharan African countries in *Energy Access Outlook 2017* (IEA, 2017).

For this *Special Focus on Africa*, the IEA refined its analysis using up-to-date datasets and the latest version of the Open Source Spatial Electrification Tool (OnSSET)[6], developed by KTH. The results provide detailed coverage of 44 countries in sub-Saharan Africa. Regional results are presented in section 10.4.2 and national results for 11 focus countries are shown in the country profiles (Chapter 12).

Overall electricity access objectives and demand projections are determined by country and region in the WEM based on population dynamics and economic growth for the Stated Policies Scenario and the Africa Case. They integrate the latest policy frameworks and national targets as well as technology and energy prices. Demand related to access is initially assumed at 250 kWh a year for rural and at 500 kWh for urban households, before growing over time to reach the national average.

Demand and other key drivers (e.g. technology and fuel costs) retrieved from WEM are then used in OnSSET in combination with several open access geospatial datasets. These include demographic indicators (e.g. population density and distribution), infrastructure (e.g. existing and planned transmission and distribution networks, roads), resources availability (e.g. solar, wind, hydro) and derivative layers (e.g. distance to the grid, to the closest road or city, diesel transportation cost) among others. The geospatial model runs a least-cost analysis mainly taking into account techno-economic factors and yields electrification investment outlooks. While grid densification (connecting areas close to the existing network) is prioritised, the geospatial model does not necessarily mirror the detail of government electrification plans (where they exist) or account for the financial and technical capacities of utilities.

10.4.1 On-grid supply

On-grid electricity supply dominates in urban areas and rural communities close to transmission lines and accounts for a majority of electricity consumption in sub-Saharan

---

[5] For the full WEM methodology, see www.iea.org/weo/weomodel/.

[6] For more details on the Open Source Spatial Electrification Tool, see www.onsset.org; for the latest OnSSET methodology update refer to Korkovelos, A. et al. (2019).

Africa (excluding South Africa) over the outlook period. The evolution and growth of grid supply by energy source varies across countries in sub-Saharan Africa, reflecting the differences in resources, costs and policies of each (Figure 10.9).

In the Stated Policies Scenario, total on-grid installed capacity in sub-Saharan Africa (excluding South Africa) increases to 270 GW by 2040, a threefold increase from the 80 GW of installed capacity in 2018. The power fleet steadily diversifies away from traditional sources of power and over half of the 190 GW of new plants commissioned over the period are non-hydro renewables (75% when hydropower is included). Gas-fired capacity expands, while the contribution of oil decreases.

In the Africa Case, the power sector proceeds further and faster with ensuring more reliable and affordable electricity for all (sections 10.5 and 10.6). The design of policies and the effectiveness of their implementation play a critical role in incentivising timely and adequate expansion of the physical infrastructure and in ensuring a better performing power sector. Deeper regional co-operation becomes more important. This all requires a steep increase in investment and a major reallocation of capital (sections 10.7 to 10.9). The generation fleet in sub-Saharan Africa (excluding South Africa) nearly doubles compared to the Stated Policies Scenario, reaching about 490 GW by 2040, driven by the substantial increase in electricity demand. The capacity mix diversifies further, with renewables accounting for over three-quarters of the additional installed capacity relative to the Stated Policies Scenario. Gas-fired generation takes on an increasingly important role in all areas and rises in tandem with renewables.

Hydropower remains a cornerstone of sub-Saharan Africa's power system (excluding South Africa) but its share declines as other renewable technologies and natural gas expand. Hydropower currently provides more than half of on-grid generation in sub-Saharan Africa (excluding South Africa) and over 80% of electricity supply in DR Congo, Ethiopia and Mozambique. In the Stated Policies Scenario, generation from hydropower almost triples by 2040. Over two-thirds of the 17 GW under construction today are scheduled to come online by 2025 including the Grand Renaissance Dam (6 GW) in Ethiopia. The Mambilla Dam (3 GW) in Nigeria helps alleviate local demand for fossil fuel resources and provide more reliable access to power. In Angola, the Laúca Dam (1 GW) is expected to be fully operational in early 2020 and the Caculo-Cabaca Dam (2.2 GW) in 2024. Construction of a major hydropower project (2.1 GW) was launched in Tanzania's Rufiji Basin in mid-2019.

In the Africa Case, better regional co-operation and integration of power networks is instrumental in unlocking a larger share of hydropower's huge potential. Larger markets absorb the power output from resources heavily concentrated in the Nile Basin and Congo River, making these resources more economical to develop. Generation from hydropower quadruples by 2040, led by DR Congo with (115 TWh) by 2040, with the completion of Stage V of Grand Inga and by Ethiopia with a quadrupling of output (60 TWh). Large hydropower projects are also developed in Mozambique (including the Mphanda Nkuwa Dam). These three countries become sizeable exporters to neighbouring countries and regions.

Figure 10.9 ▷ On-grid electricity generation by scenario in sub-Saharan Africa (excluding South Africa) **and selected countries, 2018 and 2040**

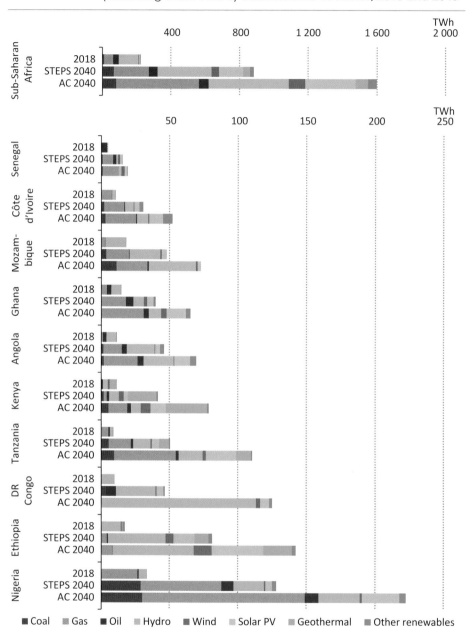

The fuel mix of on-grid electricity supply diversifies in all countries, with hydropower being increasingly complemented by gas, solar PV and geothermal

Note: STEPS = Stated Policies Scenario; AC = Africa Case.

In the Stated Policies Scenario and the Africa Case, the share of hydropower in total electricity generation declines by 15 and 20 percentage points, respectively, as alternatives become available. While hydropower remains an essential element of electricity supply, diversifying the electricity mix helps to reduce the risk of power disruptions during droughts and in the long term to strengthen resilience to changing climate conditions.

Natural gas use continues to increase: the size of the gas-fired power generation fleet in sub-Saharan Africa (excluding South Africa) more than doubles to 50 GW by 2040 as an additional 1.6 GW of capacity is added each year on average in the Stated Policies Scenario. A third of this expansion occurs among traditional gas producers, such as Nigeria, through increasing efforts to capture and make use of associated gas from oil production. Angola also expands gas-fired generation with the newly commissioned Soyo combined-cycle gas turbine plant.

In the Africa Case, gas-fired power generation overtakes hydropower in the 2030s to become the largest source of on-grid electricity generation in the region and accounts for nearly a third of the electricity mix. This expansion is also driven by additional growth in countries such as Senegal, Mozambique and Tanzania, which capitalise on newly developed domestic supplies of natural gas. These countries become pivotal actors in gas-fired generation as power sector governance improves and enhanced regional co-operation leads to the development of wider and deeper markets (see Chapter 11). Gas also plays an increasingly important role in providing back-up power during dry spells in countries that continue to depend on hydropower.

The deployment of **non-hydro renewables** accelerates to nearly 5 GW of new capacity per year between 2019 and 2040 in the Stated Policies Scenario. Solar PV represents 40% of all new capacity additions over the period. Geothermal resources play a central role in East African countries and particularly in Kenya where geothermal becomes the largest source of electricity in terms of both installed capacity and electricity production. The share of electricity from wind also increases in the Stated Policies Scenario although some of the best resources remain far from major load centres.

The uptake of these new alternative renewable sources is projected to keep pace with the higher electricity demand growth in the Africa Case and deployment accelerates to 10 GW of additional capacity each year over the period. The majority – over 70% – comes from solar PV. Installed solar PV capacity increases across the entire region to reach about 160 GW in 2040, overtaking hydropower and gas to become the largest source in terms of installed capacity (and the third-largest in terms of generation output).

Coal-fired power capacity also increases over the outlook, from around 3 GW today to 12 GW in 2040 in the Stated Policies Scenario and 17 GW in the Africa Case, as projects gradually come online in Zimbabwe, Senegal, Nigeria and Mozambique.

New **oil units** contribute to only 1% of total additions across the region in the Stated Policies Scenario, but almost 5% in Nigeria. This reflects the fact that only a few small projects are currently planned. In the Africa Case, the oil share shrinks further as programmes to convert oil-fired units to burn domestic gas accelerate, notably in Angola and Senegal.

— SPOTLIGHT —

Hydropower in Africa: strengthening resilience to a changing climate

Hydropower is the most important source of low-carbon electricity globally, accounting for 16% of total electricity generation. In Africa, hydropower plays an even bigger role, accounting for 22% of electricity generation on average (excluding South Africa). The full technical potential for African hydropower is far greater – new analysis points to a total potential of around 1 120 TWh in just 12 countries (taking into account environmental constraints), which is over eight-times today's level of hydropower generation in all of Africa.

Increasing reliance on hydropower may pose risks for the power sector due to impacts of a changing climate. Changes in rainfall patterns and temperature may lead to changes in river flows, and in evaporation and transpiration, altering the resource potential for hydropower. More frequent and intense extreme weather events such as droughts and floods may also lead to more variability in generation output. While impacts are likely to vary by region and even locally, climate-related events have already had noticeable effects on power systems, for example in Zambia where a severe drought in 2015-16 led to a drop in usable capacity of the largest hydropower plant and to power blackouts.

New analysis carried out for this *World Energy Outlook* assessed future climate change impacts on hydropower outputs and potential in 12 African countries under various climate change scenarios to 2099. The analysis linked global circulation models with hydrological models to examine changes of hydropower availability at precise locations using high-resolution discharge and elevation data (Gernaat et al, 2017). Two Intergovernmental Panel on Climate Change (IPCC) climate scenarios were compared: one leading to a global temperature rise likely to be below 2 degrees Celsius (°C) by 2100 (Representative Concentration Pathway [RCP] 2.6), implying a peak in emissions in 2020 and subsequent decline; and the second leading to a global temperature rise of around 3 °C by 2100 (RCP 6.0), implying a continuing gradual rise in emissions before they peak well into the second-half of the century (IPCC, 2014).

The annual availability of hydropower (measured by capacity factors) becomes more uncertain in both scenarios, but year-to-year variability is higher in RCP 6.0, the scenario with more climate impacts (Figure 10.10). Average annual capacity factors decline by some 2 percentage points by 2099 in both scenarios. However, hydropower capacity factors show stronger fluctuation in RCP 6.0 than in RCP 2.6 for most of the plants analysed (55 out of 64). Several Nile Basin countries (notably Sudan, Uganda, Egypt and Kenya) experience more than 50% relative increase of annual variability in RCP 6.0 compared to RCP 2.6, as do Zambia, Mozambique and Morocco. Without planning to improve resilience, this increased variability could have critical impacts on the reliability of power systems that are heavily and increasingly reliant on hydropower.

Figure 10.10 ▷ Variability of annual hydropower capacity factors for selected African countries by climate scenario, 2020-2099

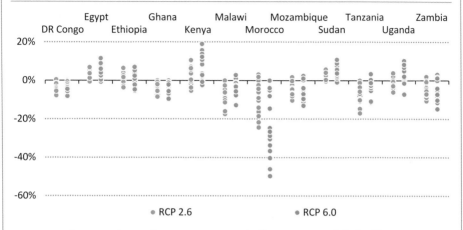

Increased greenhouse gas concentrations are associated with an intensification of year-to-year variability in hydropower generation

Note: RCP = Representative Concentration Pathway (IPCC, 2014).

Regional differences in hydropower availability also become more marked under the scenario with the higher global temperature rise (RCP 6.0). For example, hydropower in Morocco is projected to see a 9% decrease in capacity factors under RCP 2.6, while capacity factors in Nile Basin countries (Egypt, Kenya, Sudan and Uganda) would increase by 0-2%. In RCP 6.0, these differences are accentuated, with drier conditions leading to Morocco's outputs declining by 24% relative to today, compared to 4-8% increases in the Nile Basin. These striking regional differences underline the importance of developing enhanced interconnections and power pools that link countries and sub-regions together.

The sensitivity of African hydropower to rises in global temperature points to the importance of integrating climate resilience – the capacity to absorb, accommodate, adapt to and recover from climate change impacts – into planning for the construction and operation of future hydropower plants. For example, most planned hydropower projects do not currently take into account projected hydrological changes, relying instead on historical conditions (CDKN, 2015). This could lead to suboptimal operation of plants, at a time when many African countries are increasingly expecting hydropower to satisfy rapidly increasing power demand and to be a source of flexibility to support the integration of variable renewables such as wind and solar PV.

10.4.2 Role of decentralised systems to reach universal access to electricity

Geospatial analysis indicates that the least-cost way to reach full access by 2030 and to meet demand from newly connected households is to deploy mini-grids and stand-alone systems while also extending the main grid (Box 10.2). Providing electricity for all in sub-Saharan Africa would require an additional investment of around $25 billion per year above the level mobilised in the Stated Policies Scenario over the period to 2030.

In the Stated Policies Scenario, grid connections constitute the least-cost option for around 70% of the 230 million new connections that are expected to be achieved by 2030, mainly in areas that are close to a grid. A high proportion of the population lives close to a network (see Chapter 8), and grid densification connects around 70 million people, mainly in urban areas, while grid extension reaches more than 90 million, almost all living in rural areas. The number of people who gain access from decentralised solutions increases to almost 70 million over the period as technology costs continue to decline.

Figure 10.11 ▷ Solutions to provide electricity access by area and scenario in sub-Saharan Africa, 2019-2030

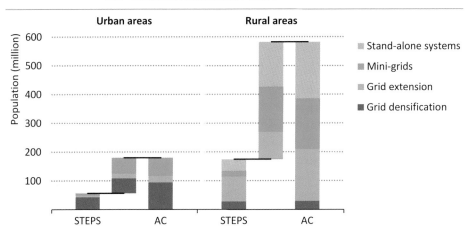

Access in urban areas will largely be via grid connections, while decentralised solutions are the least-cost option for about 370 million people in rural areas to reach full access

Sources: IEA analysis; KTH-dESA.

Decentralised systems are even more important to bring electricity to the 530 million additional people who need to be reached in the Africa Case in order to provide access to electricity for all. They represent the least-cost solution for more than two-thirds of these additional connections. Mini-grids play a major role in closing the gap in urban areas that cannot be reached by the grid before 2030, accounting for almost half of the additional urban connections. In rural areas, decentralised solutions provide more than three-quarters of additional connections, with mini-grids and stand-alone systems both having a role to play depending on population density (Figure 10.11). As a result, decentralised systems connect in total almost 450 million in the Africa Case by 2030. They can be installed quickly, providing valuable basic energy services to households who would otherwise need to rely on polluting and inefficient fuels. If deployed carefully, such systems can complement the grid, providing services immediately and preparing the way for later grid extension.

The best way to determine the optimal mix of solutions to provide access to all is to prepare integrated plans based on geospatial mapping. Such plans allow governments to develop a precise strategy, assess the investment needed, design adapted policies to reach all populations, and clarify the roles of different actors (government stakeholders, donors, private sector and non-governmental organisations). Turning such plans into actual investment flows and concrete progress on the ground raises some challenges (section 10.9), but they remain the best way to develop an integrated approach and to facilitate private sector participation.

Several countries, including Ghana, Senegal, Ethiopia, Nigeria and Rwanda, have developed long-term comprehensive strategies. As an example, the Ethiopian government announced plans in its 2019 National Electrification Plan to connect 100% of households by 2025 by connecting to the grid those 65% of households located less than 2.5 kilometres (km) from the existing network and putting in place decentralised solutions for the remaining 35%. By 2030, the government plans to extend the grid to reach households located between 2.5 km and 25 km from the existing grid. The 5% of households living farther than 25 km from the grid would have decentralised solutions over the long term.

10.5 Reliability

The provision of high quality electricity services is essential to economic growth. An electricity supply that is unreliable acts as a brake on overall economic activity and welfare, and inhibits the output of individual firms. The provision of low quality or unreliable electricity supplies forces firms to manage gaps in supply or to turn to more polluting and expensive alternatives such as diesel generators. Both choices have detrimental effects on firm efficiency and undermine competitiveness.

Poor electricity infrastructure in low-income countries is a major cause of unreliability (Figure 10.12). Under-investment in existing transmission and distribution assets and the inability to meet peak load due to installed capacity deficit result in frequent service disruptions (unscheduled outages or regular load shedding), ranging from a few hours to a few days. Between 2006 and 2018, around 80% of sub-Saharan African firms suffered frequent electricity disruptions, typically six hours in length, imposing losses of around 8% of annual sales on average (World Bank, 2018). Outages tend to be most frequent and prolonged in Nigeria (see Chapter 8). By contrast, firms in Organisation for Economic Co-operation and Development (OECD) countries experience interruptions of around one hour per month on average.

Investment in power systems, combined with improvements in the performance of utilities, results in a decline in the number of outages in sub-Saharan Africa by the end of the projection period. In the Stated Policies Scenario, the number of hours lost as a result of

Figure 10.12 ▷ Electricity outages and GDP per capita in selected regions, 2017

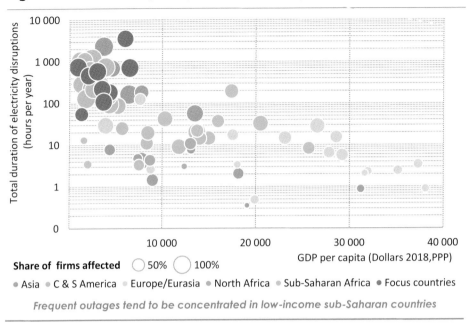

Frequent outages tend to be concentrated in low-income sub-Saharan countries

Notes: C & S = Central and South America. PPP = purchasing power parity. Focus countries are Angola, Côte d'Ivoire, DR Congo, Ethiopia, Ghana, Kenya, Mozambique, Nigeria, Senegal and Tanzania.
Source: IEA analysis based on World Bank (2019a).

Figure 10.13 ▷ Reliability indicators by scenario in sub-Saharan Africa (excluding South Africa), **2018 and 2040**

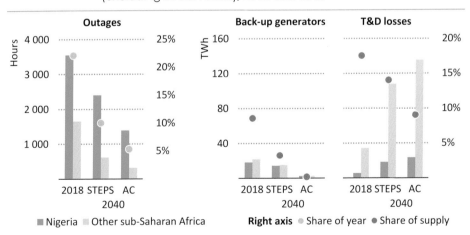

Major improvements in reliability in the Africa Case reduce the incidence of power outages by 60% in Nigeria and over three-quarters elsewhere; network losses shrink to below 10%

Note: T&D = transmission and distribution; STEPS = Stated Policies Scenario; AC = Africa Case.

outages declines to around 900 hours a year on average across the region; this ranges from below 1% of the year in Mozambique, Senegal, Kenya and Côte d'Ivoire up to nearly 30% in Nigeria, despite significant progress there. The number of outages declines even further in the Africa Case, falling to less than 500 hours a year on average and to about 15% of the year in Nigeria (Figure 10.13). As a result, the output of back-up generation declines in the Stated Policies Scenario from 40 TWh a year to around 30 TWh a year and to less than 5 TWh a year the Africa Case, reducing fuel and maintenance costs, noise and air pollution.

Inefficiencies arising from network losses can be very costly for utilities. Network losses are very high in sub-Saharan Africa (excluding South Africa) and at 18% are higher than in other developing regions (see Chapter 8). Investments in power systems result in losses falling to around 14% in the Stated Policies Scenario and to as low as 9% in the Africa Case.

Box 10.3 ▷ Improving grid reliability: a pathway for lower cost electrification

Providing access to electricity is essential, but access has to bring with it a reliable supply of electricity if households, businesses and public services are to reap the full benefits. A lack of reliable electricity supply from the grid disrupts daily lives and activities, lowers trust and use of the grid, and increases costs for consumers and utilities. Grid reliability also influences the best mix of solutions to provide universal access to electricity, by improving the cost-effectiveness of extending the grid to connect more potential consumers. This in the end affects the overall cost of electrification.

To shed light on the relations between the least-cost pathway to universal access and grid reliability, we developed a new analysis in collaboration with the MIT-Comillas Universal Energy Access Lab. Using the Reference Electrification Model (REM), building level geospatial analysis informs network and mini-grid deployment and design to optimise electrification planning (MIT-IIT, 2019). Taking an excerpt from the National Electrification Plan of Rwanda as a test case, we considered a rural and peri-urban area of 30×60 km in the Nyagatare region, with some 48 000 buildings that represent about 22 different consumer profiles (from 100 Watts [W] to 300 kW peak demand). Through the REM, we examined the least-cost electrification solutions for these consumers at various levels of grid reliability, defined as the percentage of demand served. The results highlight the complementarity of on-grid and decentralised solutions at all grid reliability levels in the area analysed (Figure 10.14). Poor grid performance, similar to the situation currently observed in many countries, contributes to the attractiveness of decentralised solutions to connect up to two-thirds of those gaining electricity access. These solutions remain attractive even with reliability improvements, a trend accentuated by expected declines in costs of decentralised systems. Nonetheless, improving the reliability of the grid could facilitate optimising the infrastructure by connecting more consumers and increasing average consumption of electricity, in addition to removing a major obstacle to development of businesses and community services. The additional investments to improve reliability by installing sufficient generation capacity to cover peaks would be offset by a decline in the cost per unit of demand served.

Investing in better grid reliability to optimise grid utilisation, while deploying decentralised systems to reach populations distant from networks, appears to be the best way to provide improve access to electricity at the lowest cost.

Figure 10.14 ▷ Connections by type to reach universal access to electricity with different grid reliability levels in a region of Rwanda

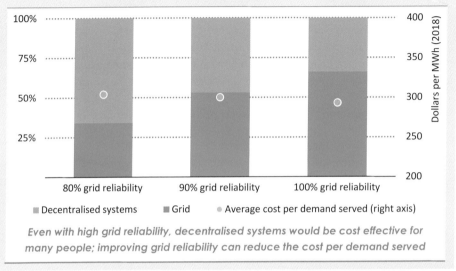

Even with high grid reliability, decentralised systems would be cost effective for many people; improving grid reliability can reduce the cost per demand served

Power sector regional integration

Increased power sector integration in sub-Saharan Africa can help with the goal of providing more affordable and reliable power. Affordability can be improved by reducing the need for investment: access to other markets allows countries to reduce the amount of installed capacity needed to meet peak demand, and sharing reserves between balancing areas means each can maintain less reserve capacity. Closer integration also enhances reliability by allowing the system to respond better to seasonal imbalances and unexpected shocks. Sub-Saharan African countries enjoy a diverse range of natural resources and have scope to benefit from the complementary nature of those resources.

Economies of scale achieved at a regional level may also enable countries to proceed with large projects that would not be justified based only on domestic power demand levels. Completion of Stage V of Grand Inga and associated interconnection projects in Southern African countries, for example, would allow the export of hydropower from DR Congo in the Africa Case and would significantly reduce average electricity generation costs in the region.

To realise these gains, governments and utilities across the region need to step up co-ordination in order to increase investment in transmission infrastructure, establish regional markets and improve regulation for cross-border trading (for example by defining and implementing regional transmission tariffs).

10.6 Affordability

The affordability of energy is a primary concern for policy makers, businesses and consumers. As discussed in Chapter 8, many households cannot afford the often very high upfront costs of grid connection, and current electricity tariffs make even basic energy services unaffordable for a large share of the population connected. However, the tariffs that constrain affordability for consumers are often set too low for utilities to be able to recoup their costs of supply. The risk is that this locks the power sector into a cycle of low revenue, high debt, inadequate maintenance, under-investment and poor quality of service. One of the biggest challenges of achieving universal access to electricity relates to the cost of providing power, which increases dramatically to supply sparsely populated and remote areas compared with households close to an existing grid. Our geospatial analysis shows that the least-cost option to provide access increases by a factor of four from easily accessible areas to the most remote ones (Figure 10.15). It is therefore inevitable that ensuring access to all requires government policies, subsidies and tax exemptions in one way or another. Much is already being done. In Togo, for example, the government recently announced its CIZO Plan to electrify 555 000 households with solar kits; the "CIZO solar cheque" will subsidise the hardware costs for households with a monthly payment, in partnership with a few licenced companies.

Supporting energy efficiency and productive uses can also help to improve affordability. Energy efficient appliances can enable consumers to access higher levels of energy services at lower costs, and so reduce the size (and the cost) of the system needed to support these services (IEA, 2017). Broadening the scope of electricity access plans to include the provision of energy for productive uses, such as agriculture or industry, can support the ability of end-users to pay while at the same time bringing down the cost of supply by increasing the load factor. Providing support for the acquisition of efficient equipment along with access to electricity can bring multiple benefits. A recent study from the World Bank (ESMAP, 2019) indicates that many productive tools and equipment appear to have a pay-back period of less than 12 months. Private companies including providers of mini-grid and solar home systems are starting to consider how best to support the development of commercial activities among electrified communities to ensure the sustainability of their projects. Success on this front will require cross-sectoral planning and co-ordination (for example between energy, water and agriculture ministries) as well as financial support.

Fossil fuel consumption subsidies have been used by a number of countries as a way of making electricity more affordable for citizens and companies (potentially helping them become more competitive). Some sub-Saharan countries, for example Ghana, subsidise certain fossil fuels as part of a strategy to promote switching from the use of traditional biomass. While fossil fuel subsidies – relatively more prevalent in North Africa – can help to support the use of energy services by the poorest households, they also create a substantial fiscal burden on what are often overstretched government budgets. We estimate the value of fossil fuel consumption subsidies in 2018 to have been $2.9 billion in Nigeria, $5 billion in Libya, $17 billion in Algeria and $27 billion in Egypt.

Figure 10.15 ▷ Levelised cost of electricity (LCOE) to achieve universal access to electricity by 2030 in sub-Saharan Africa, in the Africa Case

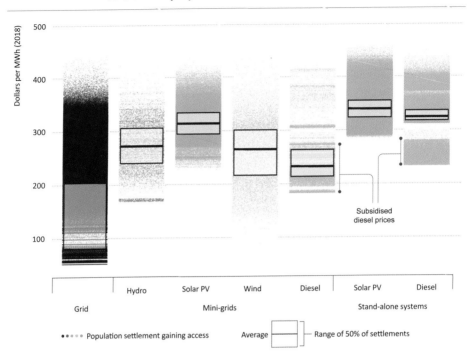

The cost of supplying electricity varies dramatically depending on household location; decentralised solutions are often the cheapest option for remote households

Notes: Each point represents an individual settlement in sub-Saharan Africa. It shows the LCOE of the least-cost solution determined for each settlement through our geospatial analysis (see Box 10.2).

Source: IEA analysis; KTH-dESA.

Amending current fossil fuel subsidy schemes is desirable for a number of reasons. These include: the need to reflect the true cost of electricity and remove distorted incentives (with implications for investment decisions); the need to reduce consumption of electricity from emissions-intense sources and encourage the use of more efficient and low-carbon sources; and the need to reduce the resultant fiscal burden that such subsidies cause.

Successful reform programmes broadly share the same key design and implementation features. They tend to focus in particular on being clear about the amount of the subsidy and the different categories of consumers who benefit from it. Obtaining wide understanding of, and support for, proposed reforms is essential: gradual implementation and assistance to the poorest households may be needed. There is plenty of experience in other countries to draw on and learn from. International development finance institutions can provide technical and financing assistance to help with fossil fuel subsidy reform.

10.7 Investment needs for reliable, sustainable and affordable power

The amount of investment needed for the provision of electricity in sub-Saharan Africa is substantial and well above the level of the current flows of capital into the region's power sector. Achieving the outcomes projected in the Stated Policies Scenario would require annual power sector investment in sub-Saharan Africa to more than double to around $46 billion per year, and would mean a cumulative total of more than $1 trillion in investment between 2019 and 2040 (1.6% of the regional GDP over the period). The Stated Policies Scenario would see the electricity access rate rise to 64% in 2030.

Reaching full access by 2030 and maintaining it to 2040, as in the Africa Case, would require multiplying current investment levels by five. The cumulative investment in this case would reach more than $2 trillion between 2019 and 2040 (2.7% of the regional GDP in the Africa Case over the period), or over $100 billion per year, more than doubling the capital needed under the Stated Policies Scenario (Table 10.2). Half of the investment needs would be spent on grid expansion, reinforcement and maintenance. Most of the rest would be for low-carbon power capacity, where solar PV takes an important role, reaching almost $25 billion per year on average. Cumulative investments in solar PV by 2040 in the Africa Case reach $535 billion. Decentralised solutions (mini-grids and stand-alone systems) would take an even more central role in this scenario, capturing a fourth of all investment in new capacity over the period to 2040.

Table 10.2 ▷ Average annual power sector investment in sub-Saharan Africa by scenario, 2019-2040 (Billion dollars, 2018)

	Stated Policies Scenario			Africa Case		
	On-grid	Mini-grid and stand-alone	Total	On-grid	Mini-grid and stand-alone	Total
Total power plants	19.3	1.7	21.0	34.1	16.8	50.8
Coal	2.0	0.0	2.0	1.7	0.0	1.7
Natural gas	1.3	0.0	1.3	2.9	0.0	2.9
Oil	0.0	0.1	0.2	0.4	0.5	0.8
Hydro	4.4	0.0	4.4	7.7	0.0	7.7
Solar PV, wind, other low-carbon	11.6	1.6	13.2	21.4	16.3	37.7
T&D	25.3	0.2	25.5	49.0	2.5	51.5
Total	44.6	1.9	46.5	83.1	19.3	102.3

Note: T&D = transmission and distribution; Other low-carbon = bioenergy, nuclear and other renewables.

In addition to higher investment levels, a reallocation of capital would be needed across countries and technologies in both scenarios. In South Africa, a major reallocation of capital away from coal-fired power (currently around 35% of the investment) towards electricity networks and low-carbon generation would need to happen (Figure 10.16). Nigeria,

Ethiopia and DR Congo are the other countries with the highest annual investment needs. Together, these four economies account for around 40% of the investment needs in both outlooks. In addition, investment in natural gas generation would also need to pick up and maintain the current investment pace. Average annual investment in natural gas in the 2019-40 period in the Africa Case is more than twice that of the Stated Policies Scenario and four-times the 2018 investment level. With more renewable plants and higher access levels, gas helps maintain security of supply at lower emission levels than other fossil fuels.

Figure 10.16 ▷ Annual average power sector investment by scenario in sub-Saharan Africa

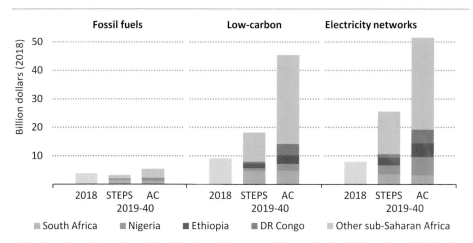

The investment gap is particularly large for renewables and electricity networks in both outlooks, and needs to accommodate more capital for natural gas in the Africa Case

Notes: STEPS = Stated Policies Scenario; AC = Africa Case. Low-carbon generation includes renewables and nuclear (nuclear only projected in South Africa).

Who will supply the capital needed to enable this investment? While investments will inevitably be funded from a variety of sources and types of funds – international and local; private and public; equity and debt – the choice of capital provider and financing vehicle makes a big difference to the pace and affordability of Africa's shift towards more reliable, sustainable and affordable power. The approach taken needs to be informed by an analysis of the ways in which power sector investments have been financed in sub-Saharan Africa, the drivers for investment decisions, and the priority areas necessary to tackle investment risks. Such an analysis can help with the design or the re-evaluation of policies and regulations to ensure their ability to reduce the cost of capital, especially for renewables where financing costs account for around half of the LCOE. The first part of this section addresses this by presenting an overall picture of the financing of the power sector in sub-Saharan Africa and describes the role of private financing in particular. The second part identifies four priority areas that require further policy and regulatory interventions to reduce investment risks and scale up the funds needed to finance investments.

10.8 Sources of finance for power investment in sub-Saharan Africa

The majority of the power sector investment in sub-Saharan Africa has been financed by public funds, mainly from domestic governments or state-owned utilities, development finance institutions (DFIs) and export credit agencies (ECAs). Of the new projects with final investment decisions in the period 2014-18, two-thirds of the new generation capacity was publicly funded. The level of reliance on public funds was highest for large, conventional generation projects and lowest for renewable projects, in part because in South Africa all new renewable capacity since 2011 has been procured via a competitive tender programme (Figure 10.17). The role of DFIs and ECAs as financiers has been important across the board, but particularly so for large coal-fired generation and hydropower projects where they accounted for around 60% of funds raised. Chinese DFIs have played an especially visible role: between 2013 and 2017 over $10 billion of Chinese funds financed 80% of the total investment for ten hydropower projects (over 6 GW) and over $6 billion for five coal-fired plants.

Figure 10.17 ▷ Financing sources for power generation investment by share, type and capacity in sub-Saharan Africa, 2014-2018

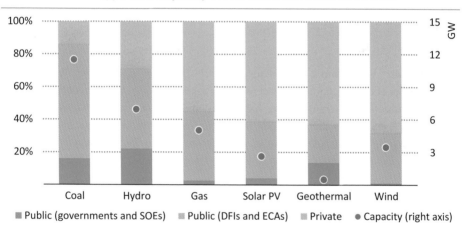

Large-scale generation projects have been more reliant on public sources of finance, while renewables were financed more with private finances

Notes: DFIs = development finance institutions; ECAs = export credit agencies; SOEs = state-owned enterprises. Based on utility-scale projects that reached financial close between 2014 and 2018.

Sources: IEA analysis based on World Bank (2019) and IJ Global (2019).

10.8.1 Investment framework and market structure

Private sector financing has been focused on generation (Figure 10.18), mainly through projects developed by independent power producers (IPPs).[7] In contrast, most of the

---
[7] IPPs are generation projects owned and operated by entities other than utilities, e.g. private developers.

investment in electricity networks has come from the public sector. To date, 16 of 43 sub-Saharan African countries do not allow for private sector participation in electricity generation or networks, and 18 others only allow it in power generation.

Figure 10.18 ▷ Private sector participation in electricity supply in sub-Saharan Africa by activity

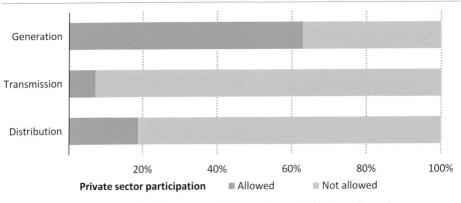

In many countries there is no private sector participation allowed. Where it is allowed, it is mainly in the generation activity.

Notes: In the distribution category, decentralised solutions are not included. Based on 43 countries in sub-Saharan Africa.

Unlike power generation, transmission and distribution grids have monopolistic characteristics and are generally subject to strong regulation. In sub-Saharan Africa, private sector participation in transmission has come about mainly through "whole-of-grid concessions"[8], but these did not result in much investment and two-out-of-three were cancelled (World Bank, 2017). Private participation in distribution networks is more common, also under concessions, but it is still far from usual: fewer than ten countries in Africa allow it (Eberhard et al., 2016).

Supportive policies and regulations as well as maturing markets have helped attract private sector investment into mini-grids and stand-alone systems. The World Bank estimates that there has been almost $4 billion of cumulative investment in Africa to date in almost 1 500 mini-grids (ESMAP, 2019). Although the majority of mini-grids were financed and are operated by state-owned utilities (some installed long ago), the privately-financed market has been growing – there are about 480 mini-grid developers in the African market today. Estimates based on another study of the global solar market of stand-alone systems show that 75% of the total funds raised by top developers between 2012 and 2017 (almost $700 million) went to developers operating in East and West Africa (IFC, 2018).

---

[8] Whole-of-grid concessions are long-term contracts where a private company is responsible to operate and maintain the existing grid, as well as investing in new lines and ensuring quality of supply. The company's annual revenues are set by a regulatory authority and subject to periodic revisions.

10.8.2 Private financing is concentrated in IPPs, mostly in South Africa

For almost 90 utility-scale IPP projects that obtained financing between 2014 and 2018 in sub-Saharan African countries, more than 60% of the funds were from private sources (World Bank, 2019a). Improved policy frameworks helped reduce perceptions about risks and increase project bankability. Lenders were willing to lend more money and the average share of costs that they were willing to cover rose from an average of 67% in 2014 to 79% in 2018.

Figure 10.19 ▷ Sources of financing for independent power producers in sub-Saharan Africa financed between 2014 and 2018

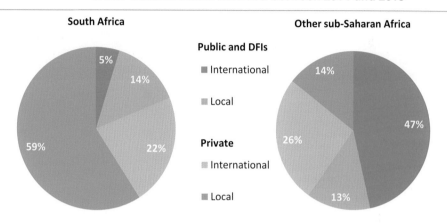

Policy frameworks to underpin IPP projects are less developed in sub-Saharan Africa, other than South Africa, which limits the ability of public and development finance to catalyse private investment

Note: DFIs = development finance institutions; IPP = independent power producers.
Sources: IEA analysis based on World Bank (2019a) and IJ Global (2019).

South Africa alone attracted two-thirds of the private finance for IPPs, or almost $7 billion over the 2014-18 period. In South Africa, IPPs were less reliant on public funds and required lower shares of equity than in other countries, with the private sector providing more than 80% of funds for IPPs (Figure 10.19). A good enabling environment, combined with a well-developed financial sectoral and clear sector policies were critical factors. A notable example is the Renewable Energy Independent Power Producer Procurement Programme (REIPPPP), a competitive programme to tender all new renewable capacity introduced in 2011, which was instrumental in enhancing the bankability of renewable projects. However, delays in the recent rounds of the programme – driven by political uncertainty as well as the deteriorating financial performance of state-owned utility Eskom – have raised questions over the positioning of the REIPPPP in South Africa's overall power sector development.

The financing picture is very different in other parts of sub-Saharan Africa. In most other countries there has been a high degree of reliance on public funds to finance IPP projects, and public and development finance has not been as effective in catalysing private financing, though this has varied between countries. Between 2014 and 2018, every dollar funded by public and development finance was matched by one or more dollars of private funds in Namibia, Nigeria, Mozambique and Ghana, but slightly above half a dollar in Zambia, and less than that in other countries. On average, every dollar funded by public and development finance was matched by half a dollar of private funds, whereas in the case of South Africa each dollar was matched by almost four dollars of private finance.

Table 10.3 ▷ Selected large development finance initiatives supporting sub-Saharan Africa's power sector

Initiative	Main financiers	Committed funds (billion $)	Preparation	Financing	Implementation
Green Climate Fund	Mainly developed countries	5.2	●	●	●
Africa Renewable Energy Initiative	France, Germany and European Commission	10.0	●	●	●
Clean Infrastructure Funds	Developed countries	8.1		●	●
New Deal on Energy for Africa	African Development Bank	12.0	●	●	●

The World Bank Group, the African Development Bank (AfDB), European governments and institutions, and the United States and Japanese governments provided most of the public funds used in the sub-Saharan African power sector between 2008 and 2017. The majority of this went to transmission and distribution projects, then to renewable-based generation and last to non-renewable power. The three main recipient countries were Kenya, Tanzania and Ethiopia (OECD, 2019). Separately, DFI funding from China has been growing rapidly (Horn, Reinhard and Trebesch, 2019). Funding has come from other sources too: a diverse array of organisations have established initiatives and committed funds to support power infrastructure development or help with project preparation, financing and implementation support (Table 10.3). Some initiatives, like US-led Power Africa or the AfDB's New Deal on Energy for Africa, expect their commitments to bring in significant additional funds. For example, the AfDB's New Deal on Energy for Africa expects to leverage $45-50 billion in co-financing by 2020. Similarly, the Power Africa programme has supported power sector investments in Africa that, if fully realised, would total more than $50 billion.

Development finance support has been substantial across sub-Saharan Africa and, to various degrees, has helped to catalyse private funds. In many cases, the presence of DFIs, providing financing and risk mitigation measures, has been critical to obtain financing. Further commitments are expected in the coming years.

However, while public financing looks set to continue to play an important role, closing the very large investment gap requires a much bigger role for private financing: there are limits to what governments can do, given their fiscal constraints, and state-owned utilities are mostly in a weak financial position. Attracting larger amounts of private funds requires policy and regulatory improvements in the region, as well as project-specific measures to reduce investment risks. Such improvements could increase the catalytic effect and allow for a more effective use of public funds.

10.9 Closing the investment and financing gap

This section highlights four priority areas that will be vital to address to reduce risk perceptions, to obtain more and cheaper financing, and to bring new actors to the power sector.

10.9.1 Improve the financial and operational performance of utilities

Utilities in sub-Saharan Africa have high transmission and distribution losses (Figure 10.20). This, combined with tariffs below costs and low collection rates, results in utilities being generally short of cash and hampers their ability to raise funds. It is estimated that only 19 out of 39 utilities in sub-Saharan Africa earned enough revenues to cover operational expenses, and only four of these covered at least half of their capital expenses (Kojima, 2016). As utilities are the main counterpart to private investors in generation, this situation can raise concerns on the part of those investors about future payment and makes it more difficult to secure financing at low cost.

Figure 10.20 ▷ Electricity network losses versus cost recovery ratio for major utilities in selected markets

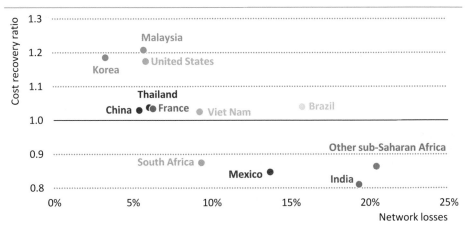

The weak financial and operational performance of utilities in sub-Saharan Africa hampers affordable financing in the power sector

Note: IEA analysis with calculations for cost recovery based on financial statements of reference utilities in each market.

Strengthening the governance framework of utilities is critical to improve operational and financial efficiencies. The World Bank study estimated that 11 utilities in the region could become financially viable if network losses dropped to 10%, cash collection rates increased to 100% and staff ratios matched those of efficient utilities in other regions (Kojima, 2016). Reducing network losses requires setting feasible targets, robust planning and a clear action plan to invest in infrastructure. Getting this done may come with challenges but there are positive examples in the region (Box 10.4). Moving towards cost recovery, which may require subsidies or cross-subsidisation, is also necessary to increase the pace of investment by easing perceived counterparty risks and allowing for increased financing at lower cost.

Box 10.4 ▷ **Improving performance of a distribution utility: lessons from Uganda**

The Government of Uganda initiated reforms in 1999 to unbundle the state-owned utility, Uganda Electricity Board. IPPs were introduced and the government awarded a concession to operate and maintain the generation assets. A state-owned transmission company was created and made responsible for planning, procurement and operation. Umeme, a private consortium, won a 20-year concession to operate and maintain the network.

Improvements in operations were slow during the first few years of the concession, but Umeme was able to reverse the situation. Network losses halved from 38% in 2005 to 17% in 2018, driven by an increase in annual investment from an average of $16 million in the 2005-09 period to $81 million in the 2014-18 period, and an improvement in the power supply. In addition, the number of customers multiplied by four while collection rates increased by 20 percentage points (Figure 10.21).

Key factors included:

- Contract-based performance indicators. Indicators included loss reduction targets and investment obligations ($65 million by the end of the fifth year).
- Regulatory independence. The government agreed that the regulator would set annual tariff adjustments based on a methodology defined in the concession contract.
- Commercial efficiencies. Increased channels to pay bills (including mobile payments) and the roll-out of prepayment metering made a difference (which in 2018 represented 24% of revenues).
- Technical support from international donors.
- Various risk mitigation mechanisms established with DFI support. The security package included a payment guarantee and insurance to cover termination and other political risks.

The process was not always smooth. In 2006, the government and Umeme renegotiated the contract after a power crisis (one of the two original investors left the concession at this point). Improvements in electricity access were slow and, although access has increased recently, the percentage of those with access remains low (around 11% in rural areas). Quality of electricity supply also remains an issue for customers despite the reduction in losses.

Figure 10.21 ▷ **Transmission and distribution network losses, collection rates and annual investments in Uganda, 2005-2018**

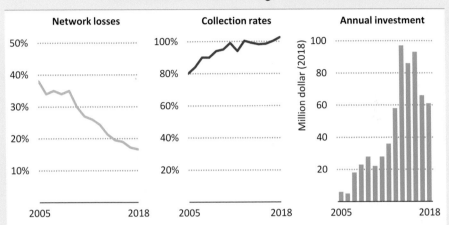

After a period of slow progress, the private distribution company reduced network losses and increased collection rates, which supported increased investment

Note: Collection rates surpass 100% due to pre-paid metering.
Sources: IEA analysis based on World Bank (2014) and Umeme (2018).

10.9.2 Enhance policy and regulatory frameworks to improve bankability

Robust procurement frameworks and well-designed contracts are crucial for project bankability. Competitive procurement is picking up in sub-Saharan Africa and is attracting strong interest from investors. Excluding South Africa, which largely acquires independent power projects through competitive tenders, half of the privately-financed IPP projects that reached financing in 2014-18 were competitively awarded (Figure 10.22).

Designing and conducting competitive tenders and auctions requires technical expertise and can take longer than direct negotiations. However, well-designed auctions bring a high degree of transparency and predictability, enhance market confidence and facilitate price discovery. Bid prices for solar PV under the REIPPP decreased by 80% between 2014 and 2018, while a programme to procure utility-scale solar PV in Zambia and Senegal together attracted almost 100 applicants and brought record low prices for the region of $48/MWh in 2017 and $43/MWh in 2018.

A key feature to ensure that procurement programmes translate into investments at scale is the bankability of the underlying contracts. Successful procurement programmes are accompanied by power purchase agreements that clearly define risks and responsibilities, e.g. clauses on dispute resolution, force majeure and termination. Credit enhancement mechanisms such as escrow accounts or public guarantees may also be necessary to mitigate payment risks and increase utilities creditworthiness. For example, the tenders in Zambia and Senegal had both sovereign and DFI-backed guarantees. Maintaining predictable, clear policies throughout the process is also important to preserve interest from private investors and lenders.

Figure 10.22 ▷ IPP capacity awarded by type in sub-Saharan Africa (excluding South Africa), 2014-2018

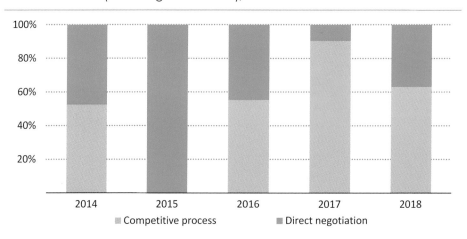

Competitive procurement is increasing in sub-Saharan African countries, though almost half of privately-financed capacity was awarded by direct negotiation

Sources: IEA analysis based on World Bank (2019a).

10.9.3 Create supportive enabling environments for rural electricity access

As discussed in section 10.4, reaching access targets requires a combination of on-grid and decentralised solutions. Scaling up investments in decentralised solutions may come with a variety of challenges, many of which evolve around revenue and regulatory uncertainty.

Supportive policies, as well as low-cost financing from DFIs, foundations and impact investors i.e. those that invest in projects that have development benefits, have fostered private-led projects, but revenue uncertainty still presents a major challenge, especially as much of future electrification will take place in more rural and generally poorer areas (Table 10.4). Mini-grid developers cannot recover the high upfront investments if customers consume little, while retailers of solar home systems may need to anticipate longer repayment periods and higher default rates from the customers they provide loans. Consumers may be restricted by their ability to pay for electricity.

Table 10.4 ▷ **Main risks and their underlying causes in deploying mini-grids and solar home systems**

	Mini-grids	Solar home systems
Revenue risk		
Low demand	Inability/delay to recover high upfront investment due to electricity demand lower than expected (over-sized mini-grid).	Low sales due to limited ability to pay, restricting demand to low-end solar home systems.
Low affordability	Customers with low and unpredictable income.	
Tariff level and subsidies	Dependence and uncertainty regarding subsidies (especially if tariffs required to be set at national uniform levels); difficulty to maintain support for and collect cost recovery tariffs (high compared to the grid), if allowed tariffs not capped.	Prices of solar home systems are generally unregulated, but developers may face uncertainty regarding regulation of subsidies (when applied), dependence on mobile services and regulation of interest rates of loans.
Regulatory risk		
Registration and licensing	Unclear rules on licensing and registration of assets and delays to obtain such permits.	Generally none.
Tariff setting	Incomplete/unpredictable tariff setting methodology. Delays to obtain tariff approvals.	N/A
Interaction with central grid	Weak/incomplete specifications of what happens when the central grid arrives to an area where a mini-grid operates (e.g. mini-grid becomes SPP or SPD; financial conditions in case of asset buy-out by utility).	N/A

Note: N/A = risk does not arise given technological and commercial characteristics; SPP = small power producer; SPD = small power distributor.

A study in rural Rwanda shows that households that received a free small PV kit used it intensively and reduced their kerosene and energy consumption: it also found that children studied longer (Grimm et al., 2016). Other studies also support the hypothesis that consumers are cash and credit constrained, and that social benefits, when fully internalised, exceed the investment costs. This points to the need for some sort of government or public support to realise these benefits.

Electricity subsidies, or similar financing mechanisms, could help rural households overcome the affordability constraint. They could also increase the sustainability of the decentralised electricity sector and encourage private companies to expand to more rural areas. Subsidies could be provided to households in the form of lower tariffs or help with connection costs or they could go to developers in the form of grants for capital expenses or concessional financing. Whatever option is chosen, subsidies need to be clear, predictable and well targeted.

Even if subsidies are expanded successfully, increased consumption will still be critical. Annual electricity demand per person in sub-Saharan Africa currently stands at around 190 kWh (excluding South Africa). This is one of the lowest levels of demand in the world. In some countries, increasing electricity access has led to increasing consumption as well, but at lower rates. For example, while access grew at an annual rate of 3.5% in Ghana between 2000 and 2018, per capita consumption grew by 1.5% per year. In Kenya, the growth in the access rate was four-times the growth in per capita consumption. Policies to increase productive uses and higher industrialisation could help reverse this trend.

Those countries where private mini-grid developers are most active, such as Tanzania, Nigeria, Kenya and Rwanda, are also those that have the best-developed regulatory frameworks. A strong and well-articulated regulatory framework that is clear about the most important issues will help to attract private investors. The issues that need to be covered include tariffs levels, subsidies and tariff setting; regulation of entry; and what happens when the central grid arrives. Tanzania's mini-grid regulation provides four alternatives on this last point: mini-grids can become small power producers (SPPs) selling electricity to the grid; they can become small power distributors (SPDs) buying electricity from the grid; they can combine the two (SPP+SPD); or they can sell the mini-grid assets to the utility. Lack of clarity over compensation issues, and concerns about enforcement of the regulation still appear to be causing concern to developers.

10.9.4 Strengthen provision of long-term finance

Given the long-lived and capital-intensive nature of power projects, the availability of long-term finance is crucial for power sector investment. In most of sub-Saharan Africa, however, access to long-term finance is severely constrained. It relies heavily on international development finance. Given the substantial risks associated with currency fluctuations, the local financial sector has a central part to play in ensuring a steady flow of long-term financing to power projects: a mismatch between revenues in local currency and costs in foreign currency (for equipment, borrowing, prices for power purchase agreements) can weigh heavily on the finances of state-owned utilities. While access to local banks in sub-Saharan Africa expanded over the past decade, it still compares unfavourably with access in other developing economies, with the exception of South Africa; and the majority of loans are still for short-term investments (Figure 10.23).

Developing the local financial sector and its ability to extend long-term finance has the potential to make a big difference to secure private investment in the power sector. DFIs can help by acting as a catalyst, for example by providing guarantees, refinancing or on-lending mechanisms. The refinancing of the Kenya Power and Lighting Company, the company that owns and operates the majority of the electricity network in Kenya, is a good example of how DFIs can strengthen the role of local banks and help utilities access cheaper and longer term finance ($500 million of commercial debt was restructured for longer term and lower cost commercial debt). Domestic pension and sovereign wealth funds could also play a more important role in financing power investments. Senegal's

Sovereign Fund for Strategic Investments, FONSIS, which has provided equity for solar PV plants in Senegal, is leading the way.

Figure 10.23 ▷ Level of private credit and loan maturity of the local banking sector in sub-Saharan Africa, 2016-2017

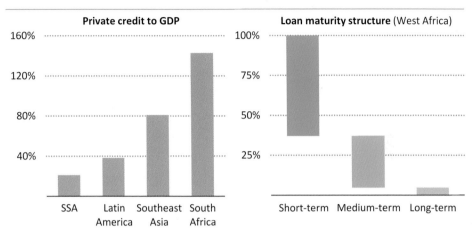

The local financial sector is playing a limited role in power sector financing other than in South Africa; the provision of long-term finance is particularly constrained

Notes: SSA = sub-Saharan Africa (excluding South Africa). Long-term loan has a maturity more than five years; medium-term loan between one and five years; and short-term loan less than one year.

Sources: IEA analysis based on World Bank (2019b) and BCEAO (2018).

Domestic policies outside the power sector also matter. Policies on issues such as the repatriation of funds, tax incentives and the regulation of public-private partnerships all affect the overall enabling environment and the regulatory framework for financiers. Clear economic policies that are conducive to private sector participation have an important role to help scale up power sector investment in Africa.

Chapter 11

Natural gas and resource management
New horizons for Africa's resources?

SUMMARY

- Africa is endowed with abundant oil, gas and mineral resources and these have provided an important source of income for the continent's economic growth. But changing energy market dynamics offer different prospects for each fuel. Oil production reaches 8.2 million barrels per day (mb/d) by 2040 in the Stated Policies Scenario, just shy of today's level (after a dip in the 2020s); gas production doubles between 2018 and 2040; and coal production remains at today's level.

- Africa also has many of the metals and minerals that are critical for clean energy technologies. It accounts for two-thirds of global cobalt production, 80% of platinum and half of manganese production. Responsible development of these resources is crucial to support the continent's economic prosperity and global energy transitions.

- Natural gas is facing a potential turning point in Africa. Outside North Africa, natural gas has not so far played a major role in energy development – at 5%, the share of gas in the energy mix in sub-Saharan Africa is one of the lowest in the world. The future looks likely to be different: recent discoveries across the continent could fit well with Africa's push for industrial growth and its need for reliable electricity supply. Developing gas infrastructure however will be a major challenge given generally small market sizes and concerns about affordability.

- Four case studies are presented to illustrate the differing dynamics for gas across the continent. Nigeria is an incumbent producer that has struggled to develop domestic gas consumption. Egypt is a resurgent producer with extensive domestic infrastructure that has successfully managed to revive upstream activity. Mozambique and Tanzania are emerging producers, thanks to recent major discoveries, but face the challenge of getting domestic value from gas. Ghana is short of gas and seeking access to liquefied natural gas (LNG) imports.

- In the Stated Policies Scenario, gas demand in sub-Saharan Africa triples to over 100 billion cubic metres (bcm) by 2040. Production across the region grows even more rapidly as sub-Saharan Africa becomes a major supplier of gas to international markets. In our Africa Case, both production and demand rise further.

- Developing domestic markets is likely to see traditional pipelines complemented by small-scale and distributed approaches in some areas. Oil use equivalent to some 10 bcm (30% of today's gas demand in sub-Saharan Africa) can already be displaced by small-scale LNG with today's costs and prices. Bringing gas into the energy system is a challenging task, and getting these new value chains up and running would require a concerted effort from Africa's decision makers.

- Many resource-holders in Africa have not yet capitalised on their resource endowments in a way that supports economic and social growth. Changing global energy dynamics are likely to put further pressure on production and on development models that rely on hydrocarbon revenues. This underscores the need for transparent resource management as well as efforts to reform and diversify the economies. Given the importance of resources to Africa's development plans, we highlight three strategic responses that could help mitigate the risks and support responsible development.

- The first is to ensure competitive investment frameworks. Recently there has been a global shift towards upstream investment with shorter lead times and a more rapid return of capital. Although there are some examples of investments of this kind in Africa, in many cases the conditions for investors are out of step with this trend, and this could weigh against decisions to put capital into projects in Africa. This makes it all the more important that clear regulatory frameworks are in place and that the role of national oil companies is clearly defined.

Figure 11.1 ▷ Annual average net income from oil and natural gas production by scenario in sub-Saharan Africa

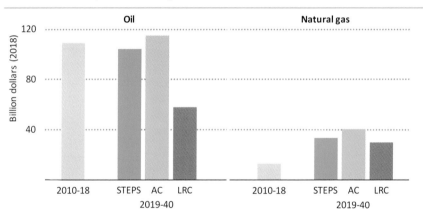

Africa faces significant challenges in sustaining net income from oil production, while natural gas offers more stable sources of revenue

Note: STEPS = Stated Policies Scenario, AC = Africa Case, LRC = Lower Revenue Case.

- The second is to develop the infrastructure required to bring resources to African consumers. This requires strategic choices on refining capacity and distribution networks. The third is to ensure that the revenue garnered from oil and gas is managed wisely. Average annual net income from oil and gas in sub-Saharan Africa amounts to $140 billion over the outlook period, but reduces sharply by 36% in a Lower Revenue Case in which global oil demand and prices are substantially lower as a result of faster energy transitions. In all cases there is a major shift towards less lucrative (per unit of energy) but potentially more stable revenues from gas.

11.1 Introduction

Africa has abundant oil, gas and mineral resources, and the revenue that these resources have generated has been an important source of income and economic growth. Oil and gas exports brought around $1.7 trillion of net income to African producers during the commodity boom in the 2000s, and the resource industry contributed around a quarter of Africa's economic growth between 2000 and 2008 (McKinsey, 2010). However, this income has also been a source of corruption and in many cases it has hindered broad-based development. Africa's resource producers remain among the least diversified economies in the world, and their economic prospects remain tied to the volatile movement of global commodity prices. The plunge in oil prices in 2014 brought these structural weaknesses into sharp relief. Many producers experienced a significant drop in export revenue (and fiscal capacity) and investment into Africa's upstream sector was severely curtailed. Oil prices have started to rise again in recent years, but changing energy dynamics make it risky to assume that ample resources will translate into reliable future revenues.

The challenges are particularly formidable for oil. Producers face both rising competition in global markets and growing uncertainty over long-term demand, and issues such as weak regulatory environments, political and social instability and the lack of local supply chain and technical expertise tend to undermine the competitiveness of African resources. Producers need to make considerable efforts to resume output growth and position themselves competitively in global export markets. There is also an imperative for both incumbent and emerging producers to take a hard look at how best to use the revenue from resource development to support the broader development of their economies. In the Stated Policies Scenario, the reduction in upstream investment since the oil price downturn puts the continent's oil production on a downward trajectory until the mid-2020s. Production resumes afterwards, but remains below today's levels through to 2040. In the Africa Case, higher domestic demand and improved resource governance lead to higher production, but net exports of oil remain well below today's levels (Table 11.1).

The story is different for natural gas. Gas takes a growing share in the global energy system in all of the IEA scenarios. Helped by a series of major discoveries in recent years, gas has a potentially important role to play in Africa's energy mix as a source of reliable baseload energy and as a companion for the rapid growth of renewables. It is however likely to be challenging to build infrastructure that makes the gas available and affordable in domestic markets. In the Stated Policies Scenario, Africa's gas production increases twofold in the period to 2040 and it rises further in the Africa Case where the potential for using gas in power and industry is exploited. Gas production is relatively more resilient in a Lower Revenue Case than oil production, offering a more stable source of income over the period.

Table 11.1 ▷ Fossil fuels demand, production and net trade in Africa by scenario

	2000	2018	Stated Policies				Africa Case	
			2025	2030	2035	2040	2030	2040
Oil (mb/d)								
Demand	2.2	3.9	4.9	5.5	6.2	7.0	6.4	7.8
Production	7.8	8.4	7.9	8.0	8.1	8.2	8.4	8.8
Nigeria	2.2	2.1	2.1	2.1	2.2	2.3	2.3	2.6
Libya	1.5	1.0	1.0	1.1	1.3	1.5	1.2	1.6
Angola	0.7	1.5	1.3	1.3	1.2	1.2	1.3	1.4
Algeria	1.4	1.5	1.3	1.3	1.3	1.3	1.3	1.2
Net exports	5.4	4.2	2.8	2.2	1.5	0.8	1.8	0.6
Natural gas (bcm)								
Demand	58	158	185	221	265	317	238	349
Production	124	240	287	372	435	508	407	561
Algeria	82	96	96	104	112	125	103	111
Mozambique and Tanzania	0	5	21	61	82	105	70	130
Egypt	18	59	81	92	95	98	93	101
Nigeria	12	44	41	45	56	65	56	84
Net exports	67	81	102	151	169	190	169	212
Coal (Mtce)								
Demand	129	160	166	161	160	161	153	148
Production	187	225	217	199	210	221	196	198
South Africa	181	209	198	174	176	173	171	148
Mozambique	0	11	13	18	24	32	18	34
Net exports	59	66	52	39	50	60	43	51

Notes: mb/d = million barrels per day; bcm = billion cubic metres; Mtce = million tonnes of coal equivalent. Net exports = domestic production + refinery processing gains – demand including international bunkers.

Against this backdrop, this chapter explores two key issues that are critical to shaping the future of African resources:

- **The role of natural gas in Africa's energy mix:** We look at the strategically important role of natural gas on the continent with the help of four case studies featuring Nigeria, Egypt, East Africa (Mozambique and Tanzania) and Ghana. The section also explores how recent discoveries could affect the position of gas in Africa's energy mix.

- **Maximising the value of Africa's resources**: We examine the outlook for resource development and related revenues in Africa under different scenarios. Despite their vast resource endowment, many African governments have been unable to capitalise on opportunities to support growth. This section discusses the strategies available to resource-holders to mitigate risks and ensure that resources are used to support broad economic and social goals.

11.2 The role of natural gas in Africa's energy mix

The position of natural gas in Africa's energy mix today varies widely across the continent. In North Africa, gas is a mainstream fuel that meets around half of the region's energy needs. In 2017, with strong growth in Algeria and Egypt, gas consumption in North Africa overtook oil use for the first time in history. However, the picture is very different in much of sub-Saharan Africa, where gas has been a niche fuel. The share of gas in sub-Saharan Africa's energy mix is around 5% – the lowest regional share in the world.

There are a number of reasons for the low penetration of gas in sub-Saharan Africa. In many cases, there is a considerable distance between production and consumption centres, necessitating large-scale, capital-intensive infrastructure. Except in Nigeria and Angola, there has been relatively little domestic production, and the commercial case for importing gas has generally been weak. Even in countries with significant gas resources, like Nigeria, it has proved difficult to align interests, build infrastructure and maintain reliable supply along the value chain from producer to end-user: existing gas-fired plants have been underutilised, and there have been periodic power outages.

There are nevertheless reasons to believe that the future of natural gas in Africa may be different from the past:

- **Africa's evolving energy needs**: Gas could be a good fit for Africa's push for industrial growth as well as a suitable partner for a rapid expansion in the role of solar photovoltaics (PV) in electricity generation.

- **Scope to displace costly oil products**: In 2018, some 15% of electricity in sub-Saharan Africa (excluding South Africa) was generated using oil products such as diesel or heavy fuel oil.[1] These are mostly back-up generators used in industry to avoid the risk of unreliable power supply. At current prices there is scope in many cases for gas to offer a cheaper alternative.

- **Major gas discoveries in every part of the continent in recent years** (Figure 11.2): The immense finds in East Africa (Mozambique and Tanzania) in recent years have been followed by further discoveries in Egypt and off the coast of West Africa on the maritime border of Mauritania and Senegal, and by the discovery of gas condensate resources in South Africa in 2019. These discoveries could have a significant influence on the outlook, especially since – with the exception of southern Africa – the continent does not have large resources of coal.

- **Favourable outlook for gas importers**: A wave of new liquefied natural gas (LNG) export capacity is coming online and exporters are keen to find a new market for their gas. Gas prices in the key importing regions have plummeted, while growing liquidity and flexibility in LNG markets are helping ease concerns over security of supply.

---

[1] In 2017, the share of oil in power generation was around 35% in Angola and almost 90% in Senegal.

- **Innovation in LNG technology**: The development of new technologies such as floating storage regasification units (FSRU) and small-scale technologies offers more flexibility for current and potential future gas users, and this could boost demand for gas. FSRUs can reduce the need for more costly onshore gas infrastructure and there is scope for them to be redeployed when no longer needed. Small-scale LNG technologies can help to provide new sources of demand and thus make gas less dependent on the realisation of complex onshore infrastructure projects (see Chapter 4, section 4.6).

In addition to helping to provide energy in producing countries, natural gas can contribute to economic growth by providing a sizeable source of fiscal revenue, although the extent to which it contributes to growth depends on the revenues being used effectively and transparently (see section 11.3).

Figure 11.2 ▷ Natural gas resource discoveries, demand and production in selected countries in Africa

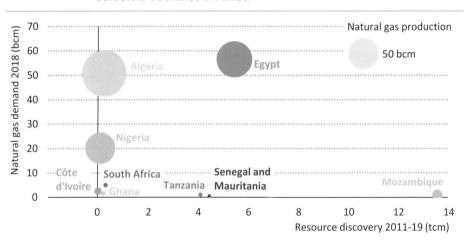

With the exception of Egypt, recent gas discoveries in Africa have been in countries with very small gas markets

Notes: bcm = billion cubic metres; tcm = trillion cubic metres. Bubble size represents production volume in 2018.

The outlook for natural gas to 2040 in the Stated Policies Scenario varies widely according to resources, market conditions and policy preferences in different countries. In this analysis, we consider four sets of circumstances:

Incumbent producers: This designation applies to Nigeria and Algeria, both of which are exporting around half of their gas production to global markets. The policy priority for these countries is to sustain production levels and stay competitive in export markets. Unlike Algeria, where gas plays a significant role in the energy mix, the penetration of gas in Nigeria is low compared to its population and resources. Nigeria therefore faces the longstanding question of whether it can develop domestic gas demand.

Resurgent producers: Egypt is a good example of a country that has well-developed gas infrastructure, but where gas production is recovering from a prolonged downturn. The policy priority in this case is to accelerate production growth to serve domestic demand and develop strategies to maximise the value of its resources.

Emerging producers: This applies to countries where there is significant resource potential but where both production and demand are small today. Following their recent major gas discoveries, Mozambique and Tanzania are at the forefront of this group, but it also includes Senegal and Mauritania. For these countries the first task is to turn discoveries into viable and successful commercial projects. Given the strong export orientation of the planned projects, they need to be competitive against other sources in global LNG markets if they are to succeed. A second task is to develop domestic gas markets where it makes sense for them to do so.

Gas importers: There are many countries with limited resources and so – if gas is to play a role in their energy future – they need to rely on imports either in the form of gas or gas-based electricity. While many African countries belong to this group, Côte d'Ivoire and Ghana are the ones that are most actively exploring import options (although Ghana has some domestic production notably from the Jubilee, Sankofa and TEN fields). South Africa has traditionally belonged to this group and imports gas from Mozambique, but this might change if recent offshore discoveries prove to be sufficiently promising. For these countries the key task is to obtain the gas they need at competitive prices while ensuring security of supply.

In this section, we look at the prospects for gas in each of these segments.

11.2.1 Prospects for gas in key regions

Incumbent producer: the case of Nigeria

Nigeria has an estimated 15 trillion cubic metres (tcm) of natural gas resources, which is more than any other African country except for Algeria. Gas accounts for around 10% of total primary energy demand today, mostly for power generation and own use in the oil and gas industry. However, the current outlook for Nigeria's gas industry is far from bright. Rapid production growth after 2000 tailed off due in part to regulatory uncertainties and the country is now struggling to arrest a decline in output. A shortage of domestic gas supply has severely affected the reliability of power supply, leading to load shedding and growing reliance on private (diesel-based) generators. Exports via the West African Gas Pipeline have been subject to frequent interruptions, and that has caused difficulties for neighbouring countries such as Ghana, Benin and Togo. Meanwhile, almost 15% of gross production is wasted through flaring, incurring both economic and environmental costs.

Upstream activity is far below a level commensurate with the country's resource base. The slump in investment, related in part to lower oil prices since 2014, is damaging for the medium-term outlook – gas production stalls through to the mid-2020s in the Stated Policies Scenario. One of the key issues is uncertainty over the fiscal conditions, exemplified by nearly two decades of uncertainty around the key provisions of the Petroleum Industry

Bill, which is designed to overhaul the legislation governing the operation of the oil and gas industries in Nigeria. Without clarity on governance and regulation, many companies will continue to rein in spending on new projects, with implications not only for Nigeria but also for adjacent countries that depend on exports from Nigeria for their gas supplies.

Reducing gas flaring would help to increase production. Nigeria has made notable progress in reducing flaring – the amount of gas flared has fallen by 70% since 2000 – but the country remains the sixth-largest gas flaring country in the world, and the value of the gas flared is estimated to be some $1.8 billion in 2018. Flaring is also a significant source of carbon dioxide (CO_2) emissions and incomplete flaring releases methane to the atmosphere as well. Despite the government's goal of eliminating flaring by 2020, the progress has stalled since 2016. Strengthened regulations and adequate gathering, processing and transportation infrastructure to bring associated gas to markets are essential to realise the government's target. In the Stated Policies Scenario, it is assumed that the envisaged reforms are gradually implemented, that output starts growing again from the late-2020s, and that Nigeria's annual gas production in 2040 rises 50% above today's level to 65 bcm.

The outlook for gas production also depends on reform of the electricity sector. Producers in Nigeria have domestic supply obligations that require them to supply a certain volume of gas to domestic markets. These obligations have typically not been met in full. Below-cost power tariffs and the precarious financial situation of electricity generation and distribution companies have led to frequent non-payment. The poor condition of the gas transmission and distribution system is also a major constraint. Without reforms to the power sector as well as the upstream, there is no guarantee that higher production would lead to improved supply of gas (and power) to the country.

Figure 11.3 ▷ **Natural gas production and use by sector and by scenario in Nigeria**

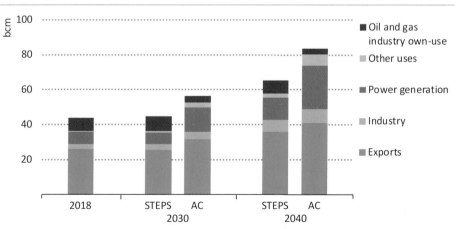

Accelerated upstream reforms and industrial growth mean that in 2040 gas use and production are 40% and 30% higher in the Africa Case than in the Stated Policies Scenario

Notes: STEPS = Stated Policies Scenario, AC = Africa Case.

While the challenges are formidable, realising the potential of gas could bring significant benefits. In the Stated Policies Scenario, electricity demand in Nigeria grows at rates in excess of 6% per year in the period to 2040, necessitating a significant increase in generation. Given the high share of gas in both existing and planned power generation capacity, using additional gas looks to be a cost–effective way of providing much-needed electricity, alongside the anticipated ramp-up of investments in renewables (from a very low base). Attracting more private capital to the sector will be essential if this is to happen: the recent 460 megawatt (MW) Azura-Edo project, Nigeria's first project-financed independent power producer project, could be an important signal in this context. Nigeria is also aiming to foster gas-based industries (for example fertiliser, methanol) to substitute domestic production for imports and to position the country as a regional hub for manufacturing, a key element in our projections for gas demand in both the Stated Policies Scenario and the Africa Case. Gas demand in Nigeria rises rapidly in the latter part of the projection period, reaching 30 billion cubic metres (bcm) in 2040 in the Stated Policies Scenario and over 40 bcm in the Africa Case (Figure 11.3).

Resurgent producer: the case of Egypt

Gas production in Egypt has undergone dramatic changes since the early 2000s. Domestic output grew by 17% per year on average between 2000 and 2008 (when production reached a peak of 62 bcm) and Egypt became an exporter of gas. However, a significant reduction in investment resulted in a 40% drop in production between 2008 and 2015. The country became a net importer of gas again in 2015, chartering two FSRUs in order to be able to bring in LNG imports. The Egyptian economy is heavily dependent on gas: more than 80% of the country's power generation capacity is gas-fired. Declining domestic output therefore caused repeated power outages and weighed heavily on industrial competitiveness. LNG export facilities were idled, and more polluting oil products started to take market share.

The discovery of the large Zohr offshore gas field in 2015 – one of the biggest finds worldwide over the last decade – changed the situation. With favourable upstream policies to expedite development, production from the Zohr field started in late 2017 and is set to reach around 30 bcm in 2019. This growth is now being supplemented by production from several other fields, notably Nooros, Atoll and the first and second phases of the West Nile Delta complex, leading to a major turnaround in the country's production. Gas production in 2018 returned to the level of the previous peak in 2008 and Egypt achieved self-sufficiency later in the year.

With sustained upstream reforms and efforts to reduce arrears to international operators, gas production in Egypt grows to around 100 bcm by 2040 in the Stated Policies Scenario. The upbeat production outlook, coupled with the country's underutilised LNG export infrastructure, opens the possibility of Egypt going well beyond self-sufficiency and acting as a regional export hub, although this would require the resolution of various political and commercial issues. However, question marks remain as to Egypt's net export position in the longer term. On the supply side, there would be need for continued upstream investment as the outputs from the Zohr and adjacent fields reach a plateau.[2] But the bigger issue is that gas demand in Egypt may rise very rapidly. Egypt is already the largest gas consumer in Africa and there is strong potential for further growth, especially in the power sector where gas is the dominant fuel.

Gas pricing is a key variable in determining the outlook for consumption. Although the government made a notable upward revision to domestic gas prices in 2014 in response to tightening supply, prices paid by power generators (around $3 per million British thermal units [MBtu]) remain below the levels to incentivise new supply investment, estimated to lie in a range between $4-6/MBtu (MEES, 2018). Prices for the petrochemical industry are within that range ($4.5/MBtu) (OIES, 2018). In contrast, prices for steel and cement producers are set at $7- 8/MBtu, and this has triggered fuel switching away from gas to coal. Earlier this year the government announced a plan to implement automatic price indexation for oil products, but gas prices for power generation are still being subsidised. A further reform of gas pricing is needed to maximise the value of the country's resources. To be effective this should cover both the upstream and the end-user, and it would be worth giving consideration as part of any reform package to the case for transparent end-user pricing schemes including clear rules of price discrimination by sector.

Emerging producers: Mozambique and Tanzania cases

East Africa's gas resources are impressive, with up to 5 tcm in Mozambique and nearly 2 tcm in Tanzania. Commercialising this significant resource base is a priority for both countries but this would require unprecedented capital investment (in Mozambique it would far exceed the country's annual gross domestic product [GDP]). This points to the need for a regulatory framework that serves domestic development priorities while remaining attractive for foreign investors.

Mozambique is taking the lead in East African natural gas development. Several projects are moving forward with a strong export orientation given the importance of economies of scale. Coral LNG, the first floating LNG project in Africa, is being built with an aim of starting operations in 2022: BP has committed to take all of the LNG that it produces into its export portfolio. The Anadarko-led Mozambique LNG, the country's first proposed onshore LNG terminal, would involve larger volumes of gas: a final investment decision (FID) was reached on this project in 2019, with a number of offtake agreements oriented toward Asian markets.[3] Another project, Rovuma LNG led by Eni and ExxonMobil, has secured sufficient offtake commitments from affiliated buyers to move ahead. Its development plan has received government approval and it is approaching a FID. The LNG project in Tanzania, however, has been held up by regulatory delays: the government now expects the project to start construction in 2022 and come into operation from the late-2020s.

---

[2] Trade with other East Mediterranean countries (e.g. imports from Israel, exports to Jordan) can also impact the trade balance.

[3] Total will replace Anadarko as a leader for the Mozambique LNG project following the merger.

In spite of the gas resources being offshore, development costs for the relevant fields and export facilities are potentially modest. Proximity to India and other fast growing Asian markets is another plus. However, the projects are expected to start operation at a time when global gas markets are very competitive, with many new and established exporters looking to gain or strengthen a foothold in the market. The delivered costs of gas to Asia for these projects are estimated to be broadly similar to those of projects in other emerging gas exporters (e.g. Canada, Australia). However the remote location, lack of established infrastructure and unfavourable security conditions create a risk of cost overruns which could significantly undermine the competitiveness of the projects. Strict control of costs and schedules is therefore very important (Figure 11.4). In the Stated Policies Scenario, gas exports from Mozambique reach nearly 75 bcm by 2040 (including exports to South Africa via pipeline) while those from Tanzania reach around 20 bcm.

Figure 11.4 ▷ Required levels of liquefaction capital cost and feedstock gas costs in East Africa by delivered prices to Asia

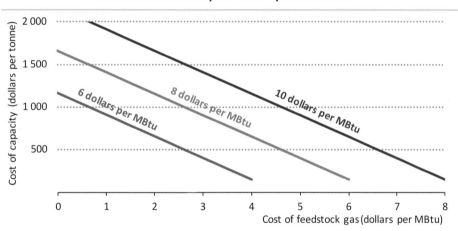

Strict cost and schedule control is indispensable for East African LNG to secure its place in global LNG markets

Notes: MBtu = million British thermal units. Delivered prices to Asia are the sum of the costs of developing gas resources, building liquefaction terminals and shipping the LNG volumes to Asia. Assumed asset lifetime is 30-years with a cost of capital in the range of 8%.

Although Mozambique and Tanzania's gas resources are primarily earmarked for export, governments in both countries are keen on developing a domestic gas industry. Domestic market obligations have been an element of discussion between the gas industry and the government with the design of the policy (how much and at which price) being a critical factor for both sides. The biggest hurdles for domestic gas consumption growth are lack of infrastructure and relatively low purchasing power of end-users.

Mozambique's gas resources are in the remote and sparsely populated north, and bringing the gas to the cities in the south would require costly pipeline infrastructure. This is under consideration (as are short-haul LNG shipments), but there is a major question mark on the economics. The development of local gas demand hubs served via FSRUs would require less capital expenditure and may prove more flexible. "Gas-by-wire" options[4] could also lower the costs of infrastructure investment, while large gas consuming "anchor" industries could play an important role in underpinning the economics of infrastructure projects. Many projects (including projects to build fertiliser, methanol and gas-to-liquids plants) have already been proposed as candidates to support an initial commercial case for developing the infrastructure.

Tanzania is in a slightly more favourable position in terms of getting gas into its domestic market: some of its gas fields are onshore and relatively close to Dar es Salaam, and there is already a pipeline which could potentially be expanded. There is a strong economic case for using gas to displace oil-fired generation: Tanzania's Gas Master Plan sets out a strategy to promote the use of gas to displace traditional uses of biomass, spur industrial growth and nurture gas-based industries. A plan to export Tanzanian gas to Kenya is also being explored. There is however more uncertainty on the upstream side about whether and when projects might move to a FID (and construction).

In the Stated Policies Scenario, domestic gas consumption reaches around 15 bcm in Mozambique and 6 bcm in Tanzania in 2040, roughly 20% of each's country's total output. In the Africa Case, gas makes further inroads into the energy system, with the combined gas demand in the two countries reaching almost 35 bcm by 2040 (Figure 11.5).

Figure 11.5 ▷ Natural gas consumption and export by scenario in Mozambique and Tanzania

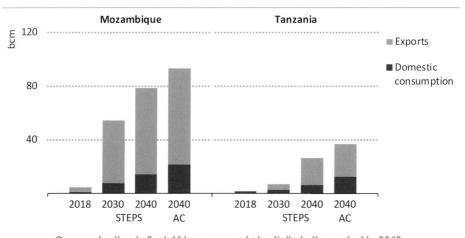

Gas production in East Africa grows substantially in the period to 2040: the vast majority is exported as LNG

Note: STEPS = Stated Policies Scenario; AC = Africa Case.

---

[4] Gas plants are built near coastal areas that have good access to LNG and electricity networks are extended to wider areas.

East Africa's gas industry, if successfully developed, has the potential to be a major contributor to long-term economic growth: increased exports would yield cumulative net fiscal incomes amounting to over $200 billion for Mozambique and nearly $50 billion for Tanzania in the period to 2040. This would help governments to address growing debt accumulation and ensure macroeconomic stability. However, it is vital that there should be transparency about the use of these revenues and that they should be used to bring tangible economic benefits for the country. History suggests that, after major discoveries, some countries have quickly scaled up public investment and accumulated excessive debts based on unrealised future revenues, damaging their economic performance in the process. Avoiding this so-called "pre-source curse" is essential if income from gas exports is to provide maximum benefit to Mozambique and Tanzania.

Senegal and Mauritania are also potential producers. Major offshore discoveries were quickly followed by a FID on the Tortue LNG project. The bulk of the gas produced is destined for export, but there is a plan to build a pipeline to the shore to serve domestic markets. As both countries currently rely heavily on oil for power supply, there is significant scope for gas to displace oil and meet growing electricity demand. In Senegal, the share of gas in power generation increases from less than 2% today to nearly half in 2040 in the Stated Policies Scenario, rising to 60% in the Africa Case.

Gas importers: the case of Ghana

Despite major gas discoveries in several parts of the continent, a large number of African countries still do not have direct access to large resources, and so need to rely on imports via LNG, pipeline or gas-by-wire. Ghana is one example, and it has been meeting its gas needs through a combination of domestic production and pipeline imports from Nigeria via the West Africa Gas Pipeline (WAGP). Domestic gas is mainly delivered to the Takoradi area in the west: imports from Nigeria serve the Tema area in the east. However, near-term domestic production prospects are uncertain and gas supply from Nigeria through the WAGP has been erratic.[5] These have led the country to explore the prospects for LNG imports. Tema LNG, the first FSRU, has been much delayed but now aims to start supplying imported gas from the early-2020s. At a delivered price of around $8/MBtu, imported gas is likely to displace the prevalent use of oil in the power and industry sectors (OIES, 2019).[6] Ghana has also recently put in place an interconnector between its two separate gas transmission systems to take surplus gas from the Takoradi system to the Tema system.

Further increases in LNG imports will depend on how demand for gas evolves. In the Stated Policies Scenario, gas demand in Ghana continues to be concentrated in the power sector. With generation from hydropower remaining stable, the growth in electricity demand is mostly met by gas and solar PV in the period to 2040, pushing up total gas demand to 4 bcm in 2040. In the Africa Case, gas plays a larger role in meeting higher electricity

---

[5] Physical attacks on the pipeline, diverting gas towards Nigeria's domestic market and the payment dispute by the Volta River Authority – Ghana's off-taker – all contributed to unstable supply of gas via the WAGP.

[6] With oil prices around $50 per barrel.

demand growth and is also used in the industry sector. Gas demand in 2040 is over 70% higher in this scenario, reaching 7 bcm, which points to a need for additional LNG imports. The requirements for LNG could be higher if Ghana pursues an option to export gas-based electricity to neighbouring countries via the West Africa Power Pool, for example to Benin, Burkina Faso, Côte d'Ivoire and Togo.

A number of other countries including Benin, Côte d'Ivoire, Kenya, Mauritius and Namibia are also reviewing options for LNG imports, but little progress has been made to date. Côte d'Ivoire was considering a plan to build an import terminal, but the project has been delayed following lower-than-expected demand growth and disputes over pricing. One of the challenges in developing smaller LNG projects focused on one country's gas demand is that small changes in either domestic demand or production outlook can easily reduce economies of scale and therefore affect the viability of the project. Concerns over the creditworthiness of off-takers and the reliability of payments can also constrain financing options for LNG projects.

For countries with little domestic production and limited access to LNG, gas has difficulty making inroads into the energy mix as small market size often does not justify investments in pipeline infrastructure. Some countries in this position are exploring a gas-by-wire option which would involve importing gas-generated electricity from countries within the same regional power pool. This option also offers an opportunity for exporting countries to increase economies of scale. It does however depend on a well-functioning power pool system (see Chapter 10).

11.2.2 Outlook for natural gas demand, production and infrastructure developments in Africa

Gas production in Africa is set to rise significantly in both the Stated Policies Scenario and the Africa Case as the development of recent new discoveries moves sub-Saharan Africa into the higher ranks of global gas producing regions.

Gas production increases in all North African countries in the Stated Policies Scenario. In Algeria, the largest gas producer in Africa today, production edges down to the early-2020s due to the lack of new developments to compensate for the decline in existing fields, but rebounds over time as additional output from legacy fields is joined by new projects. The country remains the largest gas producer in Africa through to 2040. Production in Egypt rises to some 100 bcm in 2040, reflecting developments in the Zohr and Nooros fields and plans to evaluate the Nour prospect. Libya also sees a significant increase in gas production, and it accounts for around 30% of the overall increase in North Africa.

Projected production in sub-Saharan Africa more than triples, reaching over 240 bcm by 2040. Recent discoveries mean that a range of new countries, notably Mozambique, Tanzania and Senegal, join the club of major producers. These three countries account for almost two-thirds of the increase in gas production in sub-Saharan Africa over the next two decades. Mozambique becomes the largest gas producer in sub-Saharan Africa by 2030. In the Stated Policies Scenario, production in sub-Saharan Africa in 2040 approaches the level in North Africa. In the Africa Case, where higher demand coexists with reduced regulatory risks, production rises by an additional 60 bcm, pushing it above 300 bcm by 2040.

The upbeat outlook for gas production in the Stated Policies Scenario underpins a threefold increase in demand in sub-Saharan Africa over the period to 2040. The share of gas in the energy mix rises from 5% today to just under 10% in 2040. In the Africa Case, with greater efforts to expand gas (and electricity) infrastructure, the share of gas in the energy mix rises to nearly 20% in 2040 (close to the average share of developing economies today). In this scenario, gas demand in sub-Saharan Africa climbs to around 180 bcm in 2040 (Figure 11.6). Exports from North Africa rise at a rate of 1% per year in the Stated Policies Scenario, while those of sub-Saharan Africa grow at a much faster pace. Today, Africa as a whole exports a similar volume of gas as Australia. By 2040, growing LNG exports from East Africa mean that sub-Saharan Africa is a major force in global gas markets.

Figure 11.6 ▷ Natural gas production and demand in Africa by scenario

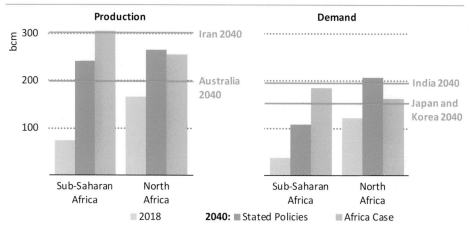

Africa strengthens its position in global gas markets in terms of both production and demand; gas makes strong inroads into the energy mix in the Africa Case

While the prospects and constraints vary by country, gas (or electricity) infrastructure is an essential prerequisite if gas is to thrive. Recent signs indicate that approaches to gas infrastructure development are likely to become more diverse. In addition to major long-distance pipelines to supply large-scale power plants, companies are increasingly looking at reaching industrial and commercial customers via small-scale LNG delivery or through distribution networks around industrial hubs. For example, Greenville LNG is operating 300 LNG trucks to deliver gas to a range of industrial customers, small-scale power plants and logistics companies in Nigeria and could potentially extend this to large mines and power plants in Burkina Faso (Africa Energy, 2019). Shell is developing distribution pipeline networks in the areas around major industrial hubs in Nigeria. Companies are also seeking better use of existing fuel distributors to distribute gas.

These small-scale, distributed approaches (akin to the value chain of liquefied petroleum gas) are likely to complement the traditional distribution channels (large pipelines) and help unlock new markets for gas. For example, many industrial and commercial customers today are paying higher costs for oil products, especially diesel, to operate their facilities: we estimate that some 10 bcm of this oil demand (30% of today's gas demand in sub-Saharan Africa) could be displaced economically given today's prices (Figure 11.7). However, given the relative ease with which oil products can be transported, this switch would happen only if policies and infrastructure allow gas suppliers to reach potential customers.

Figure 11.7 ▷ **Potential for oil displacement by small-scale LNG in power and industry in sub-Saharan Africa, 2018**

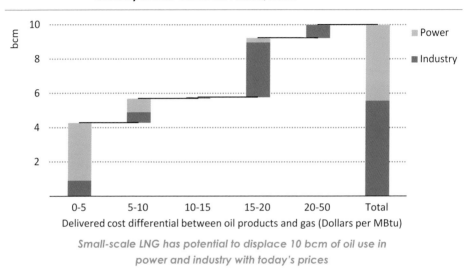

Small-scale LNG has potential to displace 10 bcm of oil use in power and industry with today's prices

11.2.3 Conclusions

Recent major discoveries present a renewed opportunity for gas to have a larger role in supporting Africa's energy and industrial development, but it is difficult to say "this time will be different" given that the challenges posed by small market size, infrastructure constraints and affordability remain considerable. Nonetheless, successful industrialisation hinges upon the stable provision of energy, and gas is well suited to providing this, whether directly where medium- and high-temperature heat is needed or indirectly where a source of relatively clean electricity is needed.

The idea of Africa leapfrogging directly from an energy system dominated by bioenergy to a fully decarbonised system is attractive, and it makes sense to develop renewables as rapidly as possible. However, it is hard to see how a fully decarbonised system can be achieved cost effectively in the coming decades while also meeting the continent's stated goals for industrialisation and economic development. Reliable electricity supply is an important element of the solution and this is set to be accompanied and supported by rapid deployment of renewables. However, in this sector there is also scope for gas to play an important role by providing a flexible and dispatchable source of electricity, helping to constrain the expansion of oil-based generators or coal-fired plants. In addition, there are other energy uses that cannot easily be electrified, including many industrial processes. The choice for these uses is occasionally between gas and renewables, but much more often between gas and more polluting fuels. The challenges for gas relate to infrastructure, affordability and business models: bringing gas into the energy system is inherently a challenging task, and cannot be taken for granted. Getting these new value chains up and running would require a concerted effort from Africa's decision makers.

11.3 Maximising the value of Africa's resources

Africa is endowed with abundant oil, gas and mineral resources. However, in many cases resource-rich African countries have not been able to capitalise on these resources in a way that supports their economic and industrial growth. A World Bank study suggests that, following a major resource discovery, countries on average not only failed to meet high expectations for economic growth, but actually registered lower growth rates in the short-run than before a discovery, mostly as a result of weak governance (Cust and Mihalyi, 2017).

Fossil fuels and minerals account for over a third of exports in roughly 60% of African countries, which together account for 80% of the region's GDP. Countries in which these commodities make up more than two-thirds of their exports account for almost half of the region's GDP (Figure 11.8). Since demand and prices for commodities tend to be highly variable depending on market circumstances, this means that the export revenues of many African countries are subject to large swings, with knock-on effects on their economies. There are many examples of the dangers of over-dependence on this narrow and volatile source of revenue. Recent market developments – and uncertainties over the future – underline the risks of a high reliance on resource revenues. The rollercoaster ride in oil prices since 2014 has exerted severe fiscal and economic strains on many producers in Africa, while the near-term impact of buoyant US tight oil production on global supply and the longer term impact of energy transitions on demand make a compelling case for improved resource and revenue management and for economic diversification and reform.

African producers are responding to these challenges. After a major slump in activity post-2014, there are signs that interest and activity in parts of Africa's upstream are picking up again. Against this backdrop, how can Africa make the most of its vast resources to spur inclusive development and growth? To answer this question, we take a look at the outlook for fossil fuel production and resource revenue under different scenarios, examine what strategies are open to African resource-owners to mitigate risks, and assess what the potential impact of abating these risks might be on production and revenue.

Figure 11.8 ▷ Share of commodities in total merchandise exports and GDP in African countries, 2017

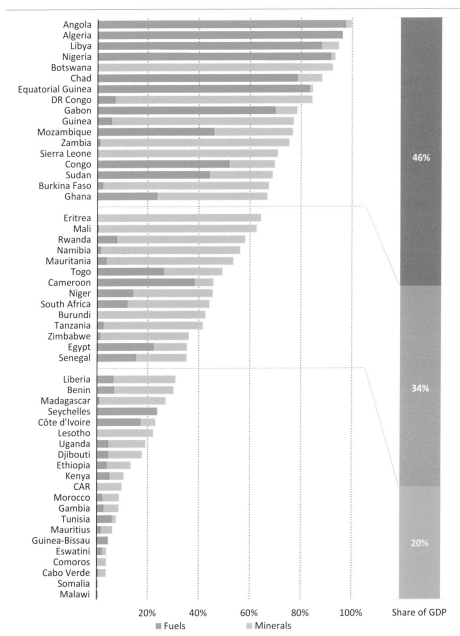

African economies are highly exposed to commodity exports; those relying on commodities for more than a third of their exports account for 80% of the continent's GDP

Notes: CAR = Central African Republic. Minerals include ores, metals, precious stones and gold.
Source: IEA analysis based on UNCTAD Stats.

11.3.1 Outlook for fossil fuel production

The oil price plunge in 2014 severely curtailed spending in Africa's upstream and this continues to weigh on the near-term production outlook. A combination of market conditions, lower revenues, ageing fields and regulatory uncertainties means that investment across Africa's upstream oil and gas sector has fallen by 40% since the high point reached in 2011, and the fall in sub-Saharan Africa has been particularly steep. More recently, there have been some signs of a pickup in activity with new licensing rounds, some major discoveries (primarily of gas) and FIDs. In the Stated Policies Scenario, after a dip in the 2020s, overall oil production almost returns to today's level by 2040.

Figure 11.9 ▷ Oil and natural gas production outlook in Africa in the Stated Policies Scenario

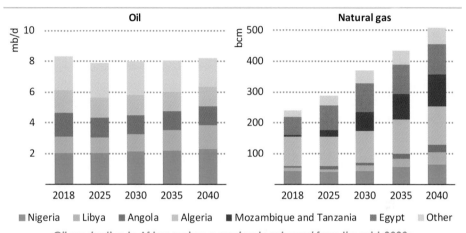

■ Nigeria ■ Libya ■ Angola ■ Algeria ■ Mozambique and Tanzania ■ Egypt ▫ Other

Oil production in Africa makes a moderate rebound from the mid-2020s, while gas production doubles as incumbent producers are joined by new players

The oil outlook is driven by a handful of major producers, all of whom face significant question marks and challenges. Libya has the largest oil reserves in Africa and remains the main source of production in North Africa. Its production continues to rebound and by 2040 almost reaches the level seen in 2010 (when the recent volatility in production levels started), although downside risks from political instability and civil unrest remain. Algeria's output continues to trend downwards due to the depletion of existing fields, dropping to 1.2 mb/d in 2040. The production outlook in sub-Saharan Africa depends heavily on two major producers, Nigeria and Angola, who have managed recent challenges in contrasting ways. In Nigeria, continued uncertainty over upstream regulation and governance mean that near-term output is projected to plateau until the mid-2020s, before improved market and (by then) domestic reforms allow for a modest resumption of production growth. In Angola, the government has accelerated reforms in an effort to stimulate investment. Assuming this effort is sustained, we project that Angola's output decline is slowed somewhat but that production nonetheless gradually falls to 1.2 mb/d in 2040 (Figure 11.9). A new group of producers also emerges in sub-Saharan Africa in our projections, including Uganda, Kenya and Senegal: their contribution to growth is relatively minor although the associated revenue promises to make a significant difference to their fiscal balances.

The gas outlook is quite different. Gas production in Africa grows by 270 bcm between 2018 and 2040, a doubling of output that is comfortably ahead of the rate at which oil production grows (see section 11.2.2). Thanks to a number of new discoveries and also the growing role of gas in the global energy system, there is a clear shift towards gas in upstream activities. Since 2010, some 20% of upstream oil and gas investment in Africa has been directed towards gas, and this increases to 36% between 2018 and 2040 in this scenario.

Coal production in Africa declines through to 2030 before making a modest rebound to 220 million tonnes of coal equivalent (Mtce) in 2040 in the Stated Policies Scenario. South Africa remains the dominant coal producer in Africa. Despite the steady decline in output, it still accounts for three-quarters of Africa's coal production in 2040, down from over 90% today. Coal output in Mozambique, mainly coking coal for export, increases threefold to around 30 Mtce in 2040, offsetting some of the decline in South African output. While coal mining is not as significant for the continent as oil and gas, the production of other minerals and metals is a key component of Africa's economy and is expected to remain so; many of its minerals and metals play a critical role in helping the global energy transition.

SPOTLIGHT

What does the clean energy boom mean for mineral production in Africa?

The rapid rise of clean energy technologies is not only changing the landscape for the power sector, but also for mineral and metal producers, including those in sub-Saharan Africa (Figure 11.10). For example, the greater need for batteries for energy storage and electric vehicles is set to supercharge demand for lithium, cobalt and manganese. What might this mean for sub-Saharan Africa's mineral production?

In 2017, net income from mineral production accounted for 2% of GDP in sub-Saharan Africa, with a majority of Africa's mineral reserves and production located in the south part of the continent. The Democratic Republic of Congo (DR Congo) is rich in cobalt; it accounts for almost two-thirds of the world's production and has half of the world's known reserves. South Africa produces 70% of the world's platinum (used both in internal combustion engines and fuel cells), 45% of the chromium (used in wind turbines) and a third of the world's manganese (a vital element for steel and advanced batteries). Rwanda and DR Congo both produce tantalum (some 30% and 40% of the global supply respectively) which is critical for electronics. Namibia and Niger were the world's fourth- and fifth-largest producers of uranium, critical for nuclear power plants (World Nuclear Association, 2019). It is also highly likely that Africa is home to rare earth elements, though no comprehensive survey has been undertaken to determine the potential (World Bank, 2017).

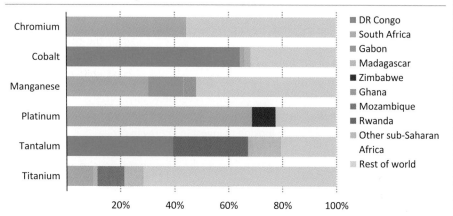

Figure 11.10 ▷ Share of sub-Saharan African countries in global production of key metals and minerals, 2018

Sub-Saharan Africa provides a significant share of the world's key metals and minerals critical to clean energy technologies

Source: USGS (2019).

As with oil and gas, sub-Saharan Africa has experienced boom-and-bust cycles of mineral exports and associated revenues. For example, a rapid recent rise in demand for cobalt caused prices to surge between 2016 and mid-2018, unleashing an opposite imbalance in the market that caused prices to fall by more than 60% from their peak. Demand for other minerals is also subject to volatility depending on the pace at which a cleaner energy future unfolds and on which technologies are in high demand, something strongly influenced by policy making. The projected growth of electric vehicles in the Stated Policies Scenario means that significant new sources of cobalt supply will need to come on line in a timely manner. Depending on the chemistry of the batteries produced, meeting the demand (230 kilotonnes per year in 2040) requires production to almost double in the Stated Policies Scenario.

There remains significant uncertainty regarding the nature and extent of sub-Saharan Africa's mineral resources. Further exploration and quantification is critical to a better understanding of the potential for extraction as well as recycling. There is also likely to be strong scrutiny of how these materials are sourced and what standards are in place all along the supply chain.[7] As with oil and gas, robust regulatory and oversight mechanisms are needed to ensure that impacts on local environments and communities are minimised and that revenues are used efficiently.

---

[7] Companies are paying increasing attention to responsible sourcing of African minerals. For example, BMW, BASF, Samsung SDI and Samsung Electronics recently launched a joint initiative to support sustainable and fair cobalt mining in the DR Congo. The initiative plans to explore ways to improve living and working conditions for the local small-scale mining operations over a three-year period.

11.3.2 Strategic responses for resource-holders in Africa

Africa's resources provide an opportunity for accelerated economic development, but the historical track record underlines the potential downside risks. Highly volatile income streams can upend economic stability and often do not benefit wider society, while changing global energy dynamics can undercut the size and reliability of this revenue. With this in mind, we explore three areas where policies can make a major difference to future outcomes: ensuring competitive frameworks for investment; investing strategically in associated infrastructure; and managing resource revenue wisely. Success in all these areas will need to be underpinned by improvements in governance to provide assurances of predictability and stability for those looking to invest in energy in Africa.

Ensure transparent and competitive frameworks for investment

In response to an uncertain and changing energy system over the last few years, company strategies and investment profiles have moved away from more capital-intensive projects towards shorter cycle projects that recoup investment more quickly. There have been shorter cycle projects in Africa as well. For example, the Zohr field in Egypt started production less than three years after its discovery in 2015, but it is the exception rather than the rule. On the other side of the ledger, there have been many incidences of project delays due to regulatory uncertainty, complex bureaucracy and politics, weak infrastructure and gaps in technical capabilities. Such project delays tend to count against African upstream investment opportunities when they are compared by potential investors with opportunities in other parts of the world. Analysis of the time lags between a FID and first production show that sub-Saharan Africa has not been following the global trend towards a shorter time-to-market for new projects (Box 11.1). In addition, many projects get stuck in planning and permitting processes before getting to a FID, as illustrated by the difficulty in proceeding with oil discoveries in Kenya and Uganda.

Box 11.1 ▷ Time-to-market: the watchword for African oil and gas?

The average size and time-to-market for oil and gas development globally has steadily declined as energy companies respond to lower revenues and market uncertainties: in 2018 the average time required to bring a project to market was 20% lower than it was in 2010 (IEA, 2019). The time-to-market for projects in sub-Saharan Africa, which tend to be smaller than the global average, however, has not changed noticeably over this period (Figure 11.11). A myriad of above-ground challenges, especially for onshore projects, bring a risk of further delays between project planning and start-up.

In response, oil and gas producers in various countries have taken steps to ensure that licensing processes and fiscal terms, including taxes, royalties, production sharing and dividends from resource extraction, are fair and transparent. Angola is an example of how a revision to a contract scheme can unlock investment decisions: following reforms to its regulatory and fiscal regime for offshore projects in 2018, Total approved the Zina

II project after many years of delay. Mozambique, which revised its Petroleum Law in 2014, scores well compared to other countries in sub-Saharan Africa on the Natural Resource Governance Institute's Resource Governance Index in terms of transparency and accountability regarding licensing and taxation (Natural Resource Governance Institute, 2019). Ghana, which has recently updated a range of upstream regulations, also scores well though it is much stronger on taxation than licensing. However, uncertainty about regulatory frameworks and a lack of transparency remain a widespread problem, and there are sometimes long delays in enacting legislation designed to address problems. Prominent examples include Nigeria's longstanding struggle to pass new petroleum industry legislation, and the as-yet unratified 2014 Hydrocarbon Law in Gabon, which has delayed the licensing round for shallow water and deepwater blocks opened in 2018.

Figure 11.11 ▷ Average time-to-market trends in the upstream oil and gas sector for sub-Saharan Africa and the world

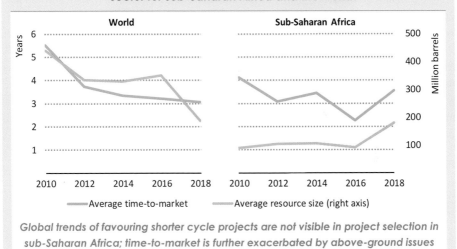

Global trends of favouring shorter cycle projects are not visible in project selection in sub-Saharan Africa; time-to-market is further exacerbated by above-ground issues

Notes: Number of years on Y-axis indicates time from FID to operation. Year on X-axis indicates FID year. Average size and time-to-market are for onshore, offshore and deepwater projects.

How a national oil company (NOC) performs can also have a significant impact on competitiveness and how the local hydrocarbon industry evolves. NOCs in Africa have grown in number, but there are questions about how effectively they are fulfilling their mandates. The role and governance of each NOC varies by country, but NOCs in sub-Saharan Africa tend to account for a smaller share of upstream investment than most of their counterparts elsewhere in the world (Figure 11.12). Many have been hampered in attracting investment by the lack of a clear mandate and by limited financing capabilities and technical expertise. Lack of oversight and accountability may also reduce the incentives for NOCs to operate efficiently.

Ensuring that the mandate of a NOC is aligned with the resource base and the needs of a changing energy sector is a crucial challenge for many African countries. While NOC models vary, some countries with modest reserves structure their NOCs to focus mostly on the downstream sector, taking on a role of product purchaser (for example, the National Oil Corporation of Kenya) or refiner (South Africa's PetroSA). For resource-rich countries the role of NOCs is more likely to focus on the development of resources and associated sectors (Ghana National Petroleum Corporation and Mozambique's National Hydrocarbon Company) and management of mature operations (Angola's Sonangol). Another model might see a focus on facilitating resource development, for example by conducting geological surveys and exploration activities where private companies are reluctant to step in.

No matter what the model is, clarity of purpose together with regulatory certainty, consistency and transparency are vital to attract investment, develop partnerships and allow a NOC to fulfil its obligations. Currently, sub-Saharan Africa ranks second to last among global regions in terms of transparent operation of its NOCs (Heller and Mihalyi, 2019). Ensuring prudent fiscal management and accountability of expenditures is another very important responsibility for NOCs, particularly given the scale of the resources involved (for example, Sonangol's debt has increased to more than 20% of Angola's GDP). NOCs also need to position themselves for broader changes in the energy sector, including national emissions reduction goals, by focusing on their environmental performance and minimising flaring and indirect emissions (especially methane leaks). In some cases, they may have the expertise and means to broaden their horizons by supporting the deployment of a range of low-emissions technologies.

Figure 11.12 ▷ Share of investment in upstream oil and gas by type of companies in selected regions, 2018

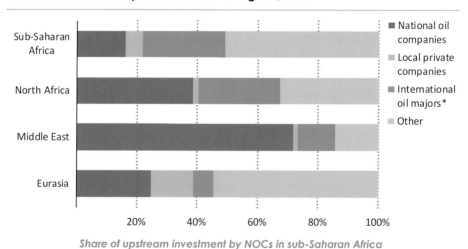

Share of upstream investment by NOCs in sub-Saharan Africa is smaller than in other regions

* Includes BP, Chevron, ConocoPhillips, ENI, ExxonMobil, Shell and Total.

Invest strategically in associated infrastructure

Once resources are extracted, two other crucial issues come into play: whether infrastructure exists to use these resources productively in Africa's domestic energy markets (discussed in this section), and whether the revenues are used prudently to promote a broader development agenda (discussed in the next section).

In the Stated Policies Scenario, oil demand in sub-Saharan Africa more than doubles between 2018 and 2040, driven by strong growth in transport and buildings. The annual rate of growth (at 3.5% per year) is one of the world's highest, and oil demand grows faster than in developing economies in Asia. An increasing number of (mostly energy inefficient) cars on the road pushes up demand for gasoline and diesel, and progress towards clean cooking raises demand for liquefied petroleum gas (LPG). Although sub-Saharan Africa produces around 5 mb/d of crude oil every year, its weak refining system means that it continues to rely on imports to meet its needs for oil products (Figure 11.13). The extent of the shortfall is even greater for low-sulfur products as many countries are moving to introduce more stringent regulations on fuel quality which local refineries struggle to meet. For example, Ghana reduced the allowed sulfur content in diesel from 3 000 parts per million (ppm) to 50 ppm in 2017 and many other countries plan to follow suit. In East Africa, where Kenya is the only country with a refinery, almost all oil products are imported.

Figure 11.13 ▷ Oil production and demand, refinery runs and net oil product imports in sub-Saharan Africa in the Stated Policies Scenario

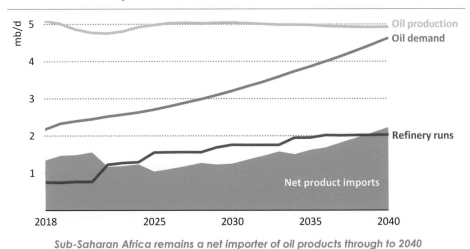

Sub-Saharan Africa remains a net importer of oil products through to 2040

The persistent deficit in oil products indicates that there is a case for new refining investments to upgrade existing facilities or to build new ones. Many projects are on the drawing board, but only one small refinery in Cameroon has come into operation in recent years. The reasons why more capacity has not been built include the relatively small size of individual country markets, high upfront investment costs, lack of local crude oil supply (except in major producing countries in West Africa), subsidised fuel prices, widespread smuggling and regulatory hurdles. Some projects are, though, making progress, with Nigeria's 650 thousand barrels per day (kb/d) Dangote project and smaller projects in Angola, South Sudan and Uganda now looking likely to be commissioned. Nigeria is also pursuing an option for modular mini-refineries which can be built and transported relatively easily and which may be well suited to small and relatively remote markets. Both the Stated Policies Scenario and the Africa Case assume that a number of refineries are built in the latter part of the projection period, but the region still remains a net importer of products through to 2040.

A growing need for imports of oil products does not necessarily pose a problem given rising international competition in refining and the possible growth of gasoline (and diesel) volumes seeking an export outlet in the global refining system. This is especially the case in East Africa where there is ample modern and cost-efficient refining capacity in relatively near reach, notably in the Middle East and India. However, a different risk is created by the lack of distribution pipelines and storage capacity in much of Africa. The lack of distribution pipelines means that imported products often have to travel thousands of kilometres by trucks from port to consumers, adding costs and accompanying a considerable risk of accidents. Plus the lack of storage undermines the ability to respond to volatile market conditions.

There is a strong case for strengthened infrastructure to bring oil products to consumers. This would require countries to harmonise varying regulations on product pricing, fuel quality and eligibility of importers, and to put in place adequate safety regulations for distributors. Given relatively smaller upfront capital requirements compared to refining, investments are already being made in distribution, storage, import facilities and service stations in Ghana, Tanzania, Uganda and elsewhere.

In the light of recent gas discoveries in Africa, midstream gas infrastructure – pipelines, gas processing and transportation – is also gaining importance (see previous section). Gas processing plants could provide an additional revenue stream for the planned gas projects and make LPG more readily available. While large-scale pipelines connecting prospective producing regions with major demand centres might well bring benefits large enough to justify their costs, such pipelines are expensive and it is likely to make sense to combine them with other options such as gas-by-wire or small-scale compressed natural gas or LNG solutions to reach small and more remote and widely dispersed markets that do not currently justify larger-scale pipeline developments.

Hydrocarbon investment can also catalyse the development of other infrastructure. For example, upstream oil and gas require electricity and could serve as an anchor load to increase the viability of new power infrastructure. Mining facilities also typically require new transport links, and could help spur new road or rail infrastructure. Major projects can also be an opportunity to develop new local supply chains and industries, an aim often pursued by local content clauses, with the potential to bring long-lasting value to local economies.

Manage hydrocarbon revenues to finance sustainable economic development

As discussed in Chapter 8, there are considerable risks for producers when national budgets are heavily dependent on hydrocarbon revenues. The extent of these risks depends on which pathway the energy sector follows, in particular with regard to future demand levels and prices (Figure 11.14).

Figure 11.14 ▷ Average annual net income from oil and natural gas production by scenario in sub-Saharan Africa

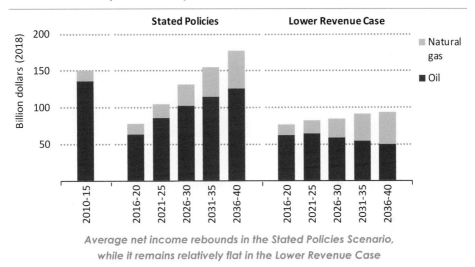

Average net income rebounds in the Stated Policies Scenario, while it remains relatively flat in the Lower Revenue Case

Notes: Net income from oil and gas is defined as the difference between the costs of oil and gas production, including a normal return on capital, and the value realised from its sale on either domestic or international markets. This net income changes over time and between various scenario projections, depending on the cost and volume of production, as well as both the international and domestic price, including any applicable energy subsidies.

Looking at the trends for individual countries and regions, projected net oil and gas income for producers in sub-Saharan Africa is generally higher by the end of the outlook period in the Stated Policies Scenario, but the trend is not a linear one and net income projections for oil and gas ultimately follow different trajectories (Table 11.2). In the near term to 2025, oil producers in sub-Saharan Africa suffer for two main reasons. First, the oil market remains well-supplied due in large part to US shale, which exerts downward pressure on prices. Second, the slowdown in investment since 2014 and rapid declines in maturing fields begins to bite, resulting in lower production. As a result, Nigeria and Angola receive on average 20% and 30% less annual income from oil output to 2025 than they did on average from 2010 to 2018. Small producers are hit much harder – producers in Central Africa (e.g. Cameroon, Chad, Congo and Equatorial Guinea) receive 40% less revenue.

By the mid-2020s, net oil income for producers in sub-Saharan Africa starts to rebound in the Stated Policies Scenario, as the effects of an anticipated near-term pickup in investment come through, US oil production plateaus and the market once again has more room for conventional supply. However, production in Nigeria and Angola remains relatively low and this is only offset in part by the rise of emerging African producers, meaning that net oil income for sub-Saharan Africa as a whole is just below its 2010-18 levels over the *Outlook* period. In the Lower Revenue Case (where global oil demand and prices are substantially lower because of stronger action in support of climate goals, in line with the objectives of the Paris Agreement) oil production in sub-Saharan Africa declines by almost 40% over the projection period, and the average net income of oil producers in sub-Saharan Africa is 45% lower than in the Stated Policies Scenario.

Table 11.2 ▷ Average annual net income from oil and natural gas by scenario and by country in Africa ($2018 billion)

	Oil & gas	Oil				Natural gas			
		Stated Policies Scenario		Lower Revenue Case		Stated Policies Scenario		Lower Revenue Case	
	2010-2018	2019-2025	2026-2040	2019-2025	2026-2040	2019-2025	2026-2040	2019-2025	2026-2040
Sub-Saharan Africa	123	83	114	67	54	19	40	18	35
Angola	36	25	31	20	14	2	3	2	3
Mozambique	1	0	2	0	1	3	13	3	11
Nigeria	54	37	52	29	24	10	13	10	11
Tanzania	0	0	1	0	1	0	3	0	3
South Africa	3	3	5	3	3	0	1	0	1
North Africa	61	33	57	28	30	16	21	19	37
Algeria	38	17	20	14	12	10	11	11	17
Libya	16	14	33	11	13	1	6	1	4
Africa	184	116	171	95	85	35	61	37	72

Natural gas is less lucrative than oil but our projections suggest that it may offer more stable revenues. Although pressure from US shale affects gas as well as oil, robust global demand for gas means that, overall net income in sub-Saharan Africa grows steadily over the projection period in the Stated Policies Scenario. The downside in the Lower Revenue Case is also less pronounced, offering some comfort to emerging producers such as Mozambique. In general, those with more gas in their portfolio fare better than those countries dependent on oil. For big oil producers such as Nigeria and Angola, cumulative net income declines by roughly 40% relative to the Stated Policies Scenario, among gas producers, in Mozambique cumulative net income declines by 20%, while it declines by 15% in Tanzania.

There are two other important points in the context of our projections for net income. The first is that there is a risk – especially for oil – of substantial volatility in market conditions and prices, which implies continued difficulties for fiscal management. The second is that producer economies in sub-Saharan Africa will need to contend with a growing population and the need to create more jobs (see Chapter 8, section 8.2.2). By 2040, there will be almost 750 million more people in sub-Saharan Africa than today. Nigeria alone accounts for almost 20% of the increase. Beyond considerations of employment, population growth also affects net income from oil and gas when calculated on a per-capita basis (Figure 11.15). This underlines the importance of ensuring that revenues from oil and gas are well managed if sub-Saharan Africa is to attain its development objectives.

Figure 11.15 ▷ Population and oil and gas net income per capita by scenario for selected African countries, 2018 and 2040

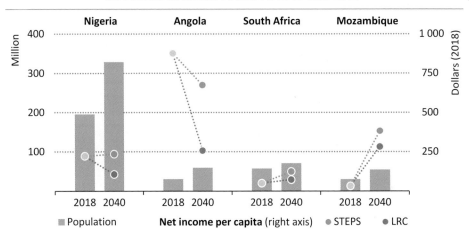

Growing population dwarfs the increase in net income from oil and gas production; the impacts are amplified in the Lower Revenue Case

Note: STEPS = Stated Policies Scenario; LRC = Lower Revenue Case.

Deciding how and when to allocate revenues generated from oil and gas is a challenge: many countries have pressing and immediate investment needs alongside the need to diversify the economy. To manage vulnerabilities and avoid inefficient spending, each country will need to develop a revenue management plan tailored to its own situation. Such plans could include fiscal rules and a sovereign wealth fund (SWF) (Box 11.2), and should include arrangements to ensure oversight of and transparency about resource revenue flows and expenditure. Assessments of oil and gas revenue management among sub-Saharan African economies indicate that most remain very vulnerable to commodity market cycles, although the examples of Cameroon and Ghana highlight what can be done.[8]

---

[8] Scoring based on 2017 Resource Governance Index from Natural Resource Governance Institute.

Public debt remains a significant problem despite the fact that almost half of the countries in sub-Saharan Africa have fiscal rules in place to manage spending and borrowing, and balance the budgets (Natural Resource Governance Institute, 2019).

Box 11.2 ▷ Role of sovereign wealth funds in sub-Saharan Africa

One often-used tool for revenue management is a sovereign wealth fund which can help counter-balance volatile commodity prices and diversify revenue streams. In general, these funds are designed as saving accounts to put money aside for future use; as buffers to support public sector financing during times of reduced oil and gas income; and as vehicles to help support economic diversification by investing in domestic infrastructure and sectors. For example, Senegal's Fonds Souverain d'Investissements Strategiques (FONSIS) has played an important role in scaling up solar PV projects in the country

In Africa, the rise of SWFs is a relatively recent phenomenon; half of them have been established in the last six years (Hove and Ncube). Some countries set up a fund as soon as they discover resources to help manage expected but yet unrealised revenue (although there is also a tendency for some countries to run into difficulties by increasing expenditure well in advance of future revenue, a phenomenon which has been called the "pre-source curse").

But a SWF may not always be an optimal solution, and there is a risk of "premature funds" in countries with limited savings and high levels of debt. Some countries have funds that are small in comparison to public debts. For example, Ghana's public debt in 2014 was 40-times the size of its savings fund, with the rate of return on savings insufficient even to service the interest rate on its debt (Bauer and Mihalyi, 2018). Others, such as Kenya, Tanzania and Uganda have either set up funds or are considering it despite the fact that revenues are unlikely to provide a significant share of fiscal revenues in the future. This raises the question of whether the money would be better spent to reduce debts, invest in infrastructure or address economic and structural deficiencies. The effectiveness of any such spending will however depend on the administrative capacity to identify and implement the projects that can bring the best economic and social returns.

11.3.3 Conclusions

Africa's resource wealth represents both an opportunity and a risk. Faced with multiple social and economic challenges and a major infrastructure deficit, resource-rich African countries are unlikely to forego the chance to develop valuable domestic resources. But extracting value from Africa's resource wealth requires the establishment of sound, stable and transparent regulatory and fiscal regimes, and the right institutions and practices to implement them. This is all the more important in an environment where there are significant uncertainties over prices and future market conditions.

In the Africa Case, it is assumed that many of the above-ground challenges currently impeding production are eased. This, alongside higher domestic demand and economic growth, helps make projects in sub-Saharan Africa more competitive globally and encourages investment from prospective investors. As a result, oil and gas production in sub-Saharan Africa is over 10% and 25% higher in 2040 than in the Stated Policies Scenario. While a higher volume of produced hydrocarbons is directed towards domestic markets, the average annual net income for oil and natural gas from 2019-40 is still 10% higher than in the Stated Policies Scenario. Translating higher income into sustained economic growth requires continued efforts to ensure transparent revenue management, along with prudent investment in infrastructure aimed at increasing and diversifying economic growth.

Chapter 12

Implications for Africa and the world

Introduction

How the African energy system evolves over the next two decades, and what it will look like in 2040, are vitally important questions not only for Africa but also the rest of the world. The future pathway is far from certain but, whatever the policy choices, the implications of those choices will resonate throughout Africa and beyond. We have outlined possible pathways for the continent's energy development to 2040 as described in detail in Chapters 9, 10 and 11. These pathways are based on an in-depth, sector-by-sector and country-specific analysis of Africa's energy sector opportunities: to the best of our knowledge, this is the most comprehensive such analysis undertaken to date.

The chapter consists of two sections:

- **A discussion of the policy implications and outcomes of the analysis in the global context**: This section provides a brief summary of what the future might hold for Africa's energy sector, and what it might mean for global energy and emissions trends, looking in particular at two scenarios. The first is the Stated Policies Scenario, which takes account of existing plans and announced intentions, and the second is the Africa Case, which is based on the Agenda 2063 vision agreed by African leaders (see Box B.2 in the introduction to Part B).

- **Detailed regional and country energy profiles**: The second part presents the results of the Stated Policies Scenario and Africa Case for sub-Saharan Africa as a whole as well as for eleven countries in this region: Angola, Côte d'Ivoire, Democratic Republic of the Congo, Ethiopia, Ghana, Kenya, Mozambique, Nigeria, Senegal, South Africa and Tanzania. The countries covered represent three-quarters of sub-Saharan Africa's gross domestic product (GDP) and energy demand today, and two-thirds of population. The profiles aim to provide decision makers with a data-rich set of information on the potential energy pathways for each country, considering their unique energy demand and supply needs and stages of development.

Implications for the world

Africa's population is among the fastest growing and youngest in the world, and this trend is set to continue in the period to 2040. One-in-two people added to the world population by 2040 are African, and a third of global urban population growth occurs in Africa (Figure 12.1).

Over the next 20 years, total population growth in Africa is more than double the combined population growth of China, India and Southeast Asia. In the coming years, Africa overtakes both China and India in terms of total population. This large increase (mostly occurring in cities) will be a major force driving Africa's energy demand growth.

Figure 12.1 ▷ Total and urban population in Africa, China and India, and share in global growth, 2018-2040

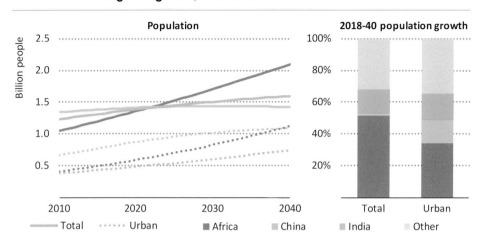

One-in-two people added to the world population and one-in-three people added to urban populations in the period to 2040 are African

Africa as a key driver for global energy demand growth

A rapidly rising population and growing pace of urbanisation make Africa a key driver of global energy demand growth. In the Stated Policies Scenario, total primary energy demand in Africa grows by 2% per year between 2018 and 2040, double the pace of global demand growth. At the same time, the composition of energy consumption in Africa increasingly moves away from the traditional use of biomass to modern and more efficient energy sources, notably electricity, natural gas and oil products.

Effective energy policy choices are essential not only to bring to fruition the continent's growth ambitions (including those contained in Agenda 2063), but also to support other economic and developmental goals. These goals include building a sustainable energy system, managing the rapid pace of urbanisation, scaling up industrial capacity and maximising the value of the continent's natural resources. As a tangible representation of the Agenda 2063 vision, the Africa Case incorporates policies to build the African energy sector in a way that allows higher economic growth to be sustainable and inclusive. It shows that achieving the goals of Agenda 2063 does not necessarily require more energy-intensive economies, compared with the Stated Policies Scenario. There is a considerable reduction of bioenergy use in the Africa Case, and growth in demand for other sources of energy is moderated by strong efficiency improvements. There is also a significant increase in electricity demand, but additional demand is mostly met by renewables. As a result, overall primary energy demand in 2040 in the Africa Case is 10% less than in the Stated Policies Scenario (Figure 12.2).

Figure 12.2 ▷ Total primary energy demand by fuel and scenario in Africa

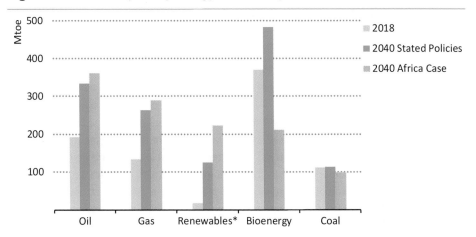

Achieving the outcome of the Africa Case adds only marginal amounts to demand for oil and gas relative to the Stated Policies Scenario while reducing the use of bioenergy

* Excludes bioenergy.

Africa emerges as a key source of **global oil demand** growth in our projections. At present, car ownership levels in Africa – especially in sub-Saharan Africa – are very low (in Ethiopia, for example, less than 2-out-of-1 000 people own a car). Oil demand grows as the size of the car fleet expands, and as liquefied petroleum gas (LPG) is increasingly used for clean cooking.

In the Stated Policies Scenario, the size of the car fleet in Africa more than doubles by 2040. This contributes to an increase of oil demand by 3.1 million barrels per day (mb/d) over the period, higher than the projected growth in China and second only to that of India. However, average car ownership levels in sub-Saharan Africa (excluding South Africa) in 2040 are still equivalent only to 60% of the level in India today. The lack of policies both for new and second-hand vehicles means that most cars have low fuel efficiency and are not subject to emissions standards that are common in many parts of the world.

In the Africa Case, the number of cars increases further to nearly 80 million by 2040, but improved vehicle efficiency offsets the expansion of car stocks and the numbers of kilometres driven. An increase in oil demand in this scenario relative to the Stated Policies Scenario is rather driven by the residential sector, where progress towards clean cooking creates additional demand for LPG.

Africa's growing weight is also felt in **natural gas markets**. The combination of renewables and natural gas provides a good fit for the development vision that African leaders signed up to in Agenda 2063. In the power sector, natural gas can help satisfy the growing appetite for baseload electricity and complement the rapid expansion of renewables, especially in those countries with large gas resources. There are also many energy uses that are hard to electrify (for example, industrial processes such as steel making) that are likely to see demand growth as African industry supports the continent's growth in urbanisation and infrastructure. In many cases, the choice for these uses is between gas and other (more polluting) fossil fuels rather than between gas and renewables.

The challenges for natural gas development relate to infrastructure, affordability and business models. A number of major gas discoveries (representing over 40% of global gas discoveries between 2011 and 2018) have been made in recent years, but the extent to which they will provide fuel for African development, as well as revenue from export, is uncertain. Making the most of these resources would require new pipeline infrastructure, although small-scale liquefied natural gas (LNG) technologies are allowing a new approach to distribute gas to consumers. Much will depend on the strength of Africa's policy push to displace polluting fuels from its energy mix, or to prevent them gaining a stronger foothold, and on the availability of finance to support the expansion of gas infrastructure.

In the Stated Policies Scenario, the share of gas in sub-Saharan Africa's energy mix rises from 5% today (one of the world's lowest) to just under 10% in 2040. In the Africa Case, it reaches almost 20% in 2040. In both scenarios, Africa becomes the third-largest source of additional gas demand globally between today and 2040, following China and the Middle East (Figure 12.3). Thanks to the emergence of new producers, notably Mozambique, Tanzania, Senegal and Mauritania, Africa also strengthens its position in global export markets.

Figure 12.3 ▷ Growth in oil and natural gas demand by region in the Stated Policies Scenario and Africa Case, 2018-2040

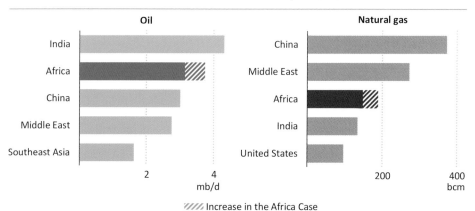

Africa emerges as a key source of demand growth for oil and natural gas. The growth in oil demand is second only to India; the growth in gas demand is the third-largest in the world

Reliable electricity supply plays a central role in meeting rising energy demand in Africa. Electricity demand in Africa is set to increase strongly, more than doubling from 700 terawatt-hours (TWh) today to over 1 600 TWh in 2040 in the Stated Policies Scenario and to 2 300 TWh in the Africa Case. Renewables make a major contribution to the additional generation required. Falling costs drive the fast deployment of utility-scale and distributed solar photovoltaics (PV), and deployment of geothermal and wind also picks up sharply: in the Stated Policies Scenario, the combined contribution of these non-hydro renewable resources increases from less than 5% today to around 30% of Africa's total power generation in 2040. Hydropower also remains a cornerstone of sub-Saharan Africa's power system – notably in the Democratic Republic of the Congo (DR Congo), Ethiopia and Mozambique – and generation almost triples by 2040. Better regional co-operation and integration of power networks is instrumental in unlocking hydropower's huge potential.

The scale of deployment of non-hydro renewables is even more significant in the Africa Case. A large part of this comes from solar PV, which overtakes hydropower and natural gas to become the largest electricity source in Africa in terms of installed capacity (and the second-largest in terms of generation output). Solar PV deployment between today and 2040 amounts to almost 15 gigawatts (GW) a year, equivalent to the amount of solar PV capacity the United States adds every year over the same period. Wind also expands rapidly in several countries benefiting from high quality wind resources, notably Ethiopia, Senegal and South Africa, while Kenya is at the forefront of geothermal deployment. The growth in overall renewable-based electricity generation in African countries is higher than in the European Union (Figure 12.4).

Figure 12.4 ▷ Growth in renewables-based electricity generation in selected regions, 2018-2040

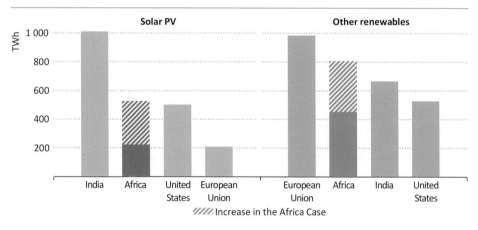

Renewables account for three-quarters of additional electricity generation in the Africa Case, and bring Africa to centre stage in global renewables markets

Note: Other renewables include hydro, wind, geothermal, concentrating solar power and biomass.

Achieving this level of deployment would require the development of efficient supply chains and the physical infrastructure necessary to facilitate smooth trade in goods and technologies between countries (as envisaged in the African Continental Free Trade Area). A favourable regulatory environment which reduces risks and the cost of finance would also be essential, as would the technical capacity to underpin a large-scale installation and maintenance sector.

Mobilising investment for reliable power supply: challenging but achievable

Africa needs to expand its energy infrastructure, especially in the power sector, to serve its growing population. Despite being home to 17% of the world's population, Africa currently accounts for just 4% of global power supply investment. On a per capita basis, power supply investment in Africa ranks among the lowest in the world (Figure 12.5). In sub-Saharan Africa, power generation capacity per capita has shown little or no growth since 1990 while that of India and Southeast Asia has grown fourfold.

Figure 12.5 ▷ Per capita power supply investment by region, 2018

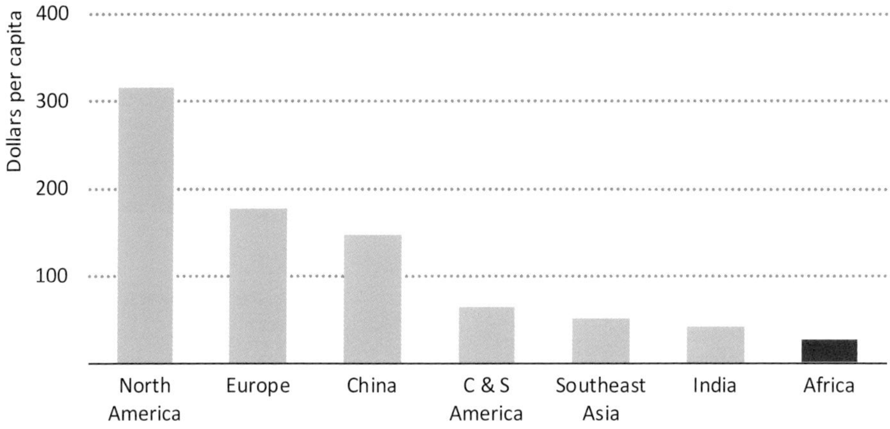

Africa's per capita investment in power supply ranks among the lowest in the world

Note: C & S America = Central and South America.

Addressing the deficit of power infrastructure in Africa will require a significant ramp-up in spending. Investments in power supply need to double through to 2040 in the Stated Policies Scenario to around $65 billion per year. The Africa Case requires a further doubling to around $120 billion per year to ensure reliable and affordable power for all and to serve an economy growing at over 6% a year. Nigeria, South Africa, DR Congo and Ethiopia are among the countries with the highest investment needs. Half of the investment is needed to expand and upgrade electricity networks – including mini-grids – and most of the rest is needed to increase low-carbon generation capacity where solar PV plays an important role. Investment needs in solar PV in sub-Saharan Africa amount to almost $25 billion per year on average in the Africa Case – almost double the level of investment in the European Union today.

The cumulative investment in Africa's power supply between 2019 and 2040 reaches $1.4 trillion in the Stated Policies Scenario (1.6% of the continent's GDP over the same period) and $2.6 trillion in the Africa Case (2.4% of GDP). Mobilising these levels of investment is a significant undertaking, but it is achievable if concerted efforts are made by African governments and the global community. There are some precedents. India, for example, has invested the equivalent of 2.6% of GDP in the power sector since 2000 and China has invested 1.9% of GDP over the same period (Figure 12.6).

Figure 12.6 ▷ Average annual power supply investment in Africa by scenario and historical power sector investment in selected regions

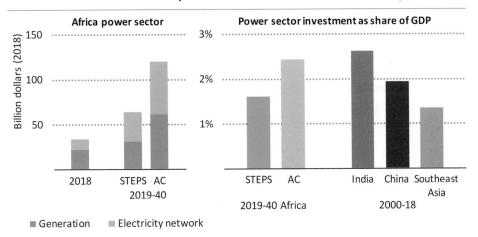

Scaling up power supply investment is challenging but achievable if concerted efforts are made to establish a favourable investment climate and reduce investment risks

Notes: STEPS = Stated Policies Scenario; AC = Africa Case.

To date, investment in power supply in Africa has relied largely on state budgets with significant contributions from development finance institutions (DFIs). The prominent role of these public sources is likely to continue. Against a backdrop of growing fiscal deficits in many countries and tightening donor resources, however, it is critical that public spending is supplemented by private capital and that funding from DFIs is used to catalyse private financing.

Mobilising private capital requires concerted efforts from both African governments and international DFIs. A large number of countries in Africa limit private participation in the power sector: 16-out-of-43 sub-Saharan African countries do not allow private participation in both generation and electricity networks. Establishing a framework for private capital is clearly a necessary first step. Many of the utilities are loss-making and have low operational efficiency: 19-out-of-39 utilities in sub-Saharan Africa are not able to recover enough cash to cover operational expenses (Kojima, 2016). Together with below-cost tariffs and low collection rates, this raises investment risks and makes it difficult to secure financing at affordable costs. Improving the financial and operational performance of utilities and moving towards cost-recovery are therefore essential to attract financing. Robust procurement frameworks (using competitive auctions, for example) and well-designed contracts are also crucial to enhance project bankability.

There is scope for international DFIs to help scale up investment and catalyse more private capital. Between 2013 and the first half of 2018, power sector investments based on private participation in infrastructure models in sub-Saharan Africa amounted to around $4.5 billion per year on average (less than 10% of the annual power sector investment needs between today and 2040), with South Africa accounting for more than half. Outside South Africa, each dollar of public funding (from DFIs and state budgets) attracted $0.6 of private capital either directly (via equity and direct loan) or indirectly (via guarantee) – the figure is $0.4 for renewables. This compares unfavourably with $0.9 for Southeast Asia and more than $4 for South Africa. It is therefore important for international DFIs not only to scale up direct investments but also to encourage private sector investment through targeted interventions (such as risk sharing, liquidity support and take-out financing). There is also a need to nurture the local financial sector to provide a sustained flow of long-term financing to infrastructure projects.

The prospects for Africa's power supply investment will be stronger if governments in African countries take account of what have worked well (and what have not) in other countries. India provides some instructive lessons. In the 2000s, the Indian government introduced a number of measures to establish a policy and regulatory framework to attract private capital, including model architecture for public-private partnerships (PPP) and financial instruments (such as an on-lending facility) to induce local financial institutions to invest in infrastructure. This contributed to a significant scale-up of private investment in power infrastructure and India was recognised as the highest recipient of PPP investments worldwide (World Bank, 2015). However, scrutiny of the commercial viability of projects was sometimes insufficiently rigorous, there were frequent construction delays, and the availability of fuel was often limited: this led to many projects performing less well than expected, and emphasises that there are potential pitfalls to manage even where the overall framework is a strong one.

Not a major emitter, but climate change matters greatly for Africa

Africa has not been a significant contributor to global greenhouse gas (GHG) emissions during the age of industrialisation. Energy-related carbon dioxide (CO_2) emissions in Africa accounted for only 2% of global cumulative emissions from 1890 to today (Figure 12.7). Although Africa experiences rapid economic growth in the Stated Policies Scenario (by two-and-a-half times from today to 2040), its contribution to global energy-related CO_2 emissions increases to just 4% over the period to 2040.

Realising the outcomes in the Africa Case would increase Africa's total CO_2 emissions over the period to 2040 by 6% compared with the Stated Policies Scenario (around 2 gigatonnes (Gt) or 100 million tonnes (Mt) per year). Although this is not a major increase globally, it is highly desirable – and in line with the vision in Agenda 2063 – that they are attained in a way which takes full account of the importance of sustainability, with a very strong role for clean energy sources.

Figure 12.7 ▷ Cumulative energy-related CO_2 emissions by region and scenario

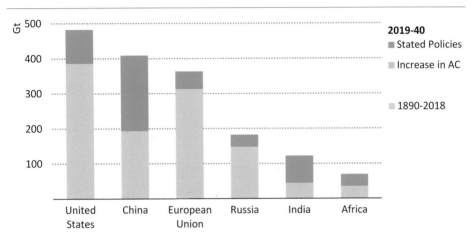

Africa has accounted for a very small share of global CO_2 emissions to date, and that does not change to 2040

Note: AC = Africa Case.

Today bioenergy accounts for over 40% of Africa's energy mix. China and India had a similar share around 1970 and 1990 respectively. In the twenty years after 1970, China relied heavily on coal (and oil to a lesser extent) to replace bioenergy and meet rapidly growing energy demand, and this resulted in cumulative emissions of around 28 Gt CO_2. This means that China incurred around 1 290 grammes of carbon dioxide (g CO_2) to generate one dollar of GDP over this period. India similarly relied on coal, oil and (to a lesser extent) natural gas to serve its expanding economy in the twenty years after 1990. This was accompanied by cumulative emissions of around 19 Gt CO_2 or 280 g CO_2 per dollar of GDP.

In our projections, however, Africa follows a different pathway, with much stronger shares of renewables and natural gas in the energy mix. In the Stated Policies Scenario, the share of renewables (excluding bioenergy) and natural gas grows significantly to 10% and 20% respectively by 2040, while the reliance on traditional uses of bioenergy and coal diminishes. As a result, only 130 g of CO_2 emissions are incurred to generate one dollar of GDP between today and 2040, while the economy grows at a rate of 4% per year.

In the Africa Case, the size of the economy almost quadruples in the period to 2040, but the continent consumes less energy overall with a higher share of cleaner energy sources. The shares of renewables and natural gas expand further to 20% and 25% by 2040, while the share of traditional uses of bioenergy declines. The emissions associated with economic growth are 15% lower in this case, at around 110 g CO_2 per dollar of GDP (Figure 12.8). These emissions relative to economic growth are lower than the figures observed in Organisation of Economic Co-operation and Development (OECD) countries between 1990 and 2018.

With the appropriate policies to support a strong expansion of clean energy and sufficient emphasis on energy efficiency improvements, Africa could be the first continent to achieve a significant level of industrialisation with cleaner energy sources playing a prominent role, requiring much less energy and emissions to deliver economic growth than other economies in the past.

Figure 12.8 ▷ Energy mix evolution and associated emissions per GDP in China, India and Africa

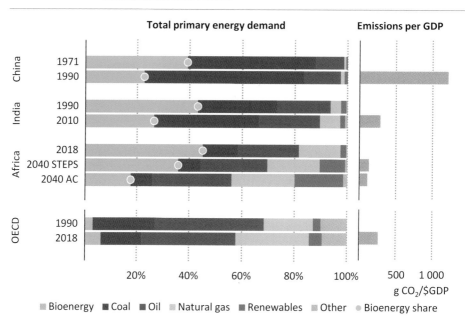

Africa could be the first continent where renewables and gas play a prominent role in supporting a shift away from bioenergy and underpinning economic and industrial growth

Notes: STEPS = Stated Policies Scenario, AC: Africa Case. Emissions per GDP = cumulative CO_2 emissions / cumulative GDP during the indicated period. Renewables exclude bioenergy. The starting year of each region is the year when bioenergy held around 40% share in the energy mix, similar to that of Africa today.

While Africa is responsible for a relatively small portion of global CO_2 emissions, its ecosystems already suffer disproportionately from global climate change, and future impacts are expected to be substantial. The continent therefore not only needs to adapt to the warming already experienced but also to prepare for the intensification of climate change impacts (World Bank, 2018). Temperatures in Africa are likely to rise faster than the global average during the 21st century. Climate change and climate variability are likely to multiply existing threats and pose increased risks to food, health and economic security in Africa (IPCC, 2014).

This underlines the importance of ensuring that new infrastructure in Africa is climate-resilient. For example, only 30% of the buildings that are likely to exist in 2040 have already been built. If building codes are implemented for new buildings to optimise the use of natural light and ventilation for passive cooling, this could reduce the need for cooling systems (fans and air conditioners) and avoid the potential heat islanding effects that could occur in cities (see Spotlight in Chapter 9). Today, a quarter of the global population living in areas that are hot enough to require cooling systems live in Africa, and this share increases to 30% by 2040 in all scenarios. As new cities are built or existing cities grow larger, smart planning is essential to ensure that buildings are highly energy efficient and to facilitate sustainable modes of public transport.

Climate change is also likely to affect the availability of hydro resources. Detailed new analysis in this report shows the negative impacts of climate change on the availability and variability of hydropower outputs in a number of countries. While hydropower remains an essential element of sub-Saharan Africa's electricity supply, diversifying the electricity mix would help to mitigate the risk of power disruptions during droughts and strengthen resilience to changing climate conditions.

In contrast to many other regions, the energy sector is not the biggest contributor to total GHG emissions in Africa. It represents around a third of total GHG emissions (compared to more than two-thirds at the global level). In sub-Saharan Africa (excluding South Africa), land use and forestry[1] (LULUCF), agriculture and waste contribute most towards total GHG emissions. The reduction in the size of Africa's forests, which are natural carbon sinks, is the primary reason for the growth of GHG emissions in Africa: some countries have seen their forest area decrease by more than half over the last 25 years (Box 12.1), highlighting the importance of deforestation for climate policies.

While the ecological and environmental toll of reliance on fuelwood for cooking cannot be exactly quantified, the traditional use of solid biomass for cooking comes at a large cost to human health and wellbeing. Air pollution in Africa is one of the leading causes of premature deaths. Around 500 000 premature deaths are attributed to smoky indoor air arising from the use of solid biomass for cooking while 300 000 premature deaths are linked to outdoor pollution in cities.

In the Stated Policies Scenario, premature deaths owing to household air pollution decrease slightly over the outlook period as a consequence of efforts to bolster access to clean cooking through LPG stoves, improved biomass stoves or biodigesters. There is a much greater adoption of these cleaner technologies in the Africa Case: over 1.1 billion people move away from traditional use of solid biomass by 2030, and the number of premature deaths from household air pollution falls by two-thirds.

---

[1] LULUCF refers to land use, land-use change and forestry.

Box 12.1 ▷   Implications of unsustainable bioenergy use

Since 1990, the total forest area in Africa has fallen by 85 million hectares (ha), which is more than the total land area of Mozambique (Figure 12.9). Some countries have been more affected than others. For instance, Nigeria has lost 60% of its forest cover since 1990, while Tanzania and Ethiopia have lost almost 20% of their forest areas (FAO, 2019).

Conversely, fuelwood consumption (directly used by households for cooking or to produce charcoal) has doubled in sub-Saharan African countries (excluding South Africa) over the same period. While the relationship between deforestation and growing demand for fuelwood is difficult to quantify, efficiency improvements across the various bioenergy value chains could play a significant role in protecting forests, biodiversity and carbon sinks.

Figure 12.9 ▷   Fuelwood consumption in the Stated Policies Scenario and the Africa Case, and forest area in selected African countries

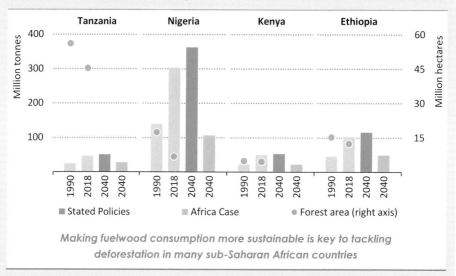

Making fuelwood consumption more sustainable is key to tackling deforestation in many sub-Saharan African countries

A number of countries have already made commitments to address deforestation in their updated Nationally Determined Contributions. Nigeria acknowledges the need to halt deforestation, conserve remaining natural forests and reverse forest degradation. Others, including DR Congo, have pledged to commit efforts to reforestation activities. Converting these ambitions into actions and extending them across the continent would make the African biomass industry more sustainable.

The increase in demand for energy services brought about by the fast-growing and rapidly urbanising population across the continent will have significant implications for air quality in cities. The increase in the overall level of air pollutant emissions in the Stated Policies

Scenario is not a surprise, given the exceptionally low baseline for current energy consumption. The mix of technologies and fuels chosen by consumers can however play an important part in mitigating the increase of pollutant emissions, which will ultimately have wide-ranging implications for the health and wellbeing for millions of people.

In the Stated Policies Scenario, sulfur dioxide (SO_2) emissions decrease by a quarter across Africa by 2040. There is an increase in industrial emissions but this is more than offset by a significant decrease in emissions from coal-fired power plants, mainly in South Africa. Emissions of nitrogen oxides (NO_X) increase by one-quarter, mainly from the incomplete combustion of fuels in cars, despite a significant fall in emissions in the power sector during the period to 2040. In the Africa Case, improved emissions standards for passenger vehicles result in emissions from this segment falling, despite the increased number of cars on the road.

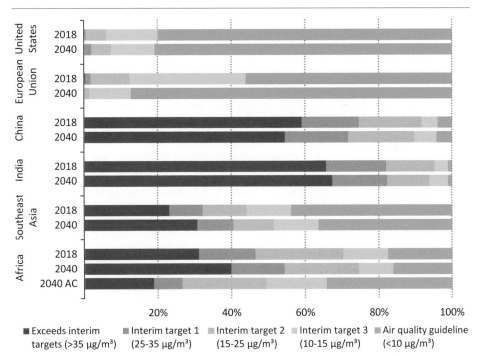

Figure 12.10 ▷ Population exposure to fine particulate pollution (PM2.5) in selected regions in the Stated Policies Scenario and Africa Case

Proportion of the population in Africa exposed to high levels of PM2.5 pollution drops in the Africa Case and remains lower than in some Asian countries in the Stated Policies Scenario

Notes: AC = Africa Case; µg/m³ = micrograms per cubic metre. Interim targets and Air Quality Guideline refer to World Health Organization exposure thresholds.

Source: International Institute for Applied Systems Analysis

Higher emissions of NOx and PM$_{2.5}$ also take a considerable toll on health and wellbeing. In the Stated Policies Scenario, the increasing concentration of PM$_{2.5}$ by 2040 means that the number of premature deaths associated with outdoor air pollution increases by almost 60%, reaching 480 000 in 2040. In the Africa Case, emissions of the three major air pollutants decline sharply from the current levels. Reduced exposure to PM$_{2.5}$ is particularly important: despite a significant increase in energy services, the number of premature deaths associated with outdoor air pollution in 2040 is almost 30% lower than in the Stated Policies Scenario.

Achieving global sustainable development goals requires the success of Africa

In many areas, global energy transition goals are closely linked to successful growth and development in Africa. The continent's economic and social prosperity are in turn closely linked to successful global energy transitions. Two examples highlight the interlinkages between the world and Africa: first, access to modern energy services; and second, Africa's role as a major supplier of the minerals necessary to achieve the global energy transition.

More than two-thirds of the world's population without access to electricity and around a third of the population without access to clean cooking live in Africa. By 2030 in the Stated Policies Scenario, most of the remaining population without access to electricity and clean cooking remain concentrated in Africa. Addressing energy access in Africa is therefore of paramount importance to solving this global concern.

Boosting energy access rates in Africa brings huge benefits in terms of reduced poverty, lower air pollution and increased economic prosperity. Access to electricity is crucial to the provision of essential services: in health centres, for instance, it is vital for the use of efficient modern equipment, the storage and preservation of vaccines and medicines, and the ability to conduct emergency medical procedures, for example during child birth. Access to clean cooking is essential to reduce the health impacts and the number of premature deaths related to household air pollution.

In the Stated Policies Scenario, around 20 million people are connected to the electricity network each year, which is less than a third of what would be needed to reach full access by 2030. By 2030, 85% of all people without access to electricity live in Africa (Figure 12.11). In DR Congo, for example, the projected number of people without access to electricity increases by 30% in this scenario, as policies fail to keep pace with population growth. Reaching full electricity access by 2030 as envisaged in the Africa Case requires a tripling of efforts to extend connection to over 60 million people each year. Reaching this level of access would need an additional push for decentralised renewables in the context of a comprehensive set of policies and investments that makes use of all available solutions, with mini-grids and stand-alone systems providing power to more than half of those gaining access by 2030. Energy efficiency also has an important part to play.

Delivering access to clean energy in an integrated way would also support economic growth and overall development. Research suggests that access could bring new avenues of productive employment to remote populations, particularly for women. In addition to freeing up time by speeding up domestic chores and giving women more time to engage in paid jobs, access to electricity can have a particular impact on female-owned businesses, helping them to transition from extreme poverty to near middle-class status, as shown recently in Ghana (Power Africa, 2019).

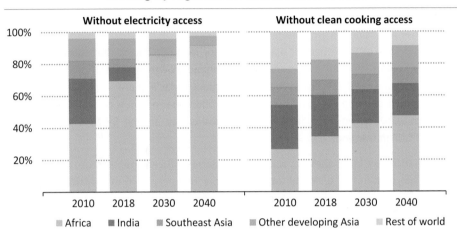

Figure 12.11 ▷ Share of population without access to electricity and clean cooking by region in the Stated Policies Scenario

Those without access to electricity and clean cooking are increasingly concentrated on the African continent

Moreover, electricity can also play an important role in improving agricultural productivity through advanced irrigation techniques, as several successful examples of stand-alone solar water pumps have demonstrated. Cold storage powered by renewable electricity could also reduce post-harvest losses of agricultural outputs, which are currently estimated at 20-50% of the food produced in sub-Saharan Africa (depending on the food type).

In the Stated Policies Scenario, Africa is one of the few regions where the number of people without access to clean cooking increases, as the expansion of clean cooking is unable to keep pace with rapid population growth, and around half of the global population without access to clean cooking in 2040 lives in Africa. There are exceptions: Ghana sees a visible improvement in this area, but many other countries are not set to emulate this example. While urbanisation increases the use of alternative options such as LPG and natural gas in many regions, solid biomass (in the form of charcoal) remains the preferred option for cooking in African cities. The Africa Case sees all households across the continent gain access to clean cooking by 2030. This reduces significantly the number of premature deaths linked to indoor air pollution.

Resource development, minerals in particular, is another area where Africa and the world share a common interest. From cobalt and manganese for batteries to chromium and neodymium for wind turbines, and to platinum for hydrogen fuel cells, minerals are a critical component in many clean energy technologies. As energy transitions accelerate, demand for minerals is set to grow significantly. For example, demand for cobalt from deployment of electric vehicles increases to around 170 kilotonnes per year (kt/year) in 2030 in the Stated Policies Scenario, higher than today's supply capacity, and to almost 360 kt/year in the case of higher electric vehicle uptake (IEA, 2019). Africa is a major producer of many of these minerals: DR Congo accounts for two-thirds of global cobalt production and South Africa produces 70% of the world's platinum.

In 2017, net income from mineral production was equivalent to around 2% of sub-Saharan Africa's GDP and minerals accounted for some 20% of total merchandise exports in Africa (77% in the case of DR Congo). Rising demand for minerals means that successful global energy transitions offer an opportunity for economic growth in mineral-rich countries in Africa. For example, if DR Congo were to maintain today's share in global production, growing global demand for cobalt would bring additional revenue of $4-8 billion to the country in 2030 (based on today's prices), equivalent to around 3-6% of the country's projected GDP in that year.

However, there are large question marks over whether African countries can keep up with rising global demand in a timely and sustainable manner. Current practices are often inefficient, polluting and subject to social protests. Given that African countries account for a large proportion of the global production of key minerals, failure to keep up with demand could not only hamper Africa's economic outcomes but also hold back the pace of global energy transitions (Figure 12.12).

Figure 12.12 ▷ Composition of Africa's merchandise exports, 2017, and key minerals production, 2018

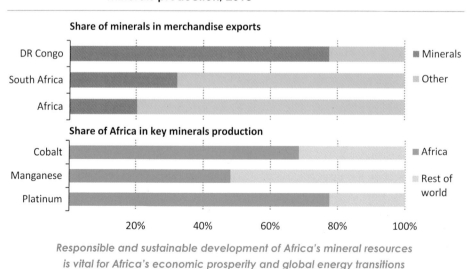

Responsible and sustainable development of Africa's mineral resources is vital for Africa's economic prosperity and global energy transitions

Source: IEA analysis based on UNCTAD Stats (2019) and USGS (2019).

Putting in place structures and governance arrangements to ensure responsible minerals development would help guard against a range of potential problems. Robust regulatory and oversight mechanisms would be needed to ensure that impacts on local environments are minimised and that revenues are used in a transparent manner. There is also a need for careful scrutiny of how minerals are sourced and how supply chains are managed. Those who use minerals can play a helpful role, as can international financial institutions. For example, BMW, BASF and Samsung recently launched a pilot initiative to support sustainable and fair cobalt mining in DR Congo, which aims to improve working and living conditions for small-scale mining operations and surrounding communities. The World Bank has launched the Climate-Smart Mining Facility to help minimise the environmental and climate impacts of mining activities. As in so many other areas, the future of Africa's development and the prospects for global sustainable growth are closely interlinked.

Regional and country energy profiles

Introduction

The following section presents the results of the Stated Policies Scenario and Africa Case for the sub-Saharan region as a whole as well as for the following eleven countries:

12.1	Sub-Saharan Africa	p526
12.2	Angola	p530
12.3	Côte d'Ivoire	p534
12.4	Democratic Republic of the Congo	p538
12.5	Ethiopia	p542
12.6	Ghana	p546
12.7	Kenya	p550
12.8	Mozambique	p554
12.9	Nigeria	p558
12.10	Senegal	p562
12.11	South Africa	p566
12.12	Tanzania	p570

Together these eleven countries accounted for three-quarters of sub-Saharan Africa's gross domestic product (GDP), two-thirds of its population and three-quarters of its energy demand in 2018. The profiles presented in this section aim to provide decision makers with a data-rich set of information on potential energy pathways that reflect each country's unique energy demand and supply needs and regional characteristics. The policy, technology and economic assumptions that underpin both the Stated Policies Scenario and the Africa Case are described in the introduction to the Special Focus on Africa and discussed on a regional basis in Chapters 9, 10 and 11, with the implications presented in the first part of this chapter.

How to read the profiles:

We use a standard format to present the country and regional profiles. Each profile contains a set of figures and tables corresponding to the following categories:

- Key characteristics of the country's energy system.
- Major macroeconomic indicators, including GDP and population growth, carbon dioxide (CO_2) emissions and data for electricity access and clean cooking access by scenario.
- Description of energy-related policy initiatives, including specific performance targets.
- Outlook of how primary energy demand and GDP (based on GDP expressed in year-2018 dollars in purchasing power parity [PPP] terms) evolve to 2040 and the role of each fuel in delivering the alternative energy futures.
- View of how the electricity mix changes over time to meet growing electricity demand.

- Final energy consumption by scenario, showing the potential efficiency gains achieved by implementing more stringent fuel economy standards, building codes, equipment and appliance efficiency requirements.
- Fuel and technology mix[2] used in cooking in 2018 and in 2030 by scenario.[3]
- The trajectory for demand and production of major fossil fuels, highlighting trade balances.
- Cumulative investment by sector required to meet the growth in energy demand and supply in both the Stated Policies Scenario and the Africa Case.[4]

Notes to profiles:

Scenarios: **AC** = Africa Case; **STEPS** = Stated Policies Scenario.

End-use sectors are industry (including manufacturing and mining), transport, buildings (including residential and services) and other (including agriculture and non-energy use).

Traditional use of solid biomass refers to the use of solid biomass with basic technologies, such as a three-stone fire, often with no or poorly operating chimneys.

Productive uses refers to energy used towards an economic purpose. This includes energy used in agriculture, industry, services and non-energy use. Some energy demand from the transport sector (e.g. freight-related) could be considered as productive, but is treated separately.

GIS maps for each country or region contained in these profiles were developed in collaboration with the Royal Institute of Technology (Sweden) – Division of Energy Systems Analysis (KTH-dESA). The maps detail the least-cost pathway to deliver universal electricity access by means of a combination of on-grid, mini-grid and stand-alone systems.[5]

Units and terms: GDP = gross domestic product; CAAGR = compound average annual growth rate; PPP = purchasing power parity; Mt CO_2 = million tonnes of carbon dioxide; Mtoe = million tonnes of oil equivalent; GW = gigawatt; TWh = terawatt-hour; kV = kilovolt; LPG = liquefied petroleum gas; mb/d = million barrels per day; Mtce = million tonnes of coal equivalent; bcm = billion cubic metres, PV = photovoltaics, GHG = greenhouse gas.

Investment data are presented in real terms in year-2018 US dollars.

---

[2] *Other clean* includes electricity, natural gas, biogas and biofuels. *Charcoal and other solid biomass* refer to the combustion of these fuels in inefficient stoves.

[3] More detailed information on the methodology can be found in Chapter 9.

[4] Investment in electricity networks and generation excludes investment in electricity access, which is counted separately in this figure.

[5] More detailed information on the methodology can be found in Chapter 10, Box 10.2.

12.1 Sub-Saharan Africa[1]

Fastest growing population   **Strong** economic growth   **Major** commodities exporter

Table 12.1A ▷ **Sub-Saharan Africa key indicators and policy initiatives**

			Stated Policies		Africa Case		CAAGR 2018-40	
	2000	2018	2030	2040	2030	2040	STEPS	AC
GDP ($2018 billion, PPP)	1 375	3 536	6 161	10 346	8 381	16 683	5.0%	7.3%
Population (million)	626	1 034	1 404	1 761	1 404	1 761	2.5%	2.5%
with electricity access	20%	43%	62%	66%	100%	100%	2.0%	4.0%
with access to clean cooking	6%	13%	31%	51%	100%	100%	6.3%	9.6%
CO_2 emissions (Mt CO_2)	130	312	534	843	762	1 154	4.6%	6.1%

Policy	Key targets and measures
Regional Strategies	▪ Agenda 2063: A prosperous Africa based on inclusive growth and sustainable development. ▪ African Continental Free Trade Area: accelerating intra-African trade and boosting Africa's trading position in the global market by strengthening Africa's common voice and policy space in global trade negotiations.

- Drastic efficiency improvements, in part due to the accelerated move away from solid biomass, result in primary energy demand being lower in the AC than in the STEPS even though GDP is 60% higher in the AC.
- Supply from natural gas and renewable sources expand in both scenarios to meet rising demand for energy as the sub-Saharan economy expands.
- Electricity access and clean cooking facilities for all are achieved by 2030 in the AC.

Figure 12.1A ▷ **Sub-Saharan Africa primary energy demand and GDP**

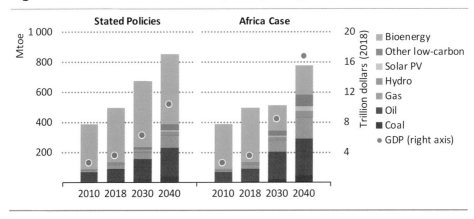

[1] Excluding South Africa.

Figure 12.1B ▷ Sub-Saharan Africa electricity generation by technology

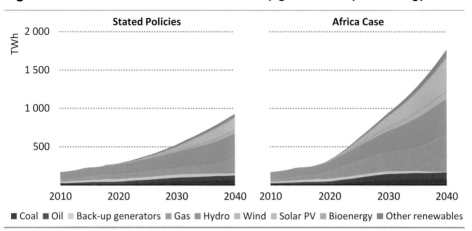

■ Coal ■ Oil ■ Back-up generators ■ Gas ■ Hydro ■ Wind ■ Solar PV ■ Bioenergy ■ Other renewables

- Today's power mix, dominated by hydro, gradually diversifies as solar PV and natural gas increasingly make inroads into the power system. In the STEPS, the combined share of solar PV and natural gas reaches the level of hydro by 2040.

- In the AC, natural gas (27%) passes hydropower (26%) as the largest source of power supply by 2040 while the share of solar PV rises to 24%.

Figure 12.1C ▷ Sub-Saharan Africa electricity access solutions by type in the Africa Case

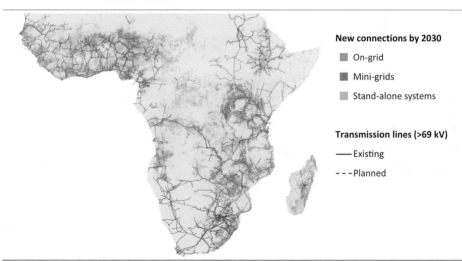

- In the STEPS, the main grid connects around 70% of the 230 million people gaining electricity access by 2030, alongside decentralised options for the remainder in more remote areas. In 2030, 530 million people remain without access.

- In the AC, decentralised solutions are the least-cost option for more than two-thirds of the 530 million additional people connected by 2030 to reach full access.

Figure 12.1D ▷ Sub-Saharan Africa final energy consumption

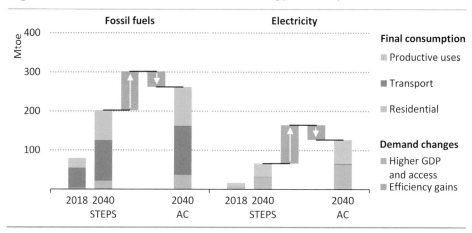

- Growing trends of urbanisation and industrialisation drive strong energy consumption growth for transport and productive uses in both the STEPS and the AC, increasing oil demand the most, especially in the AC, which sees faster economic growth.
- Electricity consumption is very low today, but quadruples through to 2040 in the STEPS, with demand growth led by light industry, appliances and cooling systems. Demand rises further in the AC.

Figure 12.1E ▷ Sub-Saharan Africa fuels & technologies used for cooking

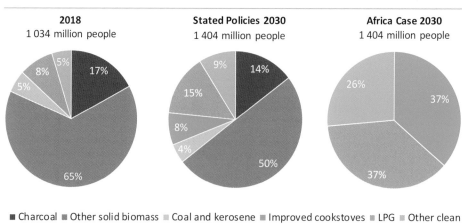

- In the STEPS, more people gain access to clean fuels and technologies for cooking by 2030, but 70% of the population still lack access.
- To bridge the gap and achieve full access to clean cooking for all in the AC, liquefied petroleum gas (LPG) is the most scalable solution for urban settlements, with improved biomass cookstoves doing most to provide access in rural areas.

Figure 12.1F ▷ Sub-Saharan Africa fossil fuel demand and production

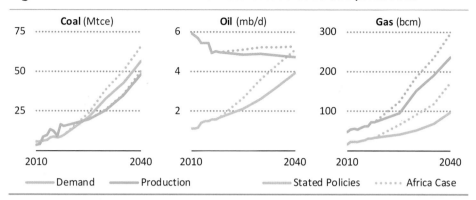

- Rapidly growing oil demand and stagnating domestic oil production reduce net oil exports in the STEPS; exports are further reduced by faster economic growth in the AC.
- Gas demand and production increase by 2040 in the STEPS, but both grow more rapidly in the AC and the region becomes a major supplier of gas to global markets.

Figure 12.1G ▷ Sub-Saharan Africa cumulative investment needs, 2019-2040

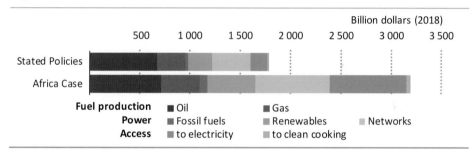

- In the STEPS, $1.8 trillion of cumulative energy supply investment is needed, with upstream oil and gas and power each accounting for around half of this.
- The AC requires 80% more capital with a stronger emphasis on power sector investments, including a doubling of spending in renewables and electricity networks.

Sub-Saharan Africa policy opportunities

- Enhanced power sector integration in sub-Saharan Africa could help to deliver more affordable and reliable power and reduce average electricity generation costs.
- Challenges relating to infrastructure, affordability and business models must be overcome if the region is to capitalise on the potential of natural gas.
- More efficient use of energy across end-use sectors such as fuel economy standards for cars and two/three-wheelers, building codes for new buildings, and more efficient industrial processes and efficiency standards for appliances and cooling systems would support wider economic development and offset growth in energy demand.

12.2 Angola

Second-largest oil producer   Oil accounts for 90% of exports   Luanda: future megacity

Table 12.2A ▷ Angola key indicators and policy initiatives

	2000	2018	Stated Policies 2030	Stated Policies 2040	Africa Case 2030	Africa Case 2040	CAAGR 2018-40 STEPS	CAAGR 2018-40 AC
GDP ($2018 billion, PPP)	72	199	287	404	349	625	3.3%	5.3%
Population (million)	16	31	45	60	45	60	3.1%	3.1%
with electricity access	12%	44%	57%	65%	100%	100%	1.7%	3.8%
with access to clean cooking	37%	50%	58%	66%	100%	100%	1.3%	3.2%
CO_2 emissions (Mt CO_2)	5	17	20	33	27	48	3.1%	4.8%

Policy	Key targets and measures
Performance targets	• Establish targets for renewable energy sources to 2025: 100 MW of solar PV; 370 MW of small and medium hydro; 500 MW of biomass; 100 MW of wind. • Up to 35% (unconditional) to 50% (conditional) reduction in GHG emissions by 2030 as compared to the business-as-usual scenario.
Industrial development targets	• National Development Plan of Angola 2018-2022: Lessen economic dependence on oil and natural gas revenues, strengthen the business environment, increase energy efficiency and achieve middle-income status by 2022.

- Angola could supply an economy three-times larger than today's in the AC with only twice the amount of energy.

- Oil remains an important energy source, but end-use tariffs that are more reflective of costs reduce its share of the overall energy mix and help diversification towards natural gas and renewables in the AC.

Figure 12.2A ▷ Angola primary energy demand and GDP

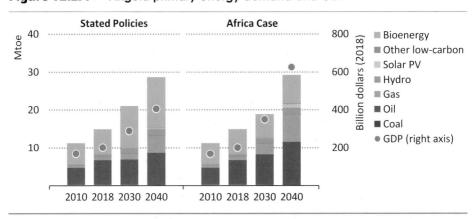

Figure 12.2B ▷ Angola electricity generation by technology

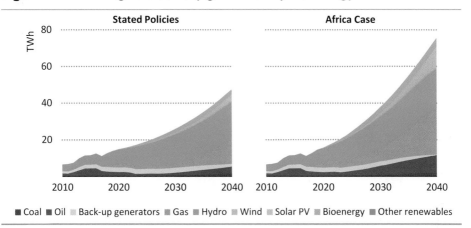

■ Coal ■ Oil ■ Back-up generators ■ Gas ■ Hydro ■ Wind ■ Solar PV ■ Bioenergy ■ Other renewables

- Angola currently relies mostly on hydropower and oil (including diesel) for power generation.
- Providing access to all increases electricity demand sevenfold in the AC. Gas and comparatively cheap hydropower play key roles in meeting this growth along with solar PV.

Figure 12.2C ▷ Angola electricity access solutions by type in the Africa Case

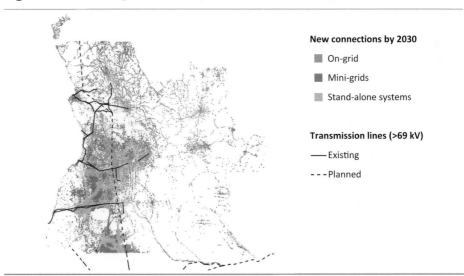

- The electricity access rate in Angola is 44% today, with most of the population currently without access located in the west of the country.
- The least-cost path to full access to electricity in the AC is mini-grids (46%), alongside grid connections for a large part of the population (38%) living near the existing and planned grids; stand-alone systems help to reach the most remote areas.

Figure 12.2D ▷ **Angola final energy consumption**

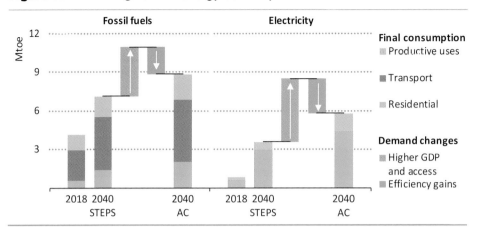

- The number of cars expands from 0.4 million in 2018 to 1.1 million in 2040 in the STEPS, and rises further in the AC. The associated increase in oil demand can be mitigated to an extent by improving fuel economy standards.
- Angola could meet nearly all of its cement demand domestically before 2040 in both scenarios provided a reliable supply of gas and electricity is available.

Figure 12.2E ▷ **Angola fuels and technologies used for cooking**

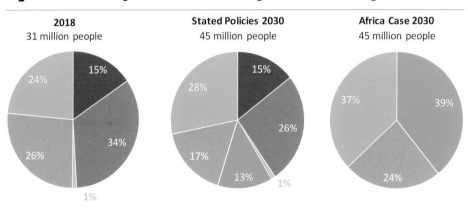

- Angola has one of the highest shares of access to clean cooking in sub-Saharan Africa, thanks to government policies supporting LPG and natural gas.
- A further push on access policies adapted to rural conditions could help provide clean cooking to 90% of people in rural areas through improved biomass cookstoves in the AC.

Figure 12.2F ▷ Angola fossil fuel demand and production

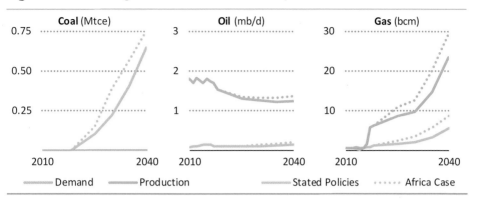

- Recent policy reforms in the oil and gas sector help stabilise the outlook for oil production in both scenarios.
- Growing population and stagnant oil production reduce per capita net income from oil and gas production in both scenarios, increasing the need for economic diversification.

Figure 12.2G ▷ Angola cumulative investment needs, 2019-2040

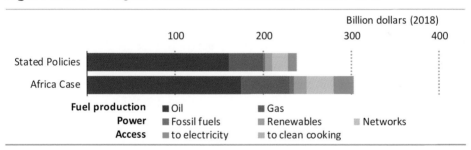

- Around $240 billion of cumulative energy supply investment is needed in the STEPS, of which over 80% goes to upstream oil and gas.
- The AC requires around 25% more capital than the STEPS, with a strong emphasis on investments in upstream gas, electricity access and networks.

Angola policy opportunities

- Angola's natural gas resources could underpin a domestic industrial base that would have the added benefit of diversifying the economy away from oil exports.
- Angola would benefit from sustaining and strengthening recent reforms in the oil and gas sector, including efforts to streamline investment procedures and restructure the role of the national oil company.
- The availability of domestic natural gas presents a significant opportunity for efficient and dependable electricity generation.

12.3 Côte d'Ivoire

Largest cocoa exporter	Rapidly expanding economy	Diversity of resources

Table 12.3A ▷ Côte d'Ivoire key indicators and policy initiatives

			Stated Policies		Africa Case		CAAGR 2018-40	
	2000	2018	2030	2040	2030	2040	STEPS	AC
GDP ($2018 billion, PPP)	56	107	221	330	264	569	5.3%	7.9%
Population (million)	17	25	34	42	34	42	2.4%	2.4%
with electricity access	50%	63%	94%	100%	100%	100%	1.1%	1.1%
with access to clean cooking	18%	30%	59%	71%	100%	100%	4.0%	5.7%
CO_2 emissions (Mt CO_2)	6	11	19	24	21	33	3.7%	5.1%

Policy	Key targets and measures
Performance targets	▪ Increase electricity generation to 4 000 MW by 2020 and 6 000 MW by 2030. ▪ *Programme National d'Électrification Rurale*: connect all localities with more than 500 inhabitants by 2021 and all areas by 2025.
Industrial development targets	▪ Achieve emerging economy status by end of 2020 while ensuring that industry accounts for 40% of the economy. ▪ Accelerate the structural transformation of the economy through industrialisation, develop infrastructure throughout the country and protect the environment. ▪ Rise in investment rate from 19.3% in 2015 of GDP to 24.5% in 2020 with significant contribution from private sector.

- Rapid industrialisation in the AC could yield an economy that is five-times larger than today but with energy efficiency the country consumes only twice as much energy.
- Natural gas has a key role to play in electricity generation in the AC. Promoting its use could see its share of the energy mix rising by eleven percentage points more than STEPS.

Figure 12.3A ▷ Côte d'Ivoire primary energy demand and GDP

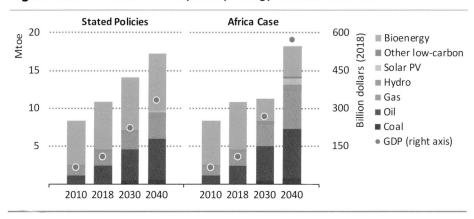

Figure 12.3B ▷ Côte d'Ivoire electricity generation by technology

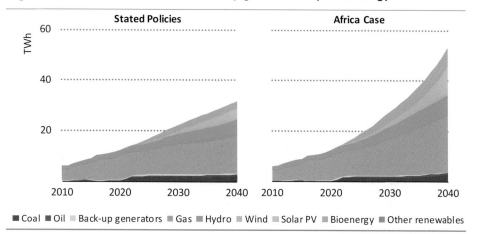

- Providing access to all and increasing industrialisation in the AC raises electricity demand sixfold compared to today.
- Gas continues to play a large role in power generation, but its share is reduced from three-quarters today to around 45% as solar and bioenergy increase in both scenarios.

Figure 12.3C ▷ Côte d'Ivoire electricity access solutions by type in the Africa Case

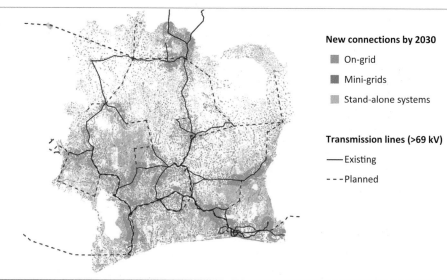

- Effective programmes supporting electrification of villages and households connect more than 90% of the population by 2030 in the STEPS.
- Given the current coverage of the grid network, grid densification and grid extension are the least-cost solution for around 40% of the population in the AC.

Chapter 12 | Implications for Africa and the world

Figure 12.3D ▷ Côte d'Ivoire final energy consumption

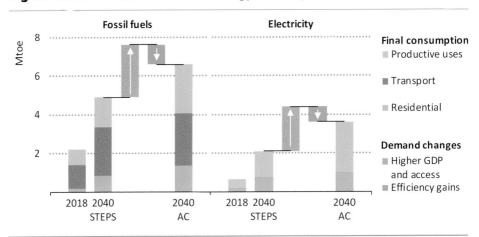

- The number of cars grows fivefold and related oil consumption fourfold in the AC, but the growth could be almost 20% larger without fuel economy standards.

- Côte d'Ivoire electrifies much of industry, with electricity displacing oil to become the major fuel. Electricity demand for residential cooling increases by almost 2 TWh in the AC.

Figure 12.3E ▷ Côte d'Ivoire fuels and technologies used for cooking

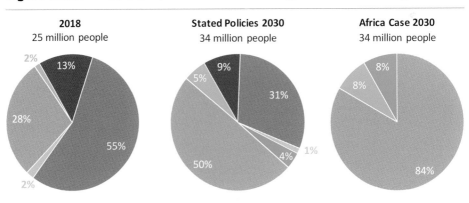

- With strong policy support, LPG is the preferred solution to improve access to clean cooking, reducing the use of traditional stoves with charcoal and other solid biomass.

- The AC sees further use of LPG and improved biomass cookstoves, particularly in rural areas, to bring access to clean cooking to all.

Figure 12.3F ▷ Côte d'Ivoire fossil fuel demand and production

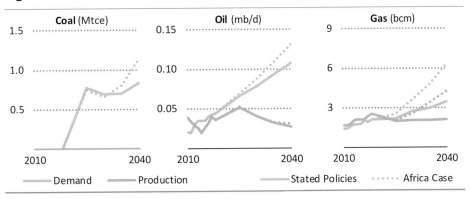

- Rapidly rising passenger car stocks and declining production lead to expanding import requirements for oil in both scenarios.
- Given the important share of gas in the power mix, strong growth in electricity demand underpins rapid growth in gas demand, especially in the AC.

Figure 12.3G ▷ Côte d'Ivoire cumulative investment needs, 2019-2040

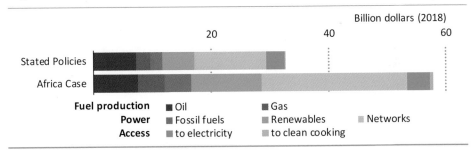

- Energy investment needs amount to $33 billion through to 2040 in the STEPS, with spending on electricity networks representing almost 40% of the total.
- The AC requires investment to increase by a further 80%, with more emphasis on gas-related spending (upstream and generation), renewables and electricity networks.

Côte d'Ivoire policy opportunities

- Increased production of natural gas provides a significant opportunity for it to be used extensively in power generation and industry.
- Prioritising energy efficiency is essential to helping Côte d'Ivoire make the most of its limited resources. With the anticipated increase in demand for cooling and other household uses, efficiency standards for appliances would materially impact the rate of energy demand growth.
- Expanding power generation capacity is crucial.

12.4 Democratic Republic of the Congo

Largest producer of cobalt   **Kinshasa:** an African megacity   **Most ambitious** hydro plan

Table 12.4A ▷ DR Congo key indicators and policy initiatives

	2000	2018	Stated Policies 2030	Stated Policies 2040	Africa Case 2030	Africa Case 2040	CAAGR 2018-40 STEPS	CAAGR 2018-40 AC
GDP ($2018 billion, PPP)	29	73	132	238	195	455	5.5%	8.7%
Population (million)	47	84	120	156	120	156	2.8%	2.8%
with electricity access	7%	9%	16%	21%	100%	100%	4.1%	11.7%
with access to clean cooking	3%	3%	4%	5%	100%	100%	2.0%	17.1%
CO₂ emissions (Mt CO₂)	1	3	10	16	18	18	8.7%	9.2%

Policy	Key targets and measures
Performance targets	• Complete the construction of Inga 3 Basse-Chute dam. • Reduce GHG emissions by 17% by 2030 compared to the business-as-usual scenario (430 Mt CO_2-equivalent), equivalent to slightly more than a 70 Mt CO_2 reduction. • Plant about three million hectares of forest by 2025.
Industrial development targets	• Achieve high-income status by 2050 by means of rigorous implementation of the National Strategic Plan for Development.

- In the AC, DR Congo supports an economy six-times larger than today's with only 35% more energy by diversifying its energy mix away from one that is 95% dependent on bioenergy.

- The power sector sees more growth than any other sector; a big increase in the use of hydropower leads to its share of the overall energy mix increasing to 23% in the AC.

Figure 12.4A ▷ DR Congo primary energy demand and GDP

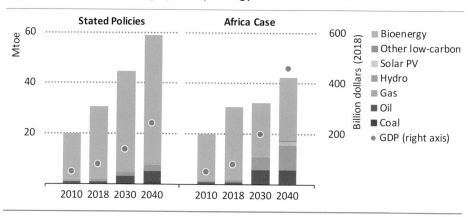

Figure 12.4B ▷ DR Congo electricity generation by technology

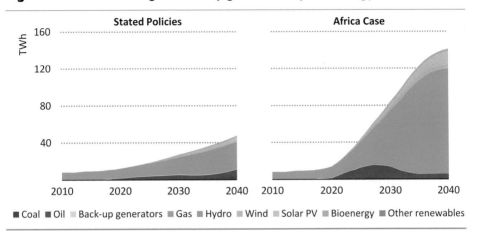

- Almost all electricity generation today comes from hydropower and the Inga project has the potential to provide much more. If network constraints are addressed, DR Congo could become an electricity exporter.
- In the AC, Phase 5 of the Inga project enables DR Congo to meet an eleven-fold increase in electricity demand; this increase is the result of achieving full access to electricity and of the growing electrification of productive uses.

Figure 12.4C ▷ DR Congo electricity access solutions by type in the Africa Case

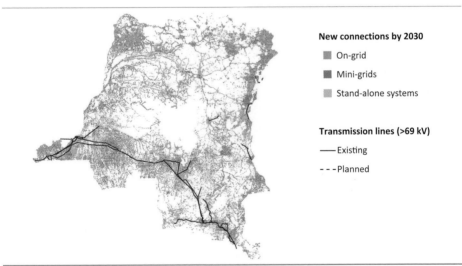

- Less than 10% of the population has access to electricity today, making DR Congo the country with the largest number of people without access in Africa after Nigeria.
- Mini-grids account for more than half of all new connections in the AC.

Figure 12.4D ▷ DR Congo final energy consumption

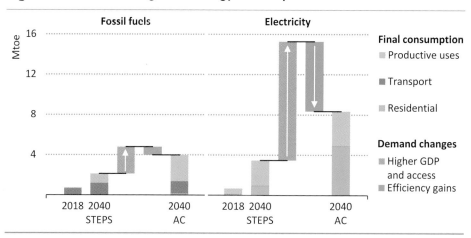

- Oil use in industry increases significantly in the AC with manufacturing and mining chiefly responsible for this growth.
- Electricity consumption is low today but is set to increase significantly in the AC as household incomes rise, access to electricity improves and mining activities increase.

Figure 12.4E ▷ DR Congo fuels and technologies used for cooking

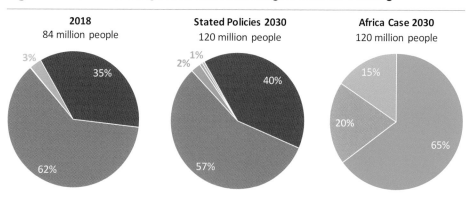

- Given the availability of fuelwood in rural areas and the affordability of charcoal in urban areas, almost all people cook with traditional stoves in 2030 in the STEPS.
- In the AC, improved cookstoves are the preferred option to provide clean cooking access in both urban and rural areas. In parallel, kilns for making charcoal are improved to increase their efficiency.

Figure 12.4F ▷ DR Congo fossil fuel demand and production

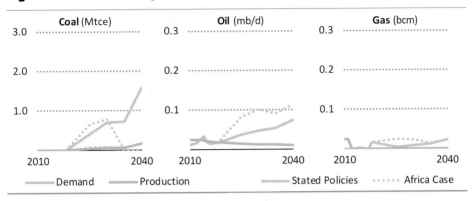

- Fossil fuel consumption is at a low level, but growing, and almost reliant on imports in both scenarios. Further industrial development depends on a large increase in imports.
- DR Congo is a major producer of minerals. It accounts for almost two-thirds of global cobalt production; this gives it a crucial role in global clean energy transitions.

Figure 12.4G ▷ DR Congo cumulative investment needs, 2019-2040

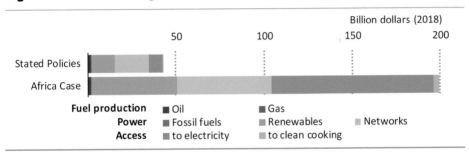

- The AC requires a quadrupling of investment compared with the STEPS, with emphasis on renewables, power networks and access to electricity and clean cooking.
- Investment opportunities in the AC are likely to be realised only if sound structures to regulate the sector and manage revenues from mineral production are in place.

DR Congo policy opportunities

- Cobalt mining activities will drive an increase in electricity demand. Meeting this through renewable hydropower would help to develop low-carbon electricity for DR Congo and a low-carbon value chain for the global electric vehicle fleet.
- Given the country's dispersed population centres, decentralised solutions offer the lowest cost way to overcome grid limitations and provide electricity access to the huge share of the population currently without it.
- Increased regional co-operation could help realise the potential of Inga, which has the potential to provide large quantities of reliable low-carbon electricity to DR Congo and its neighbours.

12.5 Ethiopia

Geothermal potential   Large hydro capacity   Strong progress on access

Table 12.5A ▷ Ethiopia key indicators and policy initiatives

	2000	2018	Stated Policies 2030	Stated Policies 2040	Africa Case 2030	Africa Case 2040	CAAGR 2018-40 STEPS	CAAGR 2018-40 AC
GDP ($2018 billion, PPP)	47	220	493	870	610	1 445	6.5%	8.9%
Population (million)	67	108	143	173	143	173	2.2%	2.2%
with electricity access	5%	45%	100%	100%	100%	100%	3.7%	3.7%
with access to clean cooking	1%	7%	34%	56%	100%	100%	9.7%	12.6%
CO_2 emissions (Mt CO_2)	3	14	29	46	32	52	5.5%	6.2%

Policy	Key targets and measures
Performance targets	▪ Increase generating capacity by 25 000 MW by 2030: 22 000 MW of hydro; 1 000 MW of geothermal; and 2 000 MW of wind by 2030. ▪ National Electrification Program (2017): 100% electrification in 2025, with 35% off-grid and 65% grid, while extending the grid to reach 96% grid connections by 2030.
Industrial development targets	▪ Achieve an annual average real GDP growth rate of 11% within a stable macroeconomic environment and become a lower middle-income country by 2025. ▪ Focus on ensuring rapid, sustainable growth by enhancing the productivity of the agriculture and manufacturing sectors, and stimulating competition in the economy.

- Ethiopia could supply a much larger economy than today in the AC, using only twice the energy, were it to diversify its energy mix and implement efficiency standards.

- In the AC, this diversification comes about as a result of a substantial expansion of geothermal energy along with increased use of oil within industry and for cooking.

Figure 12.5A ▷ Ethiopia primary energy demand and GDP

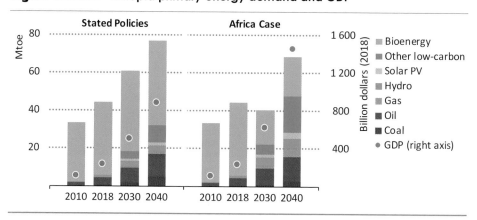

Figure 12.5B ▷ **Ethiopia electricity generation by technology**

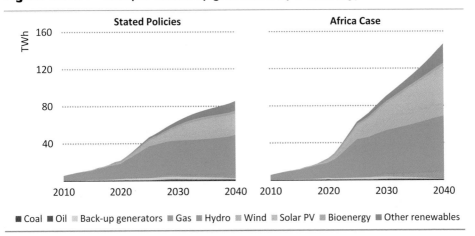

■ Coal ■ Oil ■ Back-up generators ■ Gas ■ Hydro ■ Wind ■ Solar PV ■ Bioenergy ■ Other renewables

- Ethiopia is currently heavily reliant on hydropower; plans to increase capacity to 13.5 GW by 2040 would make Ethiopia the second-largest hydro producer in Africa.
- Providing electricity access to all and electrifying productive uses will lead to a fivefold increase in generation in the STEPS, and an even bigger increase in the AC; solar PV and geothermal account for almost 45% of the power mix by 2040 in the AC.

Figure 12.5C ▷ **Ethiopia electricity access solutions by type in the Africa Case**

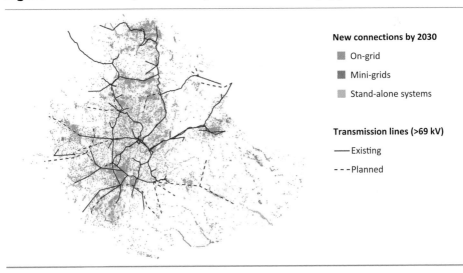

- Ethiopia currently has an electricity access rate of 45%, 11% of its population already have access through decentralised solutions. Strong government commitment to reach full access before 2030 in the STEPS.
- In both scenarios, around 80% of new connections are cost effectively delivered by grid densification and extension as a large part of the population lives close to the grid.

Figure 12.5D ▷ Ethiopia final energy consumption

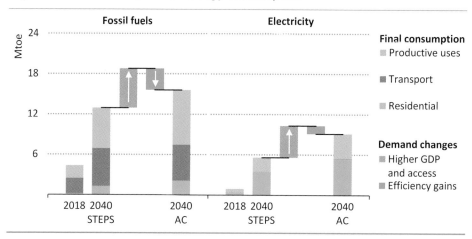

- Increased affluence in the STEPS results in a more than fourfold increase of the private vehicle stock with the number of cars reaching 700 000 by 2040. This results in a 300% increase in related oil consumption.
- To meet the needs of its growing population, Ethiopia remains a large producer of cement causing energy demand to increase significantly in both scenarios.

Figure 12.5E ▷ Ethiopia fuels and technologies used for cooking

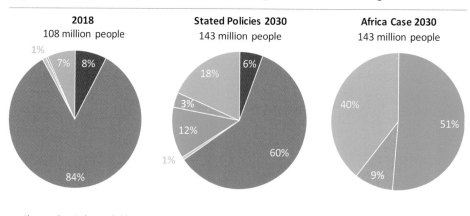

- In the STEPS, a push on improved and advanced biomass cookstoves alongside more access through LPG and electricity increases the population with access by 40 million by 2030, with 60% of this increase takes place in rural areas.
- In the AC, increased efforts using the same solutions bring access to clean cooking by 2030 to the remaining 95 million people that rely on the traditional use of biomass.

Figure 12.5F ▷ Ethiopia fossil fuel demand and production

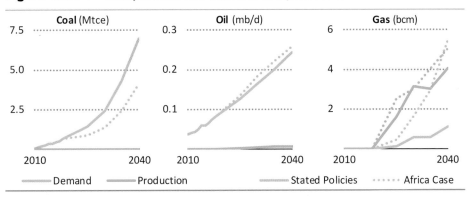

- Growing fossil fuel consumption is met almost entirely by imports in both scenarios.
- A high degree of dependency on imported fuels in both scenarios and a range of infrastructure development challenges underline the case for the development of hydropower and other renewables.

Figure 12.5G ▷ Ethiopia cumulative investment needs, 2019-2040

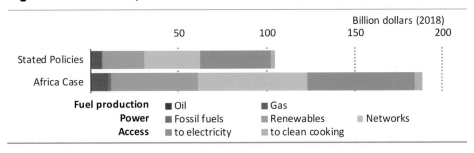

- Cumulative energy investment of $100 billion is needed in the STEPS, with electricity access and networks taking the majority.
- The AC needs around 80% more capital, including a doubling of investments in renewables and electricity networks compared with the STEPS.

Ethiopia policy opportunities

- Ethiopia will remain heavily dependent on fossil fuel imports. In both scenarios, imports of oil and coal increase; a significant increase in gas consumption (and imports) would help the country to make the most of its industrial potential.
- The need for energy imports could be reduced by a determined push to develop the country's formidable hydro resources and accelerate electrification, as well as by development of its more limited natural gas reserves.
- Continuing progress on access means that fully achieving SDG 7 is well within Ethiopia's reach. Most of the additional connections to 2025 can be made through extending the current grid.

12.6 Ghana

Oil and gas producer   Strong access record   Steel producer

Table 12.6A ▷ Ghana key indicators and policy initiatives

	2000	2018	Stated Policies 2030	Stated Policies 2040	Africa Case 2030	Africa Case 2040	CAAGR 2018-40 STEPS	CAAGR 2018-40 AC
GDP ($2018 billion, PPP)	63	191	322	438	403	728	3.9%	6.3%
Population (million)	19	29	37	44	37	44	1.9%	1.9%
with electricity access	45%	84%	100%	100%	100%	100%	0.8%	0.8%
with access to clean cooking	6%	25%	58%	73%	100%	100%	5.0%	6.5%
CO_2 emissions (Mt CO_2)	5	15	25	33	32	49	3.6%	5.4%

Policy	Key targets and measures
Performance targets	• Accelerate the displacement of light crude oil by natural gas in electricity generation. • Achieve 10% renewable energy in the national energy mix and 20% solar energy in agriculture by 2020. • 15% (unconditional) to 45% (conditional) reduction in GHG emissions by 2030 compared to the business-as-usual scenario (around 74 Mt CO_2-equivalent).
Industrial development targets	• Produce and process estimated reserves of 300 million barrels of oil and gas by 2040. • In accordance with the *One District, One Factory Initiative*, build a factory in each of the 216 districts across the country.

- Supplying an economy that is four-times the size of today's could require only three-times more energy with the implementation of efficiency standards in the AC.
- Oil remains the largest energy source in both scenarios, with nearly two-thirds of it consumed in the transport sector.

Figure 12.6A ▷ Ghana primary energy demand and GDP

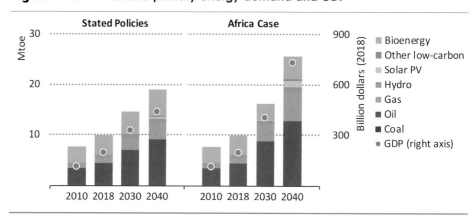

Figure 12.6B ▷ Ghana electricity generation by technology

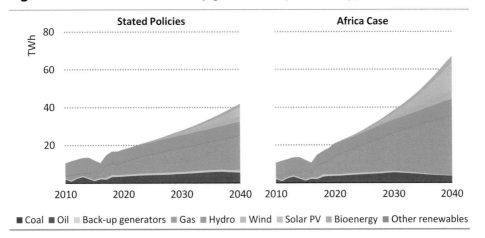

■ Coal ■ Oil ■ Back-up generators ■ Gas ■ Hydro ■ Wind ■ Solar PV ■ Bioenergy ■ Other renewables

- Almost half of today's electricity comes from hydropower; the rest comes from domestically produced gas (30%) and oil (23%).
- The 350% increase of electricity demand in the STEPS is met by increasing generation from gas, which accounts for nearly half of the power mix by 2040, and from solar PV.

Figure 12.6C ▷ Ghana electricity access solutions by type in the Africa Case

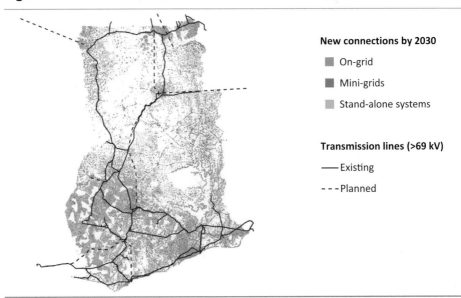

- Thanks to strong government leadership since the 1990s, Ghana had an electricity access rate of 84% in 2018, one of the highest in sub-Saharan Africa.
- To reach the remaining population, grid densification (58% of the new connections) and stand-alone systems (27%) are the two main least-cost solutions in both scenarios.

Figure 12.6D ▷ Ghana final energy consumption

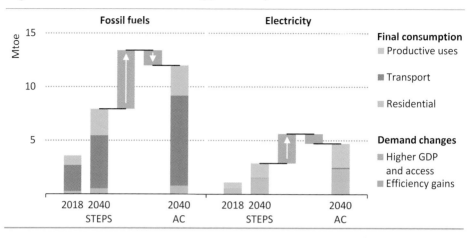

- Two/three-wheelers remain important for passenger transport; in the STEPS, an increasing number of buses accounts for nearly 70% of oil demand growth for transport.

- Millions of additional appliances and cooling systems together with the further development of bauxite mining, and steel and aluminium industries are responsible for two-thirds of the additional 45 TWh of electricity demand in the AC; around 10 TWh are avoided thanks to efficiency standards.

Figure 12.6E ▷ Ghana fuels and technologies used for cooking

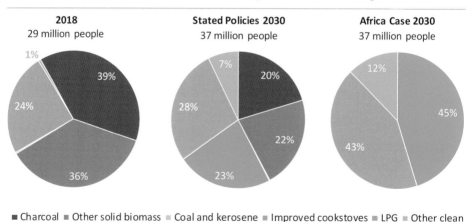

- In the STEPS, strong policies support the provision of clean cooking fuels to more than half of the population mainly through deployment of LPG and improved cookstoves.

- In the AC, 16 million people who still lack access to electricity in 2030 under the STEPS gain access through LPG, biogas or improved cookstoves.

Figure 12.6F ▷ Ghana fossil fuel demand and production

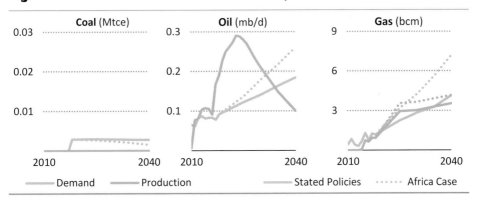

- Ghana remains a relatively minor producer of oil and gas in Africa.
- Gas demand grows strongly in the AC, lowering oil use in the power and industry sectors; this increases the need for imports of gas.

Figure 12.6G ▷ Ghana cumulative investment needs, 2019-2040

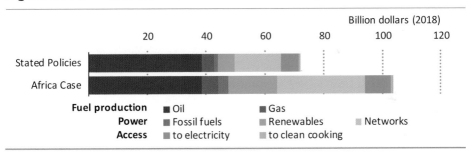

- Around $70 billion of cumulative energy supply investment is needed in the STEPS, 60% of which is for upstream oil and gas.
- Investment ramps up by nearly 45% in the AC, with a strong emphasis on renewables and electricity networks.

Ghana policy opportunities

- Thanks to notable efforts on electrification, the goal of full access is within grasp in Ghana. A mix of grid extension and stand-alone solutions would be the least-cost way to reach the decreasing share of the population that remains without access.
- Taking action to arrest (and reverse) declining oil output would reduce Ghana's reliance on imports to meet its growing future demand while a renewed push on developing domestic natural gas resources would help Ghana meet its accelerating industrial power needs.
- The government needs to develop and implement stronger efficiency policies if the potential savings identified in the AC are to be realised.

12.7 Kenya

Major access improvements   Large wind power producer   Largest geothermal producer

Table 12.7A ▷ Kenya key indicators and policy initiatives

	2000	2018	Stated Policies		Africa Case		CAAGR 2018-40	
			2030	2040	2030	2040	STEPS	AC
GDP ($2018 billion, PPP)	76	177	358	627	453	1 176	5.9%	9.0%
Population (million)	31	51	66	79	66	79	2.0%	2.0%
with electricity access	8%	75%	100%	100%	100%	100%	1.3%	1.3%
with access to clean cooking	3%	15%	46%	70%	100%	100%	7.2%	9.0%
CO_2 emissions (Mt CO_2)	8	16	27	40	33	60	4.3%	6.2%

Policy	Key targets and measures
Performance targets	▪ National Electrification Strategy: achieve universal electricity service to all households and businesses by 2022 at acceptable quality of service levels. ▪ Produce 100 000 barrels of oil per day from 2022 and develop 2 275 MW of geothermal capacity by 2030.
Industrial development targets	▪ Increase the contribution of the manufacturing sector share of GDP to 15% by 2022. ▪ Develop domestic iron and steel industries by 2030. ▪ Achieve middle-income status by 2030.

- In the AC, Kenya could supply an economy six-and half times larger than today using little more than twice its current energy consumption, if it were to move away from bioenergy and improve energy efficiency.

- Two-thirds of Kenya's energy currently comes from bioenergy. This share shrinks to 15% by 2040 in the AC thanks to increased use of geothermal resources and oil.

Figure 12.7A ▷ Kenya primary energy demand and GDP

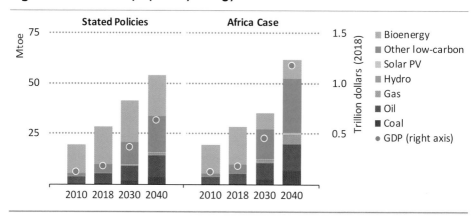

Figure 12.7B ▷ Kenya electricity generation by technology

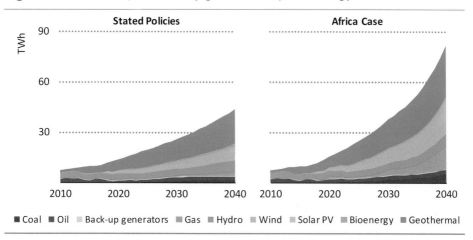

- Kenya is one of the few countries to develop geothermal energy: by 2040, it accounts for almost 50% of Kenya's power generation in the STEPS.
- The sevenfold increase in electricity demand in the AC relies on expansion of geothermal production (an increase to 4 GW) and new solar PV and gas capacity.

Figure 12.7C ▷ Kenya electricity access solutions by type in the Africa Case

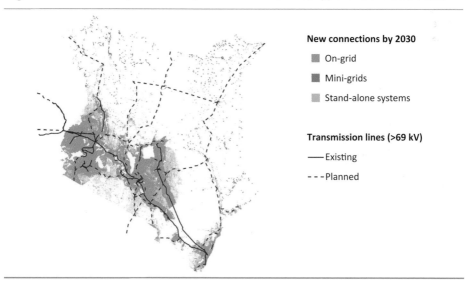

- Kenya has seen one of the fastest increases in electrification rates within sub-Saharan Africa since 2013: by 2018, 75% of the population had access.
- Kenya aims to reach full access by 2022; the grid would be the principal least-cost solution for the majority of the population (mainly in the south) still lacking access.

Figure 12.7D ▷ Kenya final energy consumption

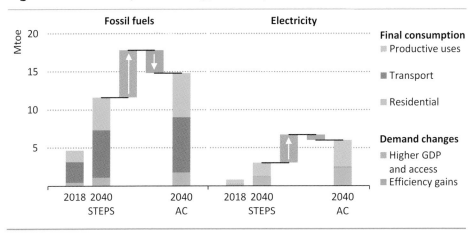

- Oil remains by far the dominant fuel in end-use sectors, and its use triples in road transport in the AC, with five million additional vehicles being added to the fleet.

- Electricity demand reaches nearly 70 TWh in the AC, as light industry grows and as ownership of household appliances and cooling systems increases; efficiency standards avoid a further 8 TWh of demand.

Figure 12.7E ▷ Kenya fuels and technologies used for cooking

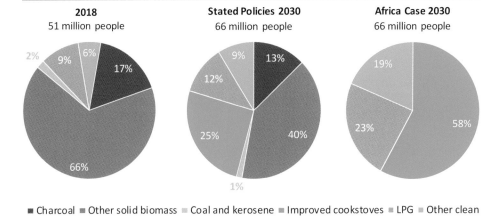

- Today three-stone fires are still used for most cooking, fuelled mostly by charcoal in urban areas and by wood in rural areas. In the STEPS, government initiatives lead to 26% of the population having access to clean cooking by 2030.

- In the AC, everybody gains access to clean cooking by 2030. Most of the 25 million people otherwise without access in rural areas gain access primarily through improved and advanced cookstoves; LPG is the least-cost fuel for most of the urban population.

Figure 12.7F ▷ Kenya fossil fuel demand and production

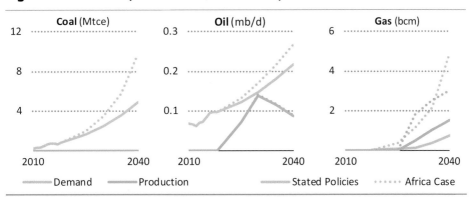

- Kenya is not a notable oil and gas producer today, but it takes some steps to develop its relatively modest resources.
- Higher economic growth underpins strong growth in fossil fuel demand in the AC. Oil demand almost triples as it expands its share of the overall energy mix.

Figure 12.7G ▷ Kenya cumulative investment needs, 2019-2040

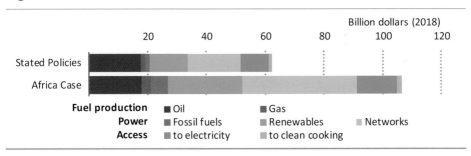

- Energy investment amounts to around $60 billion through to 2040 in the STEPS, with renewables and electricity networks accounting for half of this.
- Investments in renewables and electricity networks need to double in the AC.

Kenya policy opportunities

- Kenya is on the cusp of reaching universal access to electricity. Concerted government policy could help reach this aim through grid and stand-alone connections in roughly equal measure.
- Kenya has made notable progress in deploying renewables in large part because it has successfully attracted the necessary private investment for renewables projects. Further development of these resources would help it meet demand growth.

12.8 Mozambique

Large hydro potential   Large aluminium producer   Important gas discoveries

Table 12.8A ▷ Mozambique key indicators and policy initiatives

	2000	2018	Stated Policies 2030	Stated Policies 2040	Africa Case 2030	Africa Case 2040	CAAGR 2018-40 STEPS	CAAGR 2018-40 AC
GDP ($2018 billion, PPP)	11	39	74	140	98	219	6.0%	8.1%
Population (million)	18	31	43	55	43	55	2.7%	2.7%
with electricity access	6%	29%	60%	72%	100%	100%	4.2%	5.8%
with access to clean cooking	4%	6%	11%	22%	100%	100%	5.7%	13.4%
CO_2 emissions (Mt CO_2)	1	8	21	43	38	66	6.8%	9.9%

Policy	Key targets and measures
Performance targets	▪ Promote the construction of electricity infrastructure that is resilient to climate change. ▪ Ensure sustainable and transparent management of the natural resources and the environment.
Industrial development targets	▪ Five-Year Plan 2015–2019 focuses on empowering women and men for gender equity and equality, poverty reduction, economic development, and food security and nutrition. ▪ Expand and modernise roads, bridges, water, ports and other key infrastructure.

- Mozambique could supply an economy more than five times larger than today in the AC with four-times the energy demand if it were to diversify away from bioenergy and improve energy efficiency.

- Bioenergy, including the traditional use of biomass, currently accounts for more than 60% of primary energy supply, but recent discoveries of gas enable the energy mix to be diversified with gas accounting for 45% of the primary mix by 2040 in the AC.

Figure 12.8A ▷ Mozambique primary energy demand and GDP

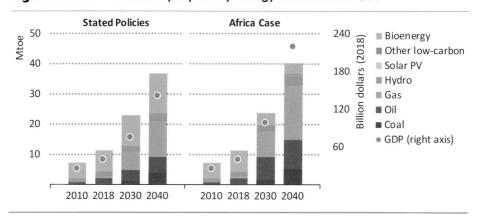

Figure 12.8B ▷ Mozambique electricity generation by technology

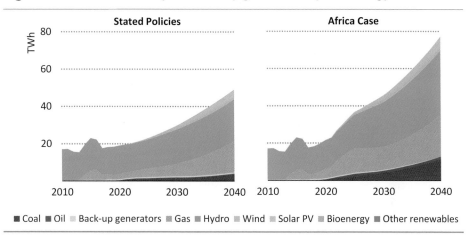

■ Coal ■ Oil ■ Back-up generators ■ Gas ■ Hydro ■ Wind ■ Solar PV ■ Bioenergy ■ Other renewables

- Providing access to all and increasing electrification of productive uses almost quadruples electricity demand in the AC.

- Hydropower remains an important source of electricity in each scenario, but its share of generation declines from four-fifths today to more than 40% in the AC; gas grows in importance as increasing use is made of domestic resources.

Figure 12.8C ▷ Mozambique electricity access solutions by type in the Africa Case

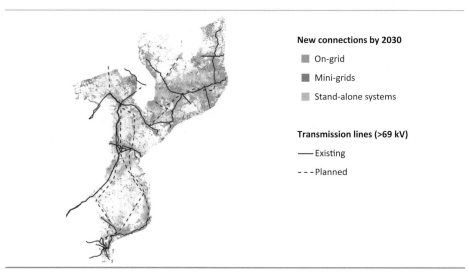

- Currently 71% of the population lacks access to electricity; decentralised solutions are the least-cost option for 55% of the new connections in the AC.

- In the AC, grid connections are the least-cost solution for the remaining 45% of new connections: a large share of the population lives close to existing and planned grids.

Figure 12.8D ▷ **Mozambique final energy consumption**

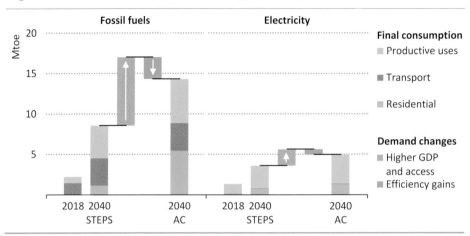

- Oil remains the major fuel used in end-use sectors, with demand growing as a result of increased use of LPG for cooking, while gas consumption exceeds electricity by 2040 in both scenarios.

- In the AC, recent gas discoveries trigger a massive increase in overall industrial gas demand: Mozambique increases production of aluminium more than fivefold by 2040, becoming a significant exporter.

Figure 12.8E ▷ **Mozambique fuels and technologies used for cooking**

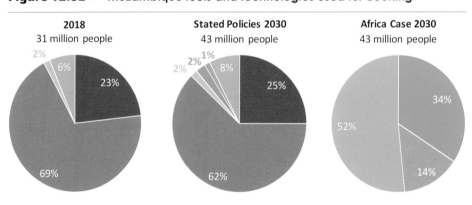

- In the STEPS, the proportion of the population relying on traditional uses of biomass decreases from 92% to 87% by 2030.

- In the AC, natural gas is the least-cost option for a quarter of the 38 million people without access in 2030; with improved cookstoves and LPG providing access for others.

Figure 12.8F ▷ Mozambique fossil fuel demand and production

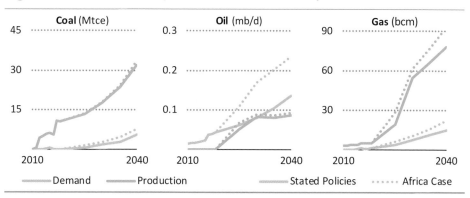

- The recent massive gas discoveries in Mozambique becoming the largest gas producer in sub-Saharan Africa by 2040 in the AC.
- While the bulk of the production is destined for export, domestic demand also grows as a result of efforts to foster gas-based industries and expand infrastructure in the AC.

Figure 12.8G ▷ Mozambique cumulative investment needs, 2019-2040

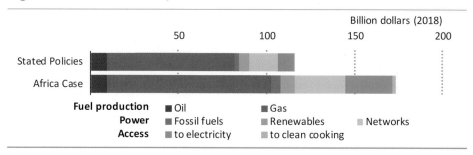

- Energy investment needs amount to $115 billion through to 2040 in the STEPS, more than 60% of which goes to gas production and infrastructure.
- The AC requires nearly 50% more capital to promote renewables in tandem with gas infrastructure.

Mozambique policy opportunities

- Mozambique's ambitions for economic and social development depend in large measure on its ability to develop its large natural gas resources.
- In addition to providing valuable export revenue, its abundant gas resources could be used to generate electricity, act as a catalyst for domestic industrial development, and to support clean cooking.
- Large industrial consumers of gas could act as anchors for smaller industries looking to increase their use of gas. The aluminium industry could be one such anchor consumer. The success of a domestic aluminium export business will depend heavily on its ability to secure affordable gas feedstock.

12.9 Nigeria

Largest population Largest economy Large chemicals producer

Table 12.9A ▷ Nigeria key indicators and policy initiatives

	2000	2018	Stated Policies 2030	Stated Policies 2040	Africa Case 2030	Africa Case 2040	CAAGR 2018-40 STEPS	CAAGR 2018-40 AC
GDP ($2018 billion, PPP)	392	1 169	1 636	2 420	2 258	3 678	3.4%	5.3%
Population (million)	122	196	263	329	263	329	2.4%	2.4%
with electricity access	40%	60%	80%	85%	100%	100%	1.6%	2.3%
with access to clean cooking	1%	10%	28%	38%	100%	100%	6.4%	11.2%
CO_2 emissions (Mt CO_2)	37	83	134	191	181	257	3.8%	5.3%

Policy	Key targets and measures
Performance targets	▪ 20% (unconditional) to 45% (conditional) reduction in GHG emissions by 2030 compared to the business-as-usual scenario. ▪ Increase oil production to 2.5 mb/d and become a net exporter by 2020, and end gas flaring by 2030.
Industrial development targets	▪ Dedicate at least 30% of the federal budget to capital expenditure. ▪ Achieve GDP growth of 7% and create over 15 million jobs by 2020 and double manufacturing output to 20% of GDP by 2025.

- Nigeria remains Africa's largest economy: in the AC, supplying an economy three-times larger than today would require less energy demand if the energy mix were to be diversified.
- In the AC, gas meets a growing share of energy demand, supported by the implementation of the government's gas masterplan.

Figure 12.9A ▷ Nigeria primary energy demand and GDP

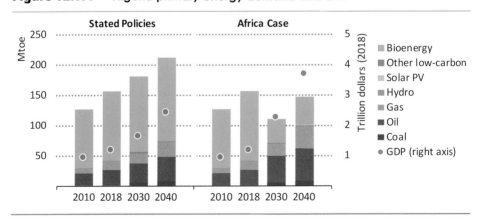

Figure 12.9B ▷ Nigeria electricity generation by technology

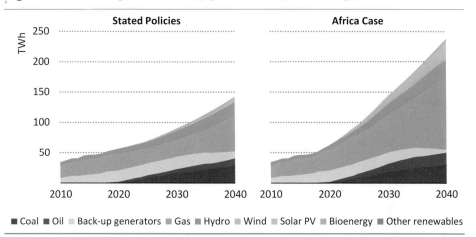

■ Coal ■ Oil ■ Back-up generators ■ Gas ■ Hydro ■ Wind ■ Solar PV ■ Bioenergy ■ Other renewables

- Today, 80% of power generation comes from gas; most of the remainder comes from oil, with Nigeria the largest user of oil-fired back-up generators on the continent.
- Natural gas remains the main source of power in the AC, although there is a shift towards solar PV as the country starts to exploit its large solar potential.

Figure 12.9C ▷ Nigeria electricity access solutions by type in the Africa Case

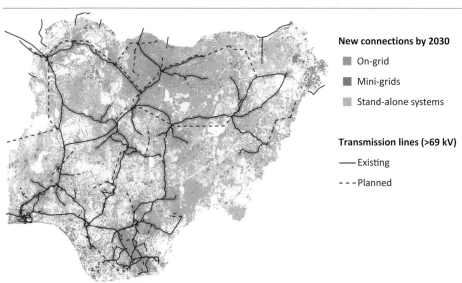

- Provided that reliability and supply improve, the grid could become the optimal solution to provide almost 60% of people with access to electricity in each scenario.
- In the AC, Nigeria achieves universal access by stepping up efforts to provide off-grid solutions to those populations that live far from a grid.

Chapter 12 | Implications for Africa and the world

Figure 12.9D ▷ Nigeria final energy consumption

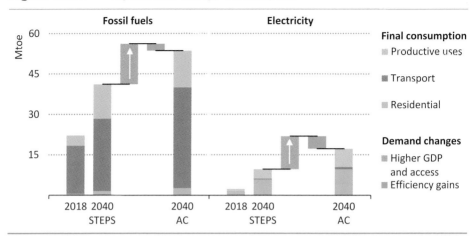

- Nigeria is a major industrial producer and large chemical exporter. In the AC, it triples chemicals production by 2040 with new gas-based methanol and ammonia plants.
- Nigeria has the second-largest vehicle stock in sub-Saharan Africa: the number of vehicles could grow from 14 to 37 million in the AC by 2040 with only two-times more oil consumption if more stringent fuel economy standards were introduced.

Figure 12.9E ▷ Nigeria fuels and technologies used for cooking

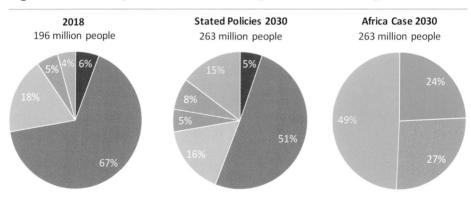

- In the STEPS, there is progress on access to clean cooking services but almost three-quarters of the population still lack access in 2030.
- In the AC, universal access is achieved through greater household access to gas networks and LPG in the main cities, and to improved cookstoves in rural areas.

Figure 12.9F ▷ Nigeria fossil fuel demand and production

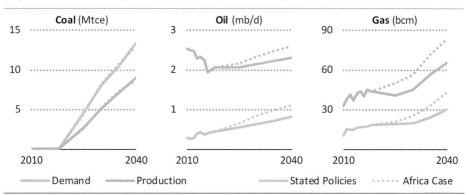

- Delayed reforms and growing competition in international oil markets means that it takes time for oil production to revive.
- In both scenarios, gas demand grows strongly in the industry and power sectors, leading to action to increase production and reduce gas flaring.

Figure 12.9G ▷ Nigeria cumulative investment needs, 2019-2040

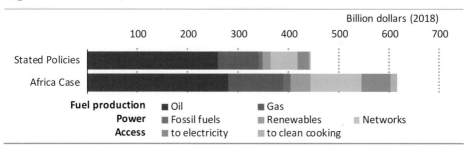

- Cumulative energy supply investment of $445 billion is needed in the STEPS, almost 80% of which goes to upstream oil and gas.
- The AC requires a significant ramp up in power sector investment. Spending on electricity networks and renewables increases by 85% and 165% respectively, compared to STEPS.

Nigeria policy opportunities

- Oil sector reforms would help to revive oil production while successful implementation of the gas masterplan would foster gas-to-power, industrial development and expansion of the gas network to industrial hubs.
- Improved power sector management and governance would help to reduce outages and transmission losses. Failure to do so would impede industrial growth and would mean continued high levels of use of polluting back-up generation.
- Reducing bioenergy use across all sectors would bring a number of benefits, not least because its use is strongly linked to deforestation and air pollution.

12.10 Senegal

Emerging local gas market   New oil and gas discoveries   Ambitious renewables plan

Table 12.10A ▷ Senegal key indicators and policy initiatives

	2000	2018	Stated Policies 2030	Stated Policies 2040	Africa Case 2030	Africa Case 2040	CAAGR 2018-40 STEPS	CAAGR 2018-40 AC
GDP ($2018 billion, PPP)	27	60	154	237	176	370	6.5%	8.7%
Population (million)	10	16	22	28	22	28	2.5%	2.5%
with electricity access	31%	69%	100%	100%	100%	100%	1.7%	1.7%
with access to clean cooking	32%	30%	47%	52%	100%	100%	2.6%	5.7%
CO$_2$ emissions (Mt CO$_2$)	4	9	19	30	19	32	5.7%	5.9%

Policy	Key targets and measures
Performance targets	■ Start producing 100 000 barrels of oil per day from 2022. ■ Achieve 200 GWh hydropower production in electricity generation output. ■ Reach universal access to electricity by 2025, with 95% of rural connections provided by the grid.
Industrial development targets	■ The overall goal of *Le Plan Sénégal Emergent* 2023 is to achieve, through structural transformation of the economy, strong, inclusive and sustainable growth for the well-being of the people and reach middle-income status by 2035.

- Senegal's economy could grow six-times larger in the AC while limiting growth in energy demand to three-times its current level by utilising new gas resources and boosting the use of renewables in power.
- In the AC, gas meets a growing share of energy demand while traditional use of biomass starts to decline in rural areas.

Figure 12.10A ▷ Senegal primary energy demand and GDP

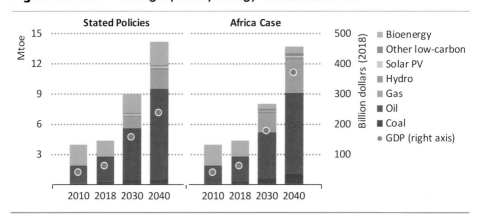

Figure 12.10B ▷ Senegal electricity generation by technology

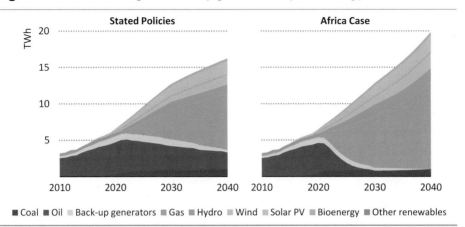

■ Coal ■ Oil ■ Back-up generators ■ Gas ■ Hydro ■ Wind ■ Solar PV ■ Bioenergy ■ Other renewables

- Electricity demand increases sharply in both scenarios, while the power mix changes, with gas playing an increasingly important role and investments in wind and other renewables bringing more diversification.
- Plans to phase out heavy fuel oil in the AC hinge on successful implementation of new gas-to-power plans.

Figure 12.10C ▷ Senegal electricity access solutions by type in the Africa Case

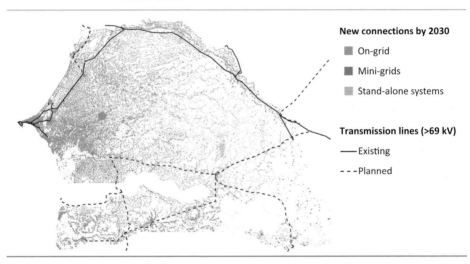

- Thanks to successful access policies, almost 70% of the population is connected today; with adoption of a comprehensive integrated plan full access is achieved by 2025.
- The grid represents the least-cost option for the majority of the population currently without electricity access today, with decentralised solutions reaching the most remote populations.

Figure 12.10D ▷ Senegal final energy consumption

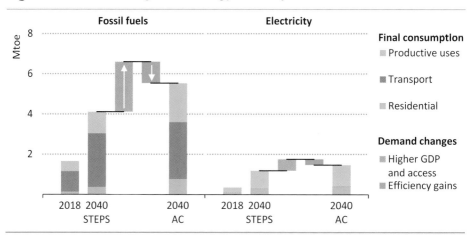

- Senegal's stock of two/three-wheelers is set to grow strongly in both scenarios and its electrification would help to free oil for other productive uses.
- In the AC, cement production could more than double to 2040, although the availability of fuels, including domestic gas, will be critical for this and for wider future industrial development.

Figure 12.10E ▷ Senegal fuels and technologies used for cooking

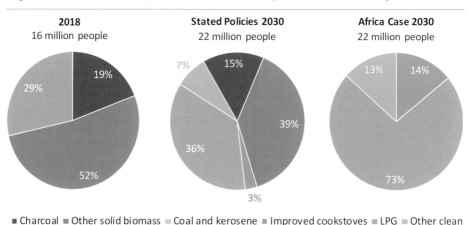

- LPG is used for cooking by almost 30% of the population today, one of the highest shares in sub-Saharan Africa. It is expected to remain the main clean cooking fuel in 2030.
- In the AC, LPG is the least-cost option in both rural and urban areas for more than 70% of the population currently still lacking access.

Figure 12.10F ▷ Senegal fossil fuel demand and production

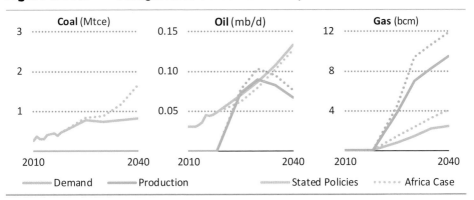

- Senegal is not a fossil fuel producer today, but the major gas discoveries are expected to change the picture and to lead to gas production of 9.5 bcm in 2040 in the STEPS.
- The greater availability of gas helps displace oil use in power generation in domestic markets while also bringing considerable export revenues.

Figure 12.10G ▷ Senegal cumulative investment needs, 2019-2040

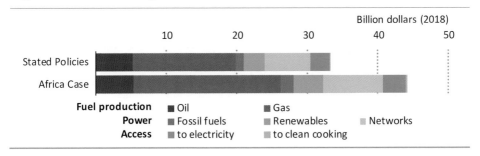

- Energy investment needs amount to $33 billion through to 2040 in the STEPS, mainly to unlock the potential for gas, expand power networks and increase electricity access.
- The AC sees this level of investment increase by a third, with more emphasis on gas and renewable generation.

Senegal policy opportunities

- Implement a robust and transparent framework for resource management and design of local content rules would help Senegal to make the most of its natural resources.
- The development of natural gas strategies that cover the entire value chain, including end-uses (gas-to-power or gas-to-industry), would help Senegal to maximise the benefits of its natural gas.
- Senegal's power sector would be strengthened by continued diversified investment in power, including renewables and natural gas, while phasing out heavy fuel oil.

12.11 South Africa

Africa's only nuclear power   Major commodity exporter   Coal-fired power sector

Table 12.11A ▷ South Africa key indicators and policy initiatives

	2000	2018	Stated Policies 2030	Stated Policies 2040	Africa Case 2030	Africa Case 2040	CAAGR 2018-40 STEPS	CAAGR 2018-40 AC
GDP ($2018 billion, PPP)	491	789	1 010	1 348	1 174	1 600	2.5%	3.3%
Population (million)	46	57	66	71	66	71	1.0%	1.0%
with electricity access	77%	95%	100%	100%	100%	100%	0.2%	0.2%
with access to clean cooking	81%	91%	93%	95%	100%	100%	0.2%	0.4%
CO_2 emissions (Mt CO_2)	280	420	321	279	289	187	-1.8%	-3.6%

Policy	Key targets and measures
Performance targets	▪ The National Development Plan 2030 envisages that adequate investment in energy infrastructure will promote economic growth and development. ▪ Decommission 35 GW (of 42 GW currently operating) of coal-fired power capacity and supply at least 20 GW of the additional 29 GW of electricity needed by 2030 from renewables and natural gas.
Industrial development targets	▪ Secure primary steel production capability and support the downstream steel sector. ▪ Automotive Masterplan 2020: raise domestic vehicle production to 1% of global output including building 20% hybrid electric vehicles by 2030.

- The economy could double in the AC with less primary energy demand compared to today by increasing the share of renewables and gas in the energy mix.
- In the AC, the role of coal in South African industry and power generation is already decreasing, while that of gas and renewables is increasing.

Figure 12.11A ▷ South Africa primary energy demand and GDP

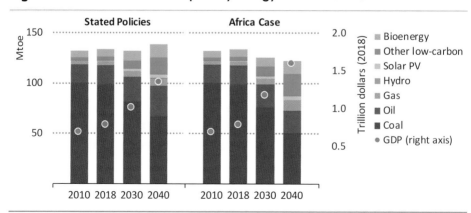

Figure 12.11B ▷ **South Africa electricity generation by technology**

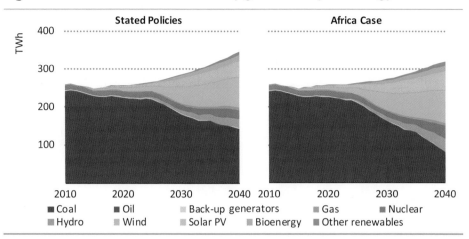

- Coal
- Oil
- Back-up generators
- Gas
- Nuclear
- Hydro
- Wind
- Solar PV
- Bioenergy
- Other renewables

- South Africa is reliant on coal but is making efforts to diversify as its coal-fired fleet is ageing; new projects will not fully compensate for the decline of the existing fleet.
- The government is focussing on diversifying the power mix by introducing natural gas and renewables, including concentrating solar power (CSP); South Africa has excellent natural resources for CSP development.

Figure 12.11C ▷ **South Africa electricity access solutions by type in the Africa Case**

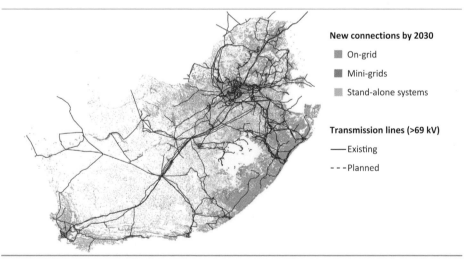

New connections by 2030
- On-grid
- Mini-grids
- Stand-alone systems

Transmission lines (>69 kV)
— Existing
--- Planned

- South Africa has a well-developed electricity network and one of the highest rates of electricity access in sub-Saharan Africa.
- The least-cost way to connect those without access is in most cases via the grid (81%) with the residual population served by mini-grids (12%) and stand-alone systems.

Figure 12.11D ▷ South Africa final energy consumption

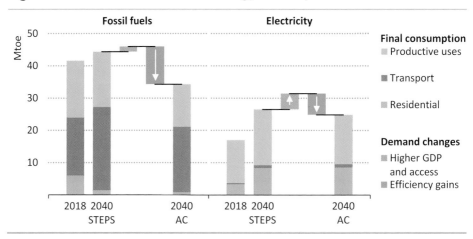

- Oil is the largest fuel in the end-use sectors; more stringent fuel economy standards would mean that a 25% increase in demand could be met with a slight increase in the amount of oil used.
- The role of coal in South African industry dwindles in the AC as gas and bioenergy are increasingly used, especially in steel production and in light industries.

Figure 12.11E ▷ South Africa fuels and technologies used for cooking

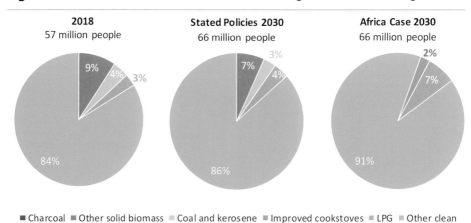

- In both urban and rural areas, electricity is the favourite option for cooking in South Africa, but more than 4 million living mainly in rural areas continue to use fuelwood for heating and cooking in 2030 in the STEPS.
- Improved cookstoves and LPG would help close the gap between the STEPS and the AC and eliminate the use of traditional biomass, reducing household premature deaths by 80% in 2030.

Figure 12.11F ▷ South Africa fossil fuel demand and production

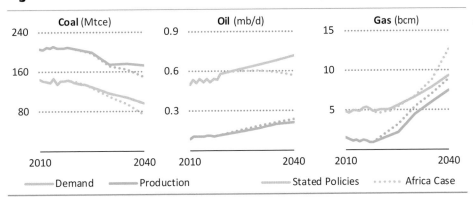

- South Africa continues to dominate coal production in Africa. Despite declining production, falling domestic demand boosts export volumes in both scenarios.
- South Africa also relies on oil and gas for its energy needs, but recent gas discoveries could reduce its import needs.

Figure 12.11G ▷ South Africa cumulative investment needs, 2019-2040

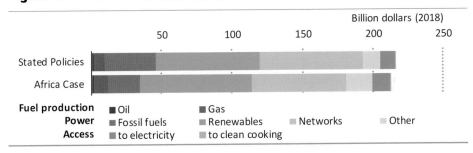

- Nearly $220 billion of cumulative energy investment is needed in the STEPS, with renewables and electricity networks accounting for the majority.
- The AC requires more investment in gas production, but overall efficiency improvements moderate the additional spending needs.

South Africa policy opportunities

- Diversifying energy supply away from coal would have many benefits, including a reduction in the number of premature deaths from pollution, but the social implications of changes would need careful management.
- Reforming and restructuring ESKOM would strengthen the reliably of the power system, support increased industrialisation and help efforts to diversify the energy mix.
- Strengthening efficiency throughout the economy would reduce demand for both materials and energy, while the implementation of minimum energy performance standards for electric motors in the industry and mining sectors would be an important first step towards unlocking further efficiency gains.

12.12 Tanzania

Strong economic growth   Minerals exporter   Plans for urban transport

Table 12.12A ▷ Tanzania key indicators and policy initiatives

	2000	2018	Stated Policies 2030	Stated Policies 2040	Africa Case 2030	Africa Case 2040	CAAGR 2018-40 STEPS	CAAGR 2018-40 AC
GDP ($2018 billion, PPP)	57	176	314	585	475	1 233	5.6%	9.3%
Population (million)	34	59	83	108	83	108	2.8%	2.8%
with electricity access	11%	37%	70%	80%	100%	100%	3.6%	4.7%
with access to clean cooking	2%	6%	46%	76%	100%	100%	12.2%	13.7%
CO_2 emissions (Mt CO_2)	3	12	24	41	36	74	5.9%	8.9%

Policy	Key targets and measures
Performance targets	▪ Reduce GHG emissions by 10-20% by 2030 compared to the business-as-usual scenario (138-153 Mt CO_2-equivalent gross emissions). ▪ Increase electricity generation capacity from 1 500 MW in 2015 to 4 910 MW and achieve 50% energy from renewable energy sources by 2020.
Industrial development targets	▪ Raise annual real GDP growth to 10% by 2021. ▪ Build a semi-industrialised country by 2025 in which the contribution of manufacturing to the national economy reaches at least 40% of GDP.

- With annual GDP growth of more than 9% in the AC, Tanzania's economy could be seven-times larger in 2040 than today, but with an increase in energy demand limited to 150% driven by fuel efficiency gains.
- In the AC, diversifying the energy mix and improving energy efficiency are the keys to achieving economic growth while limiting growth in energy demand, with oil, gas and geothermal reducing the share of bioenergy in the energy mix.

Figure 12.12A ▷ Tanzania primary energy demand and GDP

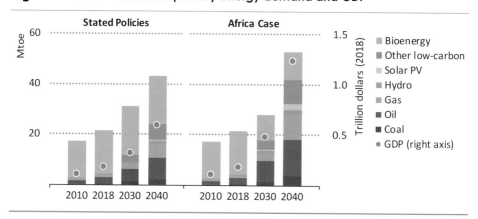

Figure 12.12B ▷ Tanzania electricity generation by technology

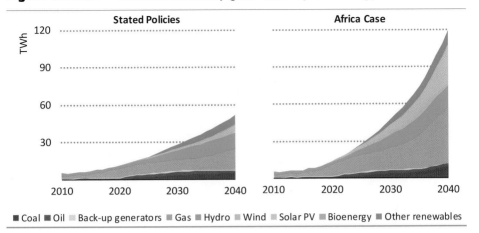

■ Coal ■ Oil ■ Back-up generators ■ Gas ■ Hydro ■ Wind ■ Solar PV ■ Bioenergy ■ Other renewables

- Gas accounts for more than half of current power generation, with the remainder coming from hydropower and oil, the latter used mostly for back-up generators.
- Providing access for all and a growth in productive uses lead to a thirteen-fold increase of electricity demand by 2040 in the AC: this is met with an expansion of gas, hydropower and solar PV.

Figure 12.12C ▷ Tanzania electricity access solutions by type in the Africa Case

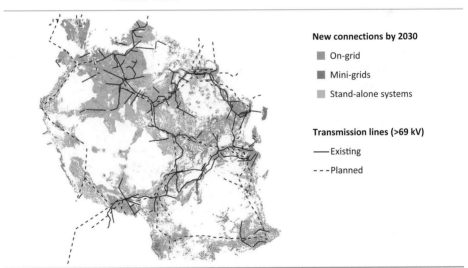

- Despite the low access rate (37%) today, the grid represents more than half of new connections by 2030 in the AC given its existing and planned coverage.
- In the AC, around one-third of the remaining population, mainly located in sparsely populated areas far from the grid, would be best reached by stand-alone systems.

Figure 12.12D ▷ Tanzania total final consumption

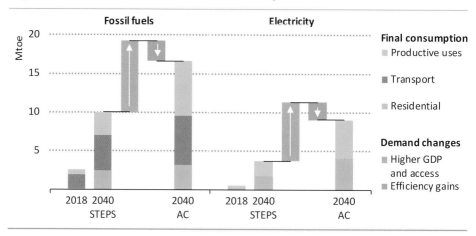

- Oil continues to play an important role in end-use sectors, not least as a result of its use by the increasing number of buses on the road as Tanzania has a large bus fleet.
- Gas and electricity use in industry is growing strongly, especially in manufacturing industries, but in the AC, energy efficiency measures have prevented consumption from being 20% higher than current levels.

Figure 12.12E ▷ Tanzania use of fuels and technologies for cooking

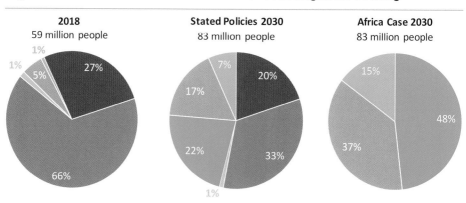

- Despite policies to promote clean cooking solutions, the number of people relying on traditional use of biomass for cooking declines from 55 million people today to 44 million in 2030 as efforts to improve access outrun by high population growth in STEPS.
- In the AC, LPG and biogas are the least-cost options for almost half of the population, with improved cookstoves the main way to extend access in rural areas.

Figure 12.12F ▷ Tanzania fossil fuel demand and production

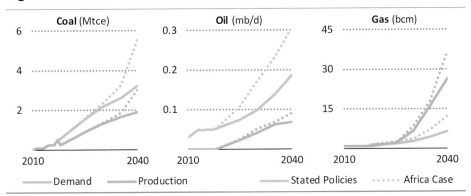

- Recent large discoveries push up gas production to almost 30 bcm by 2040 in the STEPS. Existing infrastructure helps Tanzania to increase domestic gas consumption.
- Gas demand in 2040 is twice as high in the AC, helped by efforts to promote the use of gas to displace traditional biomass and by support for gas-based industries.

Figure 12.12G ▷ Tanzania cumulative investment needs, 2019-2040

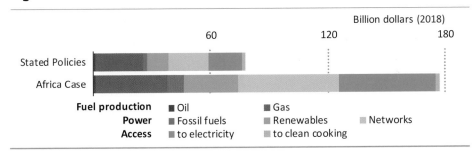

- Almost $80 billion of cumulative energy supply investment is needed in the STEPS, with most of it being used to widen access to gas and electricity.
- This level of investment doubles in the AC, with higher amounts of capital allocated to electricity access and networks.

Tanzania policy opportunities

- A rapid development of offshore resources would help to ensure greater availability of gas, and a robust framework to use export revenues in an effective manner would help to ensure that the country makes the most of those revenues.
- Maintaining investment in public transport, notably in Dar es Salaam, but also in other cities and between cities and rural areas, would help to facilitate economic growth. Government should also ensure public transport is affordable for all.

PART C
WEO INSIGHTS

The costs and potential for two low-carbon gases: **hydrogen** and **biomethane** are examined in Chapter 13. We provide the first global estimate of the sustainable technical potential of biomethane, and discuss how injecting hydrogen into gas pipeline networks can help to scale up hydrogen supply technologies. We examine the outlook for both gases in our scenarios and provide suggestions for policy makers to consider in developing long-term strategies for gas infrastructure and low-carbon gases.

Chapter 14 provides a deep dive into **offshore wind power**, a rapidly emerging renewable energy technology that can help drive clean energy transitions by decarbonising electricity and by producing low-carbon fuels. Based on a new geospatial analysis, we provide an assessment of the technical potential for offshore wind, incorporating the latest available satellite weather data and technology innovations. We explore the global and regional outlook for offshore wind, the potential for faster growth including from rising demand for renewable-based hydrogen and highlight key challenges to be addressed by policy makers, regulators and the offshore wind industry.

Chapter 13

Prospects for gas infrastructure
Is there a low-carbon future in the pipeline?

SUMMARY

- The next decade is a critical one for gas infrastructure. Short-term decisions on whether to invest in gas grids will have major long-term implications. Questions about the relative importance, and respective roles, of electricity and gas networks are central to the design of energy transitions to a low emissions future. Low-carbon electricity has huge potential to play a greater direct role in future energy systems, but there are limits to how quickly and extensively electrification can occur. Well-established gas grids can deliver twice as much energy as electricity grids today and they are a major source of flexibility. Decisions on the future of gas networks need to consider their potential to deliver different types of gas in a low emissions future, as well as their role in ensuring energy security.

- In the Sustainable Development Scenario, natural gas use grows globally to the late 2020s, to a level almost 10% higher than 2018, although advanced and developing economies follow divergent pathways. But as part of the drive for deep emissions reductions in this scenario, gas grids are gradually repurposed or retooled over time to deliver low-carbon energy. We focus here on two options: low-carbon hydrogen and biomethane.

Figure 13.1 ▷ Change in global gas demand in the Sustainable Development Scenario, 2010-2040

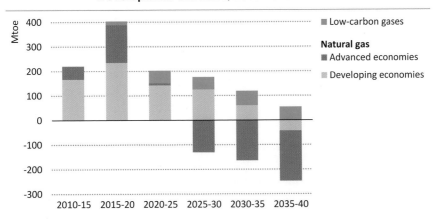

Trajectories of natural gas demand diverge between advanced and developing economies. Low-carbon gases play an increasingly important role worldwide.

- Low-carbon hydrogen has seen a recent surge of interest, and could help deliver deep emissions reductions across a wide range of hard-to-abate sectors. It is currently expensive to produce: the lowest cost options are between $12-25 per million British thermal units (MBtu). Injecting low-carbon hydrogen into gas pipeline networks would not only reduce the emissions from gas consumption but also offer the possibility of scaling up hydrogen supply technologies to bring down costs through economies of scale. This would facilitate expansion of the use of hydrogen into the buildings, industry and power sectors.

- Biomethane is also attracting increased interest. We estimate that more than 730 million tonnes of oil equivalent (Mtoe) – equivalent to over 20% of annual natural gas demand globally – could be produced today in a sustainable manner. Meeting 10% of today's gas demand with biomethane would cost $10-22/MBtu. Biomethane potential is widely spread geographically though some of the lowest cost options are available in developing economies in Asia.

- In the Stated Policies Scenario, annual consumption of biomethane is just under 80 Mtoe in 2040. Most of this takes place in China and India as a result of explicit policy support, motivated in part by a push to limit growing reliance on imports. In this scenario, there is limited blending of low-carbon hydrogen into gas networks.

- There is much greater uptake of both biomethane and low-carbon hydrogen in the Sustainable Development Scenario. Biomethane use rises to over 200 Mtoe in 2040, and more than 25 Mtoe of low-carbon hydrogen is injected into gas networks. Low-carbon gases make up 7% of total gas supply globally in 2040 and they are on a steep upward trajectory at the end of the *Outlook* period. Over 15% of total gas supply in China and the European Union is low-carbon gas in 2040.. Globally, low-carbon hydrogen and biomethane blended into the gas grid in the Sustainable Development Scenario avoid around 500 million tonnes (Mt) of annual CO_2 emissions that would have occurred in 2040 if natural gas had been used instead.

- Blending hydrogen into gas grids accelerates cost reductions in low-carbon hydrogen production, which encourages its wider use for other purposes. Over 80 Mtoe of low-carbon hydrogen is also used directly in end-use sectors in 2040. There is also around 150 gigawatts (GW) of combined-cycle gas turbine (CCGT) capacity equipped with carbon capture, utilisation and storage (CCUS) in 2040 in this scenario: this avoids 300 Mt of annual CO_2 emissions.

- We highlight some issues and provide suggestions for policy makers to consider in developing long-term strategies for gas infrastructure and low-carbon gases. These include: assessing carefully the levels of investment in gas infrastructure that are consistent with both energy security and environmental goals; the importance of regulation of gas networks to help maintain gas infrastructure during the transition to low and zero carbon energy; the need for low-carbon gas standards and incentives to encourage their use; how biomethane production can create jobs in rural locations; and how to manage distributional issues that may arise in energy transitions.

13.1 Introduction

The ability to deliver large quantities of energy to consumers flexibly and reliably is the foundation of energy markets today. Nearly all countries have an extensive grid for delivering electricity to consumers, but in many economies the gas grid provides a larger and more flexible energy delivery mechanism. Questions about the relative importance and the respective roles of these two networks in a low emissions future are central to the design of energy transitions.

The answers will vary from country to country, depending on the extent of existing infrastructure, resource endowments, the structure of the economy and the demand outlook. They will also depend on policy choices. At one end of the spectrum, there is the "electrify-everything" philosophy of decarbonisation, in which electricity becomes the main vector to meet final energy consumption directly. This route implies a massive expansion of low-carbon electricity generation and transmission infrastructure. The role of existing gas grids in this vision of the future is marginal; the core policy issue is how to manage their gradual decline.

However, most countries that have considered how to realise rapid and wholesale emissions reductions are looking instead at a future in which electricity and gas networks play complementary roles. Electricity consumption gains ground under this kind of approach as more end-uses are electrified, but developing or maintaining gas networks to deliver energy to power stations, factories and buildings also remains important. Maintaining gas grids could, for example, avoid additional investment in electricity networks and grid flexibility measures that would only be needed to meet short periods of high demand. Indeed, a number of recent studies have shown that co-ordinated policies across gas, electricity and heat, using different networks, can help maximise energy security and minimise the overall costs of decarbonisation (Navigant, 2019).

The Sustainable Development Scenario embodies this approach. The share of electricity in final consumption rises, but gaseous fuels[1] remain a central element of the global energy mix in this scenario, even though natural gas demand falls in many countries between 2018 and 2040.

The coming years are a particularly important period for policy makers, industry leaders and others considering the future of gas infrastructure: short-term decisions will have long-term implications. Gas infrastructure can take a long time to develop, and it has a long lifetime once constructed. Countries therefore need to consider "what comes next?" when planning for the future. On one hand, an investment decision could lock in gas use for a prolonged period and, if the gas delivered is natural gas, this could impact the achievement of long-term emissions reduction goals, even while delivering near-term gains where natural gas replaces more polluting fuels. On the other hand, opting out of gas networks

---

[1] We differentiate between "natural gas", which is gas of fossil origin, and "gas" or "total gas" which includes all gaseous fuels (i.e. natural gas, biomethane, hydrogen, synthetic methane).

could narrow the options available to realise future emissions reductions and introduce some energy security risks if other infrastructure does not develop quickly enough to compensate.

This chapter discusses how and why it could be useful to have gas infrastructure during energy transitions and how much investment might be required. We examine the extent to which gas networks might be used to provide low-carbon energy from hydrogen, biomethane and gas with carbon capture, utilisation and storage (CCUS). Some policy considerations that help frame the role of gas grids in energy transitions are highlighted.

13.1.1 Role of gas infrastructure today

In Europe and the United States, gas infrastructure delivers between 50-100% more energy on average to end-consumers than electricity grids (Figure 13.2).[2] Gas grids also provide a major source of flexibility. In Europe, for example, gas storage capacity today is 1 000 terawatt-hours (TWh), which is more than 50 000-times current global battery storage capacity.[3] In some countries this storage helps meet peak gas demand in the winter, which can be more than twice as large as peak gas demand in the summer. Even assuming major efficiency gains from a switch from gas to electricity, replacing this level of energy delivery with electricity would be extremely challenging. Gas grids are particularly important for satisfying energy demand in buildings and industry. In advanced economies, gas provides around a third of final energy consumption in these two sectors today.

Figure 13.2 ▷ Monthly electricity and natural gas use in Europe and the United States

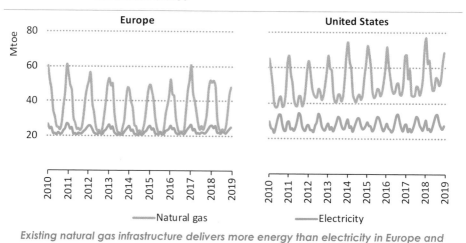

Existing natural gas infrastructure delivers more energy than electricity in Europe and the United States and is critical to meet seasonal demand fluctuations

---

[2] Infrastructure and networks are used here to refer to mid-stream assets such as LNG and transmission and distribution pipelines that transport gas from its point of production to its point of consumption.

[3] Additional storage is provided through the gas that is present within transmission and distribution pipelines.

Gas networks are also well developed in Russia, the Caspian region and parts of Latin America and the Middle East. However, gas currently plays a much smaller role in many of the developing economies in Asia that are the main sources of energy demand growth in the period to 2040. This is mainly because gas historically has not been as readily available at scale or as economically attractive as alternatives. However a number of Asian countries are actively seeking to expand their use of gas, particularly in place of coal, as a means to diversify their energy sources, reduce emissions and improve air quality. The role of gas grids in many developing economies is different from their role in advanced economies since they do not have as large a need for winter heating and so seasonal fluctuations in gas demand are far smaller (although China is one notable exception).

From a broader international perspective, transmission pipelines and liquefied natural gas (LNG) provide around 650 million tonnes of oil equivalent (Mtoe) of gas for interregional trade and are a crucial source of energy imports for a number of countries, for example those with few domestic energy resources.

13.1.2 Role of natural gas in energy transitions

Promise and limits of direct electrification

While a gas grid may play a critical role today in satisfying energy services demanded by consumers, is this role needed in the future? Global electricity demand has risen 60% faster than gas demand in final energy consumption since 2000 and the potential of electricity to provide modern energy services with no emissions at the point of use has generated interest in an "electrify-everything" approach to clean energy transitions (Jacobson et al., 2017). Indeed, it is possible to envisage a low-carbon energy system in which decarbonisation of electricity generation is accompanied by widespread electrification of industrial processes, electric heating is used rather than gas in buildings and electric transport is ubiquitous. This is already the clear direction in some Nordic countries, where electric heating is prevalent and there has been a rapid uptake of electric vehicles.

Are there limits to what can be achieved by the electrification of end-use sectors? This was one of the key questions explored in the "Future is Electric" Scenario in the *World Energy Outlook-2018* (IEA, 2018). Among other elements, this scenario assumed:

- Policies that remove non-economic barriers to the deployment of electric end-use technologies; this means that the direct use of electricity comes closer to achieving its maximum technical potential across all end-use sectors.
- More rapid adoption of connected devices leading to more electricity demand in buildings and data centres.
- Achievement of universal electricity access by 2030 and an accelerated uptake of electric appliances among households which have recently gained access.

In the Future is Electric Scenario, average electricity demand grows by around 3% per year to 2040, and electricity accounts for over 30% of final energy consumption in that year (compared with 20% today). Increased direct electricity use is largest in buildings: while today less than 15% of global space heating needs are met by electricity (whereas cooling is almost entirely electric), this grows to nearly 35% by 2040. There are also major increases in direct electricity use in industry (Figure 13.3).

Figure 13.3 ▷ Global final energy consumption in industry and buildings in the Future is Electric Scenario

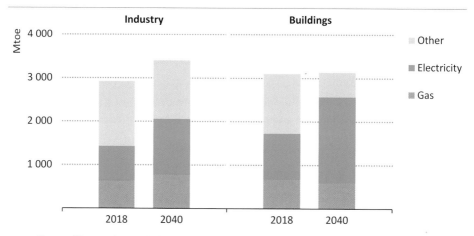

Even with a major push for electrification, natural gas would still provide around 20% of final energy consumption in industry and buildings in 2040

The Future is Electric Scenario highlighted the enormous potential of electricity to play a bigger direct role in the energy system in the future. But it also suggested that there are limits to how quickly and extensively electrification can occur:

- Electricity is not well suited to deliver all types of energy services. Even if the complete technical potential for electrification were deployed, there would still be sectors requiring other energy sources (given today's technologies). For example, most of the world's shipping, aviation, heavy-freight trucks and certain industrial processes are not yet "electric-ready". While in the future these sectors could use fuels that have been generated using electricity (such as hydrogen or synthetic fuels), these liquid or gaseous fuels would need a separate delivery infrastructure.

- Making the switch to electricity raises a number of practical issues. In the industrial sector, providing large-scale high-temperature electric heat would require significant changes in their design and operation. Further, industrial facilities tend to have long lifetimes and slow turnover of capital stock, while the highly integrated nature of industrial processes means that changing one part of a given process would often require changes to other parts. In the buildings sector, installation of an electric heat pump in place of existing gas heating can mean an intrusive retrofit with a much higher upfront capital cost than replacing an inefficient boiler with an efficient one. In addition, many high-rise urban buildings are not suitable for heat pumps.

- If electrification is to lead to significant emissions reductions, it must be produced in a low-carbon way. A massive ramp up in low-carbon electricity generation would also need to be accompanied by major additions to transmission and distribution grids, which typically face hurdles related to permitting and public acceptance.

- Wind and solar photovoltaics (PV) can be complementary to some extent but there would still need to be a huge increase in hourly, daily and seasonal electricity storage. The existing peak in winter electricity demand in Europe and some parts of North America would also be heightened by heat pumps consuming more electricity in the winter. For example, if heating in all buildings in Europe was switched to electricity using heat pumps, peak winter electricity demand would increase by more than 60%. Batteries are becoming cheaper and are well suited to manage short-term variations in electricity supply and demand; however, they are unlikely to provide a cost-effective way to cope with large seasonal swings. The expansion of hydropower, which a major source of flexibility in some countries, is limited by geography.

- The value of overlapping infrastructure can be an important consideration for policy makers. Maintaining a parallel gas infrastructure system adds a layer of resilience compared with an approach that relies exclusively on electricity. It also provides a useful hedge against the risks that electrification and the development of new electricity networks do not increase at the pace needed to displace existing fuels while meeting energy service demands.

- Electrification of the entire building stock cannot happen instantaneously. Better-off consumers are likely to be the first to make the upfront investment in new electrified heating systems. Poorer consumers would continue to rely on existing infrastructure and, under existing tariff structures, would need to shoulder the cost of maintaining this infrastructure.

Natural gas in the Sustainable Development Scenario

In the Sustainable Development Scenario, electricity demand grows by nearly 2% per year and accounts for over 30% of final energy consumption in 2040. Natural gas demand grows by around 1% per year to its peak level in the late-2020s before declining slowly. The majority of natural gas demand growth in this scenario comes from countries with energy systems currently dominated by coal (Figure 13.4). Ensuring a diverse set of sources from which to import natural gas leads to investment in long-distance transmission infrastructure. LNG terminals are particularly helpful in this regard since, by its nature, LNG is more flexible than pipelines and regasification terminals require less upfront capital investment.

In the power sector, the primary way in which coal is displaced in the Sustainable Development Scenario is through increased deployment of variable renewable electricity technologies. But it takes time for the use of these technologies to ramp up; gas-fired power plants therefore have a part to play in the interim (see Chapter 4). However, this window of opportunity for natural gas is time limited and the main role of gas-fired power in the longer term is to provide flexibility, alongside hydropower, demand-side management and batteries. There is also a significant increase in natural gas consumption in the industry sector where there are limited alternative options for the generation of high-temperature heat.

Figure 13.4 ▷ Change in natural gas demand in selected regions in the Sustainable Development Scenario

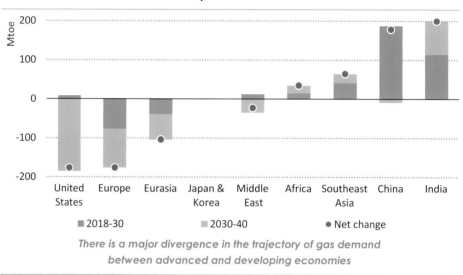

There is a major divergence in the trajectory of gas demand between advanced and developing economies

In advanced economies with existing gas grids, natural gas continues to help meet heating demand in buildings even while there is a rapid uptake in electrification. However, almost all new homes built in advanced economies in the Sustainable Development Scenario after 2030 are nearly zero-energy buildings and up to 4% of buildings are renovated each year (compared with less than 1% today). This weakens the case for purchasing new boilers or for building new gas infrastructure to deliver natural gas to buildings.

In many developing economies, the need to improve local air quality and achieve universal energy access by 2030 underpins growth in natural gas use in buildings in the Sustainable Development Scenario. The main expansion in natural gas use over the next ten years is in China, one of the few developing economies with a significant winter residential heating requirement, where it replaces coal. In general, gas pipeline expansions in developing economies are limited to urban areas. In countries that currently rely on the use of liquefied petroleum gas (LPG) in buildings, increased natural gas use in urban households means that the displaced LPG can be used in modern cookstoves in rural locations, where it can displace the traditional use of biomass. This is a major policy consideration behind the expansion of urban gas distribution networks in India.

Investment in gas infrastructure remains an important component of the energy transition envisaged in the Sustainable Development Scenario so that gas can deliver large quantities of energy and also provide a source of flexibility. Yet the trajectory of natural gas demand poses a challenge for spending levels: investment in new assets has to be sufficient to meet rising global gas demand in the near term while taking proper account of the drop in demand over the longer term.

Between 2019 and 2030, $110 billion is invested globally on average each year in LNG and pipelines in the Sustainable Development Scenario (Figure 13.5). After 2030, there is a divergence in investment trends between advanced and developing economies. In advanced economies, average annual investment drops by around 35%, with nearly all of the remaining $35 billion average annual spending used to maintain existing distribution networks. The growth in natural gas demand in some developing economies after 2030 means that investment in gas infrastructure there falls to a much smaller degree.

Figure 13.5 ▷ Average annual investment in LNG and gas pipeline infrastructure in the Sustainable Development Scenario

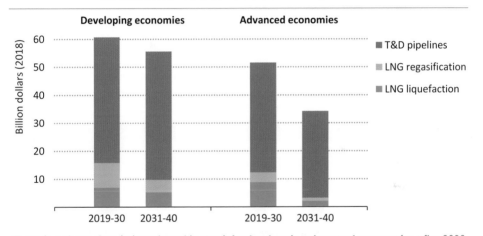

There is a sharp drop in investment in gas infrastructure in advanced economies after 2030

Note: T&D = transmission and distribution.

13.1.3 Need for gas supply to evolve

Even if natural gas brings benefits when replacing more polluting fuels, it is still a fossil fuel and there are clear environmental drawbacks to its use. Indeed, increased natural gas use by itself is far from sufficient to achieve the dramatic reductions in greenhouse gas (GHG) emissions and air pollutants that are needed in the Sustainable Development Scenario. The benefits of natural gas in this context depend critically on significant and rapid progress in eliminating methane emissions from the production, transmission and distribution of natural gas (IEA, 2018). In the Sustainable Development Scenario, global methane emissions from oil and gas operations in 2030 are 75% lower than today. Investing in gas infrastructure is important in this context: older gas pipelines generally are likely to have higher levels of fugitive methane emissions than newer pipelines, and failure to maintain them could undermine efforts to reduce methane emissions.

Box 13.1 ▷ What is a low-carbon gas?

There are many ways to produce hydrogen and biomethane for use in existing gas infrastructure, with either hydrogen or methane as the delivered energy product (Figure 13.6). However, different production methods can result in very different levels of GHG emissions. The focus here is the methods that produce low-carbon gases.

Low-carbon hydrogen: Hydrogen produced by electrolysis[4] using low-carbon electricity, or from the gasification of biomass, or from fossil fuels equipped with CCUS or, potentially, through "methane splitting" (also known as methane pyrolysis).[5]

Biomethane (also called renewable natural gas): A near-pure source of methane produced either by upgrading biogas (by removing carbon dioxide [CO_2] and other impurities) or through the gasification of solid biomass or waste followed by conversion to methane (in a process called "methanation").

Low-carbon synthetic methane: Methane produced through the methanation of low-carbon hydrogen and CO_2 from a biogenic or atmospheric source.

Figure 13.6 ▷ Alternative supply routes to produce low-carbon gases

Feedstock	Transformation	Intermediate product	Transformation	Final product
Organic matter	Anaerobic digestion	Biogas	Upgrading	Biomethane
	Biomass gasification	Bio-syngas	Methanation	Low-carbon methane
			Methanation	Low-carbon synthetic methane*
Zero-carbon electricity	Electrolysis			Low-carbon hydrogen
Fossil fuel with CCUS	Electrolysis			
	Steam reforming, methane splitting or coal gasification			

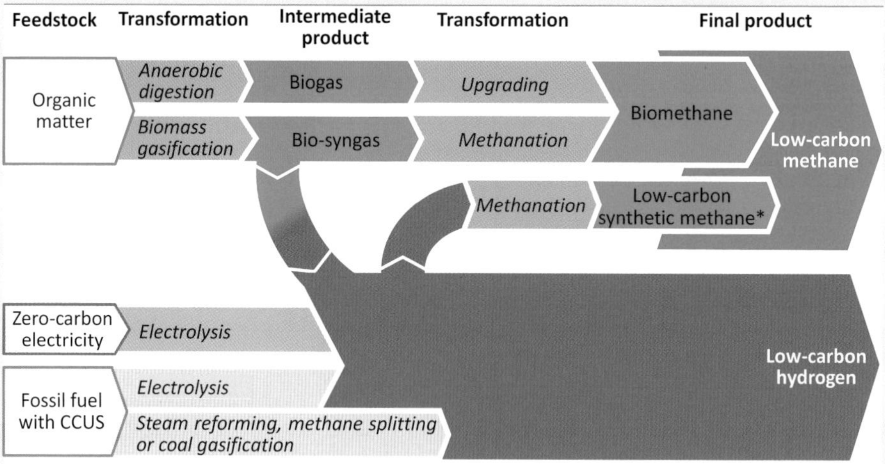

There are many pathways to produce low-carbon gases. This analysis focuses primarily on the opportunities and costs of low-carbon hydrogen and biomethane.

* Synthetic methane is only low carbon if the CO_2 originates from biogenic sources or the atmosphere.

---

[4] The conversion of electricity to another energy carrier is sometimes called "power-to-X". We avoid the use of this term as the "X" can refer to different products including chemicals, hydrogen, methane or heat.

[5] Colours are sometimes used to differentiate between production routes for hydrogen. "Blue" hydrogen is produced from fossil fuels equipped with CCUS, "green" hydrogen from renewable-based electricity, and "black/grey/brown" hydrogen for unabated fossil fuels. There are no established colours for hydrogen from biomass, nuclear, methane splitting or grid electricity and so the colour terminology is not used here.

To secure its role in a low emissions energy system, gas infrastructure will ultimately need to deliver truly low-carbon energy sources (Box 13.1). This chapter focusses on two main options. The first is hydrogen, including both blending of low-carbon hydrogen into existing natural gas pipelines and repurposing of gas grids to deliver high proportions of low-carbon hydrogen. Next, building on the biogas analysis in Chapter 7, we examine the potential and costs of biomethane. The outlook for hydrogen and biomethane in the Stated Policies and Sustainable Development scenarios are examined, along with the role of natural gas use with CCUS in these scenarios.

13.2 Low-carbon hydrogen

Low-carbon hydrogen could help deliver deep emissions reductions across a wide range of hard-to-abate sector. This includes: aviation, shipping, iron and steel production, chemicals manufacturing, high-temperature industrial heat, long-distance and long-haul road transport and buildings (especially in dense urban environments or off-grid).

The scope of the hydrogen discussion has expanded beyond an initial focus on road transport (IEA, 2010; IEA, 1995), and the stakeholder community has broadened to include renewable electricity suppliers, electricity and gas network operators, automakers, oil and gas companies and major engineering firms. Among governments, there is increasing attention from both energy exporting and importing countries, as well as city and regional authorities around the world.

Interest in hydrogen has increased sharply in recent years, reflecting the improvement in the outlook for hydrogen as a low-carbon energy carrier, especially with the declining costs of renewable electricity. Producing low-carbon hydrogen, however, is costly at the moment and investment in hydrogen and CCUS infrastructure presents significant risks in the absence of assured supply and demand.

Injecting hydrogen into gas pipeline networks has been identified as one of four key opportunities for hydrogen over the next decade (IEA, 2019). This would not only reduce the emissions from gas consumption but also offer the possibility of scaling up hydrogen supply technologies and bringing down costs through economies of scale. This in turn could facilitate the direct use of hydrogen in the buildings, transport, industry and power sectors.

13.2.1 Hydrogen use today

Hydrogen is not new to the energy system. Supplying hydrogen to industrial users is a major business globally and there are companies with extensive experience of producing and handling hydrogen. Demand for hydrogen in its pure form is currently around 70 million tonnes (Mt) per year, equivalent to around 330 Mtoe (Figure 13.7). This hydrogen is almost entirely supplied from fossil fuels: 6% of global natural gas consumption and 2% of global coal consumption goes to hydrogen production today. Most of this is used for oil refining and chemicals production. This results in around 830 Mt CO_2 per year, a large portion of which could be avoided through the use of low-carbon hydrogen. A further 45 Mt of hydrogen is used without prior separation from other gases in the industry sector.

Figure 13.7 ▷ Historic global annual demand for hydrogen

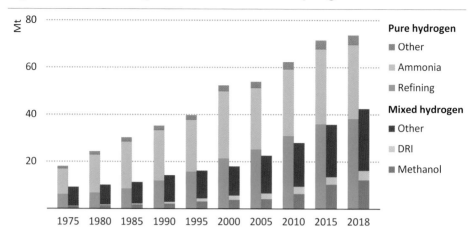

Around 70 Mt of hydrogen is used today in pure form; a further 45 Mt is used in industry without prior separation from other gases

Notes: DRI = direct reduced iron steel production. Methanol, DRI and "other mixed" represent demand for applications that use hydrogen as part of a mixture of gases, such as synthesis gas, for fuel or feedstock.
Source: IEA (2019).

Hydrogen has also been used extensively in the past in gas networks. In 1950, the United Kingdom had over 1 000 facilities producing "town gas", which was a mixture of gases produced from coal or oil that had a hydrogen content of around 50% (Arapostathis et al., 2013). Over 100 000 kilometres (km) of distribution pipelines were built to transport it to end-use sectors (Dodds and McDowall, 2013). Many of these pipelines were replaced after the discovery of natural gas in the North Sea but some are still in use. Other countries, including the United States, Canada and many northern European countries such as Austria, France and Germany, also underwent government-led transitions from town gas to natural gas from the 1950s to 1970s.

The future potential of hydrogen as a low-carbon energy source results in part from its versatility. It can be used in a wide variety of applications, such as transport and heating, or converted into electricity, or transformed into hydrogen-based fuels, such as synthetic methane, ammonia or liquid fuels. It can also support the integration of high levels of renewable-based electricity by providing a long-term storage option and dispatchable low-carbon power generation. In addition it can be produced from a wide range of low-carbon energy sources, even if less than 0.7% of hydrogen production today is low-carbon. Options include production from renewables, sustainable biomass and nuclear electricity. It can also be produced from fossil fuels, if combined with CCUS and if emissions during extraction and supply are minimised.

13.2.2 Costs and potential to blend hydrogen into gas networks

Interest in blending hydrogen into natural gas grids has risen sharply in recent years leading to several major demonstration projects. There are currently 30 such projects around the world at both the transmission and distribution levels.

The case for hydrogen blending in natural gas infrastructure is primarily that a gas supply that includes low-carbon hydrogen would mean lower CO_2 emissions, and would help scale up production of hydrogen and so reduce its costs. However blending would also facilitate the transport of hydrogen from where it is produced to where it is used, as resource availability and economies of scale dictate that individual production plants are likely to serve multiple, dispersed end-users. With the exception of around 5 000 km of hydrogen pipelines in industrial clusters, there is no established infrastructure today for hydrogen transport. The existing natural gas infrastructure in many countries is extensive and could transport hydrogen at much lower unit costs than would be the case if new dedicated hydrogen pipelines had to be built.

The energy density of hydrogen is around a third of that of natural gas and a 5% hydrogen blend by volume[6] in a natural gas pipeline would reduce the energy that the pipeline transports by around 3% (Quarton and Samsatli, 2018). As a result, end-users would need to use larger gas volumes to meet a given energy need, as would any industrial sectors that rely on the carbon contained in natural gas (e.g. for treating metal). To satisfy a given energy demand, a 5% blend of low-carbon hydrogen would reduce CO_2 emissions by 2%.

In the early phase of commercial scale up of low-carbon hydrogen, the ability to use existing infrastructure will be critically important. With only modest additional investment in infrastructure or end-use equipment, existing gas grids would be able to transport the output of new hydrogen production facilities at low marginal costs, thereby reducing the overall cost of consuming low-carbon hydrogen and building consumer confidence in the viability of hydrogen as an energy carrier. The use of existing pipelines to bring hydrogen to consumers would also help avoid the need to acquire permits for new transmission and distribution pipelines, which can take years to acquire, and the need for major new construction projects.

Blending hydrogen into the gas grid would, however, raise the costs of delivered gas because low-carbon hydrogen production costs are likely to remain higher than natural gas prices. Currently, producing low-carbon hydrogen from natural gas with CCUS costs around $12-20 per million British thermal units (MBtu), while producing hydrogen from renewable-based electricity costs around $25-65/MBtu. While "surplus" renewable electricity could be obtained at very low prices in some regions, its availability today is limited. Furthermore, electrolysis is a capital-intensive process and so using only periodic surpluses of very cheap electricity would be likely to be an expensive way to produce low-carbon hydrogen. For example, if an electrolyser had access to free electricity but operated only at a load factor

---

[6] All blend shares in this section are given in volumetric terms.

of 10%, it would cost around $50/MBtu to produce hydrogen (IEA, 2019). Nonetheless, as variable renewable electricity accounts for an increasing share of power generation in the Sustainable Development Scenario, larger hourly disparities in wholesale power prices are likely to emerge and could improve the economics of using electrolysers at times when prices are very low.

A 5% hydrogen blend in a distribution grid for a city of 3 million people today would cost around $25-50 million each year for hydrogen supply using electrolysis, injection stations, upgrades to pipelines, compressors and metering.[7] Investment in facilities to produce the low-carbon hydrogen would represent around 80% of these costs. But early stage investments of this kind would have major longer term benefits for technology learning. If 100 projects of this size were undertaken around the world, they would stimulate an additional 1 Mt of annual low-carbon hydrogen supply and require 10 gigawatts (GW) of electrolyser capacity. This would lead to a significant scale up of manufacturing and installation capabilities, promoting efficiency improvements and capital cost reductions for electrolysers of around 20%. Supplying this level of hydrogen demand from CCUS-equipped natural gas reforming would similarly promote vital experience and cost reductions.

The largest electrolyser facilities being installed today have capacities of around 10 MW, but among the larger proposed projects for coming years are facilities of 100-250 MW in Europe and North America. These will run on wind or hydropower and inject tens of thousands of tonnes of hydrogen per year into the gas network. There also are proposed projects for blending hydrogen from natural gas with CCUS in Europe, including plans in northwest England to produce around 0.1 Mt of low-carbon hydrogen per year by 2030 for injection into the gas grid, and to convert chemical plants to low-carbon hydrogen (Cadent, 2018).

Adapting the transmission grid

High concentrations of hydrogen can corrode the steel used in transmission pipelines; the performance of the equipment connected to the pipelines, such as compressors and valves, can also be adversely affected. Existing transmission grids have usually not been designed to accommodate more than minor levels of hydrogen and so upgrades to infrastructure and equipment would be needed to accommodate meaningful blending levels. With minor modifications, however, transmission networks could probably cope with hydrogen blends of up to 15-20%, depending on the local context (NREL, 2013; NATURALHY, 2009). Moving to higher concentrations of hydrogen is possible, but would require additional modifications to pipeline integrity management systems and compression stations or the replacement of some pipework. Where natural gas transmission pipeline corridors are made up of several pipelines running in parallel, the pipelines could be adapted sequentially to minimise disruption to operations.

---

[7] Assuming a ten-year project, savings from the avoided natural gas use, and no new investments are needed to upgrade end-use equipment.

The picture is mixed in terms of underground gas storage facilities. Salt cavern storage facilities are suitable for hydrogen blends of up to 100% without the need for upgrades, but depleted oil and gas reservoirs and aquifers are more permeable to hydrogen and contain contaminants that could react with it. Recent tests have indicated that hydrogen blends might be stored effectively in depleted hydrocarbon reservoirs with hydrogen losses below 20% (Underground Sun Storage, 2017), but further research is needed to understand their performance for different blend levels and time periods.

Adapting the distribution grid

Generally, distribution pipelines are expected to have a higher tolerance for hydrogen blending than transmission pipelines. Much depends on the age of different elements of the grid. Equipment that has been added recently to distribution grids can often tolerate higher shares of hydrogen (Figure 13.8).[8] Older grids and components present a bigger problem because some elements have a relatively low tolerance to hydrogen blending and would need to be replaced or adapted to accommodate higher blending levels. As hydrogen injected into the distribution system will reach all parts of the grid that are downstream, all components must be able to tolerate it to any given level.

Figure 13.8 ▷ Estimated tolerance to hydrogen blend shares of selected elements of existing gas distribution networks

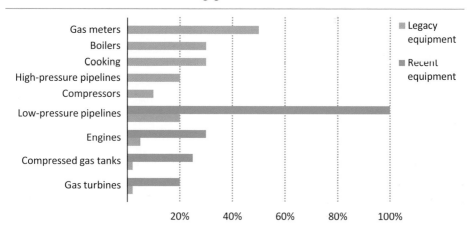

Much existing equipment would need to be adapted to tolerate meaningful levels of blending or be replaced by new equipment designed to tolerate pure streams of hydrogen

Notes: "Legacy" refers to the oldest equipment on most systems. "Recent" figures are shown where data are available.

Source: IEA analysis based on Altfeld and Pinchbeck (2013).

---

[8] In general, pipelines or equipment modified for high hydrogen blends (50% or below) can still use natural gas without added hydrogen but the combustion will not be optimised. However it is likely to be technically feasible to optimise new end-use equipment for any hydrogen blend up to 100%, as well as to produce dual-fuel appliances that can accommodate different gas supplies.

Low-pressure service pipelines that provide gas to residential buildings generally would have few problems accommodating blends of up to 20%. Indeed most service pipelines installed since the 1970s use plastic instead of steel and could accommodate a pure stream of hydrogen. Within buildings, many gas heating and cooking appliances in Europe are already certified for up to 23% hydrogen.

Other end-uses may be much more restricted. Compressed natural gas tanks, used in some vehicles, can have very low limits (although the latest "Type IV" tanks can take blends of 30%). Older natural gas engines face a similar issue and have a maximum level of blended hydrogen of 2-10%. Similarly many industrial gas users cannot accept very high blending levels: chemical producers, for example, use natural gas as a raw material and have very strict feedstock specifications. In the power sector, the control systems and seals of existing gas turbines are not designed to handle the properties of hydrogen and many are certified for less than 5% blended hydrogen. While they could accommodate higher blends if seals and safety systems were adequately adjusted, they are generally optimised for a stable gas composition rather than a blend share that varies or evolves over time.

There have been a number of successful demonstrations of hydrogen blends of up to 20% in distribution grids over the last decade. However, regulations are generally based on natural gas supply specifications or the tolerance of the most sensitive piece of equipment on the grid and in consequence only allow very low levels of blending. Currently in many countries, no more than 2% hydrogen blending is permitted (Figure 13.9).

Figure 13.9 ▷ Current limits on hydrogen blending in natural gas networks and gas demand per capita in selected locations

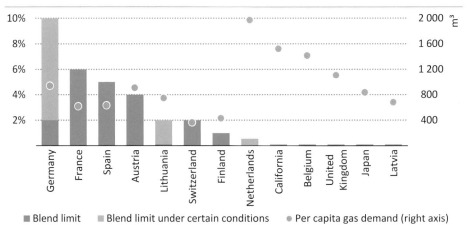

Today most natural gas networks have limits on allowable hydrogen concentrations; some with the strictest limits have the highest per capita natural gas demand

Note: The conditional limits shown reflect these parameters: in Germany if there are no compressed natural gas filling stations connected to the network; in Lithuania when pipeline pressure is greater than 16 bar; in the Netherlands for high-calorific gas.

Source: IEA analysis based on Dolci et al. (2019); HyLaw, (2019); Staffell et al. (2019).

Existing natural gas infrastructure, in theory, could also be used as a cost-effective means of hydrogen transportation to users who require a pure stream of hydrogen. Various options have been proposed and are at different stages of maturity:

- Separating hydrogen from a blend of natural gas prior to its use. There are a number of technologies that could potentially be used, but they are at an early stage of development and are relatively expensive.
- Transporting natural gas close to end-users and transforming it into hydrogen through methane splitting. Methane splitting converts methane into hydrogen and solid carbon (also called "carbon black"). The carbon black can be buried or used to produce rubber, tyres, printers or plastics. This process is still at a very early stage of development and a number of challenges still need to be resolved.
- Converting the gas grid to deliver a pure stream of hydrogen (Box 13.2).

Box 13.2 ▷ Using the gas grid to transport and deliver pure hydrogen

Delivering pure hydrogen to consumers would require much more change than just blending. A switch to 100% hydrogen supply for each part of the affected network would mean that new compressors and, in some cases, storage facilities would need to be available in advance. It would also require the replacement of meters, compressors and monitoring equipment, thorough inspection of older parts of a pipeline and replacement of current gas appliances. In addition, it would involve a temporary loss of access to the gas grid: during the conversion from town gas to natural gas in the 1960s and 1970s, households had their gas supplies cut for a day or two.

For consumers, it would require appliances that are ready to use 100% hydrogen. Although it would be feasible to produce such appliances, they would initially cost up to 20% more than current appliances (Frazer-Nash Consultancy, 2018). Consumers may need a lot of convincing to make such a change, especially given the likely disruption and additional costs. On the basis of stated policies, full conversion by 2030 is expected to be realised in fewer places than blending, and to be focussed on particular parts of national grids, such as town distribution networks and underused transmission pipelines.

The H21 project in the north of England is an example of a project to deliver pure hydrogen. It aims to use existing distribution pipes to deliver a pure stream of hydrogen to several urban areas from the late 2020s, while natural gas networks continue to operate in unconverted areas nearby, potentially using natural gas with blended hydrogen. The estimated costs of repurposing plastic distribution pipelines are $4 000/km (H21, 2016). While these costs have yet to be demonstrated in practice, they are at least a factor of ten lower than the cost of building new distribution pipelines. Other projects are under consideration. For example, the existing gas transmission network in the Netherlands, geared to handle its low-calorific natural gas, has pipelines that are likely already suitable for pure hydrogen transport and there are investigations underway to assess the opportunities to convert it to deliver 100% hydrogen.

13.3 Biomethane

Another alternative to pipeline natural gas is biomethane. Unlike hydrogen, biomethane, a near-pure source of methane, is largely indistinguishable from natural gas and so can be used without the need for any changes in transmission and distribution infrastructure or end-user equipment. There are two main biomethane production pathways: the first involves upgrading biogas and the second involves the gasification of biomass.

Upgrading biogas

The main production route for biogas is via anaerobic digestion of an organic feedstock (certain crops, animal manure, wastewater sludge and municipal solid waste are all suitable) (see Chapter 7). In its raw form, the resultant biogas is a mixture of methane, CO_2, hydrogen sulphide, nitrous oxides and other contaminants; the methane content varies from 45-75% according to the feedstock used. This is too low for biogas to be used as a direct replacement for natural gas or for it to be blended into existing gas networks. The biogas therefore has to be "upgraded" to remove most of the CO_2 and other contaminants, leaving a near-pure stream of methane.

Upgraded biogas accounts for around 90% of total biomethane produced worldwide today. Upgrading technologies make use of the different properties of the various gases contained within biogas to separate them, with water scrubbing and membrane separation accounting for almost 60% of biomethane production globally today (Cedigaz, 2019). The CO_2 that is separated from the methane is relatively concentrated and could be used for industrial or agricultural purposes or combined with hydrogen to yield an additional stream of methane: alternatively it could be a candidate for storage in geological formations.

Biomass gasification

An alternative method is to produce biomethane through the gasification of biomass. There are a few gasification demonstration plants currently in operation but the volumes of biomethane produced in this way are relatively small and the technology is at a lower level of technical maturity than anaerobic digestion. However there is arguably greater potential for gasification to generate economies of scale. There is also a higher level of interest from incumbent gas producers since gasification appears a better fit with their knowledge and technical expertise.

To produce biomethane through gasification, woody biomass is first broken down at high temperature (between 700-800 degrees Celsius [°C]) and high pressure in an oxygen-free environment. Under these conditions, the biomass is converted into a mixture of gases, mainly carbon monoxide, hydrogen and methane (sometimes collectively called syngas). To produce a pure stream of biomethane, the bio-syngas is cleaned to remove any acidic and corrosive components. A catalyst is then used to promote a reaction between the hydrogen and carbon monoxide or CO_2 to produce methane in a process called methanation (Box 13.3). Any remaining CO_2 or water is removed at the end of this process.

Box 13.3 ▷ Low-carbon synthetic methane

Synthetic methane is methane produced by chemically combining hydrogen with carbon monoxide or CO_2. If the synthetic methane is combusted, this CO_2 is again released to the atmosphere (unless the combustion process is equipped with CCUS). From a climate perspective, the source of CO_2 therefore is vitally important. For "low-carbon synthetic methane", low-carbon hydrogen is needed alongside a non-fossil source of CO_2.

Methanation is likely to be most cost effective when the hydrogen and carbon come from the same source (such as in biomass gasification). Another option is to use the CO_2 separated from methane during biogas upgrading or during bioethanol production. These processes result in a concentrated stream of CO_2, which can be captured and used with only moderate additional investment and energy. If there is production of hydrogen at the same site, then the two streams of methane can be combined to take advantage of the same infrastructure for onward transmission and distribution. This would also maximise the use of the carbon contained in the original biomass feedstock. In addition, CO_2 can be captured directly from the atmosphere. Cost estimates for direct air capture (DAC) are uncertain, but studies estimate that in the long-term costs for DAC may fall to a range of $100-230/tonne CO_2 (Keith et al., 2018; Fasihi et al., 2019).

Low-carbon synthetic methane production has been demonstrated at a scale of 1.4 million cubic metres per year (1.2 thousand tonnes of oil equivalent) in Germany, but costs remain high. With low-carbon hydrogen costing around $35/MBtu and CO_2 costing $30/tonne CO_2, low-carbon synthetic methane would cost around $60/MBtu. With DAC costing $200/tonne CO_2, costs would be higher still at around $85/MBtu. Cost reductions of around 50% are expected by 2040 from improvements in electrolysers, renewable electricity generation and methanation equipment. However these reductions are not sufficient for low-carbon synthetic methane to play a meaningful role before 2040 in the Sustainable Development Scenario.

13.3.1 Biomethane use today

There are over 700 biomethane plants in operation today producing around 2.5 Mtoe of biomethane globally. Most of these plants are in Europe and North America. Three-quarters inject biomethane into existing gas networks (Cedigaz, 2019). A further 20% of plants deliver biomethane for use in road vehicles through dedicated distribution networks. The remainder is for use in buildings and industry.

Although biomethane represents less than 0.1% of natural gas demand today, its production and use are supported by an increasing number of policies, especially in the transport and electricity sectors. For example, there are biomethane production targets and ambitions in Italy, India, China and France. Biomethane is also expected to play a role in helping to achieve Brazil's 2028 target of reducing the carbon intensity of fuels in the transport sector by 10%, and in the context of low-carbon fuel standards in the US State of California and British Colombia, Canada. Biomethane production has grown by around 30% each year on average over the past decade.

Almost all plants in operation today produce biomethane by upgrading biogas. However less than 8% of biogas produced globally today is upgraded (for the other uses of biogas, see Chapter 7). This percentage varies widely between regions: in North and South America around 15% of biogas production is upgraded; in Europe, the region that produces the most biogas and biomethane, around 10% of biogas production is upgraded; in Asia, the figure is 2% (Figure 13.10).

Figure 13.10 ▷ Biomethane production and share of total biogas production that is upgraded in selected regions, 2017

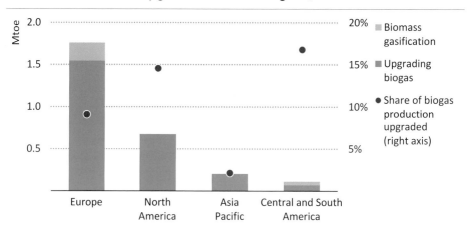

Biomethane production in 2017 is mostly concentrated in Europe and North America, although these regions upgrade only a fraction of their biogas production

Note: There are negligible levels of biomethane produced in Africa, Middle East and Eurasia.

13.3.2 Blending biomethane into gas networks: costs and potential

In this *Outlook*, we provide a first-of-a-kind assessment of the sustainable technical potential[9] and costs of biomethane supply globally, and how this might evolve in the future. (This builds on the analysis for biogas described in Chapter 7.) For biogas upgrading we include the costs and availability of 17 individual biogas feedstocks grouped into four categories: crops, animal manure, municipal solid waste (MSW) and wastewater. For biomass gasification we consider two sources of solid biomass: forestry residues and wood

---

[9] This includes feedstocks that can be processed with existing technologies, that do not compete with food for agricultural land, and that do not have any other adverse sustainability impacts (e.g. reducing biodiversity). Feedstocks grown specifically to produce biogas, such as energy crops, are also excluded.

processing residues. Each feedstock has been assessed across the 25 regions modelled within the World Energy Model.

Figure 13.11 ▷ Global sustainable technical potential of biomethane

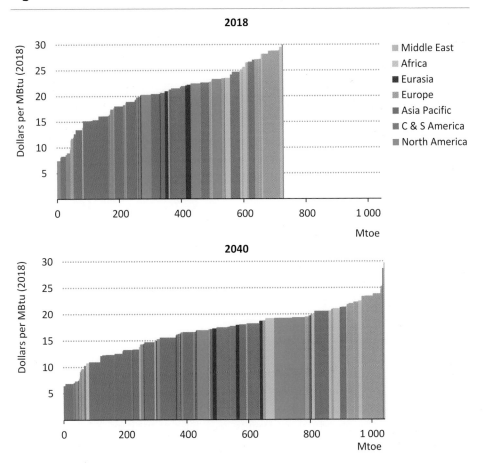

By 2040, over 1 000 Mtoe of biomethane could be produced globally.
This potential has a wide geographic spread.

Note: C & S America = Central and South America.

We estimate that around 730 Mtoe of biomethane could be produced sustainably today, equivalent to over 20% of global natural gas demand (Figure 13.11). Biogas upgrading accounts for the vast majority of this potential: crop residues (including sequential crops)[10] provide 35%, animal manure 25%, MSW 15% and wastewater less than 5%. The remaining 20% comes from biomass gasification. This potential has a wide geographic spread: the

---

[10] Sequential crops are grown between two harvested crops as a soil management measure that helps to preserve the fertility of soil and avoid erosion; they do not compete with food for agricultural land.

Chapter 13 | Prospects for gas infrastructure

largest share of the resource potential is in the United States and Europe (each with 16%), but there is also major potential in China and Brazil (each with 12%) and India (8%). The potential could be even larger if energy crops were included, but we exclude them from this assessment to avoid any potential competition between biomethane and food production (MTT Research, 2009). Nonetheless there still could be competing uses for some of the feedstocks: for example, forestry residues can be a sustainable source of heat, while crop residues can be used for animal feed or to produce advanced biofuels. Economic and population growth, changes in waste management and agricultural processes, and technological progress mean that this estimated global biomethane potential increases by over 40% between 2018 and 2040. Biomass gasification grows at a slower pace than biogas feedstocks and accounts for 15% of total potential in 2040.

We estimate that the global average cost of producing biomethane through biogas upgrading today is around $20/MBtu. Most of this cost is attributable to the collection and processing of the biogas feedstock, but biogas upgrading adds an additional $7-8/MBtu. Biomass gasification is currently the more expensive method of production in all regions: the average cost of producing biomethane by gasification is around $25/MBtu globally. In contrast to biogas upgrading, it is the gasification and methanation processes involved rather than the feedstock cost that make up most of the overall cost. There is a high degree of uncertainty over the cost reductions that could be achieved for both production routes, although constructing larger facilities should provide some economies of scale for both.[11] In 2040, we estimate that the average cost of biomethane globally will be around $4/MBtu lower than today.

Upgrading biogas captured from landfill sites is typically the cheapest way to produce biomethane in all parts of the world. Taking into account regional biomethane costs and natural gas prices, around 12 Mtoe of biomethane could be produced today for less than the domestic price of natural gas. In the Stated Policies Scenario, biomethane costs fall while natural gas prices rise over time, and so, by 2040, nearly 75 Mtoe of biomethane is fully cost competitive with natural gas. Even so, most biomethane is much more expensive than natural gas and meeting 10% of today's natural gas demand with the cheapest biomethane options available in each region would cost $2-15/MBtu more than natural gas (Figure 13.12).

One way to reduce the cost gap would be to make use of the by-products from biomethane production. Producing biomethane through biogas upgrading leaves a residue of fluids and fibrous materials called "digestate". The handling and disposal of digestate can be costly and as a result it is often considered a waste rather than a useful by-product. However in certain locations and applications digestate could be sold as a natural fertiliser and help to reduce the overall cost of biogas production. Biogas upgrading also results in a pure stream of CO_2 that could be used by other industries. In the beverage industry, for example, CO_2 is

---

[11] There is slightly more scope for cost reductions with biomass gasification than upgrading biogas, and so the cost gap between the two production routes shrinks over time.

often bought for $50-100/tonne CO_2 (Pérez-Fortes and Tzimas, 2016). The revenues that can be achieved through selling digestate or the pure CO_2 stream, however, are likely to be relatively modest and in most cases would not be sufficient to close the cost gap entirely with natural gas.

Figure 13.12 ▷ Cost of using the least expensive biomethane to meet 10% of gas demand and natural gas prices in selected regions, 2018

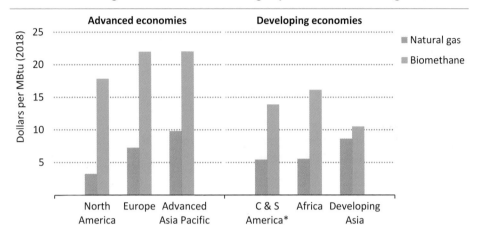

There is a large gap between the price of natural gas and the cost of biomethane production in nearly all regions today; policies will be critical to close this gap

* C & S America = Central and South America excluding Chile.

Note: Advanced Asia Pacific includes Australia, Japan, Korea and New Zealand.

Another way to reduce the cost gap would be to put a price on the GHG emissions from fossil fuels, including natural gas use. When biomethane replaces natural gas, its use reduces CO_2 emissions. Biomethane can also help to avoid large quantities of methane emissions that would otherwise be released directly to the atmosphere from feedstock decomposition (see section 13.5.2). With these avoided methane emissions taken into account, there are some countries – most notably China and India – where only a small GHG price would be required to close the cost gap between natural gas and biomethane. However, in advanced economies and other developing economies, a GHG price of $50-180/tonne CO_2-equivalent would be required. While the GHG price needed to close the cost gap falls over time in both the Stated Policies and Sustainable Development scenarios, these numbers are substantially above those in most GHG pricing systems around the world today.

Governments could take additional policy action to place a value on the additional benefits of biomethane. For example, the benefits it brings as a domestically produced low-carbon fuel that promotes rural development and enhances waste management in ways that improve water quality.

13.4 Outlook for low-carbon hydrogen and biomethane

13.4.1 Stated Policies Scenario

In the Stated Policies Scenario, just under 80 Mtoe of biomethane is consumed in 2040. Most of this occurs as a result of explicit policy measures rather than because biomethane is cost competitive with natural gas. The majority of global consumption in 2040 is in China and India (Figure 13.13): China produces over 30 Mtoe of biomethane, which is injected into its expanding natural gas grid, while India's consumption grows to 15 Mtoe of biomethane, primarily because of its target for biomethane use in the transport sector. In both cases, support for biomethane is also motivated by a desire to limit growing reliance on imported natural gas, and thereby enhance gas security. In Europe, biomethane use reaches 12 Mtoe in 2040, accounting for 2.5% of the gas used in natural gas grids. Consumption in North America increases to just under 10 Mtoe.

Figure 13.13 ▷ Biomethane consumption by sector and region in the Stated Policies Scenario

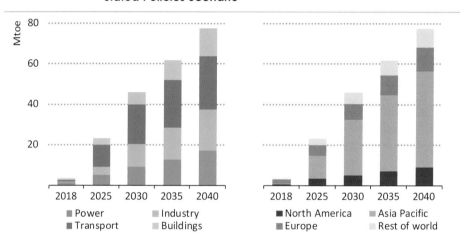

Biomethane use grows to just under 80 Mtoe in the Stated Policies Scenario, mainly because of explicit policy support in India and China

As highlighted by the case of India, the potential use of biomethane in transport is one of the main areas of focus for policy makers today. For example, the Renewable Energy Directive in the European Union requires 3.5% of fuel demand in 2030 to be met by advanced biofuels, including biomethane, while in the United States there is support from the Renewable Fuel Standard and California's Low Carbon Fuel Standard.

Because natural gas today accounts for less than 5% of total fuel demand in transport, a key element of any biomethane transport strategy is the need to increase the number of gas vehicles on the road and to build fuelling infrastructure. These conditions are likely to be easiest to satisfy in captive fleets and road freight.[12] A significant proportion of road freight activity takes place on key transport corridors, so providing fuelling infrastructure along them would provide access to a large share of demand. In Germany, for example, 60% of all freight activity occurs on roughly 2% of the road network. Legislation to support the rollout of fuelling infrastructure compatible with biomethane is already in place in the European Union and the United States.

There are a number of reasons why policy makers are interested in biomethane to reduce oil use in transport, despite the challenges of expanding gas vehicle fleets and developing refuelling infrastructure in tandem with increasing biomethane production. First, while natural gas networks are the most cost-effective means of delivering biomethane, distribution by tanker to discrete refuelling stations is one of the lowest cost ways to bring biomethane to end-use consumers in locations where pipeline infrastructure is not available. Second, biomethane has lower CO_2 and air pollutant emissions than gasoline, diesel and many conventional liquid biofuels. Third, biomethane avoids the land-use change concerns that can arise from the use of crop-based biofuels. In addition, bioethanol and biodiesel are often subject to blend share limitations (since they are not identical to existing fuels) whereas biomethane is an identical replacement and so can fully replace natural gas. Despite these benefits, the growth of biomethane in transport in the Stated Policies Scenario is modest except in India, China, Italy and the United States.

There is also some growth in the pure use of hydrogen and hydrogen-based fuels or feedstocks such as ammonia in the Stated Policies Scenario as a result of existing policy commitments and signals, although in many cases these fuels are not produced by a low-carbon route. There is also some limited blending of low-carbon hydrogen into natural gas networks.

13.4.2 Sustainable Development Scenario

There is much more significant uptake of both biomethane and low-carbon hydrogen in the Sustainable Development Scenario. Biomethane use rises to over 200 Mtoe in 2040 (Figure 13.14). The introduction of a high GHG price across advanced economies and many developing economies is a key factor underpinning this growth. In countries with existing gas infrastructure, biomethane consumption accelerates over the course of the *Outlook* period. By 2040, there is a 10% blend of biomethane in gas grids in Europe and a 5% blend in North America. While these ratios in 2040 may not seem large, natural gas consumption has peaked and is on a downward trajectory, while biomethane consumption is rising steeply, and would continue this trend beyond 2040.

In developing economies, especially those looking to expand the use of natural gas in their energy mix, the use of biomethane provides a much greater reduction in CO_2 emissions

---

[12] Captive fleets are comprised of vehicles that operate on established routes and refuel at set locations or depots. Examples include municipal buses, refuse collection vehicles, and package delivery company and supermarket fleets.

than switching from coal to natural gas. Biomethane also helps to reduce air pollution, aiding the substantial improvements in air quality required in this scenario. There is a preference for the use of biomethane in the transport sector, although in many regions biomethane use is largely shaped by the existing end-uses of grid-supplied natural gas. Globally, biomethane consumption is largest in the buildings sector.

Figure 13.14 ▷ Biomethane consumption by sector and region in the Sustainable Development Scenario

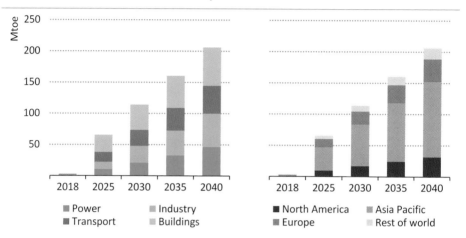

There is strong uptake of biomethane in both advanced and developing economies, and production in 2040 exceeds 200 Mtoe

Just over 25 Mtoe of low-carbon hydrogen is blended into the gas grid in 2040 in the Sustainable Development Scenario. In energy terms, this represents a much lower level of consumption than for biomethane. However hydrogen blending is a crucial building block for the development of a global industry for low-carbon hydrogen supply and for cost reduction. The European Union leads the way initially, with a 5% blend of low-carbon hydrogen in volume terms by 2030 (1.5% in energy terms). As with biomethane, blended low-carbon hydrogen is primarily used in sectors where natural gas is consumed.

By 2030, low-carbon hydrogen supply capacity for injection into the gas grid is more than 1 000-times larger than today. This is produced both through the use of low-carbon electricity and CCUS-equipped natural gas reformers. As a result, both technologies move along the learning curve and costs fall. Coupled with explicit policies that promote the use of hydrogen in hard-to-abate sectors, this encourages the direct use of low-carbon hydrogen (i.e. without blending) and hydrogen-based fuels for other purposes.

Low-carbon hydrogen blending in the gas grid continues to grow after 2030. The European Union has nearly 10% blending by 2040 while North America and China both approach 5% hydrogen blending. As the nature of gas demand shifts, the electricity sector becomes the biggest user of blended low-carbon hydrogen.

In total, more than 230 Mtoe of blended low-carbon hydrogen and biomethane are consumed in 2040. This represents around 7% of total gas demand in the Sustainable Development Scenario in energy equivalent terms (Figure 13.15). These low-carbon gases avoid more than 500 Mt CO_2 emissions that would have occurred if natural gas had been used instead. In addition to the hydrogen injected into the gas grid, just over 80 Mtoe low-carbon hydrogen is used directly in 2040 as a fuel in end-uses.

Figure 13.15 ▷ Low-carbon hydrogen and biomethane injected into gas grids in the Sustainable Development Scenario

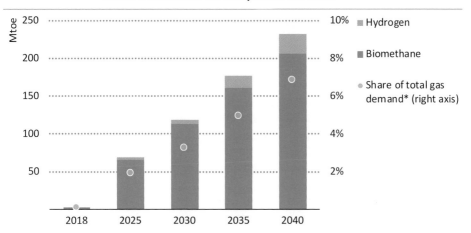

Over 230 Mtoe of low-carbon gases are delivered by the gas grid by 2040, accounting for around 7% of total gas demand

*Includes natural gas, biomethane and low-carbon hydrogen blended into gas networks in energy equivalent terms.

13.5 Implications for emissions and energy security

13.5.1 Reducing CO_2 emissions

Biomethane and low-carbon hydrogen can help to avoid CO_2 emissions by displacing the use of natural gas. In theory, a 10% volume blend of biomethane in a natural gas pipeline would reduce CO_2 emissions in the gas consumed by 10%. However, there are emissions from the harvesting, processing and transport of the biogas feedstock that need to be weighed against the CO_2 emissions that arise during the production, processing and transport of natural gas (GRDF, 2018; Giuntoli et al., 2015). These indirect emissions can vary considerably between sources of biomethane and natural gas and, unless minimised, they could reduce the CO_2 emissions savings from the use of biomethane.

For low-carbon hydrogen, a 10% volume blend in a natural gas pipeline would reduce CO_2 emissions by 3-4% (for a given level of energy). However the different potential energy inputs and conversion technologies to produce hydrogen mean a careful lifecycle emissions approach is needed to ensure that it is truly low-carbon (McDonagh et al., 2019). For electrolysis to produce low-carbon hydrogen, it has to use low-carbon electricity from dedicated renewable or nuclear facilities or from the grid. The latter faces challenges in part because the use of grid electricity is a relatively expensive way to produce hydrogen (IEA, 2019), but also because hydrogen produced in many countries using grid electricity would result in more emissions than hydrogen produced using natural gas without CCUS. It is not until 2035 in the Sustainable Development Scenario that the global average emissions intensity of electricity is low enough for hydrogen produced using grid electricity to result in fewer emissions than hydrogen from natural gas (the threshold is around 180 grammes of CO_2 per kilowatt-hour/kWh [g CO_2/kWh]). To produce low-carbon hydrogen using natural gas equipped with CCUS, it is critical both to minimise methane emissions that occur along the natural gas value chain and to maximise the volumes of CO_2 captured during the conversion process.

There is an additional opportunity for CO_2 emissions reductions from hydrogen and biomethane production. Biogas upgrading generates a highly concentrated by-product stream of CO_2 that could be captured for as little as $20/tonne CO_2 (Koornneef et al., 2013). As the CO_2 is of biogenic origin, storing it underground removes CO_2 from the atmosphere; the same is true for CO_2 captured from biomass gasification. However CO_2 transport and storage can be expensive and so an integrated CCUS system may only be cost effective for the largest biomethane and gasification plants or for plants that are close to CO_2 storage infrastructure. Another option would be to sell the captured CO_2 as a non-fossil source of carbon for synthetic fuels production or other forms of CO_2 utilisation (for example in the food industry). CCUS could also be applied to some large-scale end-uses of natural gas that are not replaced by low-carbon gases (Box 13.4).

Box 13.4 ▷ Using CCUS to reduce emissions from natural gas

Today around 30 Mt of CO_2 are captured from industrial activities in large-scale CCUS facilities. Nearly two-thirds of this is captured from natural gas processing: underground deposits of natural gas can contain significant quantities of naturally occurring CO_2 that must be removed to meet technical specifications before the gas can be sold or used. This produces a highly concentrated stream of CO_2 that is relatively easy and cost-efficient to capture. Most of the captured CO_2 is used for enhanced oil recovery (IEA, 2018).

CCUS can also significantly reduce the emissions intensity of natural gas if it is applied to end-use technologies. One option is to produce electricity from combined-cycle gas turbines (CCGTs) that are equipped with CCUS. There are currently two large-scale coal-fired power plants equipped with CCUS in operation but no large-scale CCUS gas-fired plants. The application of CCUS to end-user technologies such as power plants and the use of low-carbon gases are not mutually exclusive, but care would be needed to ensure they do not detract from each other. For example, if hydrogen were to be injected into the grid, this could reduce the capture efficiency of any CCGTs equipped with CCUS.

The lack of progress with CCUS to date stems from the significant additional capital and operating costs it entails, which are currently greater than the revenue streams that can be generated from the captured CO_2 (either from making use of the CO_2 or from policies that incentivise geological CO_2 storage). In the Sustainable Development Scenario, CCUS is a critical technology to deliver the necessary emissions reductions in both the power and industry sectors and there is extensive policy support to equip new and existing gas power plants with CCUS (see Chapter 5 for a discussion of CCUS for coal use in industry and Chapter 6 for coal-fired power generation). In the power sector, there are over 150 GW of CCGTs equipped with CCUS by 2040, just under half of which are existing plants that are retrofitted (Figure 13.16). These produce around 900 TWh electricity in 2040 with an emissions intensity of less than 40 g CO_2/kWh. In total, the use of CCUS for gas power avoids over 300 Mt CO_2 that would otherwise have been emitted in 2040.

Figure 13.16 ▷ Installed CCGTs equipped with CCUS and emissions avoided in the Sustainable Development Scenario

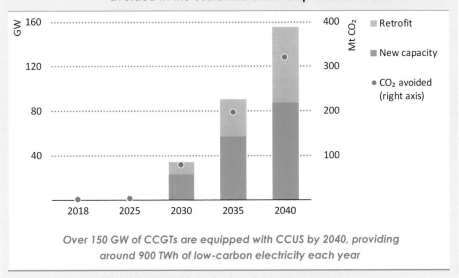

Over 150 GW of CCGTs are equipped with CCUS by 2040, providing around 900 TWh of low-carbon electricity each year

13.5.2 Avoiding methane emissions

Some of the feedstocks that are used to produce biomethane would decompose and produce methane emissions if not carefully managed. Biomethane production can avoid these emissions by capturing and processing them instead. Even if these emissions occur outside the energy sector, they should be credited to biomethane.[13] This is already the case

---

[13] The avoided methane emissions are usually associated with the waste or agricultural sectors. Our emissions analysis generally focuses on GHG emissions from the energy sector and so any avoided methane emissions would not be shown in GHG emissions reductions between scenarios. However, avoided methane emissions are included in assessing the cost effectiveness of different GHG abatement opportunities.

within California's Low Carbon Fuel Standard. Yet estimating the size of this credit is not straightforward, as it depends on a reasonable "counterfactual" case for what level of methane emissions would have occurred if the feedstock had not been converted into biomethane, which can vary according to region, feedstock type and over time.

For example, there is wide regional variation in how methane produced within landfill sites is currently handled. In Europe, most sites have capture facilities, with the captured methane (known as "landfill gas") either flared or used for power generation. In the United States, around 55% of the methane that is generated in landfill sites across the country is captured. Around 20% of what remains breaks down before reaching the atmosphere, meaning that close to 35% of the methane generated in landfills is emitted to the atmosphere.[14] There is a lack of reliable data on landfills in most developing economies but the percentage of methane that is captured is likely to be lower than in advanced economies. Animal manure can also result in methane emissions to the atmosphere. All other potential biomethane feedstock types generally degrade aerobically (and so do not result in methane emissions).

Figure 13.17 ▷ Marginal abatement costs for global biomethane potential with and without credit for avoided methane emissions, 2018

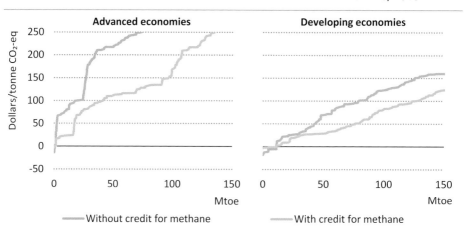

Accounting for the avoided methane emissions through the use of biomethane would greatly boost its attractiveness

Notes: Assumed methane avoided (all percentages given as share of biomethane produced): 50% from landfills in developing economies; 35% from landfills in North America; 0% from landfills in Europe; 18% from animal manure; 0% from all other feedstocks. Excludes methane leaks that occur during biomethane production since the leakage rate for natural gas and biomethane are assumed to be similar. One tonne of methane is assumed to be equal to 30 tonnes of CO_2-eq (100-year global warming potential).

---

[14] Figures are based on the 2016 US Greenhouse Gas Inventory, which is the last edition to provide a clear split between methane generation and recovery from landfills (US EPA, 2016). The percentage of methane not recovered that breaks down naturally is based on US EPA (2019).

There is also a risk that biomethane production itself results in methane emissions, notably through leaks from biodigesters. There is a large degree of uncertainty surrounding the magnitude of these leaks and they are estimated to range from close to 0% up to 5% of the biomethane that is produced (Liebetrau, 2017). The latest data suggest that the average is probably around 2%. This is broadly similar to the percentage figure for vented and fugitive emissions that are estimated to occur along the natural gas value chain, which are estimated at 1.7% on average globally (IEA, 2018). Methane leaks clearly reduce some of the climate benefit of biomethane production and so should be minimised to the fullest extent possible.

There are a number of policy frameworks for how "avoided" methane emissions should be handled or credited (e.g. the Clean Development Mechanism [UNFCCC, 2018]), but there is currently no globally agreed or universally accepted framework. Different ways of handling these emissions can have a major impact on the apparent cost effectiveness of using biomethane to reduce global GHG emissions. For example, if no credit were to be awarded for avoiding methane emissions, but a credit were to be given for the CO_2 that is avoided from displacing natural gas, then less than 45 Mtoe of biomethane potential would be economic at a $50/tonne GHG price. If avoided methane emissions were to be additionally included, then more than 90 Mtoe would be economic at a $50/tonne GHG price (Figure 13.17).

13.5.3 Energy security

One of the main attractions of biomethane and low-carbon hydrogen is that they can be produced in regions that do not have extensive fossil resources, and can help to reduce the need for natural gas imports. Even as natural gas markets become increasingly liquid, competitive and resilient (see Chapter 4), this is an important consideration.

This is especially the case for countries that see significant import growth of natural gas in our scenarios. Natural gas imports in India, for example, increase from 40 Mtoe in 2018 to 270 Mtoe in 2040 in the Sustainable Development Scenario, while in China natural gas imports increase from 150 Mtoe in 2018 to 315 Mtoe in 2040. There is widespread biomethane potential in both of these countries, a significant proportion of which is available at relatively low cost. In India and China, biomethane consumption in 2040 is around 30 Mtoe and 80 Mtoe respectively in this scenario. If this energy demand had been met instead by natural gas, imports would have been around 10% higher in India and 25% in China.

Low-carbon hydrogen and biomethane also have the potential to help all countries meet peak periods of demand, whether for electricity or heat. Indeed, one of the factors underpinning the growth of interest in hydrogen is its ability to balance the output of high shares of variable renewable electricity generation and to provide storage, including intra-day storage.

It is also possible to envisage longer distance trade of low-carbon gases as a substitute for natural gas imports. In particular, there has been growing interest recently in the potential for importing low-carbon hydrogen, although there are a number of difficulties that would need to be resolved (Box 13.5).

Box 13.5 ▷   Long-distance trade of low-carbon gases

The low energy density of hydrogen means that long-distance transmission is complex. Compression, liquefaction or incorporation of the hydrogen into larger molecules are possible options to overcome this hurdle. If hydrogen is to be shipped overseas, it can be liquefied (by cooling to minus 253 °C, much lower than the minus 162 °C required to liquefy natural gas) or transported as ammonia or in liquid organic hydrogen carriers (LOHCs).[15] Ammonia and LOHCs are generally cheaper to ship, but the costs of hydrogen conversion and reconversion are significant. There are also safety issues arising from their use (for example ammonia and some of the LOHC carrier molecules are toxic when inhaled).

Because of the need to liquefy the hydrogen or incorporate it into other molecules, hydrogen transport is relatively energy intensive, requiring energy equivalent to 15-25% of the energy contained in the hydrogen. It is also expensive: transporting hydrogen by ship is likely to increase costs by around 50-150%, depending on the hydrogen source and transport type.

This means that, in the majority of cases, domestic production of hydrogen is likely to remain cheaper than imported hydrogen. However in places where CO_2 storage is unavailable for geological or political reasons or where there are limited natural gas or renewable resources, low-carbon hydrogen imports could become cheaper than domestic production (IEA, 2019). For example, it is estimated that producing low-carbon hydrogen in Japan in 2030 using electrolysers would cost around $55/MBtu (equivalent to $6.5 per kilogramme of hydrogen or $190 per megawatt-hour [MWh]). If a production facility were to be established in Australia that combined the use of electrolysers, solar and wind generation in a resource favourable region, production would cost just over $30/MBtu. Converting this to ammonia or an LOHC, transporting it by ship and reconverting the molecules back to hydrogen would cost around $15/MBtu.[16] Importing hydrogen therefore would be less expensive than domestic production in Japan.

This $15/MBtu transport cost means that hydrogen imports are generally only cost effective if there is a difference of $30/MWh between low-carbon electricity costs in

---

[15] Making an LOHC involves "loading" a "carrier" molecule with hydrogen, transporting it and then extracting pure hydrogen at its destination. The carrier molecules are not used up when hydrogen is extracted so they can be recycled.

[16] Reconversion of ammonia back to hydrogen costs around $9/MBtu. If ammonia is required at the point of end-use rather than hydrogen, then the overall cost would be lower.

the importing and exporting regions. This is a situation that is likely to be relatively rare. Nevertheless, even if importing hydrogen is not the cheapest option, some countries may wish to consider imports to increase their energy diversity and access to low-carbon energy. In addition, countries with large swings in seasonal energy needs may find it more cost effective to import hydrogen to cover winter peaks than to build local renewable power capacity and long-term storage that will only occasionally be needed.

Because it is essentially indistinguishable from natural gas, biomethane can make use of existing cross-border trade infrastructure. It therefore can be transported by long-distance gas transmission pipelines or transformed into LNG and transported overseas in a similar way to natural gas. A number of European countries have developed biomethane registries to track the injection and extraction of biomethane from natural gas networks as well as a pan-European registry that aims to facilitate international trade.

13.6 Implications for policy makers and industry

Policy makers and the gas industry face important choices on gas infrastructure. In many gas-consuming countries, the scale and flexibility of the energy that can be supplied to houses and industry through existing gas networks is unrivalled. In many developing economies, gas offers a key opportunity to reduce reliance on more polluting fuels. In both cases, investment in gas networks will be needed. Yet there is a balance to be struck between near-term gains and long-term policy objectives, and likewise a need to align expectations and timelines between policy makers and the gas industry. The fact that natural gas is a fossil fuel means it is not easy to appraise how much needs to be invested and over what period of time: investments in the near term need to be compatible with repurposing or decommissioning networks in the long term.

Debate over the future role of gas infrastructure does not lend itself to simple conclusions, even leaving aside the point that countries have different energy systems and needs. Here we outline some possible approaches for consideration by governments and other stakeholders as they weigh the complex opportunities and risks.

- **Introduce low-carbon gas standards and incentives to encourage the use of low-carbon gases.** Significant biomethane and low-carbon hydrogen blending in existing gas grids are unlikely to happen without policy support. This could take the form of specific blending targets (along with priority access for injection), support for research, development and deployment, caps on the emissions intensity of gas consumption and GHG emissions pricing schemes.

- **Assess the level of investment needed in gas infrastructure to maintain energy security while delivering environmental goals.** While the overall volumes of natural gas consumed in many regions may fall in the future, governments will want to be sure that gas is available when it is needed. In some regions, capacity markets for power generation and demand-side response have been developed so that companies receive revenue to maintain generation capacity so that it is available when needed. Similar mechanisms may be needed to maintain gas import and transmission infrastructure.

- **Broaden the regulation of gas networks to take account of the transition to low and zero carbon energy.** The regulation of gas networks today is largely focussed on competition and liberalisation. An expanded regulatory framework may be required to ensure the compatibility of gas networks with long-term decarbonisation goals. For example, where hydrogen or biomethane are injected into gas networks, robust accounting, certification and verification for the levels and types of low-carbon gas injected and subsequently used will be essential if operators are to be paid a premium for supplying low-carbon gas.[17] Incentives should include a method to credit gases that avoid GHG emissions, wherever they may have otherwise occurred.

- **Clarify and harmonise regulations in collaboration with other governments to encourage cross-border trade of low-carbon gas.** Many countries have strict limits in place on the concentrations of hydrogen that are allowed in natural gas streams which could constrain the uptake of low-carbon gases. Changes to these limits could boost the scope for using hydrogen, particularly if new limits can be harmonised across national boundaries.

- **Stimulate hydrogen blending to promote investment and cost reductions in hydrogen infrastructure.** Hydrogen blending would boost investment in hydrogen production infrastructure, helping to lower costs and support hydrogen use in other sectors as it becomes more competitive.

- **Take a long-term strategic view of goals and consider how to manage any distributional problems that may occur.** Pursuing direct electrification of heating or district heating in parallel to hydrogen and biomethane blending could reduce the number of gas grid users at the same time as increasing the cost of gas supply. This could raise the unit price of gas for remaining users. Even within countries, different strategies may be needed for distinct parts of the network rather than a single one-size-fits-all plan.

- **Improve waste management to reduce overall biomethane costs and maximise its cost-effective use.** Biomethane provides a mechanism to extract value out of waste and residues and so can promote a more "circular" approach to handling waste. Separate collection of organic waste would help reduce overall biomethane costs.

- **Encourage the creation of biomethane production jobs in rural locations near feedstocks where this would be cost effective.** Biomethane production relies upon the use of large quantities of organic feedstocks, which can be difficult or expensive to transport over long distances. Biomethane production facilities therefore are usually

---

[17] An example is the system in California whereby some customers can purchase certificates for biomethane blended into the grid despite the gas molecules themselves being untraceable after injection.

best located close to feedstock sources. These are often in rural areas where production facilities could provide local employment and the opportunity for local communities to participate in, and benefit from, energy transitions.

- **Minimise the risk that early investments in one low-carbon option presents new challenges for other low-carbon technologies that may be needed for longer term goals.** The production and use of hydrogen, biomethane and CCGTs equipped with CCUS can be mutually supportive, but there is also scope for them to cut across each other. For example, deploying CCGTs equipped with CCUS could help reduce the cost of CCUS more generally and lower the production cost of hydrogen. Conversely, converting end-user equipment to handle high shares of hydrogen may mean that they are not optimised for biomethane. There are no easy answers here, but a systemic approach will at least help to identify the trade-offs.

- **Establish a shared understanding about the path to minimise the risk of conflicts and misunderstandings between gas market participants with differing commercial interests.** Participants at various stages of the gas value chain can have diverse interests. While a gas producer aims to maximise sales of natural gas, a transmission operator is unlikely to be concerned about whether its revenue comes from transporting natural gas, biomethane or hydrogen, while traders tend to sell both natural gas and electricity. Differences in the commercial interests of market participants could hinder the ability of the industry to act as a whole and could lead to conflicts unless all stakeholders share common expectations about the direction of change.

Natural gas results in fewer GHG and air pollutant emissions than coal and so can help to meet the world's needs during clean energy transitions. The flexibility of gas infrastructure and its ability to deliver and store large amounts of energy mean that investment in gas infrastructure remains important in the Sustainable Development Scenario. But there is a transition to be achieved in this scenario. If it is well managed, gas infrastructure can help deliver low-carbon energy sources in the longer term and serve sectors that are difficult to decarbonise through direct electrification. Low-carbon hydrogen and biomethane have a huge amount of potential in this context. There have been previous waves of enthusiasm for low-carbon gases, but today they meet only a fraction of total energy demand. It will be different in future only if there is adequate support from both industry and policy makers.

Chapter 14

Outlook for offshore wind
A strong tailwind for energy transitions

SUMMARY

- The global offshore wind market grew nearly 30% per year between 2010 and 2018. Europe has fostered the technology's development, led by the United Kingdom, Germany and Denmark. The United Kingdom and Germany currently have the largest offshore wind capacity in operation, while Denmark produced 15% of its electricity from offshore wind in 2018. China added more capacity than any other country in 2018. Offshore wind is set to pick up the pace of growth, with about 150 new projects scheduled to be completed over the next five years around the world.

- Global offshore wind is on track to be a $1 trillion business to 2040, with capacity increasing fifteen-fold. The promising outlook is underpinned by policy support in an increasing number of regions. In Europe, many countries have established policy targets to drive offshore wind growth to 2030. China has also set targets for offshore wind, and projects are now under development in Japan, Korea, Chinese Taipei and Viet Nam. State-level targets and available federal incentives are set to kick-start the offshore market in the United States.

Figure 14.1 ▷ **Growing markets and falling costs for offshore wind in the Stated Policies Scenario**

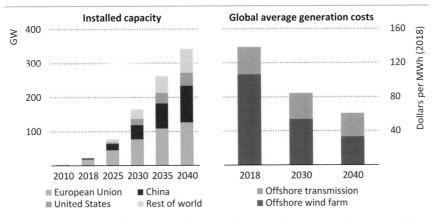

Offshore wind capacity expands fifteen-fold from 2018 as costs fall by nearly 60%

- Offshore wind is set to be competitive with fossil fuels within the next decade, as well as other renewables including solar photovoltaics (PV). By 2030 in the Stated Policies Scenario, the global average levelised cost of electricity (LCOE) from offshore wind declines towards $80 per megawatt-hour (MWh), almost 40% below the current level. This is well below the LCOE of efficient natural gas-fired power in Europe at the time (over $100/MWh). Offshore transmission infrastructure accounts for about one-third of the cost at that point, leaving the LCOE from offshore wind farms alone near to $50/MWh on average, which is in line with cost declines indicated in recent auctions. Policy makers could help to reduce these costs by nearly 30% by moderating risks to enable low-cost financing.

- New offshore wind projects have annual capacity factors of 40-50% or above, matching those of gas-fired power plants in many markets, which significantly exceed those of solar PV. Offshore wind output varies according to the strength of the wind, though its hourly variability is less than that of solar PV. These characteristics make offshore wind the only "variable baseload" technology, and one that can both reduce emissions and support electricity security.

- Innovation in offshore wind is improving performance. Individual turbines of rated capacities of 8-10 megawatts (MW) are now the standard, which is twice the average size in 2016, and rated capacities are expected to keep increasing. Innovations in foundations and grids are also evident with the development of floating turbines, high-voltage transmission lines and proposals for offshore energy "hubs" that could supply electricity or produce hydrogen.

- New geospatial analysis shows that the potential for offshore wind is enormous and widespread, and could drive clean energy transitions around the world. Our analysis examined the technical potential for all marine areas within 300 kilometres (km) of shore, with a granularity of 5-by-5 kilometres square (km^2) based on the latest available global weather data and newest turbine designs. Tapping just the most attractive potential in near shore shallow waters could provide close to 36 000 TWh globally per year, which is nearly equal to global electricity demand in 2040. Including sites in deeper waters and up to 300 km from shore raises offshore wind's technical potential to over 11-times electricity demand in 2040. On a regional basis, total potential easily exceeds aggregate electricity demand in Europe, the United States and Japan, and nearly matches demand in China and India.

- Offshore wind also faces several challenges. Absent defined long-term plans, it is difficult to establish efficient supply chains to deliver low costs. Effective system integration of offshore wind, including transmission infrastructure, also benefits from such clarity. Expanding wind turbine sizes poses logistical challenges such as the need for larger vessels and construction equipment. Marine planning practices, regulations for awarding development rights in the light of competing uses and public acceptance could also slow offshore wind development.

- After a long time on the sidelines, offshore wind is poised to make a major difference as a low-carbon technology. It accounts for more than 7 billion tonnes of carbon dioxide (CO_2) emission savings in the Sustainable Development Scenario to 2040 compared with the current power mix. It also reduces air pollution, contributes to the adequacy of electricity supply, enhances energy security by reducing reliance on imported fuels and makes clean energy transitions more affordable.

14.1 Introduction

Offshore wind is a rapidly maturing renewable energy technology. In 2018, it provided just 0.3% of global electricity supply, nevertheless its future prospects look bright. Individual offshore wind turbines have been enlarging in physical size and rated power capacity, bringing performance gains for offshore wind installations. Further technology improvements are promising steep cost reductions in the near term.

New analysis of offshore wind was performed for this *World Energy Outlook (WEO)*, with an emphasis on three areas: its technical potential; continued evolution of the technology; and its role in energy systems now and in the future. We undertook a detailed geospatial analysis of the regional and global technical potential for offshore wind, considering areas available for development on the basis of the latest wind resource data, and taking account of wind turbine technology advances. This includes an assessment of the future evolution of the relevant technology in close consultation with major industry market leaders, manufacturers, developers and other key stakeholders. The systems analysis considered the role of offshore wind with scenario-based quantifications, using our World Energy Model to provide contextual data on the wider evolution of global energy systems.

This chapter provides a deep dive into offshore wind power. It gives a snapshot of where the market and technology stand today and the outlook to 2040, in light of the policy environment and the evolving competitiveness of offshore wind. It explores the key uncertainties for the outlook, looking at both the potential for faster growth and the main challenges that could slow development. It concludes with a look at the implications of the growth of offshore wind for environmental goals, energy security and affordability.

14.2 Offshore wind power today

14.2.1 Current status

Offshore wind has emerged as one of the most dynamic technologies in the energy system. For the first time in 2010 global capacity additions of offshore wind surpassed 1 gigawatt (GW). In 2018, a total of 4.3 GW of new offshore wind capacity was completed (Figure 14.2). From 3 GW of offshore wind in operation in 2010, installed capacity expanded to 23 GW in 2018. Annual deployment has increased by nearly 30% per year, higher than any other source of electricity except solar photovoltaics (PV). By mid-2019, there were over 5 500 offshore turbines connected to a grid in 17 countries. Policy support has been fundamental to this expansion, including through technology-specific capacity tenders, progress on including offshore wind in marine planning, financial support and regulatory efforts to support grid development.

The growth of the offshore wind industry has been fostered in European countries bordering the North Seas, where high quality wind resources and relatively shallow water have provided exceptionally good conditions in which to develop offshore wind technologies and bring them to market. Stable policies supported nearly 17 GW of offshore wind capacity additions in Europe between 2010 and 2018. The United Kingdom, Germany, Belgium, Netherlands and Denmark together added 2.7 GW of capacity in 2018 alone.

China has recently taken strides forward on offshore wind and now stands among the market leaders. In 2018, China added 1.6 GW of offshore wind capacity, the most of any country. This rapid growth has been driven by the government's 13th Five-Year Plan, which called for 5 GW of offshore wind capacity to be completed by 2020, and for the establishment of supply chains to support further expansion thereafter.

Figure 14.2 ▷ Annual offshore wind capacity additions by region, 2010-2018

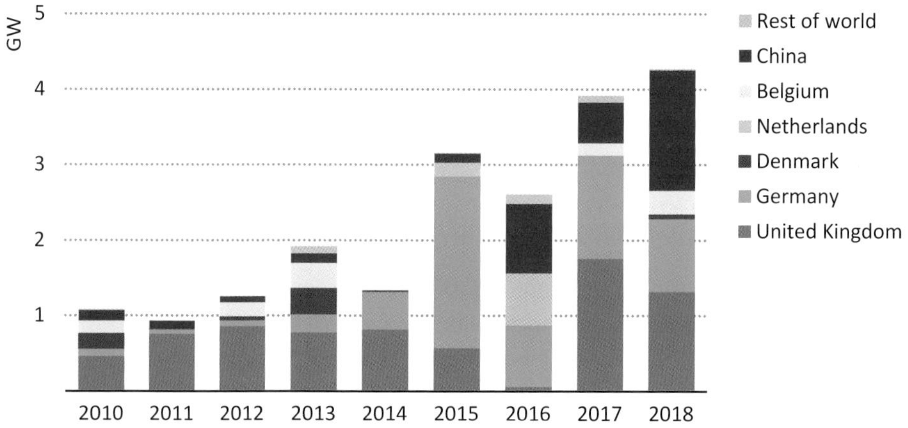

Deployment of offshore wind has increased by nearly 30% per year since 2010, second only to solar PV, as the technology and industry have matured

Note: Figure reflects date of connection to grid and power output, which may be before final commissioning.

Offshore wind is set to gain a foothold in new markets in the next five years. The current pipeline includes about 150 new offshore wind projects spread across 19 countries. Over 100 projects are scheduled to be completed by 2021, pointing to further acceleration in the rate of annual capacity additions. In the United States, there are 25 GW of offshore wind projects in the longer term pipeline (US DOE, 2019). There are also large-scale projects in Australia, Chinese Taipei, India, Japan, Korea, New Zealand, Turkey and Viet Nam.

In 2018, more than 80% of global installed offshore wind capacity was located in Europe. Around 8 GW, one-third of the total, was in the United Kingdom and 6.5 GW in Germany, with Denmark, Netherlands and Belgium providing a further 3.6 GW between them. Even as a relative newcomer, China already has 3.6 GW of offshore wind capacity (Figure 14.3).

Offshore wind power accounted for just 0.3% of global electricity supply in 2018, but played a larger role in the leading countries. It provided 15% of electricity generation in Denmark in 2018, where onshore and offshore wind together accounted for almost 50% of electricity generation. Offshore wind provided 8% of electricity generation in the United Kingdom, more than twice as much as from solar PV generation, and for 3-5% of electricity generation in Belgium, Netherlands and Germany. Despite recent growth, output from China's offshore wind fleet in 2018 represented just 0.1% of its overall power output.

Figure 14.3 ▷ **Offshore wind installed capacity and share of electricity supply by country, 2018**

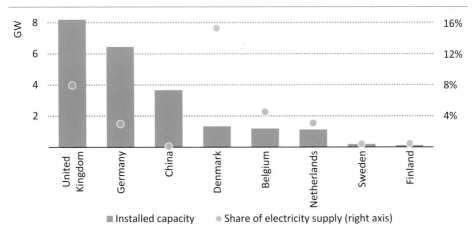

Most leading countries in offshore wind are in Europe, led by the United Kingdom, though China has quickly joined the top-three and is gaining momentum

14.2.2 Market size and key players

Today, offshore wind is a multi-billion dollar business, with developed supply chains in leading markets that span development, project construction and installation, operation and maintenance, and decommissioning activities. The offshore wind industry attracted a growing share of investment in wind and renewable energy in recent years, nearing $20 billion in 2018, up from less than $8 billion in 2010. Investment in offshore wind in 2018 accounted for nearly one-quarter of global investment in the wind sector and 6% of all investment in renewable energy. Investment was particularly pronounced in the European Union, where offshore wind accounted for about half of total investment in wind power in 2018 and one-quarter of total investment in renewable energy (IEA, 2019a).

Investment in offshore wind projects is mainly by large utilities and investment funds because the projects have relatively high upfront capital costs: a 250 megawatt (MW) project costs around $1 billion. Solar PV and onshore wind by contrast have lower upfront costs and present fewer barriers to entry for smaller players.

European companies develop and own the majority of offshore wind assets (Table 14.1). Denmark-based Ørsted owns the largest share, and is actively expanding into other markets in the United States and Asia. Germany-based RWE consolidated its share of the offshore wind market after acquiring E.ON and Innogy renewable energy assets in the North Sea and Baltic Sea, and is now the second-largest offshore wind operator in the world.

Chinese companies account for a growing share of the market. Two Chinese state-owned enterprises rank among the top-ten developers in the industry, with around 7% market share in terms of ownership. China Longyuan Power Group ranks as the largest producer of wind power across Asia, while the China Three Gorges Corporation (CTG) – previously known for its hydroelectric projects – is one of the world's largest energy companies and has become actively involved in the offshore wind industry. CTG set itself the goal of becoming the leader in offshore wind power development in China, and owns five wind installations with combined capacity of 1.2 GW, although some is under construction.

Table 14.1 ▷ Leading market players in the offshore wind industry, 2018

Organisation	Main activities	Assets (GW)			Market share	Headquarters	Ownership
		In operation	Under construction	In development			
Ørsted	DOO	2.97	2.79	5.23	12.86%	Denmark	Private
RWE	DOO	2.41	0.51	1.83	10.44%	Germany	Private
China Longyuan	DOO	1.23	0.40	1.00	5.34%	China	Public
Vattenfall	DOO	0.88	1.01	4.92	3.82%	Sweden	Public
Macquarie Capital	Investor	0.87	0.07	0.10	3.78%	Australia	Private
Northland Power	DOO	0.64	0.27	0.63	2.78%	Canada	Public
Global Infrastructure Partners	Investor	0.63	0.61	-	2.73%	United States	Private
Iberdrola	DOO	0.55	0.97	0.81	2.36%	Spain	Private
Equinor	DOO	0.48	-	2.17	2.10%	Norway	Public
Siemens Financial Services	Investor	0.46	-	-	1.98%	Germany	Private
Public Pension, Denmark	Investor	0.45	-	-	1.97%	Denmark	Public
Électricité de France	DOO	0.43	-	1.67	1.85%	France	Public
Stadtwerke München	Investor	0.41	-	-	1.79%	Germany	Public
China Three Gorges	DOO	0.40	0.88	6.87	1.74%	China	Public
Scottish and Southern Energy	DOO	0.34	0.24	0.52	1.49%	United Kingdom	Public

Notes: DOO = developer, owner and operator. Market shares are adjusted to reflect each company's equity stake across all of its projects.
Source: IEA analysis based on BNEF (2019).

Manufacturers of offshore wind turbines are mostly based in Europe, and the market is concentrated among a small number of companies (Table 14.2). Spanish-headquartered Siemens Gamesa and MHI Vestas, a joint venture between Vestas and Mitsubishi Heavy Industries, dominated the offshore wind industry, accounting for over two-thirds of the offshore wind capacity installed in 2018. Together, these two manufacturers account for over 80% of all offshore capacity commissioned from 1995 through the end of 2018. The share of turbines produced by Chinese manufacturers is expanding with its focus on the market in Asia, accounting for close to 30% of offshore wind capacity additions in 2018.

Table 14.2 ▷ Leading manufacturers of offshore wind turbines, 2018

Rank	Company	Offshore wind market share, 2018	Offshore wind market share, 1995-2018	Offshore wind capacity sold, 1995-2018 (MW)
1	Siemens Gamesa	41%	63%	13 881
2	MHI Vestas	30%	18%	3 882
3	Envision	15%	4%	804
4	Goldwind	8%	3%	574
5	Ming Yang	2%	1%	113
6	Sewind	2%	1%	306
7	GE Renewable Energy	0.4%	1%	177
8	Taiyuan	0.2%	0%	10
9	Senvion	-	6%	1 253
10	Bard	-	2%	405

Source: IEA analysis based on BNEF (2019).

Another important component in the value chain is the construction and servicing of offshore wind turbines. Between 2010 and 2018 nearly $4 billion per year was invested in the construction of offshore wind installations across Europe and China, while over $1 billion was spent annually on operations and maintenance. As the offshore wind sector expands, the opportunities to exploit synergies with oil and gas contractors and servicing companies will increase (see section 14.6.2).

14.2.3 Offshore wind technology and performance

Offshore wind technology has made impressive advances since the first turbines were installed near the shore in Denmark in 1991. Since then, equipment suppliers have focused research and development spending on developing bigger and better performing offshore wind turbines. The technology has grown dramatically in physical size and rated power output.

Technology innovation has led to an increase in turbine size in terms of tip height and swept area, and this has raised their maximum output. The tip height of commercially available turbines increased from just over 100 metres (m) in 2010 (3 MW turbine) to more than 200 m in 2016 (8 MW turbine) while the swept area increased by 230%. The larger swept area allows for more wind to be captured per turbine. A 12 MW turbine now under development is expected to reach 260 m, or 80% of the height of the Eiffel Tower (Figure 14.4). The industry is targeting even larger 15-20 MW turbines for 2030. This increase in turbine size and rating has put upward pressure on capital costs as larger turbines pose construction challenges and require larger foundations, but it has also reduced operation and maintenance costs, ultimately leading to lower levelised costs of electricity.

Figure 14.4 ▷ Evolution of the largest commercially available wind turbines

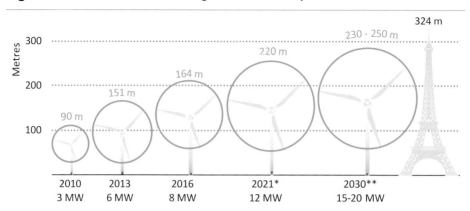

Technology advances enabled offshore wind turbines to become much bigger in just a few years and are supporting ongoing increases in scale

* Announced expected year of commercial deployments. ** Further technology improvements through to 2030 could see bigger turbines sizes of 15-20 MW.

Notes: Illustration is drawn to scale. Figures in blue indicate the diameter of the swept area.

The average turbine size used in offshore wind farms increased from 3 MW in 2010 to 5.5 MW for projects completed in 2018 (IRENA, 2019). In the same period, annual capacity factors for new projects increased from 38% to 43%.[1] New turbines of 10-12 MW promise to achieve capacity factors well over 50% (before wake losses).[2] Compared to smaller units, a bigger turbine can achieve a capacity factor improvement of two to seven percentage points given the same site conditions. However, not all projects will necessarily see a significant increase in performance as a result of using larger turbines. Capacity factors remain dependent on the quality of the wind speeds of individual sites, which may not be suitable for larger turbines. There may also be a trade-off for developers between incremental performance gains and higher costs of larger turbines.

Offshore wind provides higher capacity factors than other variable renewables. In 2018, the average global capacity factor for offshore wind turbines was 33% compared with 25% for onshore wind turbines and 14% for solar PV. Looking forward, new offshore wind projects are expected to have capacity factors of over 40% in moderate wind conditions and over 50% in areas with high quality wind resources. Other variable renewable energy technologies are also likely to see improvements, but not to match the expected capacity factors of new offshore wind projects. For example, technology improvements are raising expected capacity factors for onshore wind to between 30% and 40% in most regions (Figure 14.5).

---

[1] Capacity factor describes the average output over the year relative to the maximum rated power capacity.

[2] Wake loss refers to the effect on the space behind a turbine that is marked by decreased wind speed on a downstream wind turbine due to the fact that the turbine itself used the energy in turning the blades.

Figure 14.5 ▷ Indicative annual capacity factors by technology and region

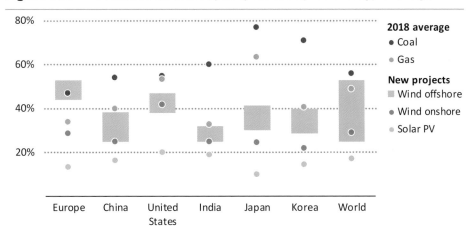

Offshore wind offers similar capacity factors to efficient gas-fired power plants in several regions, with levels well above those for other variable renewables

Simulations indicate that new offshore wind projects will produce as much electricity per unit of capacity each year as some forms of dispatchable generation, and will have capacity factors on a level with or higher than those of conventional gas-fired power plants in several regions, including Europe, China, India and Korea.[3] In Europe, new offshore wind projects are set to have an average capacity factor of over 45%, exceeding the 2018 average for Europe's coal-fired power plants. This development has implications for how offshore wind is integrated into electricity supply and its impact on grid infrastructure, as well as for its suitability for producing hydrogen, an application that could greatly expand the offshore wind market (see section on 14.4.3).

The electricity produced from offshore wind projects depends on the available wind resources at a given time, leading to variable output. High average capacity factors and variable production put offshore wind power into a category of their own as a "variable baseload" technology. Over the course of a year, offshore wind exhibits substantial variability from week-to-week. It tends to produce more electricity during the winter and less during the summer in Europe, United States and China, as indicated by the simulated output for new offshore wind projects (Figure 14.6). In India, the simulations indicate that the monsoon season from June to September would see higher output from offshore wind projects compared with other parts of the year. In all cases, the seasonal profile of offshore wind is complementary to that of solar PV, which tends to produce more electricity in the summer and less in the winter in Europe, United States and China, and less in the monsoon season in India than at other times of the year.

---

[3] Gas-fired capacity and other dispatchable power plants can readily adjust their output up to their maximum rated capacity according to market conditions and system needs.

Figure 14.6 ▷ Simulated average weekly capacity factors for new offshore wind and solar PV projects by region

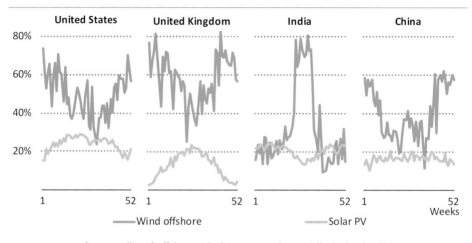

Seasonality of offshore wind can complement that of solar PV

Note: Based on weather data for 2018, 2013 and 2008.
Source: IEA analysis based on Renewables.ninja.[4]

Hour-to-hour variations in offshore wind power output tend to be comparable with those for onshore wind and lower than those for solar PV. Offshore wind typically fluctuates within a narrower band, up to 20% from hour-to-hour, than is the case for solar PV, up to 40% from hour-to-hour (Figure 14.7). Lower hourly variability offers a potential advantage for offshore wind over other variable renewables, though its impact on system flexibility needs depends on a number of factors, including the correlation of its output with electricity demand and the output from other variable renewables. At the system level, regional diversity can also moderate the impact of the variability at the project level.

Offshore wind installations are also moving further from shore and into deeper water where better quality wind resources are available. Most projects commissioned to date have been within 50 km of shore. However, several large projects in the pipeline are 100 km or more from shore. This is becoming more common as developers look to install turbines in deeper water with improved construction techniques that reflect learning from earlier projects and from the offshore oil and gas industry. The use of relatively low-cost monopile foundations has been the offshore wind industry standard for the majority of the projects installed in water depths of less than 50 m. Projects located in slightly deeper depths are also seeking to find ways to use monopile foundations rather than to have to adopt higher cost jacket and floating foundations. In order to deploy offshore wind in yet deeper waters, the industry has been deploying pilots and pre-commercial scale projects to improve the designs of floating foundations and to establish their costs (Box 14.1).

---

[4] Renewables.ninja is a publicly available tool that is validated and calibrated against real-world output in 70 countries (Staffell and Pfenninger, 2016), accessible at www.renewables.ninja.

Figure 14.7 ▷ Range of simulated hour-to-hour variations in output for new projects by technology, 2018

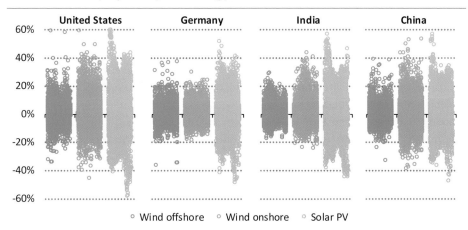

Offshore wind has similar hourly variability as onshore wind, but far less than solar PV

Note: Based on weather data for 2018.
Source: IEA analysis based on Renewables.ninja.

Box 14.1 ▷ Floating foundations – the next frontier for offshore wind

Floating offshore wind is a fast maturing technology that could harness untapped wind resources located in regions with water depths exceeding 50-60 m where traditional fixed-bottom offshore wind installations are not economically attractive. The industry is adapting various floating foundation technologies that have already been proven in the oil and gas sector, though design modifications are still required. Broadly, there are three main concepts for floating foundations: spar-buoy, semi-submersible and tension leg platforms. Other variants exist, including multiple turbines on one platform and hybrid wind/wave floating devices (IRENA, 2016; Carbon Trust, 2015).

In recent years, there have been significant developments in floating offshore wind projects including the commissioning of the world's first multi-unit installation in 2017 (30 MW Hywind in Scotland). A number of smaller demonstration projects were installed in 2018, including Floatgen in France (2 MW) and Hibiki in Japan (3 MW). In addition, at least ten new pre-commercial scale projects in Europe are in the pipeline (WindEurope, 2017), including the 30 MW WindFloat Atlantic in Portugal and the 48 MW Kincardine in Scotland. Equinor also received approval in 2019 to build a 200 MW project off the coast of the Canary Islands which is expected to be the world's largest when it begins operations in the mid-2020s. These pre-commercial and commercial-scale projects should help to establish more firmly the costs of floating foundations. Similar undertakings are under assessment for offshore substations. Both will be critical to the success and deployment of offshore wind projects in deep waters.

14.2.4 Offshore wind costs for projects commissioned in 2018

For offshore wind projects completed in 2018, the global average capital cost was $4 353 per kilowatt (kW), and the average expected capacity factor was 43% (IRENA, 2019). Applying an 8% weighted average cost of capital (WACC) in advanced economies and 7% WACC in developing economies, the global average LCOE for offshore wind in 2018 fell below $140 per megawatt-hour (MWh) (Figure 14.8).[5] Recent auction results point to rapidly falling costs for offshore wind (Box 14.2). Nearly half of the LCOE for completed projects is directly attributable to the capital investment needed for offshore wind projects, including the costs of the turbine, foundation, internal cabling, substation and offshore transmission assets. The remaining half is attributable to the project financing cost, reflecting the high capital-intensity of offshore wind projects.

Figure 14.8 ▷ Offshore wind indicative shares of capital costs by component and levelised cost of electricity for projects completed in 2018

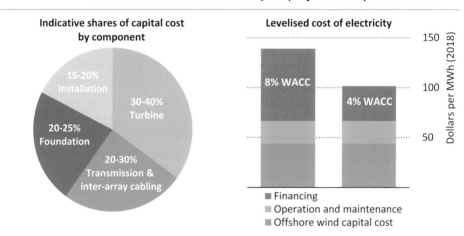

Offshore wind generation costs are heavily influenced by the cost of capital and were about $100/MWh for projects completed in 2018 based on low financing costs

Notes: WACC = weighted average cost of capital; Transmission includes offshore substations.
Sources: IEA analysis based on IRENA (2019), IJGlobal (2019) and BNEF (2019).

Improved financing terms would significantly reduce the LCOE of offshore wind, as it would the cost of other capital-intensive technologies. For example, applying a 4% WACC to 2018 costs and performance parameters yields an offshore wind LCOE of about $100/MWh, which is 30% less than the LCOE derived from the standard WACC. The sensitivity of the calculated LCOE to the cost of capital is also shown for projected costs in section 14.3.3.

---

[5] Standard assumptions are applied to all power generation technologies in the *World Energy Outlook* to represent the required return in the face of market conditions and investment risk. Operation and maintenance (O&M) costs below $100/kW per year and a 25-year economic lifespan are also assumed.

Offshore wind turbines constitute around 30-40% of total upfront capital costs. In an effort to improve the economics of this critical component, projects are getting larger, increasing the numbers of wind turbines to be delivered and enabling economies of scale, while at the same time equipment suppliers are using lighter and more resilient materials such as glass and carbon fibres for blade and nacelles[6] manufacturing, as well as improving aerodynamics. Increasing the size of turbines is also having the effect of reducing the number of foundation positions and inter-array cabling, which is reducing installation and operation and maintenance costs.

Offshore transmission, array cabling (internal wiring at a wind farm) and the offshore substation make up some 20-30% of total upfront capital costs. The costs for offshore transmission assets, in particular, are closely tied to the regional regulations for connecting the project to the onshore grid (see section 14.5.3).

Foundations account for nearly a quarter of total project costs. Monopile structures are currently the preferred technology, underpinning around 80% of deployment (WindEurope, 2019). It is becoming possible to use this type of foundation in increasingly deep water – up to 55-60 m in some cases – thus reducing the need for more expensive jacket foundations, which are suited to deeper water and are the second most used technology globally.

Box 14.2 ▷ Rapid cost reductions are on the horizon for offshore wind

Recent strike prices in Europe for offshore wind indicate significant cost reductions on the horizon (Figure 14.9). Some strike prices are at the level of wholesale electricity prices, though these are often underpinned by power purchase agreements. Exposure to wholesale prices increases market risks that can lead to higher financing costs, but competition can help drive down project costs.

The allocation of transmission costs between developer and system operator is a key factor in auction awards. Strike prices where the developer does not need to bear the cost of the grid connection to shore have generally been lower. For example, the Dunkirk project in France was awarded to EDF at $50/MWh (EUR 44/MWh) but excluded the offshore transmission asset. Some recent auctions which exclude the cost of the offshore transmission assets have not involved subsidies, including Hollandse Kust Zuid I and II and Hollandse Kust Zuid III and IV in the Netherlands, and He Dreiht, Borkum Riffgrund West 1 and 2, and OWP West in Germany.

Including the offshore transmission assets within the scope of the auctions does, however, bring competitive pressure to reduce the cost of the transmission asset and therefore could lead to lower overall prices in the future. This has been reflected in the recently awarded auctions in United Kingdom's Dogger Bank wind farm, where Scottish and Southern Energy and Equinor have joint efforts to develop 3.6 GW of total capacity. The inclusion of transmission assets under the scope of developers, but also moving

---

[6] Nacelle is the cover that houses all generating components of a wind turbine.

into deeper waters and consequently tapping into better resources, have been major drivers in bringing strike prices down to almost $50/MWh, on average. Ultimately, the responsibility to deliver these projects rest in the hands of offshore wind developers, and recent market signals indicate growing confidence from investors. This sets the stage for low-cost financing and for an increasing pipeline of future projects.

Figure 14.9 ▷ Historical LCOE of offshore wind and strike prices in recent auctions in Europe

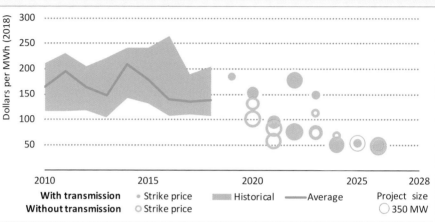

With transmission			Without transmission		
Project	Strike price ($/MWh)	Expected COD	Project	Strike price ($/MWh)	Expected COD
United Kingdom			Germany		
Beatrice	185	2019	Baltic Eagle	74	2023
East Anglia 1	152	2020	Gode wind 3	68	2024
Triton Knoll	95	2021	Gode wind 4	112	2023
Moray East	73	2022	Netherlands		
Hornsea 1	178	2022	Borssele I/II	83	2020
Hornsea 2	76	2022	Borssele III/IV	62	2021
Neart na Gaoithe	148	2023	Denmark		
Cr. Beck A Dogger Bank	51	2024	Horns Rev 3	118	2020
Cr. Beck B Dogger Bank	54	2025	Kriegers Flak	57	2021
Dogger Bank Teeside A	54	2025	Vesterhav Nord/Syd	73	2023
Seagreen	54	2025			
Muaitheabhal	51	2025	France		
Sofia	47	2026	Dunkirk	50	2026

Recent auctions in Europe set the stage for a fall in costs for new projects as the industry moves to deploy higher capacity turbines

Notes: COD = commercial operation date; LCOE = levelised cost of electricity. Historical values correspond to LCOEs including transmission. Strike prices are included for projects of over 100 MW.

Source: IEA analysis based on IRENA (2019).

14.3 Offshore wind outlook to 2040

14.3.1 Global outlook

The global offshore wind market is set to expand significantly over the next two decades, growing by 13% per year in the Stated Policies Scenario and faster still in the Sustainable Development Scenario. Bolstered by policy targets and falling technology costs in the Stated Policies Scenario, global offshore wind capacity is set to increase fifteen-fold from 2018 to 2040 (Figure 14.10). Annual offshore wind capacity additions are set to double over the next five years and increase almost fivefold by 2030 to over 20 GW per year. Beyond 2030 the cost competitiveness of offshore wind helps to maintain the pace of growth. Additional supportive policy frameworks, including designated auction schemes, drive further growth in the Sustainable Development Scenario. In this scenario, global offshore wind capacity rises to about 560 GW by 2040, a 65% increase over the Stated Policies Scenario, as part of accelerated efforts to decarbonise electricity supply. Annual capacity additions approach 30 GW by 2030 and reach 40 GW in 2040.

Figure 14.10 ▷ Projected global offshore wind capacity and share of electricity supply by scenario

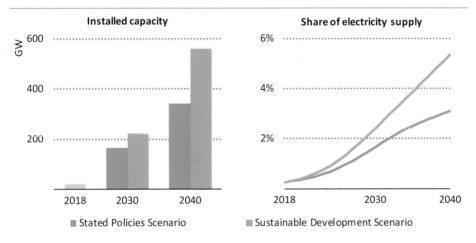

Global offshore wind installed capacity increases by fifteen-fold in the Stated Policies Scenario, raising its share of electricity supply to 3% in 2040

As the global offshore wind market expands, so does its role in supplying electricity around the world. In the Stated Policies Scenario, offshore wind accounts for 3% of global electricity supply by 2040. In the Sustainable Development Scenario, its share of global electricity supply rises to 5%.

In the Stated Policies Scenario, cumulative investment in offshore wind is about $840 billion from 2019 to 2040. Annual investment in the offshore wind power sector averages $38 billion, double the level in 2018 (Figure 14.11). This level of investment means that offshore wind accounts for 10% of investment in renewables-based power plants globally over the next two decades, roughly the same share as the total of bioenergy, concentrating solar power, geothermal and marine energy. Across the power sector more broadly, offshore wind captures almost 8% of all power plant investment, rivalling the share that goes to natural gas or to nuclear power over the period.

Figure 14.11 ▷ Cumulative capital spending on offshore wind, gas- and coal-fired capacity worldwide by scenario, 2019-2040

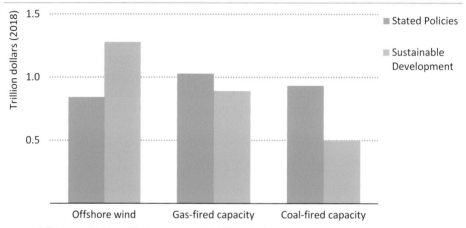

Offshore wind is set to become a $1 trillion industry over the next two decades, matching capital spending on gas- and coal-fired capacity to 2040

In the Sustainable Development Scenario, cumulative spending in the offshore wind sector rises by half to $1.3 trillion between 2019 and 2040 relative to the Stated Policies Scenario. This equates to roughly $60 billion of annual investment per year on average. This means more investment goes to offshore wind than to coal, gas or new nuclear power to 2040. Higher aggregate spending on other renewables during this period means that offshore wind attracts a similar share of total investment in renewables as in the Stated Policies Scenario (about 10%).

14.3.2 Regional outlook

In the Stated Policies Scenario, offshore wind growth is concentrated in six regions, reflecting policy ambitions, available wind resources and the improving economics of offshore wind. Europe and China lead the offshore wind market with over 70% by 2040 of installed capacity, while there is significant expansion in the United States, Korea, India and Japan, which between them capture about one-quarter of the global market (Figure 14.12).

Figure 14.12 ▷ Installed capacity of offshore wind by region and scenario

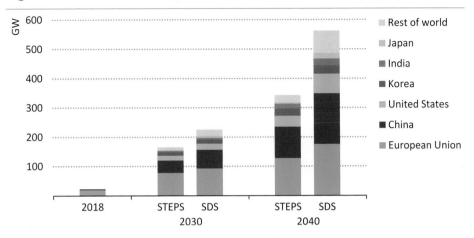

European Union and China account for 70% of the global offshore wind market to 2040, but a number of countries enter the market and increase their capacity

As of mid-2019, many regions have adopted policy targets for offshore wind to 2030 (Table 14.3). The European Union has the strongest ambitions to 2030, with targets in individual member states totalling 65-85 GW by 2030. China's Five-Year plans are encouraging provinces to expand their construction capacities for offshore wind to 2020, while state-level targets set the course for rapid growth in the United States. India, Korea and Chinese Taipei also have ambitious targets, while other countries, including Japan and Canada, are laying the groundwork for future offshore wind development.

Table 14.3 ▷ Policies targeting at least 10 GW of offshore wind by 2030

Region/country	Policy target
European Union	65-85 GW by 2030
China	5 GW by 2020 (10 GW construction capacity)
United States	22 GW by 2030
India	5 GW by 2022 and 30 GW by 2030
Chinese Taipei	5.5 GW by 2025 and 10 GW by 2030
Korea	12 GW by 2030

European Union continues to be the leader in offshore wind

Offshore wind is set for robust growth in the European Union (EU) over the next two decades. In the Stated Policies Scenario, the European Union accounts for nearly 40% of the global offshore market by 2040, increasing installed capacity to almost 130 GW by 2040. Annual investment in the offshore power sector increases from $11 billion in 2018 to an average of $17 billion per year over the outlook period. Offshore wind also plays an important role in meeting new electricity demand: output from new offshore wind projects far outpaces overall electricity demand growth to 2040, and offshore wind provides more than one-in-six kilowatt-hours generated within the European Union in 2040 (Figure 14.13).

Figure 14.13 ▷ Outlook for offshore wind in the European Union, 2018-2040

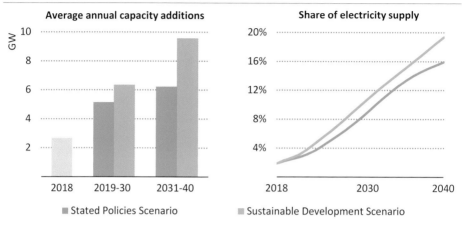

In the European Union, offshore wind is set to play a central role and has the potential to become the largest source of electricity supply, matching onshore wind

In the Sustainable Development Scenario, the European Union's offshore installed capacity increases by almost 40% relative to the Stated Policies Scenario, reaching 175 GW by 2040. Cumulative investment in offshore wind reaches $500 billion over the projection period, or $23 billion per year. Together with other renewables, nuclear power and carbon capture, utilisation and storage, offshore wind lifts the low-carbon share of generation above 90% in the European Union in 2040. Offshore wind accounts for one-fifth of electricity supply in 2040 and rivals onshore wind as the largest source of electricity in the European Union, exceeding nuclear power.

Member states have made various commitments that underpin the future development of offshore wind in the European Union (Table 14.4).[7] These targets are under regular review as offshore wind technology matures. The United Kingdom is set to be the leading EU offshore wind market to 2030, followed by Germany and the Netherlands. France, Poland and Ireland are also committed to develop offshore wind capacity.

The offshore wind industry is looking to develop 180 GW of offshore wind in the North Sea by 2050 (NSWPH, 2019). The 2016 North Seas Energy Cooperation agreement brought together a number of European countries with the aim of improving conditions for the development of offshore wind in the North Sea. It prioritises four main work areas:

---

[7] This includes countries such as Estonia, Belgium, Denmark, United Kingdom, Germany, Netherlands, France, Finland, Greece, Ireland, Latvia, and Poland, which have made technology-specific targets or permitting timelines for offshore wind technology.

maritime spatial planning; development and regulation of offshore grids; finance for offshore wind projects; and standards, technical rules and regulations for offshore wind. Alternative pathways to achieving a net-zero emissions economy may include significantly higher electricity demand due to demand for hydrogen or other fuels (see section 14.4.3) and expanded opportunities for offshore wind development, as seen in scenarios developed in support of the European Union's long-term strategy that include 240-450 GW of offshore wind in 2050 (European Commission, 2018).

Table 14.4 ▷ Policy targets for offshore wind in the European Union

Country	Policy	Capacity target	Year set
United Kingdom	UK Offshore Sector Deal	Up to 30 GW by 2030	2019
Germany	The Renewable Energies Act	15-20 GW by 2030	2017
Netherlands	The Offshore Wind Energy Roadmap	11.5 GW by 2030	2017
Denmark	Energy Agreement	5.3 GW by 2030	2019
Poland	Draft National Energy and Climate Plan	Up to 5 GW by 2030	2018
France	Multi-Annual Energy Plan	4.7-5.2 GW by 2028	2019
Belgium	Draft National Energy and Climate Plan	4 GW by 2030	2019
Ireland	Climate Action Plan 2019	3.5 GW by 2030	2019
Italy	Draft National Energy and Climate Plan	0.9 GW by 2030	2018

China moves strongly ahead with offshore wind

China has already undertaken a number of offshore wind projects and is set to play a central role in the long-term growth of offshore wind, alongside the European Union. In the Stated Policies Scenario, China has the largest offshore wind fleet of any country around 2025, surpassing the United Kingdom. Offshore wind capacity additions steadily increase throughout the period of the outlook, averaging more than 6 GW per year after 2030 (Figure 14.14). Average annual investment in the offshore wind likewise rises from $6 billion in 2018 to $9 billion per year from 2019 to 2040, accounting for nearly one-quarter of global investment in offshore wind over the period.

In the Sustainable Development Scenario, China reaches almost 175 GW of installed capacity by 2040, matching the size of the offshore wind fleet in the European Union. Annual investment increases to $13 billion on average between 2019 and 2040, meaning that 8% of total power plant investment in China is allocated to offshore wind.

Improving economics and firm policy support guide China's ambitions for the offshore wind sector. The country's 13th Five-Year Plan calls for 5 GW of capacity to be installed by 2020 and for 10 GW more to be in the construction pipeline, split among the coastal provinces so as to help develop local supply chains (Table 14.5). The National Development and Reform Commission (NDRC) recently adopted a competitive bidding scheme for offshore wind capacity in an effort to drive down costs (NDRC, 2019). Around 2030, offshore wind reaches cost parity with coal-fired generation in LCOE terms, a critical milestone that supports continued long-term growth in China. In the Sustainable Development Scenario, CO_2 pricing helps offshore wind reach cost parity with coal-fired power plants in the mid-2020s, and this leads to accelerating average capacity additions of over 10 GW per year after 2030.

Figure 14.14 ▷ Outlook for offshore wind in China, 2018-2040

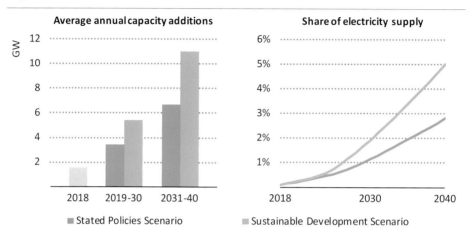

China builds as much offshore wind capacity as the European Union over the next two decades, with its growth complementing other low-carbon sources

Table 14.5 ▷ Offshore wind targets by province in China's 13th Five-Year Plan

Province	Grid-connected by 2020 (MW)	Total construction pipeline by 2020 (MW)
Tianjin	100	200
Liaoning	-	100
Hebei	-	500
Jiangsu	3 000	4 500
Zhejiang	300	1 000
Shanghai	300	400
Fujian	900	2000
Guangdong	300	1 000

Source: China National Development and Reform Commission (2016).

United States

Offshore wind gains significant ground in the United States over the next two decades. In the Stated Policies Scenario, the United States adds nearly 40 GW of offshore wind capacity by 2040, with related investment totalling $100 billion over the period. In the Sustainable Development Scenario, the United States increases its offshore wind capacity by 80%, reaching nearly 70 GW by 2040. This pace of growth calls for average investment of over $7 billion per year. Offshore wind provides over 5% of the US electricity supply in 2040, compared with the 3% level reached in the Stated Policies Scenario (Figure 14.15).

Figure 14.15 ▷ Outlook for offshore wind in the United States, 2018-2040

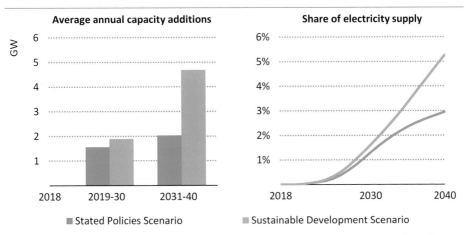

Offshore wind is set to gain traction in the United States after more than a decade of false starts, reaching 3% of generation in the Stated Policies Scenario

Note: No offshore wind capacity was completed in 2018 in the United States.

Table 14.6 ▷ Offshore wind targets and support policies in the United States

State/Jurisdiction	Policy	Target, awards or incentives	Year
New York	Climate Leadership and Protection Act	9 GW installed by 2035	2019
Massachusetts	Act to Advance Clean Energy, Act to Promote Energy Diversity	3.2 GW installed by 2035	2016
New Jersey	Offshore Wind Economic Development Act	3.5 GW installed by 2030	2018
Connecticut	Act Concerning the Procurement of Energy Derived from Offshore Wind	2 GW installed by 2030	2019
Virginia	Virginia Energy Plan	2 GW installed by 2028	2018
Maryland	Clean Energy Jobs Act	1.2 GW installed by 2030	2019
Rhode Island	'20 Clean Energy Goal	1 GW by 2025	2019
US Federal	Production tax credit (PTC)	$0.009-0.023/kWh*	1992
US Federal	Investment tax credit (ITC)	12-18%**	2002

* The exact value of the PTC for wind facilities depends on the year of construction. ** The ITC applies to projects starting construction before 2020.

A combination of federal incentives and state-level targets support the strong growth of offshore wind in the United States. The US Bureau of Ocean and Energy Management has tendered over 15 licences for offshore wind development along the east coast that are capable of supporting 21 GW of offshore capacity.[8] There have also been proposals in the US Congress to extend the tax credits available to offshore wind developers. Individual states are also setting offshore wind capacity targets totalling over 20 GW by 2035 (Table 14.6). In a nod to the important role of offshore wind in achieving its ambitious decarbonisation targets, New York revised its offshore wind target upwards from 2.4 GW to 9 GW by 2030. The technology could also play an important role in California's decarbonisation plan, but the state has not specified a capacity target for offshore wind.

Emerging markets

Offshore wind is set to establish itself in a number of new markets in the Asia Pacific region, outside of China. In the Stated Policies Scenario, countries including Korea, India and Japan account for close to 60 GW of offshore wind capacity by 2040. They account for one-sixth of cumulative global investment in offshore wind, equating to over $6 billion in average annual investment over the outlook period. In the Sustainable Development Scenario, markets in the Asia Pacific region excluding China have average annual investment of close to $10 billion over the period to 2040.

Korea becomes the largest market for offshore wind outside the European Union, China and the United States, reaching 25 GW of capacity by 2040 in the Stated Policies Scenario. Ambitious policy targets in Korea set under its Renewable Energy Plan 3020 mean that offshore wind provides more than 10% of the country's electricity by 2040, complementing 16% of electricity generation from solar PV and onshore wind. In terms of generation costs, the LCOE of offshore wind reaches parity with onshore wind and solar PV in the 2030s.

Offshore wind development also makes notable progress in India, but faces stiff competition from low-cost solar PV and onshore wind. India has set ambitious targets for offshore wind by 2030 and is expected to tender its first 1 GW wind farm in late 2019. Installed capacity reaches 16 GW by 2040 in the Stated Policies Scenario, generating more electricity than solar PV in India does today. This installed capacity increases by an additional 40% in the Sustainable Development Scenario to 23 GW by 2040, requiring approximately $2 billion in average annual investment.

Although Japan has not yet set a firm target for offshore wind by 2030, recent legislation established eleven promotion zones in five prefectures, with competitive auctions to support offshore wind deployment. In the Stated Policies Scenario, Japan has 4 GW of installed capacity by 2040. In the Sustainable Development Scenario, the total reaches 18 GW, providing nearly 7% of the country's electricity in 2040. While most of Japan's current project pipeline uses fixed-bottom type turbines, there are relatively few shallow areas available offshore, and this means that more ambitious deployment of offshore wind is likely to be tied to the successful development of floating turbines. Japan has a series of experimental floating wind farms in place.

---

[8] Number of licenses as of mid-2019.

A number of other countries outside Asia are now actively appraising their offshore wind resources. Much of this development hinges upon sustained political support and the development of local supply chains. Chinese Taipei has set offshore wind targets for 2025, while a number of countries with significant offshore potential, including Brazil, South Africa, Sri Lanka and Viet Nam, have expressed interest in the World Bank's offshore wind emerging markets fund. This programme provides upwards of $5 million in funding for offshore wind research and supply chain development (World Bank, 2019).

14.3.3 Offshore wind costs, value and competitiveness

The long-term prospects for offshore wind depend to a large extent on how competitive it is with other sources of electricity. This section considers the evolution of the cost components of offshore wind, ultimately expressed through the LCOE, the system value of offshore wind, and its competitiveness based on the value-adjusted LCOE metric.

Capital costs

The development of offshore wind is gathering momentum. The technology is in a dynamic stage, and developers are working to bring down costs in established and new markets with support from governments and regulators. Project-level upfront capital costs will continue to depend on the choice of site, and on the inevitable trade-offs between distance from shore, water depth and quality of the resource.

The evolution of capital costs for power generation technologies is heavily influenced by whether there is a sufficient pipeline of projects to create the necessary momentum for a technology to develop. Global average upfront capital costs for offshore wind (including transmission) are projected to decline to below $2 500/kW by 2030, more than 40% below today's average. This is based on the assumed learning rate that sees capital costs decline by 15% each time global capacity doubles. By 2040, global average offshore wind costs are projected to fall to $1 900/kW. Increased deployment enables industry learning about offshore wind project development and management, as well as providing opportunities to establish efficient supply chains. Wind farm operators are learning how to push costs down, and equipment manufacturers are learning how to bring bigger and more efficient components to the market.

Excluding transmission costs, the global upfront capital costs of offshore wind projects averaged some $3 300/kW in 2018 and are projected to decline to $1 500/kW in 2030 and under $1 000/kW in 2040 with the level of deployment in the Stated Policies Scenario. This would mean that in 2040, transmission capital costs would be on a similar level as the offshore wind farm. However, individual project costs varied widely reflecting project specifics and regional particularities (Figure 14.16). The availability of sites in relatively shallow waters is a critical determinant of project costs, as it is the case of the Netherlands, where projects to be commissioned in the first-half of the next decade are at the lower end of the global cost range (PBL, 2019). Capital costs are also likely to be lower in places where there are enough projects of sufficient size to achieve economies of scale, as in China.

Figure 14.16 ▷ Capital costs of offshore wind projects excluding transmission, historical and projects in development

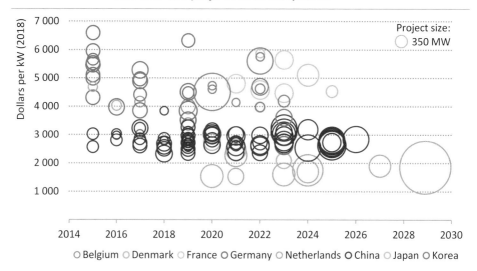

Continued industry learning propels offshore wind down the cost curve, and cost reductions can be enhanced through policy frameworks that support a healthy project pipeline

Note: Capital costs refer to the year of commissioning.
Sources: IEA analysis based on IJGlobal (2019), BNEF (2019) and company reports.

In Europe, the largest offshore wind market in terms of installed capacity, measures that harness competitive forces such as auctions are putting downward pressure on costs and generally making cost information more transparent. Continued policy support and long-term design improvements in the clustering of offshore wind farms are also helping with cost reductions. Excluding transmission costs, upfront capital costs for offshore wind farms in Europe were around $4 000/kW in 2018, though the project pipeline points to rapidly falling costs in the near term. Average offshore wind costs are projected to decline to below $2 000/kW in 2030 without transmission costs, and to about $1 500/kW in 2040.

In China, the first commissioned offshore wind projects were relatively close to shore – the average distance was less than 20 km (The Wind Power, 2019). This helped to keep total upfront capital costs (including transmission) below $2 800/kW in 2018. China is likely to see a range of project-level costs for some time as new projects adapt to specific site conditions. In the Stated Policies Scenario, capacity additions of 100 GW in China over the period to 2040 help drive down total upfront capital costs of offshore wind by nearly half.

Offshore transmission costs

There is a trend towards locating offshore wind farms further from shore. This highlights the importance of developing more advanced connection technologies and establishing appropriate regulations to govern the connection arrangements (Figure 14.17).

Figure 14.17 ▷ Offshore wind: average distance from shore by country

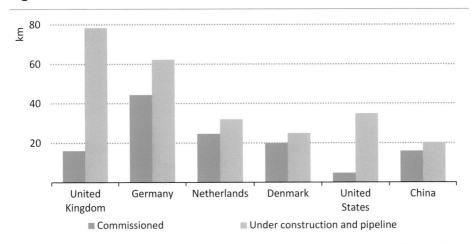

Offshore wind farms have been moving into deeper waters, amid a trend of increasing project sizes, that have impacted offshore transmission design as well as foundations

Sources: IEA analysis based on The Wind Power (2019) and BNEF (2019).

Large offshore wind farms that are more than 10 km from shore are usually attached to an offshore substation that is connected to an onshore substation. There are two main technologies behind offshore connections: alternating current (AC) systems that transport electricity directly through AC cables; and direct current (DC) systems where the current is converted from AC to DC in converter stations and back to AC in the onshore substation before the electricity is supplied to a grid. Even without considering the cost of substations, AC transmission has a cost advantage over short distances, but over longer distances high-voltage DC (HVDC) transmission can offer significant cost savings (Figure 14.18).

To date, offshore wind farms have been mostly connected to shore by radial offshore transmission assets, which imply one transmission line connecting one offshore wind farm, and the offshore transmission assets being viewed as part of the wind farm. However, where there are conflicts of use or spatial planning requirements, offshore wind farms can be designed as clusters in which several projects are connected to what is defined as an offshore "hub-and-spoke" with the aim of avoiding the impact that building several offshore transmission assets would have, while also reducing costs. That is the case with the North Sea Wind Power Hub, where Tennet together with Energinet.dk, Gasunie and the Port of Rotterdam are shaping a power hub in the North Sea. Tennet, the national transmission system operator in the Netherlands, has been playing a leading role since the installation of Borwin1. This was the first HVDC connection commissioned in Germany in 2010, with a capacity of 400 MW and 125 km offshore. Today, the longest offshore cable is located in Germany and spans 160 km, while the largest capacity of an offshore cable supports 916 MW.

Figure 14.18 ▷ Indicative upfront capital cost for high-voltage transmission cables by type and distance from shore

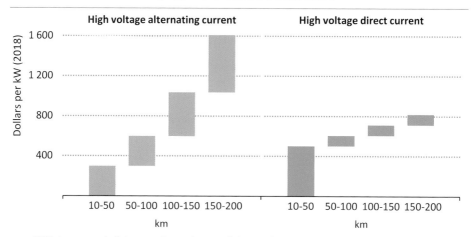

With increased distances from shore, offshore wind farms require new transmission connection technologies and regulation models to bring project costs down

Note: Installation costs for transmission cables vary based on site conditions.
Sources: IEA analysis; Xiang et al. (2016); DIW ECON (2019).

The responsibility for designing, installing and maintaining offshore transmission assets is defined by regulation, and can rest with the transmission system operator (TSO), government or project developer. This differs from arrangements for onshore wind, where transmission line liabilities are nearly always the responsibility of TSOs.

There are several models for developing offshore wind transmission in Europe. In the United Kingdom, licences for offshore transmission assets are granted through competitive auctions where offshore wind farm developers can also be in charge of building the transmission asset and then transfer it to the TSO or to a competitively appointed offshore transmission owner for operation. In other European markets such as Germany, Netherlands, France and Denmark, it is currently the system operator that provides the offshore grid connection (and in some cases the offshore substation), although Denmark has announced its intention to bring responsibility for developing offshore transmission assets into its competitive bidding framework.

There is not a one-size-fits-all solution that could guarantee the perfect regulatory framework to manage offshore transmission assets. TSOs may be able to achieve better co-ordination and standardisation of projects when the transmission connections fall under their scope. Equally it could be argued that including offshore transmission assets in auction schemes harnesses competitive market forces and could lower infrastructure costs (DIW ECON, 2019). Market transparency and long-term planning should be at the heart of whatever approach is taken, together with the need to support the development of offshore wind (Box 14.4).

Operation and maintenance costs

In much the same way as capital costs, operation and maintenance (O&M) costs are going through a phase of major development and improvement that is leading to cost reductions. Global average O&M costs for offshore wind stood at about $90/kW in 2018, and are projected to go down by one-third by 2030, before declining towards $50/kW in 2040. Regions, such as China where offshore wind markets are more developed show lower costs for O&M than others (Figure 14.19).

Figure 14.19 ▷ **Regional average annual O&M costs for new projects**

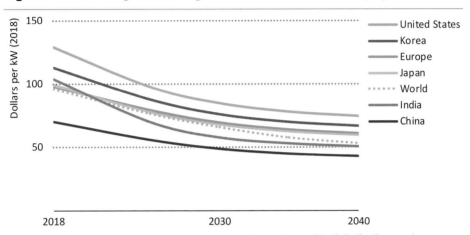

Economies of scale and industry synergies, along with digitalisation and technology development will bring current costs for O&M down by 40% in 2040

Digitalisation is bringing new techniques for monitoring that proactively identify failures not only in the turbines but also in structures and connections, helping to reduce costs. For example, the use of drones for visual inspections can enable preventative maintenance and cuts down on the need for labour-intensive inspections, along with speeding up maintenance work. In addition, synergies with the oil and gas industry have allowed offshore wind projects to draw on that industry's expertise in offshore structures in planning and carrying out maintenance activities (see section 14.6.2). The result of these improvements can help prolong the expected lifetime of projects, improving their economics and reducing the LCOE for offshore wind.

Capacity factors

As set out in section 14.2.3, offshore wind technology advancements are helping to improve performance. The size of turbines has increased and most current projects under construction in Europe involve the installation of 8-10 MW models (WindEurope, 2019). Larger turbines are able to raise wind farm electricity production by reaching a wider range of wind speeds, ultimately generating more electricity. Offshore wind farms are moving further from shore, where wind speeds tend to be higher – the maximum distance reached 90 km in recent years while distances rarely exceeded 20 km before 2006 (IRENA, 2019).

Figure 14.20 ▷ Regional average annual capacity factors for new projects

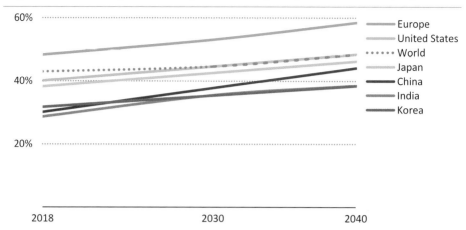

Technology advances including bigger turbine sizes and operation in areas with high quality wind resources will boost offshore wind performance in the years ahead

Source: IEA analysis based on IRENA (2019).

The ability to make use of higher quality wind resources by moving into deeper waters further from the shore has helped capacity factors reach 43% (after wake losses) for projects commissioned in 2018. Global average capacity factors are expected to increase by five percentage points by 2040 (Figure 14.20). Capacity factors move towards 60% in Europe by 2040 as a consequence of high quality resources and technology development. In China, new projects reach 45% annual capacity factors in 2040. Korea and India have lower quality wind resources overall, but new offshore wind projects are still expected to see capacity factors near 40%.

Cost of financing

Global investment in offshore wind totalled around $20 billion in 2018. Improved market maturity, underpinned by policies, regulations and the early involvement of public finance players, has facilitated lower cost financing over time from a diverse set of actors. Project finance – a financing model that requires a high degree of co-ordination and risk allocation among developers, banks and other actors – now represents the largest source of new asset financing. This suggests improved investor confidence in offshore wind developments and a greater degree of project standardisation than a few years ago, when most finance came from the balance sheets of developers and government-backed sources.

Commercial banks have increased their financing of offshore wind projects, a reflection of both the stable policy frameworks put in place in a number of countries and the successful track record of well-known investors involved in providing backing for early projects. The participation of public finance institutions such as the European Investment Bank, EKF in Denmark and KfW development bank in Germany has been instrumental in helping to manage risk and attract private capital, but their involvement has declined as the market achieved higher degrees of scale and as an array of private sector investors entered the market (IEA, 2018a). At the same time, there is a growing market for refinancing, whether by banks or through the capital markets, as well as for asset acquisitions on the part of institutional investors and pension funds.

Figure 14.21 ▷ **Offshore wind: indicative nominal cost of debt in Europe** (left) **and LCOE sensitivity analysis to cost component changes** (right)

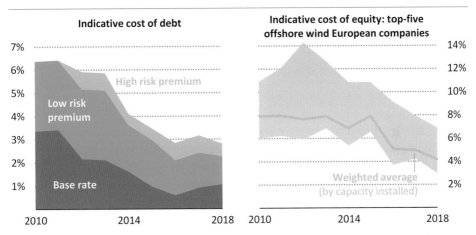

Offshore wind in Europe has benefited from low interest rates and debt risk premiums, allowing developers to decrease project WACCs, the largest component of LCOE

Notes: WACC = weighted average cost of capital; LCOE = levelised cost of electricity.

Sources: IEA analysis based on Bloomberg (2019) and company reports.

Debt financing terms continue to improve for projects in Europe. Project leverage has increased – the debt-to-equity ratios for projects achieving financial close averaged 75% in 2018, compared with 60% a decade earlier – which indicates that commercial banks are now more comfortable with offshore wind projects. Borrowing has been eased by low interest rates and lower debt risk premiums: loans are currently priced at 120 to 175 basis points above base rates (Figure 14.21) and maturities have typically increased to around 15-18 years (Green Giraffe, 2019). Nevertheless, these terms are most common for projects benefiting from long-term power purchase agreements awarded via government schemes or through contracting with a corporate off-taker: the bankability of projects relying on a high degree of revenue from wholesale markets remains less certain.

On the other side of the equation, the cost of equity has evolved in the same kind of way, with lower perceived risks from investing in offshore wind assets being underpinned by supportive policies. This has led to the cost of equity nearly halving over the last decade for companies operating in the most mature markets. The combination of lower financing costs and higher debt-to-equity ratios has directly shaped the potential WACCs that offshore wind farms have been able to achieve in recent auctions, and has led to some recent auctions being able to take place without any government subsidies.

Levelised cost of electricity

The levelised cost of electricity (LCOE) combines all the previously described elements into a single number representing the average generation cost for a technology. The cost trends described in this section taken together mean that the global average LCOE of offshore wind is set to decline from the 2018 level by nearly 40% to 2030, and by nearly 60% to 2040. Applying a standard WACC representing full market risk (7% in developing economies and 8% in advanced economies), the global LCOE falls from $140/MWh in 2018 to less than $90/MWh in 2030 and close to $60/MWh in 2040. Based on the same underlying technology costs and performance parameters but applying low-cost financing (WACC of 4%), the global LCOE of offshore wind declines from $100/MWh in 2018 to $60/MWh in 2030 and less than $45/MWh in 2040.

Offshore wind costs are set to decline in all regions. Improvements in wind turbine and foundation designs, construction processes and O&M procedures tend to have spill over effects, driving down costs in all markets as global deployment increases. Individual markets can accelerate those cost reductions further by using the development of a healthy project pipeline to help establish efficient supply chains for both equipment manufacturing and technical expertise. The LCOE of offshore wind is also influenced by the quality of wind resources and water depths in areas available for energy development.

Applying a 4% WACC on the basis of the prevailing costs of debt and equity in the region, the average LCOE of offshore wind in the European Union declines from $104/MWh in 2018 to just over $60/MWh in 2030 in the Stated Policies Scenario (Figure 14.22). Excluding transmission assets, the LCOE of European Union offshore wind farms is just $44/MWh in 2030. This is the relevant point of comparison for strike prices in markets such as Germany or the Netherlands where transmission costs are socialised by transmission system operators, paid through levies to consumers. It is also a relevant benchmark when considering dedicated projects for hydrogen and other electro-fuels. In China, where capital costs are lower but current capacity factors are lower as well, the LCOE averaged around $125/MWh in 2018 (applying a 7% WACC). Based on global and local deployment, offshore wind costs in China are projected to fall to $70/MWh in 2030 and $55/MWh by 2040, reaching cost parity with coal-fired power generation around 2030. The costs of offshore wind in nascent markets, including the United States, are linked to technology development elsewhere, but are relatively uncertain until the markets evolve.

A number of factors influence the LCOE for offshore wind, but financing costs are among the most important. As noted, a four point decline in the cost of capital – as a result of improved debt terms, but also a lower cost of equity from developers (IEA, 2019a) – translates into nearly a 30% reduction in LCOE. Improved capacity factors are another critical element for project costs, as are associated technology developments that benefit from economies of scale and standardisation. A 20% reduction in total project capital costs could reduce LCOE by about a further 17%. The LCOE benefits across the board from the economies of scale and the standardisation of processes and manufacturing supply chains that are examined more in depth in the next section.

Figure 14.22 ⊳ LCOEs for new offshore wind projects in the European Union, China and the United States, 2018-2040

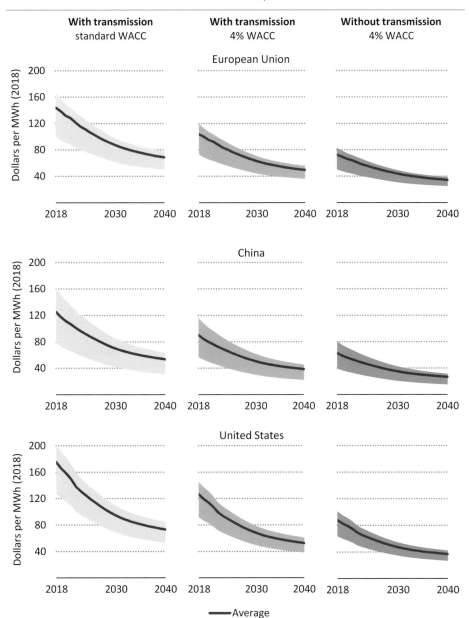

Offshore wind LCOEs are set to decline by 40% to 2030 and by over 50% to 2040, with the development of efficient local supply chains boosting global technology learning

Notes: WACC= weighted average cost of capital; LCOE = levelised cost of electricity. Standard WACC applied is 8% (real) for the European Union and United States, and 7% for China.

System value of offshore wind

Offshore wind is characterised by high capacity factors and widespread availability, making its value proposition potentially comparable to that of baseload technologies such as nuclear power and coal-fired generators. Offshore wind's value is generally higher than that of its onshore counterpart and more stable over time than that of solar PV, which has a concentrated output during daylight hours. Its energy value (equivalent to the average price received for energy sold to the market) depends on the pattern of demand and the power mix, but in most cases remains close to the average wholesale electricity price over the year (Figure 14.23), which means that in general its energy value is similar to that of baseload technologies. Even as the share of offshore wind increases, its energy value remains relatively stable, because offshore wind produces energy across all hours. Offshore wind can provide flexibility services, though incentives in most markets are insufficient to warrant the trade-off with lower energy output.

Figure 14.23 ▷ **Energy value by technology and region relative to average wholesale electricity price in the Stated Policies Scenario**

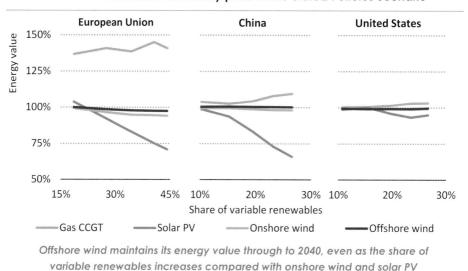

Offshore wind maintains its energy value through to 2040, even as the share of variable renewables increases compared with onshore wind and solar PV

Offshore wind can also contribute to the adequacy of electricity supply (the ability of the system to meet demand at all times from a planning perspective). The high capacity factors and seasonality of offshore wind means that 30% or more of its capacity can be counted towards reliability requirements, which is a higher percentage than for onshore wind and solar PV (Figure 14.24). This reduces the need for investment in other dispatchable capacity, including investment in combined-cycle gas turbines (CCGTs).

Pairing energy storage systems with an offshore wind project, an option which is currently being considered in connection with the North Sea Wind Power Hub, could significantly increase the capacity credit of the output from an offshore wind project. Energy storage could take a variety of forms, including batteries or thermal storage, both of which are currently being explored. The addition of energy storage to offshore wind (and other renewables) is particularly important for the Sustainable Development Scenario, where the share of variable renewables reaches 40% worldwide by 2040, and more than that in some regions.

Figure 14.24 ▷ Average capacity credit by technology and region in the Stated Policies Scenario

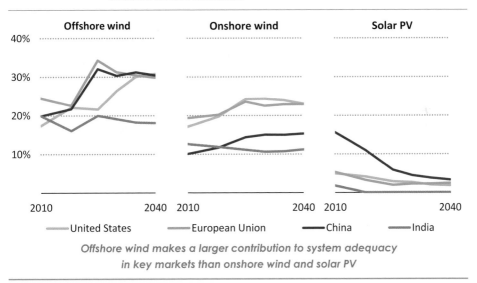

Offshore wind makes a larger contribution to system adequacy in key markets than onshore wind and solar PV

Note: In addition to fluctuating hourly profile for variable resources, other factors influence the capacity credit for a given year such as increasing capacity factors for the new unit of generation through to the total existing installed variable renewable generation in the power system.

Offshore wind could also provide flexibility services to power systems where the proper incentives are in place, as its high capacity factors mean that it could provide both upward and downward ramping. Some onshore wind projects already provide flexibility, largely in the form of downward regulation (i.e. reduced output on call). With sufficient economic incentives, onshore and offshore wind could also include upward regulation (i.e. curtailing output prior to the need to ramp upwards). To date, flexibility services have had limited attraction for most wind projects because revenues for such projects are often tied exclusively to electricity generation, but technological advances enabled by digitalisation could increase the value of flexibility for such projects in the future so as to make them economically attractive.

Offshore wind could also be used to produce low-carbon hydrogen, providing a zero-carbon fuel that could be used to provide power system flexibility, or be put to use in sectors that are hard to decarbonise, including industry, refining and transport (IEA, 2019b).

Box 14.3 ▷ Calculating the capacity credit of variable renewables

In order to quantify the expected contribution to power system adequacy (capacity credit)[9] of rising shares of variable renewables (such as offshore wind, onshore wind and solar PV), a new capacity credit tool has been developed. The tool incorporates both the annual projections generated in the IEA's World Energy Model (WEM) and the load and renewable resource profiles utilised in its hourly power module.[10]

The method employed in the capacity credit tool builds upon the approach that is currently used in the WEM and that has also been adopted by the US National Renewables Energy Laboratory (NREL) in their Regional Energy Deployment System model (NREL, 2017). It determines the contribution that a new unit of variable generation may provide to the power system during peak hours (sorted from highest to lowest) relative to a residual demand profile (actual demand less electricity production from variable renewables sorted from highest to lowest). These peak hours may be considered as a proxy for the hours with the highest risk for loss of load.

Two sorted residual demand profile are computed: one with the new unit and the other with existing variable generation. Over the peak hours, the difference between the sorted actual demand (often termed a load duration curve) and the residual demand profile without the new unit gives the capacity credit for the existing variable generation. Similarly, if we take the difference over the same period between the two sorted residual demand curves (with and without the new unit) this provides the capacity credit of the new unit of variable generation in the power system.

Value-adjusted LCOE of offshore wind

The competitiveness of power generation technologies calls for consideration of both the technology costs and the value provided to the system. The LCOE captures all the direct costs of a technology and is a commonly used metric of competitiveness. However, power generation technologies provide a number of services to the system that should be included in a comprehensive assessment of competitiveness. The services provided include contributions to the bulk energy supply, to the adequacy of the system, and to the flexibility of the system (enabling supply to match demand very closely in real-time operations).

The value-adjusted LCOE, first presented in the *World Energy Outlook-2018*, assesses the value of each of these services to the system and combines them with the LCOE to provide a single metric of competitiveness (IEA, 2018b). The metric can be applied to a wide range of power generation technologies, from variable renewables to dispatchable fossil fuels,

---

[9] Proportion of the capacity that can be reliably expected to generate electricity during times of peak demand in the network to which it is connected.

[10] For the full WEM methodology, see www.worldenergyoutlook.org/weomodel/. Wind and solar PV resource data used in the hourly modelling and capacity credit tool are sourced from Renewables.ninja, Open Power System Data (2019) and Ueckerdt et al. (2016).

and to both existing and new projects. It is applicable in a wide range of system conditions, including the share of variable renewables and available dispatchable technologies. It has similarities to other metrics applied in long-term energy modelling that combine cost and value, including the system LCOE (Ueckerdt et al., 2013) and the value-to-cost ratio applied by the US Energy Information Administration.

The value-adjusted LCOE has a basis in emerging electricity market designs around the world, incorporating and combining elements of competitive energy-only markets, ancillary service markets and capacity mechanisms (IEA, 2016; IEA, 2018c). However, the metric is applicable in all regions, though one or more value streams or the full technology costs may not be relevant to investors. The value-adjusted LCOE does not include site-specific network integration costs or externality costs that are not directly priced in the market.

Figure 14.25 ▷ Evolution of offshore wind competitiveness: value-adjusted LCOEs by technology and region in the Stated Policies Scenario

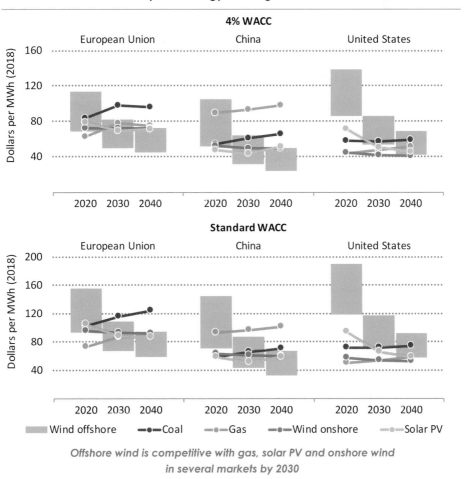

Offshore wind is competitive with gas, solar PV and onshore wind in several markets by 2030

Offshore wind closes the competitiveness gap with other low-carbon technologies in different regions over the next decade and in some cases moves ahead of fossil fuels (Figure 14.25). In the European Union, offshore wind is competitive with onshore wind, solar PV alone and solar PV paired with storage. With a rapidly rising share of variable renewables, however, the importance of flexibility in the European Union gives gas-fired power a competitive edge despite higher costs. In China, offshore wind moves within striking distance of other low-carbon technologies and coal-fired power by 2030, on the assumption of a modest carbon dioxide (CO_2) price. Offshore wind power is an attractive option in China because it can be built near population centres in the east and south of the country, complementing the extensive build-out of cross-province ultra-high transmission systems (IEA, 2017). In overall terms, the United States appears to be a challenging place for offshore wind to compete on a pure cost basis with onshore wind, solar PV and efficient gas-fired power. However, the availability of offshore wind resources in the northeast of the country and along the densely populated east coast means that it could have a role in diversifying the power mix.

14.4 Opportunities for faster growth of offshore wind

There are many factors that have a bearing on how fast the offshore wind market will expand, and some of the most important are highlighted in this section. Changes to technology, market conditions and policy decisions could accelerate offshore wind deployment in the decades to come and cause it to grow faster than projected in the Stated Policies Scenario. The upside potential is limited by the technical potential for offshore wind, not only in terms of the quality of wind resources and available turbine technologies, but also in terms of other factors that might limit its deployment including the suitability of seabed conditions for offshore wind and the nature of regulations concerning competing uses and environmental protection.

14.4.1 Global technical potential for offshore wind

A detailed assessment of the technical potential for offshore wind development was undertaken in collaboration with Imperial College London specifically for this report. Geospatial analysis was performed globally using the "Renewables.ninja" modelling tool based on the latest reanalysis of satellite data by the European Centre for Medium-Range Weather Forecasts (ECMWF) (ERA-5). Power curves corresponding to specific power output at different wind speeds were implemented for the latest turbine designs (up to 10 MW) and synthesised for designs up to 20 MW for which data are not yet available (Saint-Drenan et al., 2019). Areas available for offshore wind development excluded areas that are specified for competing uses (i.e. fishing, shipping, defence, and oil and gas exploration and production) and environmental protection (excluding Marine Projection Areas classified by the International Union for Conservation of Nature [IUCN]).[11] Areas close to submarine cables and earthquake fault lines were also excluded.[12]

Potential offshore wind performance

The quality of wind resources for energy production is best represented by the average capacity factor for new wind projects, which translates the wind speeds in a given area into the average performance over the course of a year. Based on the global assessment performed for this analysis, wind resources are generally of higher quality for energy production nearer to the poles (Figure 14.26). In Europe, the North Sea, Baltic Sea, Bay of Biscay, Irish Sea and Norwegian Sea, offshore wind has an average annual capacity factors of around 45-65%, which is higher than the comparable figures for the United States (40-55%), China (35-45%), and Japan (35-45%). The capacity factor is also high in regions off the coast of South America and New Zealand (50-65%). Moderate wind speeds resources in India translate to a 30-40% average capacity factor. The average capacity factor in general is relatively low in regions nearer to the equator for example in Southeast Asia and parts of western Africa. The detailed geospatial analysis captures varying conditions within regions, bringing out for example capacity factors in the Palk Strait between India and Sri Lanka that are well above average for the region and comparable to those found in Europe.

Figure 14.26 ▷ **Average simulated capacity factors for offshore wind worldwide**

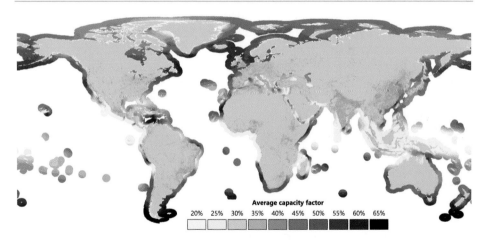

Average capacity factors reflect the quality of the wind resources available offshore around the world

Note: Inland dots depict population density of more than 500, 2 000 and 8 000 people per km² with darker shades of grey.

Source: IEA analysis developed in collaboration with Imperial College London based on Renewables.ninja.

---

[11] Classification areas Ia, Ib, II and III were excluded from the technical potential assessment (IUCN, 2013).

[12] For more information on the methodology, please refer to *Global Offshore Wind Outlook* (IEA, 2019c).

The assessed capacity factors provided the foundation of the technical potential assessment for offshore wind. Once exclusion areas and offshore wind farm designs have been factored in, the potential power capacity and annual electricity generation from each 5x5 km^2 site can be estimated and aggregated regionally and globally. Adding in estimated costs for offshore wind farms and transmission assets enables calculations to be made of the LCOE for each site, which are aggregated to provide supply curves by region.

Technical potential for offshore wind by region

Based on the assessment, we estimate that the technical potential for offshore wind worldwide is more than 120 000 GW, with the potential to generate more than 420 000 terawatt-hours (TWh) of electricity per year. This is enough to meet 11-times global electricity demand in 2040. Our study, however, does not consider other constraints such as the availability of transmission and distribution infrastructure to bring electricity generated to shore and other market related issues (see section 14.5.3). Because of their long coastlines, Russia (80 000 TWh per year, or 20% of the total), Canada (50 000 TWh per year, or 12% of the total) and the United States (over 45 000 TWh per year, or 11% of the total) together account for more than 40% of the global technical potential. Excess resources could be harnessed for export to other countries given favourable market conditions.

Figure 14.27 ▷ Ratio of technical potential to domestic electricity demand by region in the Stated Policies Scenario, 2040

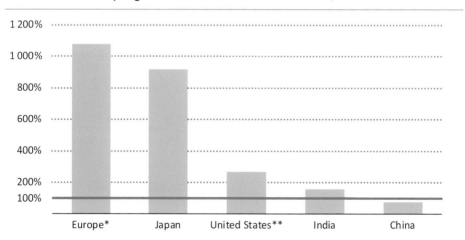

Based on technical potential, many regions could cover more than or nearly all of their domestic electricity demand from offshore wind alone

* Potential excluding Greenland and overseas territories. ** Potential available excluding Alaska and Hawaii.

Source: IEA analysis developed in collaboration with Imperial College London.

A comparison of technical potential with domestic electricity demand indicates that many countries would be able to meet their local electricity demand from offshore wind in 2040 in the Stated Policies Scenarios. For example, in Europe (excluding Greenland and overseas territories), the technical potential of offshore wind is more than ten-times demand (Figure 14.27). In the United States, excluding the offshore wind potential in Alaska and Hawaii, offshore wind could provide twice the level of total electricity demand in 2040. India and China would be able to meet the majority of electricity demand in 2040, with technical potential close to 6 000 TWh per year and 8 300 TWh per year respectively. Japan would be able to meet more than nine-times its demand based on its technical potential.

Technical potential can be divided between what is available in shallow water (i.e. < 60 m), and in deep water (i.e. 60-2 000 m). Broadly speaking, shallow water sites are suitable for established fixed-bottom foundations and are easier to access, while those for deeper water sites require floating platforms and are harder to access. The technical potential in shallow water is more than 87 000 TWh per year, which is enough to meet global electricity demand in 2040 under the Stated Policies Scenario more than twice over. The technical potential of sites located in deeper waters is more than 330 000 TWh per year. Deployment in deep water is currently costly, but the development of floating platforms technologies could bring down costs and open up enormous potential.

Within these two broad categories, the technical potential can be further subdivided: sites that are near to shore (i.e. 20-60 km) and within 50 km of existing commissioned projects; other potential sites that are close to shore; sites far from shore (i.e. 60-300 km); and sites with existing restrictions or competing uses such as oil and gas installations. Tapping just the most attractive potential in near shore shallow waters could provide close to 36 000 TWh globally per year, which is nearly equal to global electricity demand in 2040. Other near shore projects could generate a further 76 000 TWh per year. Around 70% of overall potential is in deep water and floating platform technology would be needed to develop this potential. Beyond the accessible technical potential, sites with existing competing uses and restrictions could provide around 17 000 TWh per year which could be made use of if policies and regulations made this both practicable and economic.

In Europe (excluding Greenland and overseas territories), the technical potential is close to 50 000 TWh per year, with countries bordering the North Sea and Baltic Sea such as Norway, Iceland, United Kingdom, France, Denmark, Netherlands and Germany accounting for two-thirds of this potential. Greenland adds another 14 000 TWh of potential, though its distance from large demand centres makes it less accessible. European countries are actively trying to expand their offshore wind markets to harness this potential and there are plans to develop a North Sea Wind Power Hub, to connect multiple wind farms in a hub-and-spoke configuration, with the first electricity due to come on shore in the 2030s (NSWPH, 2019). More than 1 300 TWh of Europe's technical potential is in shallow waters and close to shore, with a particularly large number of good sites located in the North Sea (Figure 14.28).

Figure 14.28 ▷ Regional technical potentials for offshore wind

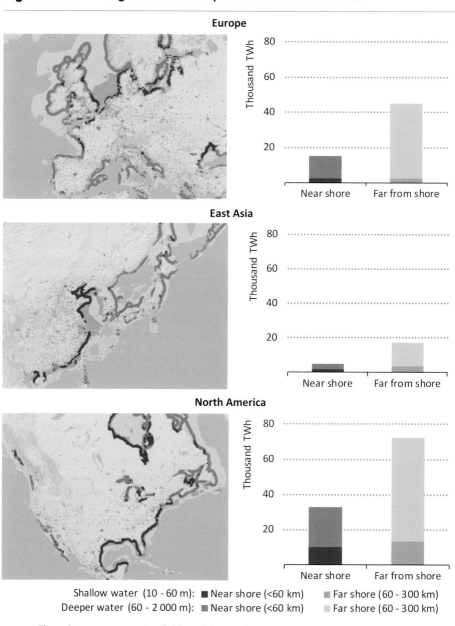

There is enormous potential for offshore wind development around the world, though the quality of wind resources varies widely

Notes: Inland dots depict population density of more than 500, 2 000 and 8 000 people per km² with darker shades of grey. Offshore regions far from shore shown in lighter shades of orange and purple respectively. For more information, refer to *Offshore Wind Outlook 2019: WEO-2019 Special Report* (IEA, 2019c).

Source: IEA analysis developed in collaboration with Imperial College London.

In East Asia, where a wave of deployment is expected in the upcoming years, the technical potential for offshore wind is over 22 000 TWh per year. In China, shallow coastal waters near major cities have the technical potential to produce close to 4 700 TWh per year, of which more than 1 800 TWh is available at sites close to the shore. This means that a good deal of the technical potential in China could be tapped at relatively low cost. In Japan, most of the technical potential of more than 9 000 TWh per year is located in deep water and would require floating platform technology in order to harness most of this potential. Even so, shallow waters could provide some 40 TWh per year, which would be enough to meet 4% of Japan's electricity demand in 2040 in the Stated Policies Scenario.

In North America, strong winds in places like Hudson Bay, Labrador Sea, Gulf of Alaska and the Atlantic seaboard together yield offshore wind potential of more than 100 000 TWh per year. The United States has technical potential of more than 46 000 TWh per year. Only around a quarter of this is in the contiguous United States, but this is still three-times more than enough to meet the total electricity demand of United States today and to meet its demand through to 2040 in the Stated Policies Scenario. Within the contiguous United States, shallow waters have the potential to provide more than 3 300 TWh per year and deep waters more than 8 700 TWh per year. Some of this potential is located off the Atlantic coast near major cities such as Washington D.C., Boston and New York. There is also technical potential of more than 900 TWh per year within the Great Lakes region.

The geospatial analysis undertaken also assessed the technical potential available at a given generation cost level, and it indicates that several hundred gigawatts of offshore wind could be deployed in leading markets at costs at or below that of coal- and gas-fired generation in 2030 (Figure 14.29). For each 5x5 km^2 site, the LCOE of a potential offshore wind project was calculated based on the estimated capacity factor combined with the project cost. The total cost of the offshore wind installation varied according to the distance from shore and the sea bed depth for each site. Deeper waters and farther distances from shore served to increase the total cost above a standard assumed cost of $2 000/kW. Based on this assessment, several regions have very large potential to develop offshore wind at low to moderate costs:

- China starts to have significant potential at about $50/MWh, despite only moderate wind conditions (shallow and close to shore).
- At around $60/MWh, an attractive cost in most cases, potential starts to become available in the United States, Europe and Australia.
- At $80/MWh, a moderate cost for a new source of electricity, China, United States and Europe each have around 1 000 GW of potential or more.
- Potential remains limited in Korea due to deep waters and middling wind conditions, while India also remains limited at these cost levels due to poor wind conditions.
- Japan also has limited potential at these costs, although this could potentially change once the costs of floating turbines are better known.

Figure 14.29 ▷ Offshore wind potential supply curves by region

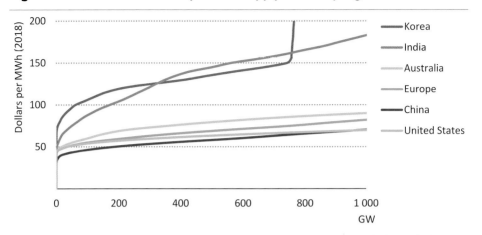

Based on near-term costs, at least 1 000 GW of offshore wind potential is available for less than $80/MWh in China, Europe and United States

Notes: LCOEs were calculated based on a standard wind farm cost of $2 000/kW, with additional costs related to greater distances from shore and deeper water. For more information, refer to *Offshore Wind Outlook 2019: WEO-2019 Special Report* (IEA, 2019c)

Source: IEA analysis developed in collaboration with Imperial College London.

14.4.2 Improved economics for offshore wind

The economic attractiveness of offshore wind to investors depends on a host of factors including technology and project costs, prevailing market conditions, the extent of government support and the evolution of other power generation technologies.

Available information for the offshore wind projects under development indicate a step change in project costs in the near term and industry expectations are for further robust cost improvements in the longer term as supply chains mature. Standardisation of project components and installation procedures could also help to reduce costs. More accurate meteorological information would help developers to choose the best sites for projects and could help make offshore wind technology more competitive with other sources of electricity. Where reductions in costs outpace those detailed in section 14.3.3, the growth of offshore wind could be accelerated further without increasing the delivered cost of electricity to consumers. Lower costs would also boost the prospects for production of renewable hydrogen.

Market conditions will play a critical role in the long-term future for offshore wind. Wholesale electricity prices seem likely to continue to experience downward pressure on the back of rising shares of variable renewables with zero or near-zero marginal costs and natural gas prices that are projected to remain at relatively low levels in several markets. Low wholesale electricity prices make it difficult to make a market-based case for investment in any power generation technology, including renewables. There are some notable exceptions for offshore wind projects in Europe, though it has yet to be seen whether this business model can be replicated for many projects. Without reforms there is a risk of inadequate signals for efficient and timely investment in new sources of power and in the power system flexibility that will be needed in the future (see Chapter 6).

Market reforms that promote such investment would improve the opportunities for market-based deployment of offshore wind. For example, capacity mechanisms are in operation or under consideration in several markets with the aim of remunerating contributions to system adequacy. Given that the characteristics of offshore wind lead to relatively high capacity credits (see section 14.3.3), capacity mechanisms seem likely to boost the competitiveness of offshore wind compared with solar PV and onshore wind. Further reforms that recognise the value of low-emission sources of electricity would also help to reinforce the business case for offshore wind and indeed other clean energy sources.

The economics of offshore wind would also be enhanced if transmissions network costs can be further reduced through innovation, economies of scale and supportive action on the part of electricity transmission grid operators. Planning that enables further expansion of the most competitive sources of electricity from a system perspective would make a significant difference to the prospects for offshore wind power in all regions.

14.4.3 Increased demand for renewable hydrogen

Demand for low-carbon hydrogen has an important role in the transition to a sustainable future (IEA, 2019b). Its versatility means that it has the potential to contribute to the decarbonisation of several sectors, as well as providing a low-carbon source of flexibility in power systems. The high capacity factors and improving cost competitiveness of offshore wind mean that it could play an important part in the production of low-carbon hydrogen.

Around 70 million tonnes (Mt) per year of hydrogen is used today, mainly in oil refining and chemical production, but it is almost all produced from fossil fuels. Using low-carbon sources of electricity to produce hydrogen could help to support a clean energy transition through sector coupling. For example, low-carbon hydrogen could be used to reduce CO_2 emissions in hard-to-abate sectors of industry such as iron, steel and chemicals. It could also be used in transport, where fuel cell electric vehicles powered by hydrogen would be well suited to heavy trucks, and where there could be other uses for low-carbon hydrogen in aviation and shipping.

Offshore wind projects dedicated to produce local, renewable-based hydrogen could offer significant cost advantages over projects using electricity direct from the grid. In part this is because dedicated offshore wind farms would benefit from cost reductions by avoiding the need for transmission. For example, the LCOE of offshore wind in Europe in 2030 is projected to be in the range of $40-70/MWh including transmission (with a 4% WACC), but just $30-50/MWh without transmission. With average European Union retail industry electricity prices projected to be around $130/MWh, offshore wind could offer 60-80%

savings on the costs of electricity input. The high upfront capital costs of electrolysers mean they need to run as often as possible to be economic, and the high annual capacity factors offered by offshore wind would fit well with this need (see section 14.3.3).

The cost of producing hydrogen is declining. Hydrogen could be produced today at a dedicated offshore wind farm without transmission costs to shore at about $9 per kilogramme (kg) of hydrogen before transportation, with 87% of this cost related to the cost of offshore wind production and 11% related to electrolyser costs. Transportation costs would add to this, but could potentially be relatively low if it were to be possible to repurpose existing oil and gas pipelines. Production costs are projected to fall below $4/kg by 2040 as a result of falling costs for offshore wind and electrolysers. In the Sustainable Development Scenario, close to 4.5 Mt of low-carbon hydrogen are used in Europe in 2040, and offshore wind accounts for a significant share of this, with 10 GW of offshore wind capacity able to produce about 1 Mt of hydrogen per year (Figure 14.30).

Figure 14.30 ▷ Offshore wind capacity needed to produce 1 Mt of hydrogen

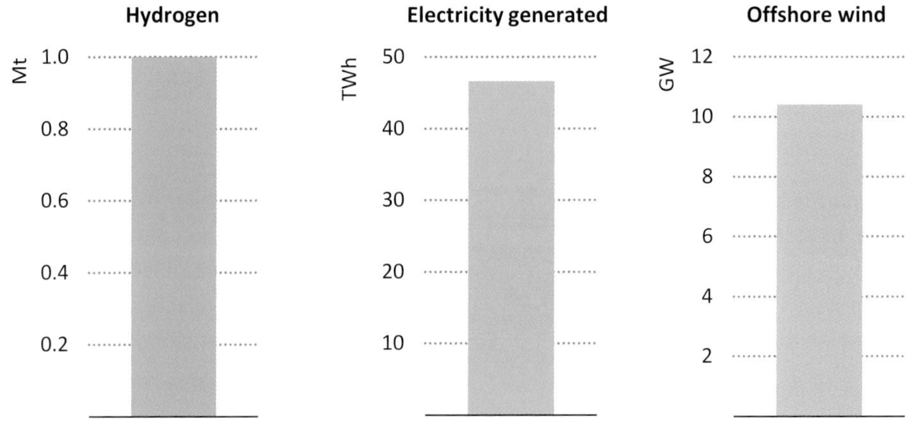

Additional demand for hydrogen offers significant upside potential for offshore wind, as 10 GW of offshore wind capacity can produce about 1 Mt of renewable hydrogen

14.4.4 Public acceptance

Offshore wind is not subject to some of the challenges faced by other technologies in terms of public acceptance precisely because of its offshore nature. Concerns about noise, visual impact and use of land do not arise in the same way that they do for onshore wind and solar PV, for example. Offshore wind does face public acceptance challenges of its own, exemplified by the decade-long challenges to the Cape Wind project in the eastern United States, and these should not be glossed over. However the challenges so far appear to be generally less widespread and intractable than for onshore wind developments.

14.5 Uncertainties that could slow offshore wind growth

As a technology poised for rapid growth, offshore wind faces a number of challenges. Developers must establish efficient supply chains, address environmental concerns, and demonstrate that the impressive cost reductions promised by recent auction bids are replicable in other markets and under a range of marine conditions. Offshore wind developers must also deal successfully with a unique set of technical and regulatory challenges, in particular in terms of marine planning and the development of supporting grid infrastructure on land to deliver electricity produced offshore to consumers. Strong growth may also bring new challenges, such as gaining access to adequate quantities of the rare earth elements that are critical to large offshore wind turbines. These various challenges will have to be overcome if offshore wind is to fulfil its potential and make a strong contribution to a clean energy future.

14.5.1 Developing efficient supply chains in new markets

One key challenge is the need to establish clear plans that enable efficient supply chains to be developed, limiting project risk and supporting low costs. In a perfect scenario for offshore wind, we would see improved turbine models and lighter and resilient foundations being delivered on time to projects, transparent planning and regulations providing clear long-term market visibility for project developers and investors, healthy competition among players to reduce costs, a good balance of international and national industry partnerships, and creating a well-trained work force.

For all these things to happen, however, markets need connections in place between all the various offshore wind supply chain links. Governments have an important part to play at the outset in demonstrating a clear commitment to national renewable energy goals that provides sufficient security to industry to procure a healthy pipeline of projects, and in backing this up by setting clear rules, for example on the handling of grid connections and the conducting auctions. This should engender the confidence in industry to develop supply chains, where the standardisation of equipment and operations help to reduce costs, and to pursue further efficiencies by making the most of synergies with the oil and gas industry (see section 14.6.2).

14.5.2 Environmental concerns

To reduce the impact on the environment, offshore wind projects are increasingly being examined carefully by government agencies at the planning stage to ensure that developers complete a detailed environmental impact assessment and if approved, put in place measures to comply with marine protection standards throughout the life of the project. For example, the Crown Estate in the United Kingdom recently blocked the expansion of Race Bank Offshore wind farm citing environmental concerns after a Habitats Regulations Assessment was conducted.

Offshore wind projects could potentially have negative impacts on fish, marine mammals and birds related to habitat change, displacement or injury during construction and operational noise, risk of collisions as well as avoidance or attraction to electromagnetic fields (WWF, 2014). On the other hand, several studies have pointed to potential benefits including enhanced biological productivity and improved ecological connectivity. Based on a long-term environmental study on the Horns Rev project in Denmark, fish species may be attracted to foundations, providing a refuge and potentially increasing the number of species in a project area (Danish Energy Agency, 2019).

While there are uncertainties about overall impact, it is important that measures are taken to protect biodiversity throughout the lifetime of the offshore wind farm. The IUCN has a set of guidelines on maritime protection areas which the industry can consult (IUCN, 2013).

14.5.3 Onshore grid development

The development of supporting onshore grid infrastructure is essential to the efficient integration of power production from offshore wind projects. Without appropriate reinforcements and perhaps some expansion of the grid, electricity from offshore wind may be curtailed. The risk posed by taking separate approaches to renewables and to grid development is larger in systems where the share of variable renewables is rising, especially where onshore wind plays a significant role and is exposed to the same kind of weather patterns as offshore wind.

Europe has set significant ambitions for the development of wind offshore, with member states together aiming for 65-85 GW of capacity by 2030. With strong and reliable wind speeds, additional development in the North Sea is of great interest. Analysis was carried out to explore the impacts of onshore grid development on integration of a wind hub in the North Sea in 2030 (Box 14.4).

Box 14.4 ▷ Tapping offshore wind potential in Europe calls for onshore transmission expansion

Along with actively deploying individual offshore wind farms, Europe is also exploring offshore "hubs" to facilitate further deployment of offshore wind. The North Sea Wind Power Hub project represents a step towards a possible 180 GW of offshore wind by 2050 (NSWPH, 2019). At that scale, offshore wind would offer the opportunity to decarbonise not only electricity supply but also other sectors through the dedicated production of hydrogen.

To illustrate the interplay between transmission network and offshore wind developments, analysis was conducted of the integration of an offshore wind hub in the North Sea. The impacts of two different grid configurations for 2030 have been simulated and analysed, based on a country-by-country representation of the European power system, and using an hourly time resolution across the whole year based on the

Stated Policies Scenario. One case includes network infrastructure that is economically optimal (from a societal point of view). It has been obtained by allowing the model to invest in the interconnection projects submitted to the ENTSO-E Ten-Year Network Development Plan 2018 as long as they are found to generate benefits that outweigh their costs. A second case includes an additional 12 GW of offshore wind power capacity, as well as including expanded interconnections with Germany (6 GW), Netherlands (4 GW) and Denmark (2 GW).

The comparison between the two cases shows that adding the offshore wind hub would trigger around 10 TWh of additional curtailment in Europe, mostly in the areas directly connected to the hub, and would increase congestion in the network (Figure 14.31). There would be significant additional congestion between the countries connected to the hub and their neighbours, notably with the Scandinavian countries, United Kingdom and Switzerland. Some interconnections would see additional congestion in 30% to 40% of the hours in the year.

Figure 14.31 ▷ Rise in the number of hours of cross-border grid congestion with the addition of a 12 GW hub in the North Sea

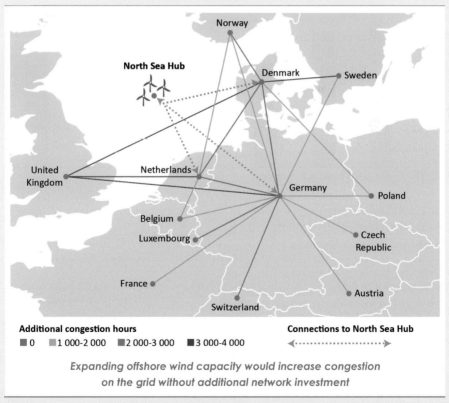

Additional congestion hours
■ 0   ■ 1 000-2 000   ■ 2 000-3 000   ■ 3 000-4 000

Connections to North Sea Hub

Expanding offshore wind capacity would increase congestion on the grid without additional network investment

Note: Other interconnections with zero additional congestion hours were not included in the figure.

To resolve these issues, an alternative case was developed by allowing the network to adapt when adding the North Sea offshore hub. The results show that an investment of EUR 750 million in 1.1 GW of interconnections from Denmark, Netherlands and Germany to their neighbours could help reduce the curtailment by around 1 TWh per year, and save EUR 130 million per year of operational costs. Furthermore, the integrated renewable energy reduces the use of thermal-based generation by the same amount (1 TWh) and the CO_2 emissions by around 0.5 Mt per year.

This comparison also shows that all onshore interconnection projects connecting Denmark, Netherlands and Germany to their neighbours and included in the ENTSO-E Ten-Year Network Development Plan 2018 are found to be cost effective (the benefits they bring outweigh their investment costs) (Figure 14.32). These new transmission capacities would relieve grid congestion, reduce overall generation costs and decrease CO_2 emissions in Europe. Additional interconnection projects might well increase the cost effectiveness of the integration of this project.

Figure 14.32 ▷ Share of potential cross-border connection projects developed by case in Europe to 2030

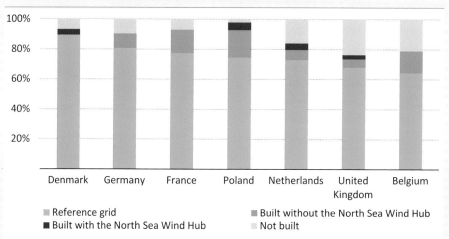

■ Reference grid   ■ Built without the North Sea Wind Hub
■ Built with the North Sea Wind Hub   ▫ Not built

Additional investment in onshore transmission is key to making the most of offshore wind

An alternative to fully utilise production from offshore wind farms would be to produce hydrogen via electrolysis, part of the long-term vision for hubs in the North Sea. At offshore wind costs of $30-40/MWh, hydrogen could be produced for $18-22 per million British thermal units in 2040 (or $2-2.5/kg of hydrogen), before transportation costs. The production of hydrogen could unlock more of the offshore wind potential.

The analysis was developed in collaboration with Artelys and carried out using Artelys Crystal Super Grid, a multi-energy systems modelling platform. A model of the European electricity system based on the IEA's Stated Policies Scenario was developed.

> The model allows for joint optimisation of investments and operations (cost-minimising criterion) for a given year using an hourly time resolution and country-level spatial granularity. The costs considered include investment and operational costs, i.e. fuel and CO_2 costs, variable O&M costs and loss of load penalties (if any) in order to ensure that electricity demand can be met at all times in the considered areas (all member states of the European Union, Norway, Switzerland, Republic of Macedonia, Montenegro, Serbia and Bosnia-Herzegovina). The model is able to simultaneously optimise investment in and the operation of all categories of assets, including different generation technologies, flexible consumption technologies, storage assets and interconnections between areas.

With offshore wind potential spread around the world, significant infrastructure development is likely to be necessary to ensure the cost-effective integration of renewable energy. This will be most effective if it is forward looking. Planning practices favouring the emergence of synergies between renewable energy deployment, onshore and offshore grids and cross-sectoral flexibility solutions are likely to result in the most cost-effective systems. It would make sense as part of this to consider hubs connected to several countries or regions and to assess the scope for synergies, in particular the possible use of offshore hubs to exchange electricity between the interconnected areas. This could reduce the overall costs across the connected countries even when the wind generation is low and increase the level of security of supply.

14.6 Implications

Offshore wind is a maturing technology that can help to achieve environmental goals while contributing to the security and affordability of electricity. The growth of offshore wind in the Stated Policies Scenario sees expansion well beyond the North Sea to markets around the world. In the Sustainable Development Scenario, offshore wind grows faster still. The falling costs of offshore wind support expansion in all scenarios, though policy makers have an important role to play in creating the necessary conditions for this to happen.

14.6.1 Achieving environmental goals

In the Stated Policies Scenario, the fifteen-fold growth of global offshore wind generation helps to decarbonise electricity supply in several key markets. Over the next two decades, the average CO_2 intensity of electricity generation declines by more than one-third. This keeps related CO_2 emissions flat in the face of demand growth of 2.1% per year. Without the growing contribution of offshore wind, cumulative global CO_2 emissions would be about 5 gigatonnes (Gt) higher over the next two decades on the assumption that its lost output was replaced by the average of the existing power generation mix. In the European Union, the central role of offshore would be particularly difficult to replace. Scaling up generation based on the current power mix in the absence of offshore wind would lead to an additional 1.4 Gt of CO_2 emissions from 2019 to 2040 in the European Union, 0.5 Gt in the United States and 1.7 Gt in China (Figure 14.33).

Figure 14.33 ▷ Avoided CO₂ emissions due to the deployment of offshore wind in the Stated Policies and Sustainable Development scenarios

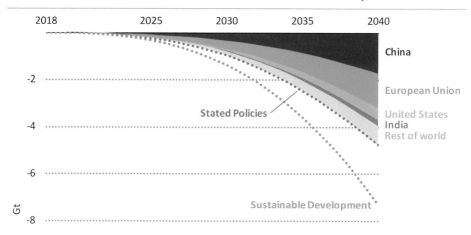

Offshore wind complements the growth of other clean energy sources, avoiding more than 7 Gt of CO_2 emissions to 2040 in the Sustainable Development Scenario

In the Sustainable Development Scenario, an additional 790 TWh of offshore wind helps avoid 2.5 Gt CO_2 emissions from 2019 to 2040. These CO_2 emissions reductions occur mainly in China, United States and India as a result of an accelerated deployment of offshore wind. They are equal to half the avoided emissions attributable to nuclear power in the transition to a more sustainable energy pathway (see Chapter 2).

The rise of offshore wind also leads to reductions of air pollutant emissions. In China, offshore wind development helps to limit the use of coal-fired power generation near large population centres in the eastern and southern parts of the country. In India and countries in Southeast Asia, offshore wind contributes to the decline in reliance on coal-fired power generation. In the United States and Europe, offshore wind displaces some gas-fired power generation and the related pollutants.

14.6.2 Synergies with oil and gas activities

The offshore wind power and oil and gas sectors share both technologies and elements of their respective supply chains, and there is scope to exploit the linkages further. Overall spending in offshore wind power reached $20 billion in 2018. Under the Stated Policies Scenario, the projected global annual spending for the next decade doubles and is two-and-a-half times higher in the Sustainable Development Scenario. This growth in offshore wind offers opportunities to those involved in providing a range of relevant offshore oil and gas services, whose skills could be very valuable to offshore wind developers in the light of the synergies between the two sectors.

In this context, it is not a surprise that oil and gas companies started investing in offshore wind projects many years ago, against the background of a trend of increasingly large projects requiring significant amounts of capital. For instance, Shell made its first entry into the offshore wind business with the Vattenfall offshore wind farm Egmond aan Zee in 2007, and is currently developing two more projects in the Netherlands that will be commissioned in the first-half of the next decade: it is also participating in the offshore market in the United Kingdom, France and United States. Equinor, a Norwegian company, has an active pipeline of projects of more than 2 GW, including 0.8 GW recently won in the first large-scale auction in New York in the United States. However, the largest portfolio of offshore wind capacity built belongs to the Danish company Ørsted (formerly DONG Energy), which recently sold its oil and gas assets to focus on renewable energy: it has an active pipeline of more than 8 GW.

Figure 14.34 ▷ Global offshore wind capital spending and potential synergies with offshore oil and gas activities

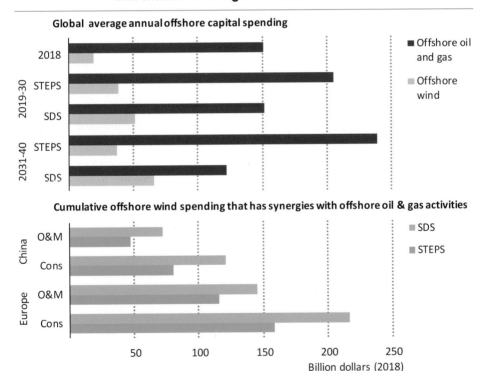

A growing pipeline of offshore wind projects opens new opportunities for traditional oil and gas companies, and these increase in the Sustainable Development Scenario

Notes: STEPS = Stated Policies Scenario; SDS = Sustainable Development Scenario. Cons = construction includes activities considered as potential synergies, e.g. foundations, installation and logistics. O&M includes potential synergies in operation and maintenance of offshore wind installations.

We estimate that about 40% of the full lifetime costs of a standard offshore wind project have significant synergies with the offshore oil and gas sector. That translates into a $400 billion market opportunity in Europe and China in the Stated Policies Scenario, and about a $550 billion opportunity in the Sustainable Development Scenario (Figure 14.34). While turbine manufacturing is specific to wind power, the construction of the foundations and subsea structures provides scope for significant learning from the oil and gas industry, especially in terms of how best to utilise vessels during installation and operation. There is likely to be much to learn from the oil and gas industry in developing floating platforms for offshore wind. Last but not least, work on the offshore substations required for large offshore wind projects and internal wiring in turbines could benefit from collaboration with the oil and gas industry.

In addition, there are a variety of equipment and support services with cross-over potential after the installation phase. Maintenance and inspection of oil and gas platforms is one area where oil and gas practices and safety standards are highly transferable.

On the other side, many processes in the offshore oil and gas platforms require electricity that is often supplied by (not very efficient) gas turbines or diesel engines that emit CO_2 and air pollutants. Nearby wind farms could supply oil and gas platforms with low-carbon electricity, reducing emissions and costs. Variable output from offshore wind could be a challenge, but batteries could help match patterns of electricity demand and supply, and limit the use of fossil fuel combustion engines.

Linkage with carbon capture, utilisation and storage would involve depleted oil and gas fields being used to store CO_2, which could be brought to the platforms using existing pipeline infrastructure. If the platforms are already electrified, they could house compression facilities that could be powered by nearby offshore wind farms.

14.6.3 Enhanced energy security and affordability

Within the power sector, the characteristics of offshore wind enable it to contribute to system adequacy in a way that is unique among variable renewables. As a result it can be a key contributor to ensuring electricity security in clean energy transitions.

Offshore wind can improve energy security in regions that rely on imported fuels for power generation by enhancing self-sufficiency and energy affordability. For example, a 1 GW offshore wind farm could replace 0.8 billion cubic metres of gas imports to produce the same amount of electricity in an efficient gas-fired power plant. At average spot prices for liquefied natural gas in 2018, such an offshore wind project would reduce annual import fuel bills by over $300 million in Japan and $220 million in Europe. On the other hand, if all electricity production from offshore wind in the Stated Policies Scenario instead were generated with the use of efficient gas-fired power generation, additional gas import bills would be over $400 million per year in Japan, close to $1.9 billion in Korea and nearly $15 billion in the European Union.

The competitiveness of offshore wind enables low-carbon electricity and hydrogen to be produced at a cost lower than otherwise would be possible. Both of these are critical to the achievement of environmental goals. Without the development of offshore wind, efforts to decarbonise electricity in the European Union, China and United States would need to depend more heavily on onshore wind and solar PV. Overall, offshore wind makes energy transitions more affordable, limits the challenges of rising flexibility needs and could, ultimately, accelerate progress towards decarbonisation.

Explore the data behind the
World Energy Outlook 2019

www.iea.org/weo

The World Energy Outlook online database provides easy access to data behind the more than 300 figures and tables in this year's Outlook, the energy balance tables as well as additional data that are not included in the book. This improved access to data reflects the priority to move towards a more "digital IEA", and our determination to remain the gold standard for long-term energy research and analysis. Please visit the database at www.iea.org/weo/weo2019secure/.

User ID: WEO2019AnnexA   password: WWC19

ANNEXES

Box A.1 ▷ *World Energy Outlook* links

WEO homepage
General information: www.iea.org/weo
WEO-2019 information: www.iea.org/weo2019

WEO-2019 figures, tables and Annex A tables
(available in Excel format)
www.iea.org/weo/weo2019secure
User ID: WEO2019AnnexA
Password: WWC19

Modelling
Documentation and methodology / Investment costs
www.iea.org/weo/weomodel

Recent analysis
Wind Offshore Energy Outlook: www.iea.org/offshorewind2019
Africa Energy Outlook: www.iea.org/Africa
Southeast Asia Energy Outlook 2019: www.iea.org/southeastasia2019
Nuclear Power in a Clean Energy System: www.iea.org/publications/nuclear
The Role of Gas in Today's Energy Transitions: www.iea.org/publications/roleofgas
Methane tracker: www.iea.org/weo/methane
Iraq's Energy Sector: www.iea.org/publications/iraqenergyoutlook

Databases
Sustainable Development Goal 7: www.iea.org/SDG
Policy Databases: www.iea.org/policiesandmeasures

Annex A

Tables for scenario projections

General note to the tables

This annex includes historical and projected data for the Stated Policies, Current Policies and Sustainable Development scenarios for the following five data sets:

- A.1. Fossil fuel production and demand by region.
- A.2. Power sector overview by region covering gross electricity generation and installed capacity; cumulative retirements and additions for 2019-2040. Global carbon dioxide (CO_2) emissions and intensity from power plants are also included.
- A.3. Energy demand, gross electricity generation and electrical capacity, and CO_2 emissions from fossil fuel combustion by region.
- A.4. Global emissions of pollutants by energy sector and fuel.
- A.5. Global average annual and cumulative energy investments by type.

Geographical coverage for Tables A.1, A.2 and A.3 include: World, North America, Central and South America, Europe, Africa, Middle East, Eurasia, and Asia Pacific. Table A.3 also covers: Brazil, China, European Union, India, Japan, Russia, South Africa, Southeast Asia and United States. The definitions for regions, fuels and sectors are in Annex C.

Both in the text of this book and in the tables, rounding may lead to minor differences between totals and the sum of their individual components. Growth rates are calculated on a compound average annual basis and are marked "n.a." when the base year is zero or the value exceeds 200%. Nil values are marked "-".

Please see Box A.1 for details on where to download the *World Energy Outlook (WEO)* tables in Excel format. In addition, Box A.1 lists the links relating to the main *WEO* website, documentation and methodology of the World Energy Model (WEM), investment costs, policy databases and recent *WEO* special reports.

Data sources

The World Energy Model (WEM) is a very data-intensive model covering the whole global energy system. Much of the data on energy supply, transformation and demand, as well as CO_2 emissions from fuel combustion, energy efficiency and prices are obtained from the IEA's own databases of energy and economic statistics (www.iea.org/statistics/).

Additional data from a wide range of external sources are also used. Historical data for gross power generation capacity are drawn from the S&P Global Market Intelligence World Electric Power Plants Database (March 2019 version) and the International Atomic Energy Agency PRIS database (www.iaea.org/pris).

The formal base year for this year's projections is 2017, as this is the last year for which a complete picture of energy demand and production is in place. However, we have used more recent data wherever available, and we include our 2018 estimates for energy production and demand in this annex (Tables A.1 to A.3). Estimates for the year 2018 are based on an update of the Global Energy and CO_2 Status Report which is derived from a number of sources, including the latest monthly data submissions to the IEA's Energy Data Centre, other statistical releases from national administrations, and recent market data from the IEA *Market Report Series* that cover coal, oil, natural gas, renewables and power. Investment estimates include the year 2018, based on the IEA's World Energy Investment 2019 (www.iea.org/wei2019/).

This annex also includes projections for primary air pollutant emissions that are emitted directly as a result of human activity. The focus is on anthropogenic emissions of sulfur dioxide (SO_2), nitrogen oxides (NO_X) and fine particulate matter ($PM_{2.5}$). Only emissions related to energy activities are reported. The base year of the projections is 2018. Base year air pollutant emissions estimates and scenario projections stem from a coupling of sectoral activity and associated energy demand of the WEM with the Greenhouse Gas and Air Pollution Interactions and Synergies (GAINS) model of the International Institute for Applied Systems Analysis (IIASA).[1]

Definitional note: A.1. Fossil fuel production and demand tables

Oil production and demand is expressed in million barrels per day (mb/d). Tight oil includes tight crude oil and condensate production except for the United States, which includes tight crude oil only (US tight condensate volumes are included in natural gas liquids). Processing gains covers volume increases that occur during crude oil refining. Biofuels and their inclusion in liquids demand is expressed in energy-equivalent volumes of gasoline and diesel. Natural gas production and demand is expressed in billion cubic metres (bcm). Coal production and demand is expressed in million tonnes of coal equivalent (Mtce). Differences between historical production and demand volumes for oil, gas and coal are due to changes in stocks. Bunkers include both international marine and aviation fuels.

Definitional note: A.2. Power sector overview tables

The power sector overview tables provide a high-level snapshot of the electricity system by region. Electricity generation expressed in terawatt-hours (TWh) and installed electrical capacity data expressed in gigawatts (GW) are both provided on a gross basis (i.e. includes own use by the generator), with more detailed data broken down by fuel and region in the A.3 tables. Power sector carbon dioxide (CO_2) emissions are expressed in million tonnes (Mt). The emission intensity expressed in grammes of carbon dioxide per kilowatt-hour (g CO_2/kWh) is calculated based on electricity-only plants and the electricity component of combined heat and power (CHP) plants.[2] For retirements and additions (both expressed in GW), the category "other" includes bioenergy, geothermal, concentrating solar power (CSP), marine, and battery storage.

---

[1] See: www.iiasa.ac.at/web/home/research/researchPrograms/air/GAINS.html for details.

[2] We assume that the heat component of a CHP plant is 90% efficient and the remainder of the fuel input is allocated to electricity to derive the associated electricity-only emissions.

Definitional note: A.3. Energy demand, electricity and CO_2 emissions tables

Total primary energy demand (TPED) is equivalent to power generation plus "other energy sector" excluding electricity and heat, plus total final consumption (TFC) excluding electricity and heat. TPED does not include ambient heat from heat pumps or electricity trade. Sectors comprising TFC include industry, transport, buildings (residential, services and non-specified other) and other (agriculture and non-energy use). While not itemised separately, hydrogen is included in total final consumption and "other energy sector". Projected gross electrical capacity is the sum of existing capacity and additions, less retirements. While not itemised separately, other sources are included in total electricity generation, and battery storage in total power generation capacity.

Total CO_2 includes carbon dioxide emissions from "other energy sector" in addition to the power and final consumption sectors shown in the tables. Total and power sector CO_2 emissions also account for captured emissions from bioenergy with carbon capture, utilisation and storage (CCUS). CO_2 emissions and energy demand from international marine and aviation bunkers are included only at the world transport level. Gas use in international bunkers is not itemised separately. CO_2 emissions do not include emissions from industrial waste and non-renewable municipal waste. For more information please visit www.iea.org/statistics/CO2emissions.

Abbreviations used: Mtoe = million tonnes of oil equivalent; CAAGR = compound average annual growth rate; Petrochem. feedstock = petrochemical feedstock.

Definitional note: A.4. Emissions of air pollutant tables

Emissions of all air pollutants (SO_2, NO_x, $PM_{2.5}$) are expressed in million tonnes (Mt) per year and are reported by sector. The energy sector is broken down into power, industry including other transformation (i.e. other energy sector excluding electricity and heat), transport, buildings and agriculture. Emissions are reported separately for all energy activities and for combustion activities; the difference between these two relates to energy processes, including, for example, cement production in the industry sector or abrasion, tyres and brakes in road transport.

Definitional note: A.5. Energy investment overview tables

The energy investment overview tables provide a high-level snapshot for the world by type of investment expressed in billion dollars, including key regional details. Total fuel investment covers oil, natural gas, coal and biofuels. Total power investment covers power plants, electricity networks and also battery storage which is not itemised separately. Electricity networks includes electric vehicle (EV) fast chargers. Renewables for end-use include solar thermal, bioenergy and geothermal applications for heating. Other end-use includes CCUS in industry sector, EVs and EV slow chargers. Abbreviations used: Renew. = renewables; Transport. = transportation.

Table A.1: Fossil fuel production

	Stated Policies Scenario							Shares (%)		CAAGR (%)
	2000	2017	2018	2025	2030	2035	2040	2018	2040	2018-40
Oil production and supply (mb/d)										
North America	14.2	20.5	23.0	28.4	29.6	29.7	28.6	24	28	1.0
Central and South America	6.8	7.2	6.6	7.5	8.1	8.7	9.7	7	9	1.8
Europe	7.1	3.7	3.7	4.0	3.3	2.8	2.6	4	3	-1.5
European Union	3.6	1.6	1.7	1.5	1.1	0.9	0.7	2	1	-3.6
Africa	7.8	8.2	8.4	7.9	8.0	8.1	8.2	9	8	-0.1
Middle East	23.5	31.2	31.7	32.2	33.6	34.6	35.6	33	34	0.5
Eurasia	7.9	14.2	14.5	14.1	13.6	13.0	12.4	15	12	-0.7
Asia Pacific	7.8	7.8	7.6	6.9	6.5	6.4	6.4	8	6	-0.8
non-OPEC	44.5	55.3	58.0	64.8	65.2	64.6	63.4	61	61	0.4
OPEC	30.6	37.5	37.4	36.2	37.6	38.7	40.1	39	39	0.3
World production	**75.1**	**92.8**	**95.4**	**101.0**	**102.8**	**103.2**	**103.5**	**100**	**100**	**0.4**
Conventional crude oil	64.5	66.9	67.1	66.1	65.1	63.0	61.9	69	58	-0.4
Tight oil	-	4.9	6.3	10.5	12.0	13.2	13.4	6	13	3.5
Natural gas liquids	9.0	16.5	17.3	19.4	20.4	21.2	21.7	18	20	1.0
Extra-heavy oil and bitumen	1.0	3.8	3.8	3.9	4.0	4.3	4.9	4	5	1.1
Other	0.6	0.7	0.8	1.1	1.3	1.5	1.6	1	1	3.1
Processing gains	1.8	2.3	2.3	2.5	2.6	2.8	2.9	2	3	1.0
World supply	**76.9**	**95.1**	**97.7**	**103.5**	**105.4**	**106.0**	**106.4**	**100**	**100**	**0.4**
Natural gas production (bcm)										
North America	763	988	1 083	1 254	1 336	1 358	1 376	27	25	1.1
Central and South America	102	177	177	188	209	244	285	5	5	2.2
Europe	338	292	277	236	206	191	188	7	3	-1.7
European Union	265	132	120	66	47	44	40	3	1	-4.8
Africa	124	233	240	287	372	435	508	6	9	3.5
Middle East	198	627	645	721	787	912	1 016	16	19	2.1
Eurasia	691	893	918	1 021	1 054	1 105	1 143	23	21	1.0
Asia Pacific	290	584	598	708	757	816	889	15	16	1.8
World	**2 507**	**3 794**	**3 937**	**4 415**	**4 720**	**5 060**	**5 404**	**100**	**100**	**1.4**
Conventional gas	2 318	2 985	3 004	3 164	3 293	3 498	3 694	76	68	0.9
Tight gas	148	253	274	276	267	244	238	7	4	-0.6
Shale gas	3	473	568	852	1 020	1 159	1 290	14	24	3.8
Coalbed methane	38	80	88	91	103	115	129	2	2	1.7
Other	-	3	3	33	36	44	54	0	1	14.2
Coal production (Mtce)										
North America	824	586	576	432	385	358	329	10	6	-2.5
Central and South America	48	89	82	75	75	62	62	1	1	-1.3
Europe	397	249	230	174	131	105	80	4	1	-4.7
European Union	307	187	163	117	79	57	30	3	1	-7.5
Africa	187	224	225	217	199	210	221	4	4	-0.1
Middle East	1	1	1	2	2	2	2	0	0	0.7
Eurasia	234	393	414	408	409	415	424	7	8	0.1
Asia Pacific	1 564	3 848	4 039	4 208	4 297	4 295	4 282	73	79	0.3
World	**3 255**	**5 391**	**5 566**	**5 515**	**5 498**	**5 446**	**5 398**	**100**	**100**	**-0.1**
Steam coal	2 504	4 164	4 342	4 349	4 393	4 394	4 394	78	81	0.1
Coking coal	449	936	955	888	857	822	790	17	15	-0.9
Lignite and peat	302	290	270	278	247	231	214	5	4	-1.0

Table A.1: Fossil fuel production

	Current Policies Scenario				Sustainable Development Scenario			
			Shares (%)	CAAGR (%)			Shares (%)	CAAGR (%)
	2030	2040	2040	2018-40	2030	2040	2040	2018-40
Oil production and supply (mb/d)								
North America	31.4	32.2	27	1.5	25.7	20.6	32	-0.5
Central and South America	8.7	11.8	10	2.7	5.7	4.5	7	-1.7
Europe	3.4	2.8	2	-1.2	2.8	1.9	3	-3.0
European Union	1.2	0.9	1	-3.0	0.9	0.5	1	-5.5
Africa	8.6	9.5	8	0.6	6.0	4.9	7	-2.4
Middle East	35.4	39.5	34	1.0	28.7	22.2	34	-1.6
Eurasia	14.2	14.1	12	-0.1	10.7	7.1	11	-3.2
Asia Pacific	6.9	7.7	7	0.0	5.4	3.9	6	-3.0
non-OPEC	68.9	72.3	61	1.0	53.8	40.8	63	-1.6
OPEC	39.8	45.4	39	0.9	31.2	24.3	37	-1.9
World production	108.7	117.7	100	1.0	85.0	65.1	100	-1.7
Conventional crude oil	68.5	70.6	58	0.2	52.7	36.9	55	-2.7
Tight oil	13.1	15.5	13	4.1	10.1	9.2	14	1.7
Natural gas liquids	21.2	23.1	19	1.3	17.7	14.8	22	-0.7
Extra-heavy oil and bitumen	4.3	6.3	5	2.3	3.3	2.9	4	-1.2
Other	1.5	2.2	2	4.8	1.2	1.2	2	2.0
Processing gains	2.8	3.3	3	1.6	2.2	1.8	3	-1.1
World supply	111.5	121.0	100	1.0	87.1	66.9	100	-1.7
Natural gas production (bcm)								
North America	1 375	1 445	25	1.3	1 209	909	24	-0.8
Central and South America	227	341	6	3.0	187	189	5	0.3
Europe	193	180	3	-1.9	189	151	4	-2.7
European Union	47	40	1	-4.8	47	42	1	-4.7
Africa	405	559	9	3.9	333	383	10	2.2
Middle East	806	1 045	18	2.2	681	651	17	0.0
Eurasia	1 145	1 315	22	1.6	921	786	20	-0.7
Asia Pacific	788	1 005	17	2.4	745	786	20	1.2
World	4 940	5 891	100	1.8	4 264	3 854	100	-0.1
Conventional gas	3 433	3 926	67	1.2	3 004	2 689	70	-0.5
Tight gas	253	232	4	-0.8	262	141	4	-3.0
Shale gas	1 113	1 532	26	4.6	863	871	23	2.0
Coalbed methane	102	143	2	2.2	101	103	3	0.7
Other	38	58	1	14.7	34	50	1	13.9
Coal production (Mtce)								
North America	430	397	6	-1.7	139	72	3	-9.0
Central and South America	90	100	2	0.9	59	11	1	-8.6
Europe	165	120	2	-2.9	63	24	1	-9.8
European Union	106	59	1	-4.6	43	17	1	-9.9
Africa	246	301	5	1.3	174	122	6	-2.7
Middle East	2	2	0	0.7	2	2	0	0.2
Eurasia	463	511	8	1.0	246	127	6	-5.2
Asia Pacific	4 539	4 968	78	0.9	2 788	1 742	83	-3.7
World	5 934	6 399	100	0.6	3 471	2 101	100	-4.3
Steam coal	4 753	5 266	82	0.9	2 672	1 515	72	-4.7
Coking coal	885	854	13	-0.5	676	497	24	-2.9
Lignite and peat	297	280	4	0.2	123	89	4	-4.9

Annex A | Tables for scenario projections

Table A.1: Fossil fuel demand

	Stated Policies Scenario							Shares (%)		CAAGR (%)
	2000	2017	2018	2025	2030	2035	2040	2018	2040	2018-40
Oil and liquids demand (mb/d)										
North America	23.5	22.3	22.8	22.5	21.5	20.3	19.1	23	18	-0.8
United States	19.6	18.0	18.5	18.4	17.4	16.3	15.1	19	14	-0.9
Central and South America	4.5	5.9	5.8	6.1	6.2	6.4	6.5	6	6	0.5
Europe	14.9	13.2	13.2	12.4	11.1	9.7	8.7	14	8	-1.9
European Union	13.1	11.2	11.1	10.1	8.8	7.4	6.3	11	6	-2.5
Africa	2.2	3.9	3.9	4.9	5.5	6.2	7.0	4	7	2.7
Middle East	4.3	7.5	7.5	8.4	8.8	9.6	10.2	8	10	1.4
Eurasia	3.1	3.8	3.9	4.3	4.3	4.2	4.2	4	4	0.4
Asia Pacific	19.4	31.0	31.6	35.8	38.0	38.9	39.2	33	37	1.0
China	4.7	12.1	12.5	14.5	15.6	15.6	15.5	13	15	1.0
India	2.3	4.5	4.7	6.4	7.5	8.4	9.0	5	8	3.0
International bunkers	5.4	8.1	8.2	9.3	10.0	10.7	11.4	8	11	1.5
World oil	77.4	95.7	96.9	103.5	105.4	106.0	106.4	100	100	0.4
Road, aviation & shipping	38.5	53.4	54.2	58.3	60.1	60.6	61.2	56	58	0.6
Industry & petrochemicals	14.4	18.4	18.3	20.5	21.5	22.4	22.9	19	21	1.0
World biofuels	0.2	1.8	1.9	2.8	3.5	4.1	4.7	2	4	4.2
World liquids	77.6	97.5	98.8	106.4	108.9	110.1	111.1	100	100	0.5
Natural gas demand (bcm)										
North America	800	974	1 067	1 163	1 183	1 195	1 221	27	23	0.6
United States	669	779	860	936	947	949	957	22	18	0.5
Central and South America	97	169	172	178	198	224	257	4	5	1.8
Europe	606	617	607	621	593	578	557	15	10	-0.4
European Union	487	487	480	477	442	416	386	12	7	-1.0
Africa	58	150	158	185	221	265	317	4	6	3.2
Middle East	186	539	535	559	646	739	807	14	15	1.9
Eurasia	471	570	598	628	639	652	674	15	12	0.5
Asia Pacific	313	758	815	1 071	1 218	1 374	1 522	21	28	2.9
China	28	240	282	454	533	598	655	7	12	3.9
India	28	59	62	103	131	166	196	2	4	5.4
International bunkers	-	0	0	11	21	34	50	0	1	34.3
World	2 530	3 778	3 952	4 415	4 720	5 060	5 404	100	100	1.4
Power	908	1 521	1 571	1 638	1 708	1 816	1 936	40	36	1.0
Industrial use	644	877	909	1 108	1 229	1 352	1 474	23	27	2.2
Coal demand (Mtce)										
North America	818	515	492	369	328	304	285	9	5	-2.5
United States	763	472	451	350	314	291	272	8	5	-2.3
Central and South America	29	46	46	48	47	47	49	1	1	0.3
Europe	578	460	447	314	263	219	203	8	4	-3.5
European Union	459	335	319	204	157	113	87	6	2	-5.7
Africa	129	157	159	165	160	160	161	3	3	0.0
Middle East	2	5	6	9	10	12	14	0	0	4.2
Eurasia	202	224	229	225	212	203	199	4	4	-0.6
Asia Pacific	1 551	4 012	4 079	4 385	4 476	4 502	4 487	75	83	0.4
China	955	2 805	2 834	2 934	2 845	2 710	2 568	52	48	-0.4
India	208	558	586	771	938	1 063	1 157	11	21	3.1
World	3 309	5 419	5 458	5 515	5 498	5 446	5 398	100	100	-0.1
Power	2 233	3 391	3 500	3 467	3 470	3 442	3 395	64	63	-0.1
Industrial use	869	1 728	1 680	1 832	1 852	1 872	1 903	31	35	0.6

Table A.1: Fossil fuel demand

	Current Policies Scenario				Sustainable Development Scenario			
			Shares (%)	CAAGR (%)			Shares (%)	CAAGR (%)
	2030	2040	2040	2018-40	2030	2040	2040	2018-40
Oil and liquids demand (mb/d)								
North America	22.5	21.6	18	-0.2	17.7	11.7	17	-3.0
United States	18.1	16.9	14	-0.4	14.2	9.1	14	-3.2
Central and South America	6.6	7.3	6	1.1	4.9	3.8	6	-1.8
Europe	12.1	10.9	9	-0.9	9.2	5.0	7	-4.3
European Union	9.6	8.2	7	-1.4	7.3	3.5	5	-5.1
Africa	5.7	7.4	6	2.9	4.9	5.2	8	1.3
Middle East	9.0	11.0	9	1.8	6.7	6.3	9	-0.7
Eurasia	4.4	4.5	4	0.6	3.8	3.1	5	-1.0
Asia Pacific	40.5	44.9	37	1.6	32.4	25.4	38	-1.0
China	17.0	18.5	15	1.8	12.7	9.3	14	-1.3
India	8.0	10.4	9	3.7	6.5	6.1	9	1.2
International bunkers	10.9	13.4	11	2.2	7.6	6.4	10	-1.2
World oil	111.5	121.0	100	1.0	87.1	66.9	100	-1.7
Road, aviation & shipping	64.7	72.7	60	1.3	47.9	32.2	48	-2.3
Industry & petrochemicals	21.5	23.0	19	1.0	18.9	18.5	28	0.0
World biofuels	2.8	3.6	3	2.9	6.3	7.7	10	6.6
World liquids	114.3	124.6	100	1.1	93.4	74.6	100	-1.3
Natural gas demand (bcm)								
North America	1 214	1 281	22	0.8	1 052	791	21	-1.4
United States	955	972	16	0.6	870	646	17	-1.3
Central and South America	217	297	5	2.5	168	169	4	-0.1
Europe	635	680	12	0.5	519	380	10	-2.1
European Union	485	502	9	0.2	387	266	7	-2.6
Africa	235	350	6	3.7	176	200	5	1.1
Middle East	674	857	15	2.2	550	507	13	-0.2
Eurasia	661	733	12	0.9	551	471	12	-1.1
Asia Pacific	1 293	1 671	28	3.3	1 234	1 322	34	2.2
China	572	730	12	4.4	508	497	13	2.6
India	154	231	4	6.2	199	303	8	7.5
International bunkers	9	23	0	29.4	14	15	0	27.0
World	4 940	5 891	100	1.8	4 264	3 854	100	-0.1
Power	1 823	2 197	37	1.5	1 580	1 248	32	-1.0
Industrial use	1 243	1 527	26	2.4	1 108	1 114	29	0.9
Coal demand (Mtce)								
North America	375	341	5	-1.7	81	50	2	-9.9
United States	351	323	5	-1.5	71	41	2	-10.3
Central and South America	55	60	1	1.2	30	23	1	-3.0
Europe	366	319	5	-1.5	129	84	4	-7.3
European Union	239	166	3	-2.9	84	59	3	-7.4
Africa	187	238	4	1.8	113	92	4	-2.5
Middle East	10	15	0	4.6	7	6	0	0.6
Eurasia	229	226	4	-0.1	136	74	4	-5.0
Asia Pacific	4 712	5 202	81	1.1	2 976	1 771	84	-3.7
China	2 949	2 838	44	0.0	2 065	1 154	55	-4.0
India	987	1 417	22	4.1	546	395	19	-1.8
World	5 934	6 399	100	0.7	3 471	2 101	100	-4.2
Power	3 789	4 156	65	0.8	1 872	858	41	-6.2
Industrial use	1 926	2 075	32	1.0	1 461	1 206	57	-1.5

Table A.2: Power sector overview

	Stated Policies Scenario						Shares (%)		CAAGR (%)
	2017	2018	2025	2030	2035	2040	2018	2040	2018-40
Electricity generation (TWh)									
North America	5 244	5 430	5 615	5 780	5 994	6 277	20	15	0.7
Central and South America	1 294	1 310	1 540	1 734	1 952	2 198	5	5	2.4
Europe	4 127	4 163	4 345	4 478	4 640	4 840	16	12	0.7
Africa	827	866	1 056	1 284	1 564	1 897	3	5	3.6
Middle East	1 126	1 147	1 306	1 570	1 876	2 169	4	5	2.9
Eurasia	1 341	1 360	1 490	1 565	1 643	1 747	5	4	1.1
Asia Pacific	11 642	12 327	15 450	17 731	20 012	22 245	46	54	2.7
China	6 639	7 170	9 037	10 177	11 192	12 071	27	29	2.4
India	1 532	1 618	2 338	3 018	3 780	4 581	6	11	4.8
World	25 601	26 603	30 803	34 140	37 682	41 373	100	100	2.0
Electrical capacity (GW)									
North America	1 415	1 429	1 535	1 654	1 787	1 934	20	15	1.4
Central and South America	346	359	434	483	532	600	5	5	2.4
Europe	1 280	1 305	1 484	1 579	1 678	1 753	18	13	1.3
Africa	228	244	315	400	501	614	3	5	4.3
Middle East	321	331	410	476	561	641	5	5	3.1
Eurasia	327	331	355	364	380	407	5	3	0.9
Asia Pacific	3 054	3 218	4 429	5 287	6 216	7 161	45	55	3.7
China	1 754	1 872	2 634	3 113	3 531	3 884	26	30	3.4
India	377	396	604	823	1 168	1 585	5	12	6.5
World	6 970	7 218	10 244	11 655	13 109		100	100	2.7
Global power sector CO_2 emissions and CO_2 intensity from electricity generation									
CO_2 emissions (Mt)	13 445	13 818	13 759	13 777	13 813	13 834	n.a.	n.a.	0.0
Intensity (g CO_2/kWh)	482	476	409	370	337	308	n.a.	n.a.	-2.0

	Stated Policies Scenario								
	Coal	Gas	Oil	Nuclear	Hydro	Wind	Solar PV	Other	Total
Cumulative retirements, 2019-2040 (GW)									
North America	127	140	64	42	32	93	39	40	578
Central and South America	7	13	19	0	9	13	3	9	74
Europe	194	59	43	77	41	158	104	43	718
Africa	32	17	22	2	3	4	3	6	90
Middle East	0	41	41	-	0	0	1	1	85
Eurasia	55	101	8	20	1	0	0	4	190
Asia Pacific	179	86	85	38	40	180	148	97	852
China	83	1	4	-	21	140	81	36	367
India	45	7	6	2	3	28	12	34	137
World	594	458	282	180	126	448	298	200	2 586
Cumulative additions, 2019-2040 (GW)									
North America	5	310	10	8	48	221	365	113	1 081
Central and South America	7	78	3	7	83	56	55	26	315
Europe	57	159	1	46	71	397	314	121	1 165
Africa	33	121	16	6	50	51	133	51	460
Middle East	6	197	23	12	7	37	92	22	395
Eurasia	33	141	0	26	20	25	5	15	265
Asia Pacific	547	358	19	139	381	952	1 980	417	4 792
China	155	128	0	90	192	620	1 041	152	2 378
India	232	17	5	27	61	211	606	169	1 327
World	687	1 365	72	243	659	1 738	2 944	765	8 473

Table A.2: Power sector overview

	Current Policies Scenario				Sustainable Development Scenario			
	2030	2040	Shares (%) 2040	CAAGR (%) 2018-40	2030	2040	Shares (%) 2040	CAAGR (%) 2018-40
Electricity generation (TWh)								
North America	5 872	6 354	15	0.7	5 527	6 186	16	0.6
Central and South America	1 811	2 351	5	2.7	1 592	1 975	5	1.9
Europe	4 622	5 012	12	0.8	4 429	5 246	14	1.1
Africa	1 338	2 029	5	3.9	1 267	1 976	5	3.8
Middle East	1 656	2 328	5	3.3	1 416	1 909	5	2.3
Eurasia	1 610	1 862	4	1.4	1 362	1 437	4	0.3
Asia Pacific	18 079	22 887	53	2.9	16 208	19 984	52	2.2
China	10 233	12 188	28	2.4	9 286	10 899	28	1.9
India	3 169	4 875	11	5.1	2 764	3 940	10	4.1
World	34 988	42 824	100	2.2	31 800	38 713	100	1.7
Electrical capacity (GW)								
North America	1 648	1 901	15	1.3	1 702	2 228	14	2.0
Central and South America	482	614	5	2.5	472	603	4	2.4
Europe	1 598	1 724	14	1.3	1 691	2 066	13	2.1
Africa	390	557	4	3.8	484	852	6	5.9
Middle East	477	631	5	3.0	508	783	5	4.0
Eurasia	367	419	3	1.1	344	424	3	1.1
Asia Pacific	5 231	6 812	54	3.5	5 841	8 522	55	4.5
China	3 074	3 818	30	3.3	3 378	4 706	30	4.3
India	834	1 386	11	5.9	993	1 741	11	7.0
World	10 194	12 658	100	2.6	11 042	15 478	100	3.5
Global power sector CO_2 emissions and CO_2 intensity from electricity generation								
CO_2 emissions (Mt)	14 951	16 594	n.a.	0.8	8 460	3 780	n.a.	-5.7
Intensity (g CO_2/kWh)	394	359	n.a.	-1.3	237	81	n.a.	-7.7

	Current Policies Scenario					Sustainable Development Scenario				
	Coal	Gas	Nuclear	Renew.	Total	Coal	Gas	Nuclear	Renew.	Total
Cumulative retirements, 2019-2040 (GW)										
North America	106	140	36	185	553	244	141	25	185	691
Central and South America	3	13	0	33	70	9	14	0	33	77
Europe	154	57	77	330	679	218	57	68	331	739
Africa	32	17	2	10	90	35	17	2	11	100
Middle East	0	41	-	2	83	0	41	-	2	84
Eurasia	55	101	20	4	189	60	102	20	4	195
Asia Pacific	179	85	41	383	842	625	87	21	386	1 317
China	83	1	-	244	363	352	1	-	245	661
India	45	7	2	44	129	128	7	2	45	228
World	529	456	176	946	2 506	1 192	460	136	952	3 204
Cumulative additions, 2019-2040 (GW)										
North America	5	319	8	614	1 023	12	187	16	1 151	1 489
Central and South America	7	94	7	210	326	3	29	8	269	321
Europe	81	223	44	693	1 097	20	161	53	1 180	1 498
Africa	62	131	6	153	404	20	63	11	537	709
Middle East	7	197	12	130	382	3	101	16	394	537
Eurasia	47	156	30	40	277	2	81	32	167	288
Asia Pacific	872	500	138	2 678	4 434	207	395	182	5 364	6 618
China	329	178	93	1 619	2 308	86	111	124	2 967	3 494
India	312	79	27	599	1 120	63	125	31	1 150	1 573
World	1 081	1 619	246	4 517	7 942	267	1 018	318	9 061	11 460

Table A.3: Energy demand – World

	Stated Policies Scenario Energy demand (Mtoe)							Shares (%)		CAAGR (%)
	2010	2017	2018	2025	2030	2035	2040	2018	2040	2018-40
Total primary demand	12 853	13 997	14 314	15 538	16 311	17 011	17 723	100	100	1.0
Coal	3 653	3 793	3 821	3 861	3 848	3 813	3 779	27	21	-0.1
Oil	4 124	4 453	4 501	4 791	4 872	4 897	4 921	31	28	0.4
Natural gas	2 749	3 128	3 273	3 638	3 889	4 167	4 445	23	25	1.4
Nuclear	719	687	709	730	801	855	906	5	5	1.1
Hydro	296	351	361	409	452	489	524	3	3	1.7
Bioenergy	1 202	1 328	1 357	1 561	1 671	1 749	1 828	9	10	1.4
Other renewables	110	256	293	548	777	1 041	1 320	2	7	7.1
Power sector	4 780	5 309	5 494	5 879	6 276	6 708	7 152	100	100	1.2
Coal	2 175	2 374	2 450	2 427	2 429	2 409	2 377	45	33	-0.1
Oil	265	225	218	197	167	148	132	4	2	-2.3
Natural gas	1 115	1 261	1 302	1 359	1 417	1 507	1 606	24	22	1.0
Nuclear	719	687	709	730	801	855	906	13	13	1.1
Hydro	296	351	361	409	452	489	524	7	7	1.7
Bioenergy	122	200	213	288	337	389	445	4	6	3.4
Other renewables	88	210	241	469	673	910	1 162	4	16	7.4
Other energy sector	1 422	1 491	1 506	1 674	1 726	1 764	1 823	100	100	0.9
Electricity	309	364	371	401	430	463	501	25	27	1.4
Total final consumption	8 841	9 744	9 954	10 997	11 607	12 145	12 672	100	100	1.1
Coal	1 050	1 019	984	991	979	963	954	10	8	-0.1
Oil	3 580	3 985	4 043	4 355	4 469	4 517	4 561	41	36	0.5
Natural gas	1 372	1 528	1 615	1 864	2 032	2 205	2 360	16	19	1.7
Electricity	1 543	1 840	1 915	2 245	2 503	2 777	3 061	19	24	2.2
Heat	274	289	296	311	313	313	312	3	2	0.2
Bioenergy	1 001	1 037	1 050	1 151	1 205	1 237	1 264	11	10	0.8
Other renewables	21	45	51	79	104	131	158	1	1	5.2
Industry	2 653	2 862	2 898	3 255	3 460	3 650	3 839	100	100	1.3
Coal	860	817	795	824	832	837	843	27	22	0.3
Oil	338	310	298	304	305	303	303	10	8	0.1
Natural gas	511	620	646	791	883	975	1 066	22	28	2.3
Electricity	639	768	803	939	1 018	1 092	1 163	28	30	1.7
Heat	123	138	143	152	152	150	147	5	4	0.1
Bioenergy	182	208	213	242	265	285	305	7	8	1.6
Other renewables	0	1	1	2	4	8	11	0	0	11.4
Transport	2 422	2 810	2 863	3 156	3 327	3 459	3 606	100	100	1.1
Oil	2 252	2 590	2 629	2 824	2 906	2 932	2 963	92	82	0.5
International bunkers	359	412	416	468	500	536	572	15	16	1.5
Electricity	25	31	32	58	88	127	173	1	5	7.9
Biofuels	56	83	89	133	167	195	223	3	6	4.3
Other fuels	89	105	113	142	166	204	247	4	7	3.6
Buildings	2 830	3 011	3 101	3 293	3 455	3 608	3 758	100	100	0.9
Coal	137	130	125	88	67	45	30	4	1	-6.3
Oil	317	325	330	308	291	273	263	11	7	-1.0
Natural gas	622	657	700	746	784	817	829	23	22	0.8
Electricity	834	974	1 011	1 165	1 307	1 462	1 625	33	43	2.2
Heat	147	148	150	155	158	160	162	5	4	0.4
Bioenergy	753	734	737	758	753	733	709	24	19	-0.2
Traditional biomass	639	619	620	628	613	582	546	20	15	-0.6
Other renewables	20	42	48	73	95	118	140	2	4	5.0
Other	937	1 061	1 092	1 292	1 365	1 428	1 470	100	100	1.4
Petrochem. feedstock	554	544	549	693	743	789	825	50	56	1.9

Table A.3: Energy demand – World

	Current Policies Scenario				Sustainable Development Scenario			
	Energy demand (Mtoe)		Shares (%)	CAAGR (%)	Energy demand (Mtoe)		Shares (%)	CAAGR (%)
	2030	2040	2040	2018-40	2030	2040	2040	2018-40
Total primary demand	16 960	19 177	100	1.3	13 750	13 279	100	-0.3
Coal	4 154	4 479	23	0.7	2 430	1 470	11	-4.2
Oil	5 174	5 626	29	1.0	3 995	3 041	23	-1.8
Natural gas	4 070	4 847	25	1.8	3 513	3 162	24	-0.2
Nuclear	811	937	5	1.3	895	1 149	9	2.2
Hydro	445	509	3	1.6	489	596	4	2.3
Bioenergy	1 626	1 736	9	1.1	1 319	1 628	12	0.8
Other renewables	681	1 042	5	5.9	1 109	2 231	17	9.7
Power sector	6 519	7 649	100	1.5	5 467	6 048	100	0.4
Coal	2 652	2 909	38	0.8	1 311	601	10	-6.2
Oil	176	158	2	-1.5	102	63	1	-5.5
Natural gas	1 513	1 823	24	1.5	1 313	1 038	17	-1.0
Nuclear	811	937	12	1.3	895	1 149	19	2.2
Hydro	445	509	7	1.6	489	596	10	2.3
Bioenergy	328	394	5	2.8	406	639	11	5.1
Other renewables	595	918	12	6.3	952	1 962	32	10.0
Other energy sector	1 831	2 062	100	1.4	1 419	1 410	100	-0.3
Electricity	454	548	27	1.8	379	423	30	0.6
Total final consumption	11 996	13 540	100	1.4	9 904	9 500	100	-0.2
Coal	1 044	1 078	8	0.4	746	533	6	-2.7
Oil	4 736	5 191	38	1.1	3 695	2 838	30	-1.6
Natural gas	2 084	2 465	18	1.9	1 816	1 719	18	0.3
Electricity	2 550	3 132	23	2.3	2 349	2 902	31	1.9
Heat	327	340	3	0.6	264	224	2	-1.3
Bioenergy	1 167	1 210	9	0.6	870	948	10	-0.5
Other renewables	86	124	1	4.1	157	269	3	7.8
Industry	3 540	4 020	100	1.5	2 949	2 904	100	0.0
Coal	866	920	23	0.7	627	470	16	-2.4
Oil	311	309	8	0.2	221	174	6	-2.4
Gas	897	1 102	27	2.5	806	813	28	1.1
Electricity	1 032	1 193	30	1.8	925	1 047	36	1.2
Heat	161	166	4	0.7	116	85	3	-2.4
Bioenergy	271	324	8	1.9	237	265	9	1.0
Other renewables	3	7	0	8.5	18	43	1	18.3
Transport	3 475	3 981	100	1.5	2 956	2 615	100	-0.4
Oil	3 129	3 512	88	1.3	2 315	1 572	60	-2.3
International bunkers	547	672	17	2.2	382	319	12	-1.2
Electricity	62	94	2	4.9	118	350	13	11.4
Biofuels	133	168	4	2.9	298	373	14	6.7
Other fuels	151	207	5	2.8	225	320	12	4.8
Buildings	3 607	4 039	100	1.2	2 735	2 709	100	-0.6
Coal	94	70	2	-2.6	54	10	0	-11.0
Oil	327	322	8	-0.1	254	175	6	-2.8
Natural gas	839	938	23	1.3	614	462	17	-1.9
Electricity	1 362	1 735	43	2.5	1 227	1 428	53	1.6
Heat	163	171	4	0.6	145	137	5	-0.4
Bioenergy	743	690	17	-0.3	306	269	10	-4.5
Traditional biomass	613	546	14	-0.6	140	75	3	-9.2
Other renewables	80	113	3	4.0	132	217	8	7.1
Other	1 374	1 500	100	1.5	1 264	1 272	100	0.7
Petrochem. feedstock	738	825	55	1.9	681	707	56	1.2

Table A.3: Electricity and CO_2 emissions – World

Stated Policies Scenario

Electricity generation (TWh)

	2010	2017	2018	2025	2030	2035	2040	Shares (%) 2018	Shares (%) 2040	CAAGR (%) 2018-40
Total generation	21 522	25 601	26 603	30 803	34 140	37 682	41 373	100	100	2.0
Coal	8 666	9 863	10 123	10 291	10 408	10 444	10 431	38	25	0.1
Oil	976	841	808	724	622	556	490	3	1	-2.3
Natural gas	4 833	5 879	6 118	6 984	7 529	8 165	8 899	23	22	1.7
Nuclear	2 756	2 636	2 718	2 801	3 073	3 282	3 475	10	8	1.1
Renewables	4 257	6 345	6 799	9 972	12 479	15 204	18 049	26	44	4.5
Hydro	3 443	4 083	4 203	4 759	5 255	5 685	6 098	16	15	1.7
Bioenergy	369	594	636	916	1 085	1 266	1 459	2	4	3.8
Wind	341	1 127	1 265	2 411	3 317	4 305	5 226	5	13	6.7
Geothermal	68	85	90	125	182	248	316	0	1	5.9
Solar PV	32	443	592	1 730	2 562	3 551	4 705	2	11	9.9
CSP	2	11	12	28	67	124	196	0	0	13.7
Marine	1	1	1	2	10	25	49	0	0	19.0

Stated Policies Scenario

Electrical capacity (GW)

	2017	2018	2025	2030	2035	2040	Shares (%) 2018	Shares (%) 2040	CAAGR (%) 2018-40
Total capacity	6 970	7 218	8 962	10 244	11 655	13 109	100	100	2.7
Coal	2 071	2 079	2 106	2 111	2 132	2 171	29	17	0.2
Oil	450	450	344	298	271	239	6	2	-2.8
Natural gas	1 693	1 745	2 070	2 254	2 448	2 651	24	20	1.9
Nuclear	413	419	417	436	459	482	6	4	0.6
Renewables	2 339	2 517	3 962	5 019	6 120	7 233	35	55	4.9
Hydro	1 269	1 290	1 448	1 586	1 707	1 822	18	14	1.6
Bioenergy	137	146	194	224	255	286	2	2	3.1
Wind	515	566	980	1 288	1 594	1 856	8	14	5.6
Geothermal	14	14	19	27	37	46	0	0	5.5
Solar PV	399	495	1 309	1 866	2 476	3 142	7	24	8.8
CSP	5	6	11	23	40	61	0	0	11.4
Marine	1	1	1	4	10	20	0	0	17.8

Stated Policies Scenario

CO_2 emissions (Mt)

	2010	2017	2018	2025	2030	2035	2040	Shares (%) 2018	Shares (%) 2040	CAAGR (%) 2018-40
Total CO_2	30 412	32 631	33 243	34 346	34 860	35 198	35 589	100	100	0.3
Coal	13 808	14 498	14 664	14 473	14 343	14 120	13 891	44	39	-0.2
Oil	10 546	11 393	11 446	11 930	12 031	11 991	12 001	34	34	0.2
Natural gas	6 057	6 740	7 134	7 944	8 486	9 087	9 697	21	27	1.4
Power sector	12 413	13 445	13 818	13 759	13 777	13 813	13 834	100	100	0.0
Coal	8 942	9 759	10 066	9 942	9 920	9 803	9 641	73	70	-0.2
Oil	844	714	692	621	526	468	418	5	3	-2.3
Natural gas	2 627	2 972	3 060	3 196	3 332	3 542	3 775	22	27	1.0
Final consumption	16 365	17 603	17 809	18 815	19 286	19 576	19 895	100	100	0.5
Coal	4 442	4 411	4 286	4 243	4 163	4 071	4 012	24	20	-0.3
Oil	9 079	10 100	10 167	10 729	10 935	10 959	11 022	57	55	0.4
Transport	6 783	7 801	7 917	8 508	8 758	8 841	8 937	44	45	0.6
Of which: bunkers	1 121	1 281	1 293	1 455	1 552	1 662	1 772	7	9	1.4
Natural gas	2 844	3 093	3 356	3 843	4 188	4 546	4 861	19	24	1.7

Table A.3: Electricity and CO₂ emissions – World

	Current Policies Scenario				Sustainable Development Scenario			
	Electricity generation (TWh)		Shares (%)	CAAGR (%)	Electricity generation (TWh)		Shares (%)	CAAGR (%)
	2030	2040	2040	2018-40	2030	2040	2040	2018-40
Total generation	34 988	42 824	100	2.2	31 800	38 713	100	1.7
Coal	11 464	12 923	30	1.1	5 504	2 428	6	-6.3
Oil	669	603	1	-1.3	355	197	1	-6.2
Natural gas	8 086	10 186	24	2.3	7 043	5 584	14	-0.4
Nuclear	3 112	3 597	8	1.3	3 435	4 409	11	2.2
Renewables	11 627	15 485	36	3.8	15 434	26 065	67	6.3
Hydro	5 171	5 923	14	1.6	5 685	6 934	18	2.3
Bioenergy	1 022	1 256	3	3.1	1 335	2 196	6	5.8
Wind	2 955	4 258	10	5.7	4 453	8 295	21	8.9
Geothermal	161	258	1	4.9	282	552	1	8.6
Solar PV	2 265	3 658	9	8.6	3 513	7 208	19	12.0
CSP	46	104	0	10.5	153	805	2	21.2
Marine	7	28	0	16.0	14	75	0	21.3

	Current Policies Scenario				Sustainable Development Scenario			
	Electrical capacity (GW)		Shares (%)	CAAGR (%)	Electrical capacity (GW)		Shares (%)	CAAGR (%)
	2030	2040	2040	2018-40	2030	2040	2040	2018-40
Total capacity	10 194	12 658	100	2.6	11 042	15 478	100	3.5
Coal	2 316	2 630	21	1.1	1 644	1 153	7	-2.6
Oil	312	268	2	-2.3	294	240	2	-2.8
Natural gas	2 387	2 908	23	2.3	2 084	2 304	15	1.3
Nuclear	436	489	4	0.7	482	601	4	1.7
Renewables	4 618	6 087	48	4.1	6 359	10 626	69	6.8
Hydro	1 555	1 758	14	1.4	1 728	2 090	14	2.2
Bioenergy	211	247	2	2.4	272	425	3	5.0
Wind	1 162	1 537	12	4.6	1 721	2 930	19	7.8
Geothermal	24	38	0	4.5	43	82	1	8.3
Solar PV	1 646	2 465	19	7.6	2 537	4 815	31	10.9
CSP	16	32	0	8.2	52	254	2	18.9
Marine	3	11	0	14.6	6	30	0	20.0

	Current Policies Scenario				Sustainable Development Scenario			
	CO_2 emissions (Mt)		Shares (%)	CAAGR (%)	CO_2 emissions (Mt)		Shares (%)	CAAGR (%)
	2030	2040	2040	2018-40	2030	2040	2040	2018-40
Total CO_2	37 379	41 302	100	1.0	25 181	15 796	100	-3.3
Coal	15 548	16 609	40	0.6	8 281	3 424	22	-6.4
Oil	12 905	14 053	34	0.9	9 436	6 433	41	-2.6
Natural gas	8 927	10 639	26	1.8	7 464	6 032	38	-0.8
Power sector	14 951	16 594	100	0.8	8 460	3 780	100	-5.7
Coal	10 839	11 813	71	0.7	5 126	1 552	41	-8.1
Oil	555	497	3	-1.5	325	200	5	-5.5
Natural gas	3 558	4 284	26	1.5	3 009	2 123	56	-1.6
Final consumption	20 507	22 561	100	1.1	15 344	11 037	100	-2.2
Coal	4 441	4 539	20	0.3	2 964	1 749	16	-4.0
Oil	11 745	12 903	57	1.1	8 684	5 964	54	-2.4
Transport	9 433	10 597	47	1.3	6 976	4 747	43	-2.3
Of which: bunkers	1 699	2 084	9	2.2	1 187	990	9	-1.2
Natural gas	4 321	5 118	23	1.9	3 696	3 324	30	-0.0

Table A.3: Energy demand – North America

	Stated Policies Scenario Energy demand (Mtoe)							Shares (%)		CAAGR (%)
	2010	2017	2018	2025	2030	2035	2040	2018	2040	2018-40
Total primary demand	2 655	2 625	2 714	2 738	2 717	2 693	2 686	100	100	-0.0
Coal	538	361	345	258	230	213	200	13	7	-2.5
Oil	997	976	997	991	947	896	851	37	32	-0.7
Natural gas	686	807	884	965	982	992	1 014	33	38	0.6
Nuclear	244	248	250	217	205	192	178	9	7	-1.5
Hydro	56	63	61	66	68	70	72	2	3	0.7
Bioenergy	111	124	126	145	160	174	187	5	7	1.8
Other renewables	24	47	52	96	125	155	186	2	7	6.0
Power sector	1 084	1 012	1 035	969	954	956	969	100	100	-0.3
Coal	490	331	315	226	199	181	168	30	17	-2.8
Oil	24	18	19	6	5	4	3	2	0	-8.7
Natural gas	226	284	317	336	334	336	348	31	36	0.4
Nuclear	244	248	250	217	205	192	178	24	18	-1.5
Hydro	56	63	61	66	68	70	72	6	7	0.7
Bioenergy	23	25	25	29	31	34	37	2	4	1.8
Other renewables	21	44	49	88	113	139	165	5	17	5.7
Other energy sector	213	236	249	293	308	315	328	100	100	1.3
Electricity	61	62	64	65	66	68	71	26	22	0.4
Total final consumption	1 821	1 839	1 908	1 969	1 962	1 946	1 938	100	100	0.1
Coal	34	23	23	23	22	21	20	1	1	-0.6
Oil	926	910	929	926	882	831	784	49	40	-0.8
Natural gas	377	409	443	473	481	484	488	23	25	0.4
Electricity	389	389	402	417	430	447	468	21	24	0.7
Heat	7	6	6	6	6	6	5	0	0	-1.0
Bioenergy	86	99	101	116	129	140	150	5	8	1.8
Other renewables	2	2	3	8	12	16	21	0	1	9.0
Industry	350	345	356	387	393	395	402	100	100	0.6
Coal	32	22	22	22	21	21	20	6	5	-0.5
Oil	45	31	30	31	30	30	31	9	8	0.1
Natural gas	134	153	161	184	187	186	187	45	46	0.7
Electricity	97	95	98	105	108	111	115	28	29	0.7
Heat	6	5	5	5	5	5	4	1	1	-0.8
Bioenergy	36	38	39	40	41	42	44	11	11	0.6
Other renewables	0	0	0	0	0	1	1	0	0	21.1
Transport	705	738	755	743	721	696	676	100	100	-0.5
Oil	661	675	689	658	620	575	535	91	79	-1.1
Electricity	1	2	2	4	7	12	21	0	3	11.9
Biofuels	25	40	42	53	61	69	75	6	11	2.7
Other fuels	19	22	22	28	33	39	46	3	7	3.3
Buildings	573	553	587	582	588	593	599	100	100	0.1
Coal	2	0	0	0	0	0	-	0	-	n.a.
Oil	51	39	40	33	27	22	16	7	3	-4.0
Natural gas	207	205	226	217	215	212	208	39	35	-0.4
Electricity	286	286	296	302	310	318	326	51	54	0.4
Heat	1	1	1	1	1	1	1	0	0	-1.8
Bioenergy	24	20	19	22	24	26	28	3	5	1.8
Traditional biomass	-	-	-	-	-	-	-	-	-	n.a.
Other renewables	2	2	3	7	11	15	19	1	3	8.5
Other	192	203	210	257	259	262	260	100	100	1.0
Petrochem. feedstock	114	79	81	109	112	113	112	39	43	1.5

Table A.3: Energy demand – North America

	Current Policies Scenario				Sustainable Development Scenario			
	Energy demand (Mtoe)		Shares (%)	CAAGR (%)	Energy demand (Mtoe)		Shares (%)	CAAGR (%)
	2030	2040	2040	2018-40	2030	2040	2040	2018-40
Total primary demand	2 806	2 878	100	0.3	2 377	2 087	100	-1.2
Coal	262	239	8	-1.7	57	35	2	-9.9
Oil	994	970	34	-0.1	766	499	24	-3.1
Natural gas	1 008	1 063	37	0.8	873	656	31	-1.3
Nuclear	208	190	7	-1.2	222	228	11	-0.4
Hydro	68	72	3	0.8	70	75	4	0.9
Bioenergy	152	178	6	1.6	214	249	12	3.1
Other renewables	115	165	6	5.4	176	345	17	9.0
Power sector	985	1 009	100	-0.1	852	889	100	-0.7
Coal	230	205	20	-1.9	34	17	2	-12.5
Oil	6	3	0	-7.6	3	1	0	-11.5
Natural gas	335	347	34	0.4	331	201	23	-2.0
Nuclear	208	190	19	-1.2	222	228	26	-0.4
Hydro	68	72	7	0.8	70	75	8	0.9
Bioenergy	31	37	4	1.9	38	62	7	4.3
Other renewables	108	154	15	5.4	155	305	34	8.7
Other energy sector	328	373	100	1.9	259	280	100	0.5
Electricity	69	75	20	0.7	58	59	21	-0.4
Total final consumption	2 007	2 051	100	0.3	1 750	1 500	100	-1.1
Coal	22	21	1	-0.3	16	11	1	-3.3
Oil	923	889	43	-0.2	717	464	31	-3.1
Natural gas	493	511	25	0.7	399	300	20	-1.8
Electricity	435	471	23	0.7	416	472	31	0.7
Heat	6	5	0	-0.7	5	4	0	-2.5
Bioenergy	121	141	7	1.5	176	186	12	2.8
Other renewables	7	12	1	6.1	21	41	3	12.3
Industry	398	413	100	0.7	346	328	100	-0.4
Coal	22	21	5	-0.3	15	11	3	-3.2
Oil	31	31	8	0.1	20	15	5	-3.1
Gas	190	193	47	0.8	160	129	39	-1.0
Electricity	108	115	28	0.7	104	117	36	0.8
Heat	5	4	1	-0.7	4	3	1	-2.8
Bioenergy	42	47	11	0.9	40	45	14	0.7
Other renewables	0	1	0	20.6	2	6	2	30.6
Transport	741	737	100	-0.1	633	463	100	-2.2
Oil	654	627	85	-0.4	481	252	54	-4.5
Electricity	3	5	1	5.0	15	57	12	17.1
Biofuels	55	67	9	2.2	103	99	21	4.0
Other fuels	29	38	5	2.4	35	55	12	4.2
Buildings	609	640	100	0.4	527	473	100	-1.0
Coal	0	0	0	-2.8	0	-	-	n.a.
Oil	34	28	4	-1.5	23	9	2	-6.7
Natural gas	228	232	36	0.1	166	102	22	-3.6
Electricity	317	344	54	0.7	293	294	62	-0.0
Heat	1	1	0	-0.6	1	1	0	-1.6
Bioenergy	22	24	4	1.0	28	33	7	2.4
Traditional biomass	-	-	-	n.a.	-	-	-	n.a.
Other renewables	6	10	2	5.5	17	33	7	11.3
Other	259	261	100	1.0	244	237	100	0.5
Petrochem. feedstock	110	111	42	1.4	100	92	39	0.6

Annex A | Tables for scenario projections

Table A.3: Electricity and CO_2 emissions – North America

	Stated Policies Scenario									
	Electricity generation (TWh)							Shares (%)		CAAGR (%)
	2010	2017	2018	2025	2030	2035	2040	2018	2040	2018-40
Total generation	5 234	5 244	5 430	5 615	5 780	5 994	6 277	100	100	0.7
Coal	2 106	1 412	1 344	985	874	810	757	25	12	-2.6
Oil	101	78	86	31	23	19	12	2	0	-8.7
Natural gas	1 216	1 585	1 793	2 026	2 074	2 124	2 240	33	36	1.0
Nuclear	935	951	959	833	788	738	683	18	11	-1.5
Renewables	868	1 212	1 242	1 738	2 019	2 302	2 583	23	41	3.4
Hydro	651	728	710	771	793	813	834	13	13	0.7
Bioenergy	84	87	87	105	114	124	133	2	2	1.9
Wind	105	296	321	538	644	764	866	6	14	4.6
Geothermal	24	25	25	29	36	44	50	0	1	3.2
Solar PV	3	72	95	291	422	543	677	2	11	9.4
CSP	1	4	4	4	7	10	17	0	0	6.9
Marine	0	0	0	0	2	5	7	0	0	26.7

	Stated Policies Scenario								
	Electrical capacity (GW)						Shares (%)		CAAGR (%)
	2017	2018	2025	2030	2035	2040	2018	2040	2018-40
Total capacity	1 415	1 429	1 535	1 654	1 787	1 934	100	100	1.4
Coal	296	276	208	181	165	154	19	8	-2.6
Oil	79	79	37	31	29	25	6	1	-5.1
Natural gas	532	546	582	622	661	715	38	37	1.2
Nuclear	122	121	107	101	94	87	8	4	-1.5
Renewables	385	406	591	699	798	892	28	46	3.6
Hydro	196	196	202	205	208	212	14	11	0.3
Bioenergy	22	23	24	26	28	29	2	2	1.2
Wind	104	113	171	198	223	242	8	13	3.5
Geothermal	5	4	5	6	7	8	0	0	2.5
Solar PV	56	68	187	260	326	394	5	20	8.3
CSP	2	2	2	2	3	5	0	0	4.4
Marine	0	0	0	1	2	2	0	0	23.3

	Stated Policies Scenario									
	CO_2 emissions (Mt)							Shares (%)		CAAGR (%)
	2010	2017	2018	2025	2030	2035	2040	2018	2040	2018-40
Total CO_2	6 297	5 735	5 887	5 544	5 322	5 102	4 953	100	100	-0.8
Coal	2 128	1 444	1 381	1 026	907	825	764	23	15	-2.7
Oil	2 575	2 492	2 543	2 395	2 256	2 099	1 962	43	40	-1.2
Natural gas	1 594	1 799	1 963	2 123	2 159	2 178	2 227	33	45	0.6
Power sector	2 579	2 055	2 074	1 721	1 595	1 520	1 485	100	100	-1.5
Coal	1 970	1 330	1 265	910	795	717	658	61	44	-2.9
Oil	78	60	63	22	15	13	9	3	1	-8.7
Natural gas	531	666	745	789	784	790	818	36	55	0.4
Final consumption	3 299	3 277	3 390	3 312	3 189	3 037	2 909	100	100	-0.7
Coal	146	104	104	103	99	96	93	3	3	-0.5
Oil	2 304	2 277	2 322	2 207	2 075	1 922	1 787	68	61	-1.2
Transport	1 955	1 996	2 039	1 947	1 834	1 700	1 582	60	54	-1.1
Natural gas	849	896	964	1 002	1 015	1 020	1 029	28	35	0.3

Table A.3: Electricity and CO_2 emissions – North America

	Current Policies Scenario				Sustainable Development Scenario			
	Electricity generation (TWh)		Shares (%)	CAAGR (%)	Electricity generation (TWh)		Shares (%)	CAAGR (%)
	2030	2040	2040	2018-40	2030	2040	2040	2018-40
Total generation	5 872	6 354	100	0.7	5 527	6 186	100	0.6
Coal	1 015	922	15	-1.7	151	70	1	-12.5
Oil	29	15	0	-7.5	13	6	0	-11.3
Natural gas	2 069	2 222	35	1.0	2 039	1 247	20	-1.6
Nuclear	798	731	12	-1.2	851	876	14	-0.4
Renewables	1 960	2 462	39	3.2	2 471	3 984	64	5.4
Hydro	796	842	13	0.8	811	869	14	0.9
Bioenergy	113	132	2	1.9	147	251	4	4.9
Wind	619	809	13	4.3	911	1 596	26	7.6
Geothermal	37	51	1	3.3	45	81	1	5.5
Solar PV	388	608	10	8.8	524	1 019	16	11.4
CSP	6	15	0	6.1	31	154	2	18.1
Marine	2	7	0	26.2	2	15	0	30.9

	Current Policies Scenario				Sustainable Development Scenario			
	Electrical capacity (GW)		Shares (%)	CAAGR (%)	Electrical capacity (GW)		Shares (%)	CAAGR (%)
	2030	2040	2040	2018-40	2030	2040	2040	2018-40
Total capacity	1 648	1 901	100	1.3	1 702	2 228	100	2.0
Coal	194	175	9	-2.0	114	44	2	-8.0
Oil	33	26	1	-4.9	29	21	1	-5.8
Natural gas	626	724	38	1.3	552	592	27	0.4
Nuclear	103	93	5	-1.2	110	112	5	-0.3
Renewables	669	835	44	3.3	865	1 372	62	5.7
Hydro	205	213	11	0.4	210	221	10	0.5
Bioenergy	26	29	2	1.1	35	56	3	4.2
Wind	191	226	12	3.2	276	441	20	6.4
Geothermal	6	8	0	2.5	7	12	1	4.7
Solar PV	238	353	19	7.8	326	591	27	10.4
CSP	2	4	0	3.5	11	45	2	15.9
Marine	1	2	0	22.9	1	5	0	27.7

	Current Policies Scenario				Sustainable Development Scenario			
	CO_2 emissions (Mt)		Shares (%)	CAAGR (%)	CO_2 emissions (Mt)		Shares (%)	CAAGR (%)
	2030	2040	2040	2018-40	2030	2040	2040	2018-40
Total CO_2	5 656	5 576	100	-0.2	3 762	2 075	100	-4.6
Coal	1 037	920	17	-1.8	192	58	3	-13.4
Oil	2 394	2 306	41	-0.4	1 728	922	44	-4.5
Natural gas	2 226	2 350	42	0.8	1 842	1 113	54	-2.5
Power sector	1 727	1 634	100	-1.1	849	303	100	-8.4
Coal	921	808	49	-2.0	116	16	5	-18.1
Oil	19	11	1	-7.6	9	4	1	-11.5
Natural gas	786	816	50	0.4	724	302	99	-4.0
Final consumption	3 353	3 292	100	-0.1	2 501	1 488	100	-3.7
Coal	102	99	3	-0.2	67	38	3	-4.5
Oil	2 200	2 103	64	-0.4	1 601	847	57	-4.5
Transport	1 934	1 855	56	-0.4	1 423	746	50	-4.5
Natural gas	1 051	1 090	33	0.6	834	604	41	-2.1

Table A.3: Energy demand – United States

	Stated Policies Scenario Energy demand (Mtoe)							Shares (%)		CAAGR (%)
	2010	2017	2018	2025	2030	2035	2040	2018	2040	2018-40
Total primary demand	2 215	2 150	2 230	2 247	2 214	2 173	2 142	100	100	-0.2
Coal	503	331	316	245	220	203	190	14	9	-2.3
Oil	806	790	812	810	766	717	671	36	31	-0.9
Natural gas	556	644	711	776	785	787	793	32	37	0.5
Nuclear	219	219	219	193	181	164	147	10	7	-1.8
Hydro	23	26	25	26	27	28	28	1	1	0.5
Bioenergy	89	101	103	118	130	143	154	5	7	1.9
Other renewables	19	40	44	79	104	131	157	2	7	5.9
Power sector	936	850	873	816	797	788	789	100	100	-0.5
Coal	463	308	293	220	195	179	166	34	21	-2.6
Oil	11	7	9	3	3	2	1	1	0	-7.9
Natural gas	185	232	264	279	273	273	280	30	35	0.3
Nuclear	219	219	219	193	181	164	147	25	19	-1.8
Hydro	23	26	25	26	27	28	28	3	4	0.5
Bioenergy	20	21	21	22	24	26	29	2	4	1.4
Other renewables	17	38	41	72	93	116	138	5	18	5.6
Other energy sector	154	162	170	199	207	208	209	100	100	0.9
Electricity	51	50	52	52	51	51	52	31	25	0.0
Total final consumption	1 513	1 520	1 584	1 638	1 622	1 600	1 582	100	100	-0.0
Coal	27	17	17	17	16	16	15	1	1	-0.6
Oil	762	748	766	766	723	676	630	48	40	-0.9
Natural gas	322	346	375	402	408	409	410	24	26	0.4
Electricity	326	321	334	344	352	362	377	21	24	0.5
Heat	7	6	6	6	5	5	4	0	0	-1.2
Bioenergy	68	80	82	96	107	117	126	5	8	2.0
Other renewables	2	2	3	7	11	15	19	0	1	9.0
Industry	274	261	272	301	303	303	306	100	100	0.5
Coal	25	17	17	17	16	16	15	6	5	-0.5
Oil	34	18	18	20	20	20	21	7	7	0.7
Natural gas	110	125	132	151	152	150	148	48	49	0.5
Electricity	71	67	70	76	77	79	81	26	27	0.7
Heat	5	5	5	5	4	4	4	2	1	-0.9
Bioenergy	29	30	31	32	33	34	35	11	11	0.6
Other renewables	-	-	-	0	0	1	1	-	0	n.a.
Transport	596	625	641	629	607	582	562	100	100	-0.6
Oil	556	568	582	552	514	472	433	91	77	-1.3
Electricity	1	1	1	3	5	10	18	0	3	13.9
Biofuels	23	38	39	50	58	66	71	6	13	2.7
Other fuels	16	18	19	25	30	35	41	3	7	3.7
Buildings	490	468	498	490	493	495	497	100	100	-0.0
Coal	2	0	0	0	0	0	-	0	-	n.a.
Oil	37	26	27	21	17	12	7	5	1	-5.8
Natural gas	182	178	197	187	185	182	178	40	36	-0.5
Electricity	251	249	260	262	266	271	276	52	55	0.3
Heat	1	1	1	1	1	1	1	0	0	-2.3
Bioenergy	15	11	10	12	14	16	17	2	4	2.4
Traditional biomass	-	-	-	-	-	-	-	-	-	n.a.
Other renewables	2	2	3	7	10	14	18	1	4	8.7
Other	153	166	172	218	219	220	217	100	100	1.1
Petrochem. feedstock	95	60	63	92	93	94	93	37	43	1.8

Table A.3: Energy demand – United States

	Current Policies Scenario				Sustainable Development Scenario			
	Energy demand (Mtoe)		Shares (%)	CAAGR (%)	Energy demand (Mtoe)		Shares (%)	CAAGR (%)
	2030	2040	2040	2018-40	2030	2040	2040	2018-40
Total primary demand	2 269	2 267	100	0.1	1 942	1 687	100	-1.3
Coal	246	226	10	-1.5	49	29	2	-10.3
Oil	800	760	34	-0.3	616	387	23	-3.3
Natural gas	791	806	36	0.6	722	535	32	-1.3
Nuclear	184	159	7	-1.4	198	197	12	-0.5
Hydro	27	28	1	0.5	28	30	2	0.8
Bioenergy	124	148	7	1.7	180	208	12	3.2
Other renewables	96	140	6	5.4	150	300	18	9.1
Power sector	816	814	100	-0.3	712	734	100	-0.8
Coal	221	200	25	-1.7	32	14	2	-12.9
Oil	3	2	0	-6.6	2	1	0	-8.9
Natural gas	268	266	33	0.0	292	174	24	-1.9
Nuclear	184	159	20	-1.4	198	197	27	-0.5
Hydro	27	28	3	0.5	28	30	4	0.8
Bioenergy	24	29	4	1.5	29	52	7	4.2
Other renewables	90	129	16	5.3	132	265	36	8.8
Other energy sector	215	227	100	1.3	179	202	100	0.8
Electricity	53	54	24	0.2	46	44	22	-0.7
Total final consumption	1 654	1 663	100	0.2	1 448	1 226	100	-1.2
Coal	17	16	1	-0.4	12	9	1	-3.1
Oil	755	712	43	-0.3	582	365	30	-3.3
Natural gas	417	427	26	0.6	338	252	21	-1.8
Electricity	354	375	23	0.5	342	386	31	0.7
Heat	6	5	0	-0.9	5	3	0	-2.6
Bioenergy	101	119	7	1.7	150	156	13	3.0
Other renewables	6	10	1	6.0	18	35	3	12.1
Industry	307	313	100	0.6	267	250	100	-0.4
Coal	16	16	5	-0.4	12	9	3	-3.0
Oil	20	21	7	0.7	13	11	4	-2.4
Gas	154	153	49	0.7	129	101	40	-1.2
Electricity	77	81	26	0.7	75	85	34	0.9
Heat	4	4	1	-0.8	4	2	1	-2.8
Bioenergy	34	38	12	0.9	32	36	14	0.7
Other renewables	0	1	0	n.a.	2	4	2	n.a.
Transport	620	606	100	-0.3	529	379	100	-2.4
Oil	540	504	83	-0.7	390	191	50	-4.9
Electricity	2	4	1	6.5	13	50	13	19.5
Biofuels	52	65	11	2.3	95	89	24	3.8
Other fuels	25	33	6	2.7	30	48	13	4.4
Buildings	509	527	100	0.3	445	396	100	-1.0
Coal	0	0	0	-2.8	0	-	-	n.a.
Oil	22	17	3	-2.1	14	3	1	-9.2
Natural gas	196	198	38	0.0	144	88	22	-3.6
Electricity	271	287	55	0.5	251	248	63	-0.2
Heat	1	1	0	-1.0	1	1	0	-2.0
Bioenergy	12	15	3	1.6	18	23	6	3.8
Traditional biomass	-	-	-	n.a.	-	-	-	n.a.
Other renewables	6	9	2	5.5	16	30	8	11.4
Other	219	217	100	1.1	207	202	100	0.7
Petrochem. feedstock	92	92	42	1.7	84	77	38	0.9

Table A.3: Electricity and CO_2 emissions – United States

	Stated Policies Scenario									
	Electricity generation (TWh)							Shares (%)		CAAGR (%)
	2010	2017	2018	2025	2030	2035	2040	2018	2040	2018-40
Total generation	4 354	4 264	4 445	4 555	4 639	4 763	4 939	100	100	0.5
Coal	1 994	1 321	1 255	960	861	802	749	28	15	-2.3
Oil	48	32	44	17	16	12	7	1	0	-7.9
Natural gas	1 018	1 338	1 542	1 711	1 714	1 734	1 803	35	37	0.7
Nuclear	839	839	841	741	696	628	564	19	11	-1.8
Renewables	452	728	758	1 125	1 350	1 585	1 814	17	37	4.0
Hydro	262	302	294	304	311	321	331	7	7	0.5
Bioenergy	73	79	79	85	93	101	110	2	2	1.5
Wind	95	257	278	458	542	643	728	6	15	4.5
Geothermal	18	19	20	21	27	35	41	0	1	3.4
Solar PV	3	67	84	252	368	473	584	2	12	9.2
CSP	1	4	4	4	7	10	16	0	0	6.6
Marine	-	-	-	0	1	3	4	-	0	n.a.

	Stated Policies Scenario								
	Electrical capacity (GW)						Shares (%)		CAAGR (%)
	2017	2018	2025	2030	2035	2040	2018	2040	2018-40
Total capacity	1 187	1 199	1 265	1 357	1 458	1 571	100	100	1.2
Coal	282	262	198	175	162	151	22	10	-2.5
Oil	57	57	22	21	20	18	5	1	-5.1
Natural gas	476	490	508	530	555	595	41	38	0.9
Nuclear	105	105	94	88	80	72	9	5	-1.7
Renewables	265	283	434	527	610	687	24	44	4.1
Hydro	103	103	104	106	107	109	9	7	0.3
Bioenergy	18	18	19	20	21	23	1	1	1.2
Wind	88	95	142	163	183	198	8	13	3.4
Geothermal	4	4	4	5	6	6	0	0	2.8
Solar PV	52	62	164	231	289	346	5	22	8.1
CSP	2	2	2	2	3	4	0	0	4.1
Marine	-	-	0	1	1	2	-	0	n.a.

	Stated Policies Scenario									
	CO_2 emissions (Mt)							Shares (%)		CAAGR (%)
	2010	2017	2018	2025	2030	2035	2040	2018	2040	2018-40
Total CO_2	5 329	4 742	4 884	4 588	4 367	4 142	3 963	100	100	-0.9
Coal	1 982	1 324	1 265	974	867	791	730	26	18	-2.5
Oil	2 060	1 981	2 035	1 914	1 784	1 637	1 505	42	38	-1.4
Natural gas	1 287	1 437	1 584	1 700	1 716	1 714	1 728	32	44	0.4
Power sector	2 329	1 807	1 828	1 551	1 435	1 358	1 313	100	100	-1.5
Coal	1 858	1 240	1 178	886	783	710	651	64	50	-2.7
Oil	37	23	30	10	9	8	5	2	0	-7.9
Natural gas	435	544	621	655	642	640	657	34	50	0.3
Final consumption	2 736	2 694	2 800	2 725	2 603	2 459	2 333	100	100	-0.8
Coal	115	78	80	80	76	73	72	3	3	-0.5
Oil	1 893	1 859	1 904	1 799	1 674	1 533	1 407	68	60	-1.4
Transport	1 641	1 682	1 723	1 632	1 521	1 395	1 281	62	55	-1.3
Natural gas	728	757	816	846	852	852	854	29	37	0.2

Table A.3: Electricity and CO$_2$ emissions – United States

	Current Policies Scenario				Sustainable Development Scenario			
	Electricity generation (TWh)		Shares (%)	CAAGR (%)	Electricity generation (TWh)		Shares (%)	CAAGR (%)
	2030	2040	2040	2018-40	2030	2040	2040	2018-40
Total generation	4 685	4 937	100	0.5	4 471	4 956	100	0.5
Coal	977	902	18	-1.5	142	59	1	-13.0
Oil	18	10	0	-6.7	11	6	0	-8.9
Natural gas	1 679	1 703	34	0.5	1 806	1 076	22	-1.6
Nuclear	706	611	12	-1.4	760	757	15	-0.5
Renewables	1 304	1 709	35	3.8	1 752	3 056	62	6.5
Hydro	311	330	7	0.5	321	351	7	0.8
Bioenergy	92	108	2	1.5	121	219	4	4.8
Wind	525	685	14	4.2	788	1 380	28	7.6
Geothermal	28	42	1	3.5	36	71	1	6.0
Solar PV	342	527	11	8.7	455	877	18	11.2
CSP	6	13	0	5.8	29	146	3	17.8
Marine	1	4	0	n.a.	1	12	0	n.a.

	Current Policies Scenario				Sustainable Development Scenario			
	Electrical capacity (GW)		Shares (%)	CAAGR (%)	Electrical capacity (GW)		Shares (%)	CAAGR (%)
	2030	2040	2040	2018-40	2030	2040	2040	2018-40
Total capacity	1 351	1 537	100	1.1	1 400	1 841	100	2.0
Coal	185	169	11	-2.0	108	39	2	-8.2
Oil	21	19	1	-5.0	19	15	1	-6.0
Natural gas	533	594	39	0.9	475	504	27	0.1
Nuclear	90	78	5	-1.4	97	97	5	-0.4
Renewables	503	640	42	3.8	672	1 107	60	6.4
Hydro	106	109	7	0.2	109	115	6	0.5
Bioenergy	20	22	1	1.1	26	47	3	4.5
Wind	158	186	12	3.1	233	373	20	6.4
Geothermal	5	7	0	2.9	6	11	1	5.3
Solar PV	212	311	20	7.6	288	515	28	10.1
CSP	2	3	0	3.1	10	43	2	15.6
Marine	0	1	0	n.a.	1	4	0	n.a.

	Current Policies Scenario				Sustainable Development Scenario			
	CO$_2$ emissions (Mt)		Shares (%)	CAAGR (%)	CO$_2$ emissions (Mt)		Shares (%)	CAAGR (%)
	2030	2040	2040	2018-40	2030	2040	2040	2018-40
Total CO$_2$	4 596	4 400	100	-0.5	3 030	1 571	100	-5.0
Coal	972	874	20	-1.7	167	46	3	-14.0
Oil	1 886	1 759	40	-0.7	1 349	672	43	-4.9
Natural gas	1 738	1 766	40	0.5	1 515	872	55	-2.7
Power sector	1 526	1 420	100	-1.1	747	237	100	-8.9
Coal	886	790	56	-1.8	109	13	6	-18.4
Oil	11	7	0	-6.6	7	4	2	-8.9
Natural gas	629	624	44	0.0	632	238	100	-4.3
Final consumption	2 730	2 628	100	-0.3	2 021	1 154	100	-4.0
Coal	79	76	3	-0.2	53	30	3	-4.4
Oil	1 770	1 650	63	-0.6	1 268	627	54	-4.9
Transport	1 597	1 492	57	-0.7	1 155	566	49	-4.9
Natural gas	881	902	34	0.5	700	497	43	-2.2

Annex A | Tables for scenario projections

Table A.3: Energy demand – Central and South America

	Stated Policies Scenario Energy demand (Mtoe)							Shares (%)		CAAGR (%)
	2010	2017	2018	2025	2030	2035	2040	2018	2040	2018-40
Total primary demand	611	657	660	727	780	844	913	100	100	1.5
Coal	26	32	32	34	33	33	35	5	4	0.3
Oil	272	271	267	285	292	299	303	40	33	0.6
Natural gas	124	142	145	150	166	188	216	22	24	1.8
Nuclear	6	6	6	6	9	16	19	1	2	5.5
Hydro	60	61	62	73	82	90	97	9	11	2.1
Bioenergy	120	135	138	156	167	178	190	21	21	1.5
Other renewables	3	10	11	22	31	41	53	2	6	7.4
Power sector	158	185	186	202	223	255	290	100	100	2.0
Coal	10	16	16	14	12	10	10	8	3	-2.0
Oil	31	24	22	19	16	16	14	12	5	-2.0
Natural gas	38	51	51	46	50	57	67	28	23	1.2
Nuclear	6	6	6	6	9	16	19	3	7	5.5
Hydro	60	61	62	73	82	90	97	33	33	2.1
Bioenergy	11	19	19	23	27	30	34	10	12	2.6
Other renewables	3	9	10	20	28	37	48	5	17	7.5
Other energy sector	87	88	88	93	100	106	114	100	100	1.2
Electricity	18	22	22	25	28	31	34	25	30	2.0
Total final consumption	463	496	498	564	606	652	699	100	100	1.5
Coal	11	12	12	14	16	17	18	2	3	1.8
Oil	215	234	231	251	259	267	272	46	39	0.7
Natural gas	67	60	62	74	85	98	112	12	16	2.7
Electricity	79	89	91	108	122	137	155	18	22	2.5
Heat	-	-	-	-	-	-	-	-	-	n.a.
Bioenergy	90	99	102	115	122	129	137	20	20	1.4
Other renewables	0	1	1	2	3	4	5	0	1	7.0
Industry	160	159	161	184	199	215	234	100	100	1.7
Coal	11	12	12	14	16	17	18	7	8	1.9
Oil	38	33	32	33	33	33	33	20	14	0.1
Natural gas	34	31	32	43	50	58	66	20	28	3.3
Electricity	35	36	36	43	47	52	57	22	25	2.1
Heat	-	-	-	-	-	-	-	-	-	n.a.
Bioenergy	42	47	49	52	54	56	59	30	25	0.9
Other renewables	-	0	0	0	0	0	0	0	0	44.9
Transport	145	174	173	194	207	220	232	100	100	1.3
Oil	123	148	146	159	163	168	171	84	73	0.7
Electricity	1	0	0	1	1	2	4	0	2	10.3
Biofuels	15	19	20	29	34	39	45	11	19	3.9
Other fuels	7	7	7	6	8	10	13	4	6	2.7
Buildings	103	116	118	128	138	149	160	100	100	1.4
Coal	0	0	0	0	0	0	0	0	0	-3.3
Oil	19	22	21	22	23	23	24	18	15	0.4
Natural gas	12	14	14	16	17	19	21	12	13	1.6
Electricity	41	50	51	60	68	77	88	43	55	2.5
Heat	-	-	-	-	-	-	-	-	-	n.a.
Bioenergy	30	29	30	29	27	26	24	25	15	-1.0
Traditional biomass	26	26	27	25	24	22	20	23	12	-1.4
Other renewables	0	1	1	2	3	3	4	1	3	6.6
Other	54	47	46	57	63	68	73	100	100	2.1
Petrochem. feedstock	24	18	17	24	27	29	32	38	44	2.8

Table A.3: Energy demand – Central and South America

	Current Policies Scenario				Sustainable Development Scenario			
	Energy demand (Mtoe)		Shares (%)	CAAGR (%)	Energy demand (Mtoe)		Shares (%)	CAAGR (%)
	2030	2040	2040	2018-40	2030	2040	2040	2018-40
Total primary demand	813	983	100	1.8	669	702	100	0.3
Coal	38	42	4	1.2	21	16	2	-3.0
Oil	310	343	35	1.1	227	178	25	-1.8
Natural gas	183	250	25	2.5	142	142	20	-0.1
Nuclear	9	19	2	5.5	9	21	3	6.0
Hydro	83	102	10	2.3	83	99	14	2.2
Bioenergy	162	181	18	1.3	150	172	25	1.0
Other renewables	28	46	5	6.7	37	73	10	8.9
Power sector	239	314	100	2.4	193	246	100	1.3
Coal	16	16	5	0.1	4	0	0	-16.1
Oil	18	16	5	-1.6	5	2	1	-10.0
Natural gas	61	87	28	2.4	33	22	9	-3.8
Nuclear	9	19	6	5.5	9	21	9	6.0
Hydro	83	102	32	2.3	83	99	40	2.2
Bioenergy	26	32	10	2.4	27	37	15	3.0
Other renewables	25	42	13	6.8	33	64	26	8.8
Other energy sector	106	131	100	1.8	87	89	100	0.0
Electricity	29	38	29	2.4	25	29	33	1.2
Total final consumption	624	740	100	1.8	525	539	100	0.4
Coal	16	19	3	2.1	13	12	2	-0.1
Oil	274	306	41	1.3	209	164	30	-1.5
Natural gas	88	116	16	2.9	82	96	18	2.0
Electricity	126	165	22	2.8	112	141	26	2.0
Heat	-	-	-	n.a.	-	-	-	n.a.
Bioenergy	118	130	18	1.1	105	116	22	0.6
Other renewables	2	4	1	5.9	4	8	2	9.8
Industry	204	243	100	1.9	169	172	100	0.3
Coal	16	19	8	2.1	12	12	7	-0.1
Oil	33	33	14	0.2	23	17	10	-3.0
Gas	52	71	29	3.7	46	49	28	1.9
Electricity	48	58	24	2.2	43	53	31	1.8
Heat	-	-	-	n.a.	-	-	-	n.a.
Bioenergy	55	61	25	1.0	44	41	24	-0.8
Other renewables	0	0	0	43.1	1	2	1	55.5
Transport	214	252	100	1.7	186	178	100	0.1
Oil	176	202	80	1.5	126	89	50	-2.2
Electricity	1	1	0	3.9	2	4	3	11.4
Biofuels	29	37	15	2.9	47	61	34	5.3
Other fuels	8	12	5	2.2	12	24	14	5.7
Buildings	144	172	100	1.8	111	125	100	0.3
Coal	0	0	0	-1.7	0	0	0	-8.4
Oil	23	25	14	0.7	22	20	16	-0.4
Natural gas	18	22	13	1.9	15	13	11	-0.4
Electricity	73	99	57	3.1	63	79	63	2.0
Heat	-	-	-	n.a.	-	-	-	n.a.
Bioenergy	27	23	14	-1.1	7	7	5	-6.5
Traditional biomass	24	20	11	-1.4	3	1	1	-12.5
Other renewables	2	4	2	5.5	3	7	5	8.6
Other	63	74	100	2.1	59	64	100	1.5
Petrochem. feedstock	27	32	43	2.8	25	29	46	2.4

Table A.3: Electricity and CO₂ emissions – Central and South America

	Stated Policies Scenario							Shares (%)		CAAGR (%)
	Electricity generation (TWh)									
	2010	2017	2018	2025	2030	2035	2040	2018	2040	2018-40
Total generation	1 128	1 294	1 310	1 540	1 734	1 952	2 198	100	100	2.4
Coal	39	67	67	63	53	44	46	5	2	-1.7
Oil	145	111	103	89	78	74	68	8	3	-1.9
Natural gas	177	248	248	257	299	348	419	19	19	2.4
Nuclear	22	22	23	24	35	61	74	2	3	5.5
Renewables	745	845	869	1 107	1 269	1 425	1 591	66	72	2.8
Hydro	694	708	717	854	950	1 041	1 126	55	51	2.1
Bioenergy	44	70	75	89	99	112	125	6	6	2.4
Wind	3	56	62	118	153	180	211	5	10	5.7
Geothermal	3	4	4	7	11	17	23	0	1	8.4
Solar PV	0	7	11	39	53	70	98	1	4	10.4
CSP	-	-	-	0	2	5	8	-	0	n.a.
Marine	-	-	-	0	0	0	0	-	0	n.a.

	Stated Policies Scenario						Shares (%)		CAAGR (%)
	Electrical capacity (GW)								
	2017	2018	2025	2030	2035	2040	2018	2040	2018-40
Total capacity	346	359	434	483	532	600	100	100	2.4
Coal	13	13	15	14	12	12	4	2	-0.3
Oil	49	49	43	37	35	33	14	5	-1.8
Natural gas	64	67	90	104	111	132	19	22	3.1
Nuclear	4	4	3	5	9	10	1	2	4.6
Renewables	217	226	282	323	363	410	63	68	2.7
Hydro	174	178	194	214	233	252	50	42	1.6
Bioenergy	19	20	23	25	27	30	5	5	1.9
Wind	17	20	37	47	55	63	6	11	5.3
Geothermal	1	1	1	2	3	4	0	1	7.6
Solar PV	5	7	26	34	44	60	2	10	9.9
CSP	-	-	0	1	2	2	-	0	n.a.
Marine	-	-	0	0	0	0	-	0	n.a.

	Stated Policies Scenario							Shares (%)		CAAGR (%)
	CO₂ emissions (Mt)									
	2010	2017	2018	2025	2030	2035	2040	2018	2040	2018-40
Total CO₂	1 090	1 167	1 158	1 206	1 249	1 308	1 382	100	100	0.8
Coal	99	126	125	128	123	120	126	11	9	0.0
Oil	723	751	737	772	782	796	801	64	58	0.4
Natural gas	268	290	295	306	343	392	454	25	33	2.0
Power sector	233	266	262	231	222	227	248	100	100	-0.2
Coal	46	70	71	64	53	45	46	27	18	-2.0
Oil	98	76	70	60	52	49	45	27	18	-2.0
Natural gas	88	121	121	107	116	133	158	46	64	1.2
Final consumption	758	806	801	878	923	970	1 012	100	100	1.1
Coal	49	51	50	59	65	70	76	6	7	1.9
Oil	584	643	635	679	695	712	721	79	71	0.6
Transport	371	444	438	476	490	505	513	55	51	0.7
Natural gas	125	112	116	140	163	188	215	14	21	2.8

Table A.3: Electricity and CO_2 emissions – Central and South America

	Current Policies Scenario				Sustainable Development Scenario			
	Electricity generation (TWh)		Shares (%)	CAAGR (%)	Electricity generation (TWh)		Shares (%)	CAAGR (%)
	2030	2040	2040	2018-40	2030	2040	2040	2018-40
Total generation	1 811	2 351	100	2.7	1 592	1 975	100	1.9
Coal	73	72	3	0.3	16	2	0	-15.9
Oil	86	76	3	-1.4	21	10	0	-10.2
Natural gas	358	538	23	3.6	203	140	7	-2.6
Nuclear	35	74	3	5.5	35	82	4	6.0
Renewables	1 258	1 591	68	2.8	1 316	1 741	88	3.2
Hydro	967	1 184	50	2.3	962	1 157	59	2.2
Bioenergy	97	118	5	2.1	100	133	7	2.7
Wind	139	189	8	5.2	170	274	14	6.9
Geothermal	11	21	1	7.8	13	29	1	9.5
Solar PV	42	72	3	8.8	65	128	6	11.8
CSP	2	7	0	n.a.	6	18	1	n.a.
Marine	0	0	0	n.a.	0	1	0	n.a.

	Current Policies Scenario				Sustainable Development Scenario			
	Electrical capacity (GW)		Shares (%)	CAAGR (%)	Electrical capacity (GW)		Shares (%)	CAAGR (%)
	2030	2040	2040	2018-40	2030	2040	2040	2018-40
Total capacity	482	614	100	2.5	472	603	100	2.4
Coal	16	17	3	1.2	10	6	1	-3.2
Oil	37	33	5	-1.8	37	33	5	-1.8
Natural gas	106	147	24	3.6	80	82	14	0.9
Nuclear	5	10	2	4.6	5	11	2	5.1
Renewables	317	403	66	2.7	338	462	77	3.3
Hydro	219	269	44	1.9	214	259	43	1.7
Bioenergy	25	28	5	1.6	25	31	5	2.0
Wind	43	56	9	4.7	54	84	14	6.7
Geothermal	2	3	1	7.1	2	4	1	8.7
Solar PV	27	44	7	8.5	41	78	13	11.3
CSP	1	2	0	n.a.	2	6	1	n.a.
Marine	0	0	0	n.a.	0	0	0	n.a.

	Current Policies Scenario				Sustainable Development Scenario			
	CO_2 emissions (Mt)		Shares (%)	CAAGR (%)	CO_2 emissions (Mt)		Shares (%)	CAAGR (%)
	2030	2040	2040	2018-40	2030	2040	2040	2018-40
Total CO_2	1 360	1 602	100	1.5	929	725	100	-2.1
Coal	145	156	10	1.0	67	42	6	-4.8
Oil	833	915	57	1.0	584	420	58	-2.5
Natural gas	382	531	33	2.7	277	263	36	-0.5
Power sector	273	325	100	1.0	108	59	100	-6.5
Coal	73	71	22	0.0	15	1	2	-16.4
Oil	57	50	15	-1.6	15	7	12	-10.0
Natural gas	143	204	63	2.4	78	51	86	-3.8
Final consumption	974	1 128	100	1.6	745	609	100	-1.2
Coal	67	80	7	2.1	48	38	6	-1.2
Oil	739	823	73	1.2	544	396	65	-2.1
Transport	529	606	54	1.5	379	266	44	-2.2
Natural gas	168	225	20	3.1	153	175	29	1.9

Table A.3: Energy demand – Brazil

	Stated Policies Scenario									
	Energy demand (Mtoe)							Shares (%)		CAAGR (%)
	2010	2017	2018	2025	2030	2035	2040	2018	2040	2018-40
Total primary demand	263	287	285	318	342	369	397	100	100	1.5
Coal	14	17	17	16	16	16	16	6	4	-0.2
Oil	105	111	108	118	123	128	130	38	33	0.9
Natural gas	24	33	30	29	32	39	48	11	12	2.1
Nuclear	4	4	4	4	7	10	12	1	3	5.0
Hydro	35	32	32	39	43	46	49	11	12	2.0
Bioenergy	82	87	89	102	109	115	122	31	31	1.4
Other renewables	1	5	5	10	13	15	19	2	5	6.0
Power sector	58	70	71	75	84	96	110	100	100	2.0
Coal	3	5	6	4	4	3	3	8	3	-2.5
Oil	4	3	3	1	1	1	1	4	1	-5.1
Natural gas	7	13	12	6	5	8	13	17	11	0.3
Nuclear	4	4	4	4	7	10	12	6	11	5.0
Hydro	35	32	32	39	43	46	49	45	45	2.0
Bioenergy	6	9	10	12	13	15	16	14	14	2.1
Other renewables	0	4	4	8	11	13	16	6	14	6.1
Other energy sector	41	42	41	46	49	52	56	100	100	1.5
Electricity	10	11	11	12	13	15	17	27	30	1.9
Total final consumption	211	229	227	260	278	298	318	100	100	1.5
Coal	7	7	7	8	8	8	9	3	3	0.7
Oil	94	103	100	112	116	121	124	44	39	1.0
Natural gas	13	13	12	14	17	20	23	5	7	2.9
Electricity	38	43	43	51	56	63	70	19	22	2.3
Heat	-	-	-	-	-	-	-	-	-	n.a.
Bioenergy	59	63	64	74	79	83	89	28	28	1.5
Other renewables	0	1	1	1	2	3	3	0	1	5.8
Industry	80	81	80	91	97	103	110	100	100	1.4
Coal	7	7	7	8	8	8	8	9	8	0.7
Oil	12	11	10	11	11	11	12	13	10	0.5
Natural gas	9	9	9	12	14	16	18	11	17	3.4
Electricity	17	17	17	20	22	24	26	21	24	1.9
Heat	-	-	-	-	-	-	-	-	-	n.a.
Bioenergy	34	36	37	40	41	43	46	46	41	1.0
Other renewables	-	-	-	0	0	0	0	-	0	n.a.
Transport	70	85	83	93	98	103	108	100	100	1.2
Oil	54	65	64	69	70	73	73	76	68	0.6
Electricity	0	0	0	0	1	1	2	0	2	10.1
Biofuels	14	17	17	23	26	29	32	21	30	2.8
Other fuels	2	2	2	1	1	1	1	3	1	-3.4
Buildings	34	38	38	43	47	52	57	100	100	1.8
Coal	-	-	-	-	-	-	-	-	-	n.a.
Oil	7	7	7	8	8	8	9	19	15	0.8
Natural gas	1	1	0	1	1	2	2	1	3	6.5
Electricity	18	23	23	27	31	35	40	61	70	2.5
Heat	-	-	-	-	-	-	-	-	-	n.a.
Bioenergy	8	7	6	6	5	4	4	17	6	-2.7
Traditional biomass	8	6	6	6	5	4	3	16	5	-3.1
Other renewables	0	1	1	1	2	2	3	2	5	5.4
Other	27	26	25	33	37	40	43	100	100	2.5
Petrochem. feedstock	9	8	8	13	15	16	18	31	42	3.9

Table A.3: Energy demand – Brazil

	Current Policies Scenario				Sustainable Development Scenario			
	Energy demand (Mtoe)		Shares (%)	CAAGR (%)	Energy demand (Mtoe)		Shares (%)	CAAGR (%)
	2030	2040	2040	2018-40	2030	2040	2040	2018-40
Total primary demand	356	427	100	1.9	299	312	100	0.4
Coal	17	18	4	0.3	10	9	3	-3.0
Oil	132	147	35	1.4	97	74	24	-1.7
Natural gas	38	60	14	3.1	26	34	11	0.5
Nuclear	7	12	3	5.0	7	12	4	5.0
Hydro	45	55	13	2.6	40	44	14	1.5
Bioenergy	106	118	28	1.3	106	118	38	1.3
Other renewables	11	17	4	5.4	14	21	7	6.6
Power sector	90	119	100	2.4	74	93	100	1.2
Coal	5	4	4	-1.2	0	-	-	n.a.
Oil	2	2	2	-1.0	1	0	0	-16.6
Natural gas	9	16	13	1.3	2	4	5	-4.4
Nuclear	7	12	10	5.0	7	12	13	5.0
Hydro	45	55	47	2.6	40	44	47	1.5
Bioenergy	13	15	12	1.7	13	15	16	1.8
Other renewables	9	14	12	5.5	11	17	19	6.5
Other energy sector	52	64	100	2.0	43	47	100	0.6
Electricity	14	19	29	2.4	12	14	29	1.0
Total final consumption	287	338	100	1.8	245	250	100	0.4
Coal	8	9	3	0.9	7	6	2	-1.3
Oil	123	139	41	1.5	91	71	28	-1.6
Natural gas	19	27	8	3.6	16	20	8	2.4
Electricity	59	75	22	2.6	52	63	25	1.7
Heat	-	-	-	n.a.	-	-	-	n.a.
Bioenergy	77	86	26	1.4	77	86	34	1.4
Other renewables	2	3	1	4.6	2	4	2	7.1
Industry	98	114	100	1.6	83	82	100	0.1
Coal	8	9	8	0.9	6	5	7	-1.4
Oil	11	12	10	0.6	9	7	9	-1.5
Gas	15	21	18	3.9	13	13	15	1.6
Electricity	22	26	23	1.9	21	25	30	1.6
Heat	-	-	-	n.a.	-	-	-	n.a.
Bioenergy	42	47	41	1.1	34	31	38	-0.8
Other renewables	0	0	0	n.a.	0	1	1	n.a.
Transport	102	118	100	1.6	89	84	100	0.0
Oil	77	87	73	1.4	50	29	34	-3.5
Electricity	0	1	0	3.9	1	2	3	11.0
Biofuels	24	29	24	2.3	37	47	56	4.6
Other fuels	1	2	2	0.3	2	6	7	4.8
Buildings	49	63	100	2.3	39	46	100	0.9
Coal	-	-	-	n.a.	-	-	-	n.a.
Oil	8	9	14	1.0	7	7	14	-0.4
Natural gas	1	2	3	7.3	0	1	2	2.1
Electricity	33	46	73	3.1	28	34	73	1.7
Heat	-	-	-	n.a.	-	-	-	n.a.
Bioenergy	5	3	5	-2.9	1	2	3	-6.1
Traditional biomass	5	3	5	-3.1	0	0	0	-19.2
Other renewables	2	2	4	4.3	2	4	8	6.2
Other	37	43	100	2.5	35	38	100	1.9
Petrochem. feedstock	15	18	42	3.9	14	17	43	3.5

Table A.3: Electricity and CO_2 emissions – Brazil

	Stated Policies Scenario							Shares (%)		CAAGR (%)
	Electricity generation (TWh)									
	2010	2017	2018	2025	2030	2035	2040	2018	2040	2018-40
Total generation	516	588	593	695	774	866	976	100	100	2.3
Coal	11	25	26	19	18	17	15	4	2	-2.5
Oil	16	16	15	5	5	5	5	2	0	-5.1
Natural gas	36	66	61	38	33	50	81	10	8	1.3
Nuclear	15	16	16	15	26	39	47	3	5	5.0
Renewables	437	465	475	618	692	755	828	80	85	2.6
Hydro	403	371	369	459	500	537	573	62	59	2.0
Bioenergy	31	51	55	63	67	71	75	9	8	1.4
Wind	2	42	47	80	103	119	139	8	14	5.0
Geothermal	-	-	-	-	-	-	-	-	-	n.a.
Solar PV	-	1	3	15	21	27	37	0	4	12.7
CSP	-	-	-	-	1	2	3	-	0	n.a.
Marine	-	-	-	-	-	-	0	-	0	n.a.

	Stated Policies Scenario						Shares (%)		CAAGR (%)
	Electrical capacity (GW)								
	2017	2018	2025	2030	2035	2040	2018	2040	2018-40
Total capacity	156	162	194	213	231	258	100	100	2.1
Coal	4	4	5	4	4	4	3	1	-0.7
Oil	8	8	7	7	7	7	5	3	-0.9
Natural gas	13	13	18	19	19	23	8	9	2.7
Nuclear	2	2	2	3	5	6	1	2	5.2
Renewables	129	135	161	179	195	217	83	84	2.2
Hydro	100	103	111	118	126	136	63	53	1.3
Bioenergy	15	16	18	19	20	21	10	8	1.3
Wind	12	14	22	28	32	36	9	14	4.3
Geothermal	-	-	-	-	-	-	-	-	n.a.
Solar PV	1	2	10	14	17	23	1	9	11.2
CSP	-	-	-	0	1	1	-	0	n.a.
Marine	-	-	-	-	-	0	-	0	n.a.

	Stated Policies Scenario							Shares (%)		CAAGR (%)
	CO_2 emissions (Mt)									
	2010	2017	2018	2025	2030	2035	2040	2018	2040	2018-40
Total CO_2	372	428	416	422	435	460	484	100	100	0.7
Coal	54	65	65	58	58	57	57	16	12	-0.6
Oil	267	292	284	302	310	320	324	68	67	0.6
Natural gas	51	71	67	62	68	83	104	16	21	2.0
Power sector	45	69	68	40	37	41	50	100	100	-1.4
Coal	18	29	31	22	21	19	18	45	35	-2.5
Oil	12	10	9	3	3	3	3	14	6	-5.1
Natural gas	16	30	28	15	13	19	30	41	59	0.3
Final consumption	302	329	319	347	361	378	390	100	100	0.9
Coal	33	32	30	32	33	34	35	9	9	0.7
Oil	242	270	263	284	292	302	307	82	79	0.7
Transport	163	198	193	208	212	219	221	60	57	0.6
Natural gas	28	27	26	31	36	42	49	8	12	2.8

Table A.3: Electricity and CO_2 emissions – Brazil

	Current Policies Scenario				Sustainable Development Scenario			
	Electricity generation (TWh)		Shares (%)	CAAGR (%)	Electricity generation (TWh)		Shares (%)	CAAGR (%)
	2030	2040	2040	2018-40	2030	2040	2040	2018-40
Total generation	811	1 053	100	2.6	706	854	100	1.7
Coal	22	20	2	-1.2	0	-	-	n.a.
Oil	12	12	1	-1.0	3	0	0	-16.6
Natural gas	55	101	10	2.3	14	28	3	-3.5
Nuclear	26	47	4	5.0	26	47	6	5.0
Renewables	696	873	83	2.8	662	778	91	2.3
Hydro	521	643	61	2.6	467	511	60	1.5
Bioenergy	66	71	7	1.2	65	73	8	1.2
Wind	93	126	12	4.5	106	148	17	5.3
Geothermal	-	-	-	n.a.	-	-	-	n.a.
Solar PV	15	29	3	11.5	23	43	5	13.5
CSP	1	3	0	n.a.	1	3	0	n.a.
Marine	-	0	0	n.a.	-	0	0	n.a.

	Current Policies Scenario				Sustainable Development Scenario			
	Electrical capacity (GW)		Shares (%)	CAAGR (%)	Electrical capacity (GW)		Shares (%)	CAAGR (%)
	2030	2040	2040	2018-40	2030	2040	2040	2018-40
Total capacity	215	271	100	2.4	202	239	100	1.8
Coal	4	4	1	-0.4	4	-	-	n.a.
Oil	7	7	3	-0.8	7	7	3	-1.0
Natural gas	21	24	9	2.9	16	16	7	1.2
Nuclear	3	6	2	5.2	3	6	3	5.2
Renewables	179	228	84	2.4	171	205	85	1.9
Hydro	125	156	58	1.9	109	119	50	0.7
Bioenergy	18	20	7	1.1	18	20	8	1.2
Wind	25	33	12	3.9	29	39	16	4.6
Geothermal	-	-	-	n.a.	-	-	-	n.a.
Solar PV	10	18	7	10.1	14	26	11	11.9
CSP	0	1	0	n.a.	0	1	0	n.a.
Marine	-	0	0	n.a.	-	0	0	n.a.

	Current Policies Scenario				Sustainable Development Scenario			
	CO_2 emissions (Mt)		Shares (%)	CAAGR (%)	CO_2 emissions (Mt)		Shares (%)	CAAGR (%)
	2030	2040	2040	2018-40	2030	2040	2040	2018-40
Total CO_2	483	568	100	1.4	310	235	100	-2.6
Coal	64	64	11	-0.0	28	19	8	-5.4
Oil	337	376	66	1.3	229	151	64	-2.8
Natural gas	82	128	23	3.0	53	65	28	-0.2
Power sector	54	68	100	0.0	8	11	100	-8.1
Coal	26	24	34	-1.2	0	-	-	n.a.
Oil	8	7	11	-1.0	2	0	2	-16.6
Natural gas	20	37	55	1.3	5	10	98	-4.4
Final consumption	387	444	100	1.5	276	204	100	-2.0
Coal	34	36	8	0.9	25	17	8	-2.5
Oil	313	350	79	1.3	217	145	71	-2.7
Transport	232	262	59	1.4	150	87	43	-3.6
Natural gas	40	58	13	3.6	34	42	21	2.2

Table A.3: Energy demand – Europe

	Stated Policies Scenario							Shares (%)		CAAGR (%)
	Energy demand (Mtoe)									
	2010	2017	2018	2025	2030	2035	2040	2018	2040	2018-40
Total primary demand	2 110	1 996	2 000	1 929	1 848	1 761	1 723	100	100	-0.7
Coal	377	322	313	220	184	154	142	16	8	-3.5
Oil	663	635	636	588	526	458	407	32	24	-2.0
Natural gas	571	504	496	510	487	474	456	25	26	-0.4
Nuclear	269	244	246	214	205	187	191	12	11	-1.1
Hydro	55	50	54	61	64	65	67	3	4	1.0
Bioenergy	146	174	182	213	226	237	247	9	14	1.4
Other renewables	30	66	73	123	155	187	213	4	12	5.0
Power sector	896	837	840	791	779	771	788	100	100	-0.3
Coal	267	226	223	134	104	79	71	27	9	-5.1
Oil	27	18	17	9	7	6	4	2	1	-6.2
Natural gas	203	170	162	177	169	169	161	19	20	-0.0
Nuclear	269	244	246	214	205	187	191	29	24	-1.1
Hydro	55	50	54	61	64	65	67	6	8	1.0
Bioenergy	49	68	72	84	90	97	103	9	13	1.6
Other renewables	25	60	66	111	140	168	190	8	24	4.9
Other energy sector	202	185	184	172	158	147	142	100	100	-1.1
Electricity	56	53	53	50	48	48	48	29	34	-0.5
Total final consumption	1 457	1 408	1 412	1 417	1 372	1 317	1 283	100	100	-0.4
Coal	66	56	53	49	45	41	39	4	3	-1.4
Oil	587	570	571	540	487	426	381	40	30	-1.8
Natural gas	330	306	306	302	291	279	269	22	21	-0.6
Electricity	299	301	304	321	334	350	367	22	29	0.9
Heat	75	65	65	67	66	65	64	5	5	-0.1
Bioenergy	95	104	108	126	134	138	141	8	11	1.2
Other renewables	5	6	6	11	15	19	22	0	2	5.9
Industry	343	341	340	346	342	338	336	100	100	-0.0
Coal	40	36	33	33	32	31	30	10	9	-0.5
Oil	43	36	35	35	34	32	31	10	9	-0.5
Natural gas	101	105	105	105	103	99	97	31	29	-0.4
Electricity	111	114	115	117	118	118	119	34	35	0.2
Heat	25	22	22	23	23	22	21	7	6	-0.2
Bioenergy	23	28	29	32	33	34	36	9	11	1.0
Other renewables	0	0	0	1	1	1	2	0	1	8.2
Transport	374	391	394	389	368	335	317	100	100	-1.0
Oil	347	362	363	345	314	272	242	92	76	-1.8
Electricity	7	7	7	12	17	26	35	2	11	7.7
Biofuels	13	16	18	26	30	30	31	5	10	2.5
Other fuels	7	6	6	6	6	8	9	2	3	1.9
Buildings	579	530	531	531	518	506	499	100	100	-0.3
Coal	23	17	16	12	9	6	5	3	1	-5.1
Oil	73	56	56	40	27	16	10	11	2	-7.7
Natural gas	197	178	177	175	166	157	147	33	29	-0.8
Electricity	176	175	176	186	193	200	206	33	41	0.7
Heat	49	42	42	43	43	43	42	8	8	0.0
Bioenergy	57	58	59	65	67	69	71	11	14	0.8
Traditional biomass	-	-	-	-	-	-	-	-	-	n.a.
Other renewables	5	5	5	9	13	16	18	1	4	6.1
Other	161	146	148	151	144	138	131	100	100	-0.6
Petrochem. feedstock	89	83	83	83	77	73	66	56	51	-1.0

Table A.3: Energy demand – Europe

	Current Policies Scenario				Sustainable Development Scenario			
	Energy demand (Mtoe)		Shares (%)	CAAGR (%)	Energy demand (Mtoe)		Shares (%)	CAAGR (%)
	2030	2040	2040	2018-40	2030	2040	2040	2018-40
Total primary demand	1 982	1 968	100	-0.1	1 689	1 470	100	-1.4
Coal	256	223	11	-1.5	90	59	4	-7.3
Oil	574	516	26	-0.9	437	231	16	-4.5
Natural gas	521	557	28	0.5	425	311	21	-2.1
Nuclear	215	201	10	-0.9	220	217	15	-0.6
Hydro	63	66	3	0.9	66	71	5	1.3
Bioenergy	219	238	12	1.2	263	285	19	2.1
Other renewables	133	168	9	3.9	188	296	20	6.6
Power sector	839	865	100	0.1	735	803	100	-0.2
Coal	172	145	17	-1.9	28	14	2	-12.0
Oil	7	5	1	-5.7	6	3	0	-7.7
Natural gas	172	199	23	0.9	147	109	14	-1.8
Nuclear	215	201	23	-0.9	220	217	27	-0.6
Hydro	63	66	8	0.9	66	71	9	1.3
Bioenergy	89	99	11	1.5	103	130	16	2.7
Other renewables	121	151	18	3.8	166	260	32	6.4
Other energy sector	168	163	100	-0.5	143	128	100	-1.6
Electricity	52	53	32	-0.0	51	64	50	0.8
Total final consumption	1 450	1 452	100	0.1	1 257	1 051	100	-1.3
Coal	48	43	3	-0.9	34	23	2	-3.6
Oil	531	481	33	-0.8	402	214	20	-4.4
Natural gas	321	329	23	0.3	254	177	17	-2.5
Electricity	343	377	26	1.0	325	385	37	1.1
Heat	69	69	5	0.3	57	50	5	-1.2
Bioenergy	128	137	9	1.1	159	153	15	1.6
Other renewables	11	16	1	4.4	22	36	3	8.3
Industry	351	354	100	0.2	295	269	100	-1.1
Coal	33	32	9	-0.2	24	20	7	-2.4
Oil	34	32	9	-0.3	23	16	6	-3.4
Gas	108	106	30	0.0	89	66	24	-2.1
Electricity	119	122	34	0.3	108	112	42	-0.1
Heat	23	23	6	0.1	16	12	5	-2.7
Bioenergy	34	38	11	1.2	30	36	13	0.9
Other renewables	1	1	0	5.0	3	5	2	13.0
Transport	392	381	100	-0.2	345	241	100	-2.2
Oil	348	327	86	-0.5	251	106	44	-5.4
Electricity	11	17	5	4.3	22	69	29	11.1
Biofuels	26	29	7	2.1	55	38	16	3.5
Other fuels	6	8	2	1.2	18	28	11	7.2
Buildings	563	585	100	0.4	484	430	100	-1.0
Coal	11	8	1	-3.3	7	1	0	-12.1
Oil	37	21	4	-4.2	24	5	1	-10.7
Natural gas	190	199	34	0.5	135	80	19	-3.5
Electricity	206	230	39	1.2	189	197	46	0.5
Heat	45	46	8	0.4	41	38	9	-0.5
Bioenergy	64	67	11	0.6	70	74	17	1.1
Traditional biomass	-	-	-	n.a.	-	-	-	n.a.
Other renewables	10	14	2	4.8	18	29	7	8.3
Other	145	133	100	-0.5	133	111	100	-1.3
Petrochem. feedstock	76	66	50	-1.0	71	55	50	-1.8

Table A.3: Electricity and CO_2 emissions – Europe

	Stated Policies Scenario									
	Electricity generation (TWh)							Shares (%)		CAAGR (%)
	2010	2017	2018	2025	2030	2035	2040	2018	2040	2018-40
Total generation	4 120	4 127	4 163	4 345	4 478	4 640	4 840	100	100	0.7
Coal	1 068	928	907	544	432	338	319	22	7	-4.6
Oil	93	65	61	31	22	19	13	1	0	-6.7
Natural gas	947	867	808	902	848	852	802	19	17	-0.0
Nuclear	1 032	936	943	822	788	717	734	23	15	-1.1
Renewables	975	1 325	1 438	2 043	2 384	2 711	2 967	35	61	3.3
Hydro	641	579	624	713	743	761	776	15	16	1.0
Bioenergy	147	217	233	280	300	326	348	6	7	1.8
Wind	153	385	417	742	968	1 189	1 357	10	28	5.5
Geothermal	11	18	21	28	33	37	40	0	1	3.1
Solar PV	23	120	138	273	325	368	394	3	8	4.9
CSP	1	6	5	7	11	18	25	0	1	7.5
Marine	0	1	1	1	4	13	27	0	1	19.7

	Stated Policies Scenario								
	Electrical capacity (GW)						Shares (%)		CAAGR (%)
	2017	2018	2025	2030	2035	2040	2018	2040	2018-40
Total capacity	1 280	1 305	1 484	1 579	1 678	1 753	100	100	1.3
Coal	227	222	145	112	88	84	17	5	-4.3
Oil	57	56	31	23	19	14	4	1	-6.0
Natural gas	262	264	307	323	356	365	20	21	1.5
Nuclear	142	142	125	118	108	111	11	6	-1.1
Renewables	590	619	867	986	1 077	1 137	47	65	2.8
Hydro	246	248	259	269	274	278	19	16	0.5
Bioenergy	46	49	58	61	65	67	4	4	1.4
Wind	178	191	299	356	402	429	15	25	3.8
Geothermal	3	3	4	4	5	5	0	0	2.9
Solar PV	114	126	243	290	320	336	10	19	4.6
CSP	2	2	3	4	6	8	0	0	5.9
Marine	0	0	0	2	6	12	0	1	19.5

	Stated Policies Scenario									
	CO_2 emissions (Mt)							Shares (%)		CAAGR (%)
	2010	2017	2018	2025	2030	2035	2040	2018	2040	2018-40
Total CO_2	4 401	3 979	3 923	3 408	3 047	2 703	2 487	100	100	-2.1
Coal	1 473	1 253	1 221	824	678	551	504	31	20	-3.9
Oil	1 671	1 596	1 589	1 443	1 282	1 095	970	41	39	-2.2
Natural gas	1 257	1 131	1 112	1 141	1 087	1 056	1 013	28	41	-0.4
Power sector	1 695	1 422	1 387	1 015	861	750	691	100	100	-3.1
Coal	1 131	967	952	570	444	335	299	69	43	-5.1
Oil	85	55	54	29	21	17	13	4	2	-6.2
Natural gas	478	399	382	416	396	397	379	28	55	-0.0
Final consumption	2 494	2 358	2 339	2 209	2 023	1 805	1 653	100	100	-1.6
Coal	296	244	229	213	195	178	168	10	10	-1.4
Oil	1 471	1 440	1 436	1 330	1 188	1 014	897	61	54	-2.1
Transport	1 050	1 099	1 101	1 047	953	823	732	47	44	-1.8
Natural gas	728	675	674	666	640	613	588	29	36	-0.6

Table A.3: Electricity and CO_2 emissions – Europe

	Current Policies Scenario				Sustainable Development Scenario			
	Electricity generation (TWh)		Shares (%)	CAAGR (%)	Electricity generation (TWh)		Shares (%)	CAAGR (%)
	2030	2040	2040	2018-40	2030	2040	2040	2018-40
Total generation	4 622	5 012	100	0.8	4 429	5 246	100	1.1
Coal	733	647	13	-1.5	101	48	1	-12.5
Oil	23	15	0	-6.1	20	9	0	-8.2
Natural gas	868	1 043	21	1.2	739	536	10	-1.8
Nuclear	825	771	15	-0.9	844	832	16	-0.6
Renewables	2 168	2 532	51	2.6	2 721	3 817	73	4.5
Hydro	733	762	15	0.9	767	824	16	1.3
Bioenergy	298	333	7	1.6	345	461	9	3.2
Wind	822	1 049	21	4.3	1 157	1 860	35	7.0
Geothermal	30	35	1	2.4	36	51	1	4.3
Solar PV	275	321	6	3.9	395	541	10	6.4
CSP	9	19	0	6.2	16	44	1	10.3
Marine	2	12	0	15.4	5	36	1	21.1

	Current Policies Scenario				Sustainable Development Scenario			
	Electrical capacity (GW)		Shares (%)	CAAGR (%)	Electrical capacity (GW)		Shares (%)	CAAGR (%)
	2030	2040	2040	2018-40	2030	2040	2040	2018-40
Total capacity	1 598	1 724	100	1.3	1 691	2 066	100	2.1
Coal	174	148	9	-1.8	65	23	1	-9.8
Oil	25	16	1	-5.5	23	15	1	-5.7
Natural gas	370	431	25	2.2	318	369	18	1.5
Nuclear	118	109	6	-1.2	128	127	6	-0.5
Renewables	891	982	57	2.1	1 132	1 468	71	4.0
Hydro	264	273	16	0.4	275	293	14	0.8
Bioenergy	60	64	4	1.2	69	88	4	2.7
Wind	313	353	20	2.8	422	590	29	5.3
Geothermal	4	5	0	2.2	5	7	0	4.1
Solar PV	245	275	16	3.6	352	461	22	6.1
CSP	3	6	0	4.7	6	14	1	8.5
Marine	1	6	0	15.3	3	16	1	20.9

	Current Policies Scenario				Sustainable Development Scenario			
	CO_2 emissions (Mt)		Shares (%)	CAAGR (%)	CO_2 emissions (Mt)		Shares (%)	CAAGR (%)
	2030	2040	2040	2018-40	2030	2040	2040	2018-40
Total CO_2	3 574	3 386	100	-0.7	2 236	1 196	100	-5.3
Coal	982	845	25	-1.7	280	112	9	-10.3
Oil	1 424	1 293	38	-0.9	1 018	443	37	-5.6
Natural gas	1 169	1 248	37	0.5	937	640	54	-2.5
Power sector	1 160	1 098	100	-1.1	471	263	100	-7.3
Coal	733	616	56	-2.0	111	16	6	-17.0
Oil	21	15	1	-5.7	19	9	4	-7.6
Natural gas	405	467	43	0.9	342	238	90	-2.1
Final consumption	2 239	2 118	100	-0.4	1 639	858	100	-4.5
Coal	207	188	9	-0.9	141	76	9	-4.9
Oil	1 323	1 203	57	-0.8	942	404	47	-5.6
Transport	1 055	992	47	-0.5	760	320	37	-5.5
Natural gas	708	728	34	0.3	556	377	44	-2.6

Table A.3: Energy demand – European Union

	Stated Policies Scenario							Shares (%)		CAAGR (%)
	Energy demand (Mtoe)									
	2010	2017	2018	2025	2030	2035	2040	2018	2040	2018-40
Total primary demand	1 726	1 617	1 613	1 514	1 414	1 313	1 254	100	100	-1.1
Coal	283	234	223	143	110	79	61	14	5	-5.7
Oil	570	531	530	478	413	345	294	33	23	-2.6
Natural gas	448	398	392	392	363	341	317	24	25	-1.0
Nuclear	239	216	216	180	166	150	155	13	12	-1.5
Hydro	32	26	29	33	35	36	36	2	3	1.0
Bioenergy	131	159	166	194	205	213	220	10	18	1.3
Other renewables	22	52	56	95	122	150	171	3	14	5.2
Power sector	725	666	662	604	578	560	562	100	100	-0.7
Coal	212	170	162	86	58	32	17	24	3	-9.6
Oil	24	16	16	8	6	5	4	2	1	-6.4
Natural gas	150	123	119	130	119	113	103	18	18	-0.7
Nuclear	239	216	216	180	166	150	155	33	28	-1.5
Hydro	32	26	29	33	35	36	36	4	6	1.0
Bioenergy	47	64	67	78	82	87	91	10	16	1.4
Other renewables	20	49	53	88	113	137	156	8	28	5.1
Other energy sector	148	138	137	122	109	100	94	100	100	-1.7
Electricity	43	41	41	37	34	33	33	30	35	-1.0
Total final consumption	1 205	1 154	1 155	1 135	1 074	1 006	958	100	100	-0.9
Coal	41	36	34	31	27	24	22	3	2	-2.0
Oil	503	479	479	441	384	321	274	41	29	-2.5
Natural gas	278	254	253	243	229	214	200	22	21	-1.0
Electricity	244	241	241	250	256	265	274	21	29	0.6
Heat	53	49	49	49	48	47	46	4	5	-0.3
Bioenergy	83	93	97	114	121	124	127	8	13	1.2
Other renewables	2	3	3	7	9	12	15	0	2	7.5
Industry	272	270	269	269	261	254	249	100	100	-0.3
Coal	24	22	21	20	19	18	17	8	7	-1.0
Oil	37	28	27	27	26	25	24	10	10	-0.6
Natural gas	85	88	88	86	82	78	75	33	30	-0.7
Electricity	88	89	89	90	88	87	87	33	35	-0.1
Heat	17	16	16	16	15	14	13	6	5	-0.9
Bioenergy	22	27	28	30	30	31	32	10	13	0.7
Other renewables	0	0	0	0	0	1	1	0	1	20.6
Transport	320	327	328	316	289	253	232	100	100	-1.6
Oil	299	303	302	279	244	199	168	92	72	-2.6
Electricity	5	6	6	9	14	22	30	2	13	7.9
Biofuels	13	15	17	25	28	29	29	5	12	2.4
Other fuels	3	3	3	3	3	4	5	1	2	1.8
Buildings	482	436	435	427	409	391	378	100	100	-0.6
Coal	14	10	10	7	5	3	2	2	0	-7.5
Oil	64	50	50	35	23	13	7	11	2	-8.5
Natural gas	172	150	149	142	132	120	109	34	29	-1.4
Electricity	147	142	142	147	149	151	153	33	40	0.3
Heat	36	32	32	33	33	33	32	7	9	0.0
Bioenergy	47	49	50	56	59	61	63	11	17	1.0
Traditional biomass	-	-	-	-	-	-	-	-	-	n.a.
Other renewables	2	3	3	6	9	11	13	1	3	7.3
Other	131	122	122	123	114	107	98	100	100	-1.0
Petrochem. feedstock	78	75	75	75	68	63	56	62	57	-1.3

Table A.3: Energy demand – European Union

	Current Policies Scenario				Sustainable Development Scenario			
	Energy demand (Mtoe)		Shares (%)	CAAGR (%)	Energy demand (Mtoe)		Shares (%)	CAAGR (%)
	2030	2040	2040	2018-40	2030	2040	2040	2018-40
Total primary demand	1 531	1 466	100	-0.4	1 311	1 101	100	-1.7
Coal	167	116	8	-2.9	59	42	4	-7.4
Oil	456	389	27	-1.4	340	161	15	-5.3
Natural gas	398	412	28	0.2	317	217	20	-2.6
Nuclear	175	168	11	-1.1	177	178	16	-0.9
Hydro	34	35	2	0.9	35	37	3	1.1
Bioenergy	199	213	15	1.1	237	240	22	1.7
Other renewables	102	132	9	4.0	146	225	20	6.6
Power sector	630	625	100	-0.3	563	613	100	-0.4
Coal	113	68	11	-3.9	19	13	2	-10.8
Oil	6	4	1	-5.9	5	3	0	-7.7
Natural gas	125	139	22	0.7	104	79	13	-1.9
Nuclear	175	168	27	-1.1	177	178	29	-0.9
Hydro	34	35	6	0.9	35	37	6	1.1
Bioenergy	82	88	14	1.2	93	104	17	2.0
Other renewables	96	122	20	3.9	131	199	33	6.2
Other energy sector	118	111	100	-0.9	101	87	100	-2.0
Electricity	37	37	33	-0.5	38	50	57	0.9
Total final consumption	1 143	1 107	100	-0.2	987	790	100	-1.7
Coal	28	24	2	-1.6	20	13	2	-4.3
Oil	423	361	33	-1.3	313	149	19	-5.2
Natural gas	256	257	23	0.1	198	126	16	-3.1
Electricity	263	282	25	0.7	252	294	37	0.9
Heat	50	49	4	0.1	42	36	5	-1.4
Bioenergy	115	123	11	1.1	143	135	17	1.5
Other renewables	6	10	1	5.5	15	26	3	10.2
Industry	269	264	100	-0.1	226	200	100	-1.3
Coal	20	18	7	-0.7	15	11	6	-3.0
Oil	26	24	9	-0.5	18	12	6	-3.6
Gas	87	84	32	-0.2	70	48	24	-2.7
Electricity	90	89	34	-0.0	83	85	42	-0.2
Heat	15	14	5	-0.7	11	7	4	-3.6
Bioenergy	31	34	13	0.9	28	32	16	0.7
Other renewables	0	1	0	15.9	2	4	2	26.0
Transport	311	288	100	-0.6	273	178	100	-2.7
Oil	274	242	84	-1.0	190	66	37	-6.7
Electricity	10	15	5	4.6	17	57	32	11.1
Biofuels	25	27	9	2.1	51	34	19	3.2
Other fuels	3	4	1	1.0	14	21	12	8.8
Buildings	448	455	100	0.2	383	329	100	-1.3
Coal	6	3	1	-5.6	3	0	0	-22.7
Oil	32	18	4	-4.5	20	4	1	-11.1
Natural gas	154	158	35	0.3	106	56	17	-4.3
Electricity	160	173	38	0.9	147	149	45	0.2
Heat	34	35	8	0.4	31	28	9	-0.6
Bioenergy	56	59	13	0.8	60	65	20	1.2
Traditional biomass	-	-	-	n.a.	-	-	-	n.a.
Other renewables	6	9	2	5.5	13	22	7	9.8
Other	115	100	100	-0.9	105	82	100	-1.8
Petrochem. feedstock	68	55	56	-1.4	62	46	56	-2.2

Annex A | Tables for scenario projections

Table A.3: Electricity and CO_2 emissions – European Union

	Stated Policies Scenario									
	Electricity generation (TWh)							Shares (%)		CAAGR (%)
	2010	2017	2018	2025	2030	2035	2040	2018	2040	2018-40
Total generation	3 336	3 269	3 279	3 351	3 396	3 468	3 565	100	100	0.4
Coal	864	710	666	351	238	126	65	20	2	-10.0
Oil	87	61	58	28	20	17	12	2	0	-6.9
Natural gas	765	663	629	698	627	591	523	19	15	-0.8
Nuclear	917	830	829	693	636	575	595	25	17	-1.5
Renewables	699	1 000	1 092	1 579	1 872	2 156	2 368	33	66	3.6
Hydro	377	301	339	383	403	414	422	10	12	1.0
Bioenergy	143	211	225	265	279	298	312	7	9	1.5
Wind	149	362	392	682	884	1 084	1 239	12	35	5.4
Geothermal	6	7	7	7	10	13	15	0	0	3.6
Solar PV	23	114	124	234	282	319	332	4	9	4.6
CSP	1	6	5	6	10	15	22	0	1	6.8
Marine	0	1	1	1	4	13	27	0	1	19.7

	Stated Policies Scenario								
	Electrical capacity (GW)						Shares (%)		CAAGR (%)
	2017	2018	2025	2030	2035	2040	2018	2040	2018-40
Total capacity	1 038	1 055	1 185	1 251	1 320	1 360	100	100	1.2
Coal	170	164	93	59	32	22	16	2	-8.7
Oil	46	45	25	19	15	11	4	1	-6.3
Natural gas	217	217	249	255	275	272	21	20	1.0
Nuclear	125	125	106	97	88	92	12	7	-1.4
Renewables	480	502	705	808	885	927	48	68	2.8
Hydro	155	155	160	166	170	173	15	13	0.5
Bioenergy	45	48	55	57	60	61	5	5	1.2
Wind	170	180	276	324	363	387	17	28	3.5
Geothermal	1	1	1	1	2	2	0	0	3.6
Solar PV	107	115	210	254	279	286	11	21	4.2
CSP	2	2	2	4	5	7	0	1	5.3
Marine	0	0	0	2	6	12	0	1	19.5

	Stated Policies Scenario									
	CO_2 emissions (Mt)							Shares (%)		CAAGR (%)
	2010	2017	2018	2025	2030	2035	2040	2018	2040	2018-40
Total CO_2	3 563	3 144	3 083	2 582	2 212	1 846	1 590	100	100	-3.0
Coal	1 116	918	874	531	394	265	192	28	12	-6.7
Oil	1 438	1 335	1 332	1 173	1 005	819	693	43	44	-2.9
Natural gas	1 009	891	878	878	812	762	705	28	44	-1.0
Power sector	1 331	1 071	1 023	699	547	417	325	100	100	-5.1
Coal	901	731	694	369	249	136	71	68	22	-9.8
Oil	76	50	49	25	18	15	11	5	4	-6.4
Natural gas	353	290	280	305	279	266	242	27	74	-0.7
Final consumption	2 058	1 915	1 903	1 745	1 547	1 324	1 167	100	100	-2.2
Coal	178	153	147	131	116	102	94	8	8	-2.0
Oil	1 259	1 199	1 197	1 076	927	753	634	63	54	-2.8
Transport	906	920	919	846	741	605	510	48	44	-2.6
Natural gas	620	562	559	537	504	470	439	29	38	-1.1

Table A.3: Electricity and CO$_2$ emissions – European Union

	Current Policies Scenario				Sustainable Development Scenario			
	Electricity generation (TWh)		Shares (%)	CAAGR (%)	Electricity generation (TWh)		Shares (%)	CAAGR (%)
	2030	2040	2040	2018-40	2030	2040	2040	2018-40
Total generation	3 515	3 702	100	0.6	3 418	4 016	100	0.9
Coal	479	293	8	-3.7	64	48	1	-11.3
Oil	21	14	0	-6.2	19	9	0	-8.3
Natural gas	663	761	21	0.9	549	401	10	-2.0
Nuclear	673	644	17	-1.1	679	685	17	-0.9
Renewables	1 677	1 987	54	2.8	2 103	2 871	71	4.5
Hydro	394	411	11	0.9	404	428	11	1.1
Bioenergy	278	303	8	1.4	317	372	9	2.3
Wind	751	962	26	4.2	1 019	1 573	39	6.5
Geothermal	9	12	0	2.6	12	20	0	5.1
Solar PV	234	271	7	3.6	333	408	10	5.6
CSP	8	16	0	5.4	13	35	1	9.1
Marine	2	12	0	15.4	5	36	1	21.1

	Current Policies Scenario				Sustainable Development Scenario			
	Electrical capacity (GW)		Shares (%)	CAAGR (%)	Electrical capacity (GW)		Shares (%)	CAAGR (%)
	2030	2040	2040	2018-40	2030	2040	2040	2018-40
Total capacity	1 271	1 344	100	1.1	1 341	1 561	100	1.8
Coal	112	66	5	-4.1	37	14	1	-10.5
Oil	20	13	1	-5.6	19	12	1	-5.8
Natural gas	304	345	26	2.1	252	262	17	0.9
Nuclear	97	92	7	-1.4	105	106	7	-0.7
Renewables	720	795	59	2.1	908	1 115	71	3.7
Hydro	163	168	13	0.4	167	175	11	0.5
Bioenergy	56	59	4	1.0	64	73	5	1.9
Wind	286	321	24	2.7	368	485	31	4.6
Geothermal	1	2	0	2.6	2	3	0	5.2
Solar PV	210	234	17	3.3	299	353	23	5.2
CSP	3	6	0	4.0	5	11	1	7.5
Marine	1	6	0	15.3	3	16	1	20.9

	Current Policies Scenario				Sustainable Development Scenario			
	CO$_2$ emissions (Mt)		Shares (%)	CAAGR (%)	CO$_2$ emissions (Mt)		Shares (%)	CAAGR (%)
	2030	2040	2040	2018-40	2030	2040	2040	2018-40
Total CO$_2$	2 661	2 327	100	-1.3	1 668	820	100	-5.8
Coal	637	424	18	-3.2	177	71	9	-10.8
Oil	1 132	975	42	-1.4	788	294	36	-6.6
Natural gas	893	929	40	0.3	702	455	55	-2.9
Power sector	795	632	100	-2.2	331	190	100	-7.4
Coal	483	292	46	-3.9	73	14	8	-16.1
Oil	19	13	2	-5.8	17	9	5	-7.6
Natural gas	294	327	52	0.7	241	167	88	-2.3
Final consumption	1 737	1 575	100	-0.9	1 243	577	100	-5.3
Coal	123	105	7	-1.6	83	41	7	-5.7
Oil	1 047	900	57	-1.3	723	263	45	-6.7
Transport	831	736	47	-1.0	578	201	35	-6.7
Natural gas	567	571	36	0.1	437	274	47	-3.2

Table A.3: Energy demand – Africa

	Stated Policies Scenario Energy demand (Mtoe)							Shares (%)		CAAGR (%)
	2010	2017	2018	2025	2030	2035	2040	2018	2040	2018-40
Total primary demand	681	817	838	992	1 100	1 202	1 318	100	100	2.1
Coal	108	110	112	116	112	112	113	13	9	0.0
Oil	161	193	194	237	266	298	333	23	25	2.5
Natural gas	90	126	133	154	184	220	263	16	20	3.2
Nuclear	3	4	4	4	7	9	11	0	1	5.2
Hydro	9	11	12	17	21	24	30	1	2	4.4
Bioenergy	308	368	378	449	471	470	471	45	36	1.0
Other renewables	2	6	7	16	39	68	97	1	7	13.0
Power sector	142	168	175	194	230	275	327	100	100	2.9
Coal	66	65	66	69	64	60	54	38	17	-0.9
Oil	17	18	19	20	19	20	19	11	6	-0.1
Natural gas	44	64	67	67	75	88	109	38	33	2.2
Nuclear	3	4	4	4	7	9	11	2	3	5.2
Hydro	9	11	12	17	21	24	30	7	9	4.4
Bioenergy	1	1	1	2	6	9	11	1	3	11.7
Other renewables	1	6	6	15	37	65	93	4	28	12.9
Other energy sector	101	126	127	164	179	183	190	100	100	1.9
Electricity	12	16	16	17	20	24	29	13	15	2.7
Total final consumption	496	594	611	727	804	881	970	100	100	2.1
Coal	17	21	22	22	23	24	26	4	3	0.7
Oil	137	165	167	207	238	271	308	27	32	2.8
Natural gas	29	42	46	61	77	96	114	8	12	4.2
Electricity	47	57	59	75	92	114	141	10	15	4.0
Heat	-	-	-	-	-	-	-	-	-	n.a.
Bioenergy	266	309	317	359	372	374	377	52	39	0.8
Other renewables	0	0	0	1	2	3	4	0	0	13.6
Industry	83	95	98	119	135	157	184	100	100	2.9
Coal	12	11	12	14	16	17	20	12	11	2.3
Oil	17	18	17	21	23	26	31	18	17	2.7
Natural gas	14	24	27	34	40	48	57	27	31	3.5
Electricity	20	23	24	28	33	38	44	24	24	2.8
Heat	-	-	-	-	-	-	-	-	-	n.a.
Bioenergy	20	18	18	22	24	27	32	19	17	2.5
Other renewables	-	-	-	0	0	0	0	-	0	n.a.
Transport	86	116	118	147	168	187	207	100	100	2.6
Oil	85	114	116	144	163	181	199	98	96	2.5
Electricity	0	1	1	1	1	2	2	0	1	6.8
Biofuels	0	0	0	1	1	1	2	0	1	20.8
Other fuels	1	1	1	2	2	3	3	1	2	3.9
Buildings	297	358	369	426	460	490	525	100	100	1.6
Coal	3	8	8	6	5	4	4	2	1	-3.7
Oil	19	19	19	23	28	37	47	5	9	4.2
Natural gas	6	12	14	20	28	37	44	4	8	5.4
Electricity	24	31	33	43	55	70	90	9	17	4.7
Heat	-	-	-	-	-	-	-	-	-	n.a.
Bioenergy	244	288	295	333	342	340	337	80	64	0.6
Traditional biomass	236	277	283	320	326	321	314	77	60	0.5
Other renewables	0	0	0	1	2	3	4	0	1	13.0
Other	30	25	26	35	41	47	54	100	100	3.4
Petrochem. feedstock	12	6	6	11	13	15	17	25	32	4.6

Table A.3: Energy demand – Africa

	Current Policies Scenario				Sustainable Development Scenario			
	Energy demand (Mtoe)		Shares (%)	CAAGR (%)	Energy demand (Mtoe)		Shares (%)	CAAGR (%)
	2030	2040	2040	2018-40	2030	2040	2040	2018-40
Total primary demand	1 126	1 395	100	2.3	698	828	100	-0.1
Coal	131	166	12	1.8	79	64	8	-2.5
Oil	274	351	25	2.7	239	247	30	1.1
Natural gas	195	291	21	3.6	146	166	20	1.0
Nuclear	7	11	1	5.2	7	20	2	7.9
Hydro	21	30	2	4.5	26	40	5	5.8
Bioenergy	471	482	35	1.1	139	123	15	-5.0
Other renewables	26	63	5	10.8	61	168	20	15.8
Power sector	246	368	100	3.4	202	307	100	2.6
Coal	80	97	26	1.8	36	18	6	-5.8
Oil	24	26	7	1.4	19	18	6	-0.4
Natural gas	84	135	37	3.2	50	40	13	-2.3
Nuclear	7	11	3	5.2	7	20	6	7.9
Hydro	21	30	8	4.5	26	40	13	5.8
Bioenergy	4	8	2	10.1	7	14	4	12.6
Other renewables	25	60	16	10.7	57	158	52	15.7
Other energy sector	189	219	100	2.5	91	100	100	-1.1
Electricity	22	32	15	3.2	18	25	25	2.1
Total final consumption	806	983	100	2.2	514	591	100	-0.2
Coal	25	31	3	1.6	19	18	3	-0.9
Oil	240	316	32	2.9	211	224	38	1.3
Natural gas	75	110	11	4.0	66	92	16	3.2
Electricity	94	143	15	4.1	91	145	24	4.1
Heat	-	-	-	n.a.	-	-	-	n.a.
Bioenergy	372	380	39	0.8	123	103	17	-5.0
Other renewables	2	3	0	12.4	4	9	2	17.5
Industry	134	184	100	2.9	115	146	100	1.8
Coal	16	23	13	3.0	12	14	9	0.5
Oil	22	26	14	1.8	19	24	17	1.5
Gas	39	55	30	3.3	34	42	29	2.1
Electricity	32	43	23	2.7	26	35	24	1.8
Heat	-	-	-	n.a.	-	-	-	n.a.
Bioenergy	26	38	20	3.3	23	29	20	2.1
Other renewables	0	0	0	n.a.	1	2	1	n.a.
Transport	169	217	100	2.8	160	175	100	1.8
Oil	166	212	98	2.8	148	147	84	1.1
Electricity	1	1	0	2.4	1	3	2	8.2
Biofuels	0	1	0	16.7	5	10	6	30.2
Other fuels	2	3	1	3.3	5	15	8	11.0
Buildings	462	528	100	1.6	201	221	100	-2.3
Coal	6	5	1	-2.3	4	2	1	-5.5
Oil	29	49	9	4.4	23	29	13	1.9
Natural gas	28	45	8	5.5	20	25	11	2.8
Electricity	57	94	18	4.9	60	102	46	5.3
Heat	-	-	-	n.a.	-	-	-	n.a.
Bioenergy	340	332	63	0.5	90	56	25	-7.3
Traditional biomass	326	314	60	0.5	78	44	20	-8.1
Other renewables	2	3	1	12.1	3	7	3	16.3
Other	41	54	100	3.4	39	49	100	3.0
Petrochem. feedstock	12	14	26	3.7	12	16	33	4.3

Table A.3: Electricity and CO_2 emissions – Africa

	Stated Policies Scenario							Shares (%)		CAAGR (%)
	Electricity generation (TWh)									
	2010	2017	2018	2025	2030	2035	2040	2018	2040	2018-40
Total generation	671	827	866	1 056	1 284	1 564	1 897	100	100	3.6
Coal	259	254	258	284	268	256	240	30	13	-0.3
Oil	64	75	79	84	79	82	78	9	4	-0.0
Natural gas	220	335	351	377	433	523	651	40	34	2.9
Nuclear	12	14	14	14	28	35	44	2	2	5.2
Renewables	115	147	163	296	474	668	882	19	46	8.0
Hydro	110	123	135	202	245	283	348	16	18	4.4
Bioenergy	1	2	2	7	22	32	40	0	2	14.9
Wind	2	12	14	39	84	120	159	2	8	11.7
Geothermal	1	5	5	9	23	41	59	1	3	12.1
Solar PV	0	5	6	34	90	172	241	1	13	18.1
CSP	-	1	2	5	9	19	34	0	2	13.9
Marine	-	-	-	-	-	-	-	-	-	n.a.

	Stated Policies Scenario						Shares (%)		CAAGR (%)
	Electrical capacity (GW)								
	2017	2018	2025	2030	2035	2040	2018	2040	2018-40
Total capacity	228	244	315	400	501	614	100	100	4.3
Coal	48	48	55	55	51	48	20	8	0.0
Oil	42	43	35	36	37	36	18	6	-0.8
Natural gas	92	103	129	146	171	207	42	34	3.2
Nuclear	2	2	2	4	5	6	1	1	5.0
Renewables	44	48	90	154	225	299	20	49	8.7
Hydro	35	36	49	57	67	83	15	13	3.9
Bioenergy	1	1	2	6	8	9	0	1	11.9
Wind	5	5	15	30	41	53	2	9	10.9
Geothermal	1	1	1	4	6	9	0	1	12.6
Solar PV	3	4	21	55	98	135	2	22	16.8
CSP	1	1	2	3	6	10	0	2	11.3
Marine	-	-	-	-	-	-	-	-	n.a.

	Stated Policies Scenario							Shares (%)		CAAGR (%)
	CO_2 emissions (Mt)									
	2010	2017	2018	2025	2030	2035	2040	2018	2040	2018-40
Total CO_2	1 017	1 181	1 215	1 357	1 464	1 621	1 797	100	100	1.8
Coal	385	391	395	382	346	332	318	32	18	-1.0
Oil	450	541	551	668	750	846	948	45	53	2.5
Natural gas	182	248	269	307	368	443	532	22	30	3.1
Power sector	420	466	480	495	490	508	530	100	100	0.4
Coal	263	257	261	275	255	239	215	54	41	-0.9
Oil	54	59	62	63	59	62	60	13	11	-0.2
Natural gas	103	150	158	157	176	208	256	33	48	2.2
Final consumption	496	620	641	776	892	1 021	1 163	100	100	2.7
Coal	66	83	85	84	85	89	96	13	8	0.5
Oil	382	472	480	590	675	767	871	75	75	2.8
Transport	257	345	351	436	495	549	603	55	52	2.5
Natural gas	48	65	76	102	131	165	197	12	17	4.4

Table A.3: Electricity and CO₂ emissions – Africa

	Current Policies Scenario				Sustainable Development Scenario			
	Electricity generation (TWh)		Shares (%)	CAAGR (%)	Electricity generation (TWh)		Shares (%)	CAAGR (%)
	2030	2040	2040	2018-40	2030	2040	2040	2018-40
Total generation	1 338	2 029	100	3.9	1 267	1 976	100	3.8
Coal	336	424	21	2.3	148	75	4	-5.5
Oil	102	113	6	1.6	76	64	3	-0.9
Natural gas	491	809	40	3.9	298	250	13	-1.5
Nuclear	28	44	2	5.2	28	76	4	7.9
Renewables	378	638	31	6.4	716	1 509	76	10.6
Hydro	243	351	17	4.5	306	464	23	5.8
Bioenergy	13	29	1	13.1	23	49	2	15.9
Wind	48	85	4	8.5	130	252	13	14.0
Geothermal	17	45	2	10.7	25	75	4	13.3
Solar PV	49	108	5	13.8	200	556	28	22.6
CSP	8	20	1	11.1	32	112	6	20.3
Marine	-	-	-	n.a.	0	1	0	n.a.

	Current Policies Scenario				Sustainable Development Scenario			
	Electrical capacity (GW)		Shares (%)	CAAGR (%)	Electrical capacity (GW)		Shares (%)	CAAGR (%)
	2030	2040	2040	2018-40	2030	2040	2040	2018-40
Total capacity	390	557	100	3.8	484	852	100	5.9
Coal	63	78	14	2.2	43	32	4	-1.8
Oil	43	48	9	0.6	40	46	5	0.3
Natural gas	161	216	39	3.4	129	149	17	1.7
Nuclear	4	6	1	5.0	4	11	1	8.0
Renewables	112	192	34	6.5	262	574	67	11.9
Hydro	57	82	15	3.9	73	115	13	5.4
Bioenergy	3	7	1	10.3	6	11	1	12.7
Wind	17	28	5	7.7	47	86	10	13.3
Geothermal	3	7	1	11.3	4	12	1	14.1
Solar PV	29	62	11	12.8	122	317	37	21.5
CSP	3	6	1	8.7	11	33	4	17.5
Marine	-	-	-	n.a.	0	0	0	n.a.

	Current Policies Scenario				Sustainable Development Scenario			
	CO₂ emissions (Mt)		Shares (%)	CAAGR (%)	CO₂ emissions (Mt)		Shares (%)	CAAGR (%)
	2030	2040	2040	2018-40	2030	2040	2040	2018-40
Total CO₂	1 590	2 117	100	2.6	1 161	1 104	100	-0.4
Coal	417	512	24	1.2	209	111	10	-5.6
Oil	774	1 001	47	2.7	668	685	62	1.0
Natural gas	400	605	29	3.7	284	307	28	0.6
Power sector	593	786	100	2.3	315	200	100	-3.9
Coal	318	385	49	1.8	137	49	25	-7.3
Oil	76	83	11	1.4	62	57	28	-0.4
Natural gas	198	318	40	3.2	117	94	47	-2.3
Final consumption	904	1 210	100	2.9	773	833	100	1.2
Coal	93	118	10	1.5	68	59	7	-1.6
Oil	681	898	74	2.9	594	620	74	1.2
Transport	503	642	53	2.8	449	446	54	1.1
Natural gas	130	195	16	4.4	110	154	19	3.3

Table A.3: Energy demand – South Africa

	Stated Policies Scenario							Shares (%)		CAAGR (%)
	Energy demand (Mtoe)									
	2010	2017	2018	2025	2030	2035	2040	2018	2040	2018-40
Total primary demand	132	133	134	135	133	135	139	100	100	0.2
Coal	101	98	99	93	82	75	68	74	49	-1.7
Oil	18	19	19	23	25	27	30	14	21	2.1
Natural gas	4	4	4	4	5	6	8	3	5	2.9
Nuclear	3	4	4	4	4	4	6	3	5	2.5
Hydro	0	0	0	0	0	0	0	0	0	9.9
Bioenergy	7	7	7	8	10	12	13	5	9	3.1
Other renewables	0	1	1	2	7	10	14	1	10	11.5
Power sector	65	62	62	60	56	57	58	100	100	-0.3
Coal	61	57	57	53	42	37	31	92	53	-2.8
Oil	0	0	0	0	0	0	0	0	0	1.8
Natural gas	-	-	-	1	1	2	3	-	6	n.a.
Nuclear	3	4	4	4	4	4	6	6	11	2.5
Hydro	0	0	0	0	0	0	0	0	1	9.9
Bioenergy	0	0	0	1	2	3	4	0	7	18.8
Other renewables	0	1	1	2	6	9	13	2	22	11.7
Other energy sector	28	24	24	25	25	26	26	100	100	0.3
Electricity	4	4	4	4	4	5	5	18	18	0.5
Total final consumption	61	68	68	73	76	81	86	100	100	1.1
Coal	15	18	19	17	17	16	15	27	18	-0.9
Oil	22	25	25	29	30	32	34	37	39	1.3
Natural gas	1	2	2	2	2	2	2	2	2	0.5
Electricity	17	17	17	19	21	23	27	25	31	2.1
Heat	-	-	-	-	-	-	-	-	-	n.a.
Bioenergy	6	6	6	6	7	7	7	8	8	1.2
Other renewables	0	0	0	0	1	1	1	0	2	9.6
Industry	25	24	24	25	26	27	28	100	100	0.8
Coal	10	9	9	9	9	9	9	36	33	0.4
Oil	1	2	2	2	2	2	2	7	6	-0.1
Natural gas	1	2	2	2	2	2	2	7	6	-0.0
Electricity	10	10	10	11	11	12	12	43	44	0.9
Heat	-	-	-	-	-	-	-	-	-	n.a.
Bioenergy	2	2	2	2	2	3	3	7	10	2.9
Other renewables	-	-	-	0	0	0	0	-	0	n.a.
Transport	16	18	18	21	23	25	28	100	100	1.9
Oil	16	18	18	20	22	24	26	98	93	1.7
Electricity	0	0	0	0	1	1	1	2	3	4.3
Biofuels	-	-	-	0	0	1	1	-	4	n.a.
Other fuels	0	0	0	0	0	0	0	0	0	29.3
Buildings	15	20	20	19	19	20	22	100	100	0.4
Coal	3	8	8	6	5	4	4	41	16	-3.8
Oil	1	2	2	1	1	1	1	8	5	-1.6
Natural gas	0	0	0	0	0	0	0	0	1	22.5
Electricity	6	6	6	7	8	10	12	30	57	3.4
Heat	-	-	-	-	-	-	-	-	-	n.a.
Bioenergy	4	4	4	4	4	4	3	20	15	-0.8
Traditional biomass	4	4	4	4	4	3	3	20	13	-1.6
Other renewables	0	0	0	0	1	1	1	1	5	9.0
Other	6	6	6	8	8	8	9	100	100	1.4
Petrochem. feedstock	1	1	1	3	3	3	3	19	39	4.8

Table A.3: Energy demand – South Africa

	Current Policies Scenario				Sustainable Development Scenario			
	Energy demand (Mtoe)		Shares (%)	CAAGR (%)	Energy demand (Mtoe)		Shares (%)	CAAGR (%)
	2030	2040	2040	2018-40	2030	2040	2040	2018-40
Total primary demand	144	160	100	0.8	112	107	100	-1.0
Coal	96	92	58	-0.3	64	39	37	-4.1
Oil	26	33	21	2.6	21	20	19	0.3
Natural gas	6	8	5	3.0	5	7	6	2.3
Nuclear	4	6	4	2.5	4	10	9	4.7
Hydro	0	0	0	8.5	0	0	0	9.9
Bioenergy	9	12	8	2.7	9	13	12	3.0
Other renewables	3	8	5	8.9	8	17	16	12.5
Power sector	65	74	100	0.8	43	40	100	-2.0
Coal	55	52	71	-0.4	28	7	18	-9.1
Oil	0	0	0	1.8	0	0	0	1.7
Natural gas	2	4	5	n.a.	1	3	6	n.a.
Nuclear	4	6	9	2.5	4	10	26	4.7
Hydro	0	0	1	8.5	0	0	1	9.9
Bioenergy	2	4	5	17.9	2	4	11	18.8
Other renewables	3	7	10	9.1	7	15	38	12.6
Other energy sector	26	28	100	0.6	25	25	100	0.2
Electricity	5	5	20	1.1	4	3	14	-1.0
Total final consumption	78	91	100	1.3	65	66	100	-0.1
Coal	18	17	19	-0.3	14	10	16	-2.6
Oil	31	37	41	1.8	26	24	37	-0.2
Natural gas	2	2	2	0.5	1	2	2	-0.4
Electricity	21	27	30	2.2	17	21	31	0.9
Heat	-	-	-	n.a.	-	-	-	n.a.
Bioenergy	6	7	8	1.0	5	7	10	1.0
Other renewables	0	1	1	7.7	1	2	3	12.1
Industry	26	29	100	0.9	21	22	100	-0.4
Coal	10	10	35	0.8	7	6	28	-1.6
Oil	2	2	6	-0.0	1	1	4	-2.7
Gas	2	2	6	0.2	1	1	4	-2.5
Electricity	11	12	42	0.8	8	9	43	-0.4
Heat	-	-	-	n.a.	-	-	-	n.a.
Bioenergy	2	3	10	2.8	3	4	19	4.5
Other renewables	0	0	0	n.a.	0	0	2	n.a.
Transport	24	30	100	2.3	21	21	100	0.7
Oil	23	29	96	2.2	19	18	84	0.0
Electricity	0	0	2	1.9	0	1	4	4.4
Biofuels	0	1	3	n.a.	1	2	9	n.a.
Other fuels	-	-	-	n.a.	0	1	3	44.2
Buildings	20	24	100	0.8	15	15	100	-1.2
Coal	6	5	20	-2.4	4	2	15	-5.6
Oil	1	1	5	-1.1	1	1	4	-4.5
Natural gas	0	0	0	21.4	0	0	0	18.4
Electricity	9	13	56	3.7	7	10	64	2.3
Heat	-	-	-	n.a.	-	-	-	n.a.
Bioenergy	4	3	14	-0.9	1	1	6	-6.4
Traditional biomass	4	3	12	-1.6	1	0	2	-10.3
Other renewables	0	1	4	7.4	1	2	11	10.9
Other	8	9	100	1.6	8	8	100	1.0
Petrochem. feedstock	3	3	38	4.8	3	3	39	4.3

Table A.3: Electricity and CO_2 emissions – South Africa

	Stated Policies Scenario							Shares (%)		CAAGR (%)
	Electricity generation (TWh)									
	2010	2017	2018	2025	2030	2035	2040	2018	2040	2018-40
Total generation	257	251	254	264	283	310	344	100	100	1.4
Coal	242	227	228	221	181	163	143	90	42	-2.1
Oil	0	0	0	0	0	0	0	0	0	1.8
Natural gas	-	-	-	4	10	17	23	-	7	n.a.
Nuclear	12	14	14	14	14	15	25	6	7	2.5
Renewables	2	10	12	25	77	114	153	5	44	12.4
Hydro	2	1	1	2	3	4	6	0	2	9.9
Bioenergy	0	0	0	2	9	13	16	0	5	19.6
Wind	0	5	7	12	42	61	81	3	24	12.0
Geothermal	-	-	-	0	0	0	0	-	0	n.a.
Solar PV	-	3	3	7	22	32	41	1	12	12.6
CSP	-	1	1	2	2	4	8	0	2	10.2
Marine	-	-	-	-	-	-	-	-	-	n.a.

	Stated Policies Scenario						Shares (%)		CAAGR (%)
	Electrical capacity (GW)								
	2017	2018	2025	2030	2035	2040	2018	2040	2018-40
Total capacity	56	56	65	82	96	112	100	100	3.2
Coal	42	41	41	36	31	27	74	24	-1.8
Oil	4	4	4	3	3	3	7	3	-0.9
Natural gas	-	-	4	7	11	17	-	15	n.a.
Nuclear	2	2	2	2	2	3	3	3	2.4
Renewables	8	8	13	31	42	54	15	48	8.9
Hydro	4	4	4	4	4	4	7	4	0.9
Bioenergy	0	0	1	2	3	4	0	3	12.6
Wind	2	2	4	13	17	22	4	19	11.2
Geothermal	-	-	0	0	0	0	-	0	n.a.
Solar PV	2	2	4	12	16	21	3	19	11.8
CSP	0	0	1	1	1	3	1	2	8.9
Marine	-	-	-	-	-	-	-	-	n.a.

	Stated Policies Scenario							Shares (%)		CAAGR (%)
	CO_2 emissions (Mt)									
	2010	2017	2018	2025	2030	2035	2040	2018	2040	2018-40
Total CO_2	419	422	420	378	321	300	279	100	100	-1.8
Coal	356	347	346	296	232	204	175	82	63	-3.0
Oil	61	71	70	77	81	86	91	17	33	1.2
Natural gas	2	4	4	5	7	10	12	1	4	5.4
Power sector	243	225	227	212	172	153	129	100	100	-2.5
Coal	243	225	227	210	168	147	121	100	94	-2.8
Oil	0	0	0	0	0	0	0	0	0	1.8
Natural gas	-	-	-	1	3	6	8	-	6	n.a.
Final consumption	116	142	143	142	143	145	148	100	100	0.1
Coal	57	70	71	64	60	57	54	50	37	-1.2
Oil	58	69	68	74	79	84	89	47	60	1.2
Transport	46	54	54	61	67	72	77	37	52	1.7
Natural gas	2	4	4	4	4	4	4	3	3	0.5

Table A.3: Electricity and CO₂ emissions – South Africa

	Current Policies Scenario				Sustainable Development Scenario			
	Electricity generation (TWh)		Shares (%)	CAAGR (%)	Electricity generation (TWh)		Shares (%)	CAAGR (%)
	2030	2040	2040	2018-40	2030	2040	2040	2018-40
Total generation	300	381	100	1.9	232	260	100	0.1
Coal	236	238	62	0.2	118	28	11	-9.1
Oil	0	0	0	1.9	0	0	0	1.8
Natural gas	12	24	6	n.a.	10	17	6	n.a.
Nuclear	14	25	6	2.5	14	39	15	4.7
Renewables	38	94	25	9.9	89	176	68	13.1
Hydro	2	4	1	8.5	3	6	2	9.9
Bioenergy	6	13	3	18.4	9	16	6	19.6
Wind	16	41	11	8.5	46	92	35	12.6
Geothermal	0	0	0	n.a.	0	0	0	n.a.
Solar PV	12	31	8	11.1	26	52	20	13.8
CSP	2	6	1	8.2	5	10	4	11.2
Marine	-	-	-	n.a.	0	0	0	n.a.

	Current Policies Scenario				Sustainable Development Scenario			
	Electrical capacity (GW)		Shares (%)	CAAGR (%)	Electrical capacity (GW)		Shares (%)	CAAGR (%)
	2030	2040	2040	2018-40	2030	2040	2040	2018-40
Total capacity	76	108	100	3.1	82	105	100	2.9
Coal	42	42	39	0.1	33	17	16	-4.0
Oil	4	3	3	-0.8	3	3	3	-0.9
Natural gas	7	14	13	n.a.	5	8	8	n.a.
Nuclear	2	3	3	2.4	2	6	6	5.1
Renewables	18	36	34	7.0	36	62	59	9.6
Hydro	4	4	4	0.9	4	4	4	0.9
Bioenergy	2	3	3	11.6	2	4	4	12.6
Wind	5	11	10	7.9	14	25	23	11.8
Geothermal	0	0	0	n.a.	0	0	0	n.a.
Solar PV	7	16	15	10.4	14	26	25	12.9
CSP	1	2	2	6.9	2	3	3	10.0
Marine	-	-	-	n.a.	0	0	0	n.a.

	Current Policies Scenario				Sustainable Development Scenario			
	CO₂ emissions (Mt)		Shares (%)	CAAGR (%)	CO₂ emissions (Mt)		Shares (%)	CAAGR (%)
	2030	2040	2040	2018-40	2030	2040	2040	2018-40
Total CO₂	379	384	100	-0.4	231	107	100	-6.0
Coal	287	271	70	-1.1	155	36	33	-9.8
Oil	84	101	26	1.7	70	63	58	-0.5
Natural gas	8	13	3	5.5	6	9	8	3.8
Power sector	223	216	100	-0.2	109	13	100	-12.1
Coal	218	208	96	-0.4	105	7	55	-14.4
Oil	0	0	0	1.8	0	0	2	1.7
Natural gas	4	8	4	n.a.	3	6	44	n.a.
Final consumption	150	166	100	0.7	118	94	100	-1.9
Coal	65	63	38	-0.6	48	30	32	-3.9
Oil	82	99	59	1.7	68	61	65	-0.5
Transport	69	86	52	2.2	58	54	58	0.0
Natural gas	4	4	3	0.5	3	3	3	-1.2

Table A.3: Energy demand – Middle East

	Stated Policies Scenario Energy demand (Mtoe)						Shares (%)		CAAGR (%)
	2010	2017	2018	2025	2030	2035	2040	2018	2040	2018-40
Total primary demand	640	776	763	848	956	1 092	1 206	100	100	2.1
Coal	2	4	4	6	7	8	10	1	1	4.2
Oil	311	324	313	354	376	409	442	41	37	1.6
Natural gas	324	443	440	466	538	615	672	58	56	1.9
Nuclear	-	2	2	11	12	18	20	0	2	11.1
Hydro	2	2	2	2	3	3	3	0	0	3.2
Bioenergy	1	1	1	2	4	7	11	0	1	12.2
Other renewables	0	1	1	7	17	31	48	0	4	21.0
Power sector	214	282	288	291	317	359	394	100	100	1.4
Coal	0	0	0	2	3	5	6	0	1	15.6
Oil	83	85	87	97	81	72	66	30	17	-1.2
Natural gas	129	193	197	173	206	236	257	68	65	1.2
Nuclear	-	2	2	11	12	18	20	1	5	11.1
Hydro	2	2	2	2	3	3	3	1	1	3.2
Bioenergy	0	0	0	1	2	4	6	0	2	35.2
Other renewables	0	0	0	4	11	22	36	0	9	22.1
Other energy sector	65	72	61	86	99	115	130	100	100	3.5
Electricity	14	21	20	22	26	31	35	32	26	2.6
Total final consumption	435	521	513	584	676	781	869	100	100	2.4
Coal	1	3	3	3	3	3	3	1	0	-0.1
Oil	210	237	231	255	286	325	359	45	41	2.0
Natural gas	163	202	198	231	269	309	337	39	39	2.4
Electricity	60	78	80	91	109	131	153	16	18	3.0
Heat	-	-	-	-	-	-	-	-	-	n.a.
Bioenergy	1	1	1	2	2	3	4	0	0	7.6
Other renewables	0	0	0	3	6	9	12	0	1	18.7
Industry	131	148	137	155	173	193	213	100	100	2.0
Coal	1	3	3	3	3	3	3	2	1	-0.1
Oil	34	23	18	18	19	20	20	13	10	0.6
Natural gas	83	107	101	116	131	147	164	73	77	2.2
Electricity	13	15	16	18	20	22	24	11	11	2.0
Heat	-	-	-	-	-	-	-	-	-	n.a.
Bioenergy	-	-	-	0	1	1	2	-	1	n.a.
Other renewables	0	0	0	0	0	0	0	0	0	24.6
Transport	116	134	131	149	165	189	209	100	100	2.2
Oil	111	127	124	139	154	177	194	95	93	2.1
Electricity	0	0	0	0	0	0	0	0	0	10.3
Biofuels	-	-	-	-	-	-	-	-	-	n.a.
Other fuels	5	7	7	10	10	12	15	5	7	3.4
Buildings	126	149	151	173	214	257	287	100	100	3.0
Coal	0	0	0	0	0	0	0	0	0	-1.7
Oil	22	17	17	15	15	15	15	11	5	-0.5
Natural gas	59	71	73	86	107	127	136	48	47	2.9
Electricity	44	59	60	69	85	104	123	40	43	3.3
Heat	-	-	-	-	-	-	-	-	-	n.a.
Bioenergy	1	1	1	1	2	2	2	1	1	4.8
Traditional biomass	0	0	0	0	1	1	1	0	0	1.5
Other renewables	0	0	0	2	5	9	11	0	4	18.3
Other	62	90	94	107	124	142	159	100	100	2.4
Petrochem. feedstock	44	71	74	84	100	117	134	78	84	2.8

Table A.3: Energy demand – Middle East

	Current Policies Scenario				Sustainable Development Scenario			
	Energy demand (Mtoe)		Shares (%)	CAAGR (%)	Energy demand (Mtoe)		Shares (%)	CAAGR (%)
	2030	2040	2040	2018-40	2030	2040	2040	2018-40
Total primary demand	989	1 279	100	2.4	802	880	100	0.7
Coal	7	10	1	4.6	5	4	1	0.6
Oil	387	484	38	2.0	270	254	29	-0.9
Natural gas	561	713	56	2.2	458	422	48	-0.2
Nuclear	13	23	2	11.8	15	33	4	13.6
Hydro	2	3	0	3.1	3	4	0	4.1
Bioenergy	4	9	1	11.5	8	16	2	14.2
Other renewables	14	36	3	19.5	44	147	17	27.3
Power sector	334	428	100	1.8	253	288	100	0.0
Coal	4	6	1	16.1	2	2	1	11.3
Oil	82	80	19	-0.4	35	16	6	-7.3
Natural gas	221	284	66	1.7	170	117	41	-2.3
Nuclear	13	23	5	11.8	15	33	11	13.6
Hydro	2	3	1	3.1	3	4	1	4.1
Bioenergy	1	5	1	33.6	4	9	3	37.1
Other renewables	9	26	6	20.3	25	106	37	28.2
Other energy sector	107	144	100	4.0	80	90	100	1.8
Electricity	28	38	26	3.0	22	27	30	1.5
Total final consumption	692	908	100	2.6	590	669	100	1.2
Coal	3	3	0	0.0	2	1	0	-4.3
Oil	294	383	42	2.3	235	235	35	0.1
Natural gas	272	343	38	2.5	231	247	37	1.0
Electricity	115	163	18	3.3	100	137	21	2.5
Heat	-	-	-	n.a.	-	-	-	n.a.
Bioenergy	2	4	0	7.8	4	7	1	10.2
Other renewables	5	10	1	17.8	19	40	6	25.4
Industry	173	217	100	2.1	148	154	100	0.5
Coal	3	3	1	0.1	1	1	1	-5.1
Oil	19	21	9	0.6	13	12	8	-1.9
Gas	131	167	77	2.3	115	116	75	0.6
Electricity	20	24	11	2.1	17	20	13	1.3
Heat	-	-	-	n.a.	-	-	-	n.a.
Bioenergy	1	2	1	n.a.	1	3	2	n.a.
Other renewables	0	0	0	23.5	0	2	1	40.4
Transport	170	226	100	2.5	136	131	100	0.0
Oil	162	215	95	2.5	118	96	73	-1.2
Electricity	0	0	0	2.7	1	3	2	20.9
Biofuels	-	-	-	n.a.	1	2	1	n.a.
Other fuels	8	10	5	1.8	16	31	24	7.0
Buildings	225	306	100	3.3	191	241	100	2.1
Coal	0	0	0	-1.6	0	0	0	-5.6
Oil	17	18	6	0.3	13	11	4	-2.0
Natural gas	112	144	47	3.1	80	80	33	0.4
Electricity	90	133	43	3.7	78	110	46	2.8
Heat	-	-	-	n.a.	-	-	-	n.a.
Bioenergy	1	2	1	4.4	2	3	1	5.3
Traditional biomass	0	1	0	1.5	0	0	0	-7.8
Other renewables	5	9	3	17.4	18	37	15	25.0
Other	123	159	100	2.4	115	143	100	1.9
Petrochem. feedstock	98	131	82	2.7	93	121	85	2.3

Table A.3: Electricity and CO_2 emissions – Middle East

	Stated Policies Scenario							Shares (%)		CAAGR (%)
	Electricity generation (TWh)									
	2010	2017	2018	2025	2030	2035	2040	2018	2040	2018-40
Total generation	833	1 126	1 147	1 306	1 570	1 876	2 169	100	100	2.9
Coal	0	1	1	11	17	23	28	0	1	18.9
Oil	286	289	294	346	303	273	251	26	12	-0.7
Natural gas	529	808	822	837	1 065	1 253	1 404	72	65	2.5
Nuclear	-	8	8	41	45	70	78	1	4	11.1
Renewables	18	21	23	71	140	257	408	2	19	14.0
Hydro	18	18	18	27	29	32	36	2	2	3.2
Bioenergy	0	0	0	2	6	14	22	0	1	35.0
Wind	0	1	1	9	28	60	109	0	5	24.7
Geothermal	-	-	-	0	0	0	0	-	0	n.a.
Solar PV	0	2	4	29	63	123	193	0	9	19.9
CSP	-	0	0	3	13	28	47	0	2	26.7
Marine	-	-	-	-	-	-	-	-	-	n.a.

	Stated Policies Scenario						Shares (%)		CAAGR (%)
	Electrical capacity (GW)								
	2017	2018	2025	2030	2035	2040	2018	2040	2018-40
Total capacity	321	331	410	476	561	641	100	100	3.1
Coal	0	0	3	4	5	6	0	1	14.9
Oil	93	96	95	86	81	77	29	12	-1.0
Natural gas	208	214	266	309	345	370	65	58	2.5
Nuclear	1	1	7	8	13	13	0	2	12.5
Renewables	18	19	40	68	115	173	6	27	10.4
Hydro	17	17	19	20	22	23	5	4	1.5
Bioenergy	0	0	0	1	2	4	0	1	31.5
Wind	0	1	3	10	21	37	0	6	19.6
Geothermal	-	-	0	0	0	0	-	0	n.a.
Solar PV	1	2	16	32	61	93	1	14	19.2
CSP	0	0	1	5	10	16	0	3	23.8
Marine	-	-	-	-	-	-	-	-	n.a.

	Stated Policies Scenario							Shares (%)		CAAGR (%)
	CO_2 emissions (Mt)									
	2010	2017	2018	2025	2030	2035	2040	2018	2040	2018-40
Total CO_2	1 493	1 776	1 805	1 936	2 104	2 317	2 481	100	100	1.5
Coal	6	15	14	23	27	32	36	1	1	4.3
Oil	809	859	829	908	913	954	994	46	40	0.8
Natural gas	678	902	961	1 006	1 165	1 331	1 450	53	58	1.9
Power sector	565	709	730	717	748	792	828	100	100	0.6
Coal	1	2	1	10	14	19	23	0	3	15.6
Oil	261	262	270	302	253	223	205	37	25	-1.2
Natural gas	303	445	459	405	481	550	600	63	72	1.2
Final consumption	826	927	934	1 043	1 174	1 331	1 446	100	100	2.0
Coal	4	12	12	12	12	12	12	1	1	-0.1
Oil	515	551	515	556	606	676	734	55	51	1.6
Transport	331	378	369	414	461	527	579	40	40	2.1
Natural gas	307	363	407	475	556	642	701	44	48	2.5

Table A.3: Electricity and CO_2 emissions – Middle East

	Current Policies Scenario				Sustainable Development Scenario			
	Electricity generation (TWh)		Shares (%)	CAAGR (%)	Electricity generation (TWh)		Shares (%)	CAAGR (%)
	2030	2040	2040	2018-40	2030	2040	2040	2018-40
Total generation	1 656	2 328	100	3.3	1 416	1 909	100	2.3
Coal	17	32	1	19.4	11	12	1	14.3
Oil	308	310	13	0.2	128	59	3	-7.1
Natural gas	1 152	1 568	67	3.0	916	703	37	-0.7
Nuclear	51	88	4	11.8	56	125	7	13.6
Renewables	128	331	14	12.9	305	1 010	53	18.8
Hydro	29	36	2	3.1	36	44	2	4.1
Bioenergy	5	17	1	33.3	14	31	2	37.0
Wind	23	65	3	21.7	119	371	19	31.8
Geothermal	-	-	-	n.a.	0	0	0	n.a.
Solar PV	62	195	8	20.0	115	362	19	23.4
CSP	8	18	1	21.3	22	201	11	35.4
Marine	-	-	-	n.a.	0	1	0	n.a.

	Current Policies Scenario				Sustainable Development Scenario			
	Electrical capacity (GW)		Shares (%)	CAAGR (%)	Electrical capacity (GW)		Shares (%)	CAAGR (%)
	2030	2040	2040	2018-40	2030	2040	2040	2018-40
Total capacity	477	631	100	3.0	508	783	100	4.0
Coal	4	7	1	15.4	3	3	0	10.5
Oil	87	88	14	-0.4	79	72	9	-1.3
Natural gas	311	370	59	2.5	279	274	35	1.1
Nuclear	8	13	2	12.5	8	17	2	13.8
Renewables	64	148	23	9.6	138	411	53	14.9
Hydro	20	23	4	1.5	23	27	4	2.3
Bioenergy	1	3	0	29.8	2	6	1	34.3
Wind	8	22	3	16.7	43	125	16	26.4
Geothermal	-	-	-	n.a.	0	0	0	n.a.
Solar PV	32	94	15	19.3	60	181	23	22.9
CSP	3	6	1	18.4	8	72	9	32.4
Marine	-	-	-	n.a.	0	0	0	n.a.

	Current Policies Scenario				Sustainable Development Scenario			
	CO_2 emissions (Mt)		Shares (%)	CAAGR (%)	CO_2 emissions (Mt)		Shares (%)	CAAGR (%)
	2030	2040	2040	2018-40	2030	2040	2040	2018-40
Total CO_2	2 191	2 699	100	1.8	1 569	1 262	100	-1.6
Coal	27	39	1	4.7	16	14	1	-0.1
Oil	944	1 119	41	1.4	593	434	34	-2.9
Natural gas	1 219	1 541	57	2.2	960	815	65	-0.7
Power sector	788	939	100	1.2	511	318	100	-3.7
Coal	14	26	3	16.1	9	10	3	11.3
Oil	257	250	27	-0.4	109	51	16	-7.3
Natural gas	517	664	71	1.7	393	257	81	-2.6
Final consumption	1 209	1 536	100	2.3	917	832	100	-0.5
Coal	12	12	1	0.1	6	3	0	-5.8
Oil	633	808	53	2.1	450	360	43	-1.6
Transport	482	642	42	2.5	352	285	34	-1.2
Natural gas	564	715	47	2.6	462	468	56	0.6

Annex A | Tables for scenario projections

Table A.3: Energy demand – Eurasia

	Stated Policies Scenario							Shares (%)		CAAGR (%)
	Energy demand (Mtoe)									
	2010	2017	2018	2025	2030	2035	2040	2018	2040	2018-40
Total primary demand	831	900	934	965	980	1 000	1 031	100	100	0.5
Coal	138	157	160	157	148	142	139	17	14	-0.6
Oil	152	182	189	198	198	193	191	20	19	0.1
Natural gas	469	477	500	524	533	544	562	54	55	0.5
Nuclear	45	54	54	52	56	64	66	6	6	0.9
Hydro	19	21	22	23	25	26	27	2	3	1.0
Bioenergy	7	9	9	10	12	16	21	1	2	4.0
Other renewables	0	0	0	2	7	15	24	0	2	22.2
Power sector	417	400	404	407	411	424	445	100	100	0.4
Coal	90	84	87	86	78	72	69	21	16	-1.0
Oil	11	6	7	6	5	5	5	2	1	-1.5
Natural gas	246	230	231	234	234	234	241	57	54	0.2
Nuclear	45	54	54	52	56	64	66	13	15	0.9
Hydro	19	21	22	23	25	26	27	5	6	1.0
Bioenergy	4	4	4	5	6	9	13	1	3	5.2
Other renewables	0	0	0	1	7	14	23	0	5	22.4
Other energy sector	143	165	170	172	171	170	172	100	100	0.1
Electricity	31	34	34	35	34	34	36	20	21	0.2
Total final consumption	534	588	614	652	670	684	702	100	100	0.6
Coal	30	43	43	43	44	44	45	7	6	0.2
Oil	120	147	152	165	167	166	165	25	24	0.4
Natural gas	182	199	219	230	237	244	252	36	36	0.6
Electricity	75	80	80	91	98	104	112	13	16	1.5
Heat	125	115	115	117	117	118	119	19	17	0.1
Bioenergy	3	4	4	5	6	7	8	1	1	2.6
Other renewables	0	0	0	0	0	1	1	0	0	18.6
Industry	156	180	185	198	203	207	212	100	100	0.6
Coal	21	36	36	37	38	40	41	20	19	0.6
Oil	13	20	20	21	20	20	19	11	9	-0.1
Natural gas	41	45	49	52	53	54	56	27	26	0.6
Electricity	34	35	35	39	40	42	43	19	20	0.9
Heat	48	43	43	47	48	49	50	23	24	0.7
Bioenergy	0	1	1	2	2	3	3	1	1	3.1
Other renewables	-	-	-	0	0	0	0	-	0	n.a.
Transport	111	114	120	130	135	139	144	100	100	0.8
Oil	68	73	76	83	86	86	87	63	60	0.6
Electricity	8	7	7	8	9	11	12	6	8	2.4
Biofuels	-	0	0	0	0	0	0	0	0	4.0
Other fuels	35	33	37	39	40	43	45	31	31	0.9
Buildings	198	216	223	226	230	233	236	100	100	0.3
Coal	9	7	7	6	5	4	3	3	1	-3.3
Oil	12	24	24	25	24	23	22	11	9	-0.5
Natural gas	70	78	85	86	87	88	89	38	38	0.2
Electricity	30	35	35	39	43	47	51	16	22	1.7
Heat	74	70	69	67	67	66	66	31	28	-0.2
Bioenergy	2	3	3	3	4	4	4	1	2	2.2
Traditional biomass	-	-	-	-	-	-	-	-	-	n.a.
Other renewables	0	0	0	0	0	0	0	0	0	16.4
Other	69	79	86	98	102	105	109	100	100	1.1
Petrochem. feedstock	48	20	19	31	33	34	36	22	33	2.9

Table A.3: Energy demand – Eurasia

	Current Policies Scenario				Sustainable Development Scenario			
	Energy demand (Mtoe)		Shares (%)	CAAGR (%)	Energy demand (Mtoe)		Shares (%)	CAAGR (%)
	2030	2040	2040	2018-40	2030	2040	2040	2018-40
Total primary demand	1 010	1 103	100	0.8	858	807	100	-0.7
Coal	160	158	14	-0.1	95	52	6	-5.0
Oil	202	201	18	0.3	175	140	17	-1.3
Natural gas	552	612	55	0.9	459	393	49	-1.1
Nuclear	56	76	7	1.5	62	79	10	1.7
Hydro	25	28	3	1.1	30	41	5	2.9
Bioenergy	11	16	1	2.6	19	51	6	8.3
Other renewables	5	12	1	18.7	18	50	6	26.4
Power sector	423	472	100	0.7	354	362	100	-0.5
Coal	89	87	18	0.0	41	12	3	-8.6
Oil	5	4	1	-1.8	5	4	1	-2.6
Natural gas	239	256	54	0.5	187	139	38	-2.3
Nuclear	56	76	16	1.5	62	79	22	1.7
Hydro	25	28	6	1.1	30	41	11	2.9
Bioenergy	5	9	2	3.4	12	42	11	10.9
Other renewables	5	12	3	18.9	17	46	13	26.3
Other energy sector	177	194	100	0.6	144	119	100	-1.6
Electricity	36	41	21	0.7	29	25	21	-1.5
Total final consumption	690	742	100	0.9	591	549	100	-0.5
Coal	44	46	6	0.2	32	24	4	-2.6
Oil	171	176	24	0.7	146	119	22	-1.1
Natural gas	248	272	37	1.0	222	209	38	-0.2
Electricity	100	117	16	1.7	86	96	18	0.8
Heat	121	124	17	0.3	97	86	16	-1.3
Bioenergy	5	6	1	1.8	7	9	2	3.6
Other renewables	0	0	0	15.4	1	4	1	27.6
Industry	207	221	100	0.8	171	161	100	-0.6
Coal	38	41	19	0.6	29	23	14	-2.0
Oil	20	18	8	-0.4	17	15	9	-1.3
Gas	57	63	29	1.1	52	51	31	0.1
Electricity	40	43	19	0.9	34	36	22	0.1
Heat	49	53	24	0.9	36	31	19	-1.5
Bioenergy	2	3	1	3.2	3	4	3	5.1
Other renewables	0	0	0	n.a.	0	2	1	n.a.
Transport	140	155	100	1.2	119	107	100	-0.5
Oil	88	94	61	1.0	74	56	52	-1.4
Electricity	9	11	7	1.8	10	13	12	2.9
Biofuels	0	0	0	3.0	0	0	0	3.8
Other fuels	43	50	32	1.4	36	38	35	0.1
Buildings	240	254	100	0.6	204	181	100	-1.0
Coal	5	4	1	-2.6	3	1	0	-10.0
Oil	26	26	10	0.3	21	16	9	-2.0
Natural gas	91	96	38	0.5	79	62	34	-1.4
Electricity	46	57	22	2.2	38	43	24	0.9
Heat	69	69	27	-0.0	58	53	29	-1.2
Bioenergy	3	3	1	0.7	4	5	3	2.5
Traditional biomass	-	-	-	n.a.	-	-	-	n.a.
Other renewables	0	0	0	12.7	1	2	1	24.7
Other	103	112	100	1.2	97	99	100	0.7
Petrochem. feedstock	33	36	32	2.9	31	33	33	2.4

Table A.3: Electricity and CO_2 emissions – Eurasia

Stated Policies Scenario

Electricity generation (TWh)	2010	2017	2018	2025	2030	2035	2040	Shares (%) 2018	Shares (%) 2040	CAAGR (%) 2018-40
Total generation	1 244	1 341	1 360	1 490	1 565	1 643	1 747	100	100	1.1
Coal	235	250	271	279	248	230	222	20	13	-0.9
Oil	9	5	5	3	2	1	1	0	0	-8.5
Natural gas	598	626	618	726	778	789	831	45	48	1.4
Nuclear	173	206	206	198	213	243	253	15	14	0.9
Renewables	229	254	260	283	324	379	440	19	25	2.4
Hydro	226	249	255	269	286	303	318	19	18	1.0
Bioenergy	3	3	3	4	9	19	34	0	2	11.9
Wind	0	1	1	7	20	41	62	0	4	22.7
Geothermal	1	0	0	1	6	13	20	0	1	19.0
Solar PV	-	1	1	3	4	4	6	0	0	7.3
CSP	-	-	-	-	-	-	-	-	-	n.a.
Marine	-	-	-	0	0	0	0	-	0	n.a.

Stated Policies Scenario

Electrical capacity (GW)	2017	2018	2025	2030	2035	2040	Shares (%) 2018	Shares (%) 2040	CAAGR (%) 2018-40
Total capacity	327	331	355	364	380	407	100	100	0.9
Coal	68	68	61	52	49	46	21	11	-1.8
Oil	9	9	6	3	1	1	3	0	-10.4
Natural gas	150	152	176	180	180	192	46	47	1.1
Nuclear	28	29	29	32	35	35	9	9	0.8
Renewables	72	73	84	97	113	130	22	32	2.7
Hydro	70	70	76	81	85	89	21	22	1.1
Bioenergy	1	1	2	3	5	8	0	2	8.1
Wind	0	0	3	9	17	25	0	6	21.0
Geothermal	0	0	0	1	2	3	0	1	17.8
Solar PV	0	1	3	3	4	5	0	1	9.3
CSP	-	-	-	-	-	-	-	-	n.a.
Marine	-	-	0	0	0	0	-	0	n.a.

Stated Policies Scenario

CO_2 emissions (Mt)	2010	2017	2018	2025	2030	2035	2040	Shares (%) 2018	Shares (%) 2040	CAAGR (%) 2018-40
Total CO_2	1 926	1 984	2 035	2 090	2 076	2 064	2 088	100	100	0.1
Coal	551	553	567	563	529	508	500	28	24	-0.6
Oil	369	421	428	446	448	440	436	21	21	0.1
Natural gas	1 005	1 010	1 039	1 082	1 099	1 116	1 151	51	55	0.5
Power sector	1 014	932	928	932	893	869	873	100	100	-0.3
Coal	389	347	360	357	322	298	286	39	33	-1.0
Oil	38	25	22	19	18	16	15	2	2	-1.5
Natural gas	587	560	546	555	554	555	571	59	65	0.2
Final consumption	791	931	979	1 012	1 028	1 036	1 048	100	100	0.3
Coal	152	201	203	201	203	205	209	21	20	0.1
Oil	294	359	369	390	396	392	388	38	37	0.2
Transport	202	217	225	245	254	255	258	23	25	0.6
Natural gas	344	371	407	420	429	440	451	42	43	0.5

Table A.3: Electricity and CO_2 emissions – Eurasia

	Current Policies Scenario				Sustainable Development Scenario			
	Electricity generation (TWh)		Shares (%)	CAAGR (%)	Electricity generation (TWh)		Shares (%)	CAAGR (%)
	2030	2040	2040	2018-40	2030	2040	2040	2018-40
Total generation	1 610	1 862	100	1.4	1 362	1 437	100	0.3
Coal	293	300	16	0.5	101	8	1	-14.8
Oil	2	1	0	-8.6	2	1	0	-8.7
Natural gas	795	894	48	1.7	572	321	22	-2.9
Nuclear	213	288	15	1.6	236	301	21	1.7
Renewables	308	378	20	1.7	452	807	56	5.3
Hydro	286	325	17	1.1	350	482	34	2.9
Bioenergy	7	20	1	9.4	32	136	9	19.2
Wind	7	16	1	15.3	48	132	9	27.0
Geothermal	5	12	1	16.3	14	39	3	22.5
Solar PV	4	5	0	6.7	7	17	1	12.9
CSP	-	-	-	n.a.	-	-	-	n.a.
Marine	0	0	0	n.a.	0	1	0	n.a.

	Current Policies Scenario				Sustainable Development Scenario			
	Electrical capacity (GW)		Shares (%)	CAAGR (%)	Electrical capacity (GW)		Shares (%)	CAAGR (%)
	2030	2040	2040	2018-40	2030	2040	2040	2018-40
Total capacity	367	419	100	1.1	344	424	100	1.1
Coal	62	60	14	-0.5	30	10	2	-8.3
Oil	3	1	0	-10.4	3	1	0	-10.4
Natural gas	180	207	49	1.4	139	132	31	-0.6
Nuclear	32	40	10	1.4	33	41	10	1.6
Renewables	90	109	26	1.8	139	237	56	5.5
Hydro	81	91	22	1.2	99	134	32	3.0
Bioenergy	2	5	1	5.7	8	29	7	14.5
Wind	3	7	2	13.8	22	50	12	24.9
Geothermal	1	2	0	15.0	2	5	1	21.2
Solar PV	3	4	1	8.7	7	17	4	15.5
CSP	-	-	-	n.a.	-	-	-	n.a.
Marine	0	0	0	n.a.	0	0	0	n.a.

	Current Policies Scenario				Sustainable Development Scenario			
	CO_2 emissions (Mt)		Shares (%)	CAAGR (%)	CO_2 emissions (Mt)		Shares (%)	CAAGR (%)
	2030	2040	2040	2018-40	2030	2040	2040	2018-40
Total CO_2	2 176	2 302	100	0.6	1 605	1 119	100	-2.7
Coal	579	577	25	0.1	315	136	12	-6.3
Oil	457	467	20	0.4	383	294	26	-1.7
Natural gas	1 140	1 258	55	0.9	907	706	63	-1.7
Power sector	951	982	100	0.3	622	355	100	-4.3
Coal	370	360	37	0.0	170	49	14	-8.6
Oil	16	15	1	-1.8	16	12	3	-2.6
Natural gas	565	607	62	0.5	436	311	88	-2.5
Final consumption	1 064	1 124	100	0.6	871	685	100	-1.6
Coal	204	212	19	0.2	142	85	12	-3.9
Oil	407	417	37	0.6	338	260	38	-1.6
Transport	261	280	25	1.0	218	167	24	-1.4
Natural gas	453	495	44	0.9	391	340	50	-0.8

Table A.3: Energy demand – Russia

	Stated Policies Scenario							Shares (%)		CAAGR (%)
	Energy demand (Mtoe)									
	2010	2017	2018	2025	2030	2035	2040	2018	2040	2018-40
Total primary demand	674	723	751	766	767	773	786	100	100	0.2
Coal	101	115	116	112	101	95	90	15	11	-1.1
Oil	125	148	153	157	155	149	145	20	18	-0.2
Natural gas	381	383	404	420	421	421	428	54	54	0.3
Nuclear	45	53	53	51	55	62	64	7	8	0.9
Hydro	14	16	16	17	18	19	19	2	2	0.8
Bioenergy	7	8	8	9	11	14	18	1	2	3.7
Other renewables	0	0	0	1	6	13	21	0	3	23.4
Power sector	364	334	336	340	342	352	365	100	100	0.4
Coal	71	60	62	61	52	48	45	18	12	-1.4
Oil	10	6	6	5	5	5	5	2	1	-1.1
Natural gas	219	195	195	200	199	196	199	58	54	0.1
Nuclear	45	53	53	51	55	62	64	16	18	0.9
Hydro	14	16	16	17	18	19	19	5	5	0.8
Bioenergy	4	4	4	5	6	9	13	1	3	5.1
Other renewables	0	0	0	1	6	13	21	0	6	23.3
Other energy sector	108	123	126	127	122	119	118	100	100	-0.3
Electricity	25	27	27	28	28	28	29	21	24	0.3
Total final consumption	433	486	508	525	531	534	539	100	100	0.3
Coal	14	28	28	27	26	26	25	5	5	-0.4
Oil	96	118	122	129	129	125	123	24	23	0.0
Natural gas	143	165	184	188	189	192	195	36	36	0.3
Electricity	62	65	65	72	77	81	85	13	16	1.2
Heat	115	106	106	106	105	105	105	21	20	-0.0
Bioenergy	2	3	4	4	4	5	5	1	1	1.6
Other renewables	-	-	-	0	0	0	0	-	0	n.a.
Industry	125	149	154	161	161	161	162	100	100	0.2
Coal	10	25	25	24	24	24	24	16	15	-0.1
Oil	10	15	15	16	16	15	14	10	9	-0.1
Natural gas	32	38	42	43	42	41	41	28	25	-0.2
Electricity	28	29	29	32	32	32	33	19	20	0.6
Heat	45	41	41	45	46	46	48	27	29	0.6
Bioenergy	0	1	1	2	2	2	2	1	1	2.0
Other renewables	-	-	-	0	0	0	0	-	0	n.a.
Transport	96	96	102	108	111	112	114	100	100	0.5
Oil	56	60	62	65	67	65	64	61	56	0.2
Electricity	7	7	7	8	9	10	11	6	10	2.3
Biofuels	-	-	-	-	0	0	0	-	0	n.a.
Other fuels	33	30	34	35	36	37	39	33	34	0.6
Buildings	149	167	173	169	168	167	166	100	100	-0.2
Coal	4	3	3	2	2	1	1	2	0	-5.9
Oil	6	16	16	16	15	14	13	9	8	-1.1
Natural gas	43	56	62	60	59	58	57	36	34	-0.4
Electricity	26	28	28	31	34	36	38	16	23	1.5
Heat	67	62	62	58	57	55	54	36	33	-0.6
Bioenergy	2	2	2	2	2	2	3	1	2	1.0
Traditional biomass	-	-	-	-	-	-	-	-	-	n.a.
Other renewables	-	-	-	0	0	0	0	-	0	n.a.
Other	63	73	79	87	91	94	97	100	100	0.9
Petrochem. feedstock	47	19	18	26	28	29	31	22	32	2.5

Table A.3: Energy demand – Russia

	Current Policies Scenario				Sustainable Development Scenario			
	Energy demand (Mtoe)		Shares (%)	CAAGR (%)	Energy demand (Mtoe)		Shares (%)	CAAGR (%)
	2030	2040	2040	2018-40	2030	2040	2040	2018-40
Total primary demand	794	844	100	0.5	680	635	100	-0.8
Coal	111	105	12	-0.5	62	35	6	-5.3
Oil	158	153	18	-0.0	139	109	17	-1.5
Natural gas	438	467	55	0.7	364	302	47	-1.3
Nuclear	55	73	9	1.5	61	74	12	1.5
Hydro	18	20	2	1.0	22	28	4	2.4
Bioenergy	10	14	2	2.4	17	44	7	7.9
Other renewables	4	11	1	19.7	15	43	7	27.5
Power sector	353	390	100	0.7	299	310	100	-0.4
Coal	63	60	15	-0.2	25	9	3	-8.6
Oil	5	4	1	-1.4	5	4	1	-2.2
Natural gas	204	213	55	0.4	161	118	38	-2.3
Nuclear	55	73	19	1.5	61	74	24	1.5
Hydro	18	20	5	1.0	22	28	9	2.4
Bioenergy	5	9	2	3.3	12	37	12	10.4
Other renewables	4	11	3	19.7	15	41	13	27.1
Other energy sector	127	132	100	0.2	104	83	100	-1.9
Electricity	29	33	25	0.9	23	20	24	-1.3
Total final consumption	549	574	100	0.6	472	427	100	-0.8
Coal	26	25	4	-0.4	20	15	3	-2.8
Oil	132	130	23	0.3	114	90	21	-1.4
Natural gas	200	214	37	0.7	176	161	38	-0.6
Electricity	79	90	16	1.5	69	76	18	0.7
Heat	108	110	19	0.2	87	76	18	-1.5
Bioenergy	4	5	1	1.3	5	6	1	2.5
Other renewables	0	0	0	n.a.	1	3	1	n.a.
Industry	165	171	100	0.5	137	125	100	-0.9
Coal	24	24	14	-0.1	18	15	12	-2.4
Oil	15	13	8	-0.5	14	12	9	-1.1
Gas	45	48	28	0.6	40	37	29	-0.7
Electricity	32	33	19	0.6	28	28	23	-0.1
Heat	47	50	29	0.9	35	30	24	-1.5
Bioenergy	2	2	1	1.9	2	3	3	3.5
Other renewables	0	0	0	n.a.	0	1	1	n.a.
Transport	115	124	100	0.9	97	84	100	-0.9
Oil	69	70	56	0.6	58	42	50	-1.8
Electricity	8	10	8	1.8	9	12	14	2.8
Biofuels	-	-	-	n.a.	0	0	0	n.a.
Other fuels	39	44	36	1.2	31	30	36	-0.5
Buildings	176	180	100	0.2	151	129	100	-1.3
Coal	2	1	0	-5.4	1	-	-	n.a.
Oil	17	16	9	-0.1	13	9	7	-2.5
Natural gas	61	62	34	-0.0	54	38	30	-2.2
Electricity	35	43	24	2.0	30	33	25	0.7
Heat	59	57	31	-0.4	50	44	34	-1.5
Bioenergy	2	2	1	0.7	2	3	2	1.6
Traditional biomass	-	-	-	n.a.	-	-	-	n.a.
Other renewables	0	0	0	n.a.	0	1	1	n.a.
Other	92	99	100	1.0	87	89	100	0.5
Petrochem. feedstock	28	30	31	2.5	26	28	31	2.1

Annex A | Tables for scenario projections

Table A.3: Electricity and CO$_2$ emissions – Russia

	Stated Policies Scenario									
	Electricity generation (TWh)							Shares (%)		CAAGR (%)
	2010	2017	2018	2025	2030	2035	2040	2018	2040	2018-40
Total generation	1 029	1 076	1 085	1 186	1 233	1 281	1 345	100	100	1.0
Coal	166	175	190	193	158	144	133	18	10	-1.6
Oil	8	2	2	2	1	1	1	0	0	-5.6
Natural gas	515	507	496	590	623	614	631	46	47	1.1
Nuclear	170	203	203	196	211	237	245	19	18	0.9
Renewables	170	189	193	205	239	285	336	18	25	2.5
Hydro	166	185	189	194	206	218	226	17	17	0.8
Bioenergy	3	3	3	4	8	18	33	0	2	12.0
Wind	0	0	0	4	17	35	55	0	4	30.1
Geothermal	1	0	0	1	6	11	19	0	1	18.5
Solar PV	-	1	1	2	3	3	4	0	0	6.8
CSP	-	-	-	-	-	-	-	-	-	n.a.
Marine	-	-	-	0	0	0	0	-	0	n.a.

	Stated Policies Scenario								
	Electrical capacity (GW)						Shares (%)		CAAGR (%)
	2017	2018	2025	2030	2035	2040	2018	2040	2018-40
Total capacity	262	265	275	279	287	304	100	100	0.6
Coal	52	52	42	33	30	27	20	9	-3.0
Oil	4	4	2	2	1	1	1	0	-7.3
Natural gas	125	126	141	140	136	141	48	46	0.5
Nuclear	28	29	28	31	33	34	11	11	0.7
Renewables	54	54	61	71	85	99	20	33	2.8
Hydro	52	52	55	58	61	63	20	21	0.9
Bioenergy	1	1	2	3	5	8	1	3	7.9
Wind	0	0	2	8	15	22	0	7	24.9
Geothermal	0	0	0	1	2	3	0	1	17.2
Solar PV	0	0	2	2	3	4	0	1	9.7
CSP	-	-	-	-	-	-	-	-	n.a.
Marine	-	-	0	0	0	0	-	0	n.a.

	Stated Policies Scenario									
	CO$_2$ emissions (Mt)							Shares (%)		CAAGR (%)
	2010	2017	2018	2025	2030	2035	2040	2018	2040	2018-40
Total CO$_2$	1 505	1 520	1 555	1 583	1 538	1 499	1 484	100	100	-0.2
Coal	405	391	400	391	351	333	317	26	21	-1.1
Oil	298	331	335	343	339	325	316	22	21	-0.3
Natural gas	803	799	820	849	847	841	851	53	57	0.2
Power sector	871	752	742	750	711	685	677	100	100	-0.4
Coal	313	252	261	258	221	204	190	35	28	-1.4
Oil	36	23	20	18	17	16	15	3	2	-1.1
Natural gas	523	477	462	475	473	465	471	62	70	0.1
Final consumption	571	707	748	750	745	731	722	100	100	-0.2
Coal	85	135	136	130	127	125	124	18	17	-0.4
Oil	229	276	283	294	293	283	275	38	38	-0.1
Transport	167	177	184	194	197	193	191	25	26	0.2
Natural gas	257	295	328	326	324	323	324	44	45	-0.1

Table A.3: Electricity and CO_2 emissions – Russia

	Current Policies Scenario				Sustainable Development Scenario			
	Electricity generation (TWh)		Shares (%)	CAAGR (%)	Electricity generation (TWh)		Shares (%)	CAAGR (%)
	2030	2040	2040	2018-40	2030	2040	2040	2018-40
Total generation	1 270	1 443	100	1.3	1 082	1 130	100	0.2
Coal	198	199	14	0.2	60	3	0	-16.9
Oil	1	1	0	-5.7	1	1	0	-5.8
Natural gas	635	683	47	1.5	448	246	22	-3.1
Nuclear	211	281	19	1.5	233	284	25	1.5
Renewables	225	280	19	1.7	339	596	53	5.3
Hydro	207	235	16	1.0	252	320	28	2.4
Bioenergy	6	19	1	9.4	30	121	11	18.9
Wind	4	10	1	20.5	40	108	10	34.1
Geothermal	4	11	1	15.9	13	36	3	22.0
Solar PV	2	4	0	6.0	4	11	1	11.3
CSP	-	-	-	n.a.	-	-	-	n.a.
Marine	0	0	0	n.a.	0	1	0	n.a.

	Current Policies Scenario				Sustainable Development Scenario			
	Electrical capacity (GW)		Shares (%)	CAAGR (%)	Electrical capacity (GW)		Shares (%)	CAAGR (%)
	2030	2040	2040	2018-40	2030	2040	2040	2018-40
Total capacity	281	313	100	0.8	264	309	100	0.7
Coal	42	40	13	-1.2	20	6	2	-9.6
Oil	2	1	0	-7.3	2	1	0	-7.3
Natural gas	140	153	49	0.9	107	90	29	-1.5
Nuclear	31	39	12	1.3	32	39	13	1.4
Renewables	65	79	25	1.8	103	171	55	5.4
Hydro	58	65	21	1.1	70	88	28	2.4
Bioenergy	2	5	2	5.6	8	27	9	14.1
Wind	2	4	1	15.9	19	41	13	28.5
Geothermal	1	2	1	14.6	2	5	2	20.7
Solar PV	2	3	1	8.7	4	10	3	14.8
CSP	-	-	-	n.a.	-	-	-	n.a.
Marine	0	0	0	n.a.	0	0	0	n.a.

	Current Policies Scenario				Sustainable Development Scenario			
	CO_2 emissions (Mt)		Shares (%)	CAAGR (%)	CO_2 emissions (Mt)		Shares (%)	CAAGR (%)
	2030	2040	2040	2018-40	2030	2040	2040	2018-40
Total CO_2	1 625	1 654	100	0.3	1 190	800	100	-3.0
Coal	395	379	23	-0.2	192	81	10	-7.0
Oil	346	338	20	0.0	296	217	27	-2.0
Natural gas	884	937	57	0.6	702	520	65	-2.1
Power sector	762	770	100	0.2	494	292	100	-4.2
Coal	264	252	33	-0.2	103	36	12	-8.6
Oil	15	14	2	-1.4	16	12	4	-2.2
Natural gas	483	504	65	0.4	374	261	90	-2.6
Final consumption	775	784	100	0.2	633	464	100	-2.1
Coal	128	124	16	-0.4	86	44	9	-5.0
Oil	302	295	38	0.2	255	186	40	-1.9
Transport	203	207	26	0.6	172	123	27	-1.8
Natural gas	345	365	47	0.5	292	234	50	-1.5

Annex A | Tables for scenario projections

Table A.3: Energy demand – Asia Pacific

	Stated Policies Scenario							Shares (%)		CAAGR (%)
	Energy demand (Mtoe)									
	2010	2017	2018	2025	2030	2035	2040	2018	2040	2018-40
Total primary demand	4 965	5 814	5 989	6 858	7 402	7 837	8 208	100	100	1.4
Coal	2 464	2 808	2 855	3 070	3 133	3 151	3 141	48	38	0.4
Oil	1 209	1 460	1 489	1 670	1 768	1 807	1 821	25	22	0.9
Natural gas	486	627	674	861	980	1 105	1 220	11	15	2.7
Nuclear	152	130	147	226	306	370	419	2	5	4.9
Hydro	95	144	150	165	190	211	229	3	3	1.9
Bioenergy	509	518	524	581	620	649	677	9	8	1.2
Other renewables	51	126	149	284	404	544	700	2	9	7.3
Power sector	1 869	2 425	2 566	3 024	3 363	3 668	3 939	100	100	2.0
Coal	1 251	1 653	1 744	1 895	1 969	2 002	1 999	68	51	0.6
Oil	72	55	47	39	34	27	22	2	1	-3.4
Natural gas	228	268	277	326	350	387	423	11	11	1.9
Nuclear	152	130	147	226	306	370	419	6	11	4.9
Hydro	95	144	150	165	190	211	229	6	6	1.9
Bioenergy	33	83	92	144	176	206	240	4	6	4.5
Other renewables	37	90	109	229	338	465	607	4	15	8.1
Other energy sector	611	618	627	693	712	728	746	100	100	0.8
Electricity	117	156	162	187	207	228	248	26	33	2.0
Total final consumption	3 277	3 885	3 981	4 604	4 990	5 301	5 574	100	100	1.5
Coal	890	860	827	836	827	813	803	21	14	-0.1
Oil	1 026	1 311	1 346	1 544	1 650	1 695	1 720	34	31	1.1
Natural gas	223	310	342	483	574	667	747	9	13	3.6
Electricity	596	846	899	1 142	1 318	1 493	1 665	23	30	2.8
Heat	67	102	110	120	124	125	124	3	2	0.6
Bioenergy	460	420	417	423	430	428	422	10	8	0.1
Other renewables	13	35	40	54	66	80	93	1	2	3.9
Industry	1 430	1 595	1 621	1 865	2 015	2 145	2 257	100	100	1.5
Coal	743	696	676	700	706	709	711	42	31	0.2
Oil	148	150	145	147	146	143	138	9	6	-0.3
Natural gas	103	155	170	256	320	383	440	10	19	4.4
Electricity	329	451	479	589	653	711	761	30	34	2.1
Heat	45	67	72	77	77	75	71	4	3	-0.1
Bioenergy	61	75	77	95	110	121	130	5	6	2.4
Other renewables	0	1	1	1	3	5	7	0	0	11.4
Transport	526	730	756	924	1 035	1 110	1 182	100	100	2.1
Oil	498	679	699	828	905	938	963	92	82	1.5
Electricity	9	15	16	32	52	74	99	2	8	8.8
Biofuels	4	8	9	21	30	38	45	1	4	7.6
Other fuels	15	29	32	42	48	61	74	4	6	3.8
Buildings	953	1 089	1 122	1 226	1 306	1 380	1 450	100	100	1.2
Coal	100	98	94	64	47	30	18	8	1	-7.2
Oil	120	149	153	151	146	138	129	14	9	-0.8
Natural gas	70	99	110	146	164	177	183	10	13	2.4
Electricity	232	339	360	466	554	646	741	32	51	3.3
Heat	22	35	38	43	47	50	53	3	4	1.6
Bioenergy	395	336	330	305	287	266	243	29	17	-1.4
Traditional biomass	376	315	309	283	262	239	212	28	15	-1.7
Other renewables	13	33	38	51	61	72	83	3	6	3.6
Other	369	470	482	588	633	666	685	100	100	1.6
Petrochem. feedstock	225	267	268	351	382	408	427	55	62	2.1

Table A.3: Energy demand – Asia Pacific

	Current Policies Scenario				Sustainable Development Scenario			
	Energy demand (Mtoe)		Shares (%)	CAAGR (%)	Energy demand (Mtoe)		Shares (%)	CAAGR (%)
	2030	2040	2040	2018-40	2030	2040	2040	2018-40
Total primary demand	7 678	8 878	100	1.8	6 232	6 085	100	0.1
Coal	3 298	3 641	41	1.1	2 083	1 240	20	-3.7
Oil	1 887	2 090	24	1.6	1 499	1 172	19	-1.1
Natural gas	1 042	1 341	15	3.2	998	1 060	17	2.1
Nuclear	303	417	5	4.8	361	551	9	6.2
Hydro	182	208	2	1.5	211	266	4	2.6
Bioenergy	606	629	7	0.8	494	644	11	0.9
Other renewables	360	551	6	6.1	585	1 153	19	9.7
Power sector	3 453	4 193	100	2.3	2 878	3 153	100	0.9
Coal	2 061	2 353	56	1.4	1 165	538	17	-5.2
Oil	34	23	1	-3.2	30	18	1	-4.2
Natural gas	401	514	12	2.9	396	410	13	1.8
Nuclear	303	417	10	4.8	361	551	17	6.2
Hydro	182	208	5	1.5	211	266	8	2.6
Bioenergy	171	204	5	3.7	216	346	11	6.2
Other renewables	301	473	11	6.9	499	1 023	32	10.7
Other energy sector	755	836	100	1.3	615	605	100	-0.2
Electricity	219	272	32	2.4	176	193	32	0.8
Total final consumption	5 171	5 973	100	1.9	4 250	4 172	100	0.2
Coal	886	914	15	0.5	631	443	11	-2.8
Oil	1 757	1 968	33	1.7	1 394	1 099	26	-0.9
Natural gas	581	766	13	3.7	551	586	14	2.5
Electricity	1 336	1 697	28	2.9	1 218	1 526	37	2.4
Heat	132	141	2	1.1	104	85	2	-1.2
Bioenergy	420	409	7	-0.1	264	284	7	-1.7
Other renewables	59	79	1	3.1	87	131	3	5.5
Industry	2 074	2 389	100	1.8	1 705	1 673	100	0.1
Coal	738	780	33	0.7	532	390	23	-2.5
Oil	153	149	6	0.1	106	76	5	-2.9
Gas	321	446	19	4.5	309	361	22	3.5
Electricity	665	789	33	2.3	592	674	40	1.6
Heat	84	86	4	0.8	59	38	2	-2.8
Bioenergy	111	135	6	2.5	95	107	6	1.5
Other renewables	2	4	0	8.1	12	25	2	17.8
Transport	1 093	1 321	100	2.6	951	891	100	0.8
Oil	988	1 162	88	2.3	735	507	57	-1.4
Electricity	37	58	4	6.2	69	200	22	12.3
Biofuels	22	33	2	5.9	56	74	8	10.0
Other fuels	46	68	5	3.5	91	110	12	5.7
Buildings	1 364	1 555	100	1.5	1 016	1 038	100	-0.4
Coal	72	53	3	-2.5	40	6	1	-11.9
Oil	160	155	10	0.0	129	88	8	-2.5
Natural gas	172	201	13	2.8	119	99	10	-0.5
Electricity	572	779	50	3.6	506	603	58	2.4
Heat	48	55	4	1.7	45	46	4	0.9
Bioenergy	285	239	15	-1.5	106	92	9	-5.7
Traditional biomass	262	212	14	-1.7	59	30	3	-10.1
Other renewables	55	73	5	3.0	71	101	10	4.5
Other	640	708	100	1.8	578	569	100	0.8
Petrochem. feedstock	382	436	62	2.2	348	361	63	1.4

Table A.3: Electricity and CO_2 emissions – Asia Pacific

	Stated Policies Scenario							Shares (%)		CAAGR (%)
	Electricity generation (TWh)									
	2010	2017	2018	2025	2030	2035	2040	2018	2040	2018-40
Total generation	8 292	11 642	12 327	15 450	17 731	20 012	22 245	100	100	2.7
Coal	4 958	6 951	7 276	8 126	8 516	8 743	8 817	59	40	0.9
Oil	278	218	180	140	116	89	67	1	0	-4.4
Natural gas	1 147	1 410	1 477	1 859	2 033	2 276	2 552	12	11	2.5
Nuclear	582	500	566	869	1 175	1 419	1 609	5	7	4.9
Renewables	1 306	2 541	2 805	4 434	5 869	7 463	9 178	23	41	5.5
Hydro	1 104	1 680	1 744	1 924	2 209	2 452	2 660	14	12	1.9
Bioenergy	91	215	237	430	534	640	757	2	3	5.4
Wind	77	377	449	959	1 419	1 952	2 462	4	11	8.0
Geothermal	28	34	36	51	73	97	124	0	1	5.8
Solar PV	6	236	338	1 062	1 605	2 271	3 096	3	14	10.6
CSP	0	0	0	8	25	44	65	0	0	26.7
Marine	0	1	1	1	4	8	14	0	0	16.4

	Stated Policies Scenario						Shares (%)		CAAGR (%)
	Electrical capacity (GW)								
	2017	2018	2025	2030	2035	2040	2018	2040	2018-40
Total capacity	3 054	3 218	4 429	5 287	6 216	7 161	100	100	3.7
Coal	1 419	1 452	1 619	1 693	1 761	1 820	45	25	1.0
Oil	121	119	97	83	68	53	4	1	-3.6
Natural gas	385	398	519	571	623	671	12	9	2.4
Nuclear	115	121	144	168	196	221	4	3	2.8
Renewables	1 013	1 125	2 008	2 692	3 428	4 192	35	59	6.2
Hydro	532	544	649	741	819	885	17	12	2.2
Bioenergy	47	52	85	103	120	138	2	2	4.6
Wind	210	235	451	638	836	1 006	7	14	6.8
Geothermal	5	5	8	11	14	18	0	0	5.7
Solar PV	219	288	813	1 190	1 623	2 120	9	30	9.5
CSP	0	0	3	8	13	19	0	0	18.5
Marine	0	0	0	1	3	5	0	0	14.2

	Stated Policies Scenario							Shares (%)		CAAGR (%)
	CO_2 emissions (Mt)									
	2010	2017	2018	2025	2030	2035	2040	2018	2040	2018-40
Total CO_2	13 066	15 528	15 928	17 329	18 005	18 357	18 532	100	100	0.7
Coal	9 165	10 715	10 960	11 528	11 732	11 752	11 642	69	63	0.3
Oil	2 827	3 452	3 473	3 843	4 049	4 099	4 118	22	22	0.8
Natural gas	1 073	1 361	1 495	1 959	2 224	2 505	2 772	9	15	2.8
Power sector	5 908	7 594	7 958	8 648	8 969	9 148	9 179	100	100	0.7
Coal	5 142	6 787	7 156	7 756	8 037	8 151	8 114	90	88	0.6
Oil	230	177	152	125	108	88	71	2	1	-3.4
Natural gas	536	630	650	767	824	909	994	8	11	1.9
Final consumption	6 581	7 403	7 431	8 111	8 464	8 648	8 793	100	100	0.8
Coal	3 729	3 716	3 602	3 571	3 505	3 422	3 359	48	38	-0.3
Oil	2 408	3 076	3 118	3 524	3 748	3 813	3 851	42	44	1.0
Transport	1 497	2 040	2 099	2 488	2 721	2 820	2 898	28	33	1.5
Natural gas	444	611	712	1 017	1 212	1 412	1 584	10	18	3.7

Table A.3: Electricity and CO_2 emissions – Asia Pacific

	Current Policies Scenario				Sustainable Development Scenario			
	Electricity generation (TWh)		Shares (%)	CAAGR (%)	Electricity generation (TWh)		Shares (%)	CAAGR (%)
	2030	2040	2040	2018-40	2030	2040	2040	2018-40
Total generation	18 079	22 887	100	2.9	16 208	19 984	100	2.2
Coal	8 998	10 525	46	1.7	4 976	2 214	11	-5.3
Oil	120	74	0	-4.0	95	49	0	-5.8
Natural gas	2 351	3 111	14	3.4	2 277	2 387	12	2.2
Nuclear	1 162	1 601	7	4.8	1 384	2 117	11	6.2
Renewables	5 426	7 554	33	4.6	7 453	13 197	66	7.3
Hydro	2 117	2 423	11	1.5	2 454	3 093	15	2.6
Bioenergy	489	608	3	4.4	675	1 136	6	7.4
Wind	1 297	2 047	9	7.1	1 918	3 810	19	10.2
Geothermal	62	94	0	4.5	148	276	1	9.8
Solar PV	1 446	2 349	10	9.2	2 207	4 585	23	12.6
CSP	12	25	0	21.3	45	275	1	35.2
Marine	3	9	0	13.9	6	22	0	18.7

	Current Policies Scenario				Sustainable Development Scenario			
	Electrical capacity (GW)		Shares (%)	CAAGR (%)	Electrical capacity (GW)		Shares (%)	CAAGR (%)
	2030	2040	2040	2018-40	2030	2040	2040	2018-40
Total capacity	5 231	6 812	100	3.5	5 841	8 522	100	4.5
Coal	1 802	2 145	31	1.8	1 380	1 034	12	-1.5
Oil	83	56	1	-3.4	84	52	1	-3.7
Natural gas	633	812	12	3.3	588	706	8	2.6
Nuclear	166	218	3	2.7	194	282	3	3.9
Renewables	2 475	3 419	50	5.2	3 485	6 102	72	8.0
Hydro	709	806	12	1.8	834	1 041	12	3.0
Bioenergy	94	112	2	3.6	126	204	2	6.4
Wind	586	844	12	6.0	857	1 554	18	9.0
Geothermal	9	14	0	4.3	23	41	0	9.7
Solar PV	1 072	1 633	24	8.2	1 628	3 171	37	11.5
CSP	4	7	0	13.5	15	84	1	26.7
Marine	1	3	0	11.6	2	8	0	16.4

	Current Policies Scenario				Sustainable Development Scenario			
	CO_2 emissions (Mt)		Shares (%)	CAAGR (%)	CO_2 emissions (Mt)		Shares (%)	CAAGR (%)
	2030	2040	2040	2018-40	2030	2040	2040	2018-40
Total CO_2	19 113	21 492	100	1.4	12 707	7 298	100	-3.5
Coal	12 360	13 559	63	1.0	7 202	2 949	40	-5.8
Oil	4 380	4 869	23	1.5	3 274	2 246	31	-2.0
Natural gas	2 373	3 063	14	3.3	2 231	2 160	30	1.7
Power sector	9 460	10 830	100	1.4	5 583	2 281	100	-5.5
Coal	8 408	9 547	88	1.3	4 567	1 410	62	-7.1
Oil	109	74	1	-3.2	96	59	3	-4.2
Natural gas	942	1 210	11	2.9	919	869	38	1.3
Final consumption	9 048	10 025	100	1.4	6 683	4 714	100	-2.0
Coal	3 756	3 831	38	0.3	2 492	1 450	31	-4.1
Oil	4 064	4 567	46	1.8	3 028	2 086	44	-1.8
Transport	2 970	3 495	35	2.3	2 208	1 527	32	-1.4
Natural gas	1 229	1 627	16	3.8	1 164	1 177	25	2.3

Table A.3: Energy demand – China

	Stated Policies Scenario							Shares (%)		CAAGR (%)
	Energy demand (Mtoe)									
	2010	2017	2018	2025	2030	2035	2040	2018	2040	2018-40
Total primary demand	2 550	3 080	3 187	3 618	3 805	3 905	3 972	100	100	1.0
Coal	1 797	1 964	1 984	2 054	1 992	1 897	1 797	62	45	-0.4
Oil	431	572	593	672	714	716	711	19	18	0.8
Natural gas	93	196	231	347	410	460	499	7	13	3.6
Nuclear	19	65	77	119	166	216	252	2	6	5.6
Hydro	61	99	104	112	122	131	138	3	3	1.3
Bioenergy	133	114	114	144	167	183	200	4	5	2.6
Other renewables	16	70	86	169	234	303	373	3	9	6.9
Power sector	957	1 368	1 479	1 740	1 893	2 019	2 117	100	100	1.6
Coal	829	1 086	1 155	1 223	1 225	1 202	1 160	78	55	0.0
Oil	9	7	7	6	5	5	4	0	0	-2.2
Natural gas	21	41	51	94	115	131	147	3	7	5.0
Nuclear	19	65	77	119	166	216	252	5	12	5.6
Hydro	61	99	104	112	122	131	138	7	7	1.3
Bioenergy	13	33	38	66	85	101	118	3	6	5.3
Other renewables	4	37	48	120	176	235	295	3	14	8.7
Other energy sector	376	383	386	419	405	387	375	100	100	-0.1
Electricity	63	91	95	107	113	118	122	25	33	1.1
Total final consumption	1 653	2 004	2 055	2 363	2 511	2 591	2 647	100	100	1.2
Coal	713	665	627	587	536	480	428	30	16	-1.7
Oil	372	518	536	628	677	681	681	26	26	1.1
Natural gas	74	132	154	236	278	318	346	7	13	3.7
Electricity	300	480	521	670	762	844	916	25	35	2.6
Heat	62	96	103	114	117	118	117	5	4	0.6
Bioenergy	121	81	76	78	82	82	81	4	3	0.3
Other renewables	12	33	38	49	58	68	78	2	3	3.3
Industry	933	993	1 000	1 099	1 131	1 142	1 139	100	100	0.6
Coal	584	519	494	468	432	390	348	49	31	-1.6
Oil	69	56	50	48	46	42	39	5	3	-1.2
Natural gas	35	56	65	108	135	160	178	7	16	4.7
Electricity	203	297	321	393	431	460	482	32	42	1.9
Heat	42	64	69	74	74	72	69	7	6	-0.0
Bioenergy	-	0	0	6	13	16	20	0	2	41.1
Other renewables	0	0	1	1	1	2	4	0	0	9.8
Transport	199	313	327	413	468	491	517	100	100	2.1
Oil	185	281	291	353	386	382	380	89	74	1.2
Electricity	6	11	12	25	39	55	71	4	14	8.6
Biofuels	2	2	3	9	15	19	23	1	4	10.2
Other fuels	7	19	22	26	28	35	42	7	8	3.0
Buildings	372	482	504	567	603	628	648	100	100	1.1
Coal	82	81	77	48	32	17	7	15	1	-10.5
Oil	36	58	60	56	51	43	36	12	6	-2.3
Natural gas	25	48	56	87	99	105	106	11	16	3.0
Electricity	79	152	168	229	270	308	341	33	53	3.3
Heat	19	32	34	40	43	46	48	7	7	1.6
Bioenergy	119	78	73	63	53	45	37	15	6	-3.1
Traditional biomass	112	71	66	56	47	37	26	13	4	-4.1
Other renewables	11	32	36	47	55	64	72	7	11	3.2
Other	148	217	224	283	309	330	343	100	100	2.0
Petrochem. feedstock	76	116	115	168	190	209	226	51	66	3.1

Table A.3: Energy demand – China

	Current Policies Scenario				Sustainable Development Scenario			
	Energy demand (Mtoe)		Shares (%)	CAAGR (%)	Energy demand (Mtoe)		Shares (%)	CAAGR (%)
	2030	2040	2040	2018-40	2030	2040	2040	2018-40
Total primary demand	3 950	4 283	100	1.4	3 226	2 915	100	-0.4
Coal	2 064	1 986	46	0.0	1 446	808	28	-4.0
Oil	784	849	20	1.6	582	421	14	-1.5
Natural gas	441	559	13	4.1	391	371	13	2.2
Nuclear	168	264	6	5.8	200	331	11	6.9
Hydro	119	133	3	1.1	131	149	5	1.7
Bioenergy	161	177	4	2.0	186	284	10	4.2
Other renewables	212	314	7	6.1	290	551	19	8.8
Power sector	1 915	2 181	100	1.8	1 628	1 678	100	0.6
Coal	1 237	1 242	57	0.3	838	428	25	-4.4
Oil	5	4	0	-2.7	6	4	0	-2.7
Natural gas	138	186	9	6.1	121	132	8	4.4
Nuclear	168	264	12	5.8	200	331	20	6.9
Hydro	119	133	6	1.1	131	149	9	1.7
Bioenergy	88	105	5	4.8	110	176	10	7.2
Other renewables	159	246	11	7.7	222	458	27	10.8
Other energy sector	428	420	100	0.4	342	293	100	-1.2
Electricity	114	126	30	1.3	95	93	32	-0.1
Total final consumption	2 625	2 878	100	1.5	2 164	1 988	100	-0.1
Coal	585	513	18	-0.9	411	223	11	-4.6
Oil	741	810	28	1.9	550	404	20	-1.3
Natural gas	282	360	13	3.9	256	225	11	1.7
Electricity	765	922	32	2.6	704	844	42	2.2
Heat	125	134	5	1.2	98	79	4	-1.2
Bioenergy	74	72	2	-0.3	76	108	5	1.6
Other renewables	53	68	2	2.7	68	93	5	4.1
Industry	1 175	1 242	100	1.0	958	834	100	-0.8
Coal	456	395	32	-1.0	325	174	21	-4.6
Oil	48	44	4	-0.6	29	16	2	-5.1
Gas	137	186	15	4.9	127	120	14	2.8
Electricity	442	514	41	2.2	394	436	52	1.4
Heat	81	84	7	0.9	57	37	4	-2.8
Bioenergy	10	18	1	40.8	20	37	4	45.2
Other renewables	1	2	0	5.4	6	12	1	15.4
Transport	503	583	100	2.7	436	399	100	0.9
Oil	437	484	83	2.3	298	173	43	-2.3
Electricity	30	47	8	6.6	52	130	33	11.6
Biofuels	10	16	3	8.4	26	33	8	11.9
Other fuels	26	36	6	2.3	60	64	16	4.9
Buildings	632	692	100	1.5	501	495	100	-0.1
Coal	55	39	6	-3.1	29	3	1	-14.0
Oil	58	46	7	-1.2	43	22	4	-4.6
Natural gas	104	119	17	3.5	63	50	10	-0.5
Electricity	269	338	49	3.2	239	265	53	2.1
Heat	44	50	7	1.8	41	42	9	1.0
Bioenergy	52	35	5	-3.3	27	33	7	-3.5
Traditional biomass	47	26	4	-4.1	5	3	1	-13.7
Other renewables	51	65	9	2.7	59	78	16	3.6
Other	315	361	100	2.2	268	261	100	0.7
Petrochem. feedstock	192	237	66	3.3	164	173	66	1.9

Table A.3: Electricity and CO_2 emissions – China

Stated Policies Scenario										
Electricity generation (TWh)								Shares (%)		CAAGR (%)
	2010	2017	2018	2025	2030	2035	2040	2018	2040	2018-40
Total generation	4 236	6 639	7 170	9 037	10 177	11 192	12 071	100	100	2.4
Coal	3 263	4 509	4 748	5 145	5 188	5 120	4 979	66	41	0.2
Oil	15	10	10	6	5	4	3	0	0	-6.1
Natural gas	92	196	255	533	662	767	877	4	7	5.8
Nuclear	74	248	294	457	636	828	969	4	8	5.6
Renewables	791	1 676	1 862	2 896	3 687	4 474	5 244	26	43	4.8
Hydro	711	1 157	1 205	1 301	1 415	1 519	1 607	17	13	1.3
Bioenergy	34	93	106	217	275	326	382	1	3	6.0
Wind	45	295	354	710	985	1 290	1 602	5	13	7.1
Geothermal	0	0	0	1	3	6	12	0	0	23.3
Solar PV	1	131	197	662	990	1 300	1 594	3	13	10.0
CSP	0	0	0	6	19	32	45	0	0	24.7
Marine	0	0	0	0	1	1	2	0	0	25.1

Stated Policies Scenario									
Electrical capacity (GW)							Shares (%)		CAAGR (%)
	2017	2018	2025	2030	2035	2040	2018	2040	2018-40
Total capacity	1 754	1 872	2 634	3 113	3 531	3 884	100	100	3.4
Coal	981	1 006	1 084	1 103	1 099	1 078	54	28	0.3
Oil	9	9	7	7	6	4	0	0	-3.0
Natural gas	74	80	155	179	194	207	4	5	4.4
Nuclear	37	46	70	93	118	136	2	4	5.1
Renewables	653	730	1 299	1 698	2 070	2 403	39	62	5.6
Hydro	344	352	416	456	493	523	19	13	1.8
Bioenergy	15	18	40	49	57	66	1	2	6.0
Wind	164	184	333	447	560	664	10	17	6.0
Geothermal	0	0	0	0	1	2	0	0	21.0
Solar PV	131	175	507	739	949	1 135	9	29	8.9
CSP	0	0	2	6	10	13	0	0	20.4
Marine	0	0	0	0	0	1	0	0	22.8

Stated Policies Scenario										
CO_2 emissions (Mt)								Shares (%)		CAAGR (%)
	2010	2017	2018	2025	2030	2035	2040	2018	2040	2018-40
Total CO_2	7 852	9 269	9 513	10 005	9 950	9 621	9 229	100	100	-0.1
Coal	6 600	7 470	7 607	7 618	7 341	6 935	6 479	80	70	-0.7
Oil	1 020	1 351	1 380	1 541	1 613	1 570	1 526	15	17	0.5
Natural gas	233	448	527	846	996	1 117	1 224	6	13	3.9
Power sector	3 483	4 585	4 893	5 268	5 310	5 238	5 091	100	100	0.2
Coal	3 402	4 465	4 749	5 025	5 021	4 913	4 729	97	93	-0.0
Oil	31	22	24	20	19	17	15	1	0	-2.2
Natural gas	50	98	120	223	270	308	347	2	7	5.0
Final consumption	4 015	4 371	4 304	4 405	4 319	4 080	3 847	100	100	-0.5
Coal	2 962	2 856	2 718	2 460	2 199	1 912	1 650	63	43	-2.2
Oil	896	1 225	1 248	1 423	1 502	1 458	1 422	29	37	0.6
Transport	559	844	874	1 061	1 161	1 147	1 143	20	30	1.2
Natural gas	157	290	338	522	618	711	774	8	20	3.8

Table A.3: Electricity and CO$_2$ emissions – China

	Current Policies Scenario				Sustainable Development Scenario			
	Electricity generation (TWh)		Shares (%)	CAAGR (%)	Electricity generation (TWh)		Shares (%)	CAAGR (%)
	2030	2040	2040	2018-40	2030	2040	2040	2018-40
Total generation	10 233	12 188	100	2.4	9 286	10 899	100	1.9
Coal	5 311	5 449	45	0.6	3 515	1 744	16	-4.4
Oil	5	3	0	-6.1	6	1	0	-11.9
Natural gas	810	1 082	9	6.8	624	643	6	4.3
Nuclear	646	1 014	8	5.8	768	1 272	12	6.9
Renewables	3 461	4 641	38	4.2	4 373	7 238	66	6.4
Hydro	1 389	1 549	13	1.1	1 526	1 733	16	1.7
Bioenergy	254	317	3	5.1	356	577	5	8.0
Wind	912	1 390	11	6.4	1 167	2 203	20	8.7
Geothermal	2	6	0	19.5	5	18	0	25.4
Solar PV	895	1 360	11	9.2	1 290	2 549	23	12.3
CSP	9	18	0	19.5	29	157	1	31.9
Marine	0	1	0	23.2	0	1	0	21.0

	Current Policies Scenario				Sustainable Development Scenario			
	Electrical capacity (GW)		Shares (%)	CAAGR (%)	Electrical capacity (GW)		Shares (%)	CAAGR (%)
	2030	2040	2040	2018-40	2030	2040	2040	2018-40
Total capacity	3 074	3 818	100	3.3	3 378	4 706	100	4.3
Coal	1 167	1 252	33	1.0	946	740	16	-1.4
Oil	7	4	0	-3.0	7	4	0	-3.0
Natural gas	202	257	7	5.4	177	190	4	4.0
Nuclear	94	139	4	5.2	108	170	4	6.1
Renewables	1 573	2 106	55	4.9	2 080	3 452	73	7.3
Hydro	446	502	13	1.6	503	576	12	2.3
Bioenergy	45	55	1	5.1	61	98	2	8.0
Wind	414	575	15	5.3	526	904	19	7.5
Geothermal	0	1	0	17.3	1	3	0	23.1
Solar PV	665	967	25	8.1	980	1 826	39	11.2
CSP	3	5	0	15.6	9	45	1	27.3
Marine	0	1	0	21.0	0	0	0	18.8

	Current Policies Scenario				Sustainable Development Scenario			
	CO$_2$ emissions (Mt)		Shares (%)	CAAGR (%)	CO$_2$ emissions (Mt)		Shares (%)	CAAGR (%)
	2030	2040	2040	2018-40	2030	2040	2040	2018-40
Total CO$_2$	10 475	10 444	100	0.4	7 053	3 093	100	-5.0
Coal	7 595	7 170	69	-0.3	4 880	1 670	54	-6.7
Oil	1 807	1 904	18	1.5	1 241	708	23	-3.0
Natural gas	1 073	1 370	13	4.4	932	752	24	1.6
Power sector	5 411	5 512	100	0.5	3 570	1 354	100	-5.7
Coal	5 068	5 060	92	0.3	3 272	1 135	84	-6.3
Oil	17	13	0	-2.7	20	13	1	-2.7
Natural gas	326	439	8	6.1	278	242	18	3.3
Final consumption	4 718	4 591	100	0.3	3 246	1 602	100	-4.4
Coal	2 403	2 002	44	-1.4	1 520	489	31	-7.5
Oil	1 687	1 782	39	1.6	1 152	654	41	-2.9
Transport	1 314	1 454	32	2.3	895	520	32	-2.3
Natural gas	628	807	18	4.0	575	459	29	1.4

Table A.3: Energy demand – India

	Stated Policies Scenario							Shares (%)		CAAGR (%)
	Energy demand (Mtoe)									
	2010	2017	2018	2025	2030	2035	2040	2018	2040	2018-40
Total primary demand	700	882	916	1 194	1 427	1 648	1 841	100	100	3.2
Coal	279	391	410	540	656	744	810	45	44	3.1
Oil	162	223	233	305	351	390	420	25	23	2.7
Natural gas	54	51	53	89	113	143	168	6	9	5.4
Nuclear	7	10	10	18	29	44	58	1	3	8.1
Hydro	11	12	13	16	21	25	28	1	2	3.8
Bioenergy	185	187	187	197	203	208	211	20	11	0.5
Other renewables	2	7	10	30	54	94	146	1	8	13.1
Power sector	239	339	354	476	585	686	776	100	100	3.6
Coal	177	262	276	350	416	450	466	78	60	2.4
Oil	7	8	6	8	8	6	5	2	1	-0.6
Natural gas	27	14	14	18	20	26	29	4	4	3.2
Nuclear	7	10	10	18	29	44	58	3	8	8.1
Hydro	11	12	13	16	21	25	28	4	4	3.8
Bioenergy	8	26	27	37	41	45	50	8	6	2.9
Other renewables	2	7	9	28	51	89	140	2	18	13.4
Other energy sector	62	84	87	108	132	158	181	100	100	3.4
Electricity	22	32	32	42	52	64	75	37	41	3.9
Total final consumption	484	591	614	812	969	1 129	1 278	100	100	3.4
Coal	87	101	105	148	184	223	260	17	20	4.2
Oil	135	196	207	278	326	367	399	34	31	3.0
Natural gas	27	36	38	69	91	114	137	6	11	6.0
Electricity	63	100	107	159	207	261	319	17	25	5.1
Heat	-	-	-	0	0	0	0	-	0	n.a.
Bioenergy	172	158	157	156	158	158	157	26	12	0.0
Other renewables	0	1	1	2	3	5	6	0	0	8.7
Industry	159	217	226	325	407	490	568	100	100	4.3
Coal	75	89	93	135	172	212	251	41	44	4.6
Oil	16	31	32	37	40	42	43	14	8	1.4
Natural gas	11	24	26	50	68	86	103	12	18	6.4
Electricity	28	40	42	63	80	97	114	19	20	4.6
Heat	-	-	-	0	0	0	0	-	0	n.a.
Bioenergy	29	32	32	40	46	51	55	14	10	2.4
Other renewables	0	0	0	0	1	1	2	0	0	17.6
Transport	65	98	104	153	190	229	262	100	100	4.3
Oil	62	94	99	141	173	204	228	95	87	3.9
Electricity	1	1	1	3	6	10	14	1	6	11.2
Biofuels	0	0	1	3	4	5	7	1	3	11.9
Other fuels	1	3	3	6	7	9	12	3	5	6.5
Buildings	207	214	218	236	259	287	317	100	100	1.7
Coal	12	12	12	12	12	10	9	6	3	-1.6
Oil	28	33	35	39	42	44	46	16	14	1.3
Natural gas	1	2	2	4	5	7	9	1	3	7.4
Electricity	23	41	44	67	91	121	155	20	49	5.9
Heat	-	-	-	-	-	-	-	-	-	n.a.
Bioenergy	143	125	124	112	108	102	95	57	30	-1.2
Traditional biomass	136	118	117	104	97	90	83	54	26	-1.5
Other renewables	0	1	1	2	2	3	4	0	1	6.9
Other	54	62	66	98	112	123	132	100	100	3.2
Petrochem. feedstock	24	19	20	36	43	49	55	30	42	4.8

Table A.3: Energy demand – India

	Current Policies Scenario				Sustainable Development Scenario			
	Energy demand (Mtoe)		Shares (%)	CAAGR (%)	Energy demand (Mtoe)		Shares (%)	CAAGR (%)
	2030	2040	2040	2018-40	2030	2040	2040	2018-40
Total primary demand	1 494	2 055	100	3.7	1 143	1 294	100	1.6
Coal	691	992	48	4.1	382	277	21	-1.8
Oil	377	486	24	3.4	306	286	22	0.9
Natural gas	132	198	10	6.2	171	260	20	7.5
Nuclear	28	56	3	7.9	29	65	5	8.7
Hydro	19	22	1	2.7	23	32	2	4.4
Bioenergy	200	202	10	0.3	142	163	13	-0.6
Other renewables	48	99	5	11.1	89	211	16	15.0
Power sector	620	903	100	4.3	457	547	100	2.0
Coal	440	620	69	3.8	195	60	11	-6.7
Oil	8	5	1	-0.2	7	5	1	-0.8
Natural gas	41	63	7	7.0	74	114	21	9.9
Nuclear	28	56	6	7.9	29	65	12	8.7
Hydro	19	22	2	2.7	23	32	6	4.4
Bioenergy	39	42	5	2.1	46	75	14	4.8
Other renewables	46	94	10	11.4	82	197	36	15.2
Other energy sector	145	208	100	4.0	118	148	100	2.5
Electricity	60	90	43	4.7	44	59	40	2.8
Total final consumption	1 002	1 364	100	3.7	806	939	100	1.9
Coal	192	282	21	4.6	138	151	16	1.7
Oil	349	461	34	3.7	282	269	29	1.2
Natural gas	89	132	10	5.8	94	140	15	6.1
Electricity	212	329	24	5.3	193	280	30	4.5
Heat	0	0	0	n.a.	0	0	0	n.a.
Bioenergy	157	155	11	-0.0	92	84	9	-2.8
Other renewables	3	5	0	7.2	7	15	2	12.9
Industry	413	582	100	4.4	342	421	100	2.9
Coal	179	271	47	5.0	129	149	35	2.2
Oil	43	48	8	1.9	32	27	6	-0.8
Gas	64	94	16	5.9	73	112	27	6.8
Electricity	78	110	19	4.4	71	94	22	3.7
Heat	0	0	0	n.a.	0	0	0	n.a.
Bioenergy	48	57	10	2.6	35	33	8	0.1
Other renewables	1	2	0	16.1	3	7	2	23.8
Transport	203	300	100	4.9	168	190	100	2.8
Oil	189	274	91	4.7	144	136	71	1.4
Electricity	3	6	2	6.6	7	28	15	14.6
Biofuels	2	4	1	8.8	8	14	8	15.8
Other fuels	9	17	6	8.2	9	12	6	6.4
Buildings	273	346	100	2.1	189	213	100	-0.1
Coal	13	12	3	-0.3	9	2	1	-7.5
Oil	46	56	16	2.2	40	37	17	0.2
Natural gas	5	8	2	7.0	2	4	2	3.2
Electricity	99	173	50	6.4	87	129	61	5.0
Heat	-	-	-	n.a.	-	-	-	n.a.
Bioenergy	107	94	27	-1.3	46	33	16	-5.8
Traditional biomass	97	83	24	-1.5	32	16	8	-8.6
Other renewables	2	3	1	5.4	4	8	4	9.9
Other	113	137	100	3.4	106	116	100	2.6
Petrochem. feedstock	43	55	40	4.7	43	53	46	4.6

Table A.3: Electricity and CO$_2$ emissions – India

Stated Policies Scenario
Electricity generation (TWh)

	2010	2017	2018	2025	2030	2035	2040	Shares (%) 2018	Shares (%) 2040	CAAGR (%) 2018-40
Total generation	982	1 532	1 618	2 338	3 018	3 780	4 581	100	100	4.8
Coal	658	1 134	1 194	1 558	1 861	2 036	2 133	74	47	2.7
Oil	24	25	17	25	24	18	16	1	0	-0.3
Natural gas	113	71	71	94	105	143	164	4	4	3.9
Nuclear	26	38	40	68	111	170	224	2	5	8.1
Renewables	160	264	295	592	916	1 412	2 045	18	45	9.2
Hydro	125	142	146	187	241	286	328	9	7	3.8
Bioenergy	15	45	47	81	93	107	123	3	3	4.5
Wind	20	51	58	146	250	399	525	4	11	10.5
Geothermal	-	-	-	0	1	1	1	-	0	n.a.
Solar PV	0	26	44	178	328	611	1 053	3	23	15.5
CSP	-	-	-	1	3	7	14	-	0	n.a.
Marine	-	-	-	-	0	0	1	-	0	n.a.

Stated Policies Scenario
Electrical capacity (GW)

	2017	2018	2025	2030	2035	2040	Shares (%) 2018	Shares (%) 2040	CAAGR (%) 2018-40
Total capacity	377	396	604	823	1 168	1 585	100	100	6.5
Coal	223	227	269	307	360	414	57	26	2.8
Oil	8	8	10	10	8	7	2	0	-0.4
Natural gas	29	29	30	31	36	39	7	2	1.4
Nuclear	7	7	9	16	25	31	2	2	7.2
Renewables	111	125	273	425	666	975	32	62	9.8
Hydro	48	49	64	80	94	106	12	7	3.6
Bioenergy	11	12	16	18	21	23	3	1	3.0
Wind	33	35	76	119	176	219	9	14	8.6
Geothermal	-	-	0	0	0	0	-	0	n.a.
Solar PV	19	28	118	207	373	622	7	39	15.1
CSP	-	-	0	1	3	5	-	0	n.a.
Marine	-	-	-	0	0	0	-	0	n.a.

Stated Policies Scenario
CO$_2$ emissions (Mt)

	2010	2017	2018	2025	2030	2035	2040	Shares (%) 2018	Shares (%) 2040	CAAGR (%) 2018-40
Total CO$_2$	1 583	2 160	2 267	3 011	3 621	4 120	4 500	100	100	3.2
Coal	1 088	1 507	1 581	2 079	2 517	2 846	3 092	70	69	3.1
Oil	399	580	578	748	869	975	1 056	25	23	2.8
Natural gas	96	73	108	184	234	299	352	5	8	5.5
Power sector	789	1 099	1 146	1 459	1 722	1 868	1 935	100	100	2.4
Coal	704	1 040	1 095	1 391	1 651	1 789	1 853	96	96	2.4
Oil	23	26	18	25	23	18	15	2	1	-0.6
Natural gas	62	33	33	43	48	62	67	3	3	3.2
Final consumption	762	1 028	1 086	1 513	1 854	2 201	2 509	100	100	3.9
Coal	382	464	482	683	859	1 047	1 228	44	49	4.3
Oil	349	526	531	693	813	921	1 001	49	40	2.9
Transport	190	285	301	429	526	621	693	28	28	3.9
Natural gas	32	38	73	137	183	232	279	7	11	6.3

Table A.3: Electricity and CO_2 emissions – India

	Current Policies Scenario				Sustainable Development Scenario			
	Electricity generation (TWh)		Shares (%)	CAAGR (%)	Electricity generation (TWh)		Shares (%)	CAAGR (%)
	2030	2040	2040	2018-40	2030	2040	2040	2018-40
Total generation	3 169	4 875	100	5.1	2 764	3 940	100	4.1
Coal	1 974	2 822	58	4.0	891	266	7	-6.6
Oil	25	18	0	0.1	23	15	0	-0.6
Natural gas	234	384	8	8.0	443	729	19	11.2
Nuclear	106	215	4	7.9	113	251	6	8.7
Renewables	830	1 436	29	7.5	1 294	2 678	68	10.5
Hydro	218	261	5	2.7	268	375	10	4.4
Bioenergy	86	97	2	3.4	111	217	5	7.2
Wind	225	402	8	9.2	364	740	19	12.3
Geothermal	0	1	0	n.a.	2	5	0	n.a.
Solar PV	300	671	14	13.2	539	1 236	31	16.4
CSP	2	3	0	n.a.	11	105	3	n.a.
Marine	-	1	0	n.a.	0	1	0	n.a.

	Current Policies Scenario				Sustainable Development Scenario			
	Electrical capacity (GW)		Shares (%)	CAAGR (%)	Electrical capacity (GW)		Shares (%)	CAAGR (%)
	2030	2040	2040	2018-40	2030	2040	2040	2018-40
Total capacity	834	1 386	100	5.9	993	1 741	100	7.0
Coal	332	495	36	3.6	223	162	9	-1.5
Oil	10	8	1	0.0	9	6	0	-1.0
Natural gas	61	101	7	5.9	85	147	8	7.7
Nuclear	16	31	2	7.2	17	36	2	7.9
Renewables	390	680	49	8.0	618	1 230	71	11.0
Hydro	72	84	6	2.5	89	121	7	4.2
Bioenergy	17	19	1	2.0	21	40	2	5.6
Wind	108	172	12	7.5	170	309	18	10.4
Geothermal	0	0	0	n.a.	0	1	0	n.a.
Solar PV	193	404	29	12.8	333	723	42	15.9
CSP	1	1	0	n.a.	4	36	2	n.a.
Marine	-	0	0	n.a.	0	0	0	n.a.

	Current Policies Scenario				Sustainable Development Scenario			
	CO_2 emissions (Mt)		Shares (%)	CAAGR (%)	CO_2 emissions (Mt)		Shares (%)	CAAGR (%)
	2030	2040	2040	2018-40	2030	2040	2040	2018-40
Total CO_2	3 875	5 476	100	4.1	2 516	2 052	100	-0.5
Coal	2 651	3 807	70	4.1	1 416	874	43	-2.7
Oil	943	1 245	23	3.6	731	650	32	0.5
Natural gas	281	424	8	6.4	370	549	27	7.7
Power sector	1 868	2 627	100	3.8	967	425	100	-4.4
Coal	1 747	2 463	94	3.8	771	173	41	-8.0
Oil	25	17	1	-0.2	23	15	3	-0.8
Natural gas	97	147	6	7.0	173	257	61	9.7
Final consumption	1 959	2 785	100	4.4	1 512	1 592	100	1.8
Coal	897	1 331	48	4.7	638	691	43	1.6
Oil	884	1 184	43	3.7	683	616	39	0.7
Transport	575	831	30	4.7	437	412	26	1.4
Natural gas	179	269	10	6.1	191	284	18	6.4

Table A.3: Energy demand – Japan

	Stated Policies Scenario Energy demand (Mtoe)							Shares (%)		CAAGR (%)
	2010	2017	2018	2025	2030	2035	2040	2018	2040	2018-40
Total primary demand	501	432	434	400	387	368	353	100	100	-0.9
Coal	115	116	115	99	87	80	75	27	21	-2.0
Oil	203	176	172	145	127	109	94	40	27	-2.7
Natural gas	86	101	100	86	76	75	74	23	21	-1.3
Nuclear	75	9	15	31	55	55	57	3	16	6.4
Hydro	7	7	8	8	8	8	8	2	2	0.4
Bioenergy	12	15	15	19	20	22	22	4	6	1.7
Other renewables	3	8	9	12	15	19	22	2	6	4.4
Power sector	233	189	190	180	184	183	182	100	100	-0.2
Coal	64	72	69	57	49	46	43	37	23	-2.2
Oil	20	14	13	8	5	3	1	7	1	-9.5
Natural gas	58	70	68	52	41	39	37	36	20	-2.7
Nuclear	75	9	15	31	55	55	57	8	31	6.4
Hydro	7	7	8	8	8	8	8	4	5	0.4
Bioenergy	6	8	9	12	13	14	15	4	8	2.7
Other renewables	3	7	8	12	13	17	20	4	11	4.1
Other energy sector	57	42	43	38	35	32	30	100	100	-1.6
Electricity	11	8	8	8	8	7	7	19	24	-0.6
Total final consumption	312	293	293	273	259	245	234	100	100	-1.0
Coal	23	22	23	21	19	17	15	8	7	-1.8
Oil	164	151	148	129	115	101	90	51	38	-2.3
Natural gas	29	30	31	33	34	35	35	11	15	0.5
Electricity	89	83	84	82	83	83	84	29	36	0.0
Heat	1	1	1	1	1	1	1	0	0	0.0
Bioenergy	6	6	6	7	7	7	7	2	3	0.3
Other renewables	1	0	0	1	1	2	2	0	1	8.2
Industry	94	87	88	83	78	73	69	100	100	-1.1
Coal	22	21	22	20	18	16	15	25	21	-1.9
Oil	25	21	20	17	15	14	12	23	17	-2.4
Natural gas	11	11	12	13	13	13	12	14	18	0.2
Electricity	33	30	30	29	28	27	26	34	37	-0.7
Heat	-	-	-	-	-	-	-	-	-	n.a.
Bioenergy	3	4	4	4	4	4	4	5	6	0.2
Other renewables	0	0	0	0	0	0	0	0	1	33.9
Transport	76	71	69	59	53	48	44	100	100	-2.0
Oil	74	69	67	56	49	42	38	97	85	-2.6
Electricity	2	2	2	2	3	4	5	2	10	5.1
Biofuels	0	0	0	1	1	1	1	1	2	3.5
Other fuels	0	0	0	0	0	1	1	0	3	15.4
Buildings	103	96	97	96	95	95	94	100	100	-0.1
Coal	0	0	0	-	-	-	-	0	-	n.a.
Oil	27	23	23	22	20	17	15	24	16	-2.0
Natural gas	18	18	19	19	20	21	22	20	24	0.7
Electricity	55	51	52	51	52	53	53	53	57	0.1
Heat	1	1	1	1	1	1	1	1	1	0.0
Bioenergy	2	2	2	2	2	2	2	2	2	-0.6
Traditional biomass	-	-	-	-	-	-	-	-	-	n.a.
Other renewables	1	0	0	1	1	1	2	0	2	7.2
Other	40	39	38	35	32	29	27	100	100	-1.7
Petrochem. feedstock	29	23	23	22	20	19	17	59	64	-1.3

Table A.3: Energy demand – Japan

	Current Policies Scenario				Sustainable Development Scenario			
	Energy demand (Mtoe)		Shares (%)	CAAGR (%)	Energy demand (Mtoe)		Shares (%)	CAAGR (%)
	2030	2040	2040	2018-40	2030	2040	2040	2018-40
Total primary demand	399	374	100	-0.7	349	300	100	-1.7
Coal	101	93	25	-1.0	49	30	10	-6.0
Oil	131	105	28	-2.2	107	57	19	-4.9
Natural gas	77	85	23	-0.7	77	52	17	-2.9
Nuclear	50	45	12	5.3	60	72	24	7.5
Hydro	8	8	2	0.4	9	12	4	2.1
Bioenergy	19	20	5	1.2	22	24	8	2.1
Other renewables	13	18	5	3.3	25	53	18	8.5
Power sector	192	192	100	0.0	171	177	100	-0.3
Coal	61	59	31	-0.7	17	6	3	-10.7
Oil	6	2	1	-7.7	3	1	0	-11.9
Natural gas	42	48	25	-1.6	47	25	14	-4.4
Nuclear	50	45	24	5.3	60	72	40	7.5
Hydro	8	8	4	0.4	9	12	7	2.1
Bioenergy	12	12	6	1.7	14	16	9	2.8
Other renewables	13	16	9	3.1	21	46	26	8.0
Other energy sector	36	30	100	-1.6	31	28	100	-1.9
Electricity	8	8	27	-0.2	7	6	22	-1.4
Total final consumption	266	250	100	-0.7	233	185	100	-2.1
Coal	19	16	6	-1.6	16	11	6	-3.1
Oil	119	99	40	-1.8	98	54	29	-4.5
Natural gas	34	37	15	0.7	28	22	12	-1.5
Electricity	86	89	36	0.3	78	80	43	-0.2
Heat	1	1	0	0.5	0	0	0	-1.1
Bioenergy	7	7	3	0.5	8	8	5	1.2
Other renewables	1	1	1	6.3	4	7	4	14.3
Industry	79	71	100	-1.0	69	57	100	-2.0
Coal	18	15	21	-1.7	15	11	19	-3.3
Oil	16	12	17	-2.3	12	8	14	-4.2
Gas	13	13	18	0.4	11	9	16	-1.4
Electricity	28	26	37	-0.6	26	24	42	-1.0
Heat	-	-	-	n.a.	-	-	-	n.a.
Bioenergy	4	4	6	0.4	4	4	7	0.1
Other renewables	0	0	0	29.3	1	1	2	40.3
Transport	55	49	100	-1.6	48	31	100	-3.6
Oil	52	45	92	-1.8	42	18	59	-5.7
Electricity	2	3	6	3.0	3	8	26	7.9
Biofuels	1	1	2	2.6	2	2	6	7.0
Other fuels	0	0	0	4.9	1	3	9	19.2
Buildings	100	102	100	0.3	84	74	100	-1.2
Coal	-	-	-	n.a.	-	-	-	n.a.
Oil	20	16	15	-1.8	14	5	7	-6.8
Natural gas	21	24	23	0.9	16	12	17	-2.0
Electricity	56	60	58	0.6	49	48	65	-0.4
Heat	1	1	1	0.5	0	0	1	-1.1
Bioenergy	2	2	2	0.2	2	2	3	0.4
Traditional biomass	-	-	-	n.a.	-	-	-	n.a.
Other renewables	1	1	1	5.6	3	5	7	13.6
Other	32	27	100	-1.6	31	24	100	-2.1
Petrochem. feedstock	20	17	63	-1.3	19	15	62	-1.9

Table A.3: Electricity and CO$_2$ emissions – Japan

Stated Policies Scenario
Electricity generation (TWh)

	2010	2017	2018	2025	2030	2035	2040	Shares (%) 2018	Shares (%) 2040	CAAGR (%) 2018-40
Total generation	1 164	1 061	1 069	1 050	1 053	1 057	1 062	100	100	-0.0
Coal	311	352	338	284	244	230	218	32	21	-2.0
Oil	96	70	64	40	27	16	8	6	1	-9.3
Natural gas	326	398	385	331	268	262	251	36	24	-1.9
Nuclear	288	33	56	120	210	212	219	5	21	6.4
Renewables	122	188	206	255	283	317	346	19	33	2.4
Hydro	84	83	89	91	93	95	97	8	9	0.4
Bioenergy	28	41	41	54	61	66	71	4	7	2.5
Wind	4	6	7	19	31	43	53	1	5	9.3
Geothermal	3	2	2	3	3	5	7	0	1	4.6
Solar PV	4	55	66	88	95	105	113	6	11	2.5
CSP	-	-	-	-	-	-	-	-	-	n.a.
Marine	-	-	-	0	0	2	6	-	1	n.a.

Stated Policies Scenario
Electrical capacity (GW)

	2017	2018	2025	2030	2035	2040	Shares (%) 2018	Shares (%) 2040	CAAGR (%) 2018-40
Total capacity	337	339	356	364	370	375	100	100	0.5
Coal	50	50	49	48	45	42	15	11	-0.9
Oil	48	45	29	18	11	5	13	1	-9.1
Natural gas	85	86	82	82	83	78	25	21	-0.4
Nuclear	41	39	34	30	27	28	11	7	-1.5
Renewables	111	118	161	182	201	216	35	58	2.8
Hydro	50	50	51	52	52	52	15	14	0.2
Bioenergy	8	8	10	12	13	14	2	4	2.7
Wind	3	4	9	13	17	20	1	5	8.0
Geothermal	0	0	1	1	1	1	0	0	4.0
Solar PV	50	56	90	104	116	126	17	34	3.7
CSP	-	-	-	-	-	-	-	-	n.a.
Marine	-	-	0	0	1	2	-	1	n.a.

Stated Policies Scenario
CO$_2$ emissions (Mt)

	2010	2017	2018	2025	2030	2035	2040	Shares (%) 2018	Shares (%) 2040	CAAGR (%) 2018-40
Total CO$_2$	1 108	1 097	1 078	908	792	722	666	100	100	-2.2
Coal	423	447	441	372	324	301	280	41	42	-2.0
Oil	478	412	401	334	290	246	214	37	32	-2.8
Natural gas	207	238	236	202	178	175	172	22	26	-1.4
Power sector	489	531	511	402	329	304	282	100	100	-2.7
Coal	290	321	309	253	216	202	191	61	68	-2.2
Oil	62	45	41	25	17	10	5	8	2	-9.5
Natural gas	136	165	160	123	96	92	87	31	31	-2.7
Final consumption	574	531	533	476	436	394	362	100	100	-1.7
Coal	114	109	114	101	92	84	76	21	21	-1.8
Oil	393	353	346	299	265	230	203	65	56	-2.4
Transport	221	205	201	168	147	126	112	38	31	-2.6
Natural gas	67	70	73	76	79	81	82	14	23	0.5

Table A.3: Electricity and CO_2 emissions – Japan

	Current Policies Scenario				Sustainable Development Scenario			
	Electricity generation (TWh)		Shares (%)	CAAGR (%)	Electricity generation (TWh)		Shares (%)	CAAGR (%)
	2030	2040	2040	2018-40	2030	2040	2040	2018-40
Total generation	1 095	1 129	100	0.2	991	1 005	100	-0.3
Coal	309	305	27	-0.5	85	25	2	-11.1
Oil	30	12	1	-7.4	16	4	0	-11.8
Natural gas	278	325	29	-0.8	306	163	16	-3.8
Nuclear	191	173	15	5.3	229	275	27	7.5
Renewables	268	294	26	1.6	334	518	52	4.3
Hydro	92	96	9	0.4	108	139	14	2.1
Bioenergy	56	58	5	1.6	64	74	7	2.7
Wind	27	40	4	8.0	53	139	14	14.2
Geothermal	3	6	1	3.9	9	26	3	11.2
Solar PV	88	92	8	1.5	98	130	13	3.1
CSP	-	-	-	n.a.	-	-	-	n.a.
Marine	0	1	0	n.a.	1	11	1	n.a.

	Current Policies Scenario				Sustainable Development Scenario			
	Electrical capacity (GW)		Shares (%)	CAAGR (%)	Electrical capacity (GW)		Shares (%)	CAAGR (%)
	2030	2040	2040	2018-40	2030	2040	2040	2018-40
Total capacity	360	356	100	0.2	358	410	100	0.9
Coal	53	52	15	0.1	29	7	2	-8.4
Oil	18	7	2	-8.1	18	5	1	-9.2
Natural gas	88	90	25	0.2	72	67	16	-1.1
Nuclear	27	22	6	-2.5	33	35	9	-0.4
Renewables	169	180	51	1.9	203	290	71	4.2
Hydro	51	52	15	0.2	59	70	17	1.6
Bioenergy	11	12	3	1.9	13	15	4	2.8
Wind	11	16	4	6.7	21	49	12	12.4
Geothermal	1	1	0	3.4	2	4	1	10.6
Solar PV	94	99	28	2.6	108	148	36	4.5
CSP	-	-	-	n.a.	-	-	-	n.a.
Marine	0	0	0	n.a.	0	4	1	n.a.

	Current Policies Scenario				Sustainable Development Scenario			
	CO_2 emissions (Mt)		Shares (%)	CAAGR (%)	CO_2 emissions (Mt)		Shares (%)	CAAGR (%)
	2030	2040	2040	2018-40	2030	2040	2040	2018-40
Total CO_2	868	805	100	-1.3	568	287	100	-5.8
Coal	384	358	45	-0.9	156	69	24	-8.1
Oil	302	245	30	-2.2	233	108	38	-5.8
Natural gas	182	201	25	-0.7	179	110	38	-3.4
Power sector	391	383	100	-1.3	186	59	100	-9.3
Coal	273	264	69	-0.7	66	3	4	-19.5
Oil	18	7	2	-7.7	10	3	4	-11.9
Natural gas	99	112	29	-1.6	110	54	91	-4.8
Final consumption	451	398	100	-1.3	358	211	100	-4.1
Coal	95	80	20	-1.6	77	56	27	-3.2
Oil	275	232	58	-1.8	215	102	48	-5.4
Transport	156	135	34	-1.8	125	55	26	-5.7
Natural gas	81	87	22	0.8	66	53	25	-1.4

Annex A | Tables for scenario projections

Table A.3: Energy demand – Southeast Asia

	Stated Policies Scenario							Shares (%)		CAAGR (%)
	Energy demand (Mtoe)									
	2010	2017	2018	2025	2030	2035	2040	2018	2040	2018-40
Total primary demand	544	676	701	843	941	1 031	1 114	100	100	2.1
Coal	86	134	143	188	216	244	272	20	24	3.0
Oil	190	237	244	291	313	325	329	35	30	1.4
Natural gas	125	134	135	168	192	220	246	19	22	2.8
Nuclear	-	-	-	-	-	-	3	-	0	n.a.
Hydro	7	16	16	17	23	28	31	2	3	3.1
Bioenergy	112	123	127	131	133	135	137	18	12	0.3
Other renewables	25	32	35	48	64	79	96	5	9	4.6
Power sector	165	230	243	313	370	428	486	100	100	3.2
Coal	49	96	103	140	162	185	208	43	43	3.2
Oil	15	7	7	5	5	5	4	3	1	-2.1
Natural gas	67	69	68	87	95	106	114	28	23	2.4
Nuclear	-	-	-	-	-	-	3	-	1	n.a.
Hydro	7	16	16	17	23	28	31	7	6	3.1
Bioenergy	3	10	14	17	21	26	32	6	7	4.0
Other renewables	25	32	35	47	63	78	94	15	19	4.5
Other energy sector	61	51	52	59	64	70	76	100	100	1.7
Electricity	7	11	11	15	17	20	22	22	30	3.1
Total final consumption	377	482	496	592	653	706	754	100	100	1.9
Coal	31	37	39	47	53	57	61	8	8	2.1
Oil	165	222	230	274	291	301	305	46	40	1.3
Natural gas	29	42	44	60	76	93	111	9	15	4.3
Electricity	52	76	79	106	129	154	180	16	24	3.8
Heat	-	-	-	-	-	-	-	-	-	n.a.
Bioenergy	99	104	105	105	103	100	95	21	13	-0.4
Other renewables	0	0	0	0	1	1	2	0	0	27.6
Industry	113	157	162	199	227	254	280	100	100	2.5
Coal	29	35	36	45	50	55	59	22	21	2.2
Oil	23	29	29	31	32	31	30	18	11	0.2
Natural gas	20	33	35	49	62	76	90	22	32	4.4
Electricity	22	33	34	46	53	60	67	21	24	3.1
Heat	-	-	-	-	-	-	-	-	-	n.a.
Bioenergy	18	26	27	29	30	31	32	17	11	0.8
Other renewables	-	-	-	0	0	0	0	-	0	n.a.
Transport	89	130	135	167	182	192	200	100	100	1.8
Oil	85	123	127	155	166	172	175	94	88	1.5
Electricity	0	0	0	1	2	3	5	0	3	13.5
Biofuels	1	4	5	8	9	11	13	4	6	4.7
Other fuels	2	2	2	3	4	5	6	2	3	4.6
Buildings	127	138	140	153	165	177	189	100	100	1.4
Coal	2	2	2	2	2	2	2	2	1	-0.2
Oil	16	21	21	23	24	25	26	15	14	0.8
Natural gas	0	0	0	1	2	3	4	0	2	11.7
Electricity	30	42	43	58	72	88	106	31	56	4.2
Heat	-	-	-	-	-	-	-	-	-	n.a.
Bioenergy	79	73	73	68	63	57	50	52	26	-1.7
Traditional biomass	79	72	72	67	62	56	49	51	26	-1.8
Other renewables	0	0	0	0	1	1	1	0	1	24.9
Other	48	57	59	73	79	83	85	100	100	1.7
Petrochem. feedstock	34	42	43	53	57	61	62	72	73	1.7

Table A.3: Energy demand – Southeast Asia

	Current Policies Scenario				Sustainable Development Scenario			
	Energy demand (Mtoe)		Shares (%)	CAAGR (%)	Energy demand (Mtoe)		Shares (%)	CAAGR (%)
	2030	2040	2040	2018-40	2030	2040	2040	2018-40
Total primary demand	966	1 185	100	2.4	797	858	100	0.9
Coal	239	341	29	4.0	111	54	6	-4.3
Oil	325	365	31	1.8	272	224	26	-0.4
Natural gas	194	245	21	2.7	177	201	23	1.8
Nuclear	-	3	0	n.a.	-	4	0	n.a.
Hydro	20	25	2	2.1	28	45	5	4.9
Bioenergy	132	133	11	0.2	85	98	11	-1.2
Other renewables	55	73	6	3.3	126	232	27	8.9
Power sector	380	513	100	3.5	332	421	100	2.5
Coal	184	271	53	4.5	72	19	4	-7.4
Oil	6	5	1	-1.8	3	2	1	-4.5
Natural gas	96	109	21	2.2	81	83	20	0.9
Nuclear	-	3	1	n.a.	-	4	1	n.a.
Hydro	20	25	5	2.1	28	45	11	4.9
Bioenergy	20	29	6	3.5	26	42	10	5.3
Other renewables	55	71	14	3.2	123	225	53	8.8
Other energy sector	68	86	100	2.3	58	62	100	0.8
Electricity	17	24	27	3.4	14	17	28	1.9
Total final consumption	667	793	100	2.2	536	556	100	0.5
Coal	54	64	8	2.3	38	33	6	-0.7
Oil	301	336	42	1.7	254	206	37	-0.5
Natural gas	77	112	14	4.3	76	99	18	3.8
Electricity	131	183	23	3.9	115	162	29	3.3
Heat	-	-	-	n.a.	-	-	-	n.a.
Bioenergy	103	96	12	-0.4	50	47	8	-3.6
Other renewables	1	1	0	25.1	3	7	1	35.0
Industry	231	287	100	2.6	187	204	100	1.0
Coal	52	62	22	2.4	36	33	16	-0.5
Oil	32	31	11	0.3	23	17	9	-2.4
Gas	63	92	32	4.5	58	75	37	3.5
Electricity	53	66	23	3.0	46	56	28	2.2
Heat	-	-	-	n.a.	-	-	-	n.a.
Bioenergy	32	35	12	1.1	22	20	10	-1.5
Other renewables	0	0	0	n.a.	1	3	2	n.a.
Transport	187	220	100	2.3	168	155	100	0.6
Oil	175	202	92	2.1	142	103	66	-1.0
Electricity	1	1	1	6.1	3	21	13	21.0
Biofuels	8	11	5	3.9	15	19	12	6.5
Other fuels	4	6	3	4.3	8	13	8	7.9
Buildings	170	200	100	1.6	105	119	100	-0.7
Coal	3	3	1	0.4	2	1	1	-5.9
Oil	25	28	14	1.3	23	21	17	-0.2
Natural gas	2	4	2	11.6	2	3	2	9.5
Electricity	76	114	57	4.5	65	84	71	3.1
Heat	-	-	-	n.a.	-	-	-	n.a.
Bioenergy	63	50	25	-1.7	11	7	6	-9.9
Traditional biomass	62	49	24	-1.8	9	5	4	-11.7
Other renewables	0	1	0	23.0	1	3	3	30.6
Other	79	86	100	1.7	76	78	100	1.3
Petrochem. feedstock	57	61	72	1.7	55	58	75	1.4

Table A.3: Electricity and CO_2 emissions – Southeast Asia

	Stated Policies Scenario									
	Electricity generation (TWh)							Shares (%)		CAAGR (%)
	2010	2017	2018	2025	2030	2035	2040	2018	2040	2018-40
Total generation	684	1 014	1 045	1 400	1 693	2 009	2 345	100	100	3.7
Coal	185	386	416	597	699	812	929	40	40	3.7
Oil	60	27	26	19	19	18	15	2	1	-2.4
Natural gas	335	362	357	478	534	608	684	34	29	3.0
Nuclear	-	-	-	-	-	-	12	-	1	n.a.
Renewables	104	239	247	306	442	571	706	24	30	4.9
Hydro	78	187	184	194	268	323	362	18	15	3.1
Bioenergy	6	20	26	39	52	69	91	3	4	5.8
Wind	0	3	4	12	22	35	53	0	2	13.0
Geothermal	19	23	25	37	52	65	77	2	3	5.2
Solar PV	0	6	7	24	47	80	124	1	5	13.8
CSP	-	-	-	-	-	-	-	-	-	n.a.
Marine	-	-	-	0	0	0	0	-	0	n.a.

	Stated Policies Scenario								
	Electrical capacity (GW)						Shares (%)		CAAGR (%)
	2017	2018	2025	2030	2035	2040	2018	2040	2018-40
Total capacity	253	261	362	438	525	623	100	100	4.0
Coal	71	74	112	126	144	165	28	26	3.7
Oil	26	26	23	21	19	13	10	2	-3.0
Natural gas	93	96	122	137	158	184	37	29	3.0
Nuclear	-	-	-	-	-	2	-	0	n.a.
Renewables	63	66	102	150	198	249	25	40	6.3
Hydro	46	47	61	84	100	111	18	18	4.0
Bioenergy	8	9	11	14	16	18	3	3	3.4
Wind	1	2	5	10	15	22	1	4	12.3
Geothermal	4	4	6	8	9	11	1	2	5.0
Solar PV	4	5	19	35	58	87	2	14	14.1
CSP	-	-	-	-	-	-	-	-	n.a.
Marine	-	-	0	0	0	0	-	0	n.a.

	Stated Policies Scenario									
	CO_2 emissions (Mt)							Shares (%)		CAAGR (%)
	2010	2017	2018	2025	2030	2035	2040	2018	2040	2018-40
Total CO_2	1 038	1 371	1 429	1 779	1 986	2 182	2 355	100	100	2.3
Coal	324	537	573	757	868	981	1 090	40	46	3.0
Oil	455	575	592	686	731	755	763	41	32	1.2
Natural gas	259	259	264	337	387	447	502	18	21	3.0
Power sector	401	570	597	784	892	1 010	1 119	100	100	2.9
Coal	198	386	416	565	652	745	837	70	75	3.2
Oil	46	22	21	16	17	16	13	4	1	-2.1
Natural gas	157	162	159	203	223	249	268	27	24	2.4
Final consumption	572	749	779	939	1 033	1 106	1 166	100	100	1.9
Coal	126	151	157	192	215	235	253	20	22	2.2
Oil	394	536	553	649	690	712	722	71	62	1.2
Transport	256	367	381	463	497	515	524	49	45	1.5
Natural gas	52	63	69	98	127	158	191	9	16	4.7

Table A.3: Electricity and CO_2 emissions – Southeast Asia

	Current Policies Scenario				Sustainable Development Scenario			
	Electricity generation (TWh)		Shares (%)	CAAGR (%)	Electricity generation (TWh)		Shares (%)	CAAGR (%)
	2030	2040	2040	2018-40	2030	2040	2040	2018-40
Total generation	1 726	2 404	100	3.9	1 496	2 083	100	3.2
Coal	788	1 205	50	5.0	304	81	4	-7.2
Oil	20	16	1	-2.1	10	8	0	-5.2
Natural gas	541	661	28	2.8	464	504	24	1.6
Nuclear	-	12	0	n.a.	-	16	1	n.a.
Renewables	377	510	21	3.4	718	1 474	71	8.5
Hydro	237	293	12	2.1	321	527	25	4.9
Bioenergy	49	78	3	5.1	67	127	6	7.5
Wind	17	29	1	9.9	129	314	15	22.5
Geothermal	44	61	3	4.0	106	183	9	9.4
Solar PV	29	49	2	9.0	94	321	15	18.8
CSP	-	-	-	n.a.	-	-	-	n.a.
Marine	0	0	0	n.a.	0	2	0	n.a.

	Current Policies Scenario				Sustainable Development Scenario			
	Electrical capacity (GW)		Shares (%)	CAAGR (%)	Electrical capacity (GW)		Shares (%)	CAAGR (%)
	2030	2040	2040	2018-40	2030	2040	2040	2018-40
Total capacity	418	575	100	3.7	508	823	100	5.4
Coal	136	206	36	4.8	93	67	8	-0.4
Oil	21	13	2	-2.9	21	13	2	-3.1
Natural gas	134	178	31	2.9	127	148	18	2.0
Nuclear	-	2	0	n.a.	-	3	0	n.a.
Renewables	123	162	28	4.2	265	580	71	10.4
Hydro	74	90	16	3.0	101	164	20	5.9
Bioenergy	13	16	3	2.8	17	25	3	5.0
Wind	8	12	2	9.1	58	135	16	21.8
Geothermal	7	9	2	3.9	16	27	3	9.3
Solar PV	22	36	6	9.6	72	229	28	19.2
CSP	-	-	-	n.a.	-	-	-	n.a.
Marine	0	0	0	n.a.	0	1	0	n.a.

	Current Policies Scenario				Sustainable Development Scenario			
	CO_2 emissions (Mt)		Shares (%)	CAAGR (%)	CO_2 emissions (Mt)		Shares (%)	CAAGR (%)
	2030	2040	2040	2018-40	2030	2040	2040	2018-40
Total CO_2	2 117	2 718	100	3.0	1 395	1 012	100	-1.6
Coal	961	1 358	50	4.0	439	171	17	-5.3
Oil	765	861	32	1.7	612	457	45	-1.2
Natural gas	391	499	18	2.9	345	383	38	1.7
Power sector	981	1 363	100	3.8	488	245	100	-4.0
Coal	739	1 093	80	4.5	288	42	17	-9.9
Oil	17	14	1	-1.8	10	8	3	-4.5
Natural gas	224	257	19	2.2	190	195	80	0.9
Final consumption	1 071	1 274	100	2.3	859	730	100	-0.3
Coal	222	266	21	2.4	150	129	18	-0.9
Oil	722	816	64	1.8	583	436	60	-1.1
Transport	522	605	47	2.1	425	307	42	-1.0
Natural gas	128	193	15	4.8	126	166	23	4.1

Table A.4: Emissions of air pollutants – World

| | Stated Policies Scenario ||||| Shares (%) || CAAGR (%) |
| | By energy sector ||||| | | |
	2018	2025	2030	2035	2040	2018	2040	2018-40
SO_2 emissions from all energy activities (Mt)								
Total	62.7	51.7	47.7	47.6	48.3	100	100	-1.2
Power	21.7	16.5	12.6	12.5	13.0	35	27	-2.3
Industry*	28.2	26.0	26.3	26.9	27.4	45	57	-0.1
Transport	6.6	3.9	4.1	4.3	4.4	10	9	-1.8
Buildings	5.1	4.0	3.4	2.8	2.5	8	5	-3.2
Agriculture	1.1	1.3	1.3	1.1	1.1	2	2	-0.2
NO_X emissions from all energy activities (Mt)								
Total	105.2	97.0	92.7	91.4	91.9	100	100	-0.6
Power	15.9	13.2	12.5	12.0	11.9	15	13	-1.3
Industry*	25.1	24.7	24.7	25.8	26.8	24	29	0.3
Transport	56.5	51.8	48.2	46.6	46.3	54	50	-0.9
Buildings	4.7	4.7	4.6	4.5	4.5	4	5	-0.3
Agriculture	3.0	2.7	2.6	2.5	2.4	3	3	-0.9
$PM_{2.5}$ emissions from all energy activities (Mt)								
Total	29.6	28.3	27.4	27.4	27.3	100	100	-0.4
Power	1.9	1.5	1.3	1.3	1.3	6	5	-1.9
Industry*	7.9	8.2	8.6	9.3	9.9	27	36	1.1
Transport	3.5	2.8	2.6	2.6	2.7	12	10	-1.2
Buildings	15.5	14.9	14.1	13.3	12.5	52	46	-1.0
Agriculture	0.7	0.8	0.8	0.8	0.9	2	3	0.8

* Industry also includes other transformation.

| | Stated Policies Scenario ||||| Shares (%) || CAAGR (%) |
| | By fuel ||||| | | |
	2018	2025	2030	2035	2040	2018	2040	2018-40
SO_2 emissions from combustion activities (Mt)								
Total	46.1	35.5	31.0	30.3	30.7	100	100	-1.8
Coal	27.0	19.8	15.5	14.9	15.0	59	49	-2.6
Oil	16.6	12.9	12.5	12.1	11.9	36	39	-1.5
Natural gas	0.4	0.4	0.5	0.5	0.6	1	2	2.1
Bioenergy	2.2	2.3	2.5	2.8	3.2	5	10	1.7
NO_X emissions from combustion activities (Mt)								
Total	93.7	86.0	81.3	79.5	79.6	100	100	-0.7
Coal	15.0	11.5	10.6	9.8	9.7	16	12	-2.0
Oil	65.9	60.7	56.4	54.5	53.7	70	67	-0.9
Natural gas	9.0	9.4	9.7	10.3	11.1	10	14	0.9
Bioenergy	3.8	4.3	4.6	4.9	5.1	4	6	1.4
$PM_{2.5}$ emissions from combustion activities (Mt)								
Total	23.4	21.7	20.4	19.7	19.1	100	100	-0.9
Coal	4.1	3.1	2.6	2.3	2.1	17	11	-3.1
Oil	4.6	3.7	3.3	3.3	3.4	19	18	-1.4
Natural gas	0.1	0.1	0.1	0.2	0.2	1	1	1.8
Bioenergy	14.7	14.8	14.3	13.9	13.5	63	71	-0.4

Table A.4: Emissions of air pollutants – World

	Current Policies Scenario				Sustainable Development Scenario			
	By energy sector		Shares (%)	CAAGR (%)	By energy sector		Shares (%)	CAAGR (%)
	2030	2040	2040	2018-40	2030	2040	2040	2018-40
SO₂ emissions from all energy activities (Mt)								
Total	49.6	52.6	100	-0.8	28.7	16.8	100	-5.8
Power	13.7	15.5	29	-1.5	4.9	2.3	13	-9.8
Industry*	26.1	27.4	52	-0.1	18.2	11.5	69	-4.0
Transport	4.6	5.4	10	-0.9	2.9	2.0	12	-5.2
Buildings	3.9	3.2	6	-2.2	2.0	0.8	5	-8.3
Agriculture	1.3	1.2	2	0.2	0.7	0.2	1	-7.0
NOₓ emissions from all energy activities (Mt)								
Total	100.8	107.6	100	0.1	58.1	31.9	100	-5.3
Power	13.3	13.8	13	-0.7	6.7	3.6	11	-6.5
Industry*	25.1	27.6	26	0.4	15.3	8.2	26	-5.0
Transport	54.9	58.7	55	0.2	32.0	18.0	57	-5.1
Buildings	4.8	4.8	4	0.1	2.6	1.6	5	-4.8
Agriculture	2.7	2.7	2	-0.4	1.4	0.5	2	-7.5
PM₂.₅ emissions from all energy activities (Mt)								
Total	27.6	28.2	100	-0.2	11.0	4.5	100	-8.2
Power	1.3	1.4	5	-1.3	0.6	0.2	4	-10.2
Industry*	8.4	9.8	35	1.0	5.0	2.0	46	-5.9
Transport	3.3	3.8	13	0.3	1.8	1.1	25	-5.2
Buildings	13.8	12.3	44	-1.1	3.1	0.9	20	-12.1
Agriculture	0.8	0.9	3	0.8	0.5	0.2	5	-5.5

* Industry also includes other transformation.

	Current Policies Scenario				Sustainable Development Scenario			
	By fuel		Shares (%)	CAAGR (%)	By fuel		Shares (%)	CAAGR (%)
	2030	2040	2040	2018-40	2030	2040	2040	2018-40
SO₂ emissions from combustion activities (Mt)								
Total	33.2	35.3	100	-1.2	16.3	8.6	100	-7.4
Coal	17.0	18.1	51	-1.8	7.6	3.2	37	-9.3
Oil	13.3	13.8	39	-0.8	6.6	3.1	36	-7.3
Natural gas	0.5	0.6	2	2.6	0.5	0.5	6	1.2
Bioenergy	2.4	2.8	8	1.2	1.7	1.8	21	-0.7
NOₓ emissions from combustion activities (Mt)								
Total	89.4	95.3	100	0.1	50.4	27.7	100	-5.4
Coal	11.5	11.6	12	-1.2	5.3	2.6	10	-7.6
Oil	63.5	67.0	70	0.1	36.2	19.3	70	-5.4
Natural gas	10.0	11.8	12	1.2	6.1	3.2	12	-4.6
Bioenergy	4.4	4.8	5	1.1	2.9	2.5	9	-1.8
PM₂.₅ emissions from combustion activities (Mt)								
Total	20.8	20.1	100	-0.7	6.5	2.2	100	-10.3
Coal	2.9	2.7	14	-1.8	1.2	0.3	13	-11.5
Oil	4.1	4.5	22	-0.0	2.0	0.9	40	-7.2
Natural gas	0.2	0.2	1	2.0	0.1	0.1	4	-1.2
Bioenergy	13.7	12.7	63	-0.7	3.2	0.9	43	-11.8

Table A.5: Energy investment

World investments	2014-2018	Stated Policies Scenario					
		2019-2030	2031-2040	2019-2040	2019-2030	2031-2040	2019-2040
(billion dollars, 2018)		Average annual			Cumulative		
Fuels	930	935	989	960	11 224	9 895	21 119
Oil and natural gas	414	867	932	897	10 409	9 321	19 730
Coal	98	59	42	51	714	419	1 133
Biofuels	5	8	15	12	101	154	256
Power	775	856	999	921	10 276	9 994	20 269
Power plants	483	491	521	505	5 890	5 212	11 101
Fossil fuels	*138*	*92*	*91*	*91*	*1 102*	*908*	*2 010*
Nuclear	*41*	*52*	*48*	*50*	*619*	*482*	*1 102*
Renewables	*303*	*347*	*382*	*363*	*4 169*	*3 821*	*7 990*
Electricity networks	291	354	455	400	4 248	4 552	8 800
Fuels and power	1 706	1 792	1 989	1 881	21 500	19 888	41 388
Energy efficiency	238	445	635	531	5 338	6 345	11 684
Renewables and other	127	220	308	260	2 644	3 080	5 723
End-use	365	665	943	791	7 982	9 425	17 407
Total	2 071	2 457	2 931	2 673	29 482	29 313	58 795

Fuels	Stated Policies Scenario					
	Upstream oil & gas	Transportation		Refining oil	Oil & gas	Total
		Oil	Gas			
Cumulative investments, 2019-2040 (billion dollars, 2018)						
North America	4 547	141	665	139	5 492	5 651
Central and South America	1 558	115	103	40	1 817	1 914
Europe	1 201	18	320	80	1 618	1 634
Africa	1 623	68	167	54	1 911	1 955
Middle East	2 098	183	291	140	2 711	2 713
Eurasia	2 089	36	329	42	2 496	2 563
Asia Pacific	2 052	80	747	405	3 284	4 239
Shipping	n.a.	927	116	n.a.	1 042	449
World	15 167	927	2 737	899	19 730	21 119

Power	Stated Policies Scenario							
	Coal	Gas	Oil	Fossil fuels	Nuclear	Renew.	Power plants	Networks	Total
Cumulative investments, 2019-2040 (billion dollars, 2018)								
North America	64	274	7	344	149	1 136	1 629	959	2 654
Central and South America	8	53	2	63	29	422	514	442	959
Europe	110	127	2	238	359	1 543	2 140	1 235	3 420
Africa	48	81	8	137	23	522	682	727	1 429
Middle East	10	138	17	164	34	249	447	292	741
Eurasia	65	120	0	185	94	144	423	247	673
Asia Pacific	632	237	11	880	412	3 973	5 266	4 898	10 394
World	935	1 029	46	2 010	1 102	7 990	11 101	8 800	20 269

Table A.5: Energy investment

World investments	Current Policies Scenario					Sustainable Development Scenario				
	2019-2030	2031-2040	2019-2030	2031-2040	2019-2040	2019-2030	2031-2040	2019-2030	2031-2040	2019-2040
(billion dollars, 2018)	Average annual		Cumulative			Average annual		Cumulative		
Fuels	1 060	1 320	12 726	13 204	25 930	727	567	8 720	5 668	14 388
Oil and natural gas	986	1 250	11 834	12 504	24 337	681	505	8 177	5 051	13 227
Coal	66	54	794	544	1 338	22	20	261	200	461
Biofuels	8	16	98	157	255	24	42	283	417	700
Power	832	918	9 986	9 181	19 167	1 014	1 463	12 167	14 631	26 797
Plants	460	481	5 525	4 810	10 336	653	793	7 833	7 934	15 767
Fossil fuels	120	134	1 444	1 342	2 786	66	65	795	651	1 446
Nuclear	53	48	631	479	1 110	62	67	749	670	1 420
Renewables	288	299	3 450	2 990	6 440	524	661	6 289	6 613	12 902
Electricity networks	360	417	4 316	4 172	8 488	345	631	4 138	6 306	10 444
Fuels and power	1 893	2 239	22 712	22 386	45 097	1 741	2 030	20 887	20 299	41 185
Energy efficiency	346	534	4 156	5 341	9 497	625	916	7 498	9 156	16 654
Renewables and other	179	200	2 144	2 005	4 148	332	950	3 985	9 504	13 489
End-use	525	735	6 299	7 346	13 645	957	1 866	11 483	18 660	30 143
Total	2 418	2 973	29 011	29 732	58 742	2 697	3 896	32 370	38 959	71 329

Fuels	Current Policies Scenario					Sustainable Development Scenario				
	Upstream oil & gas	Transport. oil & gas	Refining oil	Oil & gas	Total	Upstream oil & gas	Transport. oil & gas	Refining oil	Oil & gas	Total
Cumulative investments, 2019-2040 (billion dollars, 2018)										
North America	5 771	919	200	6 890	7 073	3 068	623	85	3 777	4 094
Central and South America	2 087	292	48	2 428	2 530	859	100	26	985	1 117
Europe	1 334	379	117	1 829	1 852	909	292	61	1 263	1 271
Africa	2 014	271	64	2 349	2 411	1 003	157	38	1 198	1 244
Middle East	2 534	568	189	3 291	3 293	1 399	243	64	1 706	1 730
Eurasia	2 623	410	46	3 078	3 171	1 314	279	34	1 626	1 667
Asia Pacific	2 467	914	591	3 972	5 034	1 533	696	230	2 458	3 027
Shipping	n.a.	500	n.a.	500	566	n.a.	213	n.a.	213	238
World	18 830	4 253	1 254	24 337	25 930	10 085	2 604	538	13 227	14 388

Power	Current Policies Scenario					Sustainable Development Scenario				
	Fossil fuels	Nuclear	Renew.	Networks	Total	Fossil fuels	Nuclear	Renew.	Networks	Total
Cumulative investments, 2019-2040 (billion dollars, 2018)										
North America	364	154	1 065	934	2 576	281	204	1 952	1 203	3 732
Central and South America	77	29	429	489	1 029	23	34	513	376	954
Europe	335	353	1 232	1 081	3 050	184	407	2 114	1 748	4 519
Africa	204	23	331	779	1 359	103	45	1 050	932	2 168
Middle East	189	34	176	321	725	80	48	719	284	1 136
Eurasia	237	112	98	269	719	77	118	396	192	786
Asia Pacific	1 380	404	3 109	4 616	9 709	697	565	6 158	5 708	13 502
World	2 786	1 110	6 440	8 488	19 167	1 446	1 420	12 902	10 444	26 797

Annex B

Design of the scenarios

The *World Energy Outlook-2019* (*WEO-2019*) presents projections for different core scenarios that are differentiated primarily by their underlying assumptions about the evolution of energy-related government policies.

The **Stated Policies Scenario** (identical in design to the previous New Policies Scenario) provides a detailed sense of the direction in which today's policy ambitions would take the energy sector. The change in name to "Stated" from "New" is intended to clarify that this scenario does not speculate on how policies might evolve in the future. It incorporates policies and measures that governments around the world have already put in place, as well as the effects of announced policies, as expressed in official targets and plans.

Given that intended policies are typically not fully reflected in legislation or regulation, the prospects and timing for their realisation are based upon our assessment of relevant regulatory, market, infrastructure and financial constraints. Where policies are time-limited, they are generally assumed to be replaced by measures of similar intensity, but we do not assume future strengthening – or weakening – of future policy action, except where there already is specific evidence which bears on this. As the name indicates, this is a scenario and not a forecast. Our intention is to inform decision makers as they consider options, not to predict the outcomes of their deliberations.

The **Sustainable Development Scenario** (SDS) starts with the outcomes to be achieved and then assesses what combination of actions would deliver them. Its approach is different from that of the other scenarios, which define the starting conditions and then see where they lead. The outcomes embodied in the Sustainable Development Scenario are derived from the Sustainable Development Goals (SDGs) of the United Nations, providing an energy sector pathway that achieves: universal access to affordable, reliable and modern energy services by 2030 (SDG 7.1); a substantial reduction in air pollution (SDG 3.9); and effective action to combat climate change (SDG 13). The Sustainable Development Scenario is fully aligned with the Paris Agreement and lays out an integrated strategy to achieve climate, air quality and access objectives while also having a strong accent on energy security.

The **Current Policies Scenario** provides a baseline for the analysis by considering only the consequences of existing laws and regulation. It excludes the effects of stated ambitions and targets that have not yet been translated into operational laws and regulations. Comparisons between this scenario and the Stated Policies Scenario underline that achieving stated ambitions and targets should not be taken for granted, especially in countries and sectors where existing laws and regulations are already quite stringent.

This annex presents some framework elements of the scenarios, including population, economic growth and fossil fuel resources, which are held constant across the scenarios, and prices for fossil fuels and carbon dioxide (CO_2) emissions, which are not.

B.1 Population

Table B.1 ▷ Population assumptions by region

	Compound average annual growth rate			Population (million)		Urbanisation share	
	2000-18	2018-30	2018-40	2018	2040	2018	2040
North America	0.9%	0.7%	0.6%	490	559	82%	87%
United States	0.8%	0.6%	0.5%	328	368	82%	87%
Central and South America	1.1%	0.8%	0.6%	520	598	81%	86%
Brazil	1.0%	0.6%	0.4%	211	231	87%	91%
Europe	0.3%	0.1%	0.0%	692	695	75%	81%
European Union	0.3%	0.0%	-0.0%	513	508	76%	82%
Africa	2.6%	2.4%	2.2%	1 287	2 095	43%	54%
South Africa	1.3%	1.1%	1.0%	57	71	66%	76%
Middle East	2.2%	1.6%	1.4%	241	324	72%	78%
Eurasia	0.4%	0.4%	0.3%	234	249	65%	70%
Russia	-0.1%	-0.1%	-0.2%	145	138	74%	80%
Asia Pacific	1.0%	0.7%	0.5%	4 138	4 652	48%	60%
China	0.5%	0.2%	0.1%	1 400	1 422	59%	77%
India	1.4%	0.9%	0.7%	1 353	1 593	34%	46%
Japan	-0.0%	-0.4%	-0.5%	126	113	92%	94%
Southeast Asia	1.2%	0.9%	0.7%	654	768	49%	61%
World	1.2%	1.0%	0.9%	7 602	9 172	55%	64%

Note: See Annex C for definitions.
Sources: UN Population Division databases; IEA databases and analysis.

- As in previous editions of the *WEO*, we use the medium variant of the United Nations projections as the basis for our projections. In this variant, global population growth slows over the coming decades, but the total population nonetheless rises from 7.6 billion today to around 9.2 billion in 2040, an increase of 1.6 billion people.

- Around half of the increase in the global population to 2040 is in Africa, underlining the importance of this continent to the achievement of the world's sustainable development goals (see special focus on Africa in Part B). India accounts for 15% of the growth and becomes the world's most populous country in the near term as China's population growth stalls.

- The share of the global population living in cities and towns is assumed to rise to 64% in 2040 from 55% today. The addition of 78 million people on average each year to the urban population, predominantly in developing economies, means that urban public policies, design and infrastructure choices become crucial variables in the future of global energy. The coastal location of many of the world's largest cities also puts them in the front line when it comes to the impacts of a changing climate.

B.2 Economic growth

Table B.2 ▷ Real gross domestic product (GDP) growth assumptions by region

	Compound average annual growth rate			
	2000-18	2018-30	2030-40	2018-40
North America	2.0%	2.0%	2.1%	2.0%
United States	1.9%	1.9%	2.0%	2.0%
Central and South America	2.6%	2.7%	3.0%	2.9%
Brazil	2.3%	2.5%	3.1%	2.8%
Europe	1.8%	1.7%	1.5%	1.6%
European Union	1.6%	1.6%	1.4%	1.5%
Africa	4.3%	4.2%	4.3%	4.3%
South Africa	2.7%	2.1%	2.9%	2.5%
Middle East	3.9%	2.9%	3.6%	3.2%
Eurasia	4.0%	2.4%	2.3%	2.3%
Russia	3.4%	1.8%	1.9%	1.8%
Asia Pacific	6.0%	5.0%	3.7%	4.4%
China	8.9%	5.2%	3.3%	4.3%
India	7.3%	7.3%	5.2%	6.4%
Japan	0.8%	0.7%	0.7%	0.7%
Southeast Asia	5.2%	4.9%	3.8%	4.4%
World	**3.7%**	**3.6%**	**3.1%**	**3.4%**

Note: Calculated based on GDP expressed in year-2018 dollars in purchasing power parity terms.

Sources: IMF (2019); World Bank databases; IEA databases and analysis.

■ As in *WEO-2018*, the global economy is assumed to grow at an average rate of 3.4% to 2040, although there have been some adjustments to individual countries and regions. A key revision comes in the Middle East, where lower near-term growth is based on the more downbeat forecasts from the International Monetary Fund (IMF). As noted in a *WEO-2018* special report, some traditional oil and gas producing regions are struggling with weak fiscal and external balances caused by lower commodity prices.

■ Some other countries and regions have slightly lower near-term growth trajectories in this *Outlook* relative to the *WEO-2018*, including the United States, China, India, Southeast Asia, Russia and Europe. This reflects uncertainties over the impact of trade tensions, as well as potential financial vulnerabilities from large public and private sector indebtedness. The World Bank (2019) has also pointed to sluggish investment levels in many developing economies despite strong needs, a concern echoed in our *World Energy Investment* analysis for the energy sector (IEA, 2019).

■ The way that economic growth plays through into energy demand depends heavily on the structure of any given economy, the balance between different types of industry and services, and on policies in areas such as pricing and energy efficiency.

B.3 Fossil fuel resources

Table B.3 ▷ Remaining technically recoverable fossil fuel resources, end-2018

Oil (billion barrels)	Proven reserves	Resources	Conventional crude oil	Tight oil	NGLs	EHOB	Kerogen oil
North America	240	2 364	244	177	141	802	1 000
Central and South America	288	852	246	60	50	494	3
Europe	15	116	60	19	29	3	6
Africa	125	452	310	54	86	2	-
Middle East	836	1 138	913	29	152	14	30
Eurasia	145	956	241	85	60	552	18
Asia Pacific	52	287	129	72	67	3	16
World	1 700	6 165	2 142	496	585	1 870	1 073

Natural gas (trillion cubic metres)	Proven reserves	Resources	Conventional gas	Tight gas	Shale gas	Coalbed methane
North America	15	141	50	10	74	7
Central and South America	8	84	28	15	41	-
Europe	5	47	19	5	18	5
Africa	19	101	51	10	40	0
Middle East	81	122	102	9	11	-
Eurasia	76	170	133	10	10	17
Asia Pacific	20	138	44	21	53	21
World	225	803	426	80	247	50

Coal (billion tonnes)	Proven reserves	Resources	Coking coal	Steam coal	Lignite
North America	258	8 390	1 032	5 839	1 519
Central and South America	14	61	3	32	25
Europe	135	977	188	388	402
Africa	13	297	35	262	0
Middle East	1	41	19	23	-
Eurasia	189	4 302	731	2 191	1 380
Asia Pacific	433	8 947	1 506	6 026	1 414
World	1 043	23 014	3 514	14 760	4 740

Notes: NGLs = natural gas liquids; EHOB = extra-heavy oil and bitumen. The breakdown of coal resources by type is an IEA estimate. Coal world resources exclude Antarctica.

Sources: BGR (2018); BP (2019); Cedigaz (2019); OGJ (2018); US DOE/EIA (2018, 2019); US DOE/EIA/ARI (2013, 2015); USGS (2012a, 2012b); IEA databases and analysis.

- The *WEO* supply modelling relies on estimates of the remaining technically recoverable resource, rather than the (often more widely quoted) numbers for proven reserves. Resource estimates are inevitably subject to a considerable degree of uncertainty.

- We distinguish in the analysis between conventional and unconventional resource types, but the distinction between the two, in practice, is an inexact and somewhat artificial one (and what is considered unconventional today may be considered conventional tomorrow).

- Remaining recoverable resources of conventional oil and gas are largely unchanged from last year's *World Energy Outlook*. The main adjustment to our estimates of remaining technically recoverable oil resources comes in the numbers for US tight oil. Total US tight crude and condensate resources in the *WEO-2019* amount to 155 billion barrels, a 35% increase from the 115 billion barrels included in the *WEO-2018*. The main revision is for tight oil resources in the Permian basin following new assessments from the United States Geological Survey (USGS). Its latest assessment of the Permian Delaware basin leads to a 50 billion barrel estimate of remaining technically recoverable crude oil and condensate resources; the *WEO-2018* included 22 billion barrels. Resources in the Permian Midland basin have also been revised up by around 33%, and there have been some upward revisions in smaller plays in Oklahoma and North Dakota.

- Remaining US shale gas resources in the *WEO-2019* are also higher than last year at 43 trillion cubic metres (tcm), a 25% increase from the level in the *WEO-2018*. The largest change is for the Haynesville/Bossier shale, for which the latest USGS assessment leads to a remaining technically recoverable resource estimate of 6.8; the *WEO-2018* included an estimate of 2.9 tcm. This revision does not have a large impact on the production profile in the Stated Policies Scenario, as the Haynesville shale is relatively expensive and it is assumed that the play is not developed in earnest until towards the end of the *Outlook* period.

- The remaining technical recoverable resources of fossil fuels are comfortably sufficient to meet the projections of global demand growth to 2040 in all scenarios.

- Overall, the gradual depletion of resources (at a pace that varies by scenario) means that operators have to develop more difficult and complex reservoirs. This tends to push up production costs over time, although this effect is offset by the assumed continuous adoption of new, more efficient production technologies and practices.

- Remaining technically recoverable coal resources are huge and more widely distributed than those of oil and gas. This means that, although environmental concerns are widespread, the availability of coal supply typically is not an issue.

- World coal resources are made up of various types of coal: around 80% is steam and coking coal and the remainder is lignite.

B.4 Fossil fuel prices

Table B.4 ▷ Fossil fuel prices by scenario

Real terms ($2018)	2000	2010	2018	Stated Policies 2025	Stated Policies 2030	Stated Policies 2035	Stated Policies 2040	Sustainable Development 2030	Sustainable Development 2040	Current Policies 2030	Current Policies 2040
IEA crude oil ($/barrel)	40	90	68	81	88	96	103	62	59	111	134
Natural gas ($/MBtu)											
United States	6.1	5.0	3.2	3.2	3.3	3.8	4.4	3.2	3.4	3.8	5.1
European Union	4.0	8.6	7.6	8.0	8.0	8.4	8.9	7.5	7.5	8.9	9.9
China	3.5	7.7	8.2	9.1	9.0	9.3	9.8	8.6	8.7	9.8	10.7
Japan	6.7	12.7	10.1	10.0	9.7	9.8	10.2	8.8	8.7	11.0	11.4
Steam coal ($/tonne)											
United States	34	58	46	51	52	53	54	49	48	59	63
European Union	48	106	92	75	76	78	78	58	60	83	90
Japan	43	123	111	83	86	88	90	65	69	94	103
Coastal China	34	133	106	88	89	91	92	74	76	98	105

Notes: MBtu = million British thermal units. The IEA crude oil price is a weighted average import price among IEA member countries. Natural gas prices are weighted averages expressed on a gross calorific-value basis. The US natural gas price reflects the wholesale price prevailing on the domestic market. The European Union and China gas prices reflect a balance of pipeline and liquefied natural gas (LNG) imports, while the Japan gas price is solely LNG imports; the LNG prices used are those at the customs border, prior to regasification. Steam coal prices are weighted averages adjusted to 6 000 kilocalories per kilogramme. The US steam coal price reflects mine-mouth prices (primarily in the Powder River Basin, Illinois Basin, Northern Appalachia and Central Appalachia markets) plus transport and handling cost. Coastal China steam coal price reflects a balance of imports and domestic sales, while the European Union and Japanese steam coal price is solely for imports.

- The oil price in the Stated Policies Scenario is lower by around 10% in 2040 than in the *WEO-2018* New Policies Scenario. This is mainly due to the upward revision in estimated tight oil resources in the United States, which allows production to remain "higher for longer" and the market to find equilibrium in a lower range.

- The oil price follows a smooth trajectory to 2040. We do not try to anticipate any of the fluctuations that characterise commodity markets in practice, although near-term demand for oil remains robust in the Stated Policies Scenario.

- The risk of a price spike would be considerably reduced if oil demand were to follow the lower pathway of the Sustainable Development Scenario. In this scenario, tight oil production limits the need to develop higher cost oil and the market finds a balance at a much lower price. The risk of market volatility in this scenario remains significant, however, not least because of the strains that this scenario implies for many large producer countries in the light of their continuing dependence on hydrocarbon revenues.

Figure B.1 ▷ Average IEA crude oil price by scenario

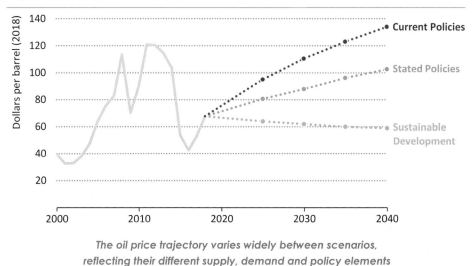

The oil price trajectory varies widely between scenarios, reflecting their different supply, demand and policy elements

- Fiscal pressures complicate this task, but nonetheless we assume, in all scenarios, that major producers maintain a strategy of market management. This means that the marginal project required to meet demand is more expensive than would be implied only by the global supply cost curve.

- Natural gas prices in the Stated Policies Scenario are also lower than in last year's edition. A downward revision in the Henry Hub price in the United States is related to ample availability of associated gas and to the implications of a higher gas resource estimate.

- The US Henry Hub price also serves as a global reference price due to a large LNG export industry actively seeking arbitrage opportunities, and this brings down prices in major importing regions as well.

- Our projections assume movement towards a more integrated global gas market, in which internationally traded gas moves in response to price signals determined by the balance between supply and demand in each region.

- The apparent oversupply in coal markets in 2019, following the high prices in 2017-18, stems from a confluence of strong supply from exporters, and policies and market forces holding down import demand in some key regions. In the Stated Policies Scenario, coal prices continue to decrease slightly from current levels until the mid-2020s as markets rebalance.

- Long-term fundamentals dictate a modest coal price increase from the mid-2020s in the Stated Policies Scenario, reflecting upward cost pressure caused by worsening geological conditions, declining coal quality in mature mining regions and the need to tap more remote coal deposits.

Table B.5 ▷ CO₂ prices in selected regions by scenario ($2018 per tonne)

Region	Sector	2030	2040
Current Policies			
Canada	Power, industry, aviation, others*	36	39
Chile	Power	5	5
China	Power, industry, aviation	20	31
European Union	Power, industry, aviation	27	38
Korea	Power, industry	28	39
Stated Policies			
Canada	Power, industry, aviation, others*	36	39
Chile	Power	12	20
China	Power, industry, aviation	23	36
European Union	Power, industry, aviation	33	43
Korea	Power, industry	33	44
South Africa	Power, industry	15	24
Sustainable Development			
Advanced economies	Power, industry, aviation**	100	140
Selected developing economies	Power, industry, aviation**	75	125

* In Canada's benchmark/backstop policies, a carbon price is applied to fuel consumed in additional sectors.
** Coverage of aviation is limited to the same regions as in the Stated Policies Scenario.

- National carbon pricing schemes are in place or planned in thirty countries around the world and this is reflected in our projections. Once China's national Emissions Trading Scheme is in place from 2020, the share of global emissions covered by carbon prices will rise to around 13% (from 7% today).

- The price of allowances in the European Union Emissions Trading Scheme rose steadily throughout 2018, averaging just under $20/tonne, and it has continued to rise so far in 2019. Future levels are uncertain, not least because the announced plans of Germany to end its use of coal-fired power plants by 2038 could lead to a large surplus of allowances unless the emissions cap is reduced by a commensurate level.

- South Africa introduced a CO_2 tax of $8.5/tonne in June 2019, although there are some tax breaks in the first phase (covering the period to 2022) that lower the effective tax rate to around $0.5-3.5/tonne.

- In the Sustainable Development Scenario, a higher and broader CO_2 price is assumed, rising to $140/tonne in 2040 in advanced economies and to $125/tonne in selected developing economies such as Brazil, China, Russia and South Africa.

- There is an interplay between the CO_2 prices assumed and a variety of other policy measures such as vehicle and building efficiency standards, renewable energy targets and support for new technology development. Further details of these policies and measures are provided in Tables B.7 – B.11.

B.5 Power generation technology costs

Table B.6 ▷ Technology costs by selected region in the Stated Policies Scenario

		Capital costs ($/kW)		Capacity factor (%)		Fuel and O&M ($/MWh)		LCOE ($/MWh)		VALCOE ($/MWh)	
		2018	2040	2018	2040	2018	2040	2018	2040	2018	2040
United States	Nuclear	5 000	4 500	90	90	30	30	105	100	105	100
	Coal	2 100	2 100	60	60	30	30	75	75	75	75
	Gas CCGT	1 000	1 000	50	50	30	35	50	60	45	60
	Solar PV	1 550	830	21	23	15	10	95	50	95	60
	Wind onshore	1 660	1 500	42	44	10	10	55	50	55	55
	Wind offshore	4 300	2 060	41	48	35	20	155	70	150	75
European Union	Nuclear	6 600	4 500	75	75	35	35	150	110	145	115
	Coal	2 000	2 000	40	40	45	45	120	145	105	125
	Gas CCGT	1 000	1 000	40	40	60	70	90	115	75	85
	Solar PV	1 090	610	13	14	15	10	110	65	105	90
	Wind onshore	1 950	1 760	28	30	20	15	95	85	95	90
	Wind offshore	4 920	2 580	49	59	20	10	140	65	135	75
China	Nuclear	2 500	2 500	75	75	25	25	65	65	65	65
	Coal	800	800	70	70	35	30	50	70	50	65
	Gas CCGT	560	560	50	50	75	85	90	110	85	100
	Solar PV	880	490	17	19	10	5	60	35	60	60
	Wind onshore	1 180	1 160	25	27	15	10	60	55	65	60
	Wind offshore	2 780	1 460	32	44	25	10	120	45	120	50
India	Nuclear	2 800	2 800	80	80	30	30	70	70	70	70
	Coal	1 200	1 200	60	60	35	35	60	55	60	50
	Gas CCGT	700	700	50	50	80	80	95	95	90	80
	Solar PV	790	430	20	21	10	5	45	30	50	50
	Wind onshore	1 200	1 160	26	29	15	10	60	50	65	55
	Wind offshore	3 400	1 720	29	38	25	15	190	65	140	70

Note: O&M = operation and maintenance; LCOE = levelised cost of electricity; VALCOE = value-adjusted LCOE; kW = kilowatt; MWh = megawatt-hour; CCGT = combined-cycle gas turbine. LCOE and VALCOEs figures are rounded. Lower figures for VALCOE indicate improved competitiveness.

Sources: IEA analysis; IRENA Renewable Costing Alliance; IRENA (2019).

- Major contributors to the LCOE include: overnight capital costs; capacity factor that describes the average output over the year relative to the maximum rated capacity (typical values provided); the cost of fuel inputs; plus operation and maintenance. Economic lifetime assumptions are 25 years for solar PV, onshore and offshore wind.

- For all technologies, a standard weighted average cost of capital was assumed (7-8% based on the stage of economic development, in real terms).

- The value-adjusted LCOE (or "VALCOE") incorporates information about both costs and the value provided to the system. Based on the LCOE, estimates of energy, capacity and flexibility value are incorporated to provide a metric of competitiveness for power generation technologies (see *WEO-2019* section 6.8). This metric provides a more robust approach to compare dispatchable technologies and variable renewables.

- Additional power generation cost information is provided online at iea.org/weo/.

B.6 Policies

The policy actions assumed to be taken by governments are a key variable in this *Outlook* and the main reason for the differences in outcomes across the scenarios. An overview of the policies and measures that are considered in the various scenarios is included in the Tables B.7 – B.11.

The policies are additive: measures listed under the Sustainable Development Scenario (SDS) supplement those in the Stated Policies Scenario (STEPS), which in turn supplement policies in the Current Policies Scenario (CPS). The tables begin with broad cross-cutting policy frameworks, followed by more detailed policies by sector: power, transport, industry and buildings. The "new policies" that are considered in the STEPS are derived from an exhaustive examination of announcements and plans in countries around the world.

Table B.7 ▷ Cross-cutting policy assumptions by scenario for selected regions

	Scenario	Assumptions
All regions	SDS	• Universal access to electricity and clean cooking facilities by 2030. • Staggered introduction of CO_2 prices (see Table B.5). • Fossil fuel subsidies phased out by 2025 in net-importing countries and by 2035 in net-exporting countries. • Maximum sulfur content of oil products capped at 1% for heavy fuel oil, 0.1% for gasoil and 10 ppm for gasoline and diesel. • Policies promoting production and use of alternative fuels and technologies such as hydrogen, biogas, biomethane and CCUS across sectors.
United States	CPS	• Extension and increase of "45Q" tax credits for carbon capture, utilisation and storage: rising to \$35/t CO_2 in 2026 for enhanced-oil or gas recovery, and to \$50/t CO_2 sequestered in saline geological formations. • State-level renewable portfolio standards. • Regional Greenhouse Gas Initiative: mandatory cap-and-trade scheme covering fossil fuel power plants in nine northeast states, and economy-wide cap-and-trade scheme in California with binding commitments.

European Union	STEPS	• NDC targets and 2030 Climate and Energy Framework:
		   ○ Reduce GHG emissions at least 40% below 1990 levels.
		   ○ Increase share of renewables to at least 32%.
		   ○ Partial implementation of goal to save 32.5% of energy use compared with business-as-usual scenarios.
		• Draft National Energy and Climate Plans (NECP) submitted in June 2019 in support of 2030 Climate and Energy Framework.
		• ETS reducing GHG emissions 43% below the 2005 level in 2030.
		• National Emission Ceilings Directive to reduce emissions of SO_2 by 79%, NO_x by 63%, $PM_{2.5}$ by 49%, NMVOC by 40% and NH_3 by 19% below 2005 levels by 2030.
		• Increase share of renewables in heating and cooling by 1% per year to 2030.

Japan	STEPS	• NDC targets: economy-wide target of reducing GHG emissions by 26% below fiscal year 2013 levels by fiscal year 2030; sector-specific targets. • The 5th Strategic Energy Plan under the Basic Act on Energy Policy.

Table B.7 ▷ Cross-cutting policy assumptions by scenario for selected regions (continued)

	Scenario	Assumptions
China	CPS	• Action Plan for Prevention and Control of Air Pollution. • ETS for the power sector.

	STEPS	• NDC GHG targets: achieve peak CO_2 emissions around 2030, with best efforts to peak early; lower CO_2 emissions per unit of GDP 60-65% below 2005 levels by 2030.
		• NDC energy target: increase the share of non-fossil fuels in primary energy consumption to 20% by 2030.
		• 13th Five-Year Plan targets for 2020:
		   o Services sector value to be increased to 56%.
		   o Non-fossil fuels to reach 15% of TPED.
		   o Energy intensity per unit of GDP limited to 15% below 2015 levels.
		   o Carbon emissions per unit of GDP limited to 18% below 2015 levels.
		   o SO_2 and NO_X emissions reduced by 15%.
		• "Made in China 2025" transition from heavy industry to higher value-added manufacturing.
		• Expand the role of natural gas.
		• ETS expansion to domestic aviation and selected industry sectors.
		• Energy price reform, including more frequent adjustments in oil product prices and reduction in natural gas price for non-residential consumers.
		• Three-year action plan for cleaner air, announced in July 2018.

India	CPS	• National Mission on Enhanced Energy Efficiency. • National Clean Energy Fund to promote clean energy technologies based on a levy of INR 400 ($6) per tonne of coal. • "Make in India" campaign to increase the share of manufacturing in the national economy.
	STEPS	• NDC GHG target: reduce emissions intensity of GDP 33-35% below 2005 levels by 2030. • NDC energy target: achieve about 40% cumulative installed capacity from non-fossil fuel sources by 2030 with the help of technology transfer and low-cost international finance. • Efforts to expedite environmental approval and land acquisition for energy projects. • Opening of coal, gas and oil sectors to private and foreign investors.

Brazil	STEPS	• NDC GHG economy-wide targets: reduce GHG emissions 37% below 2005 levels by 2025.
		• NDC energy goals for 2030:
		   o Increase share of sustainable biofuels to around 18% of TPED.
		   o Increase renewables to 45% of TPED.
		   o Increase non-hydro renewables to 28-30% of TPED and 23% of power supply.
		• Partial implementation of National Energy Efficiency Plan.

Notes: CCUS = carbon capture, utilisation and storage; NDC = Nationally Determined Contributions; GHG = greenhouse gases; LPG = liquefied petroleum gas; SO_2 = sulfur dioxide; NO_X = nitrogen oxides; $PM_{2.5}$ = fine particulate matter; NMVOC = non-methane volatile organic compounds; NH_3 = ammonia; TPED = total primary energy demand; ETS = emissions trading system. Pricing of CO_2 emissions is by emissions trading systems or taxes.

The policies and measures for the various scenarios pertaining to the regions of Africa and Southeast Asia can be found within the respective WEO 2019 special reports on Africa Energy Outlook and Southeast Asia Energy Outlook.

Table B.8 ▷ **Power sector policies and measures as modelled by scenario for selected regions**

	Scenario	Assumptions
All regions	SDS	• Increased low-carbon generation from renewables and nuclear. • Expanded support for the deployment of CCUS. • Efficiency and emissions standards preventing the refurbishment of old inefficient plants. • Stringent pollution emissions limits for industrial facilities above 50 MW$_{th}$ input using solid fuels, set at 200 mg/m^3 for SO$_2$ and NO$_X$ and 30 mg/m^3 for PM$_{2.5}$.
United States	CPS	• Extension of Investment Tax Credit and Production Tax Credit. • State renewable portfolio standards. • State 100% clean energy target by 2050. • Mercury and Air Toxics Standards. • New Source Performance Standards. • Clean Air Interstate Rule regulating SO$_2$ and NO$_X$. • Lifetimes of some nuclear plants extended beyond 60 years.
	STEPS	• Extension and strengthening of support for renewables, nuclear and CCUS. • Affordable Clean Energy Rules.
Canada	CPS	• Emissions performance standard of 420 g CO$_2$ per kWh for new coal-fired electricity generation units, and units that have reached the end of their useful life. • New Brunswick and Alberta phase out unabated coal-fired power by 2030. • Introduction of country-wide carbon pricing in 2019.
	STEPS	• Complete phase out of traditional coal-fired power in line with the Pan-Canadian Framework on Clean Growth and Climate Change. • Emissions performance standard for natural gas-fired electricity generation.
European Union	CPS	• ETS in accordance with 2020 Climate and Energy Package. • No new coal power plants post-2020 in 26 of 28 member states. • Early retirement of all nuclear plants in Germany by end-2022. • Removal of some barriers to CHP plants. • Support for renewables in accordance with overall target. • Industrial Emissions Directive.
	STEPS	• ETS in accordance with 2030 Climate and Energy Framework. • Coal phase out in a subset of member states, notably in Finland, France, Germany, Italy, the Netherlands and United Kingdom. • Extended and strengthened support to renewables-based power generation technologies in accordance with overall target taking into account draft National Energy and Climate Plans (NECP). • Support the increased use of biogas and biomethane in the power mix. • Further removal of barriers to CHP through partial implementation of the Energy Efficiency Directive. • Power market reforms to enable recovery of investments for adequacy. • New standards for Large Combustion Plants from the review of the Best Available Techniques Reference Document.
Korea	STEPS	• Third Master Energy Plan calls for 35%-40% renewables by 2040.

Table B.8 ▷ **Power sector policies and measures as modelled by scenario for selected regions** (continued)

	Scenario	Assumptions
Japan	CPS	• Air Pollution Control Law. • Retail power market liberalisation. • Support for renewables-based power generation.
	STEPS	• Power mix targets by 2030 from the 5th Strategic Energy Plan. • Lifetime of some nuclear plants beyond typical lifetime of 40 years. • Non-fossil fuels to supply 44% of power generation by 2030, corresponding to carbon intensity of 370 g CO_2/kWh. • Implementation of the feed-in tariff amendment law. • Efficiency standards for new thermal power plants (coal: 42%; gas: 50.5%; oil: 39%).
China	CPS	• Air pollutant emissions standard for thermal power plants with limits on $PM_{2.5}$: 30 mg/m³; SO_2: 100-200 mg/m³ for new plants and 200-400 mg/m³ for existing plants; NO_x: 100-200 mg/m³. • ETS for the power sector.

	STEPS	• 13th Five-Year Plan targets for 2020:
		  ○ 58 GW nuclear, 380 GW hydro, at least 210 GW wind and at least 110 GW solar.
		  ○ Retrofit of 133 GW of CHP and 86 GW of condensing coal plants in order to increase flexibility.
		  ○ Coal limited to 1 100 GW, by delaying 150 GW of new builds and retiring 20 GW of existing plants.

India	CPS	• Connect all willing households to electricity under the Pradhan Mantri Sahaj Bijli Har Ghar Yojana (Saubhagya) scheme. • Renewable Purchase Obligation and other fiscal measures to promote renewables. • Increased use of supercritical coal technology. • Restructured Accelerated Power Development and Reform Programme to finance the modernisation of transmission and distribution networks. • Pollution control rules limiting emissions from coal power plants.
	STEPS	• Environmental (Protection) Amendment Rules. • Strengthened measures such as competitive bidding to increase the use of renewables towards the national target of 175 GW of non-hydro renewables capacity by 2022 (100 GW solar, 75 GW non-solar) and 450 GW non-hydro renewables capacity target by 2050. • Expanded efforts to strengthen the national grid, upgrade the transmission and distribution network, and reduce aggregate technical and commercial losses to 15%. • Increased efforts to establish the financial viability of all power market participants, especially network and distribution companies.
Brazil	CPS	• Technology-specific power auctions for all fuel types. • Guidance on fuel mix from the Ten-Year Plan for Energy Expansion.
Chile	STEPS	• Coal phase out by 2030.
Middle East	CPS	• Partial implementation of nuclear programmes, including in Saudi Arabia and United Arab Emirates. • Partial implementation of renewable targets and programmes.
	STEPS	• Accelerated progress towards nuclear and renewables targets.

Notes: CCUS = carbon capture, utilisation and storage; MW$_{th}$ = megawatts thermal; CHP = combined heat and power; SO_2 = sulfur dioxide; NO_x = nitrogen oxides; $PM_{2.5}$ = fine particulate matter; g CO_2/kWh = grammes of carbon dioxide per kilowatt-hour; GW = gigawatts; PV = photovoltaic; ETS = emissions trading system.

Table B.9 ▷ **Industry sector policies and measures as modelled by scenario in selected regions**

	Scenario	Assumptions
All regions	SDS	• Stringent emissions limits for industrial facilities above 50 MW$_{th}$ input using solid fuels, set at 200 mg/m^3 for NO$_X$ and SO$_2$ and 30 mg/m^3 for PM$_{2.5}$. • Emission limits for facilities below 50 MW$_{th}$ based on size, fuel and combustion process. • Industrial processing plants to be fitted with the best available technologies in order to obtain operating permits. Existing plants to be retrofitted within ten years. • Enhanced minimum energy performance standards by 2025, in particular for electric motors; incentives for the introduction of variable speed drives in variable load systems, and implementation of system-wide measures. • International agreements on steel and cement industry energy intensity targets. • Mandatory energy management systems or energy audits. • Policies to support increased recycling of aluminium, steel, paper and plastics. • Policies to support increasing deployment of CCUS in various industry and fuel transformation subsectors. • Wider hosting of international projects to offset CO$_2$ emissions.
United States	CPS	• Better Buildings, Better Plants Program and Energy Star Program for Industry. • Boiler Maximum Achievable Control Technology to impose stricter emissions limits on industrial and commercial boilers, and process heaters. • Superior Energy Performance certification that supports the introduction of energy management systems. • Industrial Assessment Centers providing no-cost energy assessments to SMEs. • Permit program for GHGs and other air pollutants for large industrial installations. • Business Energy Investment Tax Credit and funding for efficient technologies.
	STEPS	• Further assistance for SME manufacturers to adopt "smart manufacturing technologies" through technical assistance and grant programs.
European Union	CPS	• ETS in accordance with 2020 Climate and Energy Package. • White certificate scheme in Italy and energy saving obligation scheme in Denmark. • Voluntary energy efficiency agreements in Belgium, Denmark, Finland, Hungary, Ireland, Luxembourg, Netherlands, Portugal, Sweden and United Kingdom. • EcoDesign Directive standards for motors, pumps, fans, compressors and insulation. • Implementation of Medium Combustion Plant Directive. • Industrial Emissions Directive.

	STEPS	• ETS in accordance with 2030 Climate and Energy Framework.
		• Implementation of Energy Efficiency Directive and extension to 2030:
		  ○ Mandatory and regular energy audits for large enterprises.
		  ○ Incentives for the use of energy management systems.
		  ○ Encouragement for SMEs to undergo energy audits.
		  ○ Technical assistance and targeted information for SMEs.

Japan	CPS	• Energy efficiency benchmarking. • Tax credits for investments in energy efficiency. • Financial incentives for SMEs to invest in energy conserving equipment and facilities. • Free energy audits for SMEs. • Mandatory energy management for large business operators. • Top Runner Programme of minimum energy standards for machinery and equipment.

	STEPS	• Maintenance and strengthening of top-end low-carbon efficiency standards:
		  ○ Higher efficiency CHP systems.
		  ○ Promotion of state-of-the-art technology, faster replacement of ageing equipment.
		  ○ Continuation of voluntary ETS.

Table B.9 ▷ Industry sector policies and measures as modelled by scenario in selected regions (continued)

	Scenario	Assumptions
China	CPS	• "Blue Skies" environmental initiative implies accelerated elimination of outdated steel and aluminium production capacity; winter production cuts across producing regions; intensified capacity control over construction industry; prohibition of establishment of new chemical parks in key regions. • Partial implementation of Industrial Energy Performance Standards. • Mandatory adoption of coke dry-quenching and top-pressure turbines in new iron and steel plants. Support of non-blast furnace in iron production. • Mechanism to incentivise energy-efficient "leaders", i.e. manufacturers and brands that exceed specific benchmarks set by the China Energy Label. • Pilot of China's ETS for some provinces and industrial sectors. • Continuation of industrial energy intensity reduction contributing to the 13th Five-Year Plan target (2016-20).
	STEPS	• Accelerated retrofit of older coal-fired industrial boilers. • Expansion of ETS to selected industry sectors. • "Made in China 2025" targets for industrial energy intensity. • Full implementation of Industrial Energy Performance Standards. • Enhanced use of energy service companies and energy performance contracting. • Clean Winter Heating Plan promoting the use of natural gas.

India	CPS	• Energy Conservation Act:
		   ○ Mandatory energy audits.
		   ○ Appointment of an energy manager in seven energy-intensive industries.
		• National Mission on Enhanced Energy Efficiency (NMEEE):
		   ○ Cycle II and III of Perform, Achieve and Trade (PAT) scheme, which benchmarks facilities' performance against best practice and enables trading of energy savings certificates.
		   ○ Income and corporate tax incentives for energy service companies, including the Energy Efficiency Financing Platform.
		   ○ Framework for Energy-Efficient Economic Development offering a risk guarantee for performance contracts and a venture capital fund for energy efficiency.
		• Energy efficiency intervention in selected SME clusters including capacity building.

	STEPS	• Further implementation of the NMEEE's recommendations including:
		   ○ Tightening of the PAT mechanism under Cycle III and continuation beyond 2020.
		   ○ Further strengthening of fiscal instruments to promote energy efficiency.
		• Strengthen existing policies to realise the energy efficiency potential in SMEs.
		• Implementation of 'New Industrial Policy' leading to a boost in domestic industrial production. 'Make in India' policy promotes manufacturing sector.
		• National steel policy target of 300 Mt annual production is achieved in the early 2030s.
		• Continuation of subsidy program to fertilizers.

Brazil	CPS	• PROCEL (National Programme for Energy Conservation). • PROESCO (Support for Energy Efficiency Projects). • Partial implementation of the National Energy Efficiency Plan, with fiscal and tax incentives for industrial upgrading, investment in training efficiency and encouragement to reuse industrial waste. • Incentives to increase biomass use in industry.
	STEPS	• Extension of PROESCO.

Notes: CCUS = carbon capture, utilisation and storage; MW$_{th}$ = megawatts thermal; mg/m^3 = milligrams per cubic metre; ETS = emissions trading system; SO$_2$ = sulfur dioxide; NO$_X$ = nitrogen oxides; PM = particulate matter; CHP = combined heat and power; SMEs = small and medium enterprises; GHG = greenhouse gases.

Table B.10 ▷ **Buildings sector policies and measures as modelled by scenario in selected regions**

	Scenario	Assumptions
All regions	SDS	• SDG 7.1: universal access to affordable, reliable and modern energy achieved by 2030. • Phase out least efficient appliances, light bulbs and heating or cooling equipment by 2030 at the latest. • Emissions limits for biomass boilers set at 40-60 mg/m^3 for PM and 200 mg/m^3 for NO$_X$. • Introduction of mandatory energy efficiency labelling requirements for all appliances. • Mandatory energy conservation building codes, including net-zero emissions requirement for all new buildings, by 2030 at the latest. • Increased support for energy efficiency measures, including building retrofits, direct use of solar thermal and geothermal, and heat pumps in certain economies. • Digitalisation of buildings electricity demand to increase demand-side response potential, through greater flexibility and controllability of end use devices.
United States	CPS	• Association of Home Appliance Manufacturers—American Council for an Energy-Efficient Economy Multi-Product Standards Agreement. • Energy Star: new appliance efficiency standards. • Steady upgrades of building codes; incentives for utilities to improve building efficiency. • Weatherisation programmes: funding for refurbishments of residential buildings. • Federal and state rebates for renewables-based heat, including Residential Renewable Energy Tax Credit for solar water heaters, heat pumps and biomass stoves.
	STEPS	• Partial implementation of the Energy Efficiency Improvement Act of 2015. • Mandatory energy efficiency requirements in building codes in some states, including California's 2019 Building Energy Efficiency Standards and recent code updates in other states. • Tightening of efficiency standards for appliances.
European Union	CPS	• Energy Performance of Buildings Directive 2010. • EcoDesign and Energy Labelling Directive including requirements for boilers to have 75-77% efficiency depending on size and to limit pollutant emissions (PM: 40-60 mg/m^3; NO$_X$: 200 mg/m^3 for biomass boilers and 350 mg/m^3 for fossil fuel boilers; CO: 500-700 mg/m^3). • Individual member state financial incentives for renewables-based heat in buildings.
	STEPS	• Partial implementation of the Energy Efficiency Directive. • 2016 update of Energy Performance of Buildings Directive mandating new buildings to be "nearly zero-energy" from 2020, and increased retrofit rates. • Implementation of proposed voluntary Smart Readiness Indicator. • Mandatory labelling for sale or rental of all buildings and some appliances. • Further product groups in EcoDesign Directive. • Enhanced renewables-based heat support in member states. • Ban of gas boilers in new buildings in certain member states.
Japan	CPS	• Building Efficiency Act for new buildings, renovations and extensions. • Top Runner Programme efficiency standards for home appliances. • Large operators to reduce energy consumption 1% per year and complete annual reports. • Energy efficiency standards for new buildings and houses larger than 300 m^2. • Capital Grant Scheme for renewable energy technologies.
	STEPS	• Extension of the Top Runner Programme. • Voluntary equipment labelling programmes. • Building Energy Efficiency Act regulations for new large-scale non-residential buildings and incentives for all new buildings. • Net zero-energy buildings by 2030 for all new construction.

Table B.10 ▷ **Buildings sector policies and measures as modelled by scenario in selected regions** (continued)

	Scenario	Assumptions
China	CPS	• Civil Construction Energy Conservation Design Standards. • Appliance standards and labelling programme. • Natural gas network extended to 57% of urban areas by 2020.

	STEPS	• Promotion of green buildings:
		  ○ New urban residential buildings to increase energy efficiency by 20% from 2015 levels to 2020.
		  ○ 50% of new urban buildings to meet energy conservation requirements.
		• Retrofit of 500 million m² of residential buildings and 100 million m² of public buildings.
		• Promotion of electricity to replace decentralised coal and oil boilers.
		• Urban gasification of 57% by 2020.
		• Solar water heaters to cover 800 million m² by 2020.
		• Mandatory energy efficiency labels for appliances and equipment.
		• Implementation of energy consumption standards for nearly-zero energy buildings.
		• Clean Winter Heating Plan: switch from coal to gas and electricity for 50 000 - 100 000 residences annually in each of the "26+2" main cities in the Beijing-Tianjin-Hebei region and surroundings. Financial support for fuel switching expanded to 43 cities.

	SDS	• Implementation of the draft standard for Building Energy Conservation and Renewable. Energy Utilization, reducing average heating and cooling energy use by 30% in residential buildings and 20% in public buildings, relative to 2016 standards.
India	CPS	• Connect all willing households to electricity under the Pradhan Mantri Sahaj Bijli Har Ghar Yojana (Saubhagya) scheme. • Promotion of clean cooking access with LPG, including free connections to poor rural households through Pradhan Mantri Ujjwala Yojana (PMUY). • Measures under the National Solar Mission. • Energy Conservation Building Code 2007 with voluntary standards for commercial buildings. • "Green Rating for Integrated Habitat Assessment" rating system for green buildings. • Promotion and distribution of LEDs through the Efficient Lighting Programme.
	STEPS	• Standards and Labelling Programme, mandatory for air conditioners, lights, televisions and refrigerators, voluntary for seven other products and LEDs. • Phase out incandescent light bulbs by 2020. • Voluntary Star Ratings for the services sector. • Measures under the National Mission on Enhanced Energy Efficiency. • Energy Conservation in Building Codes made mandatory in eight states that regulate building envelope, lighting and hot water. • Efforts to plan and rationalise urbanisation in line with the "100 smart cities" concept. • Expand PMUY LPG cooking programme to reach 80 million low-income households by 2020.
Brazil	CPS	• Labelling programme for household goods and public buildings equipment.
	STEPS	• Partial implementation of National Energy Efficiency Plan. • Mandatory certification of public lighting; ban on inefficient incandescent bulbs.

Notes: SDG = Sustainable Development Goal; mg/m³ = milligrams per cubic metre; SO_2 = sulfur dioxide; NO_X = nitrogen oxides; CO = carbon monoxide; PM = particulate matter; LED = light-emitting diodes; LPG = liquefied petroleum gas; HVAC = heating, ventilation and air conditioning.

Table B.11 ▷ Transport sector policies and measures as modelled by scenario in selected regions

	Scenario	Assumptions
All regions	CPS	• International shipping: global cap of 0.5% on sulfur content in fuel in 2020 and tightened NO$_X$ emissions standards in control areas by 2025, in line with International Maritime Organisation (IMO) regulation.
	STEPS	• Road transport: fuel sulfur standards of 10-15 ppm. • Aviation: International Civil Aviation Organization goal to improve fuel efficiency by 2% per year until 2020; aiming for carbon-neutral growth from 2020 onwards.
	SDS	• Strong support for electric mobility, alternative fuels and energy efficiency. • Retail fuel prices kept at a level similar to the STEPS, applying CO$_2$ tax across WEM regions. • PLDVs: on-road stock emissions intensity limited to 55 g CO$_2$/km in advanced economies and 70 g CO$_2$/km elsewhere by 2040. • Two/three-wheelers: phase out two-stroke engines. • Light-duty gasoline vehicles: three-way catalysts and tight evaporative controls required. • Light-duty diesel vehicles: limit emissions to 0.1 g/km NO$_X$ and 0.01 g/km PM. • Light commercial vehicles: full technology spill-over from PLDVs. • Medium- and heavy-freight vehicles: 25% more efficient by 2040 than in the STEPS. • Heavy-duty diesel vehicles: limit emissions to 3.5 g/km NO$_X$ and 0.03 g/km PM. • Aviation: fuel intensity reduced by 2.6% per year; scale-up of biofuels driven by long term CO$_2$ emissions target (50% below 2005 levels in 2050). • International shipping: annual GHG emissions trajectory consistent with 50% below 2008 levels in 2050, in line with IMO GHG emissions reduction strategy.
United States	CPS	• Renewables Fuel Standard 2. • LDVs: Phase 2 of CAFE standards until 2020 and Safer Affordable Fuel Efficient rule for model years 2021-2026. • LDVs: Tier 3 Motor Vehicle Emission and Fuel Standards, equivalent to Euro 6. • Medium and heavy-duty trucks: low range of Phase 2 of EPA/NHTSA GHG emissions and fuel efficiency standards. • HDVs: Tier 3 Motor Vehicle Emission and Fuel Standards, equivalent to Euro VI.
	STEPS	• Moderate increase of ethanol and biodiesel use after 2022 driven by state policies. • Electric cars: stock target of 4 million by 2025 across eight states. • Road freight: support for natural gas.
European Union	CPS	• Subsidy supporting biofuels blending, 7% cap on conventional biofuels blending rate. • LDVs: Euro 6 emissions and fuel sulfur standards. • HDVs: Euro VI emissions and fuel sulfur standards. • Domestic aviation: ETS.
	STEPS	• Announcements to phase out gasoline and diesel car sales including Denmark, Ireland, France, Netherlands, Norway, Slovenia, Sweden and United Kingdom. • Renewable energy share in the transport sector of 14% by 2030; as well as a cap on food-based biofuels. • Fuel Quality Directive, reducing GHG intensity of road transport fuels by 6% in 2020. • CO$_2$ targets for PLDVs and commercial LDVs with an intermediate target of 15% below 2021 levels by 2025, new cars will emit on average 37.5% less CO$_2$ and new vans on average 31% less CO$_2$ below 2021 levels by 2030. • CO$_2$ standards applied to subset of HDVs; 15% and 30% lower emissions by 2025 and 2030 respectively, assuming 2019 as a base year. • Buses: Clean vehicles directive requires local authorities to purchase at least a quarter of low/zero-emission buses by 2025 and at least a third by 2030. • Domestic aviation: ETS in accordance with 2030 Climate and Energy Framework.

Table B.11 ▷ **Transport sector policies and measures as modelled by scenario in selected regions** (continued)

	Scenario	Assumptions
Canada	STEPS	• EVs: The federal government aims for certain market shares of zero emission cars by 2040 (10% by 2025, 30% by 2030 and 100% by 2040).
Korea	STEPS	• EVs: Korea targets for 430,000 BEVs and 67,000 FCVs by 2022.
Japan	CPS	• Financial incentives for plug-in hybrid, electric and fuel cell vehicles. • PLDVs: fuel-economy target at 19.4 kilometres per litre (km/L) by 2020. • Post New Long-term Emissions Standards for LDVs and HDVs equivalent to Euro 6 and VI.
	STEPS	• Heavy-duty vehicles: New fuel efficiency standards for trucks and busses enhancing fuel efficiency by 13.4% for trucks and 14.3% for buses by 2025 compared to 2015. • PLDVs: fuel-economy target at 25.4 (km/L) by 2030. • Revitalisation strategy: target sales share of next generation vehicles of 50-70% by 2030. • EVs: stock target of 1 million by 2020, including purchase incentives and infrastructure. • Basic Strategy for Hydrogen: fleet of 800 000 fuel cell vehicles and 1 200 buses by 2030.
China	CPS	• Ethanol and biodiesel blending mandates of 10% and 7% respectively in some provinces. • Promotion of fuel-efficient/ hybrid cars and EVs; consolidation of vehicle charging standards. • PLDVs: cap on sales in some cities to reduce air pollution and traffic. • LDVs: China 6 emissions standards and Euro 6 equivalent fuel sulfur standards. • HDVs: China V (diesel) emissions standards and Euro VI equivalent fuel sulfur standards.

	STEPS	• Subsidies for alternative-fuel vehicles, mainly public buses. Policy scheme for regulating the circulation of oil-fuelled scooters and support for electric scooters.
		• PLDVs:
		  ○ Stock target of 5 million electric cars by 2020, including purchase and use incentives.
		  ○ New Energy Vehicle mandate: credit target of 12% of the car market by 2020.
		  ○ Fuel-economy target at 5 litres per 100 km by 2020, and enforcement of 4 litres per 100 km target by 2025.
		• HDVs: Stage III of National Standard targeting a 15% reduction in fuel consumption compared to 2015 from 2021 onwards.
		• Promotion of public transport in large and medium cities.
		• Targets for roll out of hydrogen refuelling stations and hydrogen vehicles by 2030.

India	CPS	• Increasing blending mandate for ethanol and support for alternative-fuel vehicles. • LDVs: Bharat IV emissions standards and Euro 4 equivalent fuel sulfur standards. • HDVs: Bharat IV emissions standards and Euro IV equivalent fuel sulfur standards.
	STEPS	• Declared intent to move to 30% electric share in vehicle sales by 2030. • Extended support for alternative-fuel two/three-wheelers, cars and public buses. • National Biofuel Policy with indicative blending share targets for bioethanol and biodiesel. • LDVs: Bharat VI emissions standards by 2020; fuel-economy standards at 113 g CO_2/km in 2022. • HDVs: Bharat VI emissions standards by 2020; fuel-economy targets for 2018 and 2021. • Dedicated rail corridors to encourage shift away from road freight. • Phase II of the FAME for promoting electrification of vehicle fleet.

Annex B | Design of the scenarios

Table B.11 ▷ Transport sector policies and measures as modelled by scenario in selected regions (continued)

	Scenario	Assumptions
Brazil	CPS	• Ethanol blending mandates in road transport of minimum 27%. • Biodiesel blending mandate of 9% in 2018 and 10% in 2019. • LDVs: PROCONVE L6 emissions standards, equivalent to Euro 5 but without limit on PM; Euro 2 (gasoline) and Euro 4 (diesel) equivalent fuel sulfur standards. • HDVs: PROCONVE P7 emissions standards, equivalent to Euro V; Euro II (gasoline) and Euro IV (diesel) equivalent fuel sulfur standards.
	STEPS	• RenovaBio: further increase of ethanol and biodiesel blending mandates to cut carbon emissions from fuels sector by 10 % through 2028. • LDVs: Rota 2030 initiative targeting fuel efficiency improvement of 11% by 2022 compared to 2017 levels. • Local renewables-based fuel targets for urban transport. • National urban mobility plan. • Long-term plan for freight transport.

Notes: ppm = parts per million; WEM = World Energy Model; NO_X = nitrogen oxides; g/km = grammes per kilometre; PM = particulate matter; CAFE = Corporate Average Fuel Economy; PLDVs = passenger light-duty vehicles; LDVs = light-duty vehicles; HDVs = heavy-duty vehicles; EVs = electric vehicles; BEVs = battery electric vehicles; FCVs: fuel cell vehicles; GHG = greenhouse gases; g CO_2/km = grammes of carbon dioxide per kilometre; FAME = Faster Adoption & Manufacturing of Electric (and hybrid) vehicles; ETS = emissions trading system; EPA = Environmental Protection Agency; NHTSA = National Highway Traffic Safety Administration; PROCONVE = *Programa de Controle da Poluição do Ar por Veículos Automotores* (Motor Vehicles Air Pollution Control Program).

Annex C

Definitions

This annex provides general information on terminology used throughout *WEO-2019* including: units and general conversion factors; definitions of fuels, processes and sectors; regional and country groupings; and abbreviations and acronyms.

Units

Area	Ha	hectare
	km^2	square kilometre
Coal	Mtce	million tonnes of coal equivalent (equals 0.7 Mtoe)
	Mtpa	million tonnes per annum
	gce	grammes of coal equivalent
Emissions	ppm	parts per million (by volume)
	Gt CO_2-eq	gigatonnes of carbon-dioxide equivalent (using 100-year global warming potentials for different greenhouse gases)
	kg CO_2-eq	kilogrammes of carbon-dioxide equivalent
	g CO_2/km	grammes of carbon dioxide per kilometre
	g CO_2/kWh	grammes of carbon dioxide per kilowatt-hour
Energy	boe	barrel of oil equivalent
	toe	tonne of oil equivalent
	ktoe	thousand tonnes of oil equivalent
	Mtoe	million tonnes of oil equivalent
	MBtu	million British thermal units
	kcal	kilocalorie (1 calorie x 10^3)
	Gcal	gigacalorie (1 calorie x 10^9)
	MJ	megajoule (1 joule x 10^6)
	GJ	gigajoule (1 joule x 10^9)
	TJ	terajoule (1 joule x 10^{12})
	PJ	petajoule (1 joule x 10^{15})
	EJ	exajoule (1 joule x 10^{18})
	kWh	kilowatt-hour
	MWh	megawatt-hour
	GWh	gigawatt-hour
	TWh	terawatt-hour
Gas	mcm	million cubic metres
	bcm	billion cubic metres
	tcm	trillion cubic metres
	scf	standard cubic foot

Mass	kg	kilogramme (1 000 kg = 1 tonne)
	kt	kilotonnes (1 tonne x 10^3)
	Mt	million tonnes (1 tonne x 10^6)
	Gt	gigatonnes (1 tonne x 10^9)
Monetary	$ million	1 US dollar x 10^6
	$ billion	1 US dollar x 10^9
	$ trillion	1 US dollar x 10^{12}
Oil	b/d	barrels per day
	kb/d	thousand barrels per day
	mb/d	million barrels per day
	mboe/d	million barrels of oil equivalent per day
Power	W	watt (1 joule per second)
	kW	kilowatt (1 watt x 10^3)
	MW	megawatt (1 watt x 10^6)
	GW	gigawatt (1 watt x 10^9)
	TW	terawatt (1 watt x 10^{12})
Water	bcm	billion cubic metres
	m^3	cubic metre

General conversion factors for energy

Convert to:	TJ	Gcal	Mtoe	MBtu	GWh
From:	multiply by:				
TJ	1	238.8	2.388×10^{-5}	947.8	0.2778
Gcal	4.1868×10^{-3}	1	10^{-7}	3.968	1.163×10^{-3}
Mtoe	4.1868×10^{4}	10^{7}	1	3.968×10^{7}	11 630
MBtu	1.0551×10^{-3}	0.252	2.52×10^{-8}	1	2.931×10^{-4}
GWh	3.6	860	8.6×10^{-5}	3 412	1

Note: There is no generally accepted definition of boe; typically the conversion factors used vary from 7.15 to 7.40 boe per toe.

Currency conversions

Exchange rates (2018 annual average)	1 US Dollar equals:
British Pound	0.75
Chinese Yuan Renminbi	6.62
Euro	0.85
Indian Rupee	68.39
Indonesian Rupiah	14 236.94
Japanese Yen	110.42
Russian Ruble	62.67
South African Rand	13.24

Source: OECD National Accounts Statistics: purchasing power parities and exchange rates dataset, Sept 2019.

Definitions

Advanced biofuels: Sustainable fuels produced from non-food crop feedstocks, which are capable of delivering significant lifecycle greenhouse gas emissions savings compared with fossil fuel alternatives, and which do not directly compete with food and feed crops for agricultural land or cause adverse sustainability impacts. This definition differs from the one used for "advanced biofuels" in the US legislation, which is based on a minimum 50% lifecycle greenhouse gas reduction and which, therefore, includes sugar cane ethanol.

Agriculture: Includes all energy used on farms, in forestry and for fishing.

Back-up generation capacity: Households and businesses connected to the main power grid may also have some form of "back-up" power generation capacity that can, in the event of disruption, provide electricity. Back-up generators are typically fuelled with diesel or gasoline and capacity can be as little as a few kilowatts. Such capacity is distinct from mini-grid and off-grid systems that are not connected to the main power grid.

Biodiesel: Diesel-equivalent, processed fuel made from the transesterification (a chemical process that converts triglycerides in oils) of vegetable oils and animal fats.

Bioenergy: Energy content in solid, liquid and gaseous products derived from biomass feedstocks and biogas. It includes solid biomass, biofuels and biogas.

Biofuels: Liquid fuels derived from biomass or waste feedstocks and include ethanol and biodiesel. They can be classified as conventional and advanced biofuels according to the technologies used to produce them and their respective maturity. Unless otherwise stated, biofuels are expressed in energy-equivalent volumes of gasoline and diesel.

Biogas: A mixture of methane, CO_2 and small quantities of other gases produced by anaerobic digestion of organic matter in an oxygen-free environment.

Buildings: The buildings sector includes energy used in residential, commercial and institutional buildings, and non-specified other. Building energy use includes space heating and cooling, water heating, lighting, appliances and cooking equipment.

Bunkers: Includes both international marine bunkers and international aviation bunkers.

Capacity credit: Proportion of the capacity that can be reliably expected to generate electricity during times of peak demand in the grid to which it is connected.

Clean cooking facilities: Cooking facilities that are considered safer, more efficient and more environmentally sustainable than the traditional facilities that make use of solid biomass (such as a three-stone fire). This refers primarily to improved solid biomass cookstoves, biogas systems, liquefied petroleum gas stoves, ethanol and solar stoves.

Coal: Includes both primary coal (including lignite, coking and steam coal) and derived fuels (including patent fuel, brown-coal briquettes, coke-oven coke, gas coke, gas-works gas, coke-oven gas, blast-furnace gas and oxygen steel furnace gas). Peat is also included.

Coalbed methane (CBM): Category of unconventional natural gas, which refers to methane found in coal seams.

Coal-to-gas (CTG): Process in which mined coal is first turned into syngas (a mixture of hydrogen and carbon monoxide) and then into "synthetic" methane.

Coal-to-liquids (CTL): Transformation of coal into liquid hydrocarbons. It can be achieved through either coal gasification into syngas (a mixture of hydrogen and carbon monoxide), combined using the Fischer-Tropsch or methanol-to-gasoline synthesis process to produce liquid fuels, or through the less developed direct-coal liquefaction technologies in which coal is directly reacted with hydrogen.

Coking coal: Type of coal that can be used for steel making (as a chemical reductant and source heat), where it produces coke capable of supporting a blast furnace charge. Coal of this quality is also commonly known as metallurgical coal.

Conventional biofuels: Fuels produced from food crop feedstocks. These biofuels are commonly referred to as first-generation and include sugar cane ethanol, starch-based ethanol, fatty acid methyl esther (FAME) and straight vegetable oil (SVO).

Decommissioning (nuclear): The process of dismantling and decontaminating a nuclear power plant at the end of its operational lifetime and restoring the site for other uses.

Decomposition analysis: Statistical approach that decomposes an aggregate indicator to quantify the relative contribution of a set of pre-defined factors leading to a change in the aggregate indicator. The *World Energy Outlook* uses an additive index decomposition of the type Logarithmic Mean Divisia Index (LMDI) I.

Demand-side integration (DSI): Consists of two types of measures: actions that influence load shape such as energy efficiency and electrification; and actions that manage load such as demand-side response.

Demand-side response (DSR): Describes actions which can influence the load profile such as shifting the load curve in time without affecting the total electricity demand, or load shedding such as interrupting demand for short duration or adjusting the intensity of demand for a certain amount of time.

Dispatchable: Dispatchable generation refers to technologies whose power output can be readily controlled - increased to maximum rated capacity or decreased to zero - in order to match supply with demand.

Electricity demand: Defined as total gross electricity generation less own use generation, plus net trade (imports less exports), less transmissions and distribution losses.

Electricity generation: Defined as the total amount of electricity generated by power only or combined heat and power plants including generation required for own-use. This is also referred to as gross generation.

Energy sector CO_2 emissions: CO_2 emissions from fuel combustion (excluding non-renewable waste). Note that this does not include fugitive emissions from fuels, CO_2 transport, storage emissions or industrial process emissions.

Energy sector GHG emissions: CO_2 emissions from fuel combustion plus fugitive and vented methane and N_2O emissions from the energy and industry sectors.

Energy services: see useful energy.

Ethanol: Refers to bio-ethanol only. Ethanol is produced from fermenting any biomass high in carbohydrates. Today, ethanol is made from starches and sugars, but second-generation technologies will allow it to be made from cellulose and hemicellulose, the fibrous material that makes up the bulk of most plant matter.

Gas-to-liquids (GTL): Process featuring reaction of methane with oxygen or steam to produce syngas (a mixture of hydrogen and carbon monoxide) followed by synthesis of liquid products (such as diesel and naphtha) from the syngas using Fischer-Tropsch catalytic synthesis. The process is similar to those used in coal-to-liquids.

High-level waste (HLW): The highly radioactive and long-lived waste materials generated during the course of the nuclear fuel cycle, including spent nuclear fuel (if it is declared as waste) and some waste streams from reprocessing.

Heat (end-use): Can be obtained from the combustion of fossil or renewable fuels, direct geothermal or solar heat systems, exothermic chemical processes and electricity (through resistance heating or heat pumps which can extract it from ambient air and liquids). This category refers to the wide range of end-uses, including space and water heating, and cooking in buildings, desalination and process applications in industry. It does not include cooling applications.

Heat (supply): Obtained from the combustion of fuels, nuclear reactors, geothermal resources and the capture of sunlight. It may be used for heating or cooling, or converted into mechanical energy for transport or electricity generation. Commercial heat sold is reported under total final consumption with the fuel inputs allocated under power generation.

Hydropower: The energy content of the electricity produced in hydropower plants, assuming 100% efficiency. It excludes output from pumped storage and marine (tide and wave) plants.

Industry: The sector includes fuel used within the manufacturing and construction industries. Key industry branches include iron and steel, chemical and petrochemical, cement, and pulp and paper. Use by industries for the transformation of energy into another form or for the production of fuels is excluded and reported separately under other energy sector. Consumption of fuels for the transport of goods is reported as part of the transport sector, while consumption by off-road vehicles is reported under industry.

International aviation bunkers: Includes the deliveries of aviation fuels to aircraft for international aviation. Fuels used by airlines for their road vehicles are excluded. The domestic/international split is determined on the basis of departure and landing locations and not by the nationality of the airline. For many countries this incorrectly excludes fuels used by domestically owned carriers for their international departures.

International marine bunkers: Covers those quantities delivered to ships of all flags that are engaged in international navigation. The international navigation may take place at sea, on inland lakes and waterways, and in coastal waters. Consumption by ships engaged in domestic navigation is excluded. The domestic/international split is determined on the basis of port of departure and port of arrival, and not by the flag or nationality of the ship. Consumption by fishing vessels and by military forces is also excluded and included in residential, services and agriculture.

Investment: All investment data and projections reflect spending across the lifecycle of a project, i.e. the capital spent is assigned to the year when it is incurred. Investments for oil, gas and coal include production, transformation and transportation; those for the power sector include refurbishments, uprates, new builds and replacements for all fuels and technologies for on-grid, mini-grid and off-grid generation, as well as investment in transmission and distribution, and battery storage. Investment data are presented in real terms in year-2018 US dollars unless otherwise stated.

Note that this investment definition is new and our methodology to assess investment has changed from the previous editions of the *World Energy Outlook*. Previously, the investment data reflected "overnight investment", i.e. the capital spent is generally assigned to the year production (or trade) is started, rather than the year when it is incurred.

Lignite: Type of coal that is used in the power sector mostly in regions near lignite mines due to its low energy content and typically high moisture levels, which generally makes long-distance transport uneconomic. Data on lignite in the *World Energy Outlook* includes peat, a solid formed from the partial decomposition of dead vegetation under conditions of high humidity and limited air access.

Lignocellulosic feedstock: Crops cultivated to produce biofuels from their cellulosic or hemicellulosic components, which include switchgrass, poplar and miscanthus.

Liquids: Refers to the combined use of oil and biofuels (expressed in energy-equivalent volumes of gasoline and diesel).

Lower heating value: Heat liberated by the complete combustion of a unit of fuel when the water produced is assumed to remain as a vapour and the heat is not recovered.

Middle distillates: Include jet fuel, diesel and heating oil.

Mini-grids: Small grid systems linking a number of households or other consumers.

Modern energy access: Includes household access to a minimum level of electricity; household access to safer and more sustainable cooking and heating fuels and stoves; access that enables productive economic activity; and access for public services.

Modern renewables: Includes all uses of renewable energy with the exception of traditional use of solid biomass.

Modern use of solid biomass: Refers to the use of solid biomass in improved cookstoves and modern technologies using processed biomass such as pellets.

Natural gas: Comprises gases occurring in deposits, whether liquefied or gaseous, consisting mainly of methane. It includes both "non-associated" gas originating from fields producing hydrocarbons only in gaseous form, and "associated" gas produced in association with crude oil as well as methane recovered from coal mines (colliery gas). Natural gas liquids (NGLs), manufactured gas (produced from municipal or industrial waste, or sewage) and quantities vented or flared are not included. Gas data in cubic metres are expressed on a "gross" calorific value basis and are measured at 15 °C and at 760 mm Hg ("Standard Conditions"). Gas data expressed in tonnes of oil equivalent, mainly for comparison reasons with other fuels, are on a "net" calorific basis. The difference between the net and the gross calorific value is the latent heat of vaporisation of the water vapour produced during combustion of the fuel (for gas the net calorific value is 10% lower than the gross calorific value).

Natural gas liquids (NGLs): Liquid or liquefied hydrocarbons produced in the manufacture, purification and stabilisation of natural gas. These are those portions of natural gas which are recovered as liquids in separators, field facilities or gas processing plants. NGLs include but are not limited to ethane (when it is removed from the natural gas stream), propane, butane, pentane, natural gasoline and condensates.

Non-energy use: Fuels used for chemical feedstocks and non-energy products. Examples of non-energy products include lubricants, paraffin waxes, asphalt, bitumen, coal tars and oils as timber preservatives.

Nuclear: Refers to the primary energy equivalent of the electricity produced by a nuclear plant, assuming an average conversion efficiency of 33%.

Off-grid systems: Stand-alone systems for individual households or groups of consumers.

Offshore wind: Refers to electricity produced by wind turbines that are installed in open water, usually in the ocean.

Oil: Oil production includes both conventional and unconventional oil. Petroleum products include refinery gas, ethane, liquid petroleum gas, aviation gasoline, motor gasoline, jet fuels, kerosene, gas/diesel oil, heavy fuel oil, naphtha, white spirit, lubricants, bitumen, paraffin, waxes and petroleum coke.

Other energy sector: Covers the use of energy by transformation industries and the energy losses in converting primary energy into a form that can be used in the final consuming sectors. It includes losses by gas works, petroleum refineries, blast furnaces, coke ovens, coal and gas transformation and liquefaction. It also includes energy used in coal mines, in oil and gas extraction and in electricity and heat production. Transfers and statistical differences are also included in this category.

Peri-urban: Peri-urban areas are zones of transition from rural to urban which often form the urban-rural interface and may evolve into being fully urban.

Power generation: Refers to fuel use in electricity plants, heat plants and combined heat and power (CHP) plants. Both main activity producer plants and small plants that produce fuel for their own use (auto-producers) are included.

Pre-salt oil and gas: These resources are referred to as such because they predate the formation of a thick salt layer, which overlays the hydrocarbons and traps them in place.

Productive uses: Energy used towards an economic purpose: agriculture, industry, services, and non-energy use. Some energy demand from the transport sector (e.g. freight) could also be considered as productive, but is treated separately.

Refining processing gains: Processing gains are volume increases that occur during crude oil refining.

Renewables: Includes bioenergy, geothermal, hydropower, solar photovoltaic (PV), concentrating solar power (CSP), wind and marine (tide and wave) energy for electricity and heat generation.

Residential: Energy used by households including space heating and cooling, water heating, lighting, appliances, electronic devices and cooking equipment.

Resistance heating: Refers to direct electricity transformation into heat through the joule effect.

Self-sufficiency: Corresponds to indigenous production divided by total primary energy demand.

Services: Energy used in commercial (e.g. hotels, offices, catering, shops) and institutional buildings (e.g. schools, hospitals, offices). Services energy use includes space heating and cooling, water heating, lighting, equipment, appliances and cooking equipment.

Shale gas: Natural gas contained within a commonly occurring rock classified as shale. Shale formations are characterised by low permeability, with more limited ability of gas to flow through the rock than is the case with a conventional reservoir. Shale gas is generally produced using hydraulic fracturing.

Solid biomass: Includes charcoal, fuelwood, dung, agricultural residues, wood waste and other solid wastes.

Steam coal: Type of coal that is mainly used for heat production or steam-raising in power plants and, to a lesser extent, in industry. Typically, steam coal is not of sufficient quality for steel making. Coal of this quality is also commonly known as thermal coal.

Tight oil: Oil produced from shales or other very low permeability formations, using hydraulic fracturing. This is also sometimes referred to as light tight oil. Tight oil includes tight crude oil and condensate production except for the United States, which includes tight crude oil only (US tight condensate volumes are included in natural gas liquids).

Total final consumption (TFC): Is the sum of consumption by the various end-use sectors. TFC is broken down into energy demand in the following sectors: industry (including manufacturing and mining), transport, buildings (including residential and services) and other (including agriculture and non-energy use). It excludes international marine and aviation bunkers, except at world level where it is included in the transport sector.

Total final energy consumption (TFEC): Is a variable defined primarily for tracking progress towards target 7.2 of the Sustainable Development Goals. It incorporates total final consumption (TFC) by end-use sectors but excludes non-energy use. It excludes international marine and aviation bunkers, except at world level. Typically this is used in the context of calculating the renewable energy share in total final energy consumption (Indicator 7.2.1 of the Sustainable Development Goals), where TFEC is the denominator.

Total primary energy demand (TPED): Represents domestic demand only and is broken down into power generation, other energy sector and total final consumption.

Traditional use of solid biomass: Refers to the use of solid biomass with basic technologies, such as a three-stone fire, often with no or poorly operating chimneys.

Transport: Fuels and electricity used in the transport of goods or persons within the national territory irrespective of the economic sector within which the activity occurs. This includes fuel and electricity delivered to vehicles using public roads or for use in rail vehicles; fuel delivered to vessels for domestic navigation; fuel delivered to aircraft for domestic aviation; and energy consumed in the delivery of fuels through pipelines. Fuel delivered to international marine and aviation bunkers is presented only at the world level and is excluded from the transport sector at a domestic level.

Useful energy: Refers to the energy that is available to end-users to satisfy their needs. This is also referred to as energy services demand. As result of transformation losses at the point of use, the amount of useful energy is lower than the corresponding final energy demand for most technologies. Equipment using electricity often has higher conversion efficiency than equipment using other fuels, meaning that for a unit of energy consumed electricity can provide more energy services.

Variable renewable energy (VRE): Refers to technologies whose maximum output at any time depends on the availability of fluctuating renewable energy resources. VRE includes a broad array of technologies such as wind power, solar PV, run-of-river hydro, concentrating solar power (where no thermal storage is included) and marine (tidal and wave).

Waste storage and disposal: Activities related to the management of radioactive nuclear waste. Storage refers to temporary facilities at the nuclear power plant site or a centralised site. Disposal refers to permanent facilities for the long-term isolation of high-level waste, such as deep geologic repositories.

Water consumption: The volume withdrawn that is not returned to the source (i.e. it is evaporated or transported to another location) and by definition is no longer available for other uses.

Water sector: Includes all processes whose main purpose is to treat/process or move water to or from the end-use: groundwater and surface water extraction, long-distance water transport, water treatment, desalination, water distribution, wastewater collection, wastewater treatment and water re-use.

Water withdrawal: The volume of water removed from a source; by definition withdrawals are always greater than or equal to consumption.

Regional and country groupings

Advanced economies: OECD regional grouping and Bulgaria, Croatia, Cyprus[1,2], Malta and Romania.

Africa: North Africa and sub-Saharan Africa regional groupings.

Asia Pacific: Southeast Asia regional grouping and Australia, Bangladesh, China, India, Japan, Korea, Democratic People's Republic of Korea, Mongolia, Nepal, New Zealand, Pakistan, Sri Lanka, Chinese Taipei, and other Asia Pacific countries and territories.[3]

Caspian: Armenia, Azerbaijan, Georgia, Kazakhstan, Kyrgyzstan, Tajikistan, Turkmenistan and Uzbekistan.

Central and South America: Argentina, Plurinational State of Bolivia (Bolivia), Brazil, Chile, Colombia, Costa Rica, Cuba, Curaçao, Dominican Republic, Ecuador, El Salvador, Guatemala, Haiti, Honduras, Jamaica, Nicaragua, Panama, Paraguay, Peru, Suriname, Trinidad and Tobago, Uruguay, Bolivarian Republic of Venezuela (Venezuela), and other Central and South American countries and territories.[4]

China: Includes the (People's Republic of) China and Hong Kong, China.

Developing Asia: Asia Pacific regional grouping excluding Australia, Japan, Korea and New Zealand.

Developing economies: All other countries not included in the "advanced economies" regional grouping.

Eurasia: Caspian regional grouping and the Russian Federation (Russia).

Europe: European Union regional grouping and Albania, Belarus, Bosnia and Herzegovina, North Macedonia, Gibraltar, Iceland, Israel[5], Kosovo, Montenegro, Norway, Serbia, Switzerland, Republic of Moldova, Turkey and Ukraine.

European Union: Austria, Belgium, Bulgaria, Croatia, Cyprus[1,2], Czech Republic, Denmark, Estonia, Finland, France, Germany, Greece, Hungary, Ireland, Italy, Latvia, Lithuania, Luxembourg, Malta, Netherlands, Poland, Portugal, Romania, Slovak Republic, Slovenia, Spain, Sweden and United Kingdom.

IEA (International Energy Agency)**:** OECD regional grouping excluding Chile, Iceland, Israel, Latvia, Lithuania and Slovenia.

Latin America: Central and South America regional grouping and Mexico.

Middle East: Bahrain, Islamic Republic of Iran (Iran), Iraq, Jordan, Kuwait, Lebanon, Oman, Qatar, Saudi Arabia, Syrian Arab Republic (Syria), United Arab Emirates and Yemen.

Non-OECD: All other countries not included in the OECD regional grouping.

Non-OPEC: All other countries not included in the OPEC regional grouping.

Figure C.1 ▷ *World Energy Outlook* main country groupings

Note: This map is without prejudice to the status of or sovereignty over any territory, to the delimitation of international frontiers and boundaries and to the name of any territory, city or area.

North Africa: Algeria, Egypt, Libya, Morocco and Tunisia.

North America: Canada, Mexico and United States.

OECD (Organisation for Economic Co-operation and Development)**:** Australia, Austria, Belgium, Canada, Chile, Czech Republic, Denmark, Estonia, Finland, France, Germany, Greece, Hungary, Iceland, Ireland, Israel, Italy, Japan, Korea, Latvia, Lithuania, Luxembourg, Mexico, Netherlands, New Zealand, Norway, Poland, Portugal, Slovak Republic, Slovenia, Spain, Sweden, Switzerland, Turkey, United Kingdom and United States.

OPEC (Organisation of the Petroleum Exporting Countries)**:** Algeria, Angola, Republic of the Congo (Congo), Ecuador, Equatorial Guinea, Gabon, Islamic Republic of Iran (Iran), Iraq, Kuwait, Libya, Nigeria, Saudi Arabia, United Arab Emirates and Bolivarian Republic of Venezuela (Venezuela), based on membership status as of November 2019.

Southeast Asia: Brunei Darussalam, Cambodia, Indonesia, Lao People's Democratic Republic (Lao PDR), Malaysia, Myanmar, Philippines, Singapore, Thailand and Viet Nam. These countries are all members of the Association of Southeast Asian Nations (ASEAN).

Sub-Saharan Africa: Angola, Benin, Botswana, Cameroon, Republic of the Congo (Congo), Côte d'Ivoire, Democratic Republic of the Congo, Eritrea, Ethiopia, Gabon, Ghana, Kenya, Mauritius, Mozambique, Namibia, Niger, Nigeria, Senegal, South Africa, South Sudan, Sudan, United Republic of Tanzania (Tanzania), Togo, Zambia, Zimbabwe and other African countries and territories.[6]

Country notes

[1] Note by Turkey: The information in this document with reference to "Cyprus" relates to the southern part of the Island. There is no single authority representing both Turkish and Greek Cypriot people on the Island. Turkey recognises the Turkish Republic of Northern Cyprus (TRNC). Until a lasting and equitable solution is found within the context of the United Nations, Turkey shall preserve its position concerning the "Cyprus issue".

[2] Note by all the European Union Member States of the OECD and the European Union: The Republic of Cyprus is recognised by all members of the United Nations with the exception of Turkey. The information in this document relates to the area under the effective control of the Government of the Republic of Cyprus.

[3] Individual data are not available and are estimated in aggregate for: Afghanistan, Bhutan, Cook Islands, Fiji, French Polynesia, Kiribati, Lao People's Democratic Republic (Lao PDR), Macau (China), Maldives, New Caledonia, Palau, Papua New Guinea, Samoa, Solomon Islands, Timor-Leste and Tonga and Vanuatu.

[4] Individual data are not available and are estimated in aggregate for: Anguilla, Antigua and Barbuda, Aruba, Bahamas, Barbados, Belize, Bermuda, Bonaire, British Virgin Islands, Cayman Islands, Dominica, Falkland Islands (Malvinas), French Guiana, Grenada, Guadeloupe, Guyana, Martinique, Montserrat, Saba, Saint Eustatius, Saint Kitts and Nevis, Saint Lucia, Saint Pierre and Miquelon, Saint Vincent and Grenadines, Saint Maarten, Turks and Caicos Islands.

[5] The statistical data for Israel are supplied by and under the responsibility of the relevant Israeli authorities. The use of such data by the OECD and/or the IEA is without prejudice to the status of the Golan Heights, East Jerusalem and Israeli settlements in the West Bank under the terms of international law.

[6] Individual data are not available and are estimated in aggregate for: Burkina Faso, Burundi, Cabo Verde, Central African Republic, Chad, Comoros, Djibouti, Equatorial Guinea, Kingdom of Eswatini, Gambia, Guinea, Guinea-Bissau, Lesotho, Liberia, Madagascar, Malawi, Mali, Mauritania, Réunion, Rwanda, Sao Tome and Principe, Seychelles, Sierra Leone, Somalia and Uganda.

Abbreviations and Acronyms

AC	Africa Case
APEC	Asia-Pacific Economic Cooperation
ASEAN	Association of Southeast Asian Nations
BEV	battery electric vehicles
CAAGR	compound average annual growth rate
CAFE	corporate average fuel-economy standards (United States)
CBM	coalbed methane
CCGT	combined-cycle gas turbine
CCUS	carbon capture, utilisation and storage
CEM	Clean Energy Ministerial
CFL	compact fluorescent lamp
CH_4	methane
CHP	combined heat and power; the term co-generation is sometimes used
CNG	compressed natural gas
CO	carbon monoxide
CO_2	carbon dioxide
CO_2-eq	carbon-dioxide equivalent
COP	Conference of Parties (UNFCCC)
CPS	Current Policies Scenario
CSP	concentrating solar power

CTG	coal-to-gas
CTL	coal-to-liquids
DER	distributed energy resources
DSI	demand-side integration
DSR	demand-side response
EHOB	extra-heavy oil and bitumen
EOR	enhanced oil recovery
EPA	Environmental Protection Agency (United States)
EU	European Union
EU ETS	European Union Emissions Trading System
EV	electric vehicle
FAO	Food and Agriculture Organization of the United Nations
FDI	foreign direct investment
FiT	feed-in tariff
FOB	free on board
GDP	gross domestic product
GHG	greenhouse gases
GTL	gas-to-liquids
HDI	human development index
HFO	heavy fuel oil
IAEA	International Atomic Energy Agency
ICE	internal combustion engine
ICT	information and communication technologies
IEA	International Energy Agency
IGCC	integrated gasification combined-cycle
IIASA	International Institute for Applied Systems Analysis
IMF	International Monetary Fund
IMO	International Maritime Organization
IOC	international oil company
IPCC	Intergovernmental Panel on Climate Change
LCOE	levelised cost of electricity
LCV	light-commercial vehicle
LED	light-emitting diode
LNG	liquefied natural gas
LPG	liquefied petroleum gas
LULUCF	land use, land-use change and forestry
MER	market exchange rate
MEPS	minimum energy performance standards
NDCs	Nationally Determined Contributions
NEA	Nuclear Energy Agency (an agency within the OECD)
NGLs	natural gas liquids
NGV	natural gas vehicle
NPV	net present value
NOC	national oil company

NO$_x$	nitrogen oxides
OECD	Organisation for Economic Co-operation and Development
OPEC	Organization of the Petroleum Exporting Countries
PHEV	plug-in hybrid electric vehicles
PLDV	passenger light-duty vehicle
PM	particulate matter
PM$_{2.5}$	fine particulate matter
PPA	power purchase agreement
PPP	purchasing power parity
PSH	pumped storage hydropower
PV	photovoltaics
R&D	research and development
RD&D	research, development and demonstration
RRR	remaining recoverable resource
SDS	Sustainable Development Scenario
SME	small and medium enterprises
SMR	steam methane reformation
SO$_2$	sulfur dioxide
STEPS	Stated Policies Scenario
SWH	solar water or solar water heaters
T&D	transmission and distribution
TES	thermal energy storage
TFC	total final consumption
TFEC	total final energy consumption
TPED	total primary energy demand
UAE	United Arab Emirates
UN	United Nations
UNDP	United Nations Development Program
UNEP	United Nations Environment Program
UNFCCC	United Nations Framework Convention on Climate Change
URR	ultimately recoverable resources
US	United States
USGS	United States Geological Survey
VALCOE	value-adjusted levelised cost of electricity
VRE	variable renewable energy
WACC	weighted average cost of capital
WEM	World Energy Model
WEO	*World Energy Outlook*
WHO	World Health Organization

Annex D

References

Part A: Global Energy Trends

Chapter 1: Overview and key findings

IEA (International Energy Agency) (2019a), *Southeast Asia Energy Outlook: World Energy Outlook Special Report*, IEA, Paris.

– (2019b), *Offshore Wind Outlook 2019: World Energy Outlook Special Report*, IEA, Paris.

– (2019c), *World Energy Investment 2019*, IEA, Paris.

– (2018), *Outlook for Producer Economies: World Energy Outlook Special Report*, IEA, Paris.

IPCC (Intergovernmental Panel on Climate Change) (2018), *Global Warming of 1.5 °C, An IPCC Special Report on the impacts of global warming of 1.5 °C above pre-industrial levels and related global greenhouse gas emission pathways, in the context of strengthening the global response to the threat of climate change, sustainable development and efforts to eradicate poverty*, [Masson-Delmotte, V., P. Zhai, H.O. Pörtner, D. Roberts, J. Skea, P. Shukla, A. Pirani, W. Moufouma-Okia, C. Péan, R. Pidcock, S. Connors, J. Matthews, Y. Chen, X. Zhou, M. Gomis, E. Lonnoy, T. Maycock, M. Tignor, and T. Waterfield (eds.), World Meteorological Organization, Geneva, https://www.ipcc.ch/sr15/.

Chapter 2: Energy and the Sustainable Development Goals

Anderson K. and G. Peters (2016), The trouble with negative emissions, *Nature*, Vol. 354, pp. 182–183.

Bento, N., C. Wilson and L. Anadon (2018), Time to get ready: Conceptualizing the temporal and spatial dynamics of formative phases for energy technologies, *Energy Policy*, Vol. 119, 08/2018.

Cárdenas Rodríguez, M., I. Haščič and N. Johnstone (2019), Commentary: Global patent applications for climate change mitigation technologies – a key measure of innovation – are trending down, IEA, https://www.iea.org/newsroom/news/2019/july/global-patent-applications-for-climate-change-mitigation-technologies--a-key-mea.html.

CEEW (Council on Energy, Environment and Water) (2019), CEEW, https://www.ceew.in/.

GCRI (Global Climate Risk Index) (2019), *Global Climate Risk Index*, https://germanwatch.org/en/16046.

CCC (Committee on Climate Change – United Kingdom) (2019), *Net Zero – The UK's contribution to stopping global warming*, https://www.theccc.org.uk/publication/net-zero-the-uks-contribution-to-stopping-global-warming/.

Gross, R. et al. (2018), How long does innovation and commercialisation in the energy sectors take? Historical case studies of the timescale from invention to widespread commercialisation in energy supply and end-use technology, *Energy Policy*, Volume 123, pp. 682-699, ISSN 0301-4215, https://doi.org/10.1016/j.enpol.2018.08.061.

IEA (International Energy Agency) (2019a), *The Future of Hydrogen*, IEA, Paris, www.iea.org/publications/reports/thefutureofhydrogen/.

– (2019b), *Nuclear Power in a Clean Energy System,* IEA, Paris, https://webstore.iea.org/nuclear-power-in-a-clean-energy-system.

– (2019c), *Material Efficiency in Clean Energy Transitions,* IEA, Paris.

– (2019d), *World Energy Investment 2019,* IEA, Paris, https://www.iea.org/wei2019/.

– (2019e), *Tracking Clean Energy Progress: Informing Energy Sector Transformations*, IEA, Paris, www.iea.org/tcep/.

– (2019f), *Clean Energy Investment Trends*, IEA, Paris, https://webstore.iea.org/clean-energy-investment-trends-2019.

– (2018a), *Global Energy and CO_2 Status Report*, IEA, Paris, www.iea.org/geco/.

– (2018b), *World Energy Investment 2018*, IEA, Paris, https://webstore.iea.org/world-energy-investment-2018.

– (2018c), *World Energy Outlook-2018,* IEA, Paris.

– (2018d), *CO_2 Emissions from Fuel Combustion 2018,* IEA, www.iea.org/statistics/.

– (2017a), *Energy Access Outlook 2017: From Poverty to Prosperity: World Energy Outlook Special Report*, IEA, Paris.

– (2017b), *Energy Technology Perspectives 2017*, IEA, Paris.

– (2017c), *Technology Roadmap Delivering Sustainable Bioenergy*, IEA, Paris.

– (2008), *World Energy Outlook-2008*, IEA, Paris.

IPCC (Intergovernmental Panel on Climate Change) (2019), *Climate Change and Land: an IPCC special report on climate change, desertification, land degradation, sustainable land management, food security, and greenhouse gas fluxes in terrestrial ecosystems (SRCCL)*, IPCC, https://www.ipcc.ch/report/srccl/.

– (2018), Global Warming of 1.5 °C, *An IPCC Special Report on the impacts of global warming of 1.5 °C above pre-industrial levels and related global greenhouse gas emission pathways, in the context of strengthening the global response to the threat of climate change, sustainable development and efforts to eradicate poverty*, [Masson-Delmotte, V., P. Zhai, H.-O. Pörtner, D. Roberts, J. Skea, P.R. Shukla, A. Pirani, W. Moufouma-Okia, C. Péan, R. Pidcock, S. Connors, J.B.R. Matthews, Y. Chen, X. Zhou, M.I. Gomis, E. Lonnoy, T. Maycock, M. Tignor, and T. Waterfield (eds.)], World Meteorological Organization, Geneva, https://www.ipcc.ch/sr15/.

– (2014), *Fifth Assessment Report*, IPCC, Cambridge University Press, Cambridge, United Kingdom and New York.

– (2006), *Guidelines for National Greenhouse Gas Inventories, Volume 3, Industrial Processes and Product Use*, IPCC, https://www.ipcc-nggip.iges.or.jp/public/2006gl/vol3.html.

IRENA (International Renewable Energy Agency)(2019), *Renewable Power Generation Costs in 2018*, IRENA, Abu Dhabi.

Masera, O., R. Bailis, and R. Drigo (2015), Environmental Burden of Traditional Bioenergy Use, *Annual Review of Environment and Resources*, 11/2015, Vol. 40, Issue 1.

Singh, D., S. Pachauri and H. Zerriffi (2017), Environmental payoffs of LPG cooking in India, *Environmental Research Letters*, 12(11).

UNFCCC (United Nations Framework Convention on Climate Change) (2019a), *Cooperative Initiatives*, https://climateaction.unfccc.int/views/cooperative-initiative-details.html?id=94.

– (2019b), *Information on COP 25, CMP 15, CMA 2*, https://unfccc.int/news/information-on-cop-25-cmp-15-cma-2, UK Committee on Climate Change, Reducing UK emissions 2019 Progress Report to Parliament, presented to Parliament pursuant to Section 36(1) of the Climate Change Act 2008, www.theccc.org.uk/publications.

Chapter 3: Outlook for oil

AAPG (American Association of Petroleum Geologists) (2000), Treatise of Petroleum Geology/ Handbook of Petroleum Geology: Exploring for Oil and Gas Traps, in Beaumont L., *Chapter 3: Petroleum Systems*, pp. 1-34, AAPG.

Automotive World Ltd. (2018), *Special Report: Africa's Auto Industry*, www.exportersec.co.za/wp-content/uploads/2018/05/special-report-africas-auto-industry.pdf, accessed 6 September 2019.

Bird (2019), *A Look at e-Scooter Safety: Examining risks, reviewing responsibilities, and prioritizing prevention*, Bird, Santa Monica, California, United States.

China Passenger Car Association (2019), *2018 Passenger Car Sales Report*, www.cpcaauto.com/newslist.asp?types=csjd&id=9345, accessed July 2019.

China Association of Automobile Manufacturers (2019), *2018 sales ranking of the top-ten Chinese car manufacturers*, www.caam.org.cn/chn/4/cate_39/con_5221287.html, accessed July 2019.

EV Volumes (2019), *EV Volumes* (commercial database), accessed 1 September 2019.

Freyman, M. (2016), *Hydraulic Fracturing and Water Stress: Water demand by the numbers – 2016 update*, https://www.ceres.org/resources/reports/hydraulic-fracturing-water-stress-water-demand-numbers, accessed 1 September 2019.

GWPC (Ground Water Protection Council) (2019), *Produced Water Report: Regulations, Current Practices, and Research Needs*, GWPC, Oklahoma City, Oklahoma, United States.

IEA (International Energy Agency) (2019a), *World Energy Investment 2019*, IEA, Paris.

– (2019b), *Global EV Outlook 2019*, IEA, Paris.

– (2019c), *The Global Fuel Economy Initiative 2019*, IEA, Paris.

– (2019d), *Oil 2019*, IEA, Paris.

– (2018a), *World Energy Outlook-2018*, IEA, Paris.

– (2018b), *Outlook for Producer Economies: What do changing energy dynamics mean for major oil and gas exporters*, IEA, Paris.

– (2012), *Golden Rules for a Golden Age of Gas: World Energy Outlook Special Report*, IEA, Paris.

– (2011), *World Energy Outlook-2011*, IEA, Paris.

IHS Markit (2018), *Vehicle Registrations and Other Characteristics at Model Level* (commercial database), accessed 20 March 2019.

Jacobs, T. (2019), Shale EOR delivers, so why won't the sector go big?, *Journal of Petroleum Technology*, 71 (5).

Kitamura et al. (1999), *Accessibility and auto use in a motorized metropolis*, Institute of Transportation Studies, University of California, Irvine, California, United States.

Lime (2019), *Year-end report 2018*, Lime, San Francisco, California, United States.

Lukash, A. (2019), *US Shale Industry Turns Cash Flow Positive*, Rystad Energy, Oslo.

Magnani et al. (2017), Discriminating between natural versus induced seismicity from long-term deformation history of intraplate faults, *Science Advances*, 3(11).

Oruganti Y. et al. (2015), *Re-Fracturing in Eagle Ford and Bakken to increase reserves and generate incremental NPV: Field Study*, Society of Petroleum Engineers, https://www.onepetro.org/conference-paper/SPE-173340-MS.

Quiroga C. et al. (2016), *Truck Traffic and Truck Loads Associated with Unconventional Oil and Gas Development in Texas, Implementation Report RR-16-01*, Texas A&M Transportation Institute, College Station, Texas, United States.

Rystad Energy (2019), *Re-Frac Market Soars Past 550 Horizontal Wells in 2018*, Rystad Energy, Oslo.

Schaller, B, (2018), *The New Automobility: Lyft, Uber and the Future of American Cities*, http://www.schallerconsult.com/rideservices/automobility.pdf.

Texas Water Development Board (2017), *2017 State Water Plan*, Austin, Texas, United States.

US EPA (United States Environmental Protection Agency) (2019), *Unconventional Oil and Gas Extraction Effluent Guidelines*, US EPA, Washington, DC.

USGS (United States Geological Survey) (2018), Assessment of Undiscovered Continuous Oil and Gas Resources in the Wolfcamp Shale and Bone Spring Formation of the Delaware Basin, Permian Basin Province, New Mexico and Texas, 2018, *Fact Sheet 2018-3073*, USGS, Reston, Virginia, United States.

Wood Mackenzie (2018), *Permian Produced Water: Slowly Extinguishing a Roaring Basin*, Wood Mackenzie, Aberdeen, United Kingdom.

World Bank (2019), *Global Gas Flaring Reduction Partnership*, World Bank, Washington DC.

Chapter 4: Outlook for natural gas

Argus Media (2019), *Global LNG Reports,* London, https://www.argusmedia.com/en/natural-gas-lng/argus-global-lng, accessed 15 July 2019

Gas Infrastructure Europe (2018), *Small-scale LNG* (database), https://www.gie.eu/index.php/gie-publications/databases/gle-lng-services-inventory, accessed 20 July 2019.

IEA (International Energy Agency) (2019a), *World Energy Investment 2019,* IEA, Paris.

– (2019b), *The Role of Gas in Today's Energy Transitions,* IEA, Paris.

– (2018), *World Energy Outlook-2018,* IEA, Paris.

– (2011), *World Energy Outlook-2011,* IEA, Paris.

IGU (International Gas Union) (2019), *Wholesale Gas Price Survey,* https://www.igu.org/, accessed 15 July 2019.

OGJ (Oil and Gas Journal) (2019), Permian gas flaring, venting reaches record high, *OGJ* 117 (6a).

Saudi Aramco (2019), *Base Prospectus*, London Stock Exchange, London, www.londonstockexchange.com, accessed 12 June 2019.

Speirs J. et al. (2019), *Can natural gas reduce emissions from transport?*, Imperial College, London, https://www.sustainablegasinstitute.org/can-natural-gas-reduce-emissions-from-transport/, accessed 5 March 2019.

White, N. and C. Brooks (2018), *Catching cold or catching fire? Current state and growth prospects for small-scale LNG,* Cedigaz, www.cedigaz.org, accessed 12 April 2019.

Chapter 5: Outlook for coal

Bloomberg (2019), Bloomberg Terminal, accessed multiple times during July-September 2019.

CRU (2019), Thermal Coal Model (database), accessed April 2019.

IEA (International Energy Agency) (2018), *World Energy Outlook-2018*, IEA, Paris.

– (2017), *World Energy Outlook-2017*, IEA, Paris.

India Ministry of Coal (2018), *Coal Directory of India, 2017-18,* Kolkata, http://www.coalcontroller.gov.in/pages/display/16-coal-directory, accessed 31 May 2019.

Miller et al. (2019), China's coal mine methane regulations have not curbed growing emissions, *Nature Communications,* 10(303).

PPCA (Powering Past Coal Alliance) (2019), https://poweringpastcoal.org/, accessed 7 October 2019.

Singh, A. K. and J .N. Sahu (2018), Coal mine gas: a new fuel utilization technique for India, *International Journal of Green Energy,* 15(12), pp. 732-743.

Thomson Reuters Eikon (2019), (database), accessed multiple times during July-September 2019.

US EPA (US Environmental Protection Agency) (2018), *Inventory of US Greenhouse Gas Emissions and Sinks, 1990-2017*, https://www.epa.gov/ghgemissions/inventory-us-greenhouse-gas-emissions-and-sinks-1990-2017.

Wang et al. (2018), Lifecycle carbon emission modelling of coal-fired power: Chinese case, *Energy, 162*, 841-852.

World Bank (2019), Worldwide Governance Indicators, Washington DC, https://info.worldbank.org/governance/wgi/, accessed 31 May 2019.

Zhu et al. (2017), An Improved Approach to Estimate Methane Emissions from Coal Mining in China, *Environmental Science and Technology, 51*(21).

Chapter 6: Outlook for electricity

Central Electricity Authority (2017), *Report of the Technical Committee on Study of Optimal Location of Various Types of Balancing Energy Sources/Energy Storage Devices to Facilitate Grid Integration of Renewable Energy Sources and Associated Issues*, CEA, New Delhi.

Climate Analytics (2017), About 80% of EU and German, virtually all Polish coal plants non-compliant with new EU 2021 air pollution regulations, Climate Analytics, Berlin, https://climateanalytics.org/media/coal_germany_briefing_15112017.pdf.

Hirth, L. (2013), The Market Value of Variable Renewables: The effect of solar wind power variability on their relative price, *Energy Economics*, Vol. 38, pp. 218-236.

IEA (International Energy Agency) (2019a), *The Future of Rail: Opportunities for Energy and the Environment*, IEA, Paris, https://webstore.iea.org/the-future-of-rail.

– (2019b), *World Energy Investment 2019*, IEA, Paris, www.iea.org/wei2019/.

– (2019c), *The Role of Gas in Today's Energy Transitions*, IEA, Paris, www.iea.org/publications/roleofgas/.

– (2019d), *Nuclear Power in a Clean Energy System*, IEA, Paris.

– (2019e), *Southeast Asia Energy Outlook: World Energy Outlook Special Report*, IEA, Paris.

– (2019f), *Status of Power System Transformation 2019*, IEA, Paris.

– (2018a), *World Energy Outlook-2018*, IEA, Paris.

– (2018b), *The Future of Cooling*, IEA, Paris, www.iea.org/futureofcooling/.

– (2018c), *World Energy Investment Report 2018*, IEA, Paris, www.iea.org/wei2018/.

– (2018d), *Status of Power System Transformation 2018*, IEA, Paris.

– (2016), The Potential for Equipping China's Existing Coal Fleet with Carbon Capture and Storage, *IEA-Insights Paper Series*, IEA, Paris.

OECD/NEA (Organisation for Economic Co-operation and Development and Nuclear Energy Agency) (2016), *Small Modular Reactors: Nuclear Energy Market Potential for Near-term Deployment*, OECD, Paris.

REN21 (2019), Renewables 2019 - Global Status Report, REN21 Secretariat, Paris.

Ueckerdt, F. et al. (2013), System LCOE: What are the costs of renewables?, *Energy*, Vol. 63, pp. 61-75.

US EIA (US Energy Information Administration) (2019), Levelized Cost and Levelized Avoided Cost of New Generation Resources, *Annual Energy Outlook 2019*, US EIA, Washington, DC.

– (2016), Today in energy: EIA electricity generator data show power industry response to EPA mercury limits, US EIA, Washington, DC, www.eia.gov/todayinenergy/detail.php?id=26972, accessed 1 September 2019.

Wanner, B. (2019), Commentary: Is exponential growth of solar PV the obvious conclusion?, IEA, Paris, https://www.iea.org/newsroom/news/2019/february/is-exponential-growth-of-solar-pv-the-obvious-conclusion.html, accessed 1 June 2019.

Chapter 7: Outlook for energy efficiency and renewables

California Air Resources Board (2019), *Low-Carbon Fuel Standard*, https://ww3.arb.ca.gov/fuels/lcfs/lcfs.htm, accessed 15 July 2019.

EBA (European Biogas Association) (2019), *EBA Statistical Report 2019*, EBA, Brussels.

– (2017), *EBA Statistical Report 2017*, EBA, Brussels.

Gibon, T. et al. (2017), Health Benefits, Ecological Threats of Low-Carbon Electricity, *Environmental Research Letters*, https://iopscience.iop.org/article/10.1088/1748-9326/aa6047.

GMI (Global Methane Institute) (2019), *Global Methane Institute*, www.globalmethane.org/partners/country.aspx?c=india, accessed 1 July 2019.

IEA (International Energy Agency) (2019a), *Energy Efficiency 2019*, IEA, Paris.

– (2019b), *World Energy Investment 2019*, IEA, Paris.

– (2019c), *Renewable Energy Market Report 2019*, IEA, Paris.

– (2019d), *Global EV Outlook 2019: Scaling up the transition to electric mobility*, IEA, Paris.

– (2018a), *The Future of Petrochemicals*, IEA, Paris.

– (2018b), *World Energy Outlook-2018*, IEA, Paris.

Material Economics (2018), *The Circular Economy a Powerful Force for Climate Mitigation*, Material Economics Sverige AB, Stockholm.

Platts (2019), *Olefins Data,* S&P Platts, London, accessed 1 September 2019.

REN21 (2019), *Renewables Status Report 2019*, REN21 Secretariat, Paris.

Suwanasri, K. et al. (2015), Biogas – Key Success Factors for Promotion in Thailand, *Semantic Scholar*, https://pdfs.semanticscholar.org/deec/9ca1d8852608aa27af3656223ea24db0b0b2.pdf.

TERI (The Energy and Resources Institute) (2019), *Reference Report for National Resource Efficiency Policy for India*, TERI, New Delhi.

Theuerl, S. et al. (2019), The Future of Agricultural Biogas Plant in Germany: A Vision, *Energies*, www.mdpi.com/1996-1073/12/3/396.

USDA (United States Department of Agriculture) (2016), *Methane Emissions from Dairy Farming*, USDA, Washington DC.

US EPA (United States Environmental Protection Agency) (2018), *Renewable Fuel Standard Program: Standards for 2019 and Biomass-based Diesel Volume for 2020*, US EPA, Washington DC.

USGS (United States Geological Survey) (2018a), *Aluminium Minerals Yearbooks and Commodities Summaries*, USGS, Washington DC.

– (2018b), *Cement Mineral Yearbooks and Commodity Summaries*, USGS, Washington DC.

Part B: Special Focus on Africa

Chapter 8: Africa today

AfDB (African Development Bank) (2018), *African Energy Portal, AfDB, Regional Profile*, https://africa-energy-portal.org/regional-profile, accessed 1 October 2019.

– (2019), *2019 African Economic Outlook*, AfDB, Tunis.

Africa Report (2019), Growth prospects hurt as Ethiopia struggles to keep the lights on, *The Africa Report*, https://www.theafricareport.com/13533/growth-prospects-hurt-as-ethiopia-struggles-to-keep-the-lights-on/, accessed 30 May 2019.

Asante, K. et al. (2018), Ghana's rural liquefied petroleum gas program scale up: A case study, *Energy for Sustainable Development*, Vol. 46, pp. 94-102.

AUC/OECD (African Union Congress/Organisation for Economic Co-operation and Development) (2018), *Africa's Development Dynamics 2018: Growth, Jobs and Inequalities*, AUC/OECD Publishing, Addis Ababa, Paris.

Baruah, B. (2010), Energy Services for the Urban Poor: NGO participation in slum electrification in India, *Environment and Planning*, Vol. 28, pp. 1011-1027.

BGR (German Federal Institute for Geosciences and Natural Resources) (2016), *Geothermal Energy: East Africa, Bundesanstalt für Geowissenschaften und Rohstoffe* (BGR) https://www.bgr.bund.de/EN/Themen/Zusammenarbeit/TechnZusammenarb/Projekte/Laufend/Afrika/2029_2016-2066-5_RegionalOstafrika_Geothermie_en.html?nn=1549142, accessed 1 July 2019.

Bloomberg Businessweek (2017), *The Army of Women Battling India's $10 Billion Power Problem,* Bloomberg Businessweek: https://www.bloomberg.com/news/features/2017-10-03/army-of-women-tackle-electricity-thieves-in-indian-slums, accessed 1 October 2019.

Bloomberg (2019), *Africa Growth at 7-Year High, No Thanks to Its Major Economies.* Bloomberg, https://www.bloomberg.com/news/articles/2019-04-03/africa-growth-at-7-year-high-no-thanks-to-its-major-economies, accessed 3 April 2019.

Carbon Africa (2015), *Kenya Market Assessment for Off-Grid*, International Finance Corporation, Washington DC.

Couture, T. and D. Jacobs (2019), *Beyond Fire: How to achieve electric cooking,* Hivos and World Future Council, https://www.worldfuturecouncil.org/wp-content/uploads/2019/05/Beyond-Fire_-How-to-achieve-electric-cooking.pdf.

Cronk, R.and J. Bartram (2018), Environmental conditions in health care facilities in low- and middle-income countries: coverage and inequalities, *International Journal of Hygiene and Environmental Health,* Vol. 221, Issue 3, pp. 409-422.

Dalberg Advisors (2018), *Scaling up clean cooking in urban Kenya with LPG & Bio-ethanol: A market and policy analysis,* https://dalberg.com/system/files/2018-06/Dalberg_Long-form%20report_FINAL_PDF_0.pdf.

ESMAP (Energy Sector Management Assistance Program) (2019), *Mini-Grids for Half a Billion People*, The International Bank for Reconstruction and Development / World Bank Group, Washington, DC.

FAO (Food and Agriculture Organization of the United Nations) (2018), *Sustainable woodfuel for food security. A smart choice: green, renewable and affordable*, FAO, Rome.

Fenix (2019), https://www.fenixintl.com/blog/fenix-reaches-500000-customers-in-6-markets-and-announces-new-leadership-team/, accessed 20 July 2019.

Geothermal Development Company (2019), *U.S.-East Africa Geothermal Partnership*, http://www.gdc.co.ke/projects_intro.php, accessed 1 July 2019.

Geothermal Energy Association (2019), *Geothermal Energy Association*, http://www.geo-energy.org/EastAfrica/EAGP.aspx?no_redirect=true, accessed 1 July 2019.

GOGLA (2019), *Global Off-Grid Solar Market Report, Semi-Annual Sales and Impact Data, July-December 2018 Public Report*, GOGLA, Utrecht, Netherlands.

Greentech Media (2019), *Africa's Wind Project Pipeline Grows to 18GW*, Greentech Media: https://www.greentechmedia.com/articles/read/africa-18-gw-wind-project-pipeline#gs.1moh9l, accessed 6 June 2019.

– (2017), *Living Under the Grid: 110 Million of Africa's Unconnected Customers Represent a Massive Opportunity,* Greentech Media*:* https://www.greentechmedia.com/articles/read/living-under-the-grid-110-million-of-africas-unconnected-customers-represen#gs.adiomx, accessed 1 October 2019.

GTM Research (2017), *Living Under the Grid: 110 Million of Africa's Unconnected Customers Represent a Massive Opportunity*, GTM Research, New York.

ICA (Infrastructure Consortium for Africa) (2017), *Infrastructure Financing Trends in Africa – 2017*, ICA, Abidjan.

IEA (International Energy Agency) (2019a), *Energy Access* (database), IEA, https://www.iea.org/sdg/electricity/, accessed 18 October 2019.

– (2019b), *Oil 2019*, IEA, Paris.

– (2019c), *Energy, Water and the Sustainable Development Goals*, IEA, https://webstore.iea.org/energy-water-and-the-sustainable-development-goals, accessed 4 October 2019.

– (2018a), *Outlook for Producer Economies: World Energy Outlook Special Report*, IEA, Paris.

– (2018b), *Renewables 2018*, IEA, Paris.

– (2018c), *Coal 2018*, IEA, Paris.

– (2017a), *Energy Access Outlook 2017: From Poverty to Prosperity: World Energy Outlook Special Report*, IEA, Paris.

– (2017b), *Digitalization and Energy*, IEA, Paris.

– (2014), *World Energy Outlook-2014*, IEA, Paris.

IEA, IRENA, UNSD, WB, WHO (International Energy Agency, International Renewable Energy Agency, United Nations Statistics Division, World Bank and World Health Organization) (2019), *Tracking SDG 7: The Energy Progress Report 2019*, International Bank for Reconstruction and Development / The World Bank, Washington DC.

IPCC (Intergovernmental Panel on Climate Change) (2018), *Global Warming of 1.5 °C, An IPCC Special Report on the impacts of global warming of 1.5 °C above pre-industrial levels and related global greenhouse gas emission pathways, in the context of strengthening the global response to the threat of climate change, sustainable development and efforts to eradicate poverty*, [Masson-Delmotte, V., P. Zhai, H. Pörtner, D. Roberts, J. Skea, P. Shukla, A. Pirani, W. Moufouma-Okia, C. Péan, R. Pidcock, S. Connors, J. Matthews, Y. Chen, X. Zhou, M.I. Gomis, E. Lonnoy, T. Maycock, M. Tignor, and T. Waterfield (eds.)], World Meteorological Organization, Geneva, https://www.ipcc.ch/sr15/.

IPU (Inter-Parliamentary Union) (2019), Percentage of women in national parliaments, IPU, https://data.ipu.org/women-ranking?, accessed 1 September 2019.

IRENA (International Renewable Energy Agency (2014), *Estimating the Renewable Energy Potential in Africa*, IRENA-KTH working paper, IRENA, Abu Dhabi.

ITU (International Telecommunication Union) (2019), ICT Statistics, https://www.itu.int/en/ITU-D/Statistics/Pages/stat/default.aspx, accessed 7 June 2019.

Kojima, M., R. Bacon and C. Trimble (2014), *Political Economy of Power Sector Subsidies: A Review with Reference to Sub-Saharan Africa*, World Bank Group, Washington, DC.

Korkovelos, et al. (2018), A Geospatial Assessment of Small-Scale Hydropower Potential in Sub-Saharan Africa, *Energies*, Vol. 11 (11), 3100 https://www.mdpi.com/1996-1073/11/11/3100.

Lee, K., E. Miguel and C. Wolfram (2016), *Experimental evidence of the demand for and costs of rural electrification*, National Bureau of Economic Research, https://www.nber.org/papers/w22292, accessed 1 October 2019.

Leo, B., V. Ramachandran and R. Morello (2014), Shedding New Light on the Off-Grid Debate in Power Africa Countries, Center for Global Development, https://www.cgdev.org/blog/shedding-new-light-grid-debate-power-africa-countries.

Luque, A. (2016), From consumers to customers: Regularizing electricity networks in São Paulo's favelas, in *Retrofitting Cities: Priorities, Governance and Experimentation*, Routledge, Abingdon.

OECD (Organisation for Economic Co-operation and Development) (2016), *The Cost of Air Pollution in Africa*, OECD Publishing, Paris.

Padam, G. et al. (2018), *Ethiopia, Beyond Connection, Energy Access Diagnostic Report Based on the Multi-Tier Framework*, World Bank Group, Washington DC.

Pilishvili, T. et al. (2016), Effectiveness of Six Improved Cookstoves in Reducing Household Air Pollution and Their Acceptability in Rural Western Kenya, *PLoS ONE*, https://journals.plos.org/plosone/article/file?id=10.1371/journal.pone.0165529&type=prin table, accessed 1 October 2019.

REN21 (2017), *Renewables 2017 - Global Status Report*, REN21 Secretariat, Paris.

RES4Africa Foundation (2019), *Africa's Future Counts: Renewables & the Water-Energy-Food Nexus in Africa*, Res4Africa, Rome, http://www.res4med.org/wp-content/uploads/2019/06/RES4Africa_flagship_2019.pdf.

SAPP (Southern African Power Pool) (2019), *Annual Report*, SAPP, http://www.sapp.co.zw/annual-reports.

Shrivastava, B. (2017, October 4). *The Army of Women Battling India's $10 Billion Power Problem*, Bloomberg Businessweek, https://www.bloomberg.com/news/features/2017-10-03/army-of-women-tackle-electricity-thieves-in-indian-slums, accessed 15 July 2019.

Solar Sisters (2019), *Our Model*, Solar Sisters, https://solarsister.org/what-we-do/our-model/, accessed 1 June 2019.

Transparency International (2019), *Corruption Perceptions Index 2018*, Transparency International, https://www.transparency.org/cpi2018.

Tusting, L. et al. (2019), Mapping changes in housing in sub-Saharan Africa from 2000 to 2015, *Nature*, 568, pp. 391–394.

United Nations (2019a), *Ending poverty*, United Nations, https://www.un.org/en/sections/issues-depth/poverty/, accessed 20 July 2019.

– (2019b), Global warming: severe consequences for Africa, *UN Africa Renewal*, https://www.un.org/africarenewal/magazine/december-2018-march-2019/global-warming-severe-consequences-africa.

UNCTAD (United Nations Conference on Trade and Development) (2019), *Foreign Direct Investment to Africa Defies Global Slump, Rises to 11%*, UNCTAD, https://unctad.org/en/pages/newsdetails.aspx?OriginalVersionID=2109, accessed 12 June 2019.

UNDESA (United Nations Department of Economic and Social Affairs) (2019), *World Population Prospects 2019 Highlights*, UNDESA, New York.

– (2018), *World Urbanization Prospects 2018: Highlights*, UNDESA, New York.

UNDP (United Nations Development Programme) (2017), *Income Inequality Trends in sub-Saharan Africa: Divergence, Determinants and Consequences*, UNDP, New York.

UNECA (United Nations Economic Commission for Africa) (2018a), *An empirical assessment of the African Continental Free Trade: Area modalities on goods,* UNECA, Addis Ababa.

– (2018b), *Clean Water and Sanitation*, UNECA, Addis Ababa, https://www.uneca.org/sites/default/files/images/arfsd_2018_-goal_6_clean_water_and_sanitation_en.pdf.

– (2017), *The Case for the African Continental Free Trade Area: the AfCFTA, Africa's trade flows and industrialization*, UNECA, Addis Ababa.

UNEP (United Nations Environment Programme) (2017), *Atlas of Africa Energy Resources*, UNEP, Nairobi.

UNESCO (United Nations Educational, Scientific and Cultural Organization) (2019), *Water as Cross-Cutting Factor in the SDGs Under Reveiw at the High-Level Policy Forum for Sustainable Development 2019 in Africa*, https://www.uneca.org/sites/default/files/uploaded-documents/ARFSD/2019/water_as_cross-cutting_factor_in_the_sdgs_under_review_at_the_high-level_panel_forum_for_sustainable_development_hlpf_2019_in_africa.pdf.

UNICEF Institute for Statistics (2019), (database), http://data.uis.unesco.org/.

– (2017), *Literacy rates continue to rise from one generation to the next*, UNESCO Institute for Statistics, Montreal.

UNICEF and WHO (United Nations Chidren's Fund and World Health Organization) (2019), *Progress on Household Drinking Water, Sanitation and Hygiene 2000-2017: Special focus on inequalities,* UNICEF and WHO, New York.

van de Valle (2015), *Africa Can End Poverty*, World Bank, https://blogs.worldbank.org/africacan/poverty-is-falling-faster-for-female-headed-households-in-africa, accessed 1 October 2019.

Wolfram, C. (2013), *Power Africa: Observations from Kenya*, Energy Institute at Haas, https://energyathaas.wordpress.com/2013/07/15/power-africa-observations-from-kenya/, accessed 15 July 2019.

World Bank (2019), *World Bank Enterprise Surveys*, www.enterprisesurveys.org/, accessed 1 September 2019.

– (2018), Worldwide Governance Indicators, World Bank, https://info.worldbank.org/governance/wgi/.

– (2017a), Infrastructure, growth, and productivity in sub-Saharan Africa, *Africa's Pulse*, https://openknowledge.worldbank.org/handle/10986/32480, accessed 1 October 2019.

– (2017b), Africa - Electricity Transmission And Distribution Grid Map, *Data Catalog*, https://datacatalog.worldbank.org/dataset/africa-electricity-transmission-and-distribution-grid-map2017https://www.enterprisesurveys.org/en/data/exploretopics/infrastructure, accessed 1 October 2019.

Wijnen, M. et al. (2018), *Assessment of Groundwater Challenges & Opportuniites in Support of Sustainable Development in Sub-Saharan Africa*, World Bank, Washington, DC.

WWAP (World Water Assessment Programme - UNESCO) (2019), *The United Nations World Water Development Report 2019: Leaving No One Behind*, United Nations Educational, Scientific and Cultural Organization, Paris.

Chapter 9: Urbanisation, industrialisation and clean cooking

ACUMEN (2018), *Accelerating Energy Access: the role of patient capital*, ACUMEN, https://acumen.org/wp-content/uploads/Accelerating-Access-Role-of-Patient-Capital-Report.pdf.

AMANHI (Alliance for Maternal and Newborn Health Improvement) (2018), Population-based rates, timing, and causes of maternal deaths, stillbirths, and neonatal deaths in south Asia and sub-Saharan Africa: a multi-country prospective cohort study, AMANHI mortality study group, *The Lancet Global Health*, Vol. 6(12), pp. 1297-1308.

Bentsen et al. (2014), Agricultural residue production and potentials for energy, material and feed services, *Progress in Energy and Combustion Science*, Vol. 40, pp. 59-73.

Biofuture Platform (2018), *Creating the Biofuture: A Report on the State of the Low-Carbon Bioeconomy*,http://biofutureplatform.org/wp-content/uploads/2018/11/Biofuture-Platform-Report-2018.pdf.

Clean Cooking Alliance (2019), *Increasing Investment in the Clean Cooking Sector*, Clean Cooking Alliance, https://www.cleancookingalliance.

Clemens et al. (2018), Africa Biogas Partnership Program: a review of clean cooking implementation through market development in East Africa, *Energy for Sustainable Development*, Vol. 46, pp. 23-31.

Demographia (2019), *World Urban Areas*, http://demographia.com/db-worldua.pdf.

FAO (Food and Agriculture Organization of the United Nations) (2017), *The Charcoal Transition*, FAO, http://www.fao.org/3/a-i6935e.pdf.

Fundira, T. and G. Henley (2017), Biofuels in Southern Africa: Political economy, trade and Policy Environment, *World Institute for Development Economic Research (UNU-WIDER) Working Paper Series 048*.

Harkouss et al. (2018), Passive design optimization of low energy buildings in different climates, *Energy*, Vol. 165, pp. 591-613.

Hofman et al. (2008), Motorcycle ambulances for referral of obstetric emergencies in rural Malawi: Do they reduce delay and what do they cost?, *International Journal of Gynecology and Obstetrics*, Vol. 102(2), pp. 191-197.

IEA (International Energy Agency) (2017), *Energy Access Outlook 2017: From Poverty to Prosperity*, IEA, Paris.

IPCC (Intergovernmental Panel on Climate Change) (2014), *Climate Change 2014: Synthesis Report Summary for Policymakers, Contribution of Working Groups I, II and III to the Fifth Assessment Report of the IPCC* [Core Writing Team, R.K. Pachauri and L.A. Meyer (eds.)], IPCC, Geneva, Switzerland.

IRENA (International Renewable Energy Agency) (2017), *Biofuel Potential in Sub-Saharan Africa: Raising food yields, reducing food waste and utilising residues*, IRENA, Abu Dhabi.

Khavari et al. (2019), *PopClusters,* https://data.mendeley.com/datasets/z9zfhzk8cr/2.

KTH-dESA (Royal Institute of Technology – Divisions of Energy Systems Analysis), *GitHub*: https://github.com/KTH-dESA/PopCluster.

Lane, J. (2019), Biofuel Mandates around the World 2019, *Biofuels Digest,* https://www.biofuelsdigest.com/bdigest/2019/01/01/biofuels-mandates-around-the-world-2019/50/.

Ministry of Energy and Petroleum, Republic of Kenya (2014), *Draft National Energy Policy*, Nairobi.

NCAR (2012), GIS Program Climate Change Scenarios GIS data portal, http://gisclimatechange.ucar.edu.

NOAA (National Oceanic and Atmospheric Administration) (2018), Global Surface Summary of the Day, 1990-2018 (dataset), US Department of Commerce: https://data.noaa.gov/dataset/dataset/global-surface-summary-of-the-day-gsod.

Odamo, L. (2019), *Closing Sub-Saharan Africa's Electricity Access Gap: why cities must be part of the solution*, World Resources Institute, https://www.wri.org/blog/2019/08/closing-sub-saharan-africa-electricity-access-gap-why-cities-must-be-part-solution.

REN21 (2017), *Renewables 2017 - Global Status Report*, REN21 Secretariat, Paris.

Sekoai. P. and K. Yoro (2016), Biofuel Development Initiatives in sub-Saharan Africa: opportunities and challenges, *Climate*, Vol 4(2), 33.

ter Heegde, F. (2019), *Technical potential for household biodigesters in Africa*, SNV (Netherlands Development Organisation), The Hague.

National Environment Management Authority, Republic of Uganda (2010), *The Potential of Biofuels in Uganda*, Kampala.

Tremeac et al. (2012), Influence of air conditioning management on heat island in Paris air street temperatures, *Applied Energy*, Vol. 95, pp. 102-110.

UNCTAD (United Nations Conference on Trade and Development) (2014), *The State of Biofuels Market: Regulatory, Trade and Development Perspectives*, United Nations Publications.

UNDESA (United Nations Department of Economic and Social Affairs), Population Division (2019), *World Population Prospects 2019*, (database), https://population.un.org/wpp/.

UNEP (United Nations Environment Programme) (2019), *Africa Used Vehicle Report*, UNEP, https://wedocs.unep.org/bitstream/handle/20.500.11822/25233/AfricaUsedVehicleReport.pdf, accessed 15 July 2019.

– (2017), *Vehicle Emissions Standards,* http://wedocs.unep.org/bitstream/handle/20.500.11822/17534/Africa_VehicleEmissions_March2017.pdf.

Chapter 10: Access to electricity and reliable power

BCEAO (Central Bank of West African States) (2018), Short-term, medium-term and long-term credits 2014-2018, Dakar, Senegal (database), accessed 25 June 2019.

CDKN (Climate & Development Knowledge Network) (2015), *Using climate information for large-scale hydropower planning in sub-Saharan Africa,* https://cdkn.org/resource/climate-change-data-hydropower-planning-sub-saharan-africa/.

Eberhard, A. et al. (2016), *Independent Power Projects in Sub-Saharan Africa: Lessons from Five Key Countries*, World Bank, Washington, DC.

ESMAP (Energy Sector Management Assistance Program) (2019), *Mini-Grids for Half a Billion People: Market Outlook and Handbook for Decision Makers*, World Bank, Washington, DC, https://esmap.org/mini_grids_for_half_a_billion_people.

Gernaat D. et al. (2017), High-resolution assessment of global technical and economic hydropower potential, *Nature Energy*, Vol. 2, pp. 821–828.

Grimm, M. et al. (2016), A first step up the energy ladder? Low cost solar kits and household's welfare in Rural Rwanda, *Policy Research Working Paper,* WPS 7859, World Bank, Washington, DC.

Horn, S., C. Reinhart and C. Trebesch (2019), China's Overseas Lending, *National Bureau of Economic Research Working Paper No. 26050*, Cambridge, Massacheutts, United States, www.nber.org/papers/w26050.

IEA (International Energy Agency) (2017), *Energy Access Outlook 2017: From Poverty to Prosperity*, IEA, Paris.

– (2014), *Africa Energy Outlook: World Energy Outlook Special Report*, IEA, Paris.

IFC (Interntional Finance Corporation) (2018), *Off-Grid Solar Market Trends Report*, IFC, Washington, DC.

IJ Global (2019), *Transactions* (database), https://ijglobal.com/data/search-transactions, accessed June-October 2019.

IPCC (Intergovernmental Panel on Climate Change) (2014), *Climate Change 2014: Synthesis Report. Fifth Assessment Report*, IPCC, Geneva, https://ar5-syr.ipcc.ch/resources/htmlpdf/WG1AR5_Chapter12_FINAL/#pf1a.

Kojima, M. (2016), *Making Power Affordable for Africa and Viable for its Utilities*, World Bank, Washington, DC.

Korkovelos, A. et al. (2019), The role of open access data in geospatial electrification planning and the achievement of SDG 7: an onsset-based case study for Malawi, *Energies*, Vol. 12/7.

MIT-IIT (2019), Optimal Electrification Planning Incorporating On- and Off-Grid Technologies: The Reference Electrification Model (REM), Proceedings of the IEEE, DOI: 10.1109/JPROC.2019.2922543.

OECD (Organisation for Economic Co-operation and Development) (2019), *Flows based on individual projects (Creditor Reporting System)*, https://stats.oecd.org/, accessed 20 September 2019.

Power Africa (2019), *Solar Mini-grids Boost Women's Entrepreneurship*, https://medium.com/power-africa/solar-mini-grids-boost-womens-entrepreneurship-8479e4c1f8f8, accessed 8 March 2019.

Power for All (2019), *Powering Jobs Census 2019: The Energy Access Workforce*, https://www.powerforall.org/application/files/8915/6310/7906/Powering-Jobs-Census-2019.pdf.

Umeme (2018), *Annual Report 2019*, https://www.umeme.co.ug/umeme_api/wp-content/uploads/2019/05/Umeme-Limited_Annual-Report-2018_compressed-1.pdf.

World Bank (2019a), *Private Participation in Infrastructure Projects* (database), https://ppi.worldbank.org/en/ppidata, accessed 20 June 2019.

– (2019b), *Global Financial Development*, Private credit by deposit money banks and other financial institutions to GDP (database), https://datacatalog.worldbank.org/dataset/global-financial-development, accessed June 2019.

– (2018), *Infrastrucure,* World Bank Group Enterprise Surveys, www.enterprisesurveys.org/data/exploretopics/infrastructure#sub-saharan-africa, accessed 20 July 2019.

– (2017), *Linking Up: Public-Private Partnerships in Power Transmission in Africa*, World Bank, Washington, DC.

– (2014), *Supplemental Implementation Completion and Results Report* (Credit Number 3411-UG), World Bank, Washington, DC.

Chapter 11: Natural gas and resource management

African Energy (2019), Small-scale LNG schemes start to gain momentum, *African Energy Newsletter*, Issue 392, https://archive.crossborderinformation.com/Article/Small-scale+LNG+schemes+start+to+gain+momentum.aspx?date=20190516&docNo=14&qid=1&page=2&from=IssueNumber/Issue+392.aspx#.

Bauer, A. and D. Mihalyi (2018), *Premature Funds: How Overenthusiasm and Bad Advice Can Leave Countries Poorer*, Natural Resource Governance Institute, https://resourcegovernance.org/sites/default/files/documents/premature-funds.pdf.

Cust, J. and D. Mihalyi (2017), *Evidence for a pre-source curse? Oil discoveries, elevated expectations, and growth disappointments*, World Bank, Washington, DC, http://documents.worldbank.org/curated/en/517431499697641884/Evidence-for-a-presource-curse-oil-discoveries-elevated-expectations-and-growth-disappointments.

Heller, P. and D. Mihalyi (2019), *Massive and Misunderstood: Data-driven Insights into National Oil Companies*, Natural Resource Governance Institute, https://resourcegovernance.org/sites/default/files/documents/massive_and_misunderstood_data_driven_insights_into_national_oil_companies.pdf.

Hove and Ncube (n.d.), *Sovereign Wealth Funds as a Driver of African Development*, Quantum Global, http://quantumglobalgroup.com/wp-content/uploads/2017/10/Sovereign-Wealth-Funds-as-a-driver-of-African-development.pdf.

IEA (International Energy Agency) (2019), *World Energy Investment 2019*, IEA, Paris.

McKinsey (2010), *What's driving Africa's growth*, McKinsey, https://www.mckinsey.com/featured-insights/middle-east-and-africa/whats-driving-africas-growth.

MEES (Middle East Economic Survey) (2018), Egypt gas turns the corner, but can it maintain investment?, *Weekly Energy, Economic and Geopolitical Outlook,* Vol. 61/7, https://www.mees.com/vol/61/issue/7.

Natural Resource Governance Institute (2019), *Resource Governance Index: From Legal Reform to Implementation in Sub-Saharan Africa*, Natural Resource Governance Institute, https://resourcegovernance.org/analysis-tools/publications/sub-saharan-africa-implementation-gap.

OIES (Oxford Institute of Energy Studies) (2019), *Opportunities for Gas in Sub-Saharan Africa,* OIES, London.

– (2018), *Egypt - A Return to a Balanced Gas Market?,* OIES, London.

UNCTAD (United Nations Conference on Trade and Development), UNCTAD Statistics, https://unctadstat.unctad.org/EN/BulkDownload.html, accessed 3 June 2019.

USGS (United States Geological Survey) (2019), *Mineral Commodity Summaries 2019*, USGS, Washington DC, https://www.usgs.gov/centers/nmic/mineral-commodity-summaries, accessed 7 October 2019.

World Bank (2017), *The Growing Role of Minerals and Metals for a Low-Carbon Future,* World Bank Publications, Washington DC, http://documents.worldbank.org/curated/en/207371500386458722/pdf/117581-WP-P159838-PUBLIC-ClimateSmartMiningJuly.pdf.

World Nuclear Association (2019), *World Uranium Mining Production*, World Nuclear Association, www.world-nuclear.org/information-library/nuclear-fuel-cycle/mining-of-uranium/world-uranium-mining-production.aspx, accessed 29 May 2019.

Chapter 12: Implications for Africa and the world

FAO (Food and Agriculture Organization of the United Nations), *FAOSTATS*, http://www.fao.org/faostat/en/#data/GF, accessed 17 July 2019.

IEA (International Energy Agency) (2019), *Global EV Outlook 2019*, IEA, Paris.

IPCC (Intergovernmental Panel on Climate Change) (2014), Africa in: *Climate Change 2014: Impacts, Adaptation, and Vulnerability. Part B: Regional Aspects. Contribution of Working Group II to the Fifth Assessment Report of the Intergovernmental Panel on Climate Change.* Niang, I., O. Ruppel, M. Abdrabo, A. Essel, C. Lennard, J. Padgham and P. Urquhart, Cambridge University Press, Cambridge, United Kingdom and New York.

Kojima, M. (2016), Making Power Affordable for Africa and Viable for its Utilities, World Bank, Washington, DC.

Power Africa (2019), *Solar Mini-grids Boost Women's Entrepreneurship...*, https://medium.com/power-africa/solar-mini-grids-boost-womens-entrepreneurship-8479e4c1f8f8, accessed 8 March 2019.

UNCTAD (United Nations Conference on Trade and Development), UNCTAD Statistics, https://unctadstat.unctad.org/EN/BulkDownload.html, accessed 3 June 2019.

USGS (United States Geological Survey) (2019), Mineral Commodity Summaries 2019, USGS, Washington DC, https://www.usgs.gov/centers/nmic/mineral-commodity-summaries, accessed 7 October 2019.

World Bank (2018), *Accelerating Climate-Resilient and Low-Carbon Development: Africa Climate Business Plan – Third Implementation Progress Report and Forward Look,* World Bank, Washington DC.

– (2015), *Innovative financing: the case of India Infrastructure Finance Company,* World Bank, Washington DC.

Part C: WEO Insights

Chapter 13: Prospects for gas infrastructure

Altfeld, K. and D. Pinchbeck (2013), Admissible hydrogen concentrations in natural gas systems, *Gas for Energy*, Vol. 3, www.gerg.eu/public/uploads/files/publications/GERGpapers/SD_gfe_03_13_Report_Altfeld-Pinchbeck.pdf.

Arapostathis S. et al. (2013), Governing transitions: Cases and insights from two periods in the history of the UK gas industry, *Energy Policy*, Vol. 52(2013), pp. 25-44.

Cadent (2018), HyNet North West, *From vision to reality.* https://hynet.co.uk/app/uploads/2018/05/14368_CADENT_PROJECT_REPORT_AMENDED_v22105.pdf.

Cedigaz (2019), *Global Biomethane Market: Green Gas Goes Global,* www.cedigaz.org/global-biomethane-market-green-gas-goes-global/.

Dodds, P. and W. McDowall (2013), The future of the UK gas network, *Energy Policy*, Vol. 60 (2013), pp. 305-316.

Dolci F. et al. (2019), Incentives and legal barriers for power-to-hydrogen pathways: An international snapshot, *International Journal of Hydrogen Energy*, Vol. 44(23), pp. 11394-11401.

Fasihi M. et al. (2019), Techno-economic assessment of CO_2 direct air capture plants, *Journal of Cleaner Production*, Vol. 224(2019), pp. 957-980.

Frazer-Nash Consultancy (2018), *Appraisal of Domestic Hydrogen Appliances*, prepared for the UK Department of Business, Energy & FNC, https://assets.publishing.service.gov.uk/government/uploads/system/uploads/attachment_data/file/699685/Hydrogen_Appliances-For_Publication-14-02-2018-PDF.pdf.

Giuntoli J. et al. (2015), *Solid and Gaseous Bioenergy Pathways: Input Values and GHG Emissions*, European Commission Joint Research Centre, Institute for Energy and Transport, Ispra, Italy.

GRDF (2018), *Perspectives gaz naturel & renouuvelables sur l'horizon 2018-2035* (Natural gas and renewables perspectives, 2018-2035), GRDF, Paris.

H21 (2016), *Leeds City Gate, Northern Gas Networks,* Leeds, United Kingdom.

HyLaw (2019), *HyLaw* (database), www.hylaw.eu/database#/database/gas-grid-issues/injection-of-hydrogen-at-transmission-level-for-energy-storage-and-enhancing-sustainabily, accessed 15 May 2019.

IEA (International Energy Agency) (2019), *The Future of Hydrogen*, IEA, Paris.

– (2018) *World Energy Outlook-2018*, IEA, Paris.

– (2010), *World Energy Outlook-2010*, IEA, Paris.

– (1995), *World Energy Outlook-1995*, IEA, Paris.

Koornneef J. et al. (2013), *Potential for Biomethane Production with Carbon Dioxide Capture and Storage*, Report 2013/11, IEAGHG, Paris.

Liebetrau J. et al. (2017), *Methane emissions from biogas plant,* IEA Bioenergy Task 37, https://www.ieabioenergy.com/wp-content/uploads/2018/01/Methane-Emission_web_end_small.pdf.

Jacobson M. et al. (2017), 100% Clean and Renewable Wind, Water, and Sunlight All-Sector Energy Roadmaps for 139 Countries of the World, *Joule*, 1(1), pp. 108-121.

Keith D. et al. (2018), A process for capturing CO_2 from the atmosphere, *Joule*, Vol. 2(8) pp. 1573-1594.

McDonagh S. et al. (2019), Are electro fuels a sustainable transport fuel? Analysis of the effect of controls on carbon, curtailment, and cost of hydrogen, *Applied Energy*, Vol. 247(1) pp. 716-730.

MTT Research (*Maa-ja elintarviketalouden tutkimuskeskus*) (Agrifood Research Centre, Finland), (2009), *Energy from field energy crops*, www.codigestion.com/fileadmin/codi/images/ENCROP/Handbook_for_energy_producers_www_version.pdf.

NATURALHY (2009), *Using the Existing Natural Gas System for Hydrogen*, www.fwg-gross-bieberau.de/fileadmin/user_upload/Erneuerbare_Energie/Naturalhy_Brochure.pdf.

Navigant (2019), *Gas for Climate: The optimal role for gas in a net-zero emissions energy system,* Navigant, Utrecht, Netherlands.

NREL (US National Renewable Energy Laboratory) (2013), *Blending Hydrogen into Natural Gas Pipeline Networks: A Review of Key Issues*, NREL, www.nrel.gov/docs/fy13osti/51995.pdf.

Pérez-Fortes, M. and E. Tzimas (2016), *Techno-Economic and Environmental Evaluation of CO_2 Utilisation for Fuel Production*, European Commission Joint Research Centre, Petten, The Netherlands.

Quarton, C. and S. Samsatli (2018), Power-to-gas for injection into the gas grid: What can we learn from real-life projects, economic assessments and systems modelling?, *Renewable and Sustainable Energy Reviews*, Vol. 98, pp. 302-316.

Staffell I. et al. (2019), The role of hydrogen and fuel cells in the global energy system, *Energy & Environmental Science*, 463-491, doi: https://doi.org/10.1039/C8EE01157E.

Underground Sun Storage (2017), *Publizierbarer Endbericht*, (Published final report), www.underground-sun-storage.at/fileadmin/bilder/SUNSTORAGE/Publikationen/UndergroundSunStorage_Publizierbarer_Endbericht_3.1_web.pdf.

UNFCCC (United Nations Framework Convention on Climate Change) (2018), *CDM Methodology Booklet,* https://cdm.unfccc.int/methodologies/documentation/meth_booklet.pdf.

US EPA (US Environmental Protection Agency) (2019), *Inventory of U.S. Greenhouse Gas Emissions and Sinks: 1990-2017*, US EPA, Washington, DC.

– (2016), *Inventory of U.S. Greenhouse Gas Emissions and Sinks: 1990-2014*, US EPA, Washington, DC.

Chapter 14: Outlook for offshore wind

Bloomberg (2019), Bloomberg Terminal, accessed 16 September 2019.

BNEF (Bloomberg New Energy Finance) (2019), *Offshore Wind Projects June 2019*, BNEF, London.

Carbon Trust (2015), *Floating Offshore Wind: Market and Technology Review*, The Carbon Trust, United Kingdom.

Danish Energy Agency (2019), *Environmental Impacts of Offshore Wind Farms*, https://ens.dk/sites/ens.dk/files/Globalcooperation/Short_materials/environmental_impacts_of_offshore_wind_farms.pdf.

DIW ECON (2019), *Market Design for an Efficient Transmission of Offshore Wind Energy*, https://diw-econ.de/en/wp-content/uploads/sites/2/2019/05/DIW-Econ_2019_Market-design-for-an-efficient-transmission-of-offshore-wind-energy.pdf.

ENTSO-E (European Network of Transmission System Operators for Electricity) (2018), *Ten-Year Network Development Plan*, ENTSO-E, Brussels.

European Commission (2018), *A Clean Planet for all: A European Long-Term Strategic Vision for a Prospectus, Modern, Competitive and Climate Neutral Economy*, European Commission, Brussels.

Green Giraffe (2019), *Recent Trends in Offshore Wind Finance*, Green Giraffe, Bilbao, Spain.

IEA (International Energy Agency) (2019a), *World Energy Investment 2019*, IEA, Paris, www.iea.org/wei2019/.

– (2019b), *The Future of Hydrogen*, IEA, Paris, www.iea.org/hydrogen2019/.

– (2019c), *World Offshore Wind Outlook: World Energy Outlook Special Report*, IEA, Paris.

– (2018a), *World Energy Investment 2018*, IEA, Paris, www.iea.org/wei2018/.

– (2018b), *World Energy Outlook-2018*, IEA, Paris.

– (2018c), *Status of Power System Transformation*, IEA, Paris.

– (2017), *World Energy Outlook-2017*, IEA, Paris.

– (2016), *Re-powering Markets*, IEA, Paris.

IJGlobal (2019), Asset Data, http://ijglobal.com, accessed 16 September 2019.

IRENA (International Renewable Energy Agency) (2019), *Renewable Power Generation Costs in 2018*, IRENA, Abu Dhabi.

– (2016), *A Supplement to Innovation Outlook: Offshore Wind*, IRENA, Abu Dhabi.

IUCN (International Union for Conservation of Nature) (2013), *Guidelines for Applying Protected Area Management Categories*, IUCN, Gland, Switzerland.

NSWPH (North Sea Wind Power Hub) (2019), *Concept Paper 4: The Benefits*, NSWPH, https://northseawindpowerhub.eu/wp-content/uploads/2019/07/Concept_Paper_4-The-benefits.pdf.

NDRC (National Development and Reform Commission) (2019), *National Development and Reform Commission on Improving Wind Power: Notice of On-Grid Tariff Policy, 21 May 2019*, NRDC, Beijing.

– (2016), *Wind Power Development 13th Five-Year Plan*, NRDC, Beijing.

NREL (US National Renewables Energy Laboratory) (2017), *8760-Based Method for Representing Variable Generation Capacity Value in Capacity Expansion Models*, NREL, www.nrel.gov/docs/fy17osti/68869.pdf.

Open Power System Data (2019), Open Power System (database), https://data.open-power-system-data.org/time_series, accessed 1 June 2019.

PBL (PBL Netherlands Environmental Assessment Agency) (2018), *Costs of Offshore Wind Energy 2018*, www.pbl.nl/en/publications/costs-of-offshore-wind-energy-2018.

Renewables.ninja, (database), www.renewables.ninja, accessed 1 July 2019.

Saint-Drenan Y. et al. (2019), *A Parametric Power Curve Model Dependent on the Main Characteristics of a Wind Turbine*, https://arxiv.org/abs/1909.13780, accessed 26 September 2019.

Staffell I. and S. Pfenninger (2016), Using Bias-Corrected Reanalysis to Simulate Current and Future Wind Power Output, *Energy*, Vol. 114, pp. 1224-1239.

The Wind Power (2019), www.thewindpower.net, accessed 16 September 2019.

Ueckerdt F. et al. (2016), Decarbonizing the Global Power Supply under Region-Specific Consideration of Challenges and Options of Integrating Variable Renewables in the REMIND Model, *Energy Economics*, Vol. 64, pp. 665-684.

Ueckerdt F. et al. (2013), System LCOE: What are the costs of variable renewables?, *Energy*, Vol. 63, pp. 61-75.

US DOE (United States Department of Energy), *2018 Offshore Wind Technologies Market Report*, US DOE, Oak Ridge, Tennessee, United States.

van Zalk, J. et al. (2018), The spatial extent of renewable and non-renewable power generation: A review and meta-analysis of power densities and their application in the United States, *Energy Policy*, Vol. 123, pp. 83-91.

WindEurope (2019), *Offshore Wind in Europe: Key Trends and Statistics 2018*, WindEurope, https://windeurope.org/about-wind/statistics/offshore/european-offshore-wind-industry-key-trends-statistics-2018/, Brussels.

– (2017), *Floating Offshore Wind Vision Statement*, Brussels, https://windeurope.org/wp-content/uploads/files/about-wind/reports/Floating-offshore-statement.pdf.

World Bank (2019), *New Program to Accelerate Expansion of Offshore Wind Power in Developing Countries*, World Bank, Washington DC.

WWF (World Wildlife Fund for Nature) (2014), *Environmental Impacts of Offshore Wind Power Production in the North Sea*, Oslo, www.wwf.no/assets/attachments/84-wwf_a4_report___havvindrapport.pdf.

Xiang X. et al. (2016), Cost Analysis and Comparison of HVAC, LFAC and HVDC for Offshore Wind Power Connection, 12th IET International Conference on AC and DC Power Transmission, *IET Conference Publications 696*, pp. 29-35, Beijing.

Annex B

BGR (German Federal Institute for Geosciences and Natural Resources) (2018), *Energiestudie 2018, Reserven, Ressourcen und Verfügbarkeit von Energierohstoffen*, (Energy Study 2018, Reserves, Resources and Availability of Energy Resources), BGR, Hannover, Germany.

BP (2019), *BP Statistical Review of World Energy 2019*, BP, London.

Cedigaz (2019), Cedigaz databases, Cedigaz, Rueil-Malmaison, France, www.cedigaz.org.

IEA (International Energy Agency) (2019), *World Energy Investment 2019*, IEA, Paris.

IMF (International Monetary Fund) (2019), *World Economic Outlook*, IMF, Washington, DC.

IRENA (International Renewable Energy Agency) (2019), *Renewable Power Generation Costs in 2018*, IRENA, Abu Dhabi.

OGJ (Oil and Gas Journal) (2018), Worldwide Oil, Natural Gas Reserves exhibit marginal increases, *OGJ* 116 (12), Pennwell Corporation, Oklahoma City, Oklahoma, United States.

United Nations Population Division, UNDESA, (database), https://www.un.org/en/development/desa/population/publications/database/index.asp.

US DOE/EIA (US Department of Energy/Energy Information Adminstration) (2019), *Assumptions to the Annual Energy Outlook 2019*, US DOE/EIA, Washington, DC.

– (2018), *U.S. Crude Oil and Natural Gas Proved Reserves, Year-end 2017,* US DOE/EIA, Washington, DC.

US DOE/EIA/ARI (US Department of Energy)/(Energy Information Administration)/(Advanced Resources International) (2013 and last updated September 2015), *Technically Recoverable Shale Oil and Shale Gas Resources: An Assessment of 137 Shale Formations in 41 Countries Outside the United States*, US DOE/EIA, Washington, DC.

USGS (United States Geological Survey) (2012a), "Assessment of Potential Additions to Conventional Oil and Gas Resources of the World (Outside the United States) from Reserve Growth", *Fact Sheet 2012–3052*, USGS, Boulder, Colorado, United States.

– (2012b), "An Estimate of Undiscovered Conventional Oil and Gas Resources of the World", *Fact Sheet 2012–3042*, USGS, Boulder, Colorado, United States.

This publication reflects the views of the IEA Secretariat but does not necessarily reflect those of individual IEA member countries. The IEA makes no representation or warranty, express or implied, in respect of the publication's contents (including its completeness or accuracy) and shall not be responsible for any use of, or reliance on, the publication.

Unless otherwise indicated, all material presented in figures and tables are derived from IEA data and analysis.

This publication and any map included herein are without prejudice to the status of or sovereignty over any territory, to the delimitation of international frontiers and boundaries and to the name of any territory, city or area.

IEA. All rights reserved.
IEA Publications
International Energy Agency
Website: www.iea.org
Contact information: www.iea.org/about/contact
Typeset in France by IEA and Printed in France by Corlet – November 2019
Cover design: IEA
Photo credits: © GraphicObsession

The paper used for this document has received certification from the Programme for the Endorsement of Forest Certification (PEFC) for being produced respecting PEFC's ecological, social and ethical standards. PEFC is an international non-profit, non-governmental organization dedicated to promoting Sustainable Forest Management (SFM) through independent third-party certification.